Xi Chen

Requirements and concepts for future automotive electronic architectures from the view of integrated safety

Requirements and concepts for future automotive electronic architectures from the view of integrated safety

von
Xi Chen

universitätsverlag karlsruhe

Dissertation, Universität Karlsruhe (TH)
Fakultät für Elektrotechnik und Informationstechnik, 2008

Impressum

Universitätsverlag Karlsruhe
c/o Universitätsbibliothek
Straße am Forum 2
D-76131 Karlsruhe
www.uvka.de

Universitätsverlag Karlsruhe 2008
Print on Demand

ISBN: 978-3-86644-225-2

Requirements and concepts for future automotive electronic architectures from the view of integrated safety

Zur Erlangung des akademischen Grades eines
DOKTOR-INGENIEURS
von der Fakultät für
Elektrotechnik und Informationstechnik
der Universität Fridericiana Karlsruhe (TH)
genehmigte
DISSERTATION
von

Diplom-Ingenieur Xi CHEN, M. Sc.

geboren in Wuhan

aus Schwäbisch Gmünd

Tag der mündlichen Prüfung: 07.02.2008

Hauptreferent: Prof. Dr.-Ing. K. D. Müller-Glaser

Korreferent: Prof. Dr.-Ing. G. Trommer

Karlsruhe, 09.01.2008

Acknowledgement

The presented work here was written during my work as a research associate from 2004 to 2007 by DaimlerChrysler (today Daimler AG) Group Research and Advanced Engineering in the department of Electric/Electronic Architecture.

I would like to sincerely express my gratitude to the supervisor of my dissertation, Mr. Prof. Dr.-Ing. Müller-Glaser, for his continuous invaluable guidance, encouragement and support. His vision and enthusiasm in advanced engineering research inspired and motivated me throughout my pursuit of the degree. I appreciate his broad knowledge, originality and insight in many areas. Working with him was a wonderful experience, and I have learned a lot from it. Many thanks to Professor Dr.-Ing. Trommer for his second expertise on this thesis.

Many thanks to Mr. Dr. Hedenetz, as the team leader of E/E-architecture of safety electronics, he supervised the technical work and gave me critical but very valuable comments on a draft of this work. Special thanks to Ms. Dr. Lauer for her support and for providing resources to perform the case study of this thesis. Acknowledgements here also to all my former colleagues in the department REI/EC at Daimler, for their interest in my work and for the professional discussions on this topic.

Also thanks to my project colleagues in WP1, WP2, WP4 and WP5 from EASIS industry consortium. They offered me many useful instructions and professional discussions. I also want to thank all my students, who helped me to build up the case study during their internship and master thesis.

I am forever indebted to my parents, Jianwei Chen, Gengxin Chen and my grandmother, Shunying Qing, for their great love, unselfish support, endless patience and strong encouragement through my life, especially the 9 years study and work in Germany, and my girlfriend, Xi Yu, for her patience and carefulness to read through my dissertation. I would dedicate all my achievements to them who I love so much.

Danksagung

Die vorliegende Arbeit entstand während meiner Tätigkeit als Ingenieur von 2004 bis 2007 bei DaimlerChrysler (heute Daimler AG) Group Research and Advanced Engineering in der Abteilung der Elektronik/Elektrik Architektur.

Mein besonderer Dank gilt

Herrn Prof. Dr.-Ing. Müller-Glaser für die Betreuung dieser Arbeit sowie die Unterstützung und Ermutigung zur Vollendung dieser Arbeit, seine Vision und Begeisterung in die Vorentwicklung und Forschung hat mich während der Arbeit ständig motiviert.

Herrn Prof. Dr.-Ing. Trommer für die Übernahme des 1. Mitberichts,

Allen Kolleginnen und Kollegen bei der Daimler Abteilung REI/EC für die gute Zusammenarbeit und hilfsbereite Unterstützung. Besonders möchte ich mich bei Herrn Dr. Hedenetz, als Teamleiter der Fahrzeug E/E-Architektur der Sicherheitselektronik, für die technische Bereuung und Korrekturlesen der Arbeit und Frau Dr. Lauer für die Unterstützung beim Aufbau des Validators bedanken.

Den Kolleginnen und Kollegen von WP1, WP2, WP4 und WP5 im EASIS-Konsortium für die erfolgreiche Zusammenarbeit und professionelle Diskussion. Den Studenten, die im Rahmen von Studien- und Diplomarbeiten zum Gelingen dieser Arbeit beigetragen haben, damit die vorgestellten Konzepte nicht nur Ideen blieben, sondern auch verwirklicht werden konnten,

Meinen Eltern Jianwei Chen, Gengxin Chen und meiner Großmutter Shunying Qing für ihre langjährige Unterstützung, unerschöpfliche Geduld insbesondere während meiner gesamten neunjährigen Studienzeit und Promotion in Deutschland, und meiner Freundin, Xu Yu, für ihre Geduld und Sorgfältigkeit bei dem Korrekturlesen meiner Dissertation.

Xi Chen

Xi Chen

Schwäbisch Gmünd, im Dezember 2007

Abbreviation list

ABS	Anti-lock Braking System
AC	Aliveness Counter
ACC	Adaptive Cruise Control
ADC	Analog Digital Converter
API	Application Programming Interface
ARC	Arrival Rate Counter
AS	Activation Status
ASAM	Association for Standardization of Automation- and Measuring Systems
ASIC	Application Specific Integrated Circuit
ASIL	Automotive Safety Integrity Level
AUTOSAR	Automotive Open System Architecture
BDHA	Basic Design Hazard Analysis
BMS	Battery Management System
BSW	Basic Software
CAN	Controller Area Network
CBC	Common Body Controller
CCA	Cycle Counter for Aliveness
CCAR	Cycle Counter for Arrival Rate
CCF	Common Cause Failures
CFC	Control Flow Checking
CFG	Control Flow Graph
CGW	Central Gateway
CMF	Common Mode Failures
COM	Communication
CPU	Central Processing Unit
CRC	Cyclic Redundancy Check
DC	Diagnostic Coverage
DLL	Dynamic Link Library
DSC	Digital Signal Controller
E/E	Electric/Electronic
EASIS	Electronic Architecture System Engineering for Integrated Safety Systems

EAST-EEA	Electronic Architecture and Software Technology – Electronic Embedded Architecture
ECM	Engine Control Module
ECU	Electronic Control Unit
EDC	Error Detection Code
EEP	EASIS Engineering Process
EEPROM	Electrically Erasable Programmable only memory
E-Gas	Engine management system of gasoline and diesel engines
EMC	Electronic Magnetic Compatibility
EMS	Energy Management System
EPS	Electronic Power Steering
ESP	Electronic Stability Program
EU	European Union
EUS	Energy Uncoupling System
FAA	Functional Analysis Architecture
FDA	Functional Design Architecture
FHA	FAA Hazard Analysis
FMEA	Failure Mode and Effect Analysis
FMEDA	Fault Mode and Effect Diagnosis Analysis
FMF	Fault Management Framework
FOU	Fail Operational Unit
FPGA	Field Programmable Gate Array
FSM	Fault State Manager
FSU	Fail Silent Unit
FT	Fault Tolerance
FTA	Fault Tree Analysis
FTCom	Fault Tolerant Communication
FTDMA	Flexible Time Division Multiple Access
FTU	Fault Treatment Unit
HA	Hardware Architecture
HGA	Hazard Graph Analysis
HiL	Hardware in the Loop
HIS	Deutsches Automobilkonsortium - Hersteller Initiative Software

HU	Head Unit
HW	Hardware
I/O	Input/Output
IC	Instrument Cluster
ID	Identity
ISR	Interrupt Service Routine
ISS	Integrated Safety System(s)
ISS EP	ISS Engineering Process
IT	Information Technology
LIN	Local Interconnect Network
MCU	Micro Controller UNIT
MiL	Model in the Loop
MMU	Memory Management Unit
MPU	Memory Protection Unit
N/A	Not available or Not applicable
NA	Not Applicable
NM	Network Management
NVM	Non volatile memory
OEM	Original Equipment Manufacturer
OIL	OSEK Implementation Language
ORC	Occupant Restraint Controller
OS	Operating System
OSEK, OSEK/VDX	„Offene Systeme und deren Schnittstellen für die Elektronik im Kraftfahrzeug", Open Systems and the Corresponding Interfaces for Automotive Electronics
PFC	Program Flow Checking
PFH	Probability of one dangerous Failure per Hour
PHA	Preliminary Hazard Analysis
PWM	Pulse Width Modulation
QoS	Quality of Service
RAM	Random Access Memory
RBTL	Reversible Belt Tensioner Left
RBTR	Reversible Belt Tensioner Right

RCP	Rapid Control Prototyping
ROM	Read-Only Memory
RTE	Run Time Environment
RTI	Real-Time Interface
RTOS	Real-Time Operating System
SBC	Sensotronic Brake Control
SC	Sensor Cluster
SER	Single Error Region
SFF	Safe Failure Fraction
SIL/ASIL	Safety Integrity Level / Automotive Safety Integrity Level
SOP	Start of Production
SotA	State of the art
SPC	Suspension Controller
SRS	Safety Requirements Specification
SSM	Standard Software Module
SW	Software
SW-C	SoftWare Component
SW-Cs	Software Components
SWM	Steering Wheel Module
SW-Watchdog	SoftWare Watchdog
TCM	Transmission Control Module
TSIU	Task State Indication Unit
TDMA	Time division multiple access
UART	Universal Asynchronous Receiver Transmitter
UML	Unified Modeling Language
V&V	Verification and Validation
VFB	Virtual Functional Bus
WCET	Worst Case Execution Time
WSM	Wheel Steering Module

Abstract

Integrated Safety System, as one of the most promising technologies of the automotive safety systems, integrates passive and active safety system, together with the interaction of cabin, chassis and powertrain electronics cross the domain board and communication networks. It provides the largest potential for future innovative safety applications. The development of ISS, compared with the current in-vehicle safety systems, puts forward more challenging requirements for the design, development, prototyping and validation process. In order to manage the complexity and fulfill high dependability requirements, new architecture concepts, engineering process and tool chains for the Integrated Safety Systems are required.

In this dissertation, requirements of future Integrated Safety Systems are identified firstly with a delta-analysis. Based on these requirements, concepts of the electronic architecture for the Integrated Safety Systems are developed as a cooperative approach of engineering process, dependable hardware architecture and software platform. In this safety justified development process, with the methodology of virtual front-loading, a distributed rapid prototyping based on the principle of early integration and validation with advanced simulation environment is suggested. The strict and well defined development steps specified here enables the correct-by-construction for the development of ISS. Dependable hardware architecture for ISS is discussed in a top-down methodology, starting with concepts to build up overall in-vehicle topology, guidelines to distribute ISS-applications, design of communication systems and concepts of fault-tolerant ECU hardware architecture are introduced. To conclude this topic, some of the most important use cases to the concepts are demonstrated. Strategy of standardized software platform has been proven to be one of the most effective levers to manage the complexity of future automotive software. Facing the challenges of higher fault tolerance towards fail-operational and increasing density of application software components on one ECU, dependability software services for the fault-tolerant communication with redundancy, end-to-end CRC and Agreement Protocol, time partitioning and space partitioning for the integration of applications, Fault management, fault treatment of dynamic reconfiguration and gateway services are designed and integrated in a layered software topology.

For the practical evaluation and validation of the concepts introduced here, a hardware-in-the-loop validator based on the Steer-by-Wire technologies was built up. On top of this validator, serial-near ISS-applications are integrated and distributed on different ECUs. The dependable software and hardware concepts are designed, prototyped and tested with fault injection on various platforms. The ISS Engineering Process and the tool-chains for the prototyping have demonstrated their viability for the development of complex ISS-applications during the validation process as well.

Zusammenfassung

Motivation:

Integrierte Sicherheitssysteme versprechen aufgrund der Interaktion und Synergieeffekten von Passiv- und Aktivsicherheitssystemen mit Innenraum, Fahrwerk und Antriebstrang, das größte Potential für die Weiterentwicklung von zukünftigen Fahrzeugsicherheitselektroniken. Die Entwicklung der integrierten Sicherheitssysteme, stellen im Vergleich zu den herkömmlichen Fahrzeugsicherheitssystemen höhere Anforderungen an den Entwurfs-, Entwicklungs-, Musteraufbau- und Validierungsprozess. Um die Komplexität und die höheren Anforderungen zu beherrschen, sind neue Architekturkonzepte, Entwicklungsprozesse und Werkzeugkonzepte für die integrierten Sicherheitssysteme notwendig.

Die Arbeit:

In der vorliegenden Arbeit werden die Anforderungen der zukünftigen integrierten Sicherheitssysteme analysiert. Auf dieser Basis wurden Konzepte der Elektronikarchitektur der integrierten Sicherheitssysteme ausgearbeitet, dabei wurde ein Ansatz gewählt, welcher den Entwicklungsprozess, die Hardwarearchitektur und die Softwareplattform mit einschließt. In dem sicherheitsgerechten Entwicklungsprozess, mit der Methodik der virtuellen Front-loading, wird ein verteiltes „Rapid-Prototyping" auf der Basis der Frühintegration und Validierung mit der Simulationsumgebung vorgeschlagen. Die strikten und vordefinierten Entwicklungsschritte ermöglichen die „correctness by construction" (frühzeitige Konzeptabsicherung) bei der Entwicklung der ISS. Verlässliche Hardwarearchitektur für ISS wurde nach dem „top-down" Ansatz entwickelt, beginnend mit dem Konzept zum Aufbau der Fahrzeugtopologie des Kommunikationsnetzes, Richtlinien zur Verteilung und Abbildung der ISS-Applikationen, Entwurf des Kommunikationssystems und Konzepte der fehlertoleranten ECU HW-Architektur werden vorgestellt. Anschließend werden ein paar wichtige Fallstudien zu Demonstrationszwecken der Konzepte erläutet. Die Strategie der standardisierten Softwareplattform ist eine der wichtigsten Ansätze um die Komplexität der zukünftigen Automobilsoftware zu beherrschen. Um den Herausforderungen der höheren Sicherheitsanforderungen der fail-operational Systeme und der steigenden Dichte von Applikationssoftware auf einem Steuergerät gerecht zu werden, wurden fehlertolerante Softwareservices zur Unterstützung der Redundanz, verlässliche Kommunikation, Ende-zu-Ende Applikations-CRC Übereinstimmungsprotokoll, zeitliche und räumliche Auftrennung bei der Applikationsintegration, Fehlermanagement und Fehlerbehandlung mit dynamischer Rekonfiguration entworfen und in einer schichtenorientierten Softwarearchitektur integriert.

Validierung der Konzepte:

Zum praktischen Nachweis des erstellten Konzeptes wurde ein auf Steer-by-Wire basierender Hardware-in-the-Loop Validator entwickelt. Auf oberster Ebene des Validators wurden die seriennahen ISS-Applikationen integriert und auf verschiedenen ECU Plattformen verteilt. Die davon abhängigen Software- und Hardwareanteile wurden entworfen, prototypisch implementiert und mit Fehlerinjektion auf verschiedenen Plattformen getestet. Die Stärke des ISS-Entwicklungsprozesses und der dazugehörenden Werkzeugkette wurde beim Prototyping und bei der Validierung der komplexen ISS-Applikationen demonstriert.

Table of contents

1 Introduction

1.1 Short introduction to future automotive safety systems

Nowadays Safety and comfort are two of the most important requirements and innovation factors for automotive OEMs and suppliers. In order to increase the safety of passengers within a vehicle and for the surrounding environment, more and more electronic systems are used in the automotive safety domain.

In the safety domain there are traditionally two safety systems: the active safety system and the passive safety system.

The following components belong to passive safety components: airbags, collapsible zones, bumpers and safety belts, etc. They help to reduce the negative effects of a traffic accident by means of the physical absorption of crash energy. On the other hand, active safety components include ABS, ESP, brake assistance and lane keeping/change assistance, which influence the powertrain, drive train and chassis system directly, in order to avoid accidents. Current active safety systems predominately directly influence the brake system only, while future active safety systems will also integrate the steering and directional stability track-holding system.

The passive safety system has already reached a relatively mature standard, in which innovation potential is almost exhausted and significant improvements are unlikely to be reached in the near future. Future safety enhancements will mainly have to rely on active safety systems to avoid accidents before they can occur and the so-called "Integrated Safety System" as a combination of both active and passive safety systems, as well as other automotive domains. An overview of the safety level reached and forecasted future development in the passive and active safety domains [GDi02] is shown in Figure 1-1, in which the potential development of ISS is demonstrated as well.

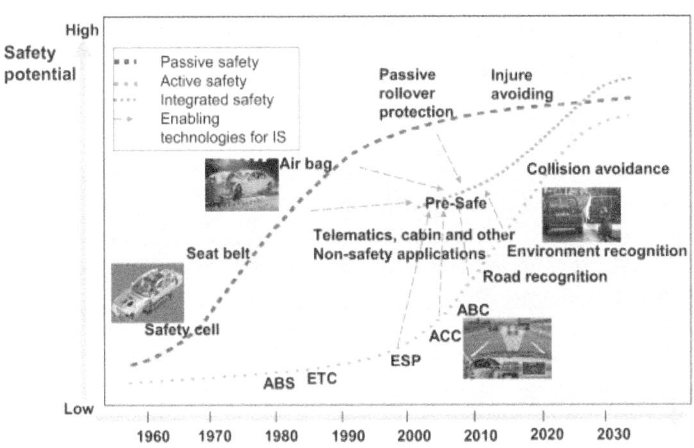

Figure 1-1: Development of active, passive and integrated safety electronic systems

The European Commission's transport policy has set new targets for 2010 with respect to road safety, as shown in Figure 1-2, where a dramatic reduction in road fatalities is expected by 2010. The traditional approach with isolated consideration for passive safety electronics, active safety electronics and other automotive domains does not provide enough potential to reach the ambitious aims of European Commission transport policy. The Integrated Safety System is the only realistic approach to the targets set by EU [EUC01].

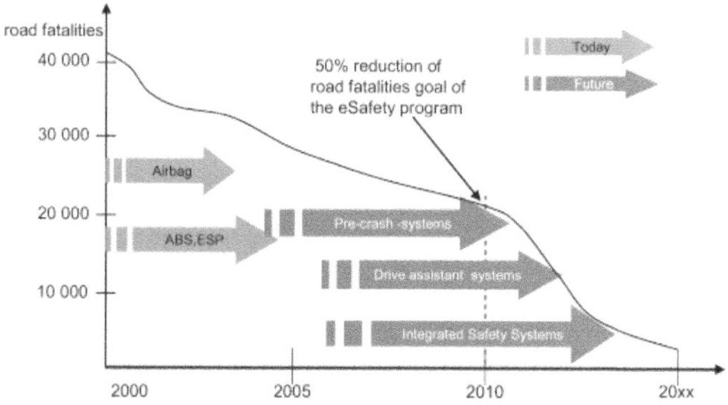

Figure 1-2: Targets set by EU of 50% reduction in the road fatalities before 2010

An Integrated Safety System is a composition of functions that enhance the level of safety not only for the passengers but also for the environment such as pedestrian, which spreads across domain borders, including passive safety components such as airbags, collapsible zones, bumpers, safety belts, etc. and active safety functions such as ABS, ESP, brake assistance, lane change assistance, etc.

The special aspects of ISS-applications compared with the traditional stand-alone safety relevant system can be summarized as follows:

- An ISS-application can spread across domain boarders, e.g., the Pre Crash System closes side windows / sunroof.
- An ISS-application can consist of functions from different domains with different safety integrity levels and safety requirements.
- An ISS-application makes use of a plethora of components, which themselves are not necessarily safety relevant (e.g. sliding roof / sunroof) in itself, thus, comfort components may inherit a high safety level.

With the interaction and integration of safety relevant / non-safety relevant applications from different domains and data sharing from environment sensors and telematics domains, the same data will not need to be measured in different domains, likewise, new applications can be created by using data which is not available from the original domain. The main benefits of Integrated Safety are listed as following:

- Information from all domains can be combined to provide a better view about the state of the vehicle and its surroundings, thereby forming a better basis for decisions taken by safety systems;
- The vehicle can be controlled in a more integrated way as control actions can be coordinated across domains.

A good example of an ISS-Function, combining passive and active safety electronics is the Pre-Safe system [ATZ05] from Mercedes-Benz (first introduced in the S-Class 2002), which detects an unavoidable crash and makes preparations to reduce the severity of the accident. The activation trigger for the Pre-Safe system is the velocity of brake pedal, vehicle velocity and plausibility of on-board electronics. Measures taken by the Pre-Safe system include activation of a reversible belt restraining system, adjusting of seat position and closing of side windows and sunroof.

1.2 Motivation for the work

As discussed in the subsection above, many potential innovations in safety electronics lie in the integration of passive safety electronics, active safety electronics and other domains. By means of sensor data fusion and information flow beyond domain borders, more new safety applications could be designed. From a technical point of view, however, current safety systems are mainly stand alone systems with a limited degree of interdependency, which can not fulfill the requirements of ISS. An integrated approach for vehicle safety systems, in particular the cooperation between the passive and active safety domains, as well as cabin, chassis and telematics domains, offers a solution to enable innovation and improve safety for both passengers and the environment. This includes the combination of active and passive safety systems, where actions will address all phases of a crash, including pre-crash and post-crash.

Dependability has always been the most important requirement for automotive safety relevant applications. For customers, it is not important whether the safety applications are implemented in traditional mechanics or with innovative electronics, but they do expect the new safety features to be at least as reliable as traditional systems.

From a hardware perspective, a safety system with redundant backups is the state of the art approach for a highly dependable system, while taking into consideration the difference in general conditions, as opposed to those specific to the automotive industry. For the future development of safety systems, however, the mechanical backup can disappear. That is to say, compared with the current automotive safety systems, which are only fail-safe or fail-silent, future integrated safety systems should be fail-operational or fail-degraded.

With the increasing number of ECUs and complexity both in hardware and software, it is a great challenge to implement complex, distributed applications, with shorter development cycles while still keeping the dependability of system at a high level. One solution is the standardization of the ECU middleware based E/E-platform to increase the modularity of software components. This trend can now be observed in different consortiums of OEMs and suppliers. With the improvement in reusability, the test depth of system components can be intensified and development time will be shortened because of the unified interface. In the automotive safety domain, the standardized ECU middleware should also support fault-tolerance mechanisms, due to the higher requirement for dependability. A fault-tolerant middleware architecture, adapted to the specific requirements of the automotive safety domain, is not available yet.

The development process for safety relevant electronic systems, compared with the general development process of automotive electronic systems, has to be adapted due to the high dependability requirements. An example of such safety measure is the introduction of preliminary hazard and safety analysis, system hazard analysis and binding to the appreciate development steps. With the derived safety integrity level from safety analysis, the development process for software should be configured as required in industry safety norms. For the development of an Integrated Safety System, the cooperation of sub-systems and their integration into the overall system is of vital importance, because even if all the sub-systems appear to work, the whole system can still fail. An early validation of the system specification is not yet fully supported by the current development process.

1.3 Overview of the dissertation

This dissertation is organized as follows:

After a short overview of current and future automotive safety systems in Subsection 1.1, motivation for the dissertation is explained in Subsection 1.2. Following an introduction to the basic terms of dependability and fault tolerance in Chapter 2, the state of the art in overall automotive electronic architecture, including an introduction to dependability, system topology, automotive software, hardware and communication systems considering safety relevant systems, their current and future requirements are explained in Chapter 3. The development process for generic electronic systems, special concepts and methods for automotive safety systems and an assessment of existing methods will be also introduced in this chapter as well.

In Chapter 4 the existing concepts, trends and current initial concepts in automotive electronic architecture and dependable electronics will be evaluated. From the challenges of ISS and known concepts, a delta analysis will be carried out. The problem is addressed from the different viewpoints of hardware, network topology, communication networks, software with FT-middleware and the development process.

In Chapter 5 EASIS (*Electronic Architecture and System Engineering for Integrated Safety Systems*) project and the approach is briefly introduced. The research work of this dissertation took place during the participation in this project.

As a basis of the design of dependability architecture framework, a dedicated for Integrated Safety System configured faulty hypothesis and fault types was carried out. The result of the analysis is specified in Chapter 6.

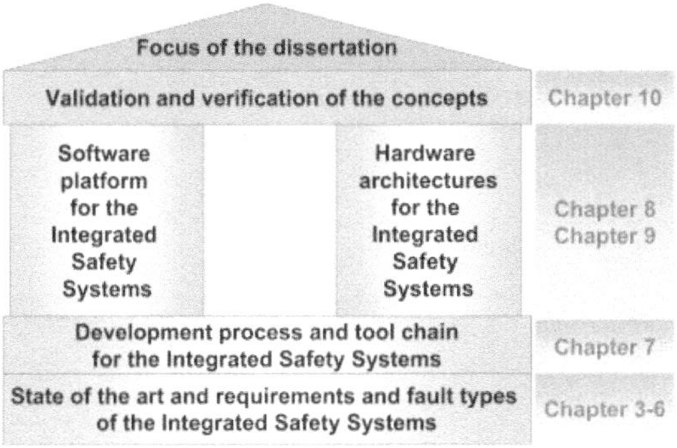

Figure 1-3: Structure of the focus in the dissertation

As depicted in Figure 1-3, emphasis of the dissertation lies in the Chapter 7 to Chapter 9. Chapter 7 introduces ISS Engineering Process for the prototyping and development of Integrated Safety System. The ISS Engineering Process takes the future development trends of safety hardware and software architecture and safety norm into consideration, follows two main principles: virtual front-loading and correct by construction approach.

Triggered by the quantitative requirements from safety norms, hardware architecture framework for the Integrated Safety System of system topologies, ECU hardware architecture and design use-cases are discussed in Chapter 8.

In Chapter 9, dependability software platform with dependability software services for ISS-communication, integration of applications with safety integrity level, fault management and gateway services are specified.

Prototyping and validation of the safety concepts proposed here are illustrated in Chapter 10. Chapter 11 concludes the work with experiences gained in the design and development of the ISS architecture framework and provides an outlook for future work. Last but not least, the appendixes include more implementation details of the validator, mathematic derivation, glossary, list of figures, literatures and author's biography.

2 Basic concepts of dependability

2.1 Definition of dependability

Since there is no common definition of dependability in existing literatures and different norms, in the following, a general definition of dependability is given with regard to the framework of this dissertation.

Dependability: Dependability is defined as the trustworthiness of a computer system, such that reliance can justifiably be placed on the service it delivers. This is a bundle of terms, as illustrated in Figure 2-1.

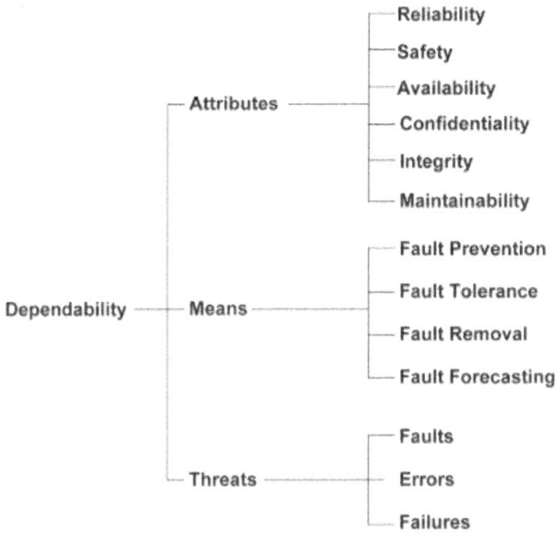

Figure 2-1: The dependability tree [Lap95]

The characteristic attributes of dependability in Figure 2-1 are explained as follows:

- **Availability** is the readiness for correct service – "Uptime" as a percentage.
- **Reliability** is the continuity of correct service, thus how likely it is that the system can complete a mission within a given duration.
- **Safety** is the absence of catastrophic consequences on the user(s) and environment. Two term concerning safety, the safety integrity and safety integrity level are explained as follows:
  - *Safety Integrity* defines the probability of a safety-related system satisfactorily performing the required safety functions under all the stated conditions within a specified period of time.

- *Safety Integrity Level* is used to describe different safety requirements. In the industry norm IEC 61508 [IEC01], SIL is specified as discrete levels of safety relevance, which require different approaches to be realized. Different safety applications involved with the same sub-systems, can have different SILs.

- **Confidentiality** is the absence of unauthorized disclosure of information.
- **Integrity** is the absence of improper system state alterations (Note: security involves malicious faults; confidentiality & integrity).
- **Maintainability** is the ability to undergo repairs and modifications.

The system dependability is a trade-off of the mentioned attributes here, e.g. the different monitoring mechanisms can improve the system safety but not necessarily the availability and maintainability since safest system is the system which doesn't work at all.

The threats of dependability are listed as follows and their relationship to the system is illustrated in Figure 2-2:

- **Fault** defines an abnormal condition that may cause a reduction in, or loss of, the capability of a functional unit to perform a required function. As shown in Figure 2-2, fault is the cause of a system failure.
- **Error** defines a discrepancy between a computed, observed or measured value or condition and the true, specified or theoretically correct value or condition. An example of an error is the occurrence of an incorrect bit caused by an equipment malfunction. Error is a system state that causes failure.
- **Failure** defines the termination of the ability of a system or functional unit to perform a required function. A failure in a sub-system can be a fault for a higher layer system. The latency time from fault to system failure is labeled as t_1, t_2 and t_3.

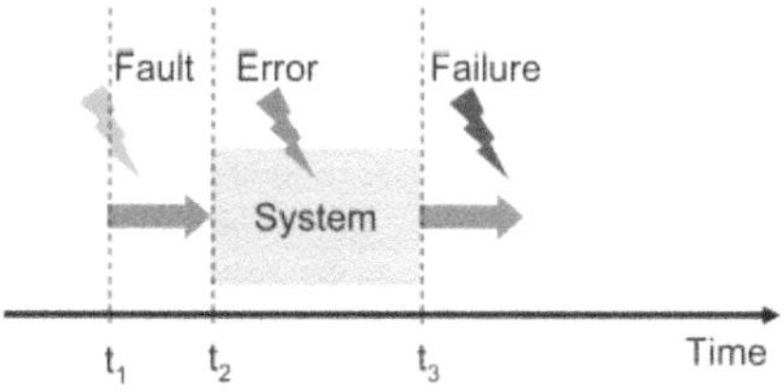

Figure 2-2: Difference between fault, error and failure

In case of failure, different strategies can be performed so as to avoid a hazard to the system, such that it continues as fail-silent, fail-safe or fail-degraded. The definition of such terms and their relationship are explained as follows:

The current semi X-by-Wire systems (e.g. SBC, EPS, E-Gas...) use a mechanical or directly cabled backup system, which exhibits the "fail-safe" and "fail-silent" characteristic.

- **Fail-silent** characterizes the property of a system or functional unit, where in the case of a fault, the output interfaces are disabled in such a way that no further outputs are made. Fail-silent is a special case of the fail-safe property.
- **Fail-safe** defines the property of a system or functional unit, where in case of a fault, the system or functional unit transits to a safe state.

For the future automotive safety systems without mechanical backup (real X-by-Wire systems), it is quite difficult or almost impossible to define a safe state for the vehicles on the road. Therefore, they must have a significantly higher dependability, which should be at least as high as the comparable mechanical system. This means that the vehicles should still offer a higher level of safety to the passenger, even if one or more sub-systems should fail. Thus fail-degraded or fail-operational systems are required.

- **Fail-degraded** defines the property of a system or functional unit, which has the ability to continue with intended degraded operation at its output interfaces, despite the presence of hardware or software faults, the "Limp home" functionality for ECU is an example of this (reduce torque to ensure arrival at home or service station).
- **Fail-operational** describes the property of a system or functional unit, which can continue normal operation at its output interfaces despite the presence of hardware or software faults or errors, for example in the braking system.

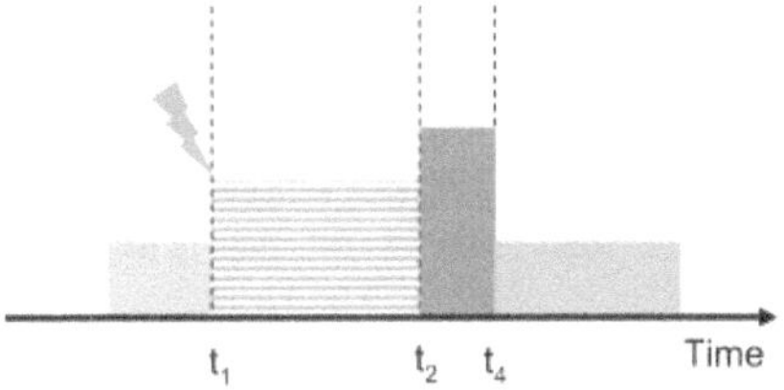

Figure 2-3: Timing requirement of fault tolerance

The timing requirement of fault tolerance is depicted in Figure 2-3. For $t \leq t_1$ the whole system is in a fault free state, $t_1 \leq t < t_2$ is an abnormal system state, in which the fault detection should take place when e.g. the monitoring threshold is exceeded. During the time period $t_2 \leq t < t_4$ a reconfiguration of the system as to the fault treatment should take place, in the time after t_4 ($t \geq t_4$) the whole system is in a reconfigured state and should exhibit a fail-safe fail-degraded or fail-operational behavior. For the sake of fault tolerance, $t_4 - t_1$ should be smaller than $t_3 - t_1$ in Figure 2-2, so that no failure should happen.

A relationship between the different operational states of an ECU, such as failure, normal, fail-operational and fail-safe (including fail-silent and fail-degraded) is illustrated in Figure 2-4.

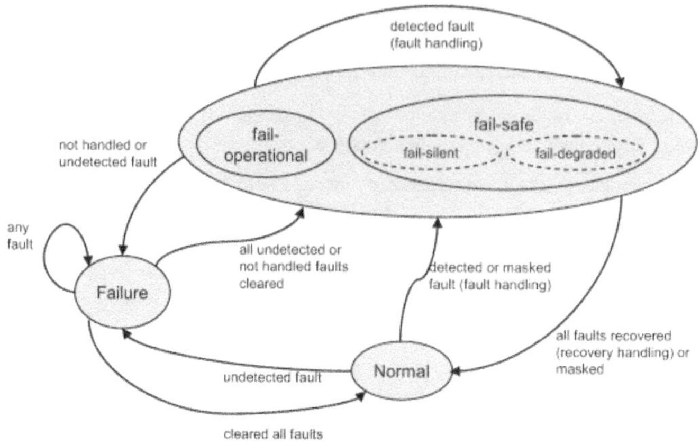

Figure 2-4: Relationship between different operational modes of an ECU

2.2 Introduction to fault tolerance

Various concepts are discussed in different literatures as to how to improve dependability, for examples by means of fault-prevention, fault-recovery, fault-tolerance and fault-forecasting, etc. Fault-tolerance is one of the most methodic of these concepts, which will be explained here in detail [Ech90].

Fault-tolerance describes the ability of a functional unit to continue to perform a required function in the presence of faults or errors. A fault tolerant system has to provide at least three activities:

- Fault diagnosis/detection
- Fault description
- Fault handling including fault isolation (damage assessment) and fault recovery

Fault tolerance removes the effects of the failure. With respect to dependability, this solution is always the best, although it may be the most expensive. In the following sub chapters the three activities/steps will be explained in detail.

2.2.1 Fault diagnosis/detection

One of the most important elements of fault-tolerance is fault diagnosis or fault detection. This describes the ability of a system, to detect and localize a particular abnormal state or state combination in a system or the environment, which has resulted in or will result in an error.

Fault diagnosis serves as the basic and the very first step, in order to avoid a fault in the global system, because measures to restrict and avoid faults in the components and system can only be carried out when unallowable states are detected.

As a conclusion, fault diagnosis can be divided into two steps. The first step is to detect the existence of a fault. The second is to localize the fault, where the error arises. This is important, in order to choose appropriate fault-handling mechanisms for the faulty part(s).

2.2.2 Fault description

In order to implement an efficient approach for fault detection, it is necessary to be aware of the potential faults. An appropriate monitoring mechanism can only be developed with the knowledge of which kinds of faults can occur at which components.

For this reason, it is necessary to define a set of potential faults, by analysis of the system and possible faults, which can be tailored in a top-down process to gain a so-called fault model. The probability of the respective faults and their hazard to the system as a whole will then be analyzed. Finally, the set of faults which can be tolerated, taking different constraints into consideration will be defined.

2.2.2.1 Fault classes

It is necessary to categorize the different faults into appreciate fault classes by analyzing the potential faults in the system. In this sub-chapter a number of basic fault classes will be specified.

There are three basic faults:

- System fault,
- Run time fault
- Other faults

System faults are faults which are embedded at the starting of the system due to conceptual problems. Run time faults, which occur during the running time, are the second most common type of fault. Run time faults can be caused by wearing faults, operating faults, coincident faults and malicious faults.

The third category of faults is other faults, which do not originate from the system nature or operational period, for example production faults or maintenance faults.

Faults can be divided further, according to their location:

- Faults in software
- Faults in hardware
- User operating faults

The final classification of faults is made according to the duration of the fault:

- Permanent faults: continuous until maintenance or other appreciate fault handling method is carried out.
- Temporary faults: last only for short time and disappear after a while.

Table 2-1 shows, as a conclusion, the faults discussed in this sub-chapter and their relationship to fault occasion, fault location and fault duration.

		software	hardware	user	permanent	temporary
system fault	specification faults	X	X		X	
	design fault	X	X		X	
	implementation fault	X			X	
	document fault	X	X		X	
runtime fault	wearing fault		X			
	operation fault			X		X
	interference fault		X			X
	coincident fault		X			X
	deliberate fault			X		X
other faults	production fault	X	X		X	X
	maintenance fault	X	X		X	X

Table 2-1: Fault classes [Ruh04]

Exclusively for distributed electronics, there are two main categories of faults [Cri91]:

- Non-specified behaviors during runtime
- Neglect of input signals and functionalities can be only restored after a restart

The first category can be divided into the following sub-categories:

- Omission fault
- Timing fault
- Response fault
- Byzantine fault

While the second category can be divided into following parts:

- Interval crash
- Partial-amnesia crash with partial history loss
- Amnesia crash with total history loss

In the following, the seven failure categories are specified in detail:

- *Omission faults*: Omission faults describe the situation when a required functionality is not performed within a certain time. A typical example is the communication between two ECUs. ECU1, according to the specification, should respond to ECU2 with a pre-defined signal. Because of interference in the communication channel the message is, however, lost or damaged. ECU2 can recognize faults by monitoring time out.
- *Timing faults*: In timing faults the required services are not delivered in the pre-defined time window. Such failure occurs very often when the communication channel is overloaded or the microprocessor is too busy.
- *Response fault*: when the required service can not be fulfilled as specified, for example faulty response/control of actuator or non-specified state change.
- *Byzantine fault*: One of the most difficult faults to handle is Byzantine fault. In this fault one "identical" message, from the point of view of the sender, will be sent out with different content to several receivers, so that the receivers, which are supposed to process identical information, have different input values. There are two different Byzantine faults: "Benign Byzantine Fault" and "Malicious Byzantine Fault". In a benign Byzantine fault a certain number of receivers can receive an identical, valid value and the rest receive nothing at all. With malicious Byzantine faults, different values will be received by the receivers.

- *Pause crash*: Pause crash is the crash with the lightest influence on the complete system, in which the functionalities of electronics resume work as normal.
- *Partial-amnesia crash with partial history loss*: compared with pause crash, only one part of the state information from before the crash remains for further operation, the rest are lost. Every ECU has some kind of history, that is to say, information regarding past actions, values and states. One example of partial data loss is a restarted ECU, which uses history information to compare computed state with real environment. A concrete example is the loss the last steering actuator angle of an EPS before the next ignition cycle.
- *Amnesia crash with total history loss* is unfortunately the most common case in automobile electronics, in which a continuation from the previously computed results is impossible.

2.2.2.2 Fault propagation behavior

When one component in the system exhibits a faulty behavior, a so called Domino Effect could occur, in which n to n+1 further components break down and this leads to a crash of the entire system. Generally speaking, there are two categories of fault propagation:

- Vertical fault propagation
- Horizontal fault propagation

The basis for analysis of fault propagation is the layered architecture model (for example [ISO7498-1]) of the electronics, a hierarchical architecture from hardware at the very bottom to the software application at the very top, as shown in Figure 2-5.

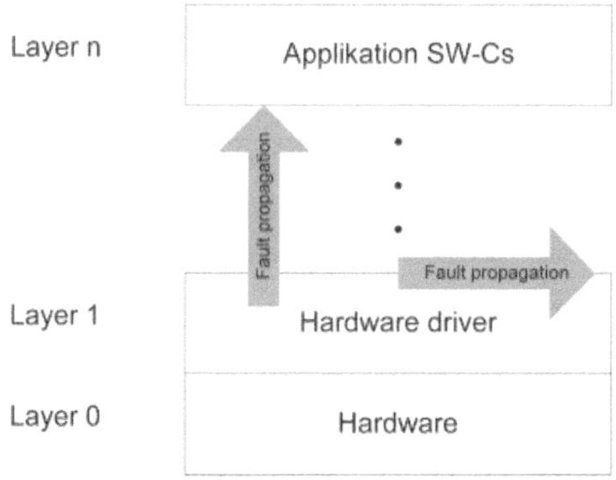

Figure 2-5: Layered architecture model of electronics

Fault propagation, which only results in other faults at the same layer, is called horizontal fault propagation. Fault propagation over the layer border is called vertical propagation.

2.2.2.3 Fault model, fault hypothesis and fault types

A fault model is a model, which describes the structure of a system, including its components and the fault possibilities (=malfunctions) of the individual components, in an at least qualitative way.

In the automobile industry there are various methods for analyzing the fault model:

- Dependability analysis
- Event Tree Analysis (ETA) und Fault Tree Analysis (FTA),
- Failure Mode and Effect Analysis (FMEA),
- Failure Mode, Effect and Diagnosis Analysis (FMEDA)
- Hazard analysis (HA)

A fault hypothesis denotes a claim about the set of faults (independent or related faults) and states an assumption about their types defined by fault number, source and temporal behavior, which are tolerated by a system within a certain time interval. [Kru98]

Detailed discussion concerning fault model analysis and fault types is not the main focus of this dissertation. Results from the fault analysis will, however, be taken into consideration by the design and reflected in the design phase of the ECU architecture and network topology.

2.2.3 Fault treatment

There are two main categories of fault treatment/handling concepts. The first is to exclude and contain the faulty components of a system and to prevent a total crash of the system. The second concept is to repair the faulty components, which implies fault detection, fault identification, fault isolation and fault recovery with a set-up of strategies to recover from the failed situation, perhaps in some degraded manner. A simple example of fault recovery could be the re-computation in an alternative way (indirect) of data when the related sensor fails or a change in the actuator's management strategy when one of the actuators in the system is partially or totally broken.

Fault location, to establish which parts of the system have been affected by the failure, is the basic step of both fault handling concepts. In order to localize the faults in a vehicle, they could be listed and analyzed and then classified on a probabilistic basis to select appropriate fault containment or fault recovery actions during the system requirements specification phase.

In the following two sub-chapters, the two fault handling concepts will be explained in detail.

2.2.3.1 Fault containment

As explained above, fault containment means deactivation and containment of faulty components, however the faults themselves will not be removed. Through fault containment the potential negative impact of faulty components on other components and system functionalities will be suppressed. The faulty components can be determined by means of, deactivated and replaced at the next maintenance. Fault region and fault containment region, as two of the most important terms in fault containment, are defined as follows:

- **Fault region**: A fault region is a set of components which are considered as either faultless or faulty as a whole. Within a fault region fault locations are not distinguished. If a fault region is faulty, its internal behavior is not of interest. Its external behavior, called malfunction, does no longer satisfy the specification. Depending on the "design philosophy", fault regions range from very small to very large. Sometimes chips are taken as fault regions in the hardware. Other examples are software layers or nodes as a whole. However, in both cases the definition of fault region reflects the design strategy of a fault-tolerant system in a comprehensive way.
- **Fault containment region**: For each malfunction a maximum set of affected components can be claimed. The fault must be kept within this set, called fault containment region. Typically this requires one of the following (or a combination of them):
    - Encapsulation reduces or even avoids interaction, to make further fault propagation impossible. Reducing the interaction may affect the exchanged values and/or the timing. Generally, "fault-critical" interaction is replaced by "less-fault-critical" interaction. Typically, this refers to the specification of interfaces.
    - Fault detection makes the existence of a failure explicit. The components can then take extra actions (such as passivation) and/or perform exception handling to some extent. Alternatively, fault detection simply turns a non-detected failure into a detected one (for a larger fault containment region).

For each malfunction, the set of affected components (outside the respective fault region) where the fault can propagate to must be defined. Typically faults can only propagate via the paths of interactions. Once the fault has propagated, the affected components become erroneous themselves, and exhibit a malfunction, which must also be specified.

2.2.3.2 Fault tolerance

As stated in the above, the second category of fault handling is fault removal, where the fault can be removed or fixed during run time, so that the local crash of a function or complete crash of the system can be prevented.

There are three different methods for fault removal:

- Fault correction
- Backwards recovery
- Forwards recovery

Fault correction includes all the existing algorithms and methods, which can correct detected faults without a state change in the applications.

The backwards recovery can bring the faulty components into a consistent, former fault-free state, from which the whole system can continue working.

Compared with backwards recovery, forwards recovery brings the faulty component into a "future" state, which is designed to be fault-free.

3 State of the art: Automotive electronic architectures

After the brief introduction to dependability and fault tolerance above, the state of the art for automotive electronic architectures, including network topology, hardware, software and development process will be introduced in the following chapters, as one of the most important concepts for improving system dependability.

The past few decades have witnessed an exponential increase in the number and sophistication of electronic systems in vehicles. More than 70% of innovations (chips instead of metal) in automotive industry [McK06] are now electronic and software related. Modern vehicles carry more computation power than Apollo spaceship that flew to the moon. Many functions, such as navigation, Infotainment or engine control can not be implemented at all without extensive use of electronics. [Bro03]

In this chapter, firstly state of the art for automotive electronic architectures will be explained, where emphases are placed on the existing hardware and software architecture for automotive ECUs and the technologies, which can be used to implement highly dependable systems. After that the communication systems and development process for automotive electronics especially for safety relevant systems will be also explained.

3.1 Building blocks of automotive electronic architectures

Generally speaking there are two views when considering automotive electronic architectures:

- Hardware architecture: From a general point of view, hardware architecture includes the following issues:
    - ECU local structure of physical components such as processor, actuator, sensor, internal communication, bus connection and hardware methods to guarantee higher dependability, including hardware redundancy and monitoring strategy.
    - In-vehicle network topology, how ECUs are connected via the communication network in a vehicle.
    - Communication system and protocol between the ECUs.
- Software architecture: A description, including specification of a standard software components, a vertical interface between the layers, a horizontal interface in one layer and functionalities of the respective software components.

Before going into detail about the state of the art for automotive electronic architectures and requirements for future automotive safety systems, a few technical terms will be explained.

- **Architecture**: The general definition of architecture means the fundamental organization of a system embodied in its components, their relationships to each other and the environment, and the principles guiding its design and evolution. The term architecture here denotes descriptions of automotive electronic systems on different abstraction levels, for example abstraction from functional analysis or logical analysis.

- **Architecture Framework**: An architectural framework is an aid which can be used for developing a broad range of different architectures, which includes
    - parts of an overall system architecture into which variable components can fit
    - a method for designing an information system in terms of a set of building blocks and for showing how the building blocks fit together
    - a set of references for supporting tools
    - providing a common vocabulary
    - a list of recommended standards
    - compliant products that can be used to implement the building blocks
- **Topology**: The term topology originates from a branch of mathematics, meaning the study of that property of geometric forms. The term network topology means the study of networks in connection with non-metric geometrical properties by investigating the interconnections between branches and nodes of networks.

  Topology here means the interconnection of ECUs (nodes) in a vehicle with the various communication networks and the E/E-architecture of the ECUs, including the hardware architecture and software architecture.

3.2 Requirements for future automotive safety systems

Nowadays automotive electronic systems are implemented in a distributed manner. This means that ECUs, sensors, actuator and networks from different suppliers are installed in various positions in a vehicle, forming a real-time distributed system, while OEMs are responsible for the system design and system integration/test of those ECUs. Theoretically the automotive electronic system can be also implemented in a central manner. The following are reasons for this distributed architecture:

- Traditionally, in the automobile industry a supply chain exists, in which the system supplier provides the OEM with the system components ready for direct installation.
- The distributed system enables an ideal modular oriented assembly system for reasons of packaging and cost.
- The dependability of the whole system can be improved by avoidance of "Single Point of Failure" with the help of the principles of fail-safe or fail-operational. The local fault of a sub-system can be tolerated so that the whole system should still work after the fault. The active redundant systems as in the steering and brake systems can be only implemented in a distributed manner.
- Additionally, a central architecture will introduce problems with cables and packaging, e.g. the number and length and cables will become critical if one ECU is designed to connect with all the other components.

The general requirements for automotive systems and specific requirements for future automotive safety systems will be addressed subsequently.

Real time with short latency period: A constant and pre-known latency period is the most important condition for a closed-loop control system. Here the latency period is defined as the period of time from the time information is produced to the time that the information is consumed, where the maximum latency period will be defined in "ms"-area according to the requirements of the application.

Fault-detection: In a distributed system it is not only important to detect the local component faults, but also faults from other nodes.

Fault-tolerant: Fault-detection alone is not enough for a safety relevant system. After the fault is detected, appropriate treatment should be taken in real-time. The system should tolerate a certain, pre-defined number of faults.

Redundant-determinisms: Active redundancy requires certain determinisms, such that the internal partner should be synchronized to a global clock, so that the result can be provided in a given time window. Here, a fault-tolerant communication channel with defined latency periods is required.

Robust: In the automotive environment, electronic components are facing numerous interferences such as high, varied temperatures, vibration, mechanical impact, dust and EMC-problems (electromagnetic compatibility). Automotive electronics should be immune to these interruptions.

Composability: In the automotive industry, even if the sub-systems from suppliers are tested and validated according to the pre-defined requirement specification and every sub-system is fault free, it is still not guaranteed that the whole system will work. OEMs, as system integrator, are legally responsible for the whole system working for the customer as promised.

In order to improve composability, the interface between the system components and the responsibilities between the sub-suppliers, system suppliers and OEMs should be well defined at the design phase. Mutual dependency should be minimized, in order to improve modularity.

Testability and Certification: With the increasing complexity and multiplied model varieties, both in hardware and software, it is a challenge to test a modern automotive electronic system. In safety electronics, every possible situation should be tested, which amounts to huge efforts in time and cost. Test automation is the solution for a rapid and reproducible way to increase the depth of automotive electronic tests in a shorter development cycle.

A large percent of source codes in modern automotive software are already generated automatically, with the help of model based software engineering and a code generator. Both development process and code generator should be certificated to the appropriate safety requirements for the applications.

Maintenance and diagnosis: One of the basic requirements for automotive ECUs is the ability to diagnose for both calibration and maintenance.

3.3 State of the art: Hardware architecture for automotive electronics

3.3.1 Requirements for hardware of safety electronic systems

According to the safety norm ISO26262-5 hardware development [ISO25], the HW safety requirements shall at least include:

- The HW safety requirements to control external failures of the HW item (i.e. failure occurring outside of the limits of the HW item under consideration), with their relevant attributes (e.g. the functional behavior required for an ECU in case of an external failure, such as an open-circuit in the input of the ECU),
- The HW safety requirements to control internal failures of the HW item, with their relevant attributes (e.g. timing and detection abilities of a watchdog),
- The HW safety requirements of monitoring mechanisms dedicated to indicate the internal or external failures to the driver (e.g. watchdog with driver warning),
- The HW safety requirements to avoid and control systematic failures (e.g. safety critical timings in normal mode of operation).

As explained in Subsection 3.1, automotive electronic hardware architectures will be examined from three different points of views: in-vehicle system topology, ECU hardware architectures and communication systems between ECUs. While system topology defines a master plan of in-vehicle ECUs, the ECU hardware architecture and communication system defines the detailed hardware implementation and the communication pattern between them.

In the following subchapter, state of the art in these three areas will be explained.

3.3.2 In-vehicle system topology

An in-vehicle system topology shows the connection of ECUs in a vehicle, including:

- The number and types of ECUs
- The number and types of communication networks
- How the ECUs are connected with the communication network

In high-end vehicles, such as the Mercedes E-Class (BR211), the system topology for the communication network is highly complex, with more than 50 ECUs, as shown in the 2-D network topology in Figure 3-1. The ECUs in this case are traditionally divided into the following 4 vehicle domains; cabin/body, chassis, powertrain and telematics, where the chassis and powertrain domains are usually combined. The term vehicles domain was introduced for the connection of ECUs with a close functional relationship and connection of ECUs which are spatial closely located.

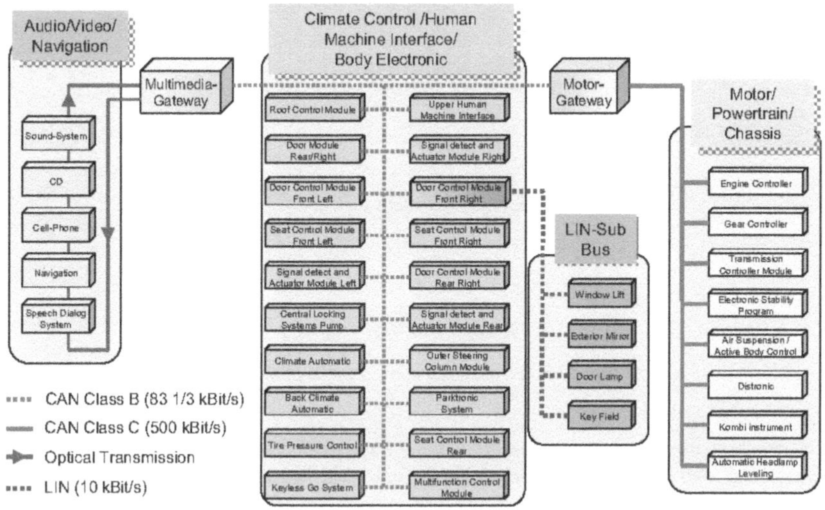

Figure 3-1: ECU topology of Mercedes E-class BR211

Modern in-vehicle network topology is primarily influenced by the concept of distributed function. Each electronic application is implemented on a single ECU as black-box. The ECUs are divided into different vehicle domains, whilst still being connected by various communication networks. With the dramatically increased number of applications in vehicles, the networking for electronic modules (sensors, actuators and ECUs) is slowly approaching the limit of mastery.

Due to reasons of cost and the requirement to reduce complexity, the current trend is towards fewer ECUs and with an increasing number of functions on each of them ([EIA04] and [ADL05]). The degree of inter-connection between ECUs will be intensified. Highly dependable applications also set new requirements for network topology, with regard to determinism and fault-tolerance.

3.3.3 ECU hardware architecture

3.3.3.1 Hardware architecture of general ECU

Generally speaking the ECU hardware can be divided into 5 main parts as shown in Figure 3-2:

- Power supply, 12 V from the in-vehicle battery
- RAM/ROM
- Interface to the communication system
- Sensors/actuators
- CPU

The other peripheries of an ECU include the analogue and digital interfaces to the sensors (A/D Converter ADC) and power semiconductor to drive the actuators and connectors.

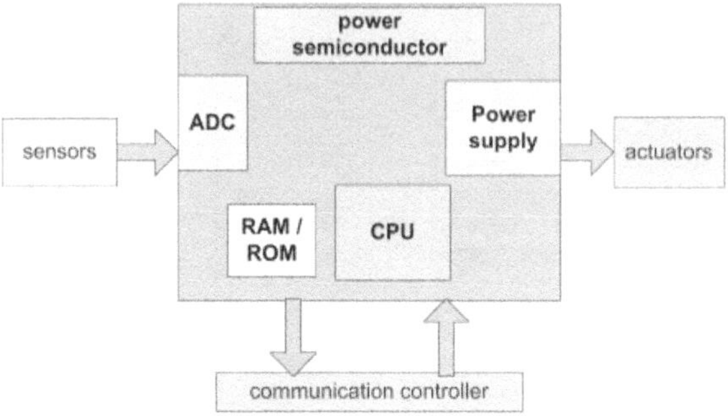

Figure 3-2: Hardware components of an ECU

Compared with domains such as chassis and powertrain, comfort and cabin electronics have fewer requirements with regards to dependability. Thus the ECUs in these domains employ the single-processor concept.

Single-Processor-ECU

The Single-Processor-ECU concept is realized with Central Processor Unit (CPU), which calculates the result of a certain algorithm and communicates with external sensors and actuators via different input and output ports. With the help of the communication driver every ECU can communicate with each other. The redundant sensors are represented from S_1 to S_n in small ellipses and redundant actuators are represented in A_1 to A_n (see Figure 3-2 und Figure 3-3).

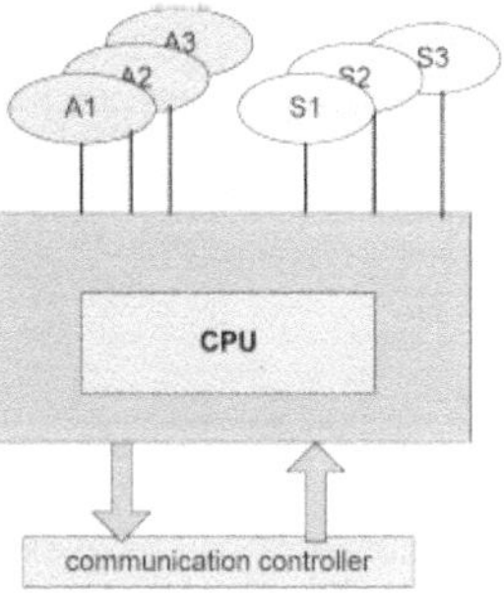

Figure 3-3: Single-Processor ECU

This kind of ECU hardware architecture, also called 1oo1 (1 out of 1 as one-channel-structure), is now most widely used in automotive electronics because of the relatively cheap material cost, considering the enormous number of deployed units. Typically, the dependability of a Single-

Processor ECU can not meet the requirements of today's active/passive safety domain and powertrain applications due to the limited fault detection capability of a single CPU, which does not have reference to a redundant unit.

3.3.3.2 Hardware architecture for safety relevant ECUs

Hardware architecture for safety relevant ECUs is characterized by high dependability, which supports fault tolerance, error detection and error handling. The traditional approach for improving hardware dependability is structural redundancy.

3.3.3.2.1 Dual-processor concept with fail-safe or fail-silent behavior

An ECU with the dual-processor concept can be implemented as fail-silent or fail-safe. The two processors work in parallel. They use the same input values, monitor each other and compare output results. Generally speaking there are two possible methods for realizing fail-silent or fail-safe ECUs, using the dual-processor concept (see Figure 3-4):

- Single processor system with a monitoring processor
- Dual-Core system with two identical processors

In single processor systems with one monitor unit, the monitoring unit, a hardware "watchdog", is usually a cheaper processor or ASIC with lower performance than the main processor. The monitoring unit runs parallel to the main processor and checks its plausibility as one channel structure with diagnosis 1oo1D-system. In case of implausibility the monitoring processor can bring the whole ECU to a safe state, for example by switching off the power supply.

Compared with single processor systems, Dual-Core systems are equipped with two identical processors with the same functionalities, which communicate via internal communication systems to compare results. The two processors communicate with an external ECU by means of a pre-defined exclusive (2oo2-system) or shared communication driver (1oo2-system). Since the same functionality can be processed on both units, a wider range of plausibility checks can be covered and therefore faults can be detected. The practical hardware realization of dual-processor-ECUs can also be different. One possibility is to realize the two processors on a common silicon chip to save costs and packaging, with the disadvantage that they can be disrupted together, while the other method is to implement the two processors in separate packages as two "independent nodes".

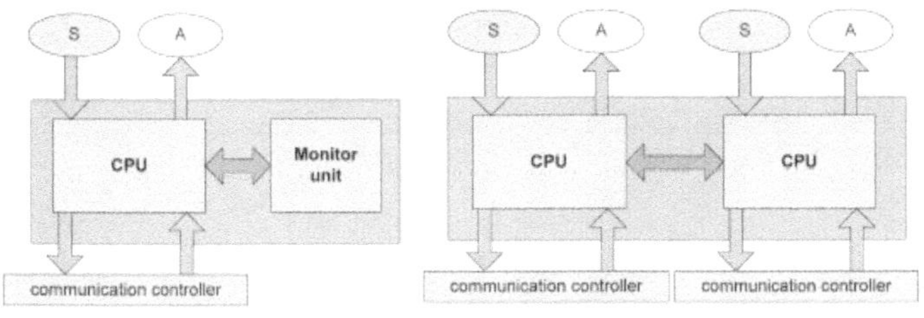

Figure 3-4: Dual-Processor ECU

The monitoring concept with a single-processor requires greater development efforts in comparison with the Dual-Core system, as two different functional entities must be implemented and tested. On the other hand, the single-processor-concept requires a processor with lower performance and thus a lower cost, meaning that production expenses can be reduced. Which of the two concepts will be chosen depends on the number of units to be produced and the safety requirements of the application. An example of the dual-processor-ECU is the modern E-Gas system or MK25 ESP-ECU [Con06] from Continental with fail-safe characteristics.

3.3.3.2.2 Fail-operational ECU (fault tolerant ECU)

The dependability realized by the dual-processor concept is fail-safe, the faulty function can return to a safe state, but can no longer be employed. For applications in X-by-Wire systems or applications in powertrain and chassis, there is no absolute safe state in many cases. Here, a fail-operational system is required, which can tolerate faults and deliver the service in spite of them. The general approach for progressing from a fail-safe system to a fail-operational system is to raise the number of redundant processor units by means of simultaneous, mutual monitoring. There is currently no real fault-tolerant automotive ECU in practice or in serial production. However, from other domains, for example in the aerospace industry, and related literatures there are a number of hardware topologies, which can realize a fault-operational ECU.

For the automotive industry only two of these will be considered, for reasons of complexity and cost:

- Triple-processor (triplex) ECU
- Dual-duplex processor ECU

Triple-Processor (triplex) ECU

The Triple-Processor ECU works in the same way as a dual-processor ECU, but with 3 processing units performing mutual comparison of data and state during data processing, instead of 2. Usually, every processing unit has its own communication controller (2oo3-system) as shown in Figure 3-5.

In order to reduce production costs, only 2 out of the 3 processing units are normally connected to the sensors, actuators and external communication system. The third microprocessor receives the data via the internal communication system from the other two processors and in case of difference during comparison an agreement protocol 2 out 3 will be performed, as shown in Figure 3-6.

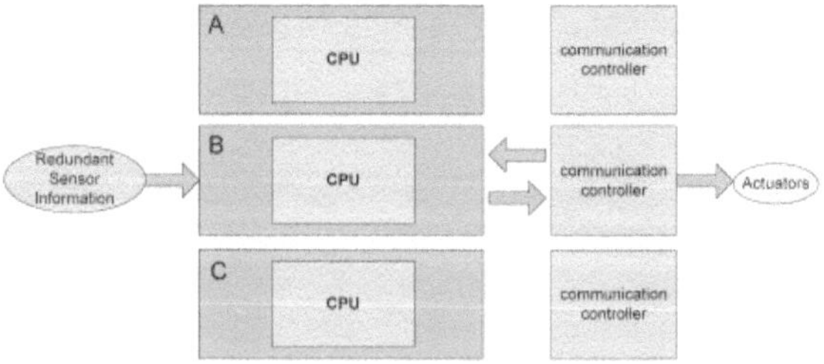

Figure 3-5: Triple-Processor ECU as a full 2oo3-system

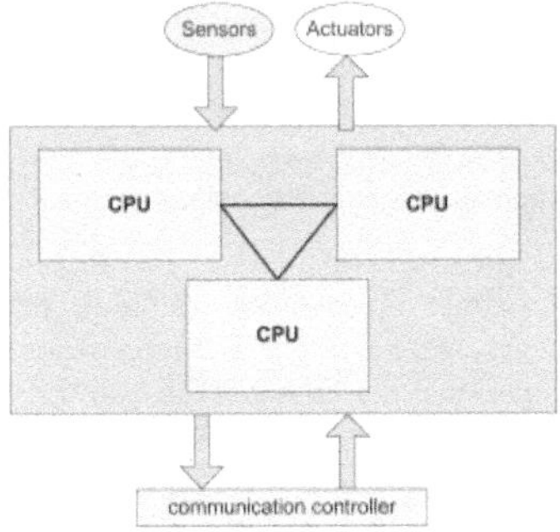

Figure 3-6: Triple-Processor ECU

In the Triple-Processor ECU, even when one processor fails to work, the other two processors can still perform fail-safe behavior, because they can still monitor each other and shut off the system in case of inconsistency (under the situation that I/O and communication channels can be redundantly controlled). The disadvantage of such a system is the greater effort required to compare the input value, results of the processor, state, etc.

In order to simplify the agreement process between the processors, a hierarchical structure, similar to the following system, can be employed.

Dual-Duplex-Processor ECU

Such hierarchical structure can be realized in a Dual-Duplex-Processor ECU. Initially, it appears to be a simple duplicated dual-processor ECU with internal communication between them, which informs the other about the state and data to compare. As shown in Figure 3-7, this initial variant is fully symmetric. By duplication of single processor with a monitoring unit, the dual-duplex architecture can be also built up as the second variant in Figure 3-7.

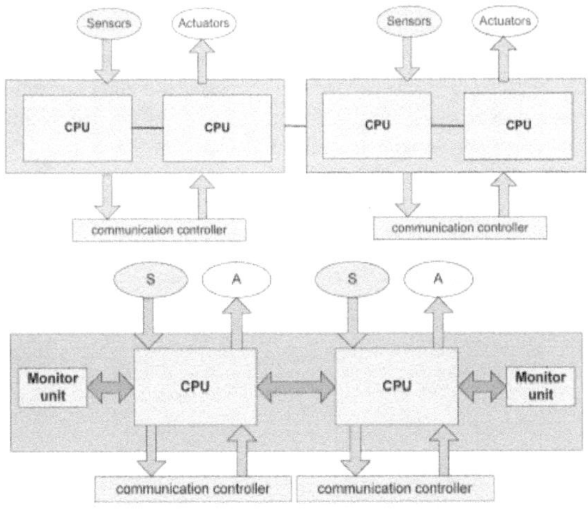

Figure 3-7: Two variant of Dual-Duplex-Processor ECU

3.3.3.2.3 Power supply for safety relevant ECUs

While dealing with safety critical functions and redundant fail-operational systems, it is important to avoid any "Common Cause" of failures. One such typical failure cause is a power supply failure. Although the topic of power supply management is not the focus of the work here, a few important points will be explained.

A fault tolerant power supply, which is essential for fail-operational systems, also requires redundancy within the power supply. Redundant power supply lines must be mutually insulated, in such a way that a failure on one must not propagate to another. Power supply strategies shall ensure fault isolation between the power supply of critical and non-critical parts in order to avoid faulty propagation. The following characteristics of the power supply system are recommended for safety relevant systems:

Power sources disconnection: disconnection of the power sources (storage units and power generation system) must be foreseen.

EMS (Energy Management System): maintains and monitors Electrical Energy Balance and ensures a proper state of charge to the storage systems (batteries)

BMS (Battery Monitoring System): accurate estimation (measurement) of the status of power storage systems.

Fuse Boxes: proper protection shall be carefully studied for each load, with particular regard to the fault isolation principle.

EUS (Energy Uncoupling System): is a component used to uncouple the Power Generator supply bus and the redundant supply buses under certain failure conditions.

3.3.3.3 Conclusion for ECU hardware architecture

State of the art in automotive hardware architecture for non-safety critical application currently remains single-processor unit with simple hardware watchdog and directly coupled actuators and sensors. With the increasing requirements of highly dependable systems, especially for future X-by-Wire applications, appreciate hardware architecture with redundant backup units is applied to increase the fault-tolerance of the system. Existing concepts of fail-operational ECUs, however, still need to be standardized and adapted for automotive applications.

3.3.4 Automotive communication systems

As shown in Figure 3-1, four different communication networks are used in the Mercedes E-Class. However, such heterorganic communication networks are no longer simply the privilege of luxury vehicles; modern, mid-range vehicles also employ several different communication networks.

3.3.4.1 Requirements for automotive communication systems

Before the various communication networks employed in modern vehicles are considered, it is necessary to consider the requirements for automotive communication systems.

- For ECU communication, configurable synchronous and asynchronous data transfers are both required.
- Many applications, especially safety relevant applications with control loops, require a deterministic manner of data transfer with QoS.

  QoS here means to the ability of a network element, such as an application, network node or gateway component, to have some level of assurance that its traffic and service requirements can be satisfied. Enabling QoS requires the cooperation of all network layers from top-to-bottom, as well as every network element from end-to-end. There four key QoS parameters:

  - Latency (end-to-end time for a service, until the time service is delivered to the application, including propagation delay through network as one part)
  - Jitter (delay variation)
  - Bandwidth
  - Packet Loss

  In time-triggered communication networks such as FlexRay, QoS is realized by static communication scheduling. In event-triggered communication systems such as CAN, important signals are sent periodically, but in particular non-periodic diagnosis messages, is sent dynamically. In telematics applications, when concerning mobile communication networks or IP-based networks, QoS is realized dynamically.
- Fault tolerance in communication: for the QoS, some kind of fault tolerant communication is also needed, to ensure the data integrity when one bus fails to transfer data due to certain physical failures.
- Scalable support for redundancy: redundant units such as actuators and sensors must also be coupled with ECUs in the network. Moreover, when redundant processor units are realized by separate packages, as independent nodes, the communication between the redundant units is also provided by the communication network.
- Global time base: a global time base is of vital importance for a time-triggered system, where the local time in each ECU is synchronized with the global time.
- Most safety relevant applications require deterministic behavior, including data transfer. The state of the art for in-vehicle communication by means of CAN-bus is based on

arbitration, in which messages with different priorities compete with each other for bus access. Each node can access the bus coincidentally, thus no peak data transfer rate can be foreseen at the development phase. This means a deterministic data transfer with deterministic bus load can not be guaranteed.

- The bus load in state of the art in-vehicle communication varies from a few Kbit/s to a few hundred Kbit/s. The transfer rate, especially in the domains of powertrain and cabin, is relative low. However, with the increasing complexity of applications and data fusion, the future demand for a higher data transfer rate of up to 10 Mbit/s is a possibility.
- Compared with the communication networks in IT, in-vehicle bus communication has a number of constraints to consider, including isolation, EMC (electromagnetic compatibility) and assembly. Therefore, the communication network should support both optical and electrical transfer mediums.
- The in-vehicle communication system should also provide support for an energy management concept for a long run battery period.

3.3.4.2 State of the art in automotive communication systems

In this section, different in-vehicle communication busses, such as, CAN, ASRB, Byteflight and FlexRay will be briefly introduced. Other in-vehicle communication networks such as LIN (Local Interconnect Network), TTCAN (Time-triggered CAN [Har02]), MOST, IEEE-1394 are not the current focus of the work and will be disregarded at this stage.

- The most widely applied automotive communication network is Controller Area Network (CAN, [ISO11898] Part 1-3). CAN is an event-triggered communication network with a transfer rate from approximately 20 Kbit/s to a maximum of 1 Mbit/s. Bus access for CAN follows the arbitration concept based on the priority of each CAN-message. This implies that when more than one node wants to access the bus medium simultaneously, only the message with the highest priority can be sent. The messages with lower priority will be withdrawn and sent at a later time, thus the time period to access the bus is not deterministic, since a message with low priority can be withdrawn many times without success and other CAN-messages could dominate the bus.
- Automotive Safety Restraints Bus (ASRB) (former Safe-by-Wire) [ASR04] is a sensor bus for the automotive passenger restraint systems. The automotive passenger restraint systems are controlled by a "Restraint Control Module" (RCM), which is connected to peripheral devices.
- First published as "SI-bus" in 1999, the Byteflight bus [Gri00] was developed to connect passive safety systems, such as airbag ECU and crash sensors. As a communication schema, Byteflight uses the FTDMA – Flexible Time Division Multiple Access algorithm. Compared with modern in-vehicle communication systems it provides a much higher bandwidth owing to the fast physical layer (10 Mbit/s) and a star-topology. The total intelligence is based on the so-called Star Net Coupler, which synchronizes the bus and controls bus access. Messages in Byteflight are sent cyclically in Mini-Slotting algorithms. The state of the art in Byteflight consists of plastic optical fiber, which is not suitable for current automotive application (heat-resistance, torsional radius, etc).

- The development of FlexRay began in 1999 shortly after the development of Byteflight and aimed to eliminate the disadvantages of Byteflight. FlexRay, ([Bel01] and [FxR02]) is the state of the art for fault tolerant communication system. It fulfills the requirements for safety relevant applications and future X-by-Wire systems. FlexRay provides two configurable transfer modes, synchronous and asynchronous. Cyclical messages with real time requirements can be transferred in the synchronous mode using a TDMA algorithm, while event triggered messages or diagnostic messages can be transferred in the asynchronous mode using an FTDMA algorithm, similar to Byteflight.

  Due to the flexibility between the dynamic and static segments, FlexRay can be used as purely time-triggered or mixed with an event-triggered communication system, as shown in Figure 3-8.

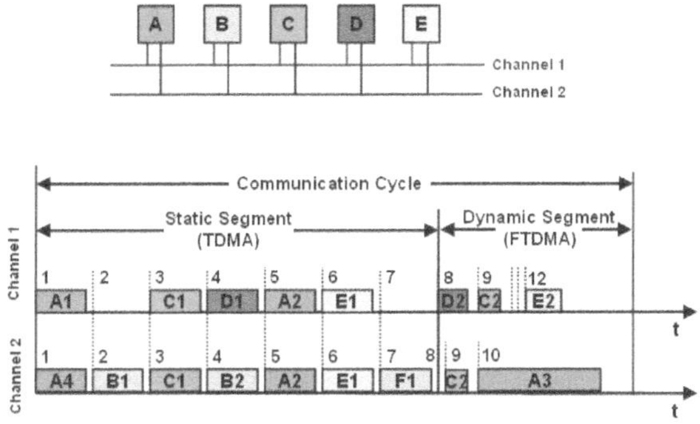

Figure 3-8: FlexRay bus access and communication cycle

One of the most important new features of FlexRay is fault tolerance, not only with the fault tolerant mechanism from the physical layer, for example CRC etc., but also directly from the protocol layer, which enables a scalable redundant system with higher dependability. A cycle counter is implemented in FlexRay which monitors the sequence of messages sent and received. Moreover, end-to-end communication safeguarding and timeout monitoring can also be implemented as extensions to the software.

FlexRay supports different network topologies with the following basic topologies:

- Point-to-point
- Passive bus
- Passive star
- Active star (capable of deactivation of faulty network branch or nodes [Ele04])

More complex topologies as depicted in Figure 3-9 are possible by combining the basic topologies as follows

- Cascading of active stars
- Combining bus- and star-topologies to a hybrid topology
- Combining dual FlexRay channel facilities with the above topologies to: dual busses, dual stars and dual hybrids

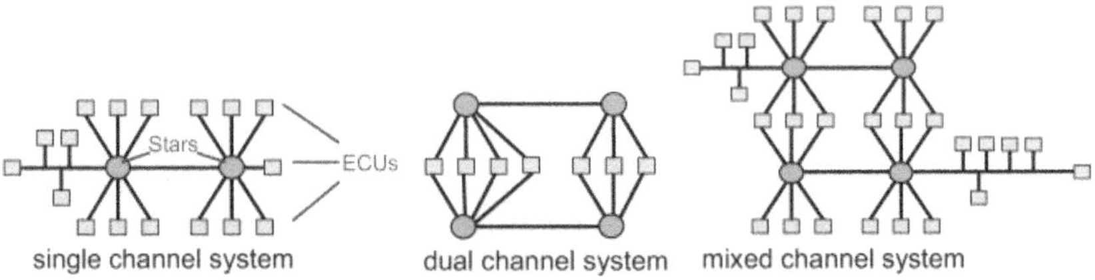

Figure 3-9: FlexRay topologies

The star-structure, compared with the passive star (ground connection as in CAN) is an active node. Physical failures, such as short-circuit, will be detected by the active star node and recovered from by turning off the damaged branches. Failures in the time domain e.g. by nodes jamming the bus as a "babbling idiot" FlexRay can prevent by the optional local or central "Bus Guardian".

The other characteristic of FlexRay (compared with Byteflight) are listed below:

- Redundant channel including bus monitoring
- Distributed time synchronization
- No intelligent Start Coupler as single point of failure
- Maximum of 2 star structures can form a daisy chaining
- Electronic physical layer with transfer rate of max. 10 Mbit/s

3.3.4.3 Conclusion for automotive communication systems

Current in-vehicle communication systems exhibit a high diversity in sub-networks, regarding hardware, transfer velocities and communication protocols. CAN, as the most common in-vehicle communication network, has established its place as a mature technology. In order to gain design correctness and deterministic communication with a higher bandwidth, to master the complexity at system integration, new communication networks, such as Byteflight and FlexRay have been implemented. For future integrated safety applications, FlexRay is considered the most suitable candidate for realizing fault-tolerant communications between distributed safety functions.

3.4 State of the art: Software architecture for automotive electronics

Besides the hardware architecture, the software architecture is one of the most important parts of an automotive electronic system. Currently, a large number of ECU functionalities in vehicles are implemented in software [McK06]; consequently according to a "post-modern" reliability theory [Kop02], software has become a serious problem for the dependability of embedded system.

3.4.1 Requirements for software in safety electronic systems

The following lists some of the general requirements for software in automotive safety electronic systems [SWR07]:

- Modern automotive software systems are highly complex and dynamic, for example every engine ECU can have a specification with more than 3,100 pages, 7,500 variables, 4,900 parameters and 500 system constants. The management of different versions of software modules on different ECU hardware alone has been a problem.
- The applications implemented in current software are mostly distributed systems with functionalities, which are implemented over different domains, making use of the synergy-effect for flow of information over domain borders.
- Compared with normal distributed systems, automotive software requires high real-time in latency, jitter and other QoS parameters. Many safety relevant applications are time triggered, thus requiring deterministic behavior both in the OS and SW-Cs, for example deterministic execution time.
- Many error detection mechanisms are now implemented in software, e.g. CRC-check. Different methods for the distribution of information concerning fault detection in real-time, for example atomic, broadcasted and asymmetric, can be employed.
- Fault-masking in redundant systems requires determinism in software, the results of which can be synchronized at the output ports of the processor units to make a unified decision with the voting-protocol.
- Resource management: Compared with the IT-industry, the embedded system has limited resources in terms of processor performance (run time of processor unit), memory (RAM/ROM), and peripheries. Services for resource management should ensure that higher level safety integrity systems always dispose of resources when resources are shared with different units, across integrity levels.
- Individual monitoring of software components: To implement fault tolerance mechanisms for safety relevant applications integrated on the same HW-platform, individual monitoring mechanisms are necessary. They should be available in an OS or implemented by other means, for example by checking whether the timing of a SW-C is within expected ranges or whether all required SW-Cs for an application started correctly.

- Reliability and robustness of automotive software is another requirement for ECU software, so that fault tolerance mechanisms can handle detected faults locally without propagation to other SW-Components, for example mechanisms for regional fault containment to avoid safe functions being contaminated by faulty functions.
- Composability of automotive software is of vital importance for system integration by OEM and system suppliers. It must show a certain level of flexibility for the configuration, allowing the user (developer) to specify certain fault-tolerance characteristics and provide the requested properties. Different safety relevant applications (and even their sub-systems) may have different safety integrity. It is important to partition the applications to manage the complexity by the integration.
- The test and certification of automotive software according to the state of the art safety norms is also an important issue, especially as the development time for automotive software becomes ever shorter.
- Maintenance and diagnosis is an important part of the software life cycle, especially for integrated applications involving more ECUs and fault-operational behavior. New diagnosis mechanisms should be researched.

3.4.2 ECU software architecture

ECU software can be divided into two general categories:

- application software
- runtime environment

The application software includes parts of the software, which vehicle users (drivers and passengers) can directly feel (for example ABS or seat adjustment).

Compared with the application software, the runtime environment is everything else in ECU software, which provides services required for the application software (for example operating system and basic software modules). The main task of a runtime environment is the abstraction of underlying hardware, so that the application can be implemented more or less independently of the ECU hardware details. This ensures that the application software can be easily transported and integrated into another ECU.

In the following, a few main concepts from the trends of development in ECU software architecture will be explained.

3.4.2.1 Concept of standard software components

The state of the art in automotive software is the standard software components (Standard Core). The software components are modulated and the runtime environment is configured statically to the time of development. The same standard software modules are used in all ECUs by an OEM.

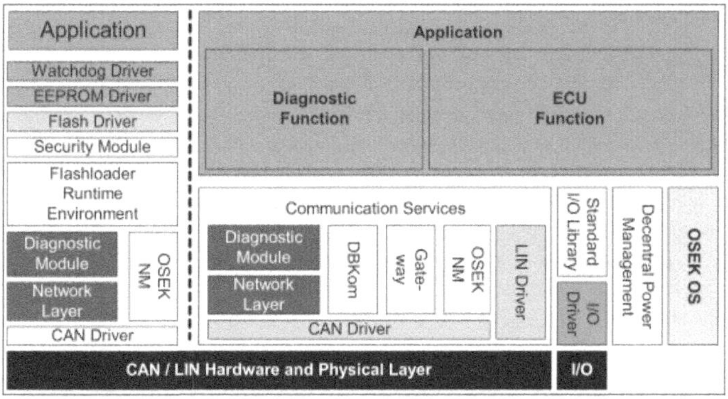

Figure 3-10: Software topology for Mercedes-Benz cabin ECU [Ruh04]

Figure 3-10 shows an example of an automotive software topology, which can be divided into two parts. The part to the left of the dashed line is for flashing; to configure and load the functional software onto the ECU by means of maintenance. The part to the right includes the application software modules and runtime environment, on top of which is the application software with ECU and diagnostic functions. Within these applications are the statically configured standard software modules, which communicate with the hardware through the defined APIs, including the following functionalities:

- Communication
- Diagnosis
- Parameterization of software
- Gateway
- Network management

3.4.2.2 Concepts of software platforms

Another possible concept for the software architecture in a distributed system is the "middleware"-concept or software platform concept. Compared with the standard software concept, middleware provides further abstraction for the underlying communication hardware and processor unit.

In the IT-industry, the current most popular concept in middleware is the "Common Object Request Broker Architecture" (CORBA). All communication mechanisms in CORBA are based on the client-server request-response concept.

A special version of middleware, modified for automotive use, was implemented in the project TITUS [Ruh04]. Here, an extra software layer for the communication was defined. The client-server concept can fulfill the requirements for event-triggered communication in a vehicle quite well. However, many functions in the control loop require a cyclical signal and a highly dependable data transfer considering time, in this case the client-server communication is less suitable.

Concept from EAST-EEA project

The EAST project (Electronics Architecture and Software Technologies) took place between mid 2001 and mid 2004, with the aim of developing a middleware based on modern standard software modules (www.east-eea.net). It defined a standardized open software architecture framework, as shown in Figure 3-11, which enables hardware-independent portability of software modules. At the very bottom of this architecture framework, there is micro-controller with the attributes of CPU power and memory size. Above it, Embedded Basic Software includes components such as operating system, hardware drivers, communication functionality system, diagnosis, hardware drivers and hardware abstraction layer. The Embedded Basic Software can not be seen directly by customers, so it is not relevant for competition in the market. With the help of EAST-middleware, different applications can be mapped onto one ECU.

This middleware offers services for transparent communication inside the classical vehicle domains as well as for cross-domain communication. Subsequently, certain selected domain oriented approaches for the body and powertrain domains and the concept for the implementation of cross-domain services are described in the EAST-project.

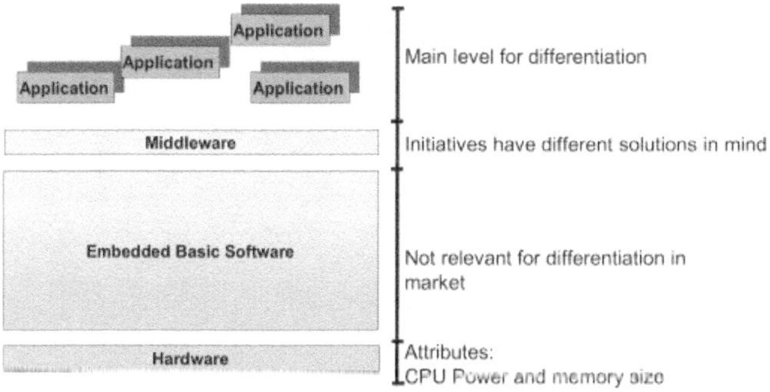

Figure 3-11: EAST architecture framework

Concept of AUTOSAR software platform

The AUTOSAR project (Automotive Open System Architecture, www.autosar.org) [AUT07], [ATO07] is a an alliance of OEMs and tier-1 automotive suppliers, who aim to develop and establish an open industry standard for an automotive E/E architecture, which will serve as the basic infrastructure for the management of functions within both future applications and standard software modules.

A general ECU software architecture, defined in AUTOSAR, is shown in Figure 3-12, which is similar in many features to the architecture framework defined in EAST.

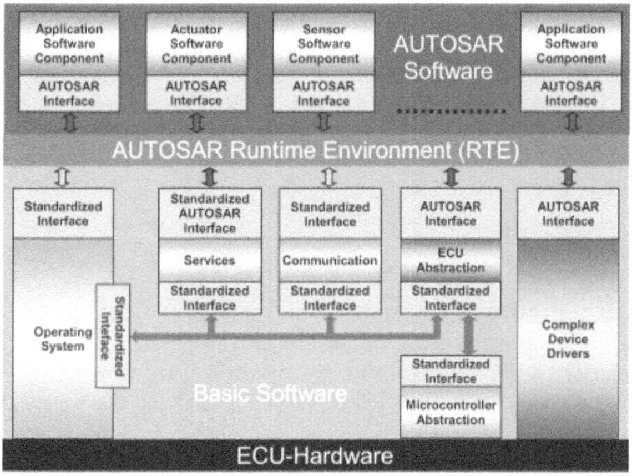

Figure 3-12: AUTOSAR architecture framework [SWA06]

In the AUTOSAR software architecture, SW-Cs of automotive functionality will be encapsulated by the AUTOSAR interface, which ensures the connectivity of software elements surrounding the AUTOSAR runtime environment.

AUTOSAR interfaces should not be considered as pieces of software dedicated to adapting any software module to an AUTOSAR environment. Rather, they are an integral part of each software module running in an AUTOSAR compliant electronic control unit.

In the place of EAST-middleware, an AUTOSAR-middleware called RTE (Runtime Environment) is defined.

- At system design level the AUTOSAR RTE acts as a communication center for inter and intra electronic control unit information exchange between architectural elements of layers connected to the AUTOSAR RTE. All communication between AUTOSAR components has to be compliant with the standardized AUTOSAR interface definitions. All other architectural elements have to provide standardized interfaces that are primarily based on standards which already exist.
- At implementation level the AUTOSAR RTE will be at least reduced to those communication channels which establish communication between SW-components allocated on different electronic control units.
- Different electronic control units will most likely run different AUTOSAR RTE. The AUTOSAR RTE will therefore be assembled with the help of a generator tool by using standard building blocks.

The software layer underlying AUTOSAR RTE is called "Basic Software". It is a standardized software layer, which provides services to the software components and is necessary for the running of the functional part of the software. It does not fulfill any functional job itself and is situated below the AUTOSAR Runtime Environment. The basic software could contain

- NVRAM management
- Transport layers for different communication technologies (e.g. CAN, LIN, etc.)
- Network management
- System services such as diagnostic protocols, etc.

Here it is to note that AUTOSAR partnership defined not only the software architecture. In AUTOSAR, UML2.0, certain standardized notation, methodologies and guidelines were applied for the design specification, where a significant effort is invested in the harmonization and unifying the languages between European OEMs and suppliers. Hence, not only the software architecture but also the developing language can be standardized and harmonized. The efficiency in the cooperation between OEM and suppliers can be improved, e.g. the tracing between requirement specification and design specification can be eased and misunderstanding can be reduced.

3.4.3 Automotive software services

Modern automotive software can already provide a large variety of services. In the following a short overview of the software services is given:

- Communication services: Asynchronous and synchronous communication services between applications running on the same or different ECUs.
- Resource management: The current ECU operating system OSEK OS 2.1 [OSK05] can already provide some resource management considering the processor running time. All tasks are assigned with a certain priority, statically in the development phase. OSEK, as a pre-emptive real-time multitasking operating system can arrange the tasks so that the tasks with higher priorities can receive processor time, without being blocked by the lower tasks.
- Network Management for energy: With the introduction of more and more ECUs with ever higher performance, energy consumption has not only been a problem for high end vehicles. Software services for network management ensure a synchronous start-up and sleep of all ECUs, thus the life cycle of the battery can be longer.
- Function management: The OS also provides mechanisms for terminating functions in other processes, in case of timing violation or demand from tasks with higher priorities.
- Memory management: Some basic access managements are provided by current automotive software, where unauthenticated access of read/write in RAM/ROM is prevented.
- Event Notification: with the flag – set und reset certain bits of state vectors. Task activation and call-back-routine can be called in the operating system.
- Diagnosis of applications and automotive middleware: With defined error symptoms, some run-time errors can be detected by the OS and an alarm will be sent. Further support for fault-tolerant systems, however, is not available.

- Calibration services for automotive software: By means of assembling/flashing or maintenance of ECUs, the software will be calibrated.
- Gateway services: As in the IT-industry, it is sometimes necessary to transport messages and signals from one domain to another. Routing, modification and rearrangement of signals are supported by current ECU software.
- Hardware abstraction: To improve transparency for sensors, actuators and micro-processors in ECUs, hardware abstraction of a certain degree is provide by state of the art ECU software.

3.4.4 Conclusion for automotive software

Automotive ECU software architecture exhibits a clear trend to the structure and abstraction of hardware, where software will be implemented for the distribution of the functionalities and the available communication drivers. This means that the mapping of ECU application software will be more flexible. Compared with standardization and high level abstraction in the IT-industry, which enables an unproblematic distribution and integration of software on any IBM compatible PC, automotive software architecture still has more potential for standardization. For safety relevant applications, especially for applications involving several ECUs across domain borders, there is still no ECU software with fault-tolerant middleware, dependability software services and deterministic behavior in automotive systems, which are all necessary for the integrated safety concept.

The project EAST-EEA, which has already been completed and ongoing projects, such as AUTOSAR, both demonstrate a clear effort on the part of automobile OEMs and suppliers towards a common automotive electronic architecture with emphasis on a middleware concept. Through the standardization of software applications and basic software services, the automotive software quality and test depth can be improved. As state of the art, there are also concepts and prototypes about dependability software services like FT-COM, OSEK-Time, HIS specification about Protected OS, etc. But they are all more or less isolated solutions without an architecture framework.

AUTOSAR, as a significant contribution towards in-vehicle embedded software platform, has the aim to design an architecture framework for all the in-vehicle domains. The original aim of AUTOSAR was to provide a common ECU electronic architecture framework for all automotive domains. However, safety domain is not the topic of AUTOSAR phase 1. The safety relevant applications implemented on ECUs in domains such as chassis and powertrain have their own specific requirements for system dependability and fault-tolerance both in data communication, software and hardware, redundancy management and diagnosis for applications spreading over a few domains, etc.

For Integrated Safety Systems, a special architecture framework with hardware, software, communication systems and a development process, taking into consideration the results of EAST-EEA, HIS and compatibility with common automotive architectures, in particular AUTOSAR, need to be defined.

3.5 State of the art: Development process for automotive electronic systems

In the above chapters, the automotive electronic architecture has been discussed, with emphasis on safety systems mainly from the perspective of hardware and software architectures. Another important topic in the architecture framework is the development process for automotive electronic systems, because electronic architectures can not be separated from the development process. It is application that drives new functions and functions create new requirements for the architecture and innovations in the architecture. The new architectures then drive changes in the development process.

3.5.1 Requirements for the development process of safety electronic systems

Before the technical details of different existing development processes and concepts are explained, a special look at the requirements for the development of electronic safety systems is meaningful.

- For safety relevant electronic systems, in the system analysis phase, safety specification and risk management for safety systems should be carried out. A Safety Requirements Specification (SRS), including a systematic, disciplined and quantifiable process for hazard analysis should be initiated in the engineering phase, for example hazard analysis and risk assessment including engineering risk management aspects of the vehicle, driver, and the environment, can be included in the engineering phase.
- Compared to the hardware development, in which the required safety integrity can be ensured by analyzing the probability of failure, the safety integrity involved in software development is reflected exclusively in higher requirements made on the development processes.
- With the increasing complexity of automotive electronic systems, the complexity of requirement specifications is also increasing explosively. The version and modification management of requirement specifications is a challenging theme. After a supplier provides an OEM with the developed product, the electronic systems will be tested against its specification. How to guarantee a traceable specification for system architecture and already take the tests into consideration in the specification phase is an important topic.
- In the design phase and system implementation phase, another problem is how to correctly transform specifications into design and design into source code.
- Early validation and verification is of vital importance for the development of safety relevant systems to improve the quality and shorten the developing time.

3.5.2 State of the art in development processes

In the state of the art development process there are a number of models, which can be used to structure and define the development stage. In the following, the most general development process, the V-model, will be briefly introduced. The general development processes will be tailored and modified for the development of automotive electronic systems with regard to the special requirements for automotive safety electronic systems.

3.5.2.1 V-Model

The development model, V-Model, is almost standard in the automotive industry. The principle of the V-Model is the description of the relationship between development and test steps, as shown in Figure 3-13. Every development step is referenced to the test step, which requires a lot effort (almost 40%) in the last phase of development. This implies a disadvantage because design failure can be only detected in the later phase of the development, which requires more time and expense.

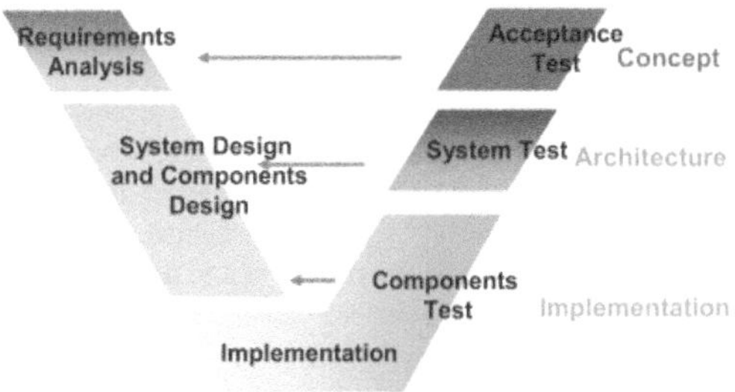

Figure 3-13: The V-Model

3.5.2.2 Development processes in automotive electronic systems

Based on the general engineering process, a specific development process for automotive electronic systems is generated and adapted for the requirements of the automobile industry.

3.5.2.2.1 Model and simulation based development process

The model based development process (sometimes also called simulation based development process) is realized on the basis of the Spiral Model, in which a sub-system will be developed iteratively. It will be tested in a pure software environment with the help of simulation tools and the functionalities of the application to be developed (the "software in the loop" test). The implementation of software on a particular hardware platform will only be initiated after a successful test of the functionalities with the help of models. In many cases, a code-generator will be used in the implementation of models, as shown in Figure 3-14 [Ruh04].

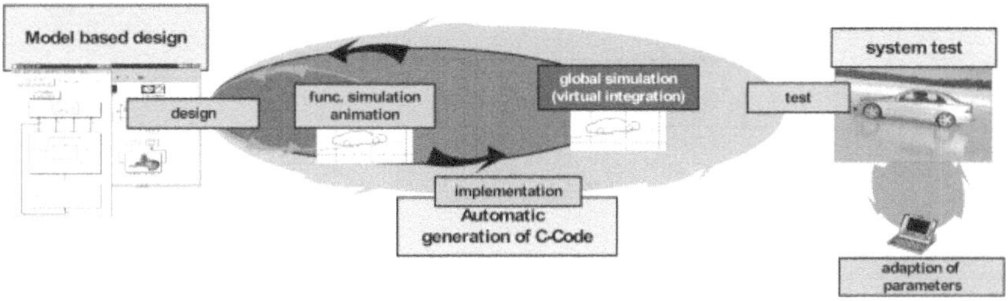

Figure 3-14: Model based development process [Ruh04]

Different development environments/tool chains for the model based development process currently exist. The most widely used simulation tool, especially in automotive electronic systems, is Matlab/Simulink. All these tools or tool chains have one common aspect; the functions/application will be verified using a simulation tool, with less consideration for the underlying hardware. In this way the conception decisions, for example mapping of functions onto the ECU and topology in the later development phase, can be evaluated in the early phase

3.5.2.2.2 Phase oriented development process

The automotive industry currently develops products in several design phases that are mainly driven by the requirements of the mechanical engineering process. This overall design process is often called the lifecycle engineering process. This lifecycle engineering process includes different management milestones (for example engineering start, styling/package freeze, release and start of production). During this process, the vehicle manufacturer produces simulation models and development car versions for different purpose (mules, varying versions of prototypes). The goal of these vehicles is to test the different components of the car. This testing includes (depending on the prototype version) component development, vehicle integration, and performance evaluation. Traditionally, the testing process steps are based on the requirements for the mechanical parts (for example engine, chassis); however, the testing of automotive electronics requires more and more attention.

To allow for vehicle development at the OEM with the most adequate electronic parts, the development process at the supplier has to be adapted to the manufacturer's process. Therefore, the building blocks according to the V model are performed not only once, but several times. The hardware result of each pass of the V cycle is called a sample (for example A-Sample, AB-Sample, B-Sample, C-Sample and D-Sample as depicted in Figure 3-15). These samples are implemented in different development stages of vehicle manufacturing. A-Samples here are defined as functional models with limited suitability for driving with constraints. There is no requirement demanding the existence of safety functions. B-Samples are samples with suitability for driving guaranteeing sufficient operational safety for first trials in the test bed an in the vehicle. The reliability of parts is dimensioned for driving operation on a test area. Safety functions are completely implemented. The C-Samples are ECUs built by production tools with constraints close-to-production. All requirements for function, reliability and interference resistance shall be complied with. The reliability of parts is dimensioned for driving operation on the road. The D-Samples are ECUs built by production tools with production constraints. The electronic control units are completely applicable.

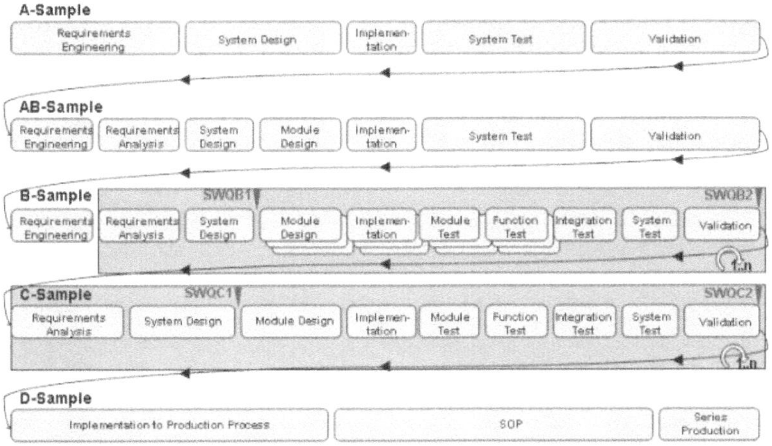

Figure 3-15: Basic V-style process model integrated in lifecycle engineering process

3.5.2.2.3 The development process for fault tolerant electronics according to Hedenetz

For the development of safety relevant systems, Dr. Hedenetz [Hed01] has modified the process described above and adapted it to safety relevant systems (as shown in Figure 3-16).

Hedenetz divided the development phase into global design and local design. In global design the whole system is considered, while in local design the details and relationships between the OEMs and suppliers are considered. In the development process of Hedenetz, simulation process is the coupled as a central part of the development, where the constraints of high dependable systems are taken into consideration.

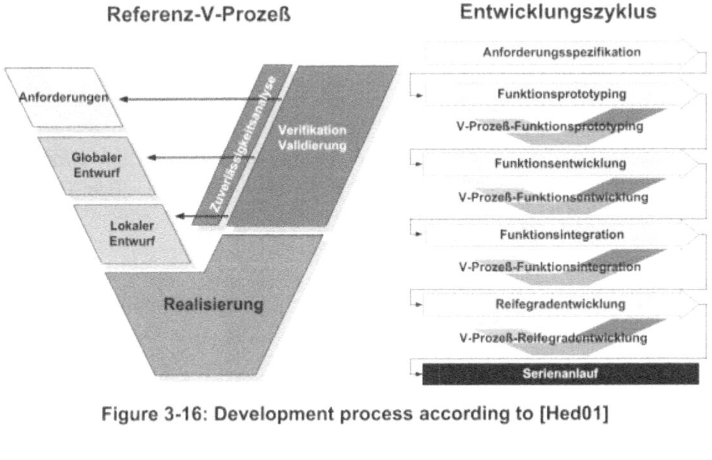

Figure 3-16: Development process according to [Hed01]

3.5.2.2.4 The development process for safety electronics according to Benz

Benz [Ben04] introduced a "Double-V-Model" for the development of automotive safety relevant electronic systems, as shown in Figure 3-17. With the help of two separate but inter-connected V-models, one for the traditional V-model, the other for the safety, the additional requirements in quality and system dependability can be fulfilled. The system safety in the development process must be given more attention compared with the existing process. Safety is a characteristic of the whole system and should be considered in each component, thus both V-Processes should be of equal important as "functional safety co-design", while the term "functional safety" is defined as following: Safety that a vehicle function does not cause any intolerable endangering states resulting from specification-, implementation or realization errors and failures while operation period reasonably foreseeable operational/users errors and reasonably foreseeable misuse.

In the Safety-V, parallel to the functional design, a functional hazard analysis will be performed, the potential impact of the safety function on its environment, including vehicle, driver and other traffic users will be analyzed and an appreciate safety level given. At the system design stage, with the help of the system safety assessment, different system architectures will be evaluated from the point of view of safely integrity. The implementation and verification/validation phase will be also evaluated as proof of fulfillment of the safety requirements.

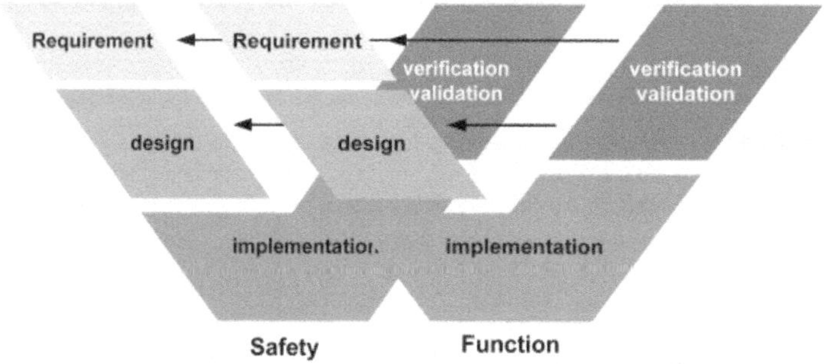

Figure 3-17: "Double-V-Model" according to [Ben04]

3.5.2.3 Safety norms in automotive industry

As have mentioned in Subsection 1.1, one of the most important factors for innovation of automotive electronics is the increasing requirements of customers for more safety and comfort, another import triggering factor is the ever strict requirements of industry norm, regulations and law makers about product liability, safety and emission to environment, etc.

In the following, state of the art of safety norms used in automotive industry will be briefly explained, since they play a significant roll in the appropriate steps in the development process.

3.5.2.3.1 Automotive SPICE

The Automotive Special Interest Group of the Procurement Forum is developing guidance on applying SPICE processes to the assessment of automotive software suppliers based on CMMI/ISO 15504 [ISO15]. A number of European automotive OEMs are already requiring that their software suppliers be subject to SPICE audits, and the intention of Automotive SPICE is to produce a common assessment framework that is acceptable to all the OEMs. Thus, if a supplier has been assessed according to Automotive SPICE and found to have a certain capability level, this is then acceptable to any OEMs, who require a supplier with a capability up to that level.

3.5.2.3.2 Automobile MISRA Guidelines

MISRA Guidelines (ISO TR 15497), better known as "Development Guidelines for Vehicle Based Software" was developed in the early 1990s by a UK-based consortium of automotive companies representing vehicle manufacturers, the supply chain and researchers. The guidelines were written against the background of the development that was taking place on IEC 61508 and acknowledged that, while the track record of the automotive industry in respect of embedded software was good, a recognized industry position on embedded software development for safety-related vehicle control systems was needed.

The notable approaches in the MISRA Guidelines are:

- The emphasis on safety activities lies in the development lifecycle.
- The use of "controllability" to classify hazards. This is a different approach from IEC 61508, although a similar process is used by the aviation industry.
- The use of (safety) integrity levels to classify systems and/or functions according to the level of risk mitigation required.

Although the MISRA Guidelines have been in use worldwide in the 10 years since their publication, a number of issues can be identified:

- There is no formal mapping for IEC 61508, despite many of the underlying principles being present.
- They do not explicitly address recent technology developments such as model-based development and automatic code generation.

3.5.2.3.3 IEC 61508

The international standard IEC 61508 relating to the functional safety of electrical, electronic and programmable electronic systems (E/E/PE) is now, also in the automotive sector, the generally recognized and most state-of-the-art basis for the development of safety-related systems. [IEC01]

Since IEC 61508 origins from the industry sector of process automation, it was developed against the model of "equipment under control", for example an industrial process in a chemical plant; and of the safety functions that need to be applied to that equipment. These safety functions may be part of the control system for that equipment, or they may form a dedicated protection system.

IEC 61508 sets requirements for the reliability of the safety functions, which is known as safety integrity. In practice, safety integrity is classified into discrete Safety Integrity Levels. Hazard analysis is used to identify the safety functions that are needed, and the SIL of the system that performs them can be determined by classifying the severity of the hazards. The SIL indicates the reliability that must be achieved – where this can be calculated, for example, for electronic hardware; or the degree of rigor that must be achieved in the design and implementation measures in order to gain adequate confidence in the system. The definition of SIL in IEC61508 is shown in Table 3-1, where the SILs are specified with two categories of threshold, one for the probability of failure to perform its designed function on demand (dimensionless, for low demand system such as airbag), the other for Probability of one dangerous Failure per Hour (PFH) (for system with high demand rate such as steering/braking system)

Safety Integrity Level (SIL)	Low demand mode of operation (Average probability of failure to perform its designed function on demand)	High demand or continuous mode of operation, PFH (Probability of one dangerous failure per hour)
1	$\geq 10^{-2} - < 10^{-1}$	$\geq 10^{-6} - < 10^{-5}$
2	$\geq 10^{-3} - < 10^{-2}$	$\geq 10^{-7} - < 10^{-6}$
3	$\geq 10^{-4} - < 10^{-3}$	$\geq 10^{-8} - < 10^{-7}$
4	$\geq 10^{-5} - < 10^{-4}$	$\geq 10^{-9} - < 10^{-8}$

Table 3-1: SIL definition in IEC 61508 [IEC01] (Part1 Subsection 7.629, pp. 76)

Another important feature for the classification of SIL is the definition of Diagnostic Coverage (DC) and Safe Failure Fraction (SFF) as shown in Figure 3-18.

Diagnostic Coverage; $DC = \frac{\sum \lambda_{DD}}{\sum \lambda_D}$

Safe Failure Fraction: $SFF = \frac{\sum \lambda_S + \sum \lambda_{DD}}{\sum \lambda_S + \sum \lambda_D} = \frac{\sum \lambda_S + \sum \lambda_D * DC}{\sum \lambda_S + \sum \lambda_D} \frac{\sum \lambda_S + \sum \lambda_{DD}}{\sum \lambda_S + \sum \lambda_{DU} + \sum \lambda_{DD}}$

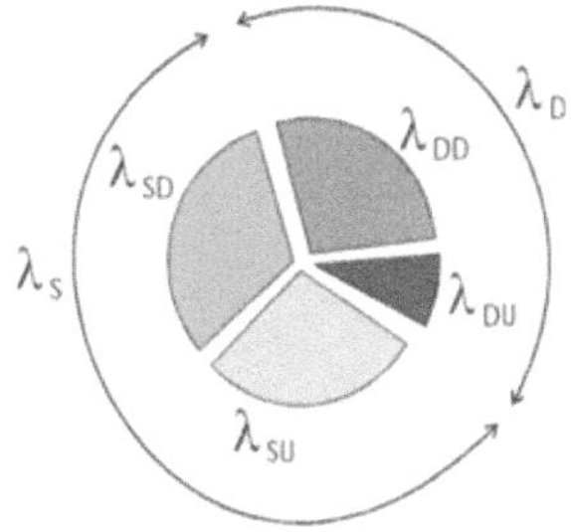

λ_D : total probabilty of dangerous failure
λ_S : total probability of safe failures
λ_{DD} : total probility of dangerous failure that are detected by the dignostic tests
λ_{DU} : total probility of dangerous failure that are undetected by the dignostic tests
λ_{SD} : total probility of safe failure that are detected by the diagnostic tests
λ_{SU} : total probility of safe failure that are undetected by the diagnostic tests

Figure 3-18: Calculation of Safe Failure Fraction according to IEC61508

As shown in Table 3-2, depending on the number of to be tolerated faults, different classification of SIL required a certain threshold of SFF to be reached. Thus SFF is a key parameter for the safety case.

Safe Failure Fraction	Number of faults to be tolerated		
	0	1	2
< 60%	Not considered	SIL1	SIL2
60% - 90%	SIL1	SIL2	SIL3
90% - 99%	SIL2	SIL3	SIL4
>= 99%	SIL3	SIL4	SIL4

Table 3-2: Mapping of SIL to the SFF according to IEC61508

As defined in Figure 3-18, it can be easily proved that $SFF \geq DC$, thus DC is an important factor to improve SFF. That is to say, if we can provide the DC of a system is for example, higher than 90%, the SFF of the system is then higher than 90%, which is require for a SIL2-application.

Safety lifecycle as an important term was also introduced in IEC61508, which means the necessary activities involved in the implementation of safety-related systems, occurring during a period of time starts at the concept phase of a project and finishes when all the systems are no longer available for use [IEC04].

IEC 61508 is already being used as the standard to apply to automotive safety systems. The IEC 61508 "frequently asked questions" [IECFAQ] suggests "automobile indicator lights, anti-lock braking and engine-management systems" as examples of safety-related systems. However there are a number of issues with the direct application of the standard in the automotive domain:

1. The term "functional safety" in IEC 61508 only refers to the safety of the equipment under control and its control system. In many automotive systems, the manufacturer needs to be concerned with the safety of the system due to its control functions, not only the "safety" functions.

2. Many of the techniques and measures recommended in IEC 61508 are very specific to the industrial process control sector.
3. The model of validation in the standard does not align with the automotive industry practices of prototype development and testing.
4. Design and testing of embedded systems are not treated sufficiently.
5. OEM and supplier relations are not described in the IEC 61508 at all.

There are many useful techniques and concepts in IEC 61508, and the importance of this standard can not be underestimated. Nevertheless, further significant work is needed to adapt it to the automotive domain.

3.5.2.3.4 ISO WD 26262 from FAKRA working group

For the constraints of IEC61508 mentioned in Subsection 3.5.2.3.3, there are now two European initiatives to implement automotive version and interpretations of IEC 61508. In Germany it is the DIN FAKRA committee AA13 and in France the BNA 0315B committee. The collaboration between the work forces of FAKRA, BNA and MISRA enables an integration and harmonization of the results into a new ISO standard ISO/WD 26262 (Working Draft, official release not expected before 2009) for the automotive functional safety.

This international standard [ISO24]:

- adopts a customer risk-based approach for the determination of the risks;
- provides a specific automotive analysis method to identify the safety integrity level of each undesirable effect (means for a vehicle function, identification of the consequences of one or some failures, leading or possibly leading to a customer claim, or a damage to the environment, up to significant damage or harm);
- uses automotive safety integrity levels (ASIL) for specifying the target level of safety integrity for the safety concept to be implemented by the E/E safety related systems;
- provides requirements for the whole lifecycle of E/E (engineering, production, operation, maintenance, decommissioning) necessary to achieve the required functional safety for E/E which are linked to the automotive safety integrity levels.

In this subsection the ongoing activities in FAKRA working group and main issues in the initial draft of ISO WD26262 will be briefly explained:

- The draft structure of ISO 26262 covers, as shown in Figure 3-19, the whole automotive life cycle, which consists of the following parts, under the general title "Road Vehicles - Functional Safety":
    - Part 1: Terms and abbreviation [ISO21]
    - Part 2: Management of functional safety [ISO22]
    - Part 3: Concept phase including system definition, hazard analysis and risk evaluation and safety concept [ISO23]
    - Part 4: Product development System [ISO24]
    - Part 5: Product development: Hardware [ISO25]
    - Part 6: Product development: Software [ISO26]
    - Part 7: Production and operation [ISO27]
    - Part 8: Supporting processes [ISO28]

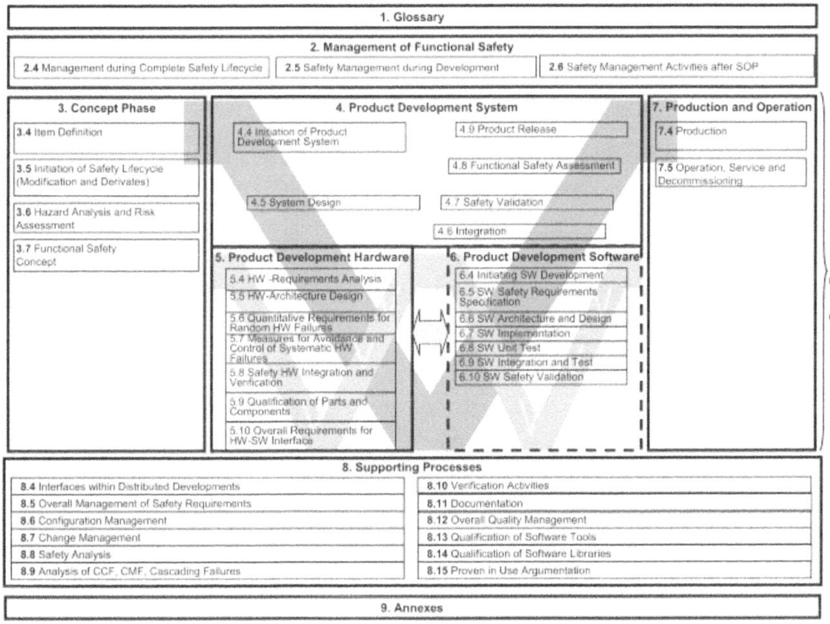

Figure 3-19: Overall framework of the safety lifecycle in ISO26262 [IEC01]

The overall framework of ISO 26262 considering the product safety lifecycle with the detailed activities in each phase is depicted in Figure 3-19, in which the shaded V-areas indicate important relations between various parts.

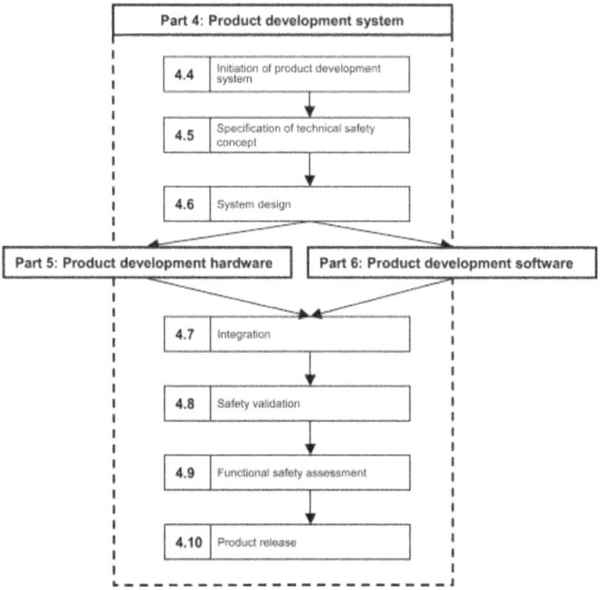

Figure 3-20: Reference phase model for the development of a safety-related item

The reference phase model of ISO26262, which is integrated in ISS Engineering Process in Chapter 7, is shown in Figure 3-20.

- Definition of ASIL: An important aspect in the activities of FAKRA is the derivation of functional safety integrity level, where the controllability of the hazard (controllability by the driver) is considered, where R = Risk is defined as

  $R = S \times f$ where S = Severity of damage; f = Frequency of occurrence

  And the frequency of occurrence f will be calculated by the formula:

  $f = E \times C \times r$ where E = Exposition [-], C = Controllability [-] and r = Failure Rate

  According to the risk level, the safety integrity level scale devised in ISO26262 has four levels {ASIL A, ASIL B, ASIL C, ASIL D}.

Figure 3-21 shows a preliminary mapping between the IEC 61508 and ISO/WD 26262 scales. The mapping of ASIL-C can be chosen to the SIL-2 and SIL-3, while ASIL-D is only mapped to SIL-3 with PFH of 10^{-8}/h to 10^{-7}/h, that is to say, SIL-3 is further classified into ASIL-C and ASIL-D in automotive industry, while ASIL-D application should have a PFH < 10^{-8}/h, so that it can fulfill the ISO-requirement anyway. In the following of this dissertation, ASIL will be used instead of SIL.

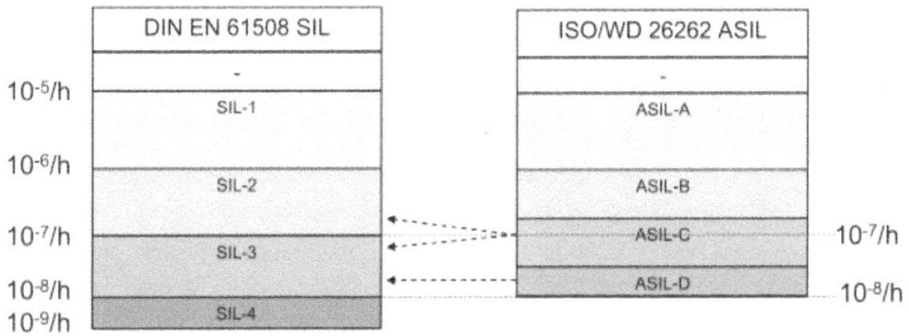

Figure 3-21: Preliminary mapping of safety scales between IEC 61508 and ISO 26262

- Safety Life Cycle: The safety life cycle in IEC61508 is not designed for vehicle as "consumer goods" but originated from process and automation industry, e.g. validation after installation etc. In consequence typical automotive test processes before SOP are not considered as for example HiL-tests, fleet tests and user oriented on-road tests. The activities of past SOP such as operation, maintenance and decommissioning are missing as well. The draft structure of ISO26262 is grouped into three large phases:
  - Concept phase
  - Vehicle development and production
  - Post SOP phase

Thus the whole safety life cycle of vehicles is considered.

3.5.3 Conclusion for development process

The current development processes for automotive electronics tend to correspond to an iteration and adaptation of V-model. The simulation and model based development process reflects the trend towards increasing requirements for an early design validation, in order to avoid additional cost and time delays in case of system modification in late phases.

The development of safety relevant systems requires a dependability-justified process. Concepts of Hedenetz and Benz reflect this trend, in which safety analysis during the development of safety electronics with fault-tolerant mechanisms and consideration of safety aspects through the product life cycle, are integrated into the V-model. According to the state of the art in the safety norms, especially the future safety norms ISO/WD26262, dedicated safety steps should be carried out according to the automotive safety levels.

The trend for automotive safety electronics is heading towards freely distributed applications with high dependability, using standardized, configurable, fault-tolerant software and hardware architecture. These new requirements of integrated safety electronics cannot be fully supported by the existing solutions of development process.

4 Assessment of state of the art approaches and conclusion for challenges

In the chapters above, a general overview of the state of the art for automotive electronic architectures has been given. Current concepts and solutions for hardware architecture, software architecture and development processes of automotive electronics, taking dependability requirements into consideration, have also been introduced. In this chapter the assessment of different approaches from state of the art against new requirement and challenges of future automotive safety electronics (the so-called gap-analysis or delta-analysis) will be concluded.

- *Challenges in higher dependability despite of increasing system complexity*

  There are already many automotive safety relevant applications based on distributed ECUs. In these applications, sub-functions or signals, which have previously been considered non-safety relevant, could become safety relevant for ISS. For example, the seat adjustment in the Pre-Safe system, the trail light control from the cabin for the brake assistant function and telematics function for calling medical rescue in the after-crash system can become also safety relevant in ISS.

  Compared with state of the art safety relevant systems in vehicles, the Integrated Safety System is characterized by an even higher complexity, in which the implementation focus shifts from traditional mechanics and hardware to SW intensive systems. The timing of the cooperation between the distributed applications, which is relatively easy on one ECU, is undertaken by means of a communication network in ISS beyond domain borders. Thus an automotive ECU topology with fault tolerant communication, which supports the requirements of ISS in a cost effective manner is needed.

  In future ISS-applications, X-by-Wire (steering, braking and engine control applications) without mechanical backup will be included. The higher dependability requirements for this (the ISS systems should be at least as reliable as traditional mechanical system, thus from fail-silent/fail-safe to fail-operational) can be only implemented with fault-tolerance characteristics.

  As a summary, current E/E-architectures are unable to handle the required system complexity with higher safety requirements introduced by the ISS.

- *Challenges in the hardware architecture of automotive safety electronics*

  With regard to ECU hardware architecture, a more and more complex in-vehicle network topology with different communication sub-networks is evolving, operating with different velocities and protocols, interconnecting ECUs and different automotive domains via gateways. With the introduction of new applications, there is a clear trend towards synergy effects by means of data fusion and more and more data flowing over domain borders. Safety relevant applications over domain borders, however, require deterministic behavior. QoS in communication is needed in both data integrity and timing. FlexRay, as the first truly fault-tolerant time triggered communication bus, will most probably be chosen for safety relevant communication.

Owing to the rapidly increasing complexity of automotive electronic systems, the existing solutions in automotive electronic architectures have almost reached their limits, especially with regard to safety relevant functions and different degrees of fault tolerance (fail-silent, fail-safe, fail-degraded and fail-operational).

Concepts for fail-operational ECUs with triplex or dual-duplex processors have already been given. However, these concepts must be adapted to the specific requirements for automotive applications, for example the optimization of complexity. The concepts for fault-tolerant ECUs should also be standardized and unitized with regard to the different requirements of safety integrity levels, to reduce development time and costs.

- *Challenges in the software architecture of automotive safety electronics*

  In the ECU software architecture, there is a clear trend towards standardization and implementation of standard underlying software architecture (middleware), for the transparency of hardware implementation and application development, which will certainly improve the testability and dependability of ECU-software.

  In the development of ECU software for safety relevant applications, the same trend is also apparent. Although software concepts concerning fault-tolerant mechanisms in ECUs already exist, there is no real fault-tolerant middleware architecture with standard dependability software modules to support common fault tolerance services (HW-transparent fault tolerance). As well as the basic standard software modules, some standardized software modules to provide common safety services shall also be defined. The different safety integrity level of distributed functions, implemented on several ECUs, requires configurable and fault tolerant software architectures to guarantee sufficient dependability while managing the system complexity and development effort. Fault model, especially in ECU software and containment region in software architectures are further important issues.

- *Challenges in the development process of automotive electronics of Integrated Safety System*

  As previously mentioned, in the state of the art, traditional development processes and current simulation-based development processes have reached their limits in terms of managing the complexity and providing a full guarantee of the dependability requirement.

  The new requirements in ISS with regard to distributed development and system integration should also be embedded or reflected during the whole development process, in which the safety requirements can be traced and validated. A certain safety integrity level should be supported during the system integration and validation phases.

  In a development process, which fulfill both future safety norms (ISO/WD26262) and manages the challenges of ISS, standardized dependability services in both ECU software and hardware architectures should be reflected during the life cycle of the Integrated Safety System.

5 Introduction to EASIS project and EASIS approaches

The EU project EASIS (*Electronic Architecture and System Engineering for Integrated Safety Systems*) (see www.easis.org) was initiated in the year 2004 and finished in the first quarter of 2007. In the form of an industry partnership, EASIS is composed of 7 European OEMs, 8 system suppliers, 4 tool producers and other research institutes. EASIS is pursuing the goal of providing enabling technologies for the introduction of integrated safety systems, by developing an open and standardized dependable in-vehicle fault tolerant electronic architecture framework (reference model in both dependable software and hardware platforms) and a standardized systems engineering approach for integrated safety systems with a dependable E/E-platform.

This modular E/E architecture consists of two main branches: i) Software platform; and ii) Hardware platform. The software branch defines a common platform for software-based functionality, upon which future applications can be built, and the hardware branch defines a cost-effective on-board electronic hardware infrastructure. In this project, requirements and needs concerning ISS have been collected which constitute the basis for the definition activities for the two branches mentioned above [TRA06].

The EASIS work group on the software architecture designed a standard software platform for the execution of ISS, satisfying both ISS and external requirements, such as standards and hardware architecture. It will provide a basis for future in-vehicle electronic systems and include principles for software topology, an interface between hardware and software, basic fault tolerance and diagnosis mechanisms and inter- and intra-domain gateway services. The main composition of layers and functional areas of the software platform will be explained in the Subsection 9.2 in details.

A hardware architecture, which supports the high ISS demands regarding dependability, computational power, high speed, accurate information exchange and the software layers, will also be produced. This architecture must be scalable for standardized usage in both safety and non-safety applications, able to adapt to different domains and vehicle classes, capable of handling various sensors and actuators, have well-defined, standardized interfaces, support fault tolerance, error detection and error handling and have optimized costs and reliability.

Achieving system dependability will be much challenging for integrated safety systems than for traditional safety-critical automotive subsystems due to higher content of safety-critical software, higher complexity, and higher interaction of subsystems from different suppliers. The state of the art of automotive dependability methodology has to improve significantly to cope with these challenges. EASIS will provide guidelines for all major dependability-related activities. These guidelines are structured with the help of a simple dependability activity framework as shown in Figure 5-1.

The introduction of ISS will increase the networking needs not only for signals. Different ISS functions may work in the car thus making functional coordination and arbitration necessary. Connecting the vehicle to surrounding information infrastructures may influence the hard real-time chassis or powertrain control systems even more. All in all, the well understood basic-function-control-loop is transformed into an ISS-control-loop, see Figure 5-2. ISS-control-loop design requires new development processes and supporting tool chains, which the work on processes and tools will provide. A software specific systems engineering process (EASIS Engineering Process, EEP) suitable for the automotive industry, will be defined, integrating the results of system dependability, and a tool chain recommendation produced.

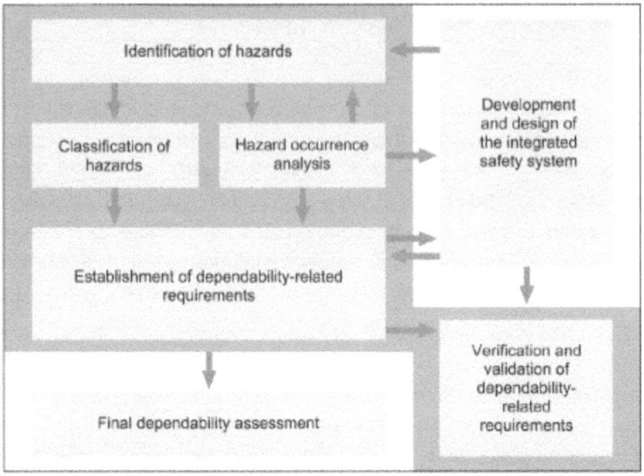

Figure 5-1: Framework of dependability activities

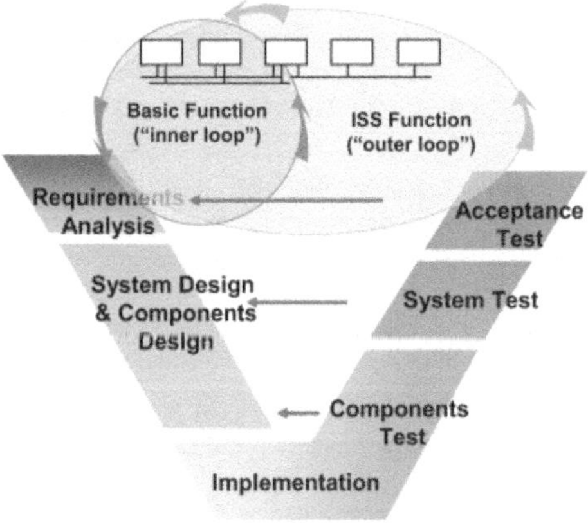

Figure 5-2: V-model indicating inner and outer loop

In the following sections the different work packages of EASIS-project will be briefly explained.

WP1: Software Platform

The primary drivers during development of the software platform are the main dependability attributes of safety, reliability, availability and security. Based on this, the work on dependability in the software platform is performed addressing the following three aspects: i) detection and handling of faulty and erroneous states, i.e., fault tolerance aspects. ii) management of information concerning the state of the system, e.g., fault codes and internal fault information; and iii) consistency and integrity of data over the telematics link. Based on these three aspects dependability services are provided by the EASIS software platform.

WP2: Hardware Platform

The work on the hardware architecture focused on the following four aspects: i) logic system architecture composed of multiple ECUs; ii) communication networks connecting the ECUs; iii) internal ECU hardware architecture; and iv) power supply network. Based on these aspects, topics as Gateway, system topology, building blocks Network technologies, ECU hardware architecture and power supply were covered.

The hardware platform and software platform have been defined to closely integrate with each other, such that the services provided to the application components of ISS will have access to a wide selection of services.

WP3: System Dependability

The goal of the work package "system dependability" is to support the system engineering of integrated safety systems by providing guidelines for all major dependability-related activities in system development. The EASIS guidelines are structured with the help of a simple framework of dependability-related activities in system development and focus on the following three aspects: i) a comprehensive approach for the establishment of dependability requirements based on hazard identification, hazard classification and hazard occurrence analysis; ii) verification and validation of complex distributed control systems; and iii) means for demonstrating that safety was sufficiently considered in system development (safety case construction).

WP4: Processes and Tools

The work on processes and tools is focused on the following three aspects: i) systems engineering processes for ISS and their functional software components; ii) tool chains supporting the engineering processes; and iii) certification and test activities. Based on these aspects, an analysis of requirements (collected within EASIS and from other projects) was conducted that revealed new challenges for the development process that arise with the introduction of ISS: The shift from traditional hardware to software intensive systems, leads to a composition of safety functions within the vehicle that may be distributed across different ECUs. In case of diverging or conflicting signals or requests to actuators, the coordination / arbitration of different ISS functions will be a challenge, while requirements of system dependability will increase.

WP5: Validation

Key concepts and methodologies were validated in this work-package in order to show the applicability of these approaches to provide solutions to the problems and requirements identified concerning the development of integrated safety systems. For this purpose, two main validator set ups will be created:

- The main principles defined in the software platform and the hardware platform will be integrated into a validator resembling an automotive electronic system, including a telematics gateway, automotive sensors and actuators and several ECUs, all connected in a FlexRay network. In this validator, the feasibility and validity of the main principles of the software and hardware platforms are demonstrated.
- A safe speed function for use in commercial vehicles, automatically reducing truck speed to the maximum safe level for the road, by limiting engine torque and, if necessary, applying brake torque, will be created.

Concepts and designs in this dissertation were developed during my work as work-task leader in WP1, WP2, WP4 and WP5 of EASIS project, while the focus of the work lies in the work-tasks of software architecture, dependability software services, EASIS engineering process, tool chains and validation of concepts and methodologies.

6 Fault type and fault hypothesis for ISS

As written in Subsection 2.2.2.3, although fault model and faulty types analysis is not the focus this dissertation, the dependability requirements of a system are determined by its fault type definition and fault hypothesis. As mentioned in Subsection 1.1 and Chapter 4, future automotive Integrated Safety System, are facing quite a few new challenges. Thus we need to define a complete fault types and fault hypothesis for the ISS, which are mapped the development concepts of ISS in hardware, software design and development process.

The following subsections start with the introduction and motivation for fault hypothesis, the HW and SW.

6.1 Fault hypothesis – a major design step

A fault hypothesis states the assumption about the types and number of errors that a fault tolerant system must tolerate. It can be considered as a set of errors as $E_{FH} = \{e \mid e\ is\ a\ single\ error\ to\ be\ tolerated\}$.

And with f_{system} denoting the functionality of a system S depending on a fault e.

$$f_{system}(S,e) := \begin{cases} error-free \\ erroneous \end{cases}$$

A fault tolerant system S is than defined as:

$$\begin{aligned} &\forall e \in E_{FH} \\ &f_{system}(S,e) = error-free \\ &\Rightarrow the\ system\ S\ is\ fault\ tolerant \end{aligned}$$

If the number of faults occurring in the real world exceeds the number assumed in the fault hypothesis, the assumption coverage will be low. Otherwise if the number of faults occurring in the real world is smaller than the number assumed in the fault hypothesis, the assumption coverage will be high. Even the best and most detailed fault model is useless if its fault hypothesis coverage is low.

A fault hypothesis can be used as a major design parameter for the system design and validation:

- Motivation for the design of fault tolerance algorithms: Without a precise fault hypothesis it is not known which fault classes must be addressed during the system design. Each introduced fault tolerance of self-checking mechanism has to be tailored according to the statements which are made in the fault hypothesis.
- Foundation for the System validation: As we know, a dependable system could not necessarily tolerate the faults, which it is not designed to tolerate, but it should be guaranteed that faults considered in the design phase should be able to be tolerated always. Thus the fault hypothesis is the key element for all kinds of validation techniques. For instances, the fault hypothesis can work as the description of the setup for fault injection experiments in an experimental validation. Different faults can result in the same or alike system errors. If the error can be tolerated by the system, it is most possible that the faults, which result in this error, can be also tolerated by the system.

It is to note that an exact fault hypothesis regarding number, source and temporal behavior with mathematic model are often not possible or realistic, esp. for the future software intensive embedded systems, taking the time constraints of the development into consideration.

In the following subsections, a general approach for the definition of a fault hypothesis suggested in will be used as a guideline [Kru98]:

- Partitioning o the system into SERs: A Single Error Region (SER) is a component of the system. It is assumed that faults will appear independent to each other and do not propagate to other SERs. The system can be partitioned into SERs by specifying these components by functionality plus the corresponding error.
- Failure semantic: Based on the previous step the failure semantic of SERs is defined. The failure semantic describes the behavior of a SER, in case of an error leading to a failure.
- Cause of failures: This step determines the maximum number of concurrently erroneous SERs and the SER internal causes. Thus, this step determines the actual requirements for the fault tolerance and self-checking mechanism of a SER.

6.2 Fault hypothesis of hardware in ISS

Starting from the fault hypothesis on a single ECU hardware structure, the fault hypothesis of communication networks, gateways and the other main hardware parts of integrated safety system will be defined in this subsection [DFP06].

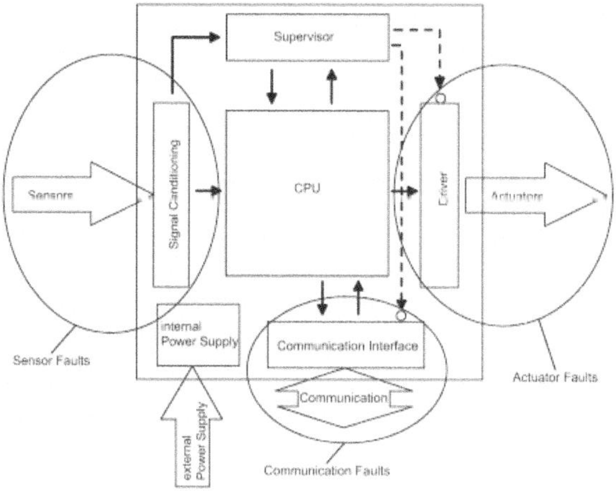

Figure 6-1: Simplified fault hypothesis of a standard ECU

As depicted in Figure 6-1, a very simplified fault hypothesis of an ECU can be divided into five categories:

1. Faults in the microprocessor/supervisor on the ECU
2. Faults by sensor coupling

3. Faults by actuator coupling
4. Faults by the power supply
5. Faults by communications

The 1st category of faults is listed as following:

Microprocessor core and supervisor:

1. Calculation errors (e.g. HW error, logic error)
2. Value errors (e.g. HW error, memory/register corruption, EMI, SEU, etc.)
3. Program flow errors (e.g. HW error)
4. Interrupt errors (sequence, frequency, delay, disregarding, etc.)
5. Algorithmic errors (= Compiler/Logic Synthesizer errors / design faults)
6. Timing errors (error group V is not fully orthogonal to the other error groups)
7. Synchronization lost between CPU and supervisor
8. Supervisor and CPU are getting different information from the extern environment
9. CPU and supervisor use different, but both valid, rules to judge the control.

RAM/ROM:

10. Errors in the RAM/ROM (memory cell defective)
11. Faulty RAM/ROM access (wrong memory address)
12. Faulty memory mapping (=Compiler or linker errors / design faults)
13. Memory overflow

I/O-Interface:

14. Interface errors (errors in ADC/digital IO/ ...)

The 2nd category of faults concerning sensor coupling:

1. The sensor delivers no value or an error signal.
2. The read value of the sensor is wrong.
3. The sensor delivers a value with a wrong timing.

The 3rd category of faults concerns actuator coupling:

1. The actuator is not driven.
2. The actuator is permanently driven (without controller command).
3. The actuator is not driven at the right time.
4. The actuator is not driven with the correct performance.

The 4th category of faults concerning power supply can be divided into:

1. Over voltage
2. Under voltage
3. Short circuit
4. Over current (due to erroneously activated actuators, defective components, etc.)
5. Leakage current too high
6. Brown out (slow decrease of the supply voltage below the minimum limit)
7. Start-up timing
8. Shutdown timing

The 5th category of faults in the communication system between different nodes can be described as message/signal faults between the nodes. Faults in data exchange between the software-components within one node (intra-ECU communication) will be considered in the context of software faults.

Faults in lower levels, e.g. physical faults on a communication bus or communication interface can be mapped to the fault types listed below. At node level the following main faults will be relevant:

1. Data values of a received message are faulty (faulty data value)
2. The message is received later than a deadline (late message or message omission)
3. The message is received too early
4. The message can not be sent out in the given time window
5. The message can not be sent out

In distributed systems it is important to consider the fault propagation by distribution of message, which may stretch over several nodes and domain border. The faults can be subdivided into symmetric and asymmetric faults. The following main fault will be relevant:

6. All receiver of the message (in a special case only one receiver exists) regard the message as faulty with respect to the same main fault type, which is one of the fault types in 1 to 3.
7. All receivers of the message regard the message as faulty with respect to one of the main fault types 1 to 3, which can be different for each receiver.
8. Some of the receivers get a correct message, while the others get a faulty message with respect to one of the main fault types 1 to 3, which is the same for each receiver of the faulty message.
9. Some of the receivers get a correct message, while the others get a faulty message with respect to one of the main fault types 1 to 3, which can be different for each receiver of the faulty message.

Remark: The asymmetric fault types (7 to 9) are special cases of the well known "Byzantine General Problem" (as long as they are undetected), which is explained in details as following:

The replication of components, called structural redundancy, is very often a way to ensure that these systems are free from single points of failure. However, the use of redundancies within an electronic system introduces some undesirable effects.

Byzantine faults, also called as Generals Problem, was initially addressed by Lamport [LSP80] and [LSP82] more than 24 years ago. It refers to totally arbitrary failures, where the node/message may give contradictory information to different observers of the system, where Byzantine failure refers to the loss of system service due to Byzantine faults. i.e., arbitrary behavior in both the value and the time domain with inconsistent redundant values could be forwarded to different nodes. Generally speaking, there are two kinds of Byzantine faults:

- *Benign Byzantine fault:* A faulty node sends a message to a set of nodes. One or more of them do not receive a message at all.
- *Malicious Byzantine fault:* A faulty node sends a message to a set of nodes. One or more of them do receive a message, but with a wrong value.

In [SIV04], the Byzantine faults in embedded systems are revisited from a practitioner's perspective, where a few typical Byzantine faults are demonstrated.

- Digital signal that is stuck at "1/2" because of the physical nature of logic circuitry (some internal faults in the transmitter) as shown in Figure 6-2.

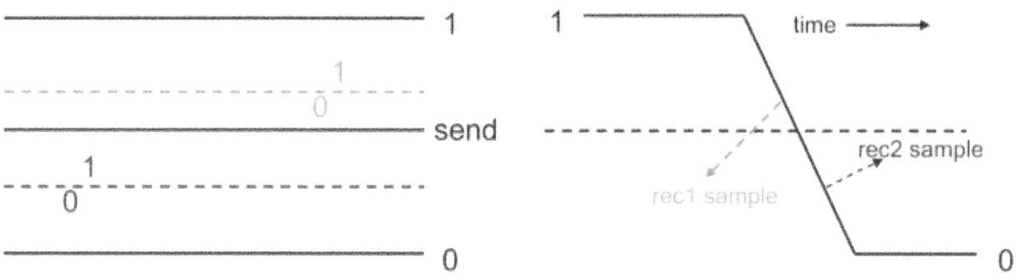

Figure 6-2: Cause of Byzantine fault with "1/2" area threshold

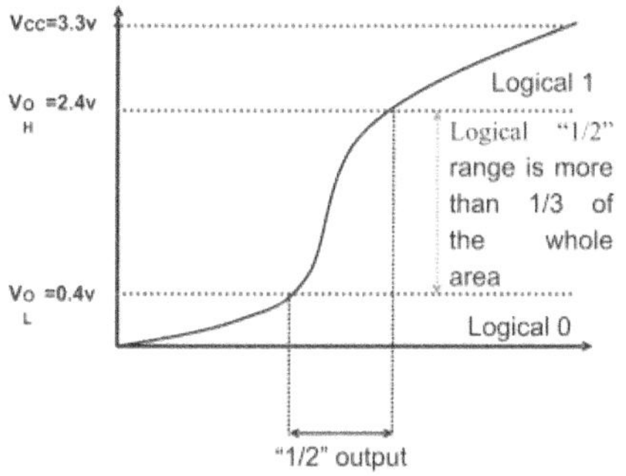

Figure 6-3: Gate transfer function with "1/2" area

Figure 6-3 shows a logic gate's transfer function for a common 3.3 volt logic circuit family. The transfer function shown here has a much less step slope than a typical logic circuits in order to show the 1/2-area.

- Byzantine faults may propagate through multiple stages of logic and still remain at an ambiguous level. Thus even if the transferred data is protected with CRC (e.g. Schrödinger's CRC), each CRC bit affected by the ½ data bit can also be ½. As shown in Figure 6-4, the resulting data received by a and b are different, but each copy has a correct CRC for its data as shown in Figure 6-4. Thus CRC can not provide a guaranteed protection against Byzantine fault propagation.

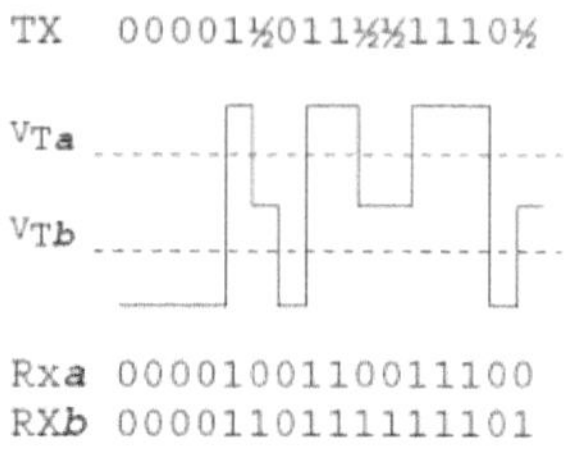

Figure 6-4: Byzantine Faults in Schrödinger's CRC [SIV04]

In time triggered communication systems with a deterministic fault tolerant communication protocol, Byzantine faults can not be excluded either. Firstly, because of EMC-problems and random radioactive decay, bit-flips in register and RAM can be induced. The dominant Byzantine faults are due to marginal transmission timing, that a message comes to the receiver at the margin of the time cycle. In this case, for example, corruptions in the time-base could lead to the situation, that a message transmitted slightly too early are accepted by the nodes with slightly fast clocks, while nodes with slightly slow clocks will reject this message.

6.3 Fault hypothesis of software in ISS

While the focus of traditional fault hypothesis of in-vehicle electronics lies in the hardware architecture of ECU, an important part of fault hypothesis of Integrated Safety Systems is software relevant. Two main challenges by the development of ISS are the implementation of software intensive system and integration of functions with different safety requirements on one ECU.

- The introduction of ISS requires by its natural consensus from different ECUs. This required consensus can be low-level (e.g. time synchronization among the distributed nodes with inter-communication) or at a higher level (e.g. a coordinated interaction of distributed applications like distributed controlling of brake force in ESP or distributed fault tolerance on several nodes in a coordinated manner).
- Taking the supply-chain of automotive industry, while the responsibility of tier1 suppliers is shifting from components to the system solutions, the focus of OEMs are shifting to the system integration. The integration of software components from different suppliers or even products third-party pure software providers on the same ECU could inevitably introduce more efforts of integration by OEMs. Although improved software requirement specification, standardized software development process and software architecture with defined interface can help to ease the problem, a specific fault hypothesis of software will surely help in the validation and debugging phase of system integration.

As introduced in the state of the art in the Subsection 9.2 software architecture for the in-vehicle electronics, an appropriate fault hypothesis of ECU software platform with layered standard software services will be introduced as following.

6.3.1 Software timing faults

As state of the art of automotive electronics, an OSEK-conform operating system with pre-emptive scheduling is used, because it is assumed that not all tasks can be scheduled as periodic and that event triggered tasks are still needed. In an operating system with pre-emptive scheduling, according to a predefined scheduling algorithm (e.g. priorities) the scheduler shall execute the parallel running schedulable entities, in which scheduling faults can take place:

1. Activation during running process of other entity: During execution of a schedulable entity, it might be possible that other schedulable entities shall be activated. Due to the reason, that parallel processing is not possible, one of the schedulable entities can not run.
2. Deadline/execution time budget violation: Deadline/execution time violation occurs, when a scheduled entity can not fulfill its job within the predefined time budget.
3. A specific schedulable entity (e.g. non periodic task) is activated too often: During runtime a schedulable entity might be activated several times in the same time window. This in general leads to inconsistencies and errors, if that routine is not prepared to do so.
4. Maximal number of entity activation is reached: During runtime more schedulable entities might be activated than the scheduler implemented on the ECU can manage, e.g. too many tasks are scheduled and activated.

In general all these scheduling concerning the operating system faults can result in two final faults: missed activation and missed termination of the OS-objects.

The scheduling faults are mainly considered to be faults, which shall be avoided during the design and development phase of the system or functional unit, but this kind of faults can not be excluded since there is no 100% fault free software. Thus the scheduling faults shall be handled (detected and tolerated) by the dependability software platform in the runtime. Detailed fault hypothesis considering timing faults in the runtime will be discussed in Subsection 9.3.2.3.2 in the dependability software service of software watchdog.

6.3.2 Communication between SW-components

As mentioned in Chapter 3, a clear trend in the embedded systems, especially in the automotive industry is the distributed control system. Distributed control system, by its natural, require consensus and coordinated behavior among the constituent parts. The required consensus might be low-level, (e.g. time synchronization among the distributed nodes) or at a higher level (e.g. in form of coordinated distribution of brake force in ABS or ESP system).

Thus the communication link is a critical chain with strict dependable requirements. At level of software-components the following main faults or errors concerning data exchange and communication between software components on one ECU (intra-ECU communication) or on different ECUs (inter-ECU communication) will be relevant: Except for the five fault types listed in Subsection 6.2 about communication, API access fault is an additional fault to be considered here.

6.3.3 Concurrent resource access

As assumed, different application SW-Cs (or redundant SW-C) can be integrated onto one ECU, limited resource on ECU may be accessed concurrently by different entities. The resources here can be CPU running time, SW-services, peripheral or other resources.

For peripheral access, one physical unit/peripheral device (like an actuator) is requested to be controlled by two or more independent software-components. But in general it is assumed that a specific physical device is controlled by only one software-component at a certain time. This assumption does not exclude e.g. a possible configuration, that two different application SW-Cs control an actuator. For example, one unit can "enable" the actuator and the other one can control it. This means that the actuator works like an AND gate, which reduces the possibility of unwanted control. Faults can occur here during a simultaneous access (timing and duration) of a peripheral device in the control of peripheral drivers. The fault in concurrent resource access can mainly be considered as a fault which has to be avoided during system design phase. A device that can be accessed by several applications would have a guarding queue. Possible problems that can occur during a simultaneous access of a peripheral device should be checked and minimized during system design and integration.

7 Engineering Process of Integrated Safety System

As required by the safety norms in Subsection 3.5.2.3, development process according to the appropriate safety integrity level is the most important lever to face the challenges summarized in Chapter 4. In the following an Engineering Process for the development of future Integrated Safety System will be discussed.

7.1 Introduction to the ISS Engineering Process

In the Subsection 3.5.2, state of the art in the development process for automotive safety electronics and challenges introduced by development of new ISS-applications are introduced. The analysis of the gap shows the necessity to improve today development processes. The widely accepted safety norm IEC 61508 and its up-coming adaptation for the automotive industry, the industry norm ISO/WD 26262 should be taken into consideration. At the same time the system dependability should be improved by managing the complexity of ISS.

At the beginning of this chapter, an overview of the ISS Engineering Process as part of the global development process is given in Figure 7-1.

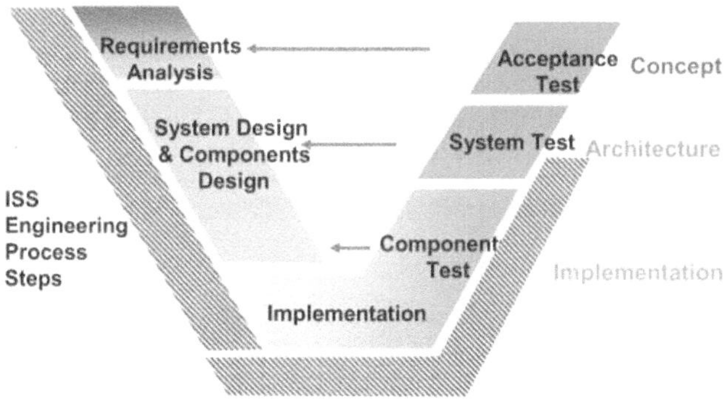

Figure 7-1: Overview of ISS Engineering Process with virtual front-loading

The complete development of ISS covers multiple phases from design to the serial production as shown in Subsection 3.5.2.2.2 as a series of V-models. The ISS Engineering Process focuses on the phases from A-sample to the B-sample, while keeping the consistent requirement engineering, safety analysis and system validation with the following phases of C-sample and D-sample.

The basic of the ISS Engineering Process is the V-Model'97. In the ISS Engineering Process, the intern V is extended with an extern V as the pink part shown in Figure 7-1. The extension of external V is based on the EASIS Engineering Process (EEP) proposed in the EASIS project [EAD41].

Compared with the EASIS EEP, the ISS Engineering Process integrates a consistent requirement engineering phase, which covers from the requirement analysis including hazard analysis to the component test and part of system test with fault injection. The requirements from ISO26262 are joined together with the detailed development steps. The dependable software and hardware architectures, defined in the Chapter 8 and Chapter 9 as a trend of platform strategy, are combined in the ISS Engineering Process as well.

From the global view however, the emphases of the ISS Engineering Process are listed as following:

- Virtual front-loading: Strict analysis and validation activities are commonplace in today's automotive development processes. However, these activities are usually performed rather later – when major design decisions have already been performed and the effort of changes is relatively high. Moreover there is often simply insufficient time left to conduct complete system test. According to a McKinsey survey [McK05], about half of automotive electronics are tested in vehicle 18 months before SOP and only 6 months before SOP does the level reach 90%. The ISS requires an even more decentralized development process, a distributed rapid prototyping based on the principle of early integration and validation with advanced simulation environment is one of the enabling technologies to solve this problem. The physical integration of serial sensors and actuators with the microcontroller in vehicles can take place in the late phase of product development; however, a lot of safety features can be already covered and tested with virtual front-loading.
- Correct-by-construction approach: correct-by-construction means that development artifacts adhere to certain quality measures by following a strict development approach. The correct by construction approach can reduce development costs by automating development steps and reducing the analysis of the results. Furthermore, it can increase the resulting quality in terms of non functional requirements like availability and safety of systems. Thus it is a promising approach especially for the development of ISS.

Main benefits of using the ISS Engineering Process are:

- This process helps to handle the complexity by splitting up the system development into manageable parts without losing overview (as a strict development approach for correct by construction).
- This process supports the identification and further handling of safety related requirements with enhanced end-to-end requirement tracing. On every development step, the necessary safety activities are identified and triggered.
- The virtual frontloading of hazard analysis activities, model-in-the-loop, software-in-the-loop and hardware-in-the-loop test support the early detection of (design) errors. Problems with the overall concept (architecture) can be detected early. This offers the possibility to make the adjustments on concept level, and not on implementation level.
- The process takes the current trends of standardization of dependable hardware and software platform as well as safety norms into consideration.

Figure 7-2 shows in detail the domain of virtual front-loading in the ISS Engineering Process. It begins with the requirement engineering. After the FAA (Functional Analysis Architecture) is mapped to the hardware architecture, the FDA (Functional Design Architecture) model derived can be validated with the rapid-prototyping. The virtual front-loading accompanies these steps and checks the results against the stepwise improved and complemented requirements.

The correct by construction approach is implemented here with the exactly defined steps in the EASIS Engineering Process as well the mapping and interconnection between the steps, which improves the quality of the designed prototype. The virtual frontloading here, in return, can find out more potential deficiencies of the design and enables a prompt and cost-effective re-engineering of construction, which contributes to the correct by construction approach as well.

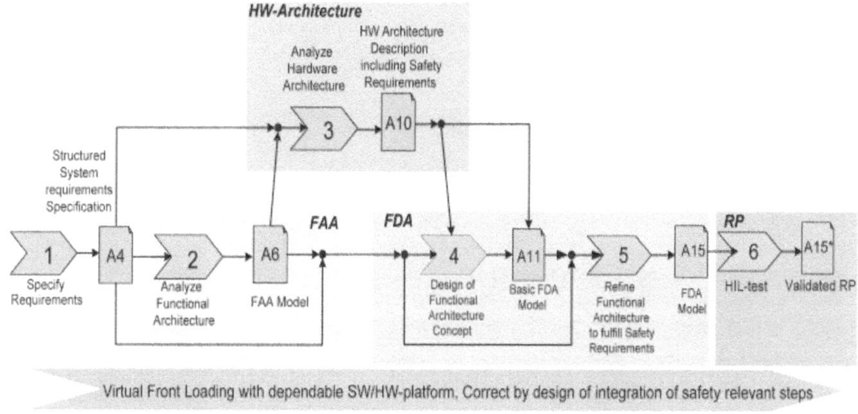

Figure 7-2: ISS Engineering Process with virtual front-loading in details

As depicted in Figure 7-2, the ISS Engineering Process can be divided into five main parts as following:

Part 1: Initial of requirement engineering (specify preliminary requirements)

All requirements of the system function under development are collected, specified and structured. Safety related requirements are analyzed from a system level point of view. At this step, it is not necessary to make assumptions regarding the E/E architecture, as mainly the intended functional behavior is considered and the possible malfunctions are analyzed.

Details on this process step are given in Subsection 7.2.1.

Part 2: Development of Functional Analysis Architecture

Based on the structured system requirements specification, the functional properties are modeled using the Functional Analysis Architecture. In this process part, firstly the functional architecture is defined (defining the building blocks) and the functional interchange mechanisms (timing and dynamic behavior) are analyzed. Secondly, the functional behavior of the single building blocks is modeled.

In this stage, the basic cooperation of the function under development with the other components of the vehicle is analyzed. This minimizes the risk of late detection of basic incompatibilities between the specified behaviors of the different functions in the vehicle. The FAA defines the possible basic entities and interaction rules usable for the considerations on this abstraction level (notation and semantics).

The FAA model as elaborated in this step is analyzed e.g. consistency, timing and formal properties. The list of the identified FAA blocks can be used as basic input for dependability related analysis (e.g. FMEA). Virtual front-loading with model based simulation of functional behavior is used for validation as to the functional requirements. Following the dependability analysis, a first Hazard Analysis is performed, possibly generating new requirements and an update to the list of hazards. Details on this process step are given in Subsection 7.2.2

Part 3: Development of hardware architecture

With the FAA Model of the function system under consideration, it is possible to tentatively allocate the defined function blocks to one or more hardware entities (hardware architecture with layout of ECU units, sensors, actuators and communication network). Thus the Hardware Architecture (HA) is integrated here as an input factor. This enables the analysis of the dependability requirements that these hardware components have to fulfill, leading to a description of necessary redundancy and of the possible failure modes. This will lead to a refinement of the hardware architecture, and give a specification of failure reaction requirements. Details on this process step are given in Subsection 7.2.3.

Part 4: Development of basic Functional Design Architecture

The hardware architecture gives the basis for the next analysis and modeling steps. First, the final allocation or mapping of functional components (as defined in the FAA model) to the HW Architecture is performed, while also taking the safety constraints into consideration. The software platform for ISS including dependability software services and its impacts on the functional architecture model are added to the design of the functional architecture. In other words, the whole dependable software platform is integrated into the design.

Details on this process step are given in Subsection 7.2.4.

Part 5: Refinement of and validation of FDA model with SiL-test

The Basic FDA Model covers the behavior of the function under the assumption that no faults (HW, SW or communication) occur. In Part 5, the handling of faults according to the system fault model is integrated, giving a final FDA model including the safety concept of the function system under development.

Details of these process steps are given in Subsection 7.2.5.

Part 6: Hazard analysis and validation of FDA model with HiL-test

Hazard analysis will be carried out for the complete FDA model with the new information. By means of hardware-in-the-loop test with consistent evaluation cases on a hybrid-system of rapid-prototyping platform, evaluation microcontroller and rest-bus simulation method, an end-to-end tracing to the test specification and requirement specification can be guaranteed.

Details on this process step are given in Subsection 7.2.6.

The enabling steps for the development of ISS, along with safety relevant steps as extension of EASIS Engineering Process, are listed as following:

- Enhanced requirement engineering phase with seamless process oriented mapping and tracing of requirement specification (seeking "end-to-end" quality assurance).
- Using UML for the functional architecture design of the embedded system and the specification of interface and specification of dynamic behavior.
- Front-loading with the virtual integration, taking the current trends of hardware architecture and software platform standardization (results from Chapter 8 and Chapter 9) into consideration (early model-in-the-loop and software-in-the-loop test).
- Mapping of app. SW-Cs to electronic architecture taking safety requirements into consideration and a guideline for the mapping work.
- Application of dependable SW-platform and hardware platform in the system integration by configuration of dependability services.
- Verification and validation with rest-bus-simulation and hardware-in-the-loop test and fault injection based on fault hypothesis.

In the following chapters, the development processes with the appropriate tool (hardware and software) are given step by step in details after the landscape of the whole development process is shown. The notations and symbols used in the landscape of the ISS Engineering Process are shown as following Figure 7-3.

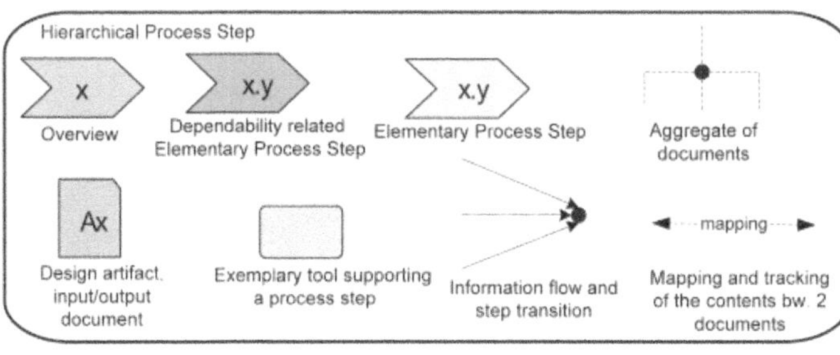

Figure 7-3: Notation in the ISS Engineering Process

7.2 Development steps of ISS Engineering Process

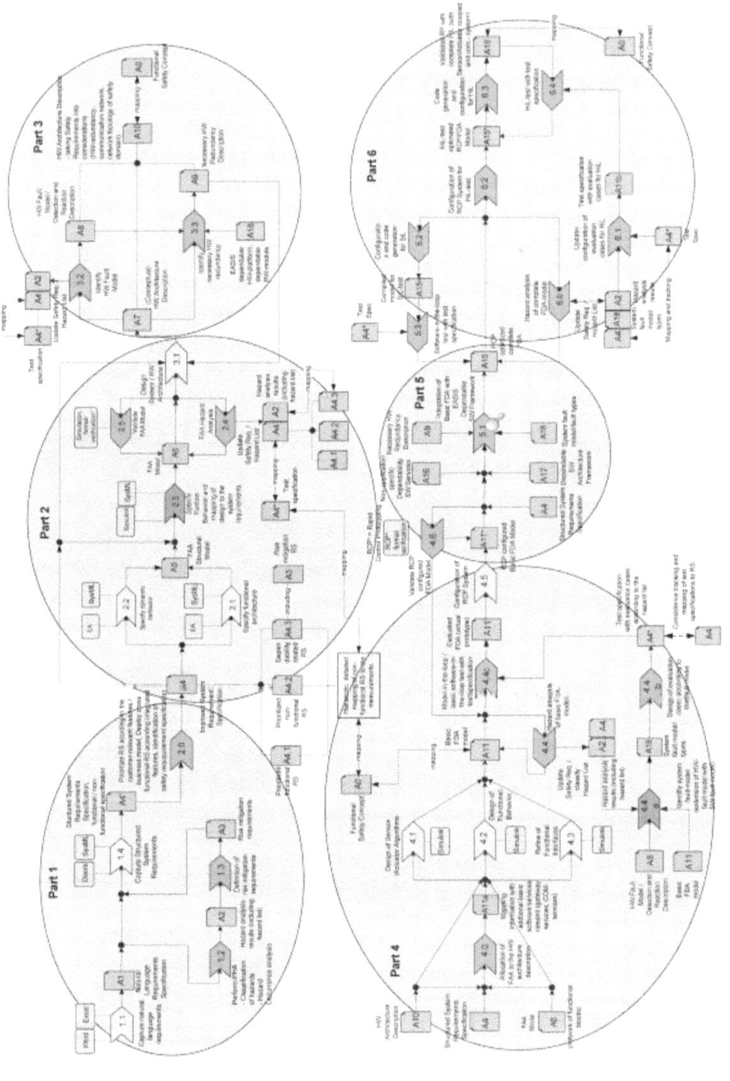

Figure 7-4: Global view of ISS Engineering Process

An overview of ISS Engineering Process is depicted in Figure 7-4, which is subdivided into 7 parts as the following subsections.

7.2.1 Part 1: Initial of requirement engineering (specify preliminary requirements)

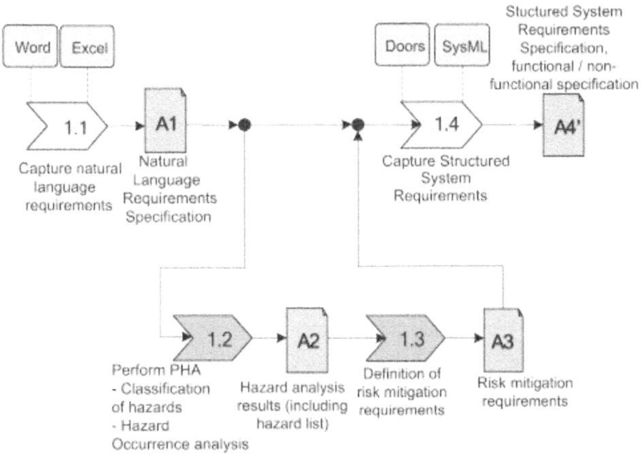

Figure 7-5: Specify preliminary requirements

Process step 1.1: Capture natural language requirements

The visions of a new system including system requirements and system concept are firstly collected in a requirement specification of natural language as depicted in Figure 7-5, where e.g. a high-volume vehicle manufacturer targeting the mass market cost-consciously strives to carryover concepts from previous projects. The next vehicle system concept is typically an adapted variant of previous one.

Already in this initial step two issues should be considered [McK06]:

- Applying right level of specification granularity – If the granularity is too high, there is the danger of misunderstanding. If too low, the efficiency can suffer.
- Deploy cross-functional feature specifications by setting up a cross-functional team to work on the specifications.

The proposed tools are more non-formal tools like Word, Excel and state-transition diagrams.

Process step 1.2: Perform Preliminary Hazard Analysis (PHA)

Preliminary Hazard Analysis (PHA) is the first execution stage of a product project specific plan, performed at the conceptualization stage of the ISS lifecycle. PHA does not require concrete designs. PHA identifies potential hazards leading to unacceptable risk, and specifies corresponding requirements to reduce the risk to within tolerable limit. As the ISS progresses towards concretion and requirements flow down, the PHA may be iteratively applied to the corresponding subsystems. PHA is iterated until residual risks are within tolerable limits.

Measures to perform the process step:

- Identify losses to be avoided & corresponding severity category
- Identify hazards based on previous experiences
- Identify hazards based on concept review
- Categorize hazards by severity
- Assess risk of malfunctions (see remark [1])
- Tabulate hazard-loss combinations & estimate risk level
- Establish tolerable risk targets
- Perform Hazard Graph Analysis (HGA)

While the results of such a Preliminary Hazard Analysis are defined as following:

- (General) list of losses to be avoided
- Database of hazard-loss combinations, corresponding risk level
- (General) preliminary hazard list, including
  - Hazards from driver overload, based on workload and cognitive load model
  - Hazards from driver confusion about mode
  - Hazards with unacceptable risk levels

The proposed tools for the early stage of the PHA can be supported some form of templates, questionnaires and tables, e.g. in Excel. The hazard graph, as mentioned above, can be integrated with the state machine model.

Process step 1.3: Definition of risk mitigation requirements

This process step specifies requirements to reduce the risk of each hazard to within tolerable limit. As mentioned above, the PHA is iterated until residual risks are within tolerable limits.

Measures to perform the process step:

- Identify requirements for risk mitigation
- Perform Hazard Graph Analysis
- Identify preliminary verification and validation (V&V) requirements

Results of the process step:

- Preliminary requirements and constraints for risk mitigation
- Additional hazards introduced due to risk mitigation measures
- Analysis of newly introduced hazards
- Identification of newly introduced hazards
- Conflicts with other requirements, including trade-off analysis and conflict resolution

[1] According to ISO26262-3 [ISO23] Section 6 hazard analysis and risk assessment, Risk of malfunction = severity * probability. In this context, "probability" is defined as "the probability that the vehicle is in a driving situation, in which the malfunction could cause a mishap with the considered severity". Example: A possible malfunction of a gearbox is that the car does not start driving when intended by the driver. This is normally no problem, as a car that is standing is the safest. But if this malfunction happens when the car is standing on a railroad crossing (for any reason), this is not safe at all. For this situation, this analysis step analyses what percentage of vehicle's lifetime that it is in such a particular situation.

- Preliminary requirements for verification and validation
- Hazard graph after applying risk mitigation requirements. The risk mitigation here means both the reduction of the hazard probability and mitigation of the effects of hazards.

The proposed tools are again the templates, questionnaires and tables as used in step 1.2.

Process step 1.4: Specify structured system requirements

In this step the system safety requirements are integrated with the discovery of adverse relationship (conflict) with other requirements, thus some kind of trade-off preliminary decision should be carried out. The proposed tools for the requirements engineering including requirements managements and tracing are DOORS and SysML [SML06].

7.2.2 Part 2: Development of Functional Analysis Architecture (FAA Model)

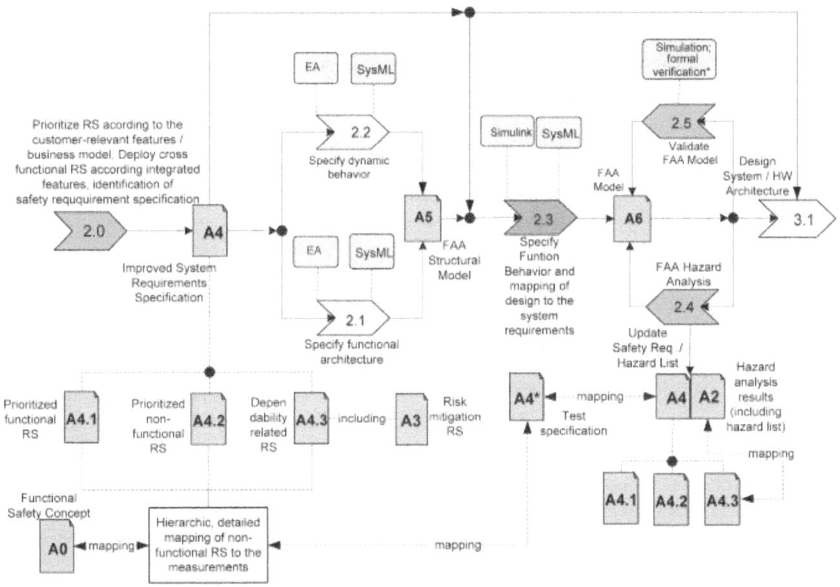

Figure 7-6: Safety integrated requirement specification and development of the FAA-model

Process step 2.0: Improvement of requirement specification

This process step (as depicted in Figure 7-6) improves the system requirement specification by further iteration of prioritization and categorization of the requirements according to different constraints.

Measures to perform the process step:

- Prioritization of the requirement specifications
    - The safety relevant requirement specifications, of course, should be given the highest priority (e.g. the allocation of ASIL according to ISO26262). The ISS, however, are mostly cross-functional integrated features. The system dependability requirement should be split to the sub-functional safety requirements.
    - Identify requirements for customer relevant features. In-vehicle electronics inject a lot of innovation in automotive industry, but not all the features can be perceived by the customers in proportion to the development efforts behind. With the introduction of vehicle platform strategy, based on one vehicle platform esp. on the vehicle E/E-platform, different vehicle variants according to the defined business model can be configured. Thus for each vehicle model, the individual customer relevant requirement specification should be selected and given a higher priority.
- Categorization of the requirement specifications
    - The easiest way is to divide the requirements into functional and non-functional requirements, where functional requirements specify what the developed software should do e.g. data, function and interfaces and non-functional requirements specify under which constraints the functional requirement should be fulfilled. The most important requirements are mostly non-functional requirements, such as performance, quality and requirements for the methods of realization, implementation requirements and organization requirements.
    - For the development of ISS, the emphasis will be put on the dependability (safety) related requirements, which are mostly non-functional requirements. Here the relevant dependencies between the system-components, the risk mitigation requirements should be included in the dependability related requirements, e.g. switch-off of ACC in case of sensor failure of ESP.

The result of this process step is an improved, prioritized and categorized requirement specification, which is the most important milestone and basis for the requirement engineering work. As a living document, the specification should be supplemented and traced during the whole ISS Engineering Process.

As proposed, the tools DOORS and SysML will be used. One trend which should be mentioned here is the so-called model based requirement engineering, in which the requirement specifications are mapped to the model based functional design. With the tools of DOORS and SysML, the structured requirement specification can be further linked and mapped to the appropriate UML-design or model-blocks with Matlab/Simulink. Further on, they can be linked to the V&V requirements and approaches,

Process step 2.1: Specify functional architecture

Specify functional architecture is a typical step of "divide and conquer". The whole system is decomposed according to domain-specific considerations. While ISS-applications are characterized with inter-domain and inter-ECU distribution, the developer should not lose the system overview and the interaction between the sub-components.

Process step 2.2: Specify dynamic behavior

Here a sequence of functions is defined. Eventually these must conform to the functional architecture defined in step 2.1. By using UML2.0 or SysML, e.g. the standardized notation of interface definition, the time sequence editors and use-case diagrams, the interaction between the sub-systems can be defined.

Process step 2.3: Specify Function Behavior

While steps 2.1 and 2.2 mainly concentrate on the "outer" view of a functional architecture, this step emphasizes the "inner" view as it cares for the implementation of the fine-grained functional behavior. The development of embedded software by means of a model-based design is already practiced in certain organizations in the automotive and other industrial domains. However, it is still not state-of-practice, esp. for the safety relevant systems. With the introduction of improved model based development tools, certified code generator and compiler, it can be foreseen that not only in the rapid prototyping phase but also in the serial development of safety relevant systems, the model based approach will be applied.

Process step 2.4: FAA Hazard Analysis (FHA)

FAA Hazard Analysis (FHA) extends and updates the results of PHA, through detailed analysis of constituent subsystems and components of the vehicular system, to identify previously unidentified hazards and to assess the associated risks.

Measures to perform the process step:

- Identify hazards and analyze their effects
- Identification of faults that can lead to the hazards, updates the hazard list and fault types/hypothesis of the system
- Identify risk mitigation measures and design criteria
- Allocate Automotive Safety Integrity Level (ASIL) to the FAA components according to ISO26262-3 - concept phase [ISO23]. With a qualitative analysis the safety integrity requirement (quantitative requirement of PFH) can be mapped to FAA-model. As shown in Figure 7-7, based on experience, the system PFH can be mapped to the FAA components of sensor/input data processing, communication, ECU computation and output to actuator.

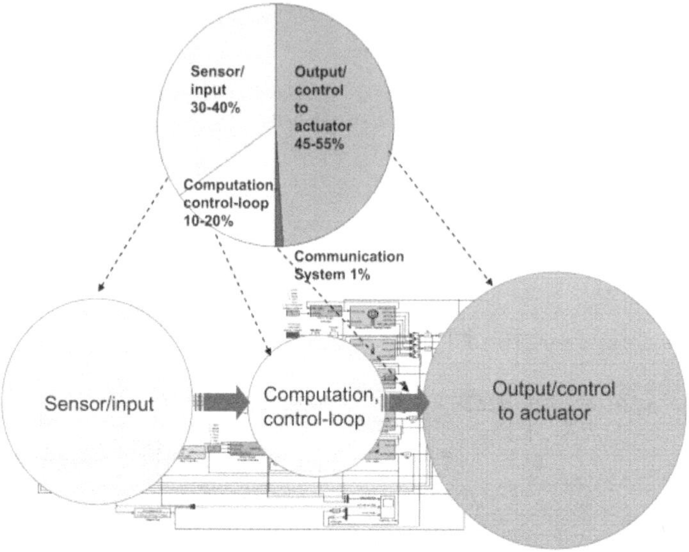

- Output/control to actuator: 49%
- Communication system: 1%
- Computation and control-loop: 15%
- Sensor/input: 35%

Figure 7-7: Mapping of system-PFH to FAA-components

A plausible mapping example is shown as above with ASIL-D system PFH as $1*10^{-8}$, the PFH of the communication system should not exceed the $1\%*1*10^{-8}$ as $1*10^{-10}$. It demonstrates on one side the possible weakest chain in the system, such as communication system. On the other side, in order to improve the system PFH, one should concentrate on the system components as sensor and actuator control.

Results of the process step:

- A supplemented list of hazards (with additional and refined hazards compared to PHA) and mapping to the defined dependability specification in A4.3.
- Assessment and refinement of severity and risk level for each hazard
- Residual risk level of each identified hazard
- Additional and refined requirements for residual risk mitigation
- Additional and refined requirements for V&V, an initial mapping here should be performed to the test specification of the identified hazards. IEC 61508-3 calls for a test specification as the result of the functional step "Module Design" (ISS Engineering Process part 4). But already in this very early phase, the software developers, who are responsible for producing the test specification, and in particular the design of test cases, should make thoughts of test specification and their mapping to the requirement specification [SMT06].

Process step 2.5: Validate FAA model

In this step an early validation of FAA model (a typical step of virtual front-loading) is carried out, in which the functional behavior of one or more functional units must have been specified. For the behavior description, an executable specification is given for each unit. The level of details in these specifications might still vary, ranging from an abstract and coarse grain specification (e.g. a simplified state-chart model) or a detail specification (i.e. definition) of the functional behavior. In either case it is assumed that a commonly used modeling language with executable semantics is used for the specification of the functional behavior, e.g. MATLAB/Simulink/Stateflow. This FAA model is still regarded to be independent from an execution platform, i.e. it does not contain mapping information resulting from mapping the model onto a specific aimed hardware platform.

In the validation process step a preliminary "model-in-the-loop" test can be carried out. It is checked whether the FAA model conforms to the structured system requirements specification. Therefore, a test model has to be derived from the requirements specification and fault types/hypothesis. The test model might be specified in the same language as the functional behavior. In this case the FAA model may be embedded in a test frame that has been triggered in a test environment by the test model, e.g. by performing an offline simulation in MATLAB/Simulink/Stateflow.

If the FAA model is given as a state-chart, formal verification techniques can also be employed. In this case the test model contains assertions and the verification which are used to check whether the FAA model fulfills these assertions.

If the FAA model reflects a control loop, also methods from control theory for stability analysis can be used. In case of a linear system there are standard methods available. To some extent there are also methods for non-linear systems, although these require additional effort.

The proposed tools for the validation of FAA models are DOORS (for specification of the system requirements), DOORS, MATLAB/Simulink/Stateflow (for the test model), test automation tools on the PC-platform (for the simulation engine and execution of tests) and dSpace Embedded Validator (for formal verification).

As depicted in Figure 7-1, there are two important mappings, the mapping of Functional Safety Concept (artifact A0) to the improved Requirement Specifications of A4 and the mapping of Test Specification of A4* to the Requirement Specifications of A4. Since the FAA-model developed in Part2 only contains the most elementary functionalities of ISS, the mapping here is only a first step for a complete Functional Safety Concept required in [ISO23], which is living-document through out the safety life-cycle. The technical safety concept here together with the Test Specification and test results form the basic parts of safety cases, which serve as the proof of the safety integrity level.

7.2.3 Part 3: Development of HW Architecture

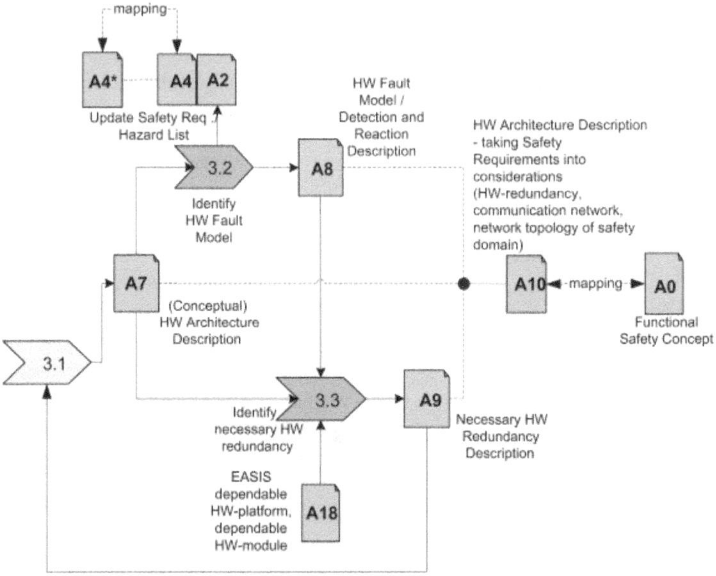

Figure 7-8: Development steps for HW architecture

Figure 7-8 shows the consideration of dependability aspects by the development of HW Architecture Description (artifact A10). A system or hardware architecture represents one of several system design alternatives that are generated to explore the design space and to find an optimized solution. This optimum is mainly driven by the continuous system optimization, but for the safety relevant applications, however, the same safety case (the same safety integrity) should be provided despite of the optimization. Thus the hardware architecture should be mapped to the Functional Safety Concept as well.

Process step 3.1: Design system HW architecture

As introduced in Subsection 3.3.3, The HW architecture describes the vehicles topology, bus systems, ECUs, gateways as well as the employed sensors and actuators. The HW architecture has to fulfill the system requirements specification, which already contains requirements to eliminate or mitigate hazards identified during early hazard analysis on system level. The analysis functions (FAA components) are tentatively allocated to the HW architecture. This includes the functional devices of the FAA model that are mapped to the HW, guidelines for the definition of the HW architecture and the distribution of the functionality. The result is a conceptual HW architecture description that can be analyzed and iteratively refined until it complies with the safety requirements. Necessary HW redundancy identified during further design steps has to be incorporated into the HW architecture. This means that either the description has to be updated or a new HW architecture has to be chosen that is able to fulfill safety requirements but may not need the redundant parts. For example, instead of two separated redundant sensors, one sensor with integrated redundancies on one silicon-board with independent data paths could be selected.

Process step 3.2: Identify HW fault model and risk mitigation measures

In order to identify possible HW faults and to determine which faults may lead to which system-level fault modes, a hazard occurrence analysis with e.g. a qualitative Fault Tree Analysis (FTA) has to be performed. The result of this activity is a starting point for the definition of error detection mechanisms and an action that has to be taken in case of a detected error Within this step, since SW related mechanisms are strongly connected with the fault-treatment, the SW mechanisms here are covered in step 5.1. The proposed tools are tools for FMEA und FTA.

Process step 3.3: Identify necessary HW redundancy

The application of dependable HW-architectures platform, e.g. redundancy concept with sensors has an impact on the HW architecture as well as on the software design. The HW architecture has to be iteratively refined for the redundancy. The HW redundancy description establishes a specification to incorporate typical ISS dependability software services, such as voting and Agreement Protocol, into the functional architecture.

7.2.4 Part 4: Development of Functional Design Architecture (FDA)

1) As depicted in Figure 7-9, the Basic FDA Model (A11) is designed as the mapping of FAA to the HW with the basis software services needed as OS, gateway-service and communication service, etc. No attention is paid here to the constraints of timing, bandwidth of communication network, etc.

2) With the new information of "basic FDA model", the HW fault model is supplemented to the system fault model based on the fault model defined here.

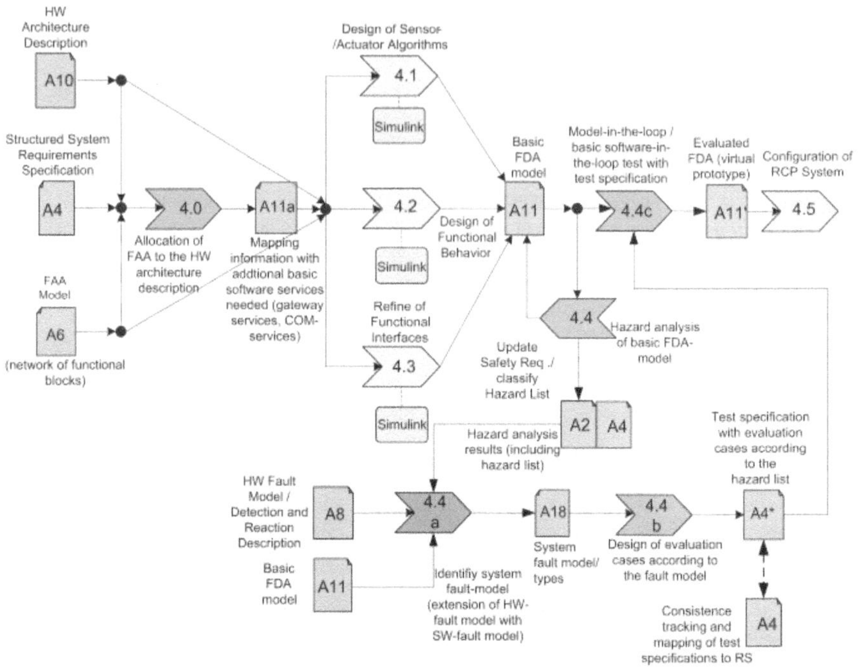

Figure 7-9: Design and validation of Functional Architecture Model with virtual front-loading

Here it is to note that one of the main differences between the FAA and the FDA is the level of details. The FAA typically does not consider domain-specific software infrastructure (e.g. diagnostics, state management and network management, etc.). The FDA with mapping on HW, on the other hand, should consider these aspects.

Process step 4.0: Allocation of FAA to the HW-architecture description

In this step, a first version of the FDA-model of the function under development is generated. The basic idea of this process step is to combine the results of part 2 and part 3. The already existing FAA-model, as a functional network designed from model based development, is allocated to the HW-architecture description including safety requirements.

The model of application software (each can consist of more than one software component, e.g. sub-systems implemented with MATLAB/Simulink) will be mapped to the HW-platform (real ECU, µC or rapid prototyping HW node), taking the system safety requirements and other constraints into consideration. That is to say, by the mapping of application software models to the rapid prototyping platform, esp. for the integration of ISS applications with different ASILs, safety requirements with regards to the reliability, bus load (communication effort) between the function models, communication delay between the functions integrated on the nodes, etc. should also be considered for the first time here. As for some of the constraints, because no 100% accurate prognoses can be made, a rough estimation is enough to define the first FAA-model.

There can be more than one mapping possibility of FAA-model to the nodes of rapid-prototyping-platform. These mapping results should be assessed to the structured system requirements A4. Mapping of FAA to HW-architecture also implies a first rough decision for HW/SW-partitioning. While this is always a situation based case and there are no golden rules, a list of criterion is given as following:

- End-to-end timing: Hard real-time requirements can be often only implemented with hardware.
- Overall costs for the hardware and the software solution.
- Changeability during lifetime is easy to be implemented with a software solution. This is nearly impossible in a hardware implementation (except for FPGA).
- Know how protection: It is more difficult to perform reverse engineering of a hardware solution than of a software solution.
- Time for development: Algorithms can often easier and faster to implement in software than in hardware. The software based solutions are esp. preferred for prototype.
- Verified legacy solutions: If a solution is already implemented in hardware or in software, it is often beneficial to reuse it. However, even if a new solution would be more efficient, it is sometimes difficult to replace it, since the old solution is established.
- Background and competences of the developing team. In real life the project members have an influence on which selections are chosen for hardware/software partitioning.

Considering the correct-by-construction approach and reuse of prototyping here for the serial development, constraints of cost should also be considered. A guideline for the mapping work regarding the constraints is discussed in Subsection 9.3.2.

For the allocation of the FAA-model to the HW architecture, there are in practice no mature tools for the design work. Vector DaVinci supports the mapping of FAA-model to the HW architecture according to the principles of runnables defined in AUTOSAR as explained in Subsection 9.3.2.3.2, while the expertise of considering the multiple constraints and trade-off between them remains the responsibility of the developers. No intelligent and automatic mapping support can be foreseen in the near-future.

Process step 4.1: Design of sensor-, actuator-algorithms (I/O incl. bus I/O)

Based on the chosen hardware (mapping to the HW), the algorithms for evaluating the respective sensors and actuators must be designed. The proposed tools are MATLAB/Simulink extended by Real-Time block-sets (dSPACE) with bus-interface and INTECRIO including ASCET-SD (ETAS).

Process step 4.2: Design of functional behavior

While the FAA provides a coarse description of the functions, this process step is charged with the design of the algorithms that carry out the function. Moreover, additional issues such as calibration and mode management must be taken into account. The proposed tools here are MATLAB/Simulink and ASCET-SD.

Process step 4.3: Refinement of functional interfaces

Stemming from algorithm design, the requirements of the input values as to resolution, minimum/maximum and error values, etc. are known. The same is true for the values provided as output of the algorithm. Thus this process step is charged with the exact definition of the interfaces in the value domain and to the dynamic behavior - also the timing domain. The proposed tools here are TargetLink (dSPACE) and ASCET-SD.

Process step 4.4: Basic Design Hazard Analysis (BDHA)

Basic Design Hazard Analysis extends and updates the results of PHA and FAA Hazard Analysis, through detailed analysis of constituent subsystems and components to identify previously unidentified hazards and to assess the associated risks. The new components introduced by the work in steps 4.0 – 4.3 have to be analyzed.

Measures to perform the process step:

- Identify hazards and analyze their effects
- Identify risk mitigation measures and design criteria
- Assess residual risk level
- Allocate ASIL to the FDA components, in step 2.4 the allocated ASIL to FAA will be verified here with the additional mapping information to HW.

Results of the process step:

- Enlarged list of hazards (with additional and refined hazards compared to PHA and FHA)
- Assessment and refinement of severity, probability and risk level for each hazard
- Residual risk level of each identified hazard
- Additional and refined requirements for residual risk mitigation
- Additional and refined requirements for V&V

Proposed tools are still tools for FMEA and FTA.

Process step 4.4a: Identify system fault-model

Step 4.4a is in fact a sub-step of step 4.4, with the complexity of ISS, the identification of system fault model is outlined. In this step, the existing HW fault-model from step 3.2 will be reconsidered with the new knowledge from the FDA model. Generally speaking, the HW faults are mostly isolated faults except for the common mode fault as the power supply, while for the ISS, the distribution and mapping of SW to the HW and communication between the components, an integrated view of system wide fault model with SW fault model as introduced in Chapter 6 should be considered here.

Process step 4.4b: Design of evaluation cases according to the fault model

The mapping of improved fault model/fault types to the evaluation case and from the design of evaluation cases to the test specification as well as front-loading of system test is a consistent evaluation process in the ISS Engineering Process.

Due to the complexity of ISS, the design of evaluation cases should follow the same principle of Fault Tree Analysis with the knowledge of FAA and FDA. Since the possibility of fault propagation is amplified in ISS, the evaluation cases with fault injection on the remote node should be considered here as an emphasis. One of the aims of designing such intelligent evaluation cases is to find out the side effects, which are not foreseen in the design phase.

The mapping between items in the fault model to the test specification and later evaluation cases is an n:m (n to m) mapping, which should be checked for the completeness. The tools suggested here are again SysML and DOORS, which enable the mapping of model based evaluation cases to the text based test specification.

Process step 4.4c: Model-in-the-loop/basic software-in-the-loop test with test specification

With the mapping information in the FDA-model, an initial separation of system model to the hardware platform can be made. The communication between the separated models (the initial nodes) can be emulated with the models of communication network using the appreciate block-sets provided by the simulation tools. The environment and driver input can be emulated now more realistically, since accuracy and hardware constraints like interface, specifications of input signals from sensors and feedback from actuators can be included in the models step-by-step. With evaluation cases by fault injection according to the fault model, not only the system functionality but also the system dependability behavior in presence of faults can be evaluated as well.

In addition to the model-in-the-loop test, instead of using the simulation tools to emulate some of the basic services in the models such as COM, TP-services or normal gateway services, these basic software services as well the applications above can be compiled together. Here it is notable that these general software services can be chosen independent from the aimed HW-platform or the specific HW are physically available. Both Matlab/Simulink and Vector CANoe or DaVinci support the binding of compiled ECU-model e.g. Dynamic Link Library file (DLL-file) in the simulation model. Thus a more near-praxis simulation and front-loading of design evaluation of risk mitigation can be carried out.

Within the framework of model-based software development in conjunction with automatic code generators, the test cases designed for the MiL-test can be also reused for the SiL test.

Process step 4.5/4.6: Configuration and validation of RCP system

As shown in Figure 7-10, the basic FDA model will be configured for the rapid prototyping process. According to the certain tool chain and requirements from rapid prototyping HW, prototyping specific parameters, blocks configuration and environments will be integrated to the FDA-models. One example here is the configuration of FlexRay scheduling for Rapid Control Prototyping (RCP) using the DECOMSYS tool chain. The separation of communication mechanisms and scheduling of communication table using DECOMSYS tools require that certain guideline of designs should be followed.

In order to customize the FDA-model for the RCP, based on the validation results of RCP, certain iterations should be carried out.

7.2.5 Part 5: Refinement and validation of FDA model with SiL-test

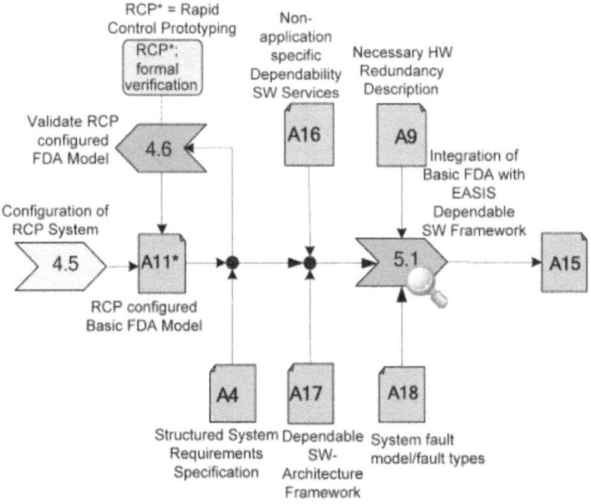

Figure 7-10: Refinement of FDA model with dependable SW-platform

Process step 5.1: Integration of Basic FDA with EASIS Dependable SW Framework

In this step, based on the RCP configured basic FDA model as depicted in Figure 7-10, HW redundancy description and HW fault detection mechanisms, appropriate dependability software services from the dependable architecture framework as introduced in Chapter 9 are chosen to support the fault detection and fault treatment mechanisms.

The application software, dependability software services and the basic software services will be configured as specified in the dependable SW architecture framework and integrated into a first complete FDA model.

As depicted in Figure 7-11, details in process steps 5.1.1, 5.1.2, 5.1.3 and 5.1.4 are defined as following.

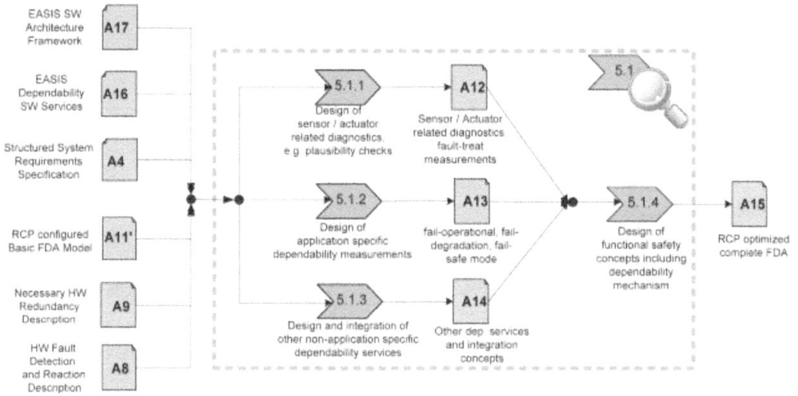

Figure 7-11: Process step 5.1 under magnifying glass

Process step 5.1.1: Design of sensor/actuator related Diagnostics

In this step, based on the specified functional system description and the characteristic behavior of the sensors/actuators in case of sensor/actuator faults, an appropriate diagnosis strategy is defined for them. In EEP step 4.1, the basic handling of (virtual) sensors has been designed. In step 5.1.1, the algorithms to detect the possible errors are integrated into the FDA model.

Process step 5.1.2: Design of application specific dependability measurements

The most plausibility checks are strongly connected with the design of applications. Based on the specific application requirements and the HW redundancy, appropriate dependability software services will be configured to perform the plausibility check of the following entities:

- Sensor signals
- Received signals from communication system, esp. the plausibility of the signal combination, e.g. based on the relationship among temperature, pressure and volume of engine, some plausibility assertion of the sensor signals can be made.
- Intermediate calculation results
- Sent signals to the communication system and computed desired value to the actuator.

Process step 5.1.3: Design and integration of other non-application specific dependability services

As introduced in Subsection 3.4.2.2, most dependability SW-services introduced here are application independent and can be configured to the application safety requirements. Such services as Agreement Protocol, fault-tolerant inter-ECU communication, Software Watchdog and Fault Management Framework can be selected and configured for the specific needs of the application with a reference to the dependability SW-architecture framework. According to ISO norm 26262 part 6 [ISO26], the application of dep. SW-architecture framework, reuse and configuration of dep. SW-services should be documented individually for each software component.

Process step 5.1.4: Design of functional safety concept including dependability mechanism

Based on the concepts in A12 (sensor/actuator related diagnostics, fault-treat measurements), A13 (different modes for fail-safe, fail-silent, fail-degraded and fail-operational, etc.) and A14 (other dependability software services and integration concepts), fault tolerance mechanism of ISS-applications are integrated in the FDA-model. The dependable software platform as well as the ISS-applications should be optimized here for the rapid prototyping in another iteration as well.

Process step 5.2: Configuration and code generation for software-in-the-loop test

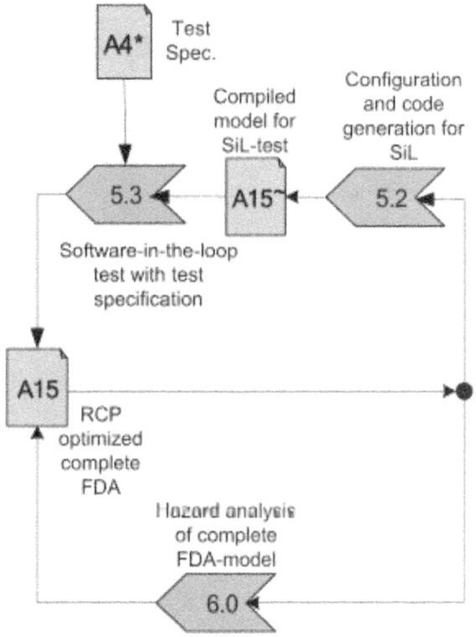

Figure 7-12: Validation of FDA-model with software-in-the-loop (SiL) test

In process step 4.4c, approaches of the model-in-the-loop-test and an initial software-in-the-loop test are integrated in the ISS Engineering Process. Compare with 4.4c, step 5.2 in Figure 7-12 is an extension with HW-specific features and system topology in FDA-model. In the software-in-the-loop test here, the generic hardware drivers and emulation of bus communication etc. can be substituted step-by-step with the µC or rapid-prototyping platform specific sensor/actuator related I/O-drivers, bus-drivers or other hardware libraries.

By applying appropriate code-generators and hardware specific compilers, the object codes can be generated and manual configured (e.g. additional check-point or break for the validation) for the aim of software-in-the-loop test.

Process step 5.3: Software-in-the-loop test with test specification

As mentioned in process step 4.4c, there are simulation tools supporting the embedding of compiled complete system model as a software based node. According to the availability of other nodes and developing phase, each node in the system topology can be chosen as the following seven variants:

1. Initial model, emulating very simple system behavior without any "intelligent" code behind.
2. Enhanced model with models computing system behavior based on the inputs without consideration of hardware constraints.
3. Software node with code generated from the enhanced model in variant 2, using generic hardware drivers, code generator and compilers.
4. Complete model as an extension of the model from variant 3, with integration of hardware specific issues like signals, timing and bus scheduling.
5. Software node with code generated from complete model in variant 4, using hardware specific drivers and code generators as well as compilers.
6. Hardware node with external bypassing, that is to say, the extended/new function is loaded to the rapid-prototyping platform, while the existing functions and I/Os are still integrated on the original ECU.
7. Hardware node, with complete code from variant 6 (bypassing and original ECU) loaded to the evaluation board or rapid-prototyping platform.

As depicted in Figure 7-13, in the different stages of development, from MiL, SiL to HiL, nodes are substituted successively with the advanced variants, while a consistent reuse of test cases and traced comparison of test results should be guaranteed.

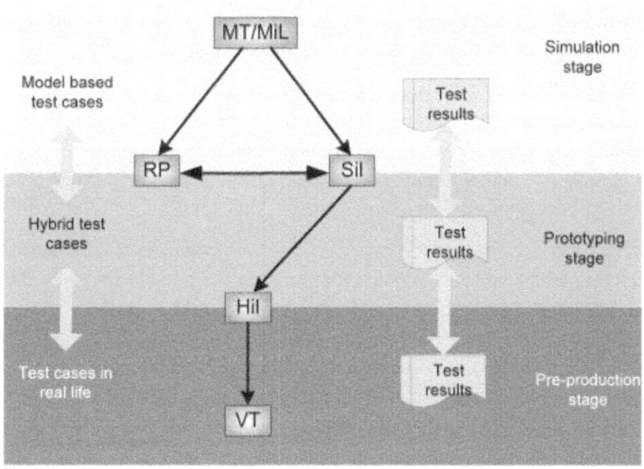

MT = model test
MiL = model in the loop
RP = rapid prototyping
SiL = software in the loop
HiL = hardware in the loop
VT = vehicle test, e.g. on-road trial

Figure 7-13: Test in different stages

A complete system topology is made up of nodes and communication system. While the nodes in the software-in-the-loop test environment can be one of the seven variants introduced here, the communication system can be also chosen from the following three variants:

1. Bus-communication connector emulated by the model-based development tools. In this case, no bus-scheduling, communication matrix or signal/frame lengths need to be considered. Most of this kind of information remains unknown in this early phase of development.

2. Emulated bus communication: Once the type of communication bus is chosen according to the bandwidth, safety requirement, complexity and cost factors, with help of rest-bus-simulation tools, the physical bus can be simulated with defined parameter as transfer speed, timing and frame length, etc.

3. Rest-bus communication: When there is the need to test real bus communication, the communication bus can be substituted piecewise with real communication bus. Once one of the node or some the nodes are available in the advanced format (from the initial model of variant 1 to the hardware nodes of variant 7), the nodes can be substituted one by one with the advanced form. Thus the whole software-in-the-loop test bench is a free-configurable hybrid system of SW-nodes and HW-nodes, which evolve towards the hardware-in-the-loop test bench.

   The advantages of the rest-bus simulation are significant for the distributed development between OEMs and suppliers in both time domain and space domain.

   - From the time domain: since traditional development cycle of a totally new modern vehicle model is about 3 to 5 years, while the development cycles of microcontroller are much shorter, most of the ECU or μC are not available or only available in form of evaluation board at the early phase of development of in-vehicle electronics.
   - From the space domain: The development of in-vehicle electronics is geographically distributed as a cooperation of OEMs and suppliers. The front-loading of concept evaluation and integration test should be carried out almost parallel to the coding and implementation work on the side of suppliers. It is much simpler and quicker to transmit software models in form of a virtual node, for example, over networks than to ship evaluation boards.

In order to overcome the bottleneck from the time and space domain, the software-in-the-loop test with the rest-bus simulation provides the proper approach for the cooperation between OEMs and suppliers, while keeping the consistent test specification and requirement tracing with the later phase of development.

The common approach of the suppliers, to show the OEM the compliance of software quality assurance measures, is the software module test. In step 5.3 it is the first time that software codes are generated and implemented, According to IEC 61508-3, the software module test in form of SiL should be performed directly after software coding. This SiL-test provides the first opportunity to detect any deviations between the specifications and the software or to find any weak points in the software and to rectify the problems at an early stage.

7.2.6 Part 6: Hazard analysis and validation of FDA model with HiL-test

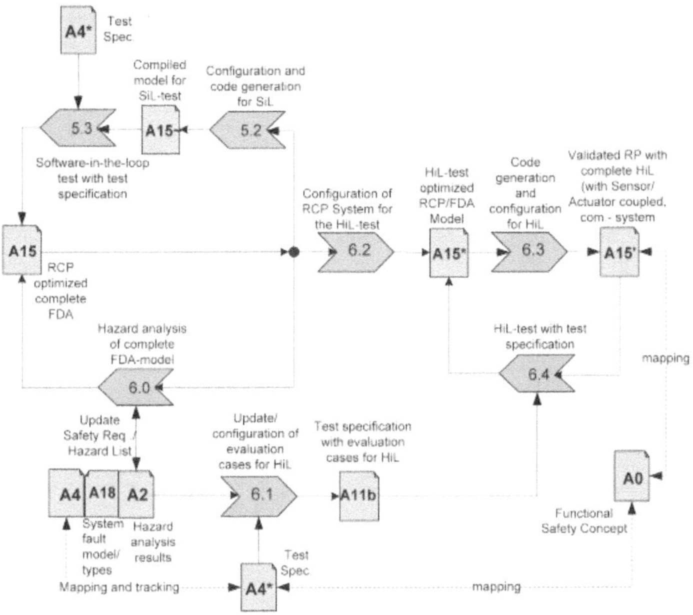

Figure 7-14: Validation of FDA-model with hardware-in-the-loop test

The term "hardware-in-the-loop" (HiL) test is widely used in the automotive industry. The hardware-in-the-loop test for the serial development usually refers to a large and sophisticated test bench with all the serial ECUs, sensors and actuators connected with the real communication bus for the preparation of in-vehicle test, in which the environment as on-road test is simulated as well.

The "hardware-in-the-loop" approach in Figure 7-14 refers to the test bench made up of rapid-prototyping platforms and evaluation boards for the evaluation of FDA models.

Process step 6.0: Hazard analysis of complete FDA-model

Hazard analysis of complete FDA-model extends and updates the results of PHA in step 1.2 and hazard analysis of FAA in step 2.4, through detailed analysis of constituent subsystems and components of the vehicular system, to identify previously unidentified hazards and to assess the associated risks. The new components introduced by the work in steps 5.x have to be analyzed. Results from step 6 work as an update of safety requirements in A4, A 18 and A2. An iteration of re-mapping and tracking between requirement specification and test specification should be performed as well.

Process step 6.1: Update and configuration of evaluation cases for HiL

The evaluation cases from model-in-the-loop test and software-in-the-loop test of legacy can be updated and reconfigured for the hardware-in-the-loop test. The benefits here are quite obvious, it prevents the wheels from being reinvented while keeping the dependability requirements traced as formal test. The evaluation results can be compared to gain more knowledge about the design.

Process step 6.2: Configuration of RCP system for the HiL-test

The RCP-system should be further configured to get ready for code generation of the hardware-in-the-loop test, for example, according to certain system requirement, communication scheduling might need to be separated from the model, some evaluation cases might be embedded in the system model for the sack of test automation and fault injection.

Process step 6.3: Code generation and configuration for HiL

With consideration of hardware resource, requirement of evaluation/test cases, interface of sensor/actuator/communication system, availability of hardware platform, cost and other constraints, different rapid-prototyping platform can be chosen. With appropriate code generator and tools, source code of the models as well as the evaluation case can be generated or implemented and compiled for the hardware-in-the-loop test.

Process step 6.4: HiL-test with test specification

As discussed in step 5.3, with the substitution of the nodes from pre-variant in the early development phase with the advanced variants, a seamless evolution of the software-in-the-loop test to the hardware-in-the-loop test can be made with consistent reusable test specification and test cases. In a complete HiL test-bench, the test environment is more realistic, the timing-constraints can be checked under real-time condition (comparison with the safety requirements, assumptions and previous testing results is possible) hardware faults according to the fault-types definition can be injected directly to the system. The results of the evaluation results from the HiL-test can be looped back to the FDA-model quickly.

For the ISS, as introduced in Process step 2.4: FAA Hazard Analysis (FHA), the allocated ASIL to the components as well as to the system should be validated in the HiL-test bench as well. The validation results from the iteration in A15 according to the test-specification A4* will be mapped and integrated to the Functional Safety Concept in A0.

7.2.7 Association of ISS Engineering Process with ISO26262

As introduced in Subsection 3.5.2.3, the ISO norm 26262 (draft version) is currently the most promising and future oriented safety norm in the automotive industry. Although the most important measures suggested in ISO 26262 are integrated implicitly in ISS Engineering Process here, it is necessary to map the safety steps from ISO 26262 to the ISS Engineering Process. Safety systems, developed by ISS Engineering Process should prove their conformance to the ISO26262 finally, to show the evidence that all reasonable safety objectives are sufficiently met.

As note in Subsection 3.5.2.3.3 the ISO 26262 is divided into eight parts, due to the limited focus of ISS Engineering Process (as shown in Figure 7-2), only the following parts of ISO 26262 are associated directly with ISS Engineering Process: Part 2 Management of functional safety [ISO22], Part 3 Concept Phase [ISO23], Part 4: Product development System [ISO24], Part 5: Product development: Hardware [ISO25], Part 6: Product development: Software [ISO26] and Part 8: Supporting processes [ISO28].

Association with ISO 26262 Part 2

As depicted in Figure 3-20, development reference phase 4.4 in ISO 26262 [ISO22] is the initialization of product development. The mapping of this part is listed in Table 7-1.

ISO WD 26262 part 2: Management of functional safety	ISS Engineering Process
Subsection 4.4.6 [ISO22], in the safety management during the development process, for all categories of safety relevant systems (ASIL A, B, C and D), the safety responsibilities of the persons, departments and organizations responsible for each phase during development should be defined.	In ISS Engineering Process notated in Figure 7-4, the dependability related process steps should be ensured to be carried out the by a safety manager, who is appointed by the project manager to take the role for functional safety management tasks.

Table 7-1: Mapping of ISO 26262 Part 2 to ISS Engineering Process

Association with ISO 26262 Part 3

The ISO26262-3 - Concept Phase [ISO23] includes three parts: system definition, hazard analysis and risk evaluation and safety concept, which are mapped to ISS EP in Table 7-2.

ISO WD 26262 part 3: Concept phase	ISS Engineering Process
Chapter 5 Initiation of the safety lifecycle, during the system definition phase the safety lifecycle should be initiated.	Because ISS Engineering Process is firstly conceived for the design and prototyping of innovative ISS-applications, the life cycle here is generally considered as new development.
Chapter 6.4.7, a safety goal (different ASIL categories) should be assigned as an attribute to the system, sub-system and components in this process step.	The development phase of hazard analysis and risk evaluation is mapped to the ISS Engineering Process 1.2, 2.0, 2.4, 4.4 and 6.0, as a continuous assessment step of system hazards according to the latest knowledge about the system design. Because the detailed risk analysis and assessment is not the emphasis of this dissertation, the detailed activities specified in [ISO23] Chapter 6 are followed in ISS EP without adaptation.
Chapter 7 Functional Safety Concept	In the ISS Engineering Process Part1 and Part2, the Functional Safety Concept is mapped to different ISS-steps. In Part6, the validation step is mapped to the Functional Safety Concept as well.
Result from ISO26262 Part 3 (Subsection 7.2) is the functional safety concept such as architectural decisions and the concepts including the warning and degradation concept as well as the redundancy concept.	The design steps of fault detection and failure mitigation specified in [ISO23] are all considered in ISS Engineering Process 5.1, while the design of driver warning is not in the scope of this dissertation.
Subsection 7.4.3: Derivation of functional safety concept is the allocation of ASIL requirements (combination/decomposition) to the system/sub-system/components.	ISS Engineering Process Step 2.0, 2.4 and 4.4. Based on the new knowledge from the design step, the overall system safety requirement (ASIL) can be decomposed to the sub-system/components, thus the dependability related RS in A4.3, as a living document, should be continuously updated. This decomposition of ASIL to the components is one of most important criterion of verification and validation for HiL-test in step 6.4.

Table 7-2: Mapping of ISO 26262 Part 3 to ISS Engineering Process

Association with ISO 26262 Part 4

The ISO 26262-4 - Product Development System [ISO24] covers the reference phase "System Design" as shown in Figure 3-20, which is mapped in ISS EP in Table 7-3.

ISO WD 26262 part 4: Product development system	ISS Engineering Process
Chapter 6 System design, Subsection 6.3 The prerequisite of this part is the system requirements specification.	Product of ISS Engineering Process 2.0 as a artifact A4
6.4.1 Process requirements: review of safety concept, safety requirements specifications, design specification	A4, A4* and mapping between them. In each development phase of the FDA-model, the appropriate design specifications will be updated.
6.4.2 Requirements for architecture	Integration of dependability architecture - ISS EP step 3.3 and A18, ISS EP step 5.1 and A17.
6.4.3 Avoidance of systematic faults, for all ASILs, separation of safety related sub-systems and other sub-systems, avoid impacts of common cause failures. 6.4.3.6 Modular system design	Definition of system fault model and fault type in ISS EP step 3.2, 4.4a and artifact A18. Other non-functional RS as design guidelines. Application of dependable HW/SW-platform
6.4.4 Methods for detection and control of random failures	Dependability services in HW and SW in ISS EP.
6.4.6 System design partitioning into hardware and software	Development approach from FAA to FDA by mapping of FAA to HW in ISS EP.
6.4.7 Verification of system design in System design inspection, simulation, prototyping, safety analyses	Virtual front-loading with RP and FMEA in ISS EP.
6.4.8 System design documentation	UML/SysML based design, FAA and FDA in model-based approach.

Table 7-3: Mapping of ISO 26262 Part 4 to ISS Engineering Process

Association with ISO 26262 Part 5

The ISO 26262-5 - Product Development Hardware [ISO25] covers the reference phase 4.6 to 4.7 (depicted in Figure 3-20) of ISO reference model, which is mapped according to Table 7-4.

ISO WD 26262 part 5: Product development Hardware	ISS Engineering Process
4 Hardware – Requirements analysis	The HW RS in ISS Engineering Process 2.0 as a artifact A4
5 Hardware architecture design and implemented measures for the control of random hardware failures during operation. 6 Quantitative Requirements for random hardware failures	ISS EP 3.2: Identify HW fault model and risk mitigation measures for systems of ASILs, while the failure rate analysis of the HW is not the focus here. The detailed activities specified in [ISO25] are followed without adaptation.
7 Measures for prevention and control of systematic hardware failures	Not covered by the ISS EP, such HW failures can not be tested by means of RP.
8 Safety hardware integration and verification 9 Qualification of parts and components, e.g. compliance of electronic components with its definition of safety	Not covered by the ISS EP
10 Overall Requirements for software hardware interface: development of a SW/HW interface of a microcontroller included in a technical safety concept.	Partly covered by ISS EP, e.g., definition and verification of timing constraints

Table 7-4: Mapping of ISO 26262 Part 5 to ISS Engineering Process

Association with ISO 26262 Part 6

The ISO 26262-6 - Product Development: Software [ISO26] is a parallel activity to the hardware development in ISO reference model. The mapping of the activities there is specified in Table 7-5.

ISO WD 26262 part 6: Product development Software	ISS Engineering Process
5 Software safety requirements specification with suggested methods of natural language or computer-aided specification tools or semi-formal methods.	ISS EP step 1.1, 1.4, 2.0, 2.3, 2.5 and 4.6. applied to all ASILs
6 Software architecture and design	Applied to all ASILs in ISS EP.

ISO WD 26262 part 6: Product development Software	ISS Engineering Process
6.4.1/6.4.2 Choosing of design methods and tools for non-functional requirements to SW	ISS EP tool chains, e.g. documentation and compute aided-designs
6.4.3/6.4.4/6.4.5/6.4.6 Hierarchic software architecture, non-functional requirements to SW, guidelines for the modeling of SW	ISS EP 5.1, A16 and A17 application of dependability layered SW-architecture, Modeling of FAA and FDA with UML
6.4.7/6.4.8 Allocation of safety integrity level to the SW-Cs and evidence of independence with ISO26262-8 Annex Concept of the independence of software components (draft)	Partly supported by ISS EP with the introduction of design step of fault model/fault region
6.4.9/6.4.10 Hazard analysis in SW and measures for detection and handling of errors	ISS EP 4.4, 5.1, A16 and A17 and iteration of the steps
6.4.11-6.4.16 Individual specification and verification of the SW-C/SW-architecture, e.g. reuse or sourcing and configuration	ISS EP 5.1 (esp. 5.1.3), A16 and A17
7 Software implementation	Applied to all ASILs, only partly covered because ISS EP with focus on RP
8 Software unit test 9 Software integration test	Partly covered by ISS EP 4.4c, 6.4 and following iteration steps, not all the testing methods suggested in the norm are included
10 Software safety acceptance test 10.4.1 A software safety acceptance test shall be planned.	Covered by ISS EP completely.
10.4.2 Appropriate test methods for the software acceptance test shall be selected, e.g. interface, HiL, HiL within ECU network and test in the test vehicles	Partly covered by ISS EP until HiL-test with ECU network
10.4.4 Test specification with test cases, tests from the preceding test phases and their results can be reused.	ISS EP follows this guideline exactly, with the consistent test case of mapping and reuse
10.4.5 The software safety validation shall be performed on the target system.	Not within the scope of ISS EP

Table 7-5: Mapping of ISO 26262 Part 6 to ISS Engineering Process

Association with ISO 26262 Part 8

The ISO 26262-8 Supporting Processes [ISO28] describes a broad scope of topics. The relevant parts to the implementation ISS EP are listed as in Table 7-6:

ISO WD 26262 part 8: Supporting processes	ISS Engineering Process
4 Interfaces within distributed developments	ISS EP itself covers more technical issues than the organizational topics; for discussions here related to ISO 26262, see Subsection 7.4.
5 Overall management of safety requirements such as communication, traceability (wherefrom, whereto), documentation and administration	Traceability and documentation are completely covered in ISS EP, while the other topics are discussed in Subsection 7.4.

Table 7-6: Mapping of ISO 26262 Part 8 to ISS Engineering Process

7.3 Tool chains for the development of ISS

A suggestion of tool chain to support the development of ISS is given here, in which the tool chain is categorized into software tools and hardware prototyping platforms.

Main categories of software tools:

- Tools to support requirements engineering process and requirement tracing as well as design process.
- Tools for development environment of embedded software, model-based tools, development environment and code generators.
- Tools to support interchange format global process, e.g. interface description and communication scheduling.
- Tools to support mapping processes of FAA to FDA.
- Scheduling design of communication system, synchronization of tasks and communication.
- Tools for test automation and design of evaluation cases with fault injection in the SiL and HiL test bench.

Main categories of hardware prototyping platforms:

- Rapid-prototyping platform of general purpose and Rapid-prototyping platform for special purpose with specific hardware interfaces.
- Evaluation board from semiconductor producers.

7.3.1 Software tools

- Tools for requirements engineering as listed in Table 7-7 [EAD42]
  - Gathering, categorization and documentation of the requirements
  - Requirement tracing in design, implementation and test phase
  - Change management of requirements to the design

Tool	Vendor	Features
Word, Excel	Microsoft	Capturing and structuring of requirements
DOORS	Telelogic	Support for requirements tracing
Requisite Pro	Rational	Full featured support for requirements engineering
CaliberRM	Borland	Full featured support for requirements engineering
IRqA	TCPSI	Full featured support for requirements engineering
ARTiSAN Studio	ARTiSAN Software	Requirements diagram with SysML

Table 7-7: Tools for requirements capture and analysis

- System Analysis and Design
  - Identification of functions (in terms of their interfaces) as part of an overall functional feature network (equipments and features of the E/E-architecture in plan).
  - Definition of the vehicle E/E-architecture (i.e. number and types ECUs and connecting bus systems including gateways).
  - Mapping of functions to ECUs.
  - Identification of the communication matrix (transmission of inter-ECU signals via bus systems, mapping of signals to frames, mapping of frames to busses, assignment of message priorities in case of event-triggered protocol, temporal scheduling of frames in case of a time-triggered protocol like FlexRay).
  - System level diagnosis, i.e. backward and forward error analysis.
  - System level analysis, e.g. feasibility of the distribution or allocation is checked, i.e., whether the requirements of the application system are satisfied, e.g., memory and processor capacity on the ECUs, capacity on the communication busses, and overall timing requirements.
  - Documentation of system design, test specification and verification results. By application of the software tools, while following certain development guidelines, a framework of such documentation can be generated automatically. Of course these documents should be reviewed and updated manually finally.

The software tools here are listed in Table 7-8.

Tool	Vendor	Features
DaVinci	Vector Informatik	Widely featured system level design for automotive software systems, software integration on single ECUs, embedded software design according to AUTOSAR principle
Preevision	Aquintos	Design and evaluation of in-vehicle E/E-architectures
INTECRIO	ETAS	Structured design of software components
DesignerPro	DECOMSYS	System level design with special focus on FlexRay as a time-triggered bus system
RT-Builder	TNI Software	System level design with special focus on real-time analysis
ARTiSAN Studio	ARTiSAN Software	Structured design of software components
Enterprise Architect[2]	Sparx Systems	Structured design of software components with UML2.0

Table 7-8: Tools for systems analysis and design

- Function Analysis and Design

  The tools listed in Table 7-9 are operated in a pure PC based environment. When the correct interaction with a real-time environment shall be validated, the functional models can be extended to operate on real signals of this environment.

Tool	Vendor	Features
MATLAB, Simulink, Stateflow	Mathworks	Graphical modeling and simulation of data and control flow models, code generation for prototyping purpose
ASCET	ETAS	Graphical programming language for data and control flow models
SCADE	Esterel Technologies	Graphical modeling of synchronous control systems
INTECRIO	ETAS	Rapid prototyping system supporting different models of computation
Real-Time Interface	dSPACE	Block-set extension for MATLAB/Simulink to support rapid control prototyping
MTest	dSPACE	Model based testing of MATLAB/Simulink models

Table 7-9: Function analysis and design

[2] For reasons of simplicity only one representative for the large family of UML tools is mentioned here. These tools are similar in scope and usually support UML 2.0 or derived UML profiles as SysML.

- Software Analysis and Design

  The function design tools in Table 7-10 are usually accompanied by code generators, which convert the abstract functional description carried out by means of the design tool to source code. The latter can be further processed and compiled to ECU-ready executable code. The usage of automatic code generators can further contribute to reliability and safety aspects of ECU software. Also the manual coding of the control software, supporting analysis tools e.g. for code coverage analysis or code style checking can also be used in combination with code generators.

Tool	Vendor	Features
TargetLink	dSPACE	Production code generation, code coverage analysis and target simulation
Real-time Workshop	dSPACE	Code generation for rapid prototyping platforms
Embedded Coder	Mathworks	Production code generation
ASCET	ETAS	Production code generation
CodeWarrior	Motorola	Development environment for the Freescale microcontroller platform
Visual Studio	Microsoft	Development environment of C
Embedded Validator	OSC	Formal model checking
Polyspace Verifier	Polyspace	Code analysis

Table 7-10: Software analysis and design

- Tools in Table 7-11 are applied for the scheduling design of communication system, synchronization of tasks and communication in the development of ISS.

Tool	Vendor	Features
DesignerPro and SIMCOM	DCOMSYS	Separation of functional design and communication design, scheduling and configuration of FlexRay [DSC05]
CANoe FlexRay extension	Vector Informatik	scheduling and configuration of FlexRay, rest-bus simulation of FlexRay, gateway design

Table 7-11: Design of communication system with FlexRay

- Tools for test automation and design of evaluation cases with fault injection in the SiL and HiL test bench are listed in Table 7-12.

Tool	Vendor	Features
ControlDesk	dSpace	Central module of dSpace experiment software to manage and instrument the experiments with integrated Simulink interface for offline management of controller models and extended options for automation.
ControlDesk failure simulation	dSpace	A software component of ControlDesk to drive electrical failure simulation in the ECU cable harness, Remote control of the failure insertion unit.
AutomationDesk	dSpace	Environment for test automation of calibration and measurement tools as well as automatic test generation.
CANoe	Vector Informatik	Rest-bus simulation with SiL-test

Table 7-12: Design of evaluation and test automation

After this brief list of software tools, another important point here worth mentioning is the interaction between tools. Different tools offered by a single vendor are usually designed in such a way that they can be combined and interact with each other to solve complex engineering tasks as an integrated process. Beyond that, long standing partnerships between different companies have been established in the past, which enables the interoperability of product offerings, e.g. the interconnection between Matlab/dSpace tool chain and DECOMSYS tool chain.

In recent years research consortia and industry partnership have contributed to this by defining standard to enable the interoperability of tools. Among these groups are ASAM, HIS and AUTOSAR. Interoperability of tools can be accomplished by diverse technical means, such as (i) run-time interaction of tools via well-defined APIs and (ii) definition of interchange formats to be read/written by tools in the global process, e.g. interface description, communication scheduling with common communication design framework with XML.

7.3.2 Hardware prototyping platforms

There are wide varieties of hardware prototyping platform in the automotive industry. Here, a short overview about typical applied products is given:

- dSpace products
  - MicroAutoBox hardware: real-time fast function prototyping platform with a wide range of applications with CAN, LIN, and FlexRay interfaces.
  - Simulator hardware: Hardware-in-the-loop simulators and simulator-specific hardware for ECU testing.
  - AutoBox and modular hardware: AUTOBOX is an expansion box to load modular hardware with wide range of I/O boards and processor boards.
- Vector-Informatik and TZM products
  - Network interfaces to CAN, LIN, FlexRay and MOST
- DECOMSYS products
  - DECOMSYS::BUSDOCTOR is a FlexRay monitoring node with measurement features connected to the FlexRay bus and visualization of bus communication.
- Mathworks products
  - xPC system: real-time rapid prototyping and hardware-in-the-loop simulation using PC hardware
- Evaluation boards from various semiconductor producers
  - Usually years before the serial microcontroller are available as mass product for the automotive serial production, the semiconductor producers will provide evaluation boards for OEMs and suppliers to test the main functionalities in the main-core and interfaces of the future products. The experiences gathered with the prototyping can be applied to the serial development directly.

7.4 The ISS Engineering Process under challenges

7.4.1 View from side of OEM

Similar to the development of in-vehicle electronics from other vehicle domains, the development of ISS is typically distributed between several partners.

Before the project initialization phase, the OEM will usually have an assessment of suppliers. According to ISO26262 [ISO28] 4.4.2, the key criteria here is the capability and experience of the supplier in the development of systems with comparable ASIL. In the initiation phase, persons in charge of safety issues (Safety Manager) should be assigned both for the OEM and suppliers. During the whole product life cycle, OEM will require a consistent practice of the ISO26262 from suppliers to their sub-suppliers as well as sub-contractors.

The OEM is usually responsible for the overall functionality of the vehicle, i.e. the OEM defines which features and sub functions are installed in a vehicle platform or in a specific variant. Looking at certification and homologation, the OEM is also responsible for the safety of the vehicle as a whole and must be able to prove that the vehicle is "safe enough" to drive on the road. Thus OEMs should have enough skills in system integration, architectures of embedded systems and making strategic decision of employing new approaches for embedded systems.

Usually, the OEM does not develop all the system functions alone, but with the help of suppliers. As shown in Figure 7-15, the OEM defines the components and the interfaces that he then gives to suppliers for implementation. Another possibility to distribute workload is the splitting of work at certain steps. E.g. it is possible that a vehicle function is developed on FAA level by the OEM. The following steps are given to other parties for detailed development or for target integration.

The relationship between OEM and first-tier suppliers can be extended as a template for the relationship between a 1.tier supplier and his (sub) suppliers, and so on.

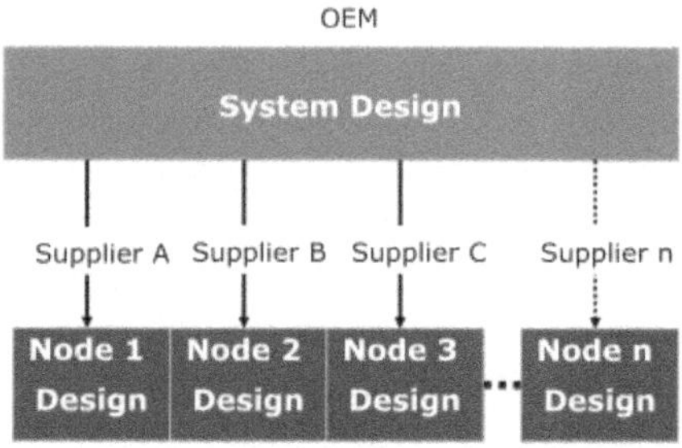

Figure 7-15: OEM - Supplier Workflow

With the challenges mentioned in the Subsection 3.5.3, the development of an inter-domain and inter-function ISS-application is characterized by an intensified distributed collaboration between OEMs and suppliers.

By the development of ISS, the OEM should cooperate much more closely with suppliers. OEMs (mostly high-end car-makers) add value by the integrating of different products from suppliers, thus the OEM should work as the initiator for the system architecture.

As depicted in Figure 7-16, in the first step, driven by the customer perceivable innovations, the features of the to-be-designed vehicle platform (as a basis of different vehicle model variants) are designed. The features are grouped to systems and prioritized to focus the efforts.

In the second step, a feature and function network (FDA in ISS Engineering Process) is defined, which is mapped to the third step hardware architecture to form the Functional Design Architecture. The ISS Engineering Process mainly covers from step 1 to step 3. The mapping process, disassembly and assembly of grouped and prioritized features to the functions in FAA, from FAA to FDA is considered as the enabling transmission between the steps, where the dependability aspects are considered with the ISS Engineering Process.

In the fourth step constraints, such as installation space in vehicles, cable harness (length, EMC), will be considered for the installation in vehicles.

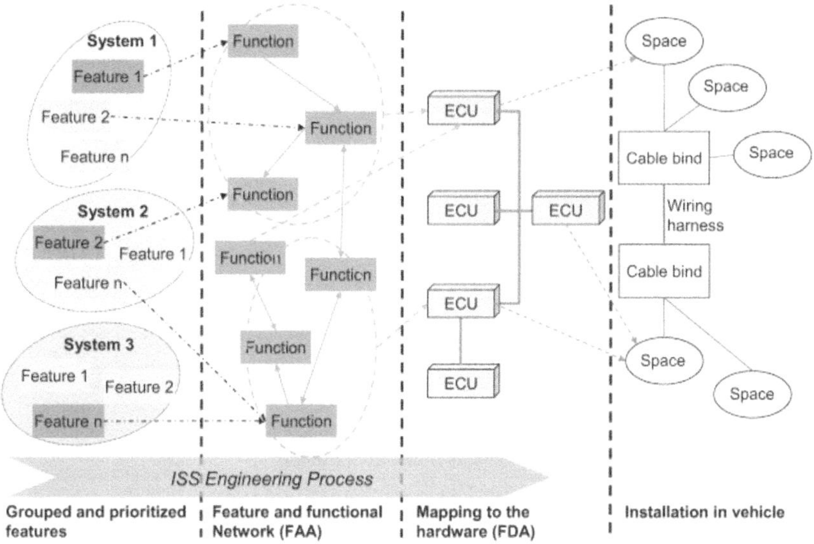

Figure 7-16: OEM view in the design of electric/electronic architecture

If the OEMs are only a "lean" branded integrator, the system suppliers should drive the architecture and provide integrated solution of the final vehicle with system know-how. In this case, however, an obvious disadvantage can be foreseen. The automotive OEM has to expose his vehicle function knowledge to the system supplier, who also supplies to the manufacturer's competitors. It is quite challenging to protect intellectual property in such an environment. Re-implementation of vehicle functions results in design cycles of several weeks (outsourced engineering). Often, the function that might return to the system integrator does not fully match the required behavior, resulting in additional iterations because of misunderstandings.

7.4.2 View from side of supplier

On the suppliers' side, a common approach is to gather the different requirement specifications from different OEMs to the comparable systems. At the beginning, the requirements from OEMs are gathered and merged together. By applying the design to cost principle, the requirements are sorted and prioritized as a whole.

In the development of ISS, the system supplier (first-tier), is responsible for the development of the system (more than one components) and must be able prove that the system meets all the requirements incl. safety requirements.

Considering the sophisticated system design, it is reasonable to build up one lean system architecture to integrate the most important functionalities, using standard dependability software architecture and standard hardware modules. In this early phase, the majority of effort can be focused on the design and validation of this lean prototypic platform instead of provision of customer specific solution. The approach of applying platform with standard software and hardware architecture enables an efficient and cost optimized configuration and extension to the additional requirements of each OEM.

Thus the "wrappers" and too much customization parallel to the end phase of finishing the validation of the lean-system can be reduced. Efforts of development teams can be transferred to configure the customer specific features and fine calibration.

Particularly for the development of safety relevant systems, confirmation measures to the safety goals according to ISO26262 [ISO28] 4.4.9 "safety assessment at supplier's premises" are more penetrative to the supplier. Except for the review of FAA, FDA as well as the safety concept suggested in ISS EP, supplier should provide "safety cases" as safety evidence of all ASILs. Additional for ASIL-C, FTA, safety assessments and safety audits should be executed by the supplier. For ASIL-D, common cause failures (CCF), common mode failures (CMF) should be considered as well.

7.4.3 Distributed cooperation between OEM and suppliers

In the following, the two most important enabling levers for improving the distributed development between OEM and suppliers are discussed:

Process:

By the development of ISS, the OEM should have a clear defined role and core competencies. The OEM should agree with the suppliers on the interface in the cooperation and responsibilities.

One good practice is that the OEM involves suppliers in the early phase of system design. OEM's define high level architecture and standardization guidelines, but leave enough freedom for the supplier to decide on the implementation approach as software libraries and modules for effective reuse. The design guide, design approach, implementation experience, lesson learned and test specification in the rapid-prototyping phase can be seamlessly transferred and reused in the serial development and production. In the later phase, a well structured development model is applied to let the supplier develop the specification further. The prioritized features and requirements are traced together, while the supplier focuses on the component test and the OEM takes the role of system integration and system dependability test. Here it is noteworthy that both OEM and supplier should agree on the integration effort provided by both sides.

Architecture

The application of standardized dependable software architecture in ISS creates benefits for a typical "win-win" business situation between OEMs and suppliers.

- OEM can share across better validated legacy platforms software with suppliers.
- The inter-domain interfaces in the ISS (e.g. engine – transmission) can be serviced better.
- The platform strategies create technical pre-conditions for business with "software options"
- It leaves room for improvements and innovation, because an ISS system based on platform provides more flexibility for adaptation.

8 Hardware architectures for the Integrated Safety System

In this chapter concepts about the hardware architecture for the Integrated Safety System are discussed as a framework from different aspects. Based on the ISS-requirements for the hardware architecture as introduced in Subsection 3.3.1, the most important building-blocks of the hardware architecture for ISS and its configurations are specified here. As introduced in Chapter 7 about ISS Engineering Process, after the system design along with the technical safety concept [ISO23] is created, the next step is the design of the hardware architecture according to the steps in ISS EP Part 3 (Subsection 7.2.3). The emphasis of this chapter is focused to the design of hardware architecture of highly safety relevant ASIL-C and ASIL-D applications (as defined in Subsection 3.5.2.3.4), while the hardware architecture for ASIL-A and ASIL-B is not considered in this chapter.

The structure of this chapter is organized as following:

As introduced in Subsection 3.3, the automotive hardware architecture refers here to the three main views: system topology, communication systems and ECU architecture. In the following subsections, the design of hardware architecture for ISS is discussed from these three aspects. The last subsection here gives to each aspect a design use-case (based on dummy ASIL-D ISS applications), which concludes the discussion as about the hardware architecture framework.

8.1 Concepts of the system topologies

Following the V-process and the discussion as shown in Figure 7-16, usually OEMs take the responsibility to define the vehicle wide system topology and communication system, while the system suppliers and sub-suppliers define the ECU local hardware architecture according to the requirements of OEMs.

Generally speaking the definition and assessment of system topology for the in-vehicle electronics is a sophisticated challenge, which is influenced by many constraints, like application real-time requirement, cost, packaging, dependability, communication bandwidth, reuse of legacy system (take-over parts), maintenance, platform strategy of different models, supplier strategy, etc. This list of various constraints can be extended even much longer and much more detailed. What makes it even complicated is the weight of each metric/constraint for each OEM is seldom the same. It is even different for different product lines within one OEM.

For the discussion in this dissertation, since the cost-analysis and packaging issues are only meaningful for the design of a certain vehicle product line and beyond the dependability discussion here, the focus was chosen on the design of ISS hardware architecture in the early phase according to functional and safety requirements of ISS-applications. The granularity was limited on the definition of system topology from a high layer.

The approach here can be started with the question of which ISS-applications/features do we need for a certain vehicle platform, which extensions are expected on this vehicle platform. The answer of this question can contain the following parts:

1. A list of prioritized features or applications
2. To realize these functions, a list of certain components is needed e.g. sensors, actuators and interfaces (also HMI)
3. Furthermore the applications can be broken up into functions, which are connected to each other via interface for information exchange. (a very first functional network on the application level)

While the answer to 1st and 3rd question is a rough FAA model (as shown in Figure 7-7), the answers to the 2nd question is like a collection of "Lego"-modules, directly resulted from the features. With the first answers to these questions, the design of system topologies can be started.

8.1.1 Future frameworks and design guidelines of system topologies

As shown in Subsection 3.3.2, the in-vehicle system topology is typically organized in different domains, while the number of domains is mainly driven by the introduction of new features, which can not be integrated into the traditional domains because of e.g. lack of communication bandwidth, quality of services or management of domain complexity. On the other hand for the cost reasons the legacy domains and communication systems are preferred for the reuse, but at one point the needs of innovation and technical complexity to keep the legacy system running may overwhelm the cost benefit of take-over.

One of the most important design criteria, regarding the choice of domains, is the technical requirement of functions, in which safety is one of most critical technical requirements. To give an example, for the ISS-applications distributed beyond the domain borders, in each domain at least one gateway is needed for the inter-domain communication. In order to fulfill the functional requirement e.g. QoS and safety requirements communication e.g. PFH, etc., one potential approach is to use one high-speed communication network and connect all ECUs together. But on the other side a high-speed communication link is more costly than a low speed link. Thus a hybrid system with domains connected with gateway might be preferred.

In consideration of the ISS-Applications:

- Small and mid size cars have today two vehicle domains, body and powertrain. These two domains will be equipped with more add-on telematics service and safety applications.
- Today's high-end vehicles have three or four vehicle domains: body, powertrain, telematics and an optional chassis-domain, while the future high end car can have five or six vehicle domains, body, powertrain, telematics, chassis and optional HMI, environment sensing and/or safety domain.

For the further discussion, three types of possible future architecture frameworks are given in the following as vertexes. The future platform, however, can be a mixture of them [HWA06].

Figure 8-1 shows the backbone architecture, as a possible system topology for the high-end luxury vehicle platform, where the time-triggered communication bus connects the traditional vehicle domains and the chassis domain. Although the backbone bus and gateways between traditional networks will introduce additional development effort, this future oriented hierarchical architecture provides more potential for the cross domain communication regarding bandwidth and service quality. If some future ISS-applications should require a fault-tolerant or fail-operational behavior for the inter-domain communication, the backbone architecture with time-triggered communication bus could provide the best preliminary condition.

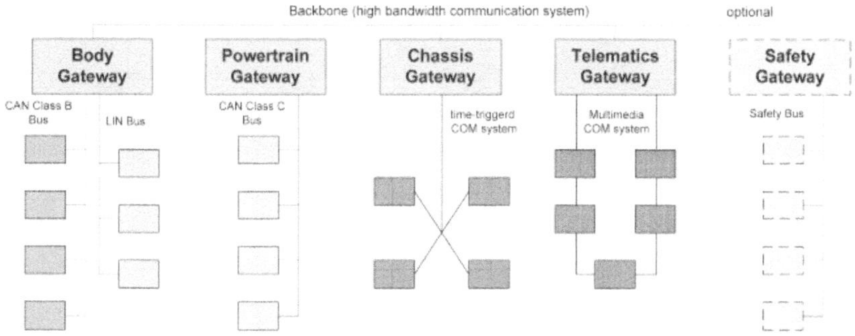

Figure 8-1: Backbone architecture

The other end of backbone architecture is depicted in Figure 8-2, in which the communication between the different sub-domains is provided by a single central gateway.

This solution requires a relative sophisticated gateway supporting a complex functionality to service all the different bus systems (e.g. bus interfaces) and to perform the routing and conversion tasks necessary. For the low-end vehicle, however, when not all the domains are required (in the simplest case only the CAN-B for the cabin/body domain and CAN-C for the chassis/powertrain domain), the central gateway architecture with simple CAN-routing is for sure a cost effective architecture solution.

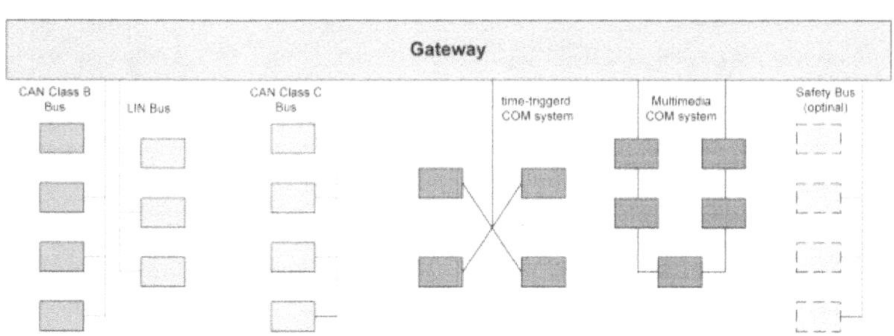

Figure 8-2: Central gateway architecture

A middle way solution is shown in Figure 8-3, where the sub-systems are provided by gateways integrated into single ECUs in a relative loose but flexible architecture. Low end and effective gateways (with 2 or maximal 3 bus-interfaces) are applied to connect the different domains.

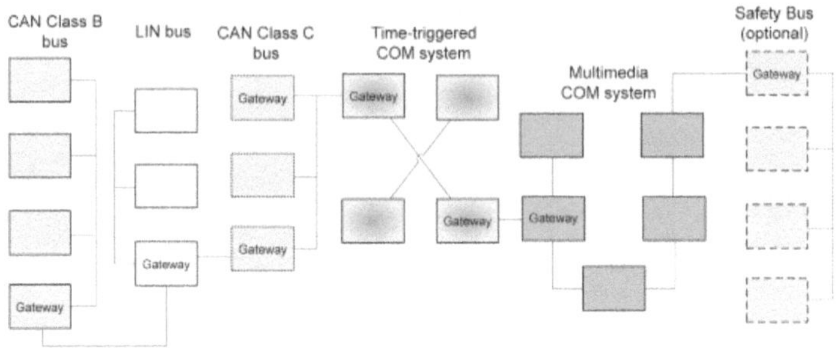

Figure 8-3: Multi gateway architecture

8.1.2 Distribution of ISS applications to the system topology

The FDA-model specified in ISS EP is derived from the mapping of FAA-model to the hardware system topology. The question how to distribute applications to the system topology is again highly application specific and there exist no general rules, taking all the constraints and guidelines into consideration. In the following, however, a few basic rules for the mapping and partitioning of ISS-applications are discussed:

1. Safety functions with hard real-time requirements to react to external events (e.g. tight control loops) must be placed on the ECUs, which are close to the event in the control loop, e.g. actuator control functions have to be placed on the ECU directly connected to the actuator.

2. Traditionally sensors are directly connected to the consumer of the information, usually to the same ECU as the actuator, thus the same vehicle dynamic data is redundant measured at different position of vehicles with appropriate resolutions and ranges for different applications. With the introduction of ISS, however, as long as timing (bandwidth) and safety requirement are fulfilled, the sensor can be displaced freely within the in-vehicle network (e.g. steering angle sensor). With distribution of sensor information, redundant sensors can be saved.

3. To keep the amount of data exchange, the network traffic under a reasonable limit, the applications, processing large blocks of raw data, such as data from camera or radar, should be mapped close to the source, e.g. by using intelligent sensors to provide information ready-to-use to the communication system.

4. Distribution of vehicle dynamic data with sufficiently high resolution (e.g. velocity, engine state, battery state and driver demand) can lead to new applications with sensor data fusion and share. But as soon as the safety relevant data is sent from one ECU to the other, the whole end-to-end data transmission path should inherit the highest safety integrity among the applications, which use this communication path. The application of

time-triggered communication is one possible solution but it alone does not solve the problem. Time triggered communication should be supported with other dependability software services for ISS such as application CRC and Agreement Protocol, etc. In this case event-triggered communication can be applied as well, if the communication can be supplemented with other dependability mechanisms as discussed in Subsection 8.2.

5. High level control functions can be usually placed on any ECU, where the necessary resources are available (processing power, memory and communication interfaces etc.), without degradation of performance. That is to say, the high level control functions could be freely shifted as long as software partitioning is guaranteed with e.g. the mechanisms discussed in Subsection 9.3.2.

6. Basic safety functions, such as brake, steering control and airbag, must not be integrated on the same ECU to avoid single point of failure. The approach of using fail-operational nodes without mechanical back-up in this case is still not suitable due to the system complexity in the near future.

7. For the same reasons, it is sometimes still necessary to keep the primary sensors of these basic functions redundantly distributed on different nodes. The distribution of sensor values via communication bus could make communication system as the weakest chain of the whole system, e.g. distribution of ASIL-D signals with CAN alone is not dependable enough for requirement of ASIL-D application. That is to say, if ASIL-D relevant signals should be transferred via CAN, additional safety services e.g. application CRC as discussed in Subsection 9.3.1 should be applied. In this case, the redundant sensors can be still applied, so that the ASIL-D application is able to work in the stand-alone mode in case of total communication failure.

8. For these redundant sensors mentioned in design rule 7, the state of the art situation is that little or less plausibility check among the sensors is applied. Such redundant signals can be used for plausibility check to gain more system safety, but the system availability might not be improved. Thus the local primary sensor should have a higher weight in the voting mechanism of plausibility check.

9. Reuse of legacy system (take-over parts) can be an important constraint by the distribution of ISS applications. Analysis shows that the warranty cost alone of ECUs (the additional development cost not considered at all) that have been significantly extended or redesigned compared to the predecessor model are up to five times that of reused ECUs [McK05]. Thus the cost/benefit-analysis of remapping should be carried out carefully.

10. If it is decided to integrate some of the safety functions on the same hardware, it is preferable to integrate only functions from the same domain (e.g. chassis-domain) with very tight communication/interaction on one ECU, because these functions usually have similar safety integrity requirements. On one side, due to the similar safety requirements, the same dependable hardware architecture can be applied. On the other side, the domain structure can be maintained, which is important for architecture simplicity.

11. The component safety integrity level can be reduced by intelligent mapping of safety functions of ISS to the hardware platform. This approach is illustrated with the following example:
    The safety relevant driver assistant function Adaptive Cruise Control (ACC) can be designed that the ACC gives the ESP-system the braking command. If the braking command will be 100% executed by ESP, the ACC should be developed as the same ASIL-D application as ESP. Not only a 99% DC (Diagnostic Coverage) of ACC should be verified but also the signal path from ACC to ESP should be kept as redundant as well. But if we can develop a sub-function in ESP (according to the ASIL-C/ASIL-D requirements), which limits the brake command from ACC, so that even the ACC gives a faulty command, the vehicle can be still controlled by the driver and the following vehicles have enough reaction time, the safety requirement of ACC can be now reduced to ASIL-A. From this example, we can see, by intelligent design and distribution of safety relevant functions (mapping of the limiting function to ESP instead of ACC-ECU), the safety requirement to the component can be reduced.

12. For reasons of simplicity and reliability (e.g. signal latency), the communication path between two functions that need to exchange information should be kept as short as possible.

13. Distribution of redundancies of ISS-applications:
    In order to meet the high dependability requirements of fail-safe and fail-operational, the conventional approach is to make use of redundancy. Typical examples of redundancies are separated power supplies, sensors, computation units, communication networks and actuators against common mode faults.

    The integration of several functions on one ECU increases the potential threat of losing several safety functions at one time. Such threat can be reduced by the following two approaches:

    a. Integrate only "related" functions on one ECU, so in case of a failure only one branch of the safety function tree is lost. This approach is best used if it is combined with a smooth transition into the degraded mode, thus the driver has the chance to react to the degraded driving behavior of the car.

    b. Integrate functions on redundant ECUs, so that a single point of error does not lead to any degradation. Redundant units, can resident on the same hardware node or on remote nodes, which fulfill the same functionality in the system runtime or are activated when the original unit fails to work. By using dynamic reconfiguration (see Subsection 9.3.3.2), the backup unit can be "switched" on.

8.2 Design concepts of the communication systems

As demonstrated in the Subsection 8.1.1 about the different system topologies for ISS, in the near future there are basically two different in-vehicle communication systems, the event triggered communication bus such as CAN and time-triggered communication bus such as FlexRay. As mentioned in Subsection 3.3.4.2, although LIN and MOST can be applied in ISS, but they are not designed for the safety relevant applications, thus they are not discussed here.

CAN, as now the most widely used in-vehicle communication network, has proved itself as a cost effective and reliable bus system. The greatest constraint of CAN is the bandwidth and non-determinism since it was originally developed for the comfort application of cabin electronics and cyclic powertrain communication.

- As an event-triggered bus, CAN guarantees no worst-case transfer delay of messages, esp. in the case when a large burst of event-triggered messages are sent onto the bus. For the transfer of safety relevant messages like the battery status (voltage and key position), light controlling signal on Cabin-CAN, etc., dedicated measures have to be taken to secure the signal.
- CAN provides no time synchronization among the participants. A typical example problem is shown as Figure 8-4, when redundant steering angle sensors are connected with CAN to an ECU (consumer of the sensor signal). Since the redundant sensors can not be synchronized to each other to start the measurement at the same time and the transmission time over CAN could be slightly different, the result is that the measurements and the sending of the signal via the CAN to the respective sensor node are not synchronized but relocated with a certain time offset. Within this time offset the steering wheel can be turned with such a large velocity of angle, that the difference between two sensor signals overruns the safety relevant threshold.

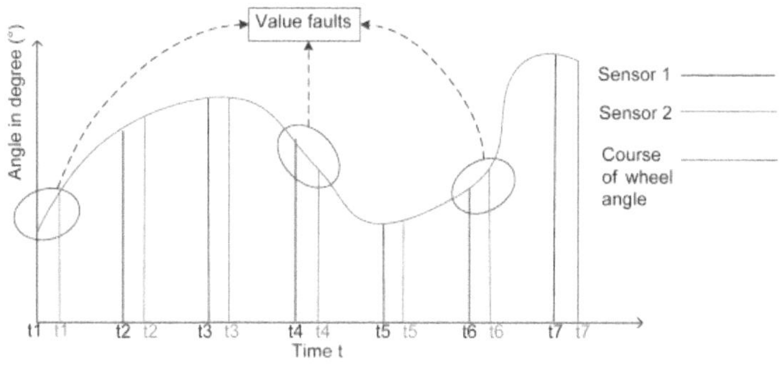

Figure 8-4: Different sampling of two sensors

- The limited bandwidth of 100 Kbit/s to 500 Kbit/s can be a bottleneck for future Integrated Safety Systems, which require deterministic communication with large volume of data exchanged between the nodes within the same domain or beyond the domain border.
- CAN provide only limited fault tolerant mechanisms such as CAN CRC-15.

Despite of these disadvantages, CAN is also applied for the safety relevant applications. The implementation of CAN for the specific application depends on the individual communication requirements regarding the quality of service. Generally speaking, there are following methods:

1. Using additional frame sequence number to monitor the transfer of CAN-messages, so that communication faults such as repeated or missing frames can be identified.
2. Using worst-case end-to-end transfer time (time-out monitoring) to monitor the message transfer on the receiver side.
3. Using acknowledgement to confirm the successfully transferred message by e.g. handshake protocol.
4. Using worst-case traffic simulation to limit the bus-load under 50% of the maximal bandwidth to guarantee the quality of service and relatively predictable end-to-end delay time. In some cases, the so-called "private CAN" – a dedicated short-range CAN is used to connect ECU with intelligent sensor and actuator.
5. Redundant CAN-communication with duplicated physical transfer paths or redundant messages on the same CAN. Here a typical trade-off between the performance and dependency should be made, since the redundant messages should be synchronized and compared on the receiver on cost of the real-time.
6. Use reliable control loop and fault detection filter to make the system more reliable against communication fault.
7. Intelligent algorithms to tolerate the timing drift of CAN messages.

While the methods 1 to 5 are relative self-explained, the methods 6 will be discussed in (process safety as discussed in the example of Subsection 9.3.1.3) as a software-based dependability service, the method 7 is demonstrated here with a detailed example.

To solve the problem in Figure 8-4, some intelligent mechanisms as following can be applied. One possible way is to synchronize the send/receive execution by trigger signal from receiver (time-synchronization with event-triggered bus). Another possible approach is to check integral of variation of sensor values $\Delta(S_1 - S_2)$ over a defined time period. With the proper choose of monitoring period T, the integral of sensor signal offset $\int_T \Delta(S_1 - S_2)$ (comparable with the integral of white noise in the telecommunication within a sufficient long period of time) should be under a limited threshold in faultless situation.

As introduced in Subsection 3.3.4.2, FlexRay, as a time-triggered protocol, is the in-vehicle communication network with the highest dependability potential. In the coming two or three years, more and more high-end cars from European OEMs will be running on road with FlexRay (the new

BMW X5 has been equipped with FlexRay for adaptive drive application in chassis since December 2006). The trend from CAN to FlexRay is triggered by the two of most superior aspects: bandwidth and dependability. FlexRay will find its way first in the safety relevant chassis applications or as backbone to provide required large bandwidth in chassis domain. The substitute of CAN-interfaces with FlexRay and to maintain a hybrid communication system of CAN and FlexRay will require additional effort, but it can be foreseen in the near future this effort will be compensated by the economy of scales.

As mentioned in the design rule 4 in Subsection 8.1.1, the safety feature of communication bus provides only the physical data integrity between the communication controllers, however, it provides no guarantee between the application-to-application data integrity service. The topic about application-CRC to ensure the application to application communication will be discussed in Subsection 9.3.1.3.

8.3 Concepts of ECU hardware architectures

Emphasis in this subsection is firstly the concrete design of dependable ECU architecture. In the second part of this chapter, hardware architecture of microcontroller components to improve the dependability will be discussed.

8.3.1 Dependable architecture driven by the safety integrity requirements

The decision for the dependable HW-architecture is mainly driven by the requirement of safety integrity level of the system. While the fail-safe and fail-operational units (1oo1D-, 2oo2-, 2oo3- and dual-duplex-system discussed in Subsection 3.3.3) are categorized according to the fault-treatment strategy, in the praxis, the decision of hardware architecture is mainly determined by the compliance with the required safety integrity level and the fulfillment of the system safety criterion.

That is to say, the desired safety level should be reached with the help of the dependable hardware architecture and SW dependability services. A quantitative safety case (e.g. SFF and PFH) should be given to the applied dependable hardware architecture to support the decision of hardware architecture (with e.g. FMEDA or quantitative FTA).

According to IEC61058 (see Table 3-2), SFF with at least 99% is required by ASIL-C and ASIL-D safety application. Experience from praxis has shown that, if such a system should be built up with one-channel system (e.g. 1oo1D system), considerable effort should be carried out to tolerate hardware transient and permanent faults (e.g. CPU computation fault, data fault in RAM as discussed in the hardware fault types in microprocessor core and supervisor of Subsection 6.2). The monitoring of such hardware faults should be implemented with software monitoring (e.g. software diversity of different algorithms). While the software diversity can not guarantee a 100% fault detection as well, because some low-level hardware faults can influence the both independent control algorithms (common cause). As a clear trend [µCI06], in the safety relevant automotive applications, architecture with dual-core (the 2nd variant of Dual-Duplex-Processor ECU as shown in Figure 3-7) has aroused the most future interest. In the following, further research of 2oo2D-system and 1oo2D-system will be carried out.

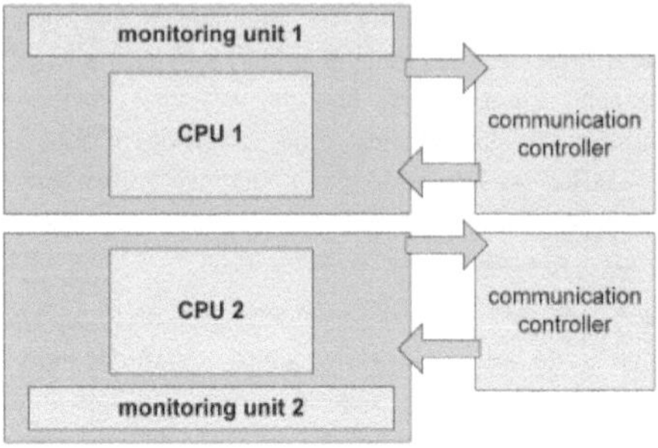

Figure 8-5: Variant of Dual-Duplex-ECU as 2oo2D-system

It can be seen that the 2oo2D.system in the simplest case as shown in Figure 8-5, is a replication of 1oo1D-structure without hardware diversity. Because the control of actuator over the communication system is redundant and no influence from one channel to the other is allowed, such a system can only guarantee an improved availability but not a high quality of system diagnosis coverage. SFF with at least 99% is difficult to reach with real components and design shown in Figure 8-5, because an unknown fault in any channel (e.g. HW-failure) can lead to a safety relevant failure of the whole system.

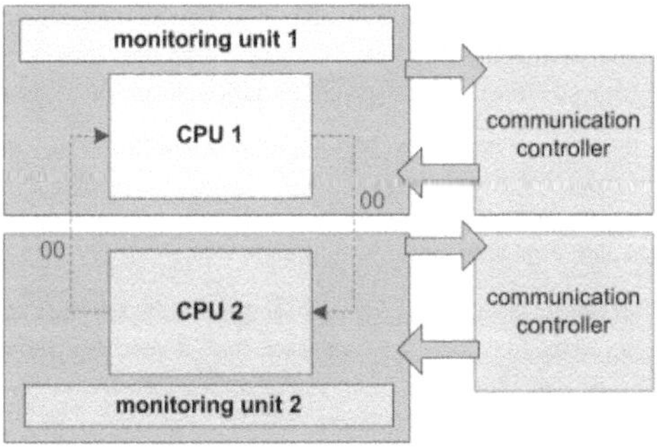

Figure 8-6: 2oo2D-system with diversity in hardware

In order to reach the ASIL-D requirements with 2oo2D hardware architecture, the two 1oo1Ds should be implemented with hardware diversities. According to [IEC02] Table A16, A17 and A19, hardware diversity is recommended as highly effective to control the systematic fault (e.g. hardware computation fault). Each subsystem should have DC of 90% as required by ASIL-B. Mutual monitoring between the two CPUs should be implemented as well.

As depicted in Figure 8-6, in this architecture for example, system 1 can be an ASIC microcontroller while system 2 can apply FPGA technology. Here design diversity can be also used, e.g. system 1 and system 2 are both ASIC but from different semiconductor producers. Because of the hardware diversely, low level hardware near software should be implemented in a diversity manner as well. The mutual monitoring and comparison of computation results are depicted with red dashed lines, which can be realized with direct µC I/O or external communication bus.

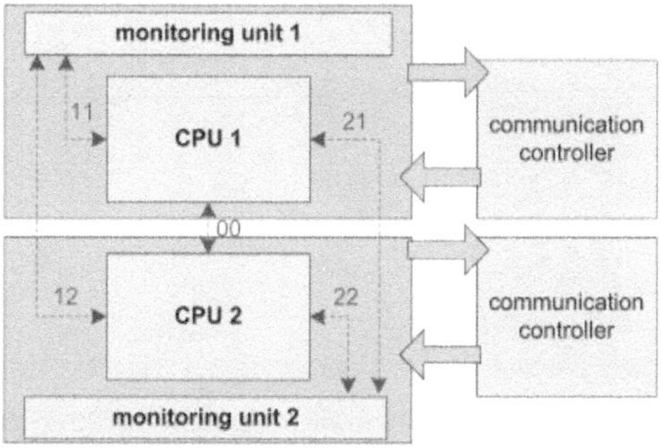

Figure 8-7: Variant of Dual-Duplex-ECU as 1oo2D-system

According to the safety norm IEC 61508-6 Annex B [IEC06] and IEC 61508-2 Subsection 7.4.3.1.6 [IEC02], the 1oo2D-system shown in Figure 8-7 with self-test is highly recommended for ASIL-C and ASIL-D systems, which requires a high SFF of 99%. Compared with 2oo2D-system, 1oo2D-system lets the monitoring units and CPUs communicate with each (as depicted with the red signal paths). The dual-core CPUs compare the intermediate results and final results with mutual agreement (signal path 00). Each CPU is not only monitored with local monitoring unit but can be monitored as an option remotely (e.g. signal paths from monitoring unit 11 to CPU1 and 12 to CPU2). If a remote fault is detected, such an additional signal path can be used to passivate/deactivate the neighbored processor unit in order to keep the whole systems running in a fail-degraded mode.

The signal paths here (00, 11, 12, 21 and 22) can be implemented with direct microcontroller I/O interfaces or with the external bus-communication or as a hybrid communication system of both. When the both CPUs are realized on the same silicon, the communication here can be implemented with shared memory or internal communication bus. In case of inconsistence the neighbored channel can be shut down. Thus the 1oo2D-system guarantees both high safety and availability.

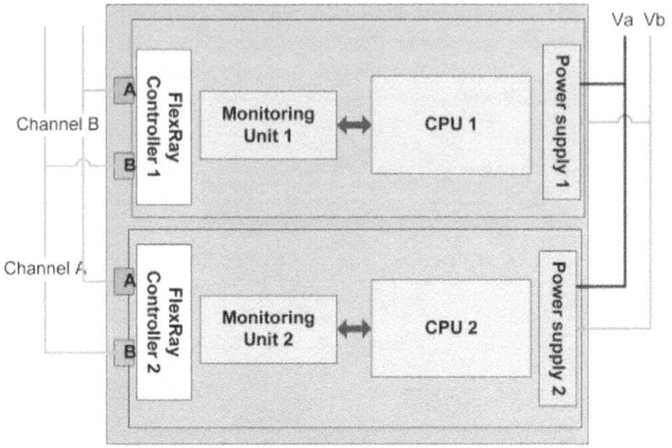

Figure 8-8: Implementation of Dual-Duplex-ECU as 1oo2D-system [HWAA06]

An implementation of Dual-Duplex-ECU as 1oo2D-system in praxis is shown in Figure 8-8, where the fail-operation unit is made up of two fail-silent units. The communication within the fail-silent unit (between the monitoring unit and main processor) is implemented on board. While the communication between the two main processors and the possible remote monitoring of main processor is implemented with dual-bus FlexRay communication. It is to mention here, the dual-bus communication is only an option, which depends on the overall in-vehicle network architecture. The single bus solution with end-to-end communication services in Subsection 9.3.1.3 can applied as well, as long as the PFH of the communication system as $1*10^{-10}$ as discussed in 7.2.2 ISS EP 2.4 is fulfilled.

It is to mention that the 1oo2D-system can be implemented with two units on one silicon (dual-core) or dual-processor (two separated CPU). Generally speaking the first approach of dual-core may share more resources, e.g. integrity checking information such as counters, CRCs, program flow information and memory management, etc. The intra-µC communication can be implemented more easily.

As a summary, in order to reach the aimed the ASIL-D requirement, the following dependable hardware architectures can be applied:

1. A 2oo2D systems with two 1oo1D (one channel) system as depicted in Figure 8-6.
2. A 1oo2D system with a dual-core system without diversities in SW (as depicted in Figure 8-7 with aimed SFF of 99%)
3. A 1oo1D system with diversity in computation units (e.g. integer and floating computation unit) and intelligent SW monitoring mechanisms of aimed SFF of 99%.

Innovative safety applications with ASIL-D requirement can be developed by the premium OEMs with the 1st approach (redundant HW of diversity concept). Such applications are seldom provided as serial product and are usually only available as an optional feature with a limited production volume. The short development cycle of the safety innovations with this hardware architecture brings the competitive advantages needed by the premium OEMs. An example here is the hardware architecture of the first generation of Active Front Steering System [Rei05], in which the safety relevant functions (main function and monitoring unit) are computed on diverse processors (MPU and NEC microcontroller) with mutual monitoring.

With the experience gained from such a pilot project and the requirement of continuous optimization of the system, when this safety application is provided as serial equipment or even considered by the middle-class OEMs, the safety architecture of the 2nd approach can be considered. The systematic HW and SW fault should be managed by safety process in the 2nd approach. The coincident common cause faults resulted from hardware or environment can be managed by the dual-core hardware architecture. With dual-core architecture it is also easier to analyze and proof functional safety. As long as the quantitative safety requirements to the SFF and DC can be reached in the safety case, no diversity in software development is required. The additional cost in hardware here can be compensated with the economy of scale production, simplified assembling and maintenance, thus the second approach can be preferred for the safety relevant systems by the premium OEMs for the volume production.

The 3rd approach is the simplest solution from the view of hardware architecture. From the hardware side, HW-Watchdog, as discussed in Subsection 8.3.4 with enhanced features such as intelligent algorithms to watch the program control flow and check the plausibility of computation results should be applied. On side of software, significant effort is required to achieve system dependability, e.g. each safety relevant signal should be monitored with diverse software mechanisms as to the value and gradient, safety relevant functions should be computed on the diverse computation unit (e.g. integer and floating) in one microcontroller. The ASIL-D safety functions should be implemented with software diversity, e.g. C and model-based approach with two independent development teams. The significant development effort in software is fortunately not proportional to the production volume. Thus the third approach can be preferred for the safety relevant systems by the volume OEMs.

8.3.2 Monitoring of sensor and actuator components

As discussed in Chapter 7 IEP Step 2.4 about mapping of system-PFH to FAA-components, about 75% to 90% system failures lie in the input (I/O, sensor and communication system with outside) and output (control of actuator). The monitoring of ECU input and output provides a significant potential to reduce the PFH of the whole system. However, the monitoring of sensors and actuators is quite application specific/dependent and there are less platform independent strategies. In the following, only a few examples from praxis are discussed.

The easiest approach is to use the redundant sensors to check the plausibility of the signals as shown in Figure 3-3. Asymmetric redundancies can be also applied, in which low-priced sensor (eventually with another principle of physical measurement) is used as a slave to monitor the main sensor. The input of the sensor should be checked in the time and value domain (with gradient/threshold check, message counter, CRC, etc). On the consumer side of the sensor signals, ECU can compare the received signals with other input values to check the physical plausibility of the sensor signals, e.g. the plausibility check of the gas pressure to its temperature.

By the monitoring of actuators we concentrate on the mechanisms of fault detection and deactivation of actuator for the fail-silent strategies. An example is depicted in Figure 8-9, in which microcontroller controls the electronic motor via power electronics. By monitoring different physical parameters of power electronics as well as motor (e.g. position and rotation speed of the rotor, voltages/current at power electronics and intern voltages/current at motor coil), microcontroller can disable the actuator via different independent "enabling" links to power electronics. The enabling-paths of actuator (to power electronics and directly to motors) should be implemented redundantly. Only at a positive output of the AND-logic of all enabling-paths, the motor can be activated. The different enabling paths should be checked at system initiation phase, the last operational mode should be saved during shut-down phase as well. By measuring the key-parameters of motors (rpm, coil current and voltage, rotor position, etc.), the microcontroller can calculate the expected parameter of the actuator for the next monitoring cycle with a simulated motor-model. By comparison of the current physical value with the results from the simulation model, the running behavior of the motor can be monitored.

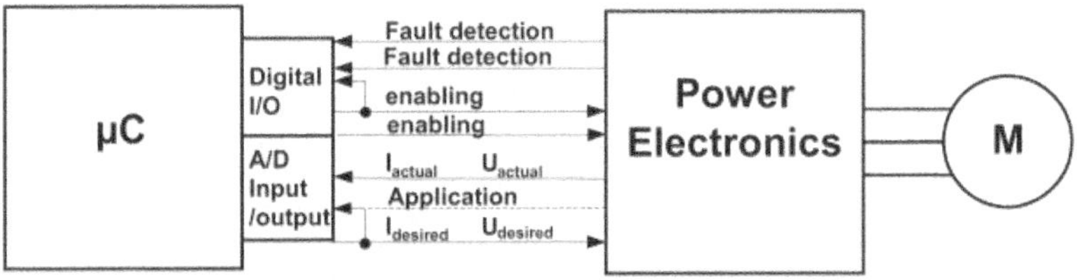

Figure 8-9: Example of the actuator monitoring (electronic motor)

8.3.3 Memory protection

As a state of the art method to protect memory area from EMV or other coincident influences from environment, there are following mechanisms to check the consistency of the memory area.

Invariable memory ranges

- CRC Signature: one word (8-bit CRC) and double word (16-bit CRC) signature are recommended in IEC 61508 – 7 A3.4 [IEC07]. The effectiveness of the signature depends on the width of the signature in relation to the block length of the information (pay-load) to be protected (at least with a hamming-distance of 2 [DIN93]).
- Block replication: In order to prevent common cause failure, safety relevant data should be saved redundantly in different areas of memory.
- In different phase of system states (e.g. start-up, operational mode and shut-down), the phase specific data along with CRC should be saved (e.g. at the end of each phase or a periodic test during the operation).
- In PROM, which contains executable code, all unused area of memory should be written with a fail-safe data value, e.g. an illegal operation code, which triggers a fail-safe interrupt.
- For the aim of diagnostics and filter of non-systematic fault, all the fault detection in the memory range should be recorded in a fault-detection counter.

- For each "write" operation to the memory, the written data should be read and compared directly after the write operation.

variable memory ranges

- With Parity-bit for RAM only a limited DC can be reached.
- RAM monitoring with a modified Hamming code or detection of data failures with error-detection-correction codes (EDC)
- Double RAM with hardware or software comparison and read/write test.

With the trend of increasing density of applications, how to prevent the undesired mutual influence of application components is an important topic for the ISS-application. While the design and software service to improve the system dependability will be discussed in Subsection 9.3.2, from the hardware side, it should be guaranteed that no non-authorized or faulty application components should allocate system resource assigned to the other applications. By means of system resource the following objects are considered: memory space, CPU time and permission to allocate system object like error-hook and interrupts. The protection of memory space brings new requirement to the ECU hardware architecture to have an additional Memory Management Unit (MMU) and Memory Protection Unit (MPU).

One of the preconditions for the memory protection is the memory portioning, so that each parallel running application can be assigned to an individual memory area. In this way, each task can be strictly controlled in the RAM memory that can be accessed. To avoid the waste of memory resource, the allocated units should be divided small enough to fit the requirement of a normal task of the automotive embedded system. The tasks themselves should be categorized as trusted task or non-trusted task. Each of them has their own private memory areas, while non-trusted task can not write or use call-back function to access the private area of other tasks. The memory protection unit controls the position and range of addressable segments in memory and the read/write operations allowed within addressable memory segments. Any illegal memory access can be detected by the memory protection hardware, which then invokes the appropriate service routine to handle the error.

By means of this, non-trusted tasks are the safe object, which are controlled under the OS all the time, while the number of privileged trusted-objects should be limit as small as possible.

8.3.4 Hardware watchdog monitoring

The hardware watchdog ([GAN03] and [LAN97]) is the last line of defense when the code collapses, in which hardware watchdog timer is the state of the art. The hardware watchdog timer should be prompted periodically by the monitored object (writing a "service pulse") to prevent a hardware reset. The intention is to bring the system recovered from the hung state into normal operation. Usually the hardware watchdog is a co-processor or external parallel running ASIC-chip to the main processor.

Traditional hardware watchdog has a separated time base as the main CPU to monitor the activities in an adaptive time window, such hardware watchdog has only a low diagnostic coverage. By the so-called question/answer mechanism, the hardware watchdog can trigger/ challenge the main CPU to check if it works logically correctly. In this case the logical sequence of program flow can be checked as well. By combination of these both mechanisms, the temporal and logical monitoring of program sequences, a high diagnostic coverage can be reached. Thus such an intelligent hardware watchdog is recommended by IEC61508-2 [IEC02] to improve the system diagnostic coverage.

8.4 Design use-cases of ISS hardware architectures

In the following subsections, the distribution and mapping of ISS-applications to the system topology is demonstrated with a few examples, following the design guideline proposed as above.

8.4.1 Distribution of ISS applications to vehicle domains

A near-future scenario of ISS-applications as specified in Subsection 1.1 is illustrated in Figure 8-10. This figure also describes the hierarchy of the different ISS-functions, starting from the elementary physical sensors and actuation devices (bottom level) up to the high level integrated application (top level, what the customer can experience as safety application).

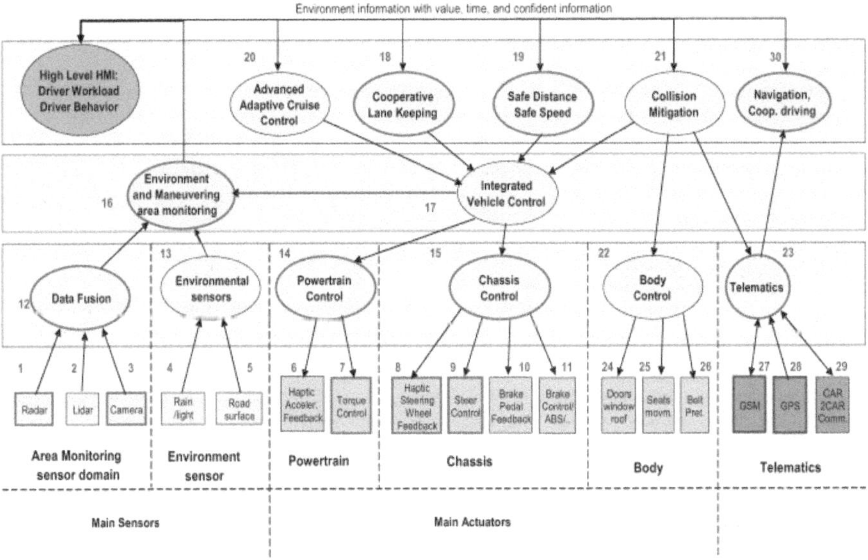

Figure 8-10: Examples of ISS application and mapping to the vehicle domains

Here we focus on two use-cases of ISS applications, the SafeLane and SafeSpeed. As highlighted with red color in Figure 8-10, SafeLane is a lane keeping assistant application while SafeSpeed is an ISS-application to automatically limit the vehicle speed to an externally commanded maximum value. The present example ISS application can be mapped into the following four different vehicle domains:

- Chassis Domain

  The SafeLane function mainly uses camera information and steering wheel sensors as environment information source. Additionally, it may use information coming from the Adaptive Cruise Control for monitoring the road in front of the vehicle and from the chassis units for understanding the vehicle dynamics. The function can provide haptic feedback through direct actuation on the steering wheel.

  Steer-by-wire can be applied here to cooperate with SafeLane, it performs the vehicle steering function by means of reading the drivers demand (steering wheel sensors) and by driving the appropriate actuators (wheel steering motor) of the steering system.

- Telematics Domain

  SafeLane and SafeSpeed consider infrastructure information elements, such as road traffic, digital road maps, precise vehicle positioning and local speed regulations.

- Powertrain Domain

  SafeSpeed may interact with the powertrain and chassis domain by requesting a reduced speed (reduce the engine torque and brake) or by providing a haptic warning to the driver through the accelerator pedal.

- Body Domain

  All the functions must interact with the driver via displays and acoustics for warnings and information about operational modes. All the functions should provide an integrated diagnostic system service.

In Figure 8-11, a possible mapping of ISS-Application SafeSpeed to the chassis and powertrain is demonstrated. SafeSpeed uses both engine- and brake-control to adapt the vehicle speed to the environment (e.g. telematics). Actually the distribution of ISS-application to vehicle domains is the mapping to of FAA to the hardware architecture, which generates the first FDA-model. Following the mapping rule in Subsection 8.1.2, the brake ECU and engine ECU are reused. At the first glance, the introduction of SafeSpeed brings a new ECU (as the prototype ECU with MABX in Figure 8-11), but for the serial development the SafeSpeed function can be integrated on an ECU for Integrated Vehicle Dynamic Control or Brake Management ECU, which has the similar safety integrity.

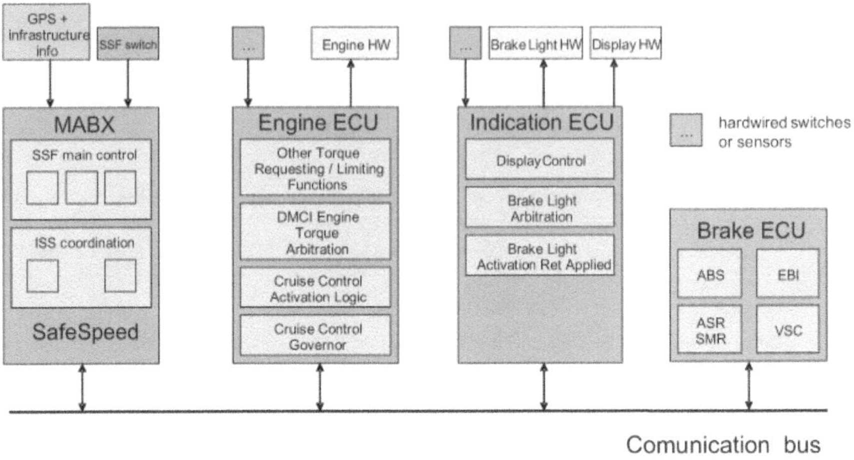

Figure 8-11: Distribution of SafeSpeed to the HW-architecture (FAA → FDA)

8.4.2 ISS system topologies and communication systems

After the discussions above about concepts of communication system, ECU hardware architectures and concepts about system topologies, esp. system design rules in Subsection 8.1.2, a few concrete examples about vehicle system topologies of ISS are given as following:

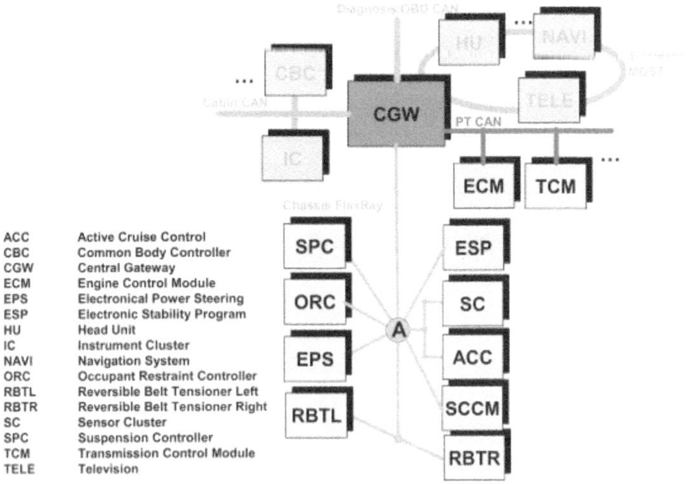

Figure 8-12: Exemplary system topology of future mid size car [HWAA06]

In-vehicle system topology shown in Figure 8-12 is a practical implementation of the central gateway concept for the mid size cars shown in Figure 8-2. The integrated safety applications are distributed among the ECUs, highlighted in yellow and connected with chassis FlexRay and powertrain CAN. A single channel FlexRay is applied here to enable the time triggered communication between the chassis nodes with a star topology with help of FlexRay active star [Ele04]. The system topology here is mainly designed for ISS, which require the fail-silent/fail-safe behavior with the mechanic back.

A good example here is a dummy "SafeSteering" ISS application. Compared with "SafeLane" as a lane keeping assistant system, SafeSteering can increase vehicle stability by interference the steering and vehicle dynamic actively. A practical example here is the superposition of an additional actively controlled steering angle to the driver input as introduced in Audi Active Dynamic Steering [ADS07]). In low speed area, it can provide the driver with park assistance support as well. SafeSteering is an ASIL-D ISS-application without any comparable system from the praxis, which is used here to demonstrate the dependable hardware architecture.

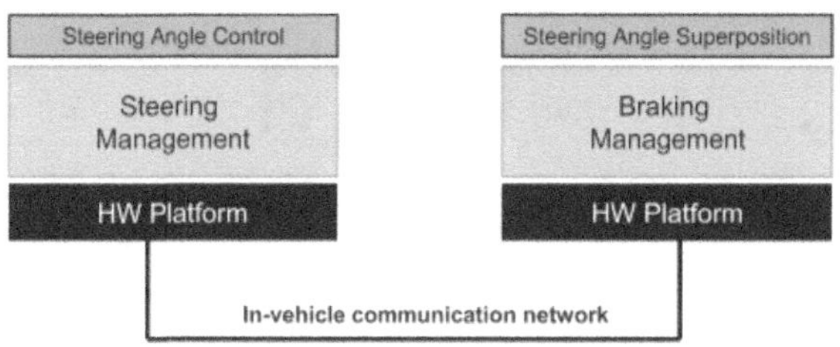

Figure 8-13: Communication and functional architecture of SafeSteering

As depicted in Figure 8-13, the SafeSteering is distributed between two main function units, the Steering Management Unit (SMU, as an electronic steering system of semi steer-by-wire technology with mechanical backup) and Braking Management Unit (BMU, as an enhanced "ESP"). The SMU provides the BMU with current wheel steering position (based on the information from actuator, e.g. rack position and rotor position of electronic motor), driver command in steering wheel angle and steering torque. BMU can provide SMU with the suggested additional steering recommendation with steering torque support or superposition angle according to the different driving situations, e.g. parking and highway.

Theoretically the computation of superposition angle and assistant steering support can be mapped to both SMU and BMU, since they are of the similar safety integrity of ASIL-D. The communication between SMU and BMU can be implemented with CAN or FlexRay as long as the ASIL-D requirement of communication is reached.

Compare the Figure 8-13 with Figure 8-12, the SMU is comparable with EPS, while the BMU, in the simplest case, is an ESP. Following the design rule 1, since most of the vehicle dynamic data needed are available on SMU, the computation of supposition angle can be mapped to ESP as shown in Figure 8-12, while the result of superposition angle can be sent to SMU. By detection of any safety relevant fault in SMU itself, the assistant steering moment /superposition angle of SMU should be switched off completely with a reduced ramp. A mechanic steering system is available in the fail-degraded situation, so that the driver can still control the vehicle. At the same time, the communication from SMU to other systems (e.g. sending-out steering-wheel-angle to ESP) can be passivated in fail-silent mode.

If the communication channel between BMU and SMU is implemented with CAN, a weak chain can be identified in the end-to-end communication for the steering recommendation. In order to reduce the safety requirement for CAN, as discussed in the rule 10, the superpositioned angle/additional steering torque should be limited in range along with the process-safety discussed in Subsection (Subsection 9.3.1.3), so that even when the signal is faulty, the vehicle steering can be still controlled by the driver.

In case of a complete communication break down, SMU should be able to work in a stand alone mode. For this reason, the steering wheel sensor and actuator are connected directly to the EPS in Figure 8-12.

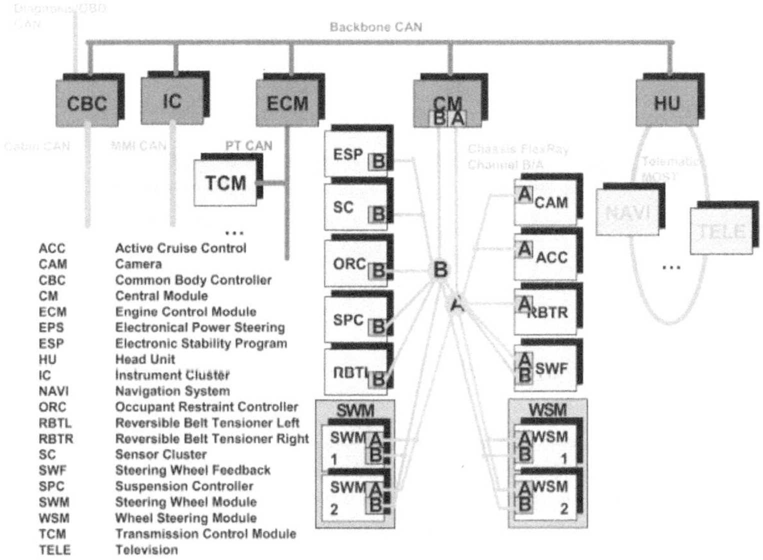

Figure 8-14: Exemplary system topology of future large size car [HWAA06]

Figure 8-14 shows an example of system topology of future large size luxury vehicles with steer-by-wire technology. Without the mechanic backup, the steering function here requires a fail-operational behavior.

Take again the dummy "SafeSteering" as an example: now this ISS is distributed on four nodes, SWM (Steering Wheel Module), ESP, CM (Central Module) and WSM (Wheel Steering Module), which are connected with a dual-channel FlexRay. One possible mapping is to map the steering management function to WSM, while the braking management function is mapped to ESP. The advanced steering/braking assistance and recommendation can be mapped to CM.

Here, the nodes with the highest safety integrity are the CM, SWM (sensor) and WSM (actuator ECU), because they are directly involved with Steer-by-Wire. These nodes are connected to both FlexRay channel A and channel B (partly redundant FlexRay with active stars). While the other less safety relevant nodes such as ESP, SC, ACC, CAM, etc. are only connected to single channel FlexRay A or B with their relevant sensors and actuators.

In the CM, different integrated safety functions and other chassis functions can be merged (sensor fusion) on one ECU to have an integrated vehicle dynamic control. The CM is responsible to give the SMU and BMU (ESP) the additional steering recommendation and braking recommendation.

It is to note that here only the SWM and WSM are redundant in 1oo2D architecture and connected with both FlexRay channels, while the CM is not redundant. In case of any problem with CM, the basic electronic steering function is still available with SWM and WSM. The SWF (Steering Wheel Feedback) is not implemented in a redundant architecture either, because the feedback force can be still provided with mechanical spring force.

CBC (Common Body Control), IC (Instrument Cluster), ECM (Engine Control Module), CM and HU (Head Unit) in Figure 8-14 are connected with a backbone CAN.

8.4.3 ECU hardware architecture for ISS

The hardware structure of SMU discussed in Figure 8-12 can be designed as a 1oo2D dual-core system. Figure 8-15 depicts one of the possible hardware architectures (adapted from the approach 2 as discussed in Subsection 8.3.1). The 1oo2D architecture is integrated on a dual-core chip. The same data input will be provided to the both dual-core processor with an intended short time delay, so that the comparing unit can compare the computation results from both CPUs step-by-step. In case of computation fault, one of the enable-path of the actuator control will be disabled.

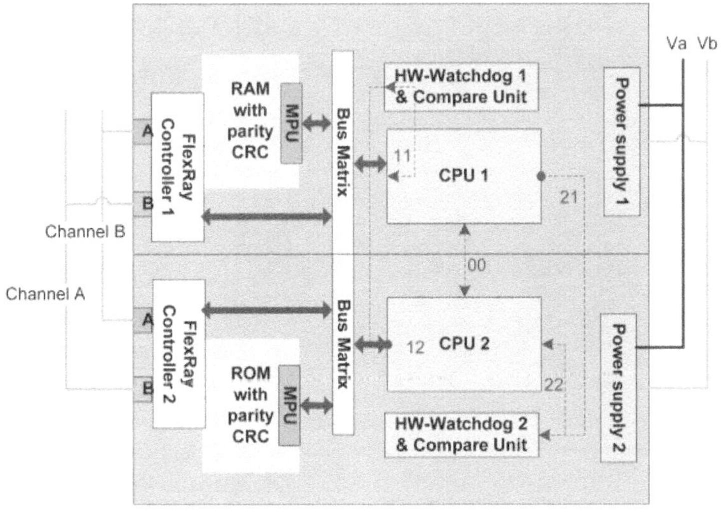

Figure 8-15: Exemplary 1oo2D dual-core ECU architecture

Since the both cores are built on one silicon board, the access to the memory is controlled by the MPU with the help of a bus matrix for the high-speed internal data exchange. In order to tolerate common-mode faults, both processors have independent memory and power-supply. RAM and ROM are cyclically checked with parity and CRC, etc. The redundant FlexRay controller to both FlexRay channel A and channel B is only an option. Single channel FlexRay with one transceiver can be applied as well as long as the safety requirement of system can be reached.

8.5 Conclusion of hardware architecture framework

Guided by the safety integrity requirements as a red line, the design of ISS HW architecture reflects again the distributed development between OEMs and suppliers.

OEM defines the overall system safety integrity requirement, topology and communication networks for ISS, while taking the distribution ISS-application into consideration. A balanced design considering different constraints and requirements (safety as the most important among all) will make the system integration more effective and improve overall system dependability significantly. The suggested system topologies and guidelines provide a framework for this decision phase.

On the other side, suppliers should provide the safety concepts in the hardware architecture, which fulfill the safety requirements from OEM. With the platform strategy and standardization, system suppliers are more and more involved in the early design phase to contribute their know-how by the definition of ECU architecture. Since the quantitative safety requirement is always a system wide term, the optimization of hardware architecture (e.g. evolution of safety concept from hardware diversity to dual-core) should be accompanied with intelligent software based dependability services.

9 Software platform for the Integrated Safety Systems

9.1 Trends of software platform – a benchmark with IT-industry

As mentioned in the state of the art about trends and challenges in the embedded in-vehicle software architecture, there is a clear trend towards standardization of a common middleware. From this point of view it is quite interesting to take a look in the history of IT-industry about the development of operating system. Similar like the OS in embedded world, OS in IT provides a set of functions needed and used by most applications and the necessary linkages to control a computer's hardware [HOS07].

The early computers lack any form of operating system and first original operating systems on computers go back to the early 1960s. Early operating systems were very diverse, with each OEM producing one or more operating systems specific to their particular hardware. Every operating system, even from the same OEM, could have radically different models of commands, operating procedures, and such facilities as debugging environment (quite similar with automotive embedded electronics now). Typically, each time the manufacturer brought out a new machine, there would be a new operating system. This state of affairs continued until the 1960s when IBM developed the System/360 series of machines which all used the same instruction architecture. With the rise of minicomputers, the UNIX operating system was developed at AT&T Bell Laboratories in the 1970s, which was conceptually the same across various hardware platforms. It was with the introduction of 8-bit home computers and game consoles of 1980s that real operating system for private customer was introduced. The decreasing cost of peripheries and processors made it practical to provide user more complicated operating systems with enhanced features.

When we compare this history with the development of embedded in-vehicle software, some similarities could be identified. The state of the art in the software platform in embedded in-vehicle software is somehow unfortunately comparable to the situation of IT-industry in 1970's. While a common operating system (OSEK, etc.) already exists, it is still not applied in each vehicle domains by each OEM. Each automotive OEM has his own adapted operating system, its own standard software modules. Automotive tier-one-suppliers esp. system suppliers are providing their ECUs along with their specific house-made software and hardware. The migration process of the systems for another customer (OEMs) requires a lot of effort. The tier-two/three suppliers (e.g. semiconductor and ECU hardware providers) are stilling developing the hardware along with their own drivers and low basic software modules, which again have a poor reusability for the tier-one suppliers and OEMs.

But why do we have this ca. twenty years' distance to the development of IT-industry? On one side the introduction of automotive electronics began only at the early of 1980's (as depicted in Figure 1-1), on the other side, because automotive industry has been traditionally strongly shaped by the mechanical engineering, which has quite different requirements and constraints compared with IT-industry. Thus the embedded in-vehicle electronics can be only a slower-follower of IT-industry.

Based on this benchmark, a conclusion can be drawn that the same development as in the IT-industry for the future development of embedded in-vehicle software will happen in the near future. This evolution could be slower than IT despite of lesson learned from IT, however, with a quite clear and steady similar direction as in IT.

Taking the Moore's Law and cost-awareness of automotive industry into consideration, the hardware devices and peripheries are quickly becoming cheaper, while more and more functions with will be implemented with software. Since ECU software platforms (including: hardware-drivers, operating system and standard software modules) can not be experienced by the customers directly, they are not really competition relevant for the OEMs, and neither for the Tier-1 suppliers. Standardized open software architecture will help to fix software bugs, improve the maturity of basic software modules and gain the experiences from integration of application software components with the underlying software platform. For the safety relevant applications, as have mentioned before, the same trend to have a software platform with dependability software services will take place. The standard library of dependability software services can be commonly applied by different OEMs and suppliers. This will surely contribute a lot to improve dependability of the whole system.

9.2 Concepts of dependability software architecture

As introduced in Chapter 5, in the EU-funded industry partnership EASIS, the first for the Integrated Safety System designed software architecture was developed. The software architecture framework for ISS is illustrated as Figure 9-1.

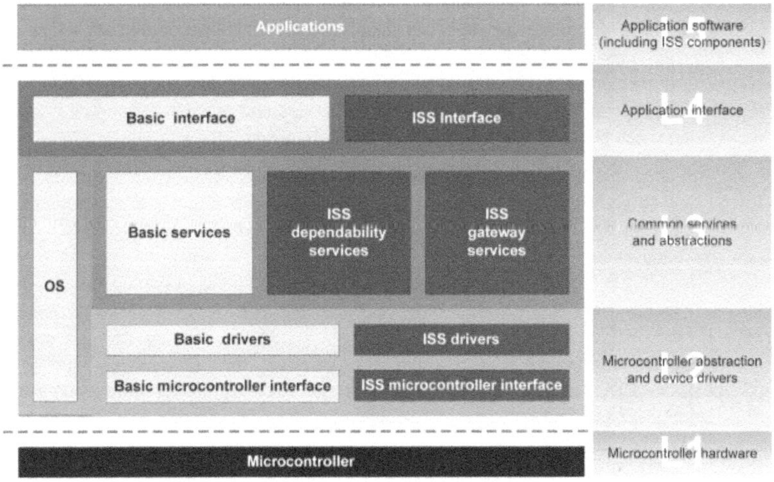

Figure 9-1: EASIS layered software topology [SWP06]

In Figure 9-2, the EASIS software topology with an overview of embedded basic software modules and dependability software services [HIL06] is given, where the software services with * on the symbol of package are optional and configurable to the safety requirements of the applications.

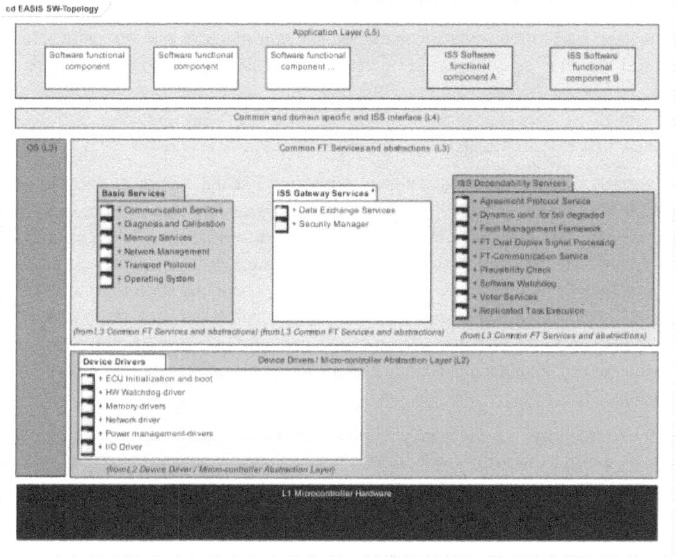

Figure 9-2: EASIS software topology –software components and services [SWT06]

In the following the layered software architecture framework depicted as Figure 9-1 is specified in details.

Layer 5 (Application Software): On the top of the EASIS layered architecture is the Application Layer. This layer harbors all application software components that provide the "actual" functionality that the system/ECU is to provide (i.e. the payload functionality). These functional components use the common/domain interface to access data and other items contained in the platform. The functional components can be domain-specific components, e.g., providing algorithms for powertrain control, or they can be cross-domain ISS components. In the application layer, software functional components with different dependability requirements can be integrated onto one ECU, including non-safety relevant functional components.

Layer 4 (Application Interface): Between the Application Layer L5 and Service Abstraction Layer L3 is the Common/Domain specific and ISS Interface Layer. It serves as application interface, which contains an interface for the functional components of the upper layers and is therefore the platform upon which all functional components are built. This is the ultimate abstraction layer of all underlying layers. It is worthy to mention that layer L4 keeps conformance to the AUTOSAR Runtime Environment (RTE) layer.

Layer 3 (Common Services and Abstractions): Layer 3 is the emphasis of EASIS work in software architecture, as shown in the Figure 9-1, the services are divided into three categories, the basic services, gateway services and ISS dependability software services (the blocks highlighted in blue in Figure 9-1).

- The first category comprises all basic software modules [BSW06], such as communication, diagnostics, calibration, network management and memory management, etc. These are gathered together as common services. All of these services are required to provide the upper functional layers with the platform they need for their functionality, e.g., the communication services provide abstract communication services and hide away all specifics arising from the physical communication systems chosen.
- The second category here is gateway services [GTD06], which support the applications distributed beyond network borders. The software services here, as well as ISS dependability services, are optional and configurable according to the requirement of the applications. The ISS gateway services consist of the services which are needed for safe and secure inter-domain as well as intra-domain communication. These services are not needed in non-gateway ECUs. In gateway ECUs some services like the Routing Service are required in any case, while others are optional depending on the required safety and security for the inter domain communication (e.g. Authentification Service and Firewall Service).
- The third category, as the focus of this dissertation, is the ISS dependability software services, which include different standardized software services to support safety relevant applications, such as Agreement protocol, Fault-tolerant Communication Services (as an add-on extension of common communication service), Resource Management and the Software Watchdog. It is notable that although the dependability software services here are categorized as parallel to the common software services, they are more like a configurable software service library compared with the basic software services, and make use of the API and services from the common software services.

Layer 2 (Microcontroller Abstraction and Device Driver): The Device Drivers and Microcontroller Abstraction Layer is an integration of hardware near software services. This layer provides the basic abstraction of the underlying hardware and the micro-controller and drivers of all devices included in the hardware platform.

The abstraction provides access to registers, I/O-ports, interrupts, etc. and hides away all specifics of the hardware and micro-controller. This interface does not provide any semantic information, meaning that data is handled exclusively as binary data, without any interpretation of the meaning of it. It also provides drivers such as network controllers, UARTs, digital I/O, AD/DA converters, TPU-functions (e.g. PWM) and so on. These device drivers provide some basic semantic information to the upper layers and may contain basic fault handling and other housekeeping functionality.

Layer 1 (Microcontroller hardware): Microcontroller hardware is layered at the bottom of this software architecture framework.

OS: In order to manage the execution of the software components included in the architecture, an operating system (OS), compatible to OSEK and AUTOSAR OS is used here. For some of the dependability services like Fault Management Framework, Software Watchdog and dynamic configuration services, extension to AUTOSAR OS will be needed. The placement of the OS between the L4 Interface Layer and L1 Microcontroller Hardware in the figure above is to indicate that it can be only accessed by layers between. In order to support dependability software services, an operating system with MPU (Memory Protection Unit) is about to be specified here in Subsection 9.3.2.2.

9.3 Concepts of the dependability software services

As stated in Subsection 1.1, two of the most distinguishing aspects of ISS are distribution of application across domain boarders and integration of functions with different safety integrity levels.

Specific for these two challenges brought by ISS, four types of dependability software services are designed, prototyped and validated in this dissertation.

- Dependability services for ISS communication
- Dependability services for the integration of applications on one HW-platform
- Dependability services for fault treatment
- Dependability software services for the gateway services

9.3.1 Dependability services for ISS communication

The requirements of communication among ISS application components come from the design regarding data exchange and relationship between the functions. As depicted in Figure 7-7 about the ISS engineering process, if we take a close look at the mapping process from FAA to FDA as depicted in Figure 9-3, the purple lines denotes the communication between application software components in FAA-model. Such related application software components can be mapped to one ECU or different ECUs or even ECUs belong to different networks.

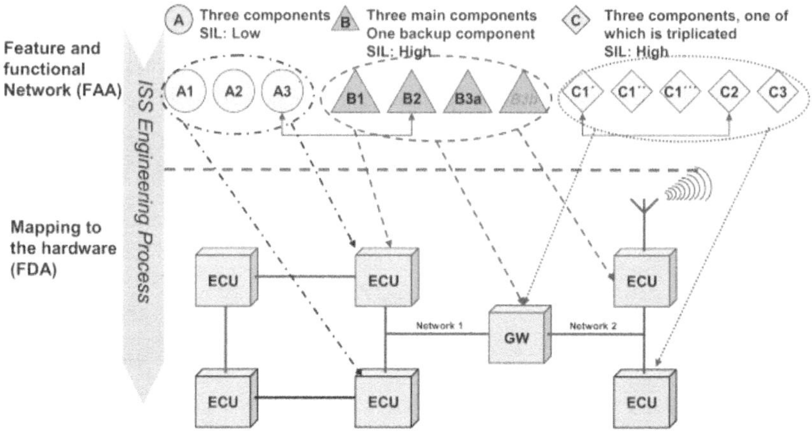

Figure 9-3: Mapping App.SW-Cs of different ASILs to the hardware topology

9.3.1.1 Categories of communication among ISS-application software components

As discussed in the mapping guidelines in Subsection 8.1.2 guideline 8, it is preferable to integrate only functions of the similar ASIL on one ECU. In practice, however, it is still possible to have applications with high ASIL integrated with applications with low ASIL. In order to guarantee the availability of the applications with high ASIL, it is also possible to allocate the replica of applications with high ASIL to be integrated onto ECUs, which originally only host the applications of low ASIL. The communication between the software components in this scenario is depicted as following:

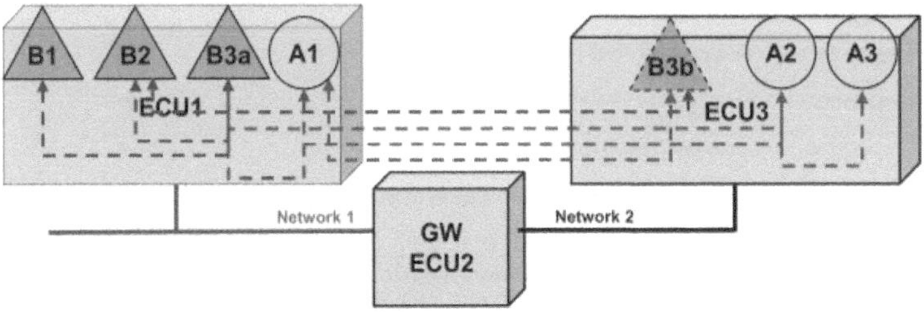

Figure 9-4: Example scenario of intra-ECU and inter-ECU communication

As denoted in Figure 9-4. ISS communication can be generally divided into intra-ECU communication denoted with the dashed red line, and inter-ECU communication, blue lines, e.g. ISS application B, highlighted in red, is made up of 3 software components B1, B2 and B3a, which are mapped to ECU1. The replica of B3a, B3b is mapped to the ECU3, where the app. SW-C A2 and A3 of lower ASIL are resident. App. SW-C A1 of low ASIL is also mapped to ECU1.

The communication between the app. SW-Cs can divided into following categories:

1. Mutual communication between App.SW-C of high ASIL with App.SW-C of high ASIL
2. Mutual communication between App.SW-C of high ASIL with App.SW-C of low ASIL
3. Mutual communication between App.SW-C of low ASIL with App.SW-C of low ASIL
4. Communication from App.SW-C of low ASIL to App.SW-C of high ASIL
5. Communication from App.SW-C of high ASIL to App.SW-C of low ASIL
6. Communication between the redundancies of App.SW-C of high ASIL

Among these 6 categories of communication:

- The category 1 refers to the intra-ECU communication between B1 and B3a or inter-ECU communication between B1 and B3b in Figure 8-4.
- The category 4 refers to the one way intra-ECU communication from A1 to B3a or inter-ECU communication from A1 to B3b, which is only a special case of communication category 2, where the App.SW-Cs of higher ASIL uses the information from App.SW-Cs of low ASIL.

- The category 6 is more or less a special case of category 1, since only the safety relevant app. SW-Cs can have replica. In order to reach a voted/agreed result among the replica, the replica should have the consistent inputs in the synchronized time window

  An integrated solution for these will be discussed from Subsection 9.3.1.2 to Subsection 9.3.1.4, in which the discussed algorithms are mapped to the communication category 1, 2, 4 and 6.

- The category 3 and 5 are not safety relevant, since the consumer of the information is not of high safety integrity, thus they will not be discussed in details here.

9.3.1.2 Redundant communication service

As introduced in the Chapter 3, Redundancy is a common approach to improve the dependency. In case of communication, three types of redundancy may be used to achieve a fault tolerant communication:

- **Information redundancy**: by information redundancy we mean redundancy of the complete information (e.g. redundant information from different memory areas by read/write) and abstract or footprint of the information (e.g. CRC).

  Within a communication frame, Error detection mechanisms already present in most protocols are a common form of information redundancy. Completely replicating a signal within a frame is a possible solution for improving error detection or even performing error correction, but as it does not protect against loss of frame it is not as efficient as the other redundancy types.

- **Time redundancy**: the signal is replicated in different frames or the same frame is sent multiple times.

- **Structural redundancy**: a signal is transmitted on multiple physical communication channels.

In case of inter-ECU, the three categories of redundant communication are depicted in Figure 9-5.

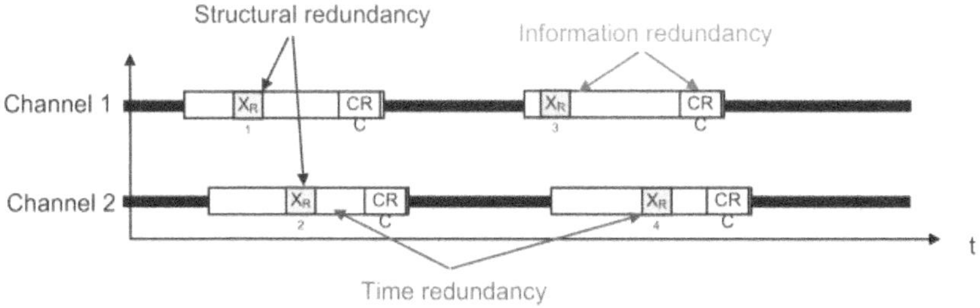

Figure 9-5: Communication redundancy mechanisms

The structure redundancy here fulfills the ASIL-D requirement of independent and redundant communication paths. In comparison, the information redundancy and time redundancy are cost effective, but require intelligent task scheduling, synchronization and plausibilisation of redundant information. While the information redundancy within one frame can be read almost once, the time redundancy can be a problem for the real-time control algorithms.

The redundant communication can be applied for the communication categories 1, 2 and 4 as specified in Subsection 9.3.1.1

9.3.1.3 End-to-end communication service with application CRC

As have discussed in the Subsection 8.2, various algorithms are applied for the fault detection of data transmission on the communication bus, but suppose that there should be some faults between the transceiver and application SW-C, e.g. memory error and application coincident fault in data read/write, such a fault can not be detected by the CRC in the communication protocol, since the protocol CRC only ensures the data integrity from bus transceiver of the sender to transceiver of the receiver. In Chapter 7 ISS EP Process Step 2.4, the mapping of system PFH to the PFH of FAA-components brings even higher safety integrity requirement of $1*10^{-10}$, which can not be guaranteed with the bus protocol alone.

Figure 9-6 shows the end-to-end communication scenario between two ECUs and the possible communication fault model. Faults, occurred in the red high-lighted area can not be detected.

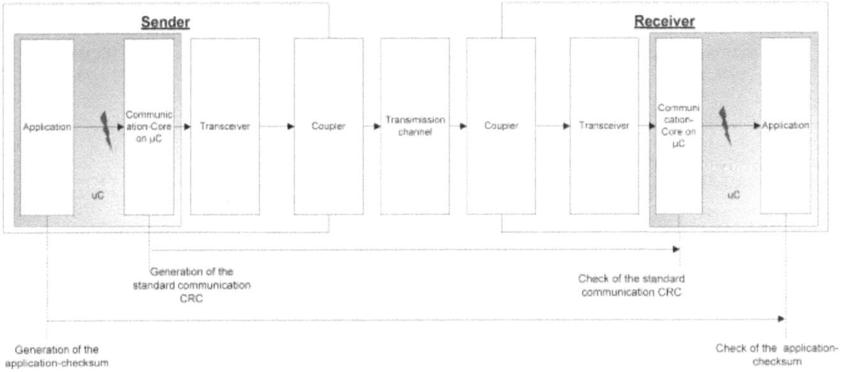

Figure 9-6: Reference model for end-to-end communication checksum

A detailed data flow from application SW-C to the transceiver is shown in Figure 9-7. On side of receiver as depicted in Figure 9-8, the same data transmission path in the inverse direction is vice-versa supposed.

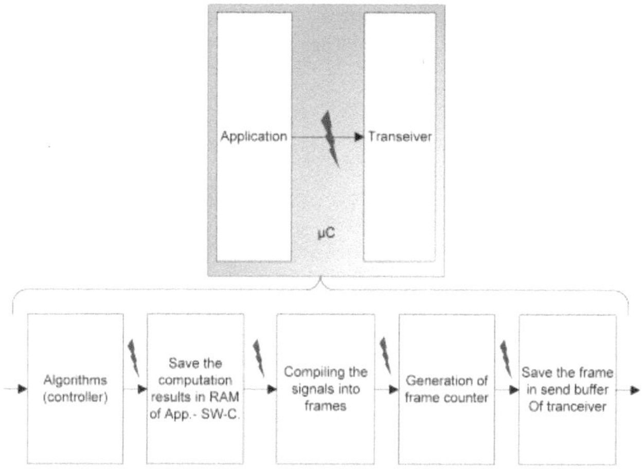

Figure 9-7: Data flow from application SW-C to transceiver (sender)

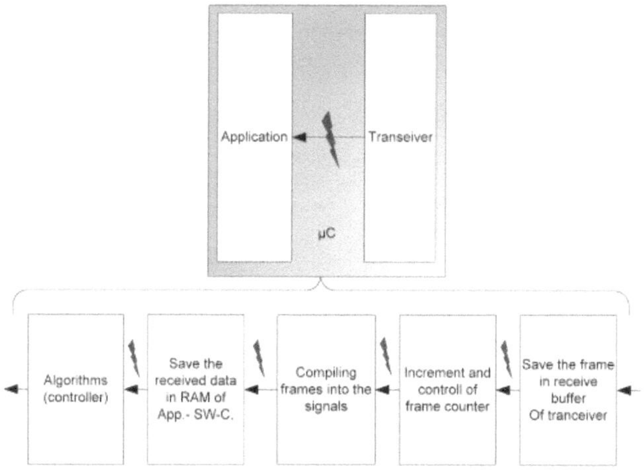

Figure 9-8: Data flow from transceiver to application SW-C (receiver)

A detailed list of fault types is given in Table 9-13, in which each chain of data flow of application SW-C to the transceiver can result an error, which is not detectable with protocol CRC.

Ref.	Fault location	Fault type	Error
1.	RAM of App. SW.-C	coincident bit-kipper	Data error
2.		DC fault model[3] of data and address	Data error
3.		Dynamic cross-over of memory cell	Data error
4.		Faulty address by read/write operation	Data error
5.	Program counter, Stack pointer	DC fault model*	Data error
6.	Program control flow	Not defined	Data error
7.	µC interrupt	Cross-over von Interrupts	Data error
8.	CPU Address	Not defined	Data error

Table 9-13: Fault types of date flow from application SW-C to transceiver

Thus it is reasonable to put an application CRC in the earliest chain of data flow of the sender (or the last chain on side of the receiver), as shown in the following Figure 9-9, so that faults in the following chains of data transmission can be also detected with application CRC.

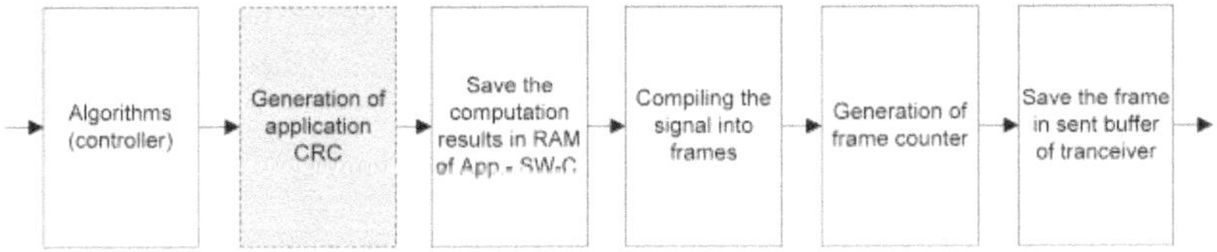

Figure 9-9: Date flow with application CRC on the sender

When decision is made to use application CRC, it is still needed to decide, which CRC-algorithm should be applied for the implementation of the safety relevant application.

Let us go back to the dummy ISS application SafeSteering as discussed in Figure 8-13. In SafeSteering, the additional wheel angle superposition is sent from BMU to SMU, and the both applications are ASIL-D with PFH requirement of $< 10^{-8}$. As discussed in ISS EP Step 2.4 Figure 7-7 about PFH proportion of communication system to the system PHF, the communication channel between the two systems, as a system component, should be implemented with a PFH of $<= 1*10^{-8} * 1\% = 1*10^{-10}$ /h (even higher than ASIL-D). That is to say, the possibility of a fault in the communication channel, which could result in a system wide safety relevant failure, should be

[3] DC (direct current) fault model includes: stuck-at faults, stuck-open, open or high impedance outputs and short circuits [IEC02]

smaller than $1*10^{-10}$ per hour. Based on the safety integrity of the application, a few questions should be clarified before the design:

- What kind of system fault tolerant behavior is required here, fail-safe or fail-operational?
- Is it enough to detect the communication fault (fail-safe) or do we need to correct it (fail-operational)?
- How sensible is the algorithms of control loop as to an undetected faulty input data? It is common that not every faulty transmitted message could result in a faulty control output (also called process safety).

In order to answer these questions, a system wide consideration should be carried out, e.g. the plausibility monitoring of input values and the controllability of the driver in case of faulty output should be considered in the tolerance of communication fault as well.

In SafeSteering, if CAN is used for the communication between Steering Management Unit and Braking Management Unit, additional dependability software services are required for the ASIL-D signals. Take the wheel steering position as an example: let us assume the pay-load of the steering angle as p bits, which are sent out every t_s s, while the total data transfer rate is v bits/s.

In order to describe the quality of physical transmission, term of Bit Error Rate (possibility of corrupted received bits) is introduced. For the automotive communication system, an assumption is made that the Bit Error Rate varies from 10^{-4} in case of an aggressive environment to 10^{-6} in the case of a benign environment. According to an experience research about Bit Error Rate of CAN [BER04], even under an aggressive environment the Bit Error Rate of CAN is only $2.6*10^{-7}$/h.

Three different integrity classes (I1, I2 and I3) are assigned in the industry norm [DIN93] to the communication path, in which I1 has a hamming-distance of at least 2, while I2 and I3 has a hamming-distance of at least 4. If we choose a relative conservative value of the Bit Error Rate of $1*10^{-4}$, as the same assumption from [DIN93], the probability of undetected error in the communication channel (denoted as R) is 10^{-6} by I1, 10^{-10} byI2 and 10^{-14} by I3 ([DIN93] diagram 1 about data integrity class). Here it is to mention that, only the undetected communication error is dangerous for the system (as dangerous failure undetected by the diagnostic test λ_{DD}).

As depicted in Figure 2-3 of Subsection 2.1, the latency time of fault tolerance $t_4 - t_1$ should be smaller than $t_3 - t_1$, during this latency time, the number of faulty messages (process safety) we can tolerate can be calculated as

$$m = \left\lfloor \frac{t_6 - t_4}{t_s} \right\rfloor$$

That is to say, only if m+1 successively faulty messages are undetected received and used by the control algorithms, a system wide dangerous failure could happen. We assume the worst-case here that each system failure will result in an uncontrollable safety relevant failure regardless of the driving situation and driver controllability.

The possibility of a system safety relevant failure resulted from the communication failure can be calculated as

$$Pcom = R^\wedge m = R^\wedge \left\lfloor \frac{t_6 - t_4}{t_s} \right\rfloor$$

In one hour, there are *n* bits sent to the receiver ($n=3.6*10^3*v$), let us take another worst-case assumption that each undetected faulty bit can result in a undetected faulty message. Since only *m* successive undetected message can result in a system failure, the number of system failure in one hour can be calculated as

$$FH = Data\_transferate * 1\_hour * possibility\_of\_system\_failure$$
$$= n * Pcom = 3.6*10\text{^}3*v*R\text{^}\left\lfloor \frac{t_6 - t_4}{t_s} \right\rfloor$$

, which should be smaller than the Dangerous Failure per Hour (PFH) required by AISL-D of $1*10^{-10}$. A further research of the formal above shows that the parameters t_s, t_6 and t_4 are system and application specific and can be less influenced by the application CRC. The decision of the CRC algorithms, however, influences the value of *R* directly.

In the following the whole approach to determine the CRC-algorithms is demonstrated with the SafeSteering, in which the parameter used are exemplary but not directly from the praxis.

Let us take high-speed CAN with 500 Kbit/s as *v*, system reconfiguration time t_6 as 10 ms, t_4 as 7 ms and steering angle superposition information is sent every 1 ms, during the reconfiguration time $t_6 - t_4$, we can tolerate $\left\lfloor \frac{t_6 - t_4}{t_s} \right\rfloor = (10-7)/1 = 3$ faulty undetected messages, thus

$$FH = 3.6*10\text{^}3*v*R\text{^}\left\lfloor \frac{t_6 - t_4}{t_s} \right\rfloor = 3.6*10\text{^}6*5*10\text{^}5*R\text{^}3$$

and $FH < 1*10^{-10}$

We could get that R should be smaller than $3.8*10^{-8}$. Again according to [DIN93] diagram 1, the integrity class 2 and class 3 should be applied, in which a hamming distance of at least 4 is required.

According to the Figure 2 and Table 3 in the paper of Koppman [Kop04], a hamming distance of 4 can be reached by 8 bits CRC as CRC8 and CRC0x97, 7 bits CRC as 0x5B and even 6 bits CRC as 0x2C, etc. In the embedded system of automotive industry, the CRC-8 is a quite commonly applied algorithm, thus the same algorithm can be applied for the application CRC with reuse of legacy codes.

9.3.1.4 Agreement Protocol

As specified in Chapter 6 about fault hypothesis and fault types for ISS, the Byzantine General problem are introduced.

A common approach to implement fault-tolerance in the safety relevant systems is the use of structural redundancies, defined in [Ech90] as the expansion of the system by additional components (software or hardware) to realize the same function redundantly. The structural redundancies within an electronic system however, introduce some problems themselves.

As discussed in [Bes02] and [Ech90], different sources of undesirable non-determinism can be identified. The simplest non-determinism is in value domain. For example, while measuring the same dimension using redundant sensors, the measured values, which theoretically should be identical, may exhibit slight deviations due to the physical nature of the sensors and are therefore non-deterministic. The use of these values leads to the computation of non-deterministic results, which is not acceptable for safety-related applications. Inconsistency in the value domain for structural redundancy can be solved by fault masking, which assumes the presence of a k-out-of-n system (see Figure 9-10). This consists of n redundant components, at least k of which must be non-faulty to ensure fault-tolerance.

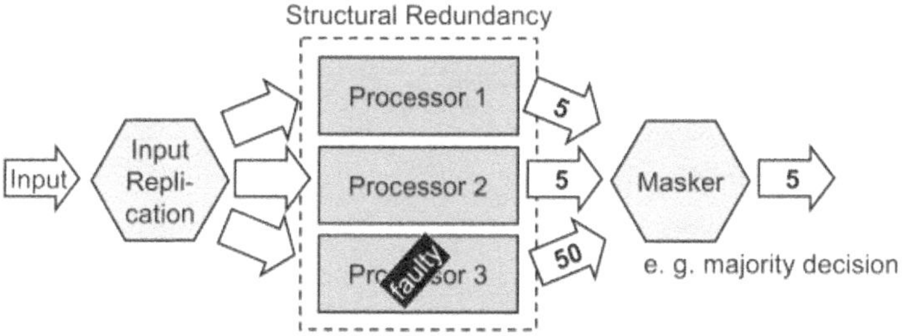

Figure 9-10: Fault masking 2-out-of-3 system [Ech90]

Given the high requirements for reliability in some applications, the masker, which constitutes a single point of failure, should also be replicated. Since additional communication is required to reach a consistent decision among maskers, the overall messages overhead is increased [CHS06].

The other form of non-determinism affects the time-domain and arises in the presence of a high latency jitter in the communication system, so that neither the arrival time of redundant messages nor their sequence in time can be predicted accurately. Such non-determinism in the time domain can be solved by the synchronization of data transmission and simultaneous data processing on redundant nodes. The introduction of a time-triggered in-vehicle communication system such as FlexRay and a predicable ECU operating system to guarantee the time constraints, simplifies the synchronization effort significantly.

Another form of non-determinism as the worst of all, is asymmetric fault (also known as the Byzantine fault, as discussed in [Lam82], Subsection 6.2) can also occur due to software and hardware coincident fault,

To ensure fault-tolerant communication even with Byzantine Faults, Lamport et al. provide different algorithms to solve the problem in [Lam80], so that non-faulty components can reach agreement on a value regardless of faults. These algorithms allow the non-faulty nodes to compute the same *interactive consistency vector* (*ICV*) containing the private value of each non-faulty node. Once interactive consistency is achieved, each non-faulty node can apply an averaging or filtering function to the interactive consistency vector, according to the needs of the application. In the following, a brief introduction to the different variants of Agreement Protocols as well as the assessment to the dependability requirements in the automotive industry is given:

Oral Messages Protocol

The notion of oral messages was introduced together with the Byzantine Generals problem ([EiK06], [Lam80], [FTD06]) and the definition of an oral message is embodied in the following assumptions with regard to the communication system:

- Every sent message is correctly received.
- The sender of a message is known to the receiver.
- The absence of a message can be detected.

As depicted in Figure 9-11:

- *P* is the set of all nodes.
- *V* is the set of all values.
- A *k*-level scenario σ is a mapping from a non-empty string of length $\leq k+1$ over *P* to *V*, thus summarizing the outcome of a *k*-phase exchange of messages. For example, let $w = p_1 p_2 \ldots p_r$. $\sigma(w) = v$ is then interpreted as the value p_2 tells p_1 that p_4 told p_3... that p_r told p_{r-1} is p_r's private value. $\sigma(p)$ simply designates the private value of node *p* and σ_p is the restriction of σ to the messages received by *p* in a scenario (i.e. only for $w = pw'$)

In the following, the algorithm OM(*m*, *n*, σ) is described in terms of *p*'s computation of the element of the Interactive Consistency Vector ICV corresponding to each node q, for a given scenario σ, an arbitrary number of faulty nodes $m \geq 0$ and a total number of nodes $n \geq 3m+1$:

(1) If for some subset *Q* of *P* of size $> (n+m)/2$ and some value *v*, $\sigma_p(pwq) = v$ for each string *w* over *Q* of length $\leq m$, *p* records *v*.

(2) Otherwise, *q* must be faulty. The algorithm OM ($m-1, n-1, \hat{\sigma}$) is then recursively applied with *P* replaced by $P - \{q\}$, and $\hat{\sigma}$ defined by $\hat{\sigma}(pw) = \sigma(pwq)$ for each string *w* of length $\leq m$ over $P - \{q\}$. This recursive call delivers an ICV with $n-1$ elements, since the node q is excluded. The *i*-th element corresponds to the value all non faulty nodes agree it has been received by the *i*-th node p' from *q* during the first phase (i.e. $\hat{\sigma}(p')$ which is in fact equal to $\sigma(p'q)$). If at least $\lfloor (n+m)/2 \rfloor$ of the $n-1$ elements agree, then *p* records the common value. Otherwise, *p* records *nil*.

The correctness of this algorithm as well as the necessity of $n \geq 3m+1$ is proved in [Lam80].

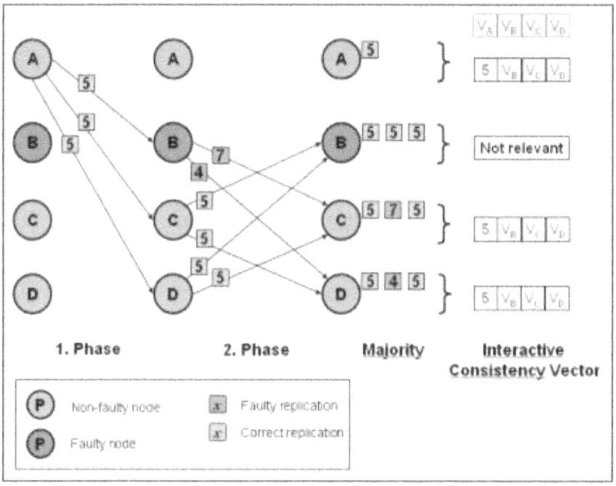

Figure 9-11: OM algorithm with n = 4 and m = 1

Signed Messages Protocol

Since oral messages can be easily forged by faulty nodes before being forwarded, reaching agreement among non-faulty nodes with the help of the OM algorithm requires at least $n = 3m + 1$ nodes to tolerate *m* faulty ones and hence induces a significant communication overhead. By using signed messages instead of oral messages, agreement can be reached for any number $n > m$ nodes and also requires $m + 1$ communications phases.

As opposed to oral messages, a signed message from a node *p* including some data item *d* cannot be forged, as this message holds an authenticator $A_p[d]$ from *p*, allowing each receiver to check the message integrity as well as the identity of the original sender. Since this authenticator cannot be falsified, faulty nodes are no longer able to alter the content of a message before forwarding it without being found out. Moreover, forwarded messages are additionally signed by the forwarding node.

Let $v = \sigma(p)$ for a node *p* (see the subsection above about Oral Message Protocol for the definition of σ). *p* communicates this value to another node *r* by sending the message consisting of the triple (p, a, v), where $a = A_p[v]$. After receiving the message, *r* checks whether $a = A_p[v]$. If the test is successful, *r* takes *v* as the value of $\sigma(p)$ (i.e. $\sigma(rp) = v$) and forwards the message $(r, A_r[(p, a, v)], (p, a, v))$ to all other nodes except *p*; otherwise $\sigma(rp)$ = *nil*.

More generally, if *r* receives a message of the form $(p_1, a_1(p_2, a_2, \ldots, (p_k, a_k, v)\ldots))$, where $a_k = A_{p_k}[v]$ and for $1 \le i \le k-1$, $a_i = A_{p_i}[(p_{i+1}, a_{i+1}, \ldots, (p_k, a_k, v))]$, then $\sigma(rp_1 \ldots p_k) = v$ and r forwards the message after signing it to every other node different than p_1, p_2... and p_k; otherwise, $\sigma(rp_1 \ldots p_k)$ = *nil*.

In the following, the signed messages algorithm SM (m, n, σ) is described in terms of the value a non-faulty node p records for a given node q and a scenario σ. m designates the number of faulty nodes ($m \geq 0$) and n the total number of nodes ($n > m$).

Let S_{pq} be the set of all non-nil values $\sigma_p(pwq)$, where w ranges over strings of distinct elements with length $\leq m$ over $P - \{p,q\}$. If S_{pq} has exactly one element v, p records v for q in its ICV; otherwise, q must be faulty and p records *nil*.

As proved in [Lam80], this algorithm allows the non-faulty nodes to compute the same ICV containing the private value of each non-faulty node and *nil* for each faulty node. Although $n > m$ nodes are sufficient to allow non-faulty nodes to reach agreement in the presence of m faulty ones, it makes more sense to use $n \geq 2m+1$ nodes, since non-faulty nodes must be in majority. This would enable a downstream masker using a majority decision to vote for the correct result.

As depicted in Figure 9-12, the ICV of each node is computed in the distribution protocol during the communication phases. Before the first phase, each node initializes its ICV to *nil* values except for its corresponding element where the node's private value is recorded. The vector is then actualized upon each communication phase as follows. When a message $(p_i, a_i(p_j, a_j...(p_k, a_k, v)...))$ is received by a node p and turns out to be valid, then

- If $ICV_p[k]$ = *nil*, then the value v is recorded for the k-th node, so that $ICV_p[k] = v$.
- If $ICV_p[k] \neq$ *nil* and $ICV_p[k] \neq v$, then a replication fault obviously occurred in the node p_k and $ICV_p[k]$ is set to *faulty*, so that any further values for p_k are ignored. Nevertheless, valid messages are always forwarded independently of the value of $ICV_p[k]$.

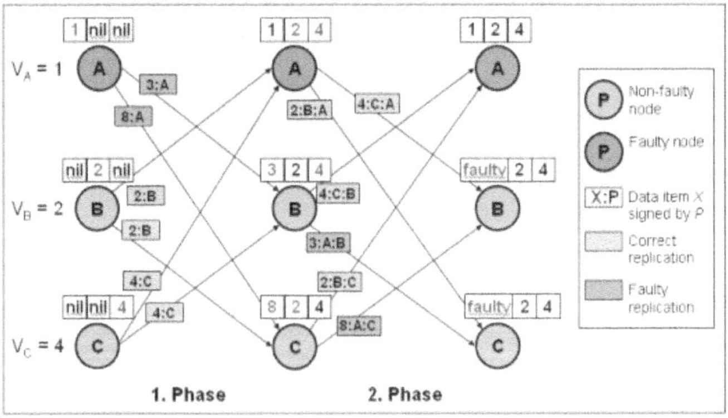

Figure 9-12: Time diagram of the Signed Message Protocol with 3 nodes

Pendulum Protocol

Independently of whether faults actually occur, the previously presented protocols always induce a worst-case communication overhead, since an interactive consistency vector is computed for all nodes before a final result is decided. The *pendulum protocol*, which is introduced in [Ech90], aims at reducing the communication overhead for the faultless scenario at the expense of fault detection by computing a single agreed value. The name of the protocol comes from its algorithm which executes in a way similar to a pendulum stroke over the nodes.

The pendulum consists of a *main protocol*, which is sufficient in the faultless case, and an *auxiliary protocol* applied to deal with faults when these occur. Moreover, it requires the use of some *distance decision* (i.e. interval decision or sphere decision) and thus the definition of δ as depicted in Figure 9-13.

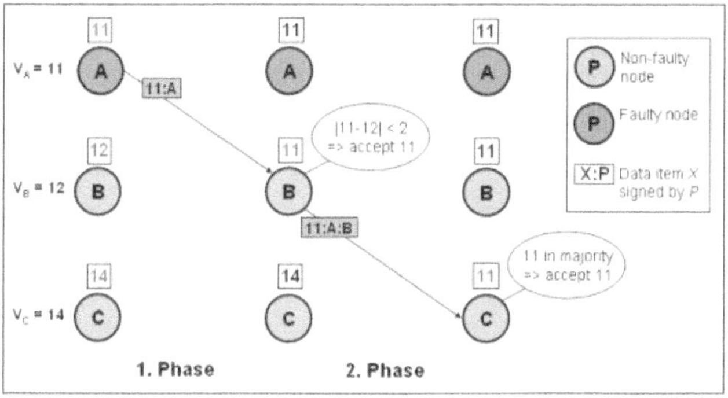

Figure 9-13: Pendulum Protocol for 3 nodes with δ = 2

Assessment of the agreement protocol variants

The following criteria were applied during the assessment of these three variants of protocols:

- Predictability: the algorithm behavior does not depend on run time events (e.g. the occurrence of faults).
- Fault-detection: the algorithm includes mechanisms that enable the detection of components providing faulty values.
- Fault-tolerance: agreement can be reached despite the presence of faults.
- Number of redundancies: the minimal number of redundancies required to tolerate *m* faulty processing units.
- Minimal communication overhead: the minimal number of messages required to reach agreement among *n* processing units, e.g. in the absence of faults.
- Maximum communication overhead: the maximal number of messages required to reach agreement among *n* processing units, e.g. in the presence of faults.

- Number of communication phases: the number of communication phases required to tolerate *m* faulty processing units.

These criteria can be classified with respect to their priorities, as depicted in Table 9-14, which categorizes the properties of the agreement protocols in three levels, Priority 1, Priority 2 and Priority 3.

Priority 1	Priority 2	Priority 3
▪ Predictability ▪ Fault tolerance ▪ Fault detection	Number of redundancies	▪ Minimal com. overhead ▪ Maximum com. overhead ▪ Number of com. phases

Table 9-14: Priorities of Criteria for the assessment of agreement protocols

Based on these criteria, the three agreement protocols are compared in Table 9-15, where *n* is the total number of processing units and *m* the number of faulty units to be tolerated. The mathematical derivation these results can be found in Appendix. 2

	OM Algorithm	SM Algorithm	Pendulum Protocol
Predictability	yes	yes	no
Fault tolerance	yes	yes	yes
Fault detection	yes	yes	no
Number of redundancies	$3m+1$	$2m+1$	$2m+1$
Minimal com. overhead	$\frac{n\cdot(n-1)\cdot((n-1)^{m+1}-1)}{n-2}$	$\frac{n!}{(n-2)!}+...+\frac{n!}{(n-m-2)!}$	$n-1$
Maximal com. overhead	$\frac{n\cdot(n-1)\cdot((n-1)^{m+1}-1)}{n-2}$	$\frac{n!}{(n-2)!}+...+\frac{n!}{(n-m-2)!}$	variable
Number of com. phases	$m+1$	$m+1$	variable

Table 9-15: Comparison of three agreement protocols [Appendix 2]

From the results of this comparison, the pendulum protocol does not seem to be adequate even if it provides the lowest communication overhead in the faultless scenario. In fact, besides failing to fulfill the predictability requirement, it does not provide 100% fault detection and could even lead to the agreement on a faulty value. Moreover, the pendulum protocol is limited to applications supporting only an interval or a sphere decision, which makes it less qualified for a standard service implementation.

Both the OM and the SM algorithms fulfill the most relevant requirements. Although the signing of messages induces an additional computational overhead, the SM protocol is most appropriate in the scope of this work, since it can guarantee fault-tolerance with less redundancies and a lower communication overhead than the OM protocol.

Based on the safety requirements in automotive industry, the Agreement Protocol service is embedded in the dependable software architecture specified in Subsection 9.2. As depicted in Figure 9-14, the application using the AP service must provide an input value as the private value of the node, the number of phases (depending on the number of faults to be tolerated), the tolerance band and additional node information required for the signature mechanism. The data exchanges are carried out using a fault-tolerant communication service via FlexRay.

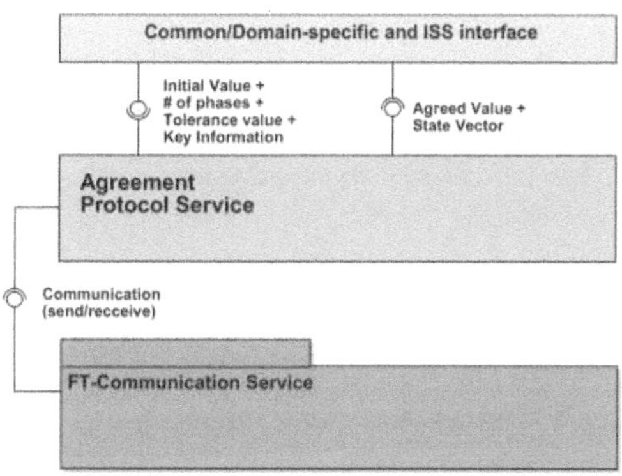

Figure 9-14: Interface to fault-tolerant communication and application

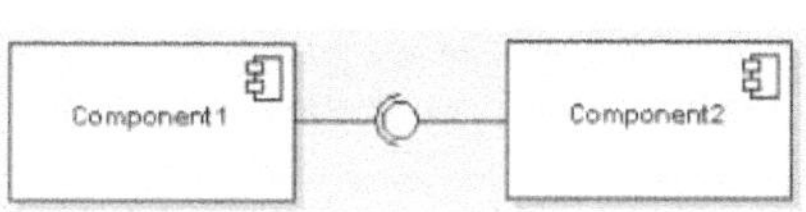

Figure 9-15: Assembly symbol in UML with Enterprise Architect

The "assembly" symbol of UML 2.0 used in Figure 9-14 is a connector between two components, which denotes that one component provides the services that another component requires. As shown in Figure 9-15 the assembly bridges a component's required interface (Component1) with the provided interface of another component (Component2).

9.3.1.5 Conclusion of ISS communication

As discussed in the subsections above, the ISS communication is characterized with the high dependability requirement (ASIL-D communication with PFH of 10^{-10}). Fault tolerance in each chain of the communication channel even with the presence of Byzantine Faults is required for the future ISS applications. The approaches presented in this subsection focus on the fault-tolerant communication services, which are integrated in the software platform as dependability software services. In the praxis, the real implementation should be tailored, configured and supplemented with application specific services such as identifier for the communication objects, aliveness counter, error detection code, acknowledgement, etc. Under the requirement that safety integrity level from the safety analysis should be met, a balanced communication concept considering system complexity can be chosen.

9.3.2 Dependability services for the integration of applications on one HW-platform

As have mentioned in Subsection 9.1, there is the trend of increasing integration density of applications on one HW-platform, in which the main challenges are as following:

1. Managing the complexity in the integration process of various functions on the same ECU during the system development
2. Increasing the system overall dependability by the partitioning of applications, which have different safety integrity levels

The first challenge is mainly considered in Chapter 7 with the help of ISS Engineering Process, while the second topic should be solved in an integrated manner with a dependable architecture framework.

9.3.2.1 Challenges and consideration from an integrated manner

As specified and required in [ISO26] Baseline 9, Chapter 6 Software design 6.4.10 and 6.4.11, freedom of interference by software partitioning is one of the most important technologies for the integration of applications with shared resources. The shared resources on ECU refer to computation time, memory and communication. While the dependable communication is discussed in Subsection 9.3.1, the partitioning topic here focuses in the computation time and memory.

The term of partitioning is meaningful when two different entities can influence each other by means of interface or shared resource. Generally speaking, partitioning is a fault containment technique which consists in precluding a failure in a partition to propagate (to control the additional hazard) and to cause other partitions to fail, which can be divided into hardware partitioning and software partitioning.

HW-partitioning is discussed in Chapter 8 by means of guideline to map applications to the HW-Architecture and hardware design. Partitioning, when applied to software, the intent is to control the additional hazard that is created when a software partition shares its or part of its resources with other partitions (shared resource of memory, computation time and peripheral access, etc,). Partitioning with clear interface definition and layered architecture is the topic of software design, while partitioning in this chapter refers to the software partitioning with shared resource and software partitioning is not intended to protect against failure of each software partition but against their propagation.

Since we are talking about partitioning, it is necessary to define the software framework to be partitioned with the help of dependable ECU software architecture as defined in Subsection 9.2.

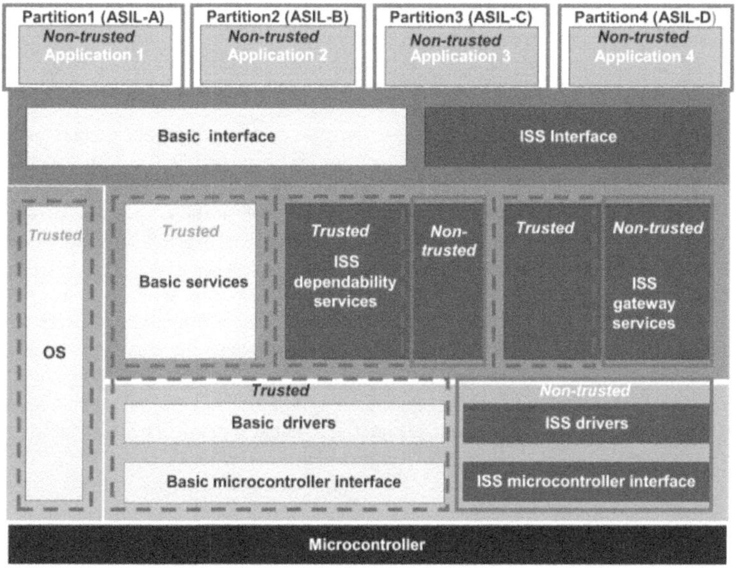

Figure 9-16: Software partitioning in the dependable software architecture

As depicted in Figure 9-16, on top of the dependable software architecture various application software components with different safety integrity levels can be integrated. As mentioned in Subsection 9.3.1.1, it can still happen, that ASIL-B functions are integrated with the ASIL-D functions if they all belong to the same ASIL-D safety application, e.g. diagnostic function in the SafeSteering application mentioned in Subsection 8.4.1. In this case, the functions with different ASILs should be partitioned to suppress the mutual influence. For same reason, the underlying software framework belong the interface layer (below EASIS SW-layer L4) could come from difference sources (different suppliers) as well. They are not all standardized and developed to the ASIL-D requirement, thus they should be partitioned to the trusted and non-trusted software modules. As the emphasis of partition, the App.SW-Cs are generally considered as non-trusted. Basic software services / drivers, OS, which have been used in many serial projects, are considered as trusted object. Part of fully validated ISS-dependability and gateway services can be identified as trusted-objects as well. Since the regulation of partition are completely or partly deactivated for trusted object. The trusted objects, which have a higher flexibility than non-trusted objects, can be the more dangerous for the system in case of error.

Software partitioning in this subsection encompasses space partitioning and time partitioning:

- Space partitioning addresses unauthorized data access and illegitimate commands of peripherals assigned to other partitions, which is discussed in detailed in Subsection 9.3.2.2 [ISO26].
- Time partitioning that focuses on disturbance of the timing of events in other partitions (scheduling, order of execution, etc.) is addressed as in Subsection 9.3.2.3.

9.3.2.2 Space partitioning with protection services

When talking about space partitioning, it is necessary to define the objects on the ECU, which are to be partitioned.

Following the ECU SW development process, the next mapping step after Figure 9-3 is the mapping of application runnables to the task on the ECU as depicted in Figure 9-17, where runnables from one App.SW-C can be mapped to different ECUs, runnables from different application SW-Cs can be mapped to the same task as well. In order to save resource esp. computation time, state of the art approach is to pack the functions (runnables are defined in Figure 9-17 as the atomic code sequences of App.SW-C), which are executed to the same time cycle, together to one frame as a task. In an embedded OS, which is both compatible with time-triggered and event-triggered scheduling, tasks could have different priorities, in which task with lower priorities can be preempted by task of higher priorities. Usually the most important tasks are assigned with the highest priority and executed with the highest frequency supported by the microcontroller. Tasks with lower priority, which do not have the hard real-time requirements, can be preempted to guarantee that, sufficient resource is available for tasks of higher priority.

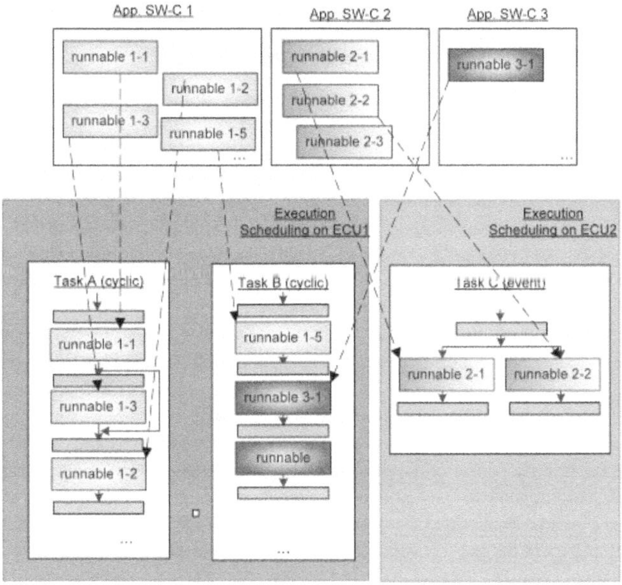

Figure 9-17: Mapping of application to ECUs and runnables to tasks

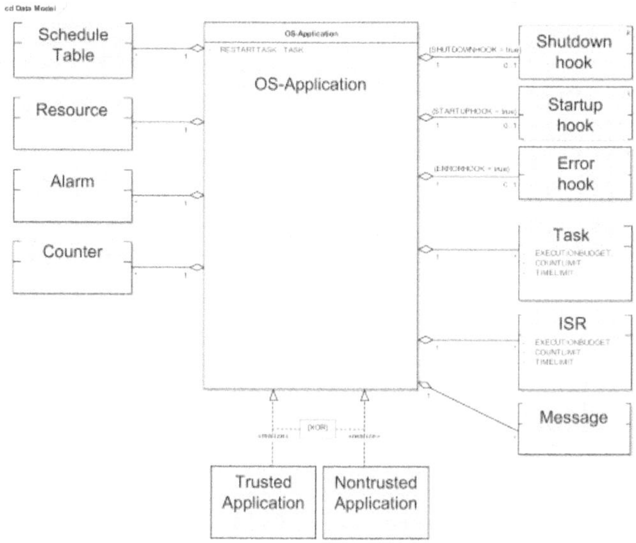

Figure 9-18: UML-model of objects in operating system

From view of operating system, all application SW-Cs, basic SW-Ms, dependability SW-Ms, etc. are implemented in OS-objects as depicted in as in Figure 9-18, such as schedule, resource, alarm, counter, hook, task, ISR and message etc. The most frequently used OS-objects are task, alarm, counter and hook. The space partitioning should be built up based on these OS-objects with help of system design, MMU/MPU and OS support. The OS-objects shown in Figure 9-18 can be restructured into two classes [AOS06]:

1. Trusted OS-objects are allowed to run with monitoring or protection features disabled at runtime. They may have unrestricted access to memory, the OS API, and need not have their timing behavior enforced at runtime. They are allowed to run in privileged mode when supported by the processor. In this sense, trusted OS-Applications are easy to use but dangerous.
2. Non-Trusted OS-objects are not allowed to run with monitoring or protection features disabled at runtime. They have restricted access to memory, restricted access to the OS API and have their timing behavior enforced at runtime. They are not allowed to run in privileged mode. Non-trusted applications are secured applications that they can not easily disturb other applications.

Especially for the non-trusted OS-objects, the first step of space partitioning is that they should be structured already in the design phase properly. This means, that they must be structured according to their function and not e. g. according to their time-frequency. Runnables with high priorities, which do not belong to the same function, should be structured to different tasks. If the given functionality of OS-objects regarding data encapsulation is not used, the problem of unintended writing in memory can of course not be solved. If it is used, an MMU enables protection even in case of programming errors that an application tries to write to the memory region of another application. When a task intends to access a memory area by write operation, by comparison of the task ID with a preconfigured task/memory allocation table, OS could trigger an error hook if the task is not authorized to perform the operation. The memory partitioning could consider the following rules:

- Each task is assigned with a memory area for all possible operation.
- For tasks, which belong to the same application, can be assigned with the application wide shared memory area.
- Eventually part of memory area can be reserved for all applications for a free access.
- As the structuring of tasks is a purely static matter and current operating systems do not allow dynamic reconfiguration in the task structure, the necessary activities have to be carried out at design time and cannot be automated. The restructuring of tasks is another example to improve the system dependability at expense of resource optimization.

So far in the space partitioning, only the direct data access by means of read/write operation is discussed. As depicted in Figure 9-19, the indirect memory access to variable can happen with function call of a runnable in task 2 in the 2nd situation. Because this runnable is originally allowed to change the variable, no error will be reported. The allocation of runnables/functions and the access to a certain driver interface should be also partitioned and protected as to the memory as well. In order to prevent the scenario in Figure 9-19, the runnable will temporally lose the unlimited authority as trusted object to the variable by inherit the status of "untrusted" from task1 when it is called by task1.

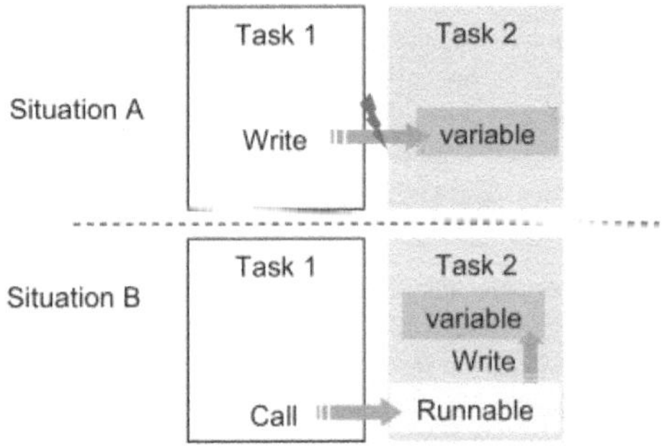

Figure 9-19: Memory protection of indirect data access

9.3.2.3 Time partitioning with OS and Software Watchdog service

To face the challenges stated in Subsection 9.3.2.1, time partitioning to monitor individual application software components is required to improve the overall system dependability [Tin03]. Time partitioning of application SW-Cs implies again an integrated approach by monitoring run-time behavior from real time and logical control flow requirements. In the following subsections, the designs for the time portioning will be discussed around these two aspects.

9.3.2.3.1 Run-time monitoring with Operating System

From the point of view in the software architecture, operating system is the best suitable basic software module to monitor the timing behavior of running entities.

- In the OSEK-compatible operating systems, as mentioned above, all the OS-objects are static configured and structured, in another word, they are all known to the operating system.
- Scheduling service is one of the basic services in the OSEK-compatible operating systems. Time services for the scheduling can be directly applied for the time monitoring.
- As discussed in Subsection 9.3.2.2, for the space portioning service all the untrusted OS-objects are associated with a look-up table in the OS for the space partition services. If this table can be extended with the timing parameter, they can be monitored by the OS as well.

In the following, a few terms used to discuss the runtime monitoring in operating system will be briefly explained:

Execution Time	As depicted in Figure 9-20, the execution time refers to the time, an OS-object spends in the RUNNING state (task state as defined in OSEK OS 2.1 [OSK05]) without entering the SUSPENDED or WAITING state. For ISRs it is the time from the first to the last instruction. For Tasks/ISRs it excludes all preemptions due to higher priority tasks/ISRs executing in preference. The execution time includes the time spent in the error, pretask and posttask hooks and the time spent making OS service calls.

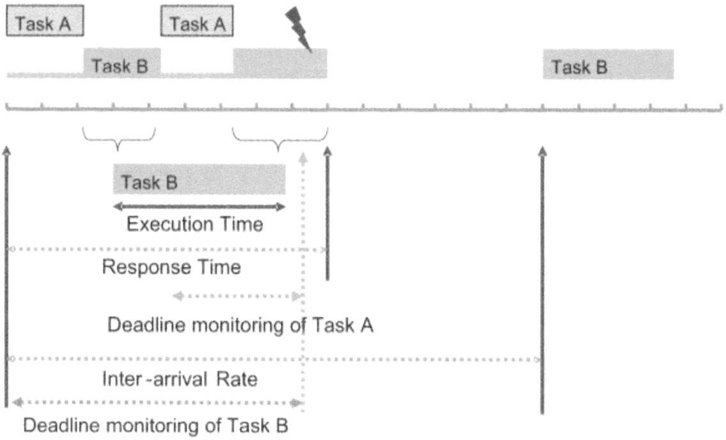

Figure 9-20: Definition of task execution time

Execution Time Budget	Maximum permitted execution time of an OS OS-object. Typically, this will be the worst case execution time (WCET) of a Task/ISR. For tasks the WCET budget counter is suspended if the task enters the WAITING or SUSPENDED state, and continued after the task is in READY state again.
Protection Error	Systematic error in the software of an OS-Application. ▪ Memory access violation: A protection error caused by access to an address in a manner for which no access right exists. ▪ Timing fault: A protection error that violates the timing protection.

In Figure 9-20, the constraints of the traditional approach of deadline monitoring in OSEKtime ([OSK01] and [HRK00]) is shown. The task B could violate the deadline, because task A, was e.g. executed too often and preempts the task B. In such a situation, the deadline monitoring based on WCET of task A will not report any error, while the task B, which is running properly, will be identified as faulty by the deadline monitoring. To solve such a problem, it is more reasonable to monitor the task execution time according to the preconfigured execution time budget as defined above to discover the real faulty task. Similar as designed in the OSEKtime, an OS interrupt can be triggered to report the timing fault for further fault treatment.

9.3.2.3.2 Run-time monitoring with Software Watchdog

Again as depicted in Figure 9-20, with execution time monitoring, faulty tasks, which take too much system resource of computation time can be detected. But as demonstrated in the example, if the task A has a priority high enough, it can preempt/block other tasks with the same or lower priority. Each time when the task A is executed, it fulfils its own requirement of task execution time, while keeping other tasks fail to reach the deadline. The faulty task A, even if it is executed much more often than normal, could stay undetected with task execution time monitoring. In the worst case, task B with lower priority can theoretically never report error with execution time monitoring but never finishes it job.

As discussed in Figure 9-17 of Subsection 9.3.2.2, tasks can be made up of runnables from different applications. So far the granularity of fault detection on the layer of task is the deadline monitoring and execution time monitoring, which are not fine enough for fault detection of runnables. As two sides of a coin, we could use a dedicated dependability software service to monitor the execution frequency of a runnable with aliveness indication to guarantee that the runnable is still running but not executed too often. In order to reduce the complexity and save the computation overhead, not all the runnables should be monitored, we can choose the runnables from the critical program flow in the following aspects

- Input signals collection
- Preparation of input data signals
- Signal processing and evaluation/plausibilisation of the results
- Control of the actuators

as the monitored objects. As mentioned above, usually these runnables are also mapped to tasks with higher priority.

Time partitioning also includes the correct logical control flow of applications, which is highly recommended in [ISO26] Table 6.5 for ASIL-D. As shown in Figure 9-21, in the faulty situation runnable R2 is branched. In case of a permanent branch fault, the aliveness fault of R2 can be detected by the aliveness monitoring of the runnable but it can be detected sooner with a runnable control flow monitoring. The control flow error might be detected with the execution time monitoring or deadline monitoring, but only to a much later time point.

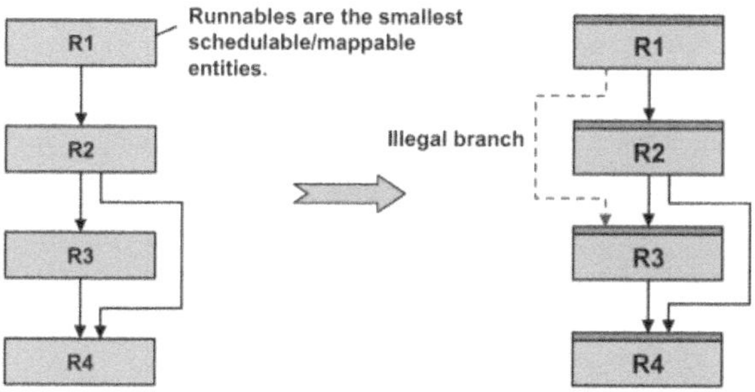

Figure 9-21: Control flow check of runnables

In the field of control flow checking, a lot of research has been carried out in the IT-industry to check the correctness of program flow. Most of the current methods of control flow checking are based on assigning signatures to blocks as introduced in [Nah02]. Such a technique suffers from high performance overhead and low flexibility with regard to modification of programs.

Software Watchdog here is supposed to ensure the real-time characteristics of the system as an extension to the hardware watchdog mentioned in Chapter 8 to face the challenges of increasing density of application SW-Cs. The following functionalities are planned to be implemented within the framework of the SW-Watchdog service:

- Monitoring the heartbeat of individual runnables including aliveness and arrival rate monitoring.
- Detecting control flow errors: control flow errors are unexpected changes in the sequence of instructions executed by the main processor. Taking the complexity into consideration, only the execution sequences of runnables are supervised by the SW-Watchdog.
- Gathering the detected error information from the two above mentioned services, compare the number of the detected errors to the threshold and derivate error indication states of the tasks and send them to fault treatment.

The interactions between the SW-Watchdog service and its runtime environment are specified as following. The dependability software architecture specified in Subsection 9.2 represents the environment, where the SW-Watchdog service is embedded and integrated. The major information, such as heartbeats and execution information of runnables will be sent to the SW-

Watchdog through Common/Domain-specific and ISS interface. And the error report from SW-Watchdog to Fault Management Framework (FMF) will be specified by FMF standard interface

In the following the functional structure of SW-Watchdog is specified in three basic blocks:

- Heartbeat monitoring: With the help of the heartbeat indication routine, runnables send their heartbeats to the SW-Watchdog. SW-watchdog supervises those heartbeats and checks if runnables are executed frequently enough (aliveness monitoring) or if they are executed too frequently (arrival rate monitoring).
- Control flow checking (CFC): The control flow checking service monitors the execution sequence of runnables by comparing real executed successors with the possible successor set of the predecessors.
- Task State Indication Unit (TSIU): To provide a proper tolerance of the errors detected in the upper two blocks, they should be sent to the TSIU. And TSIU should compare the number of the detected errors to the threshold and derivate error indication states of the tasks and send them to FMF.

The Figure 9-22 shows the functional structure of the SW-Watchdog.

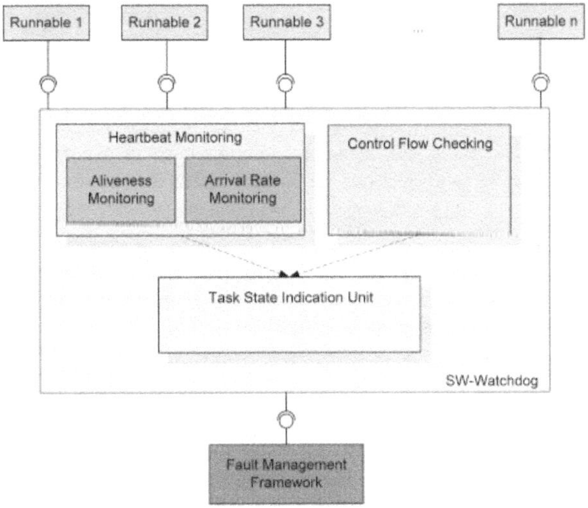

Figure 9-22: Different units constituting the SW-Watchdog service

In the following sections the functionalities of SW-Watchdog will be introduced in more details.

Heartbeat Monitoring

In order to setup Heartbeat Monitoring the attributes of each monitored runnable must be provided at the configuration phase. The attributes in this context means parameters, which describes the boundary of the correct execution. There are five parameters required, as listed in Table 9-16.

Parameters	Description
m	Minimal number of expected heartbeats within n cycles
n	Predefined number of monitoring cycles in the fault assumption for aliveness monitoring
p	Maximal number of expected heartbeats within q cycles
q	Predefined number of monitoring cycles in the fault assumption for arrival rate monitoring
runnableID	ID of runnable.

Table 9-16: Parameters for runnable in Heartbeat Monitoring

After specifying the parameters of a runnable, on the side of SW-Watchdog some data resource for the monitoring are required too, as listed in Table 9-17.

Data resource	Description
AC	Aliveness Counter, track the aliveness of monitored runnable
CCA	Cycle Counter for Aliveness, record the monitored cycles of aliveness
ARC	Arrival Rate Counter, track the arrival rate of monitored runnable
CCAR	Cycle Counter for Arrival Rate, record the monitored cycles of arrival rate

Table 9-17: Data resource for runnable in Heartbeat Monitoring

The interactions between the current runnable and heartbeat monitoring are implemented through heartbeats. Heartbeat will be sent, when the runnable is executed. After the heartbeat arrives in heartbeat monitoring, the data resource, such as AC and ARC, will be updated or in other words incremented by one.

On the other side Heartbeat Monitoring will update their own counters (e.g. *CCA* and *CCAR*) at the beginning of each monitoring cycle. It must be ensured that the monitoring cycle is independent from the cycle, of which aliveness is sent, and within one monitoring cycle all monitored runnables will be checked against their predefined boundary. According to the fault assumption two services must be implemented: Aliveness Monitoring and Arrival Rate Monitoring. Aliveness Monitoring checks, whether the runnable is out of its lower limit *m*, at the same time Arrival Rate Monitoring checks the upper limit *p* of runnable. Once the faulty runnable is detected, the ID of the faulty runnable will be reported to TSIU.

The last step of the monitoring is to decide, whether the counters (e.g. *AC* and *CCA*, *ARC* and *CCAR*) are needed to be reset. At the Aliveness Monitoring *AC* and *CCA* will be reset, when the predefined number of monitoring cycles has expired. Arrival Rate Monitoring is slightly different, the *ARC* and *CCAR* should be reset at the end of each cycle, if a fault of the arrival rate monitoring is determined at the end of this cycle. The following figures show how Aliveness Monitoring and Arrival Rate Monitoring work.

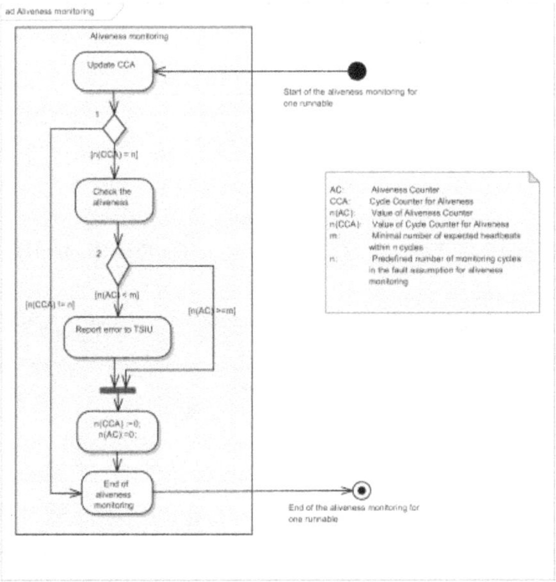

Figure 9-23: Activity diagram for Aliveness Monitoring

As demonstrated in Figure 9-23, after starting aliveness monitoring for one runnable, the *CCA* will be incremented by one, then we will compare *CCA* with predefined number of monitoring cycle *n* (see condition 1 depicted in Figure 9-23 as ◊1). If *CCA* is not equal *n*, aliveness monitoring for one runnable will be ended. In the case that *CCA* and *n* are equal, we will check the aliveness counter *AC* of the runnable (see condition 2 in Figure 9-23), whether it reached the minimal number of expected heartbeat *m*. If the expected heartbeat is not reached, an aliveness monitoring error will be reported to FMF. Before end of aliveness monitoring *CCA* and *AC* will be reset to zero.

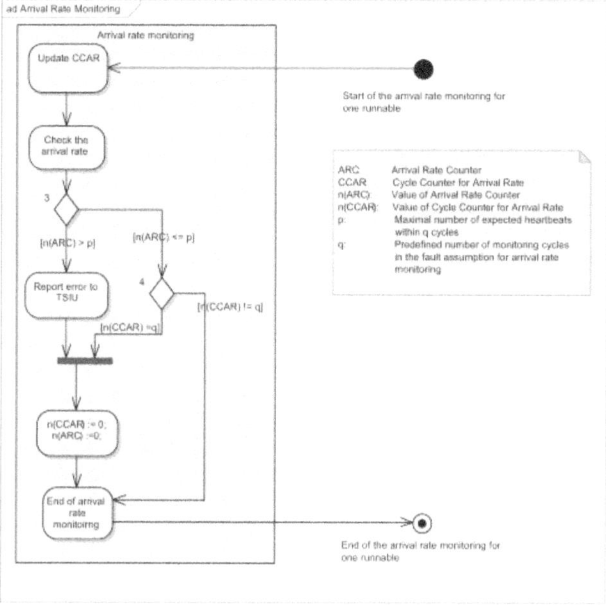

Figure 9-24: Activity diagram for Arrival Rate Monitoring

The design of Arrival Rate Monitoring is slightly different compared to the design of Aliveness Monitoring, as depicted in Figure 9-24. The first step of Arrival Rate Monitoring is also to update the cycle counter *CCAR*, but *CCAR* will not directly compare with predefined number of monitoring cycle *q*. We will check first the arrival rate counter *ARC*, whether it is greater than the maximal expected heartbeats *p*. In the case that *ARC* is greater than *p*, an error of arrival rate monitoring will be reported to TSIU. In other case if *ARC* equals or is smaller than *p*, another comparison, *CCAR* with predefined monitoring cycle *q*, will be carried out. If *CCAR* is equal to *q*, *CCAR* and *ARC* will be reset to zero. Otherwise arrival rate monitoring will be ended.

To demonstrate the concept and algorithms of Heartbeat Monitoring, an example will be given here. For a better overview, this example only demonstrates the supervision of one runnable.

For this runnable the fault assumption is:

- There are less than 2 heartbeats from the monitored runnable in 3 monitoring cycles, $m=2$, $n=3$.
- There are more than 3 heartbeats from the monitored runnable within 2 monitoring cycles, $p=3$, $q=2$.

Therefore, the examination terms of the heartbeat monitoring are:

$$n(AC) \geq 2 \text{ if } n(CCA) = 3$$

$$n(ARC) \leq 3 \text{ for all } n(CCAR) = 1...2$$

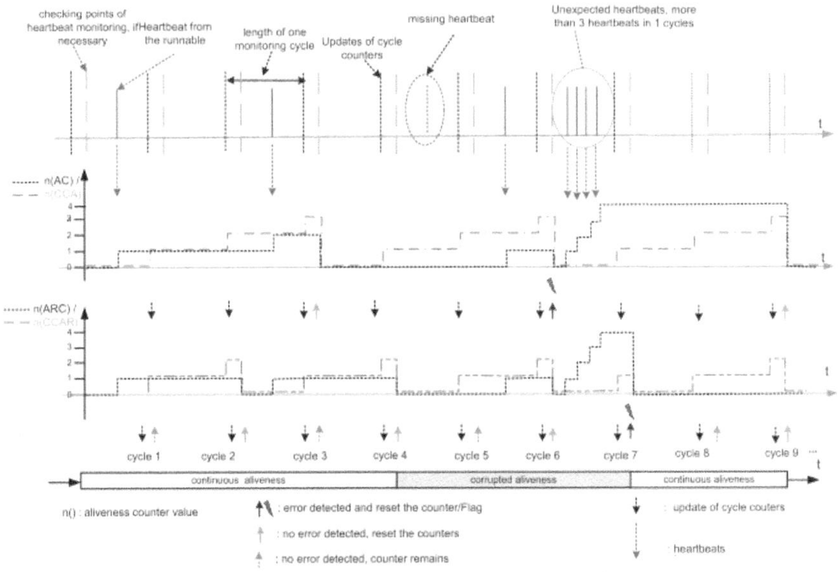

Figure 9-25: Example of Heartbeat Monitoring

In the simple example of heartbeat monitoring in Figure 9-25, it is assumed that all the counters have the initial value 0 at the beginning of the selected monitoring period. The work of SW-Watchdog in the following 9 cycles, is explained in detail:

- Cycle 1 to cycle 4, and cycle 8 to 9: There are no missing and unexpected heartbeats.

  → Checking criterion fulfilled. *AC* and *CCA* are reset at each 3 cycles, while *ARC* and *CCAR* are reset at each 2 cycles.

- Cycle 5: there is a missing heartbeat in this cycle, since the value of *CCA* has not yet reached 3.

  → No fault is detected and *AC* retains its current value.

- Cycle 6: There is only one heartbeat in cycle 4, 5 and 6. And the value of *CCA* reaches 3 at the end of cycle 6.

  → A fault in the aliveness monitoring is reported, because CCA reaches 3 while *AC*=1, *CCA* and *AC* are reset at the end of cycle 6,

- Cycle 2, 4 and 6: When *CCAR* reaches 2 (defined as q=2)

  → *CCAR* and *ARC* are reset. The *ARC* are checked at each cycles (yellow arrows), while the reset is only executed at each 2 cycles if no fault is found.

- Cycle 7: More than 3 aliveness indications are sent by the runnable.

  → *ARC* is over 3 and a fault is detected. *ARC* and *CCAR* are reset to their initial value 0.

Control Flow Checking

Another service provided by SW-Watchdog is Control Flow Checking. In comparison to traditional definition of Control Flow Checking in IT, which controls the correct execution of computer programs, here we only focus on the correct execution sequence of runnables in critical tasks. Before describing the internal mechanism of Control Flow Checking, some information about runnables contained within the critical task, must be prepared at configuration phase, as listed in Table 9-18.

Parameters	Description
successorTable	Table of runnable ID, which specify all possible successors belonged to the critical runnable.

Table 9-18: Parameters for runnable in Control Flow Checking

For each safety-critical task, Control Flow Checking will provide a predecessor indicator to record the last executed runnable of this task, so that the type of predecessor indicator must be the same type as runnable ID. Once a critical runnable moves into the running state, it will notify SW-Watchdog by using its service *checkControlFlow*. After that Control Flow Checking will examine whether the current runnable belongs to the possible successor set of last executed runnable. If the current runnable is not a possible successor, an error will be reported to Task State Indication Unit by using the service *reportError*, as depicted in Figure 9-26. After checking the control flow the predecessor indicator will be updated with the current runnable.

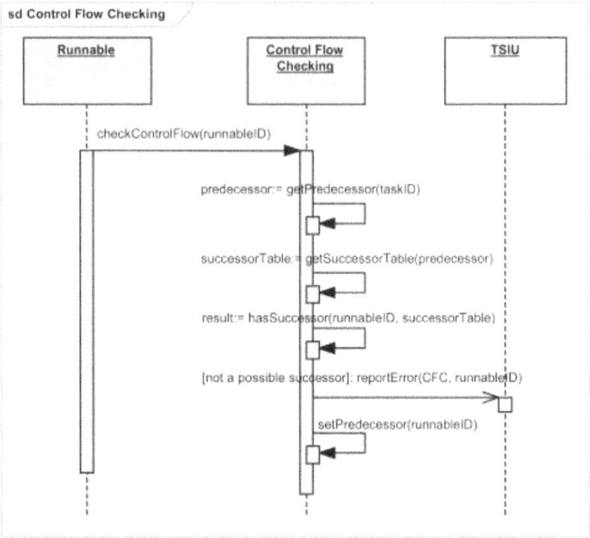

Figure 9-26: Sequence diagram for Control Flow Checking

Task State Indication Unit

The Task State Indication Unit must provide two vectors for each critical task, a condition vector and an error indication vector. The condition vector contains predefined bounds for each sort of error and it should be specified at the configuration phase. The error indication vectors are designed to record the results reported by heartbeat monitoring and control flow checking of a certain task. Each error indication vector contains three elements. The first element is the number of detected aliveness errors. It is the sum of the aliveness errors detected by the aliveness monitoring. Likewise, the second and third element should be the number of arrival rate errors and control flow errors. The determination of error indication state of a task is based on comparisons of condition vector with error indication vector. The two vectors are depicted in Table 9-19.

Name	1st element	2nd Element	3rd Element
conditionVector	Max. errors of Aliveness Monitoring	Max. errors of Arrival Rate Monitoring	Max. errors of Control Flow Checking
errorIndicationVector	Errors of Aliveness Monitoring	Errors of Arrival Rate Monitoring	Errors of Control Flow Checking

Table 9-19: Two vectors of TSIU

Once an error of one runnable in a task is reported, one element of error indication vector should be incremented depended on the error type, then the error indication vector will be compared with the condition vector of the same task. As long as one element of indication vector is greater than the one in condition vector, the error indication state of the task will be set to "`faulty`", and the faulty state will be sent to FMF by using service *indicateTaskState*. Since the state "`faulty`" is already sent to the FMF, the error indication vector must be reset to zero, in order to record new error for the following supervision, as depicted in Figure 9-27.

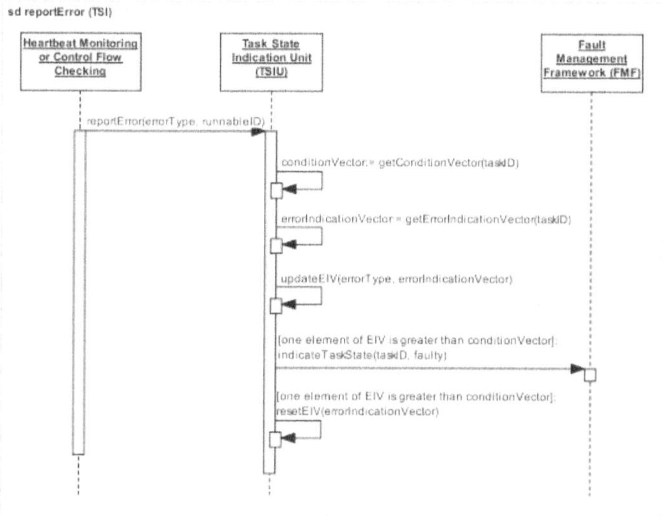

Figure 9-27: Sequence diagram for TSIU

Configuration of SW-Watchdog

In order to setup the SW-Watchdog a few parameters of SW-Watchdog must be specified at the configuration phase. Based on their application scopes, they can be split mainly into three categories.

- The first category contains the information about configuration of the Heartbeat Monitoring for each monitored runnable. The parameters *m* and *p* give the low limit and upper limit of heartbeats within predefined monitoring cycles (e.g. *n* and *q*), and this information should be provided by application designer. Another basic information is *runnableID*, which is unique in the whole system and consists of ID of the task, ID of the application and ID of runnable it self. Thus the *runnableID* will not be given by the application designer; it should be generated automatically by design tools.
- The second category contains the information about configuration of the Control Flow Checking. For Control Flow Checking each critical runnable should have a *successorTable*, which describes all the possible successors of the critical runnable. Since a runnable does not know its successors, until the mapping process of runnable on the task is finished, this information should also be generated automatically and provided by design tools.

- The last category is configuration of Task State Indication Unit (TSIU). For each task the *conditionVector* must be defined, which gives the bound of maximal tolerant faults. This bound is mostly depending on specific application design and system requirement, thus this parameter must be specified by the application designer.

As a summary to the Subsection 9.3.2.3, time partitioning services as OS with MPU, Protected Application service and SW-Watchdog service will improve on one side the system dependability with runtime monitoring. On the other side, from the view of a system integrator, they also contribute to locate and analyze the error at the system integration of functions from different suppliers, e.g. they can help OEM to determine the supplier that is actually responsible for the malfunction.

9.3.3 Dependability services of fault treatment

In the last subsections we discussed quite a few dependability software services for the individual monitoring and fault detection of application SW-Cs. Again with increasing intensity of application SW-Cs, it is required to gather these fault detection information, to management and filter out the safety relevant information. Thus an application wide system state can be pictured and an application wide coordinated fault treatment can be carried out.

9.3.3.1 Fault Management Framework

Functional design

The FMF ([FMF06] and [EiK07]) was designed for fault detection and subsequent compensation or recovery. This later aspect includes fault management (identify fault scenario) and fault treatment (identify and conduct fault mitigation strategy). Further on, a logging service was integrated into the FMF. All these aspects are described briefly in the following:

- Fault detection: Generally speaking, fault detection is often highly application specific, however, generic fault detection services (e.g. Software Watchdog) can be provided by the ISS framework. Here the task of the FMF is to handle faults and errors reported by the Software Watchdog.
- Fault management: Similar to fault detection, fault management is often application specific as well, but the FMF can offer some generic fault management mechanisms so that the handling of faults can be separated from the nominal task and the application programmer can concentrate on the application. Hereby the FMF provides general services allowing for applications (as well as the dependable software platform in Subsection 9.2) to obtain a consistent and global view of the fault state of the ECU as well as of individual applications on the ECU. This information can be used for isolation and diagnosis purposes as well as to make decisions on appropriate recovery strategies.
- Fault treatment: Based on predefined patterns from the application developers and system integrators, the FMF can combine these local (or even distributed) fault detections to derive a global view of the fault state of the ECU and the applications. Based on this global view, recovery actions could be initiated according to predefined principles. The fault-treatment measures are statically defined by the application developers and system integrators, since

a deterministic fault tolerant behavior is desired for the safety applications. Such a design brings also the constraints that only faults, which are considered in the fault hypothesis in the design phase, can be treated during runtime, since the fault tolerant system can only tolerate the faults which were considered in the design. Transient faults (e.g. transient faults from EMV) should be filtered by the FMF when the application functionalities are not affected.

- Logging service: It is also important to store appropriate information ("Logging") concerning the various faults and actions for later use in, e.g., workshops, so as to enable off-line analysis leading to fault assessment and repair, and maybe even future prevention.

In order to achieve these functionalities as discussed above, the FMF was structured into four functional units as can be seen from Figure 9-28 [FMF06].

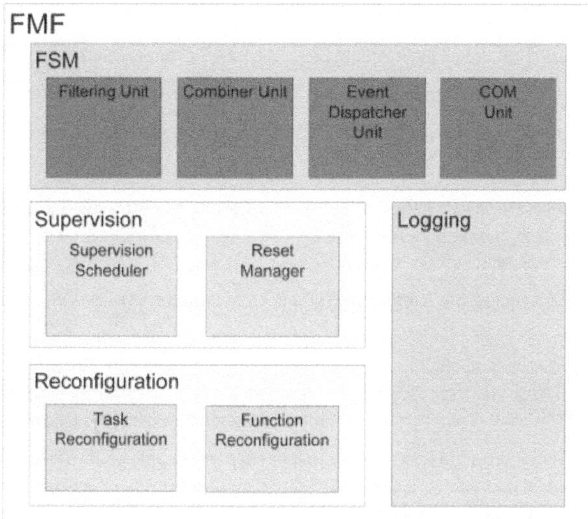

Figure 9-28: Fault Management Framework Structure

9.3.3.1.1 Fault State Manager

FSM is the core of the FMF since it receives the error reported from the software component and dispatches events to other units and software components. The FSM also manages the state of each error: present, confirmed, memorized and any combination of the three. The FSM performs error filtering, detection of error combinations, error event dispatching and distribution of error information through its four sub units.

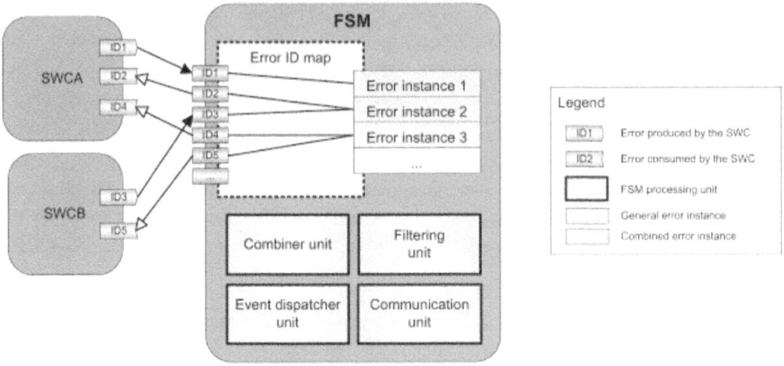

Figure 9-29: FSM structure and ID mapping

As depicted in Figure 9-29, each SW-C could have two kinds of interactions with the FSM:

- They can report error status to the FSM, those errors are called produced errors
- They can read the status of error from the FSM, those errors are called consumed errors

The mapping between the errors IDs used by the SW-Cs and the error instances is defined in the FSM configuration. This mapping allows reducing the coupling between SW-Cs, therefore enables a more flexible SW architecture.

1. The filtering unit handles the filtering algorithms (e.g. counter filter or sliding window based a stability filter)
2. The combiner unit is responsible for managing combined error instances. When an error is reported by a SW-C, the corresponding mechanism for combined errors is the evaluation of its logical expression. The evaluation of combined errors must respect the following rules:
   - A combined error has to be evaluated each time one of the error instances or its reference is updated (e.g. consecutively to a report or time based filtering), and during the same processing session.
   - A combined error is evaluated only once per execution cycle.

The evaluation of a combined error instance also includes its update, so the first rule implies that each time an error instance is updated, all combined errors, which depend on it directly or indirectly, have to be evaluated during the same execution cycle. This means that any update can trigger a chain of evaluation of combiner errors. The combiner handles those chained evaluations as following:

- Each time when an error instance is updated outside the combiner, the combiner is notified of the update and marks all the dependent combined errors as pending.
- At the end of an execution cycle, just before running the event manager, the combiner is called / run to evaluate all the pending combined errors, clearing the pending flag at each evaluation.

3. The event dispatcher in FSM as shown in Figure 9-30 is responsible for dispatching error events to SW-Cs and other FMF services. A given error event is linked to its consumers through event handlers, which define the notification mechanisms to be used and its parameters. There are three main types of event handler:

    - Synchronous event handlers which treat the event in the same execution cycle the event is detected. One example mechanism is a callback function which is called each time the event is dispatched.
    - Asynchronous event handlers which are notified of the event but do not treat it in an error processing session. One example mechanism is to increase a counter to signal that the event has occurred.
    - Supervision event handler: this is a special synchronous event handler dedicated for supervision, there can only be one such handler in the FSM event dispatcher. Its role is to run the Supervision's own scheduler which dispatches itself the events to its Fault Treatment Units.

    Similar as the combiner unit, the event dispatcher uses a delayed execution/dispatching pattern:

    - Each time the confirmed status of an error instance changes, the dispatcher is notified of the event and it marks the event and its associated handlers as pending.
    - At the end of a processing session the dispatcher will be executed, e.g. it notifies all pending handlers and after that clears all the pending flags. In this way, event handlers which are activated during a given session can be triggered in the correct order.

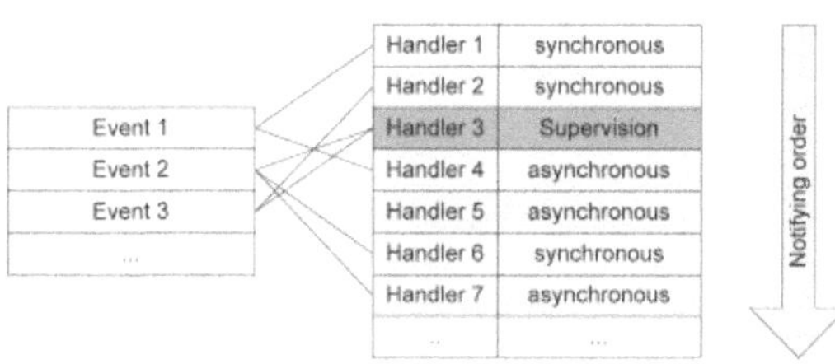

Figure 9-30: Event dispatcher example

4. The communication unit is responsible for distributing error information between the ECU local FMFs for a system wide distributed fault treatment.

    One of the aims of FMF is to manage the application wide coordinated fault treatment. Since the application SW-C can be mapped to different ECUS, the FMF should be distributed as well with each ECU runs its own local copy FMF. This means that there is no central fault management ECU. The FMF provides a mechanism to distribute error information. The FMF also allows the deployment of local and system fault treatment strategies through its Supervision and Reconfiguration units.

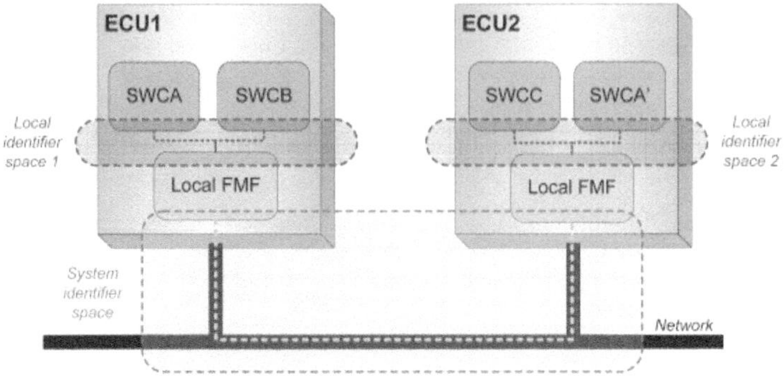

Figure 9-31: HW/SW mapping of the FMF

As depicted in Figure 9-31, there are two levels of communication in the FMF:

- Local level: communications between the SW-C and the local FMF
- System level: communications between different local FMFs

The two levels are kept separated, this means in particular that error information is never sent by a SW-C to a remote FMF, and it should always go through the local FMF instead.

For its inter ECU communication, the FMF relies on the COM-services. The communication requirements are divided in two categories:

- The requirements for error information distribution: the bandwidth requirements mostly depend on the number of system level errors, the timing requirement for each error and communication protocol used. For example the use of Agreement Protocol may require a higher bandwidth.
- The requirements for system fault treatment strategies: the strategies are completely up to the application / system designer and so are their communication requirements.

In order to avoid a modification on one ECU to impact the FMF instances on all the other ECUs, error identification is always local to an identifier space. Each local FMF has its own local identifier space to communicate with local SW-C, and there is one system identification space as shown in Figure 9-31. The identifier spaces are kept separated, so that the error identifiers of one space are not visible from the other space. The mapping between the identifier spaces is defined by the local FMF configuration. A local FMF only has the visibility on its local identifier space and the system identifier space, therefore it is not possible to map directly two local identifier spaces, and they must be linked through the system identifier space.

To clearly differentiate the local identifier spaces from the system identifier space, the FMF uses different error ID types:

- System level error identification: it is based on global identifiers which are used to identify an error uniquely within the system. Contrary to the local IDs, global IDs are not visible from the software component as they are defined internally in the FMF configuration.

- ECU level error identification: local identifiers are the only ones to be directly visible by the software components. Those local identifiers enable the local FMF to identify errors uniquely within the ECU.

9.3.3.1.2 Supervision Unit

The role of Supervision Unit is to supervise the execution of the fault treatment units. An FTU (Fault Treatment Unit as depicted in Figure 9-32) can be a sub module of a SW-C, or a completely standalone service, but it is managed / executed only by the Supervision Unit, and for that purpose it must have a special interface. Contrary to other SW-Cs, the FTUs can access the FMF reconfiguration services.

The entity responsible for managing the FTUs is the supervision scheduler. It processes error events (from the FSM), system events (startup, timer, etc.) and dispatches them to the FTUs.

The last important entity of supervision is the reset manager. The reset manager allows the FTUs to make a reset request or inhibit reset requests.

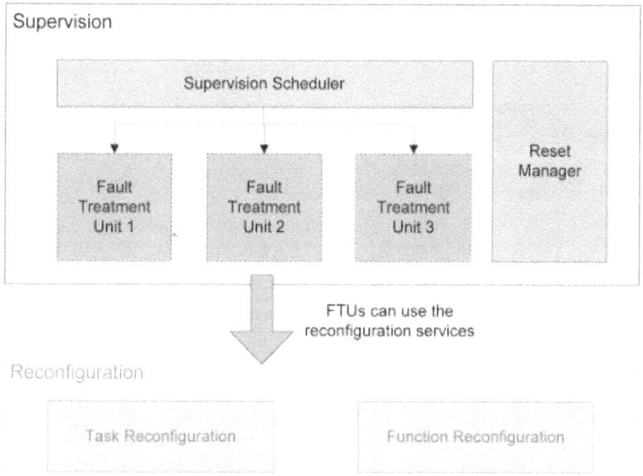

Figure 9-32: Overview of supervision

Supervision scheduler

The supervision scheduler has a similar role as the OS scheduler. The main difference is that where the OS manages the execution of tasks, the supervision scheduler manages the execution of the FTUs event handlers. Also the supervision scheduler forbids reentrancy. This centralization of the execution of the FTUs allows having a better control over the coherency and determinism of the integration of the different fault treatment strategies.

Fault Treatment Unit

The FTUs implement the fault treatment strategies. An FTU is actually implemented by the application programmer but it has specific characteristics which differentiate it from an App.SW-C. First an FTU can access the reconfiguration services and the reset manager. Secondly an FTU can only be executed through its event handlers and by the supervision scheduler. Even though the content of an FTU is application dependant, FTUs can be divided in two categories:

- Local FTUs which only interact with SW-Cs and FTUs located on the same ECU.
- System FTUs which also interact with system FTUs located on other ECUs.

9.3.3.1.3 Reconfiguration Unit

The role of the Reconfiguration Unit is to provide mechanisms to reconfigure functions, runnables and tasks of the ECU. Functions can be activated or deactivated while tasks can additionally be restarted. The services provided by this unit can only be used directly by the Supervision Unit. More details about reconfiguration can be found in Subsection 9.3.3.2.

9.3.3.1.4 Logging Unit

The logging unit is responsible for storing errors and freezing associated frame data in the non volatile memory, which is especially useful for later fault removal and system recovery after the ECU restart. The logging unit interacts with the following software services: the non volatile memory manager, the diagnosis and calibration service.

9.3.3.2 Fault treatment with dynamic configuration

Compared with the IT-Industry, where the dynamic reconfiguration is already a common practice for the system recovery and update, state of the art in the embedded in-vehicle electronics is still a statically configured system. There are various reasons for that [DyR96]:

- As its name implies, embedded electronics are embedded together with the hardware. Once a system is embedded, there might be no reason to change or update it.
- In the embedded system, resources such as processor power and memory are relative limited.
- Safety relevant systems usually require hard real-time requirement. Reconfiguration, instead, can require more than time than allowed.
- A deterministic behavior rather than a dynamic configuration is vital for the safety relevant applications.

As have mentioned in Subsection 3.4.4, the same evolution in the IT-Industry will take place in the automotive embedded electronic, where the reconfiguration will be a more and more important topic. The motivation and scope of the dynamic reconfiguration in automotive electronics, however, are quite different to the IT-industry:

- Enabling technologies such as common software architecture, model-based software design and component-based software services (esp. support from RTOS) are becoming mature technologies.
- Product platform strategy: It is possible for an entire product line to use the same software core, perhaps stored as a library in an EPROM. Based on the hardware configuration and desired functionality, the software can easily be band-end configured for each individual product model. This eliminates the need to develop and maintain separate software for each different product, esp. for the automotive supplier, dynamic reconfiguration is becoming more and more interesting to support different vehicle models on the same platform of one OEM or even for different OEMs.
- More availability with dynamic reconfiguration and fail-degradation: State of the art in the in-vehicle safety system is fail-safe and fail-silent. As introduced in Subsection 2.1, dependability is a balance of different aspects. In order to improve the availability while keeping the system safe as a trade-off, dynamic reconfiguration with fail-degradation is a proven approach, esp. for semi-X-by-Wire application.
- Remote upgrades and maintenance: Compared with the life cycle of 4 to 8 years of traditional pure mechanical vehicles, the life cycle of electronics of 1-3 years is much shorter [McK05]. "Patches" can be downloaded / updated to the customer to fix the bugs and introduce more features/extensions.

It is clear that only systems which consist of several components can be reconfigured; these components can be hardware or software. Also it is clear that reconfiguration requires some kind of redundancy, which is not an identical ECU in every case, but it can be any kind of HW or SW unit that can, at least partly, take over the duty of the faulty unit.

A reconfiguration in software usually changes the flow of information between the components and/or changes the purpose of components. With these actions the functional mapping is redefined for some or all components. The aim of the reconfiguration is a system that consists only of error-free components and that exhibits the maximum of functionality possible (fail-degradation).

Reconfiguration can be static or dynamic. Static reconfiguration appears in the development phase of a system, dynamic reconfiguration occurs during runtime. The reconfiguration mentioned here refers mainly to the dynamic reconfiguration in software, which contributes to the fault tolerance, but does not address fault repairing. Dynamic reconfiguration defined in [IEC07] C3.13 "remapping the logical architecture back onto the restricted resources left functioning" with "run-time redundancy" is not recommended for ASIL-C and ASL-D applications according to [IEC03] Table A.2 about software architecture design, thus dynamic reconfiguration with resource reallocation is not in the scope of work here.

Two types of events can trigger a dynamic reconfiguration for the fault tolerance:

1. The occurrence of a fault causing an error
2. The disappearance of an error in a component.

In the first case three actions have to take place:

- The fault has to be detected and confirmed.
- The fault has to be confined.
- The functionality has to be taken over by another component. If the functionality of the erroneous component is not replaced completely then the system operates in a degraded mode (graceful degradation). The functionality of the erroneous component can be taken over by a redundant component or another component that changes its purpose for this action.

The premise for the second case is that an error recovery has taken place. After the disappearance of the error within the component two steps may occur:

- The component is reintegrated in the system.
- The component takes over its original functionality

The reintegration of repaired components could enhance the survival probability of the system significantly.

In general, two technical implementations are possible:

- Redundant components: Two or more components with the same functionality are installed in a cold- or hot-stand-by manner. When an error is detected in one component, this component is disabled and one of the remaining components takes over its duty. Fail-silent components disable themselves, non-fail-silent components must be disabled by a supervisor. In a cold-standby system this supervisor is also responsible for activating the redundant component.
- Components that can change their purpose (functional alternatives [SCK04]): In case of an error in one component the functionality is relocated to another component. The component that takes over the functionality does not execute its original function anymore. This is only an appropriate solution for control components that are not connected to special hardware. There are already prototypes in the telematics domain of the function relocation, e.g. when the on-board display of navigation system fails to work, an enhanced audio navigation assistant can be switched on to provide the driver a degraded navigation service. In an automotive safety electronics, however, it will be a challenging task to undertake the relocation in real-time. One example for these technical realizations could be that the control of a SafeSteering can take over part of stabilizing functionalities by providing more dynamic inference with steering when the braking management System goes to a degraded mode due to an error.

After having discussed about the constraints and guidelines about dynamic reconfiguration, in the following concrete functional design of dynamic reconfiguration will be given:

The requirements of deterministic behavior and dynamic reconfiguration seem to be contradictive at the first glance. But a closer look shows that a system that is reconfigured during runtime can still be deterministic. The reason for this is that we have a fault hypothesis containing errors to be tolerated. Thus we know which errors can occur and can therefore define statically how the FMF should react to the errors. We can define e.g. several schedules or configurations which based on the detected errors are switched on the fly (see Figure 9-33).

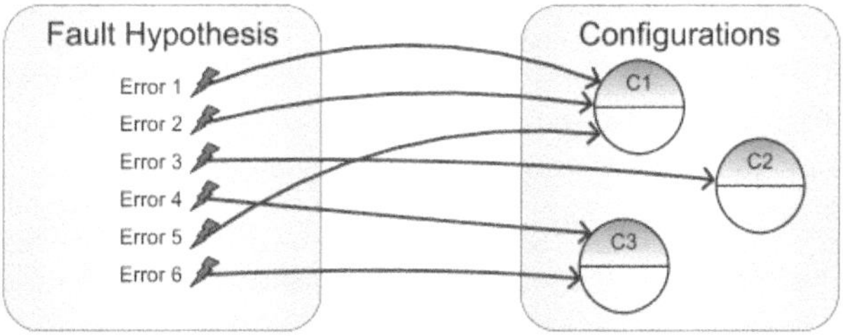

Figure 9-33: Mapping of Errors to Configurations

Based on the mapping concepts shown in Figure 9-33, dynamic reconfiguration can be divided into

- Function/runnable level: passivation or termination of faulty functions/runnables
- Task level: passivation or termination of faulty tasks
- Application level: passivation or termination of faulty application on the same ECUs or applications distributed on remote ECUs.
- ECU level: hardware reset of ECUs.

In order to perform the above reconfigurations, the OSEK-compatible OS as specified in Subsection 9.3.2.3.1 should be extended with the following functions:

Kill OS-Application	The operating system frees all system objects, e.g. kill tasks, disable counters and interrupts, etc., which are associated to the OS-Application. OS-Application (refer to Subsection 9.3.2.3.1 for detailed information) and internal variables are potentially left in an undefined state.
Kill OS-Object	The OS terminates an OS-Object (e.g. task or ISR) and releases all its held OSEK resources and re-enables all interrupts disabled by it.
Stop an OS-Application	Activate a task in the OS-Application and allow this task to stop (in a consistent manner) activities of the OS-Application.
(Re-) Start an OS-Application	An OS-Application is restarted after the self-termination (or being killed by the OS because of a protection error). If the OS activates the OS-Application it uses the configured restart task.

Function/runnable level reconfiguration

Function level reconfiguration is an indirect reconfiguration mechanism; it means that the FMF does not act directly on the function but provides the information whether the function should be activated or not. The term function is used here in an abstract way, it could be a runnable or any other element which can be activated or deactivated. The function level reconfiguration keeps track of certain events that have occurred which should prevent a function from being executed. For example, if a sensor signal is missing or faulty a function using this value should not be executed. The reconfiguration service will govern the status regarding whether a function is allowed to run or not, but it is the responsibility of the application to check the status before calling the function. In case the function is inhibited, the application may choose some backup strategy but this does not concern the FMF.

To manage the inhibition status an *inhibition mask* is used where bit *X* indicates whether event *X* has occurred or not. This mask is then compared with the *inhibition conditions* that have been specified for the function. These describe which event combinations that should cause the function to be inhibited. The reconfiguration service gets the event notices from the supervision.

To enable function level reconfiguration the following properties must be specified for each function:

- Function/runnable ID
- IDs of the events that should result in an update of the inhibition status
- Inhibition conditions

Task level reconfiguration

For task level reconfiguration the FMF offers the following possibilities:

- Termination of faulty tasks (and chained task) with help of OS
- Deactivation/Activation of tasks (and chained task) with help of OS
- Restart of a task, i.e. the init-function of the task is called.

Task level reconfiguration is more powerful than function level reconfiguration and requires support from the OS. Moreover, since deactivating/activating tasks has a strong impact on the system behavior, it is required that all activation patterns and tasks are known beforehand in order to enable a dependability analysis. That is, all fault management strategies involving task reconfiguration must be specified off-line and it is only the supervision that is entitled to issue such reconfiguration commands.

Termination of faulty task (e.g. tasks, which violate the space and time partitioning, which have a hazard to other safety relevant tasks) refers to an immediate reconfiguration to kill the running task with OS. The resulted error-hook reports the error to the OS. OS terminates the task with *KillTask<TaskId>* to free all the resource allocated by the faulty task. In case that the output of the faulty task is awaited by the following task, the so-called chained task, and the following task should be deactivated as well.

Tasks may be either time-triggered or event-triggered. Time-triggered tasks are activated solely by the OS based on timer interrupts. Event-triggered tasks are activated by having an external entity (e.g., another task) calling *ActivateTask(<TaskId>)*. The external entity can also be an interrupt routine, e.g., associated with a timer. The task type affects how the deactivation/activation is to be implemented by the reconfiguration service.

Deactivation of a task means that following invocations of the task will be disabled. However, if the task is currently running it will be allowed to finish compared with task termination. The deactivation involves modifying the timers for the task (if it is time-triggered) or disabling the events/calls that trigger the task (if it is event-triggered). For safety reasons a running task may not be killed by any other entity than itself or the OS. This is why the task deactivation activity does not involve killing the task. It might, however, be necessary to kill a task that has become corrupted and is, for example, stealing processing capacity from other tasks. Activation of a task means that the events or timers that trigger the task are enabled again. Activation may also require (re)initialization of the task, that is, the init-function of the task is called.

Restart of a task means that its init-function is called. Thus, each task must register their init-functions as callbacks with the FMF.

One critical point is if a task responsible for the acquisition of sensor signal or give the commands to actuator is involved in the task level reconfiguration, or if the output of the task is could block some other chained tasks, a fail-safe mechanism should be provided to avoid deadlock or failure propagation. During the reconfiguration, some callback function can be triggered by the error-hook to provide the "service-not-available" or "inactive" information for the chained task consumer of the task output. Another possible mechanism is to provide the consumers of the task output with time out mechanisms or signal counter to switch into the fail-safe mode. Since the measures here are most application specific, no detailed discussion will be given in this thesis.

Although it is the supervision that issues the reconfiguration commands, the reconfiguration service keeps track of the current state in a manner similar to functional level reconfiguration. This enables applications (or the rest of the FMF) to check whether or not a task is active.

Application level reconfiguration

Strictly speaking, the application level reconfiguration does not differentiate from the task level reconfiguration significantly. As depicted in Figure 9-31, application, distributed on different ECUs can be reconfigured with distributed FMF on each node. The pre-conditions for the application level reconfiguration are the synchronized configured fault-treatment on the different nodes, which requires a time-triggered communication system for the synchronization the reconfiguration and enough bandwidth to meet the real-time requirements. A concrete design and prototyping of the application level reconfiguration will be shown in Subsection 10.3.1

ECU level reconfiguration

Suppose that one or some of the most safety relevant applications on one ECU are identified as faulty, in this case, a reconfiguration to fail-degradation and fail-safe will be executed immediately. If there is the need to heal the faulty ECU in the same drive cycle, the most direct and simple way is to restart ECU. Since there are quite a few fault handling strategies linked with the ECU reset, the reset demands should be arbitrated and performed with a few pre-conditions and pre-reset procedures.

In FMF, the reset manager is responsible for arbitrating the reset demands and calling the necessary system/OS service to perform a reset. Note that the arbitration between requests has to be defined in the reset manager configuration. In order to guarantee a deterministic manner also in reset, the arbitration should be carried out with preconfigured look-up table or state-machine.

The reset manager is also responsible for keeping track of the reason why a reset occurred. The reset manager provides the following services:

- *FMF_RMInit*: This service is called by the supervision scheduler during start-up.
- *FMF_RMShutdown*: service is called by the supervision scheduler during shutdown. It takes as parameter the shutdown reason passed to FMF_Shutdown.
- *FMF_RMGetResetStatus*: This service may be called by any FTU at the start-up phase to determine the cause the shutdown.
- *FMF_RMRequestReset*: This service may be used by any FTU to request a reset of the ECU, for example for fault correction or reconfiguration purposes. A parameter identifies the request in order to allow the reset manager to perform arbitration between different multiple and maybe conflicting requests.
- *FMF_RMInhibitReset*: This service may be used by any FTU to ask the reset manager to inhibit resets requested through the *FMF_RMRequestReset* service. A parameter identifies the enable/disable request in order to allow the reset manager to perform arbitration between different multiple and maybe conflicting requests.

9.3.4 Dependability software services for gateway

Gateway, as mentioned in Subsection 3.3 and Chapter 8 about hardware architecture, is commonly applied to connect different in-vehicle domains and external of vehicles. The general definition of gateway is a functional unit that interconnects two computer networks with different network architectures, for the ISS-safety systems, gateway is an ECU, which receives, process, save or forward data from one protocol to the other. Typical examples are ECUs, which process the sensor data and forward the signal to other in-vehicle network nodes or forwarding signals from one in-vehicle communication network to the other, e.g. LIN to CAN, CAN to FlexRay, Cabin-CAN to chassis/powertrain CAN, etc.

The ISS gateway services as depicted in Subsection 9.2 are summarized as following [GTS06]:

- Routing Service
- Network Address Translation: The Network Address Translation Service allows the mapping of addresses from one domain to addresses of the other domain.
- Communication Service with security and integrity
- Firewall Service
    - Authentication Service
    - Cryptography Service
    - Access Control List

As AUTOSAR provides basic inter-domain communication mechanisms, the AUTOSAR SW-architecture, incl. the AUTOSAR PDU Router for the routing between different networks, is taken as a basis for the gateway services discussed here.

As stated in definition of dependability in Subsection 1.2, dependability includes aspect of safety, integrity and confidentiality. Although gateway services are not the focus of this dissertation, in the following, the dependability issues (as issue 3 and 4 of the list, bold font) of gateway will be discussed briefly.

Communication service with security and integrity

Gateway dependable communication service here includes read, write (store) and forward of the signals. By both processes of read and write, plausibility check as well as data integrity check (e.g. with CRC) should be performed.

Signal routing includes usually unpack from original protocol, assemble/segmentation and pack in the new protocol. There are two critical aspects here: timing requirement and data integrity. For the end to end data transmission via gateway, the transmission time on the network is more or less limited with network bandwidth, so that delay time on the gateway can become the bottle neck of the overall delay. The layered software architecture should be optimized for the data processing and forwarding. In order to guarantee the data integrity, end-to-end application CRC as specified in Subsection 9.3.1.3 can be applied. In order to accelerate the process of data forwarding, the signal frame CRC might not be checked again, since in case of any fault detection, the signal should be still forwarded to the receiver with a flag for the fault identification. In order to protect the information of assemble/segmentation, a CRC to the whole block of message can be added as a signature of the gateway at the end of the pay-load.

Security in telematics gateway

One of the most important features of ISS-applications is the interconnection with telematics service provider. As some ISS applications require communication with external agents through a wireless link, like other vehicles for cooperative driving, or with the vehicles environment and infrastructure for transmission of floating car data and reception of traffic information, the gateways are needed to provide suited communication links. Security in telematics gateway is an important step in the integration of telematics services for safety systems. As specified in the list, however, the security issue here does not differentiate a lot from the security topic in the IT-industry, esp. with the rapidly increasing computation power of embedded system.

9.4 Configuration of dependability software services

Interfaces between basic software services and dependability software services, which are specified in the EASIS projects, are summarized and depicted in Figure 9-34. The dependability software services discussed in this dissertation are:

- Dependability RTOS – Subsection 9.3.2.2 and 9.3.2.3.1
- Fault tolerant communication services – Subsection 9.3.1.2 and 9.3.1.3
- Agreement Protocol – Subsection 9.3.1.4
- Software Watchdog – Subsection 9.3.2.3.2
- Fault Management Framework – Subsection 9.3.3.1
- Dependability software services for gateway – Subsection 9.3.4

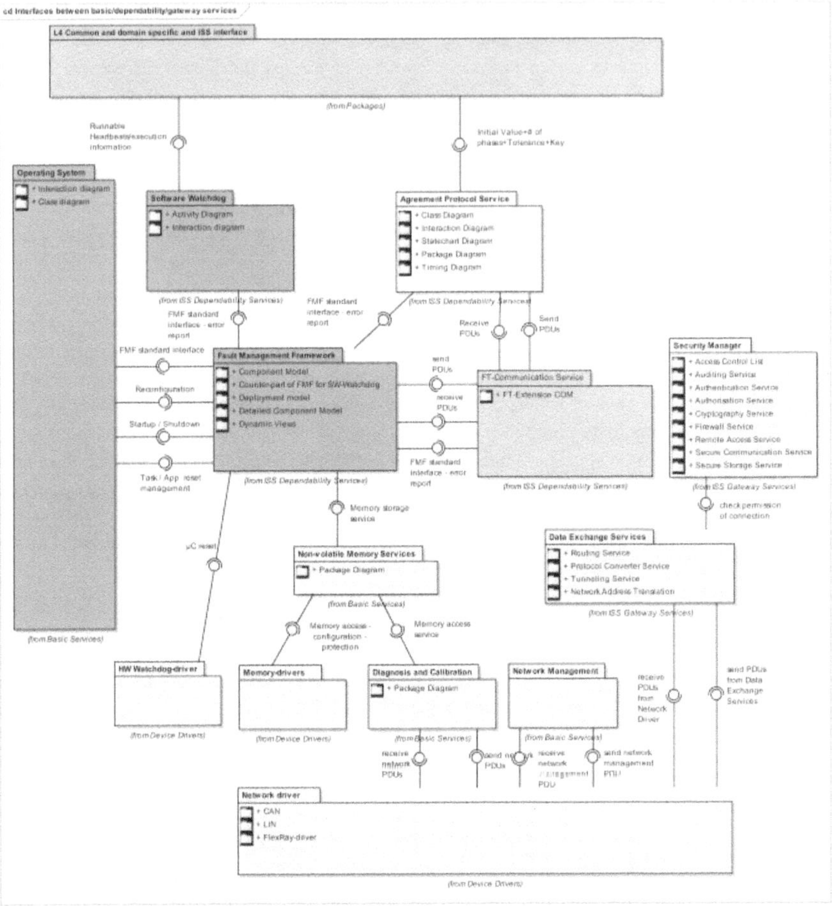

Figure 9-34: Interface of basic, dependability and gateway software services

RTOS is needed for each embedded electronic, the extended features introduced here can be categorized in different service classes as an add-on module to the core.

- Class 1: OSEK-compatible RTOS core with scheduling, basic resource management etc
- Class 2: Support for MMU and MPU (space partitioning)
- Class 3: Support for extended protected application services in time partitioning (execution time monitoring)
- Class 4: support for Fault Management Framework in dynamic reconfiguration e.g. termination, passivation of faulty tasks and task-groups

- Class 5: Support of full space partitioning and time partitioning services (e.g. runnable-call, periphery access protection, etc.)

It is to mention here that Class 2 is needed for the following two classes, while Class 3 and Class 4 are more or less independent from each other.

In a similar way the communication services can be divided into the different classes as well:

- Class 1: OSEK compatible COM with support to different in-vehicle networks
- Class 2: Fault tolerant COM with interface to application CRC
- Class 3: Support for redundancies and interfaces for additional fault tolerance mechanisms
- Class 4: Integrated fault tolerant communication services with Agreement Protocol

Class 1 and Class 2 are relative generic without preference for the communication networks, while the Class 3 and Class 4 are only required by the certain safety application with additional support.

The two main services provided by Software Watchdog, Aliveness Monitoring and Control Flow Checking, are also quite independent from each other. They can be configured to the needs of applications, the monitoring objects and threshold/counter borders can be configured in the look-up table.

The 4 sub-components of Fault Management Framework can be configured individually as well. The configuration varies from the simplest case as a fault handler to the complex variant, coordinated system wide reconfiguration with help of SW-Watchdog and OS. The reconfiguration part is the most complex component, while the reconfiguration strategy, as highly application dependent, could be chosen based on the framework defined here.

Dependability software services for gateway are only needed for gateway ECUs, the security issue is of course only interesting for the links with external communication.

9.5 Conclusion of the dependable software platform

In this chapter, started with a benchmarking with IT-industry, a dependable software platform is specified as an enable lever to improve the system dependability. Focused on the three aspects of Integrated Safety System, the communication (including gateway), the integration of application on one hardware platform and the fault treatment are discussed. Last but not least, the dependability software services are classified in different classes for the configuration and integration.

10 Prototyping and validation of the concepts

Before we can explore the prototyping and validation of the concepts for the future automotive electronic architecture in detail, a brief overview of the development process, methodology of the validation activities is given in the following subsection.

10.1 Introduction to the architecture validator

10.1.1 Prototyping approaches and validation process

As mentioned in Subsection 3.5.2, state of the art in the development of in-vehicle electronics is the model and simulation based development process, in which a sub-system is developed and tested in a pure software environment with the help of simulation tools (Model-in-the-Loop and Software-in-the-Loop test). The implementation of software on a particular hardware platform will only be initiated after a successful test of the functionalities with the help of simulation models.

Here this approach was adopted and extended – using rapid prototyping systems as real-time execution platforms, which allow for short design cycles while evaluating effects of different realization choices including the flexible variation of hardware setups.

Due to the networked architecture, the development process needed to cover system and node level design as depicted in Figure 10-1. The tools used for early development and validation are mainly MATLAB/Simulink together with Real-Time Interface Block-sets and experimentation tools, which enabled the validation of the dependability software services on rapid prototyping platforms with FlexRay interfaces from dSPACE [Eck06].

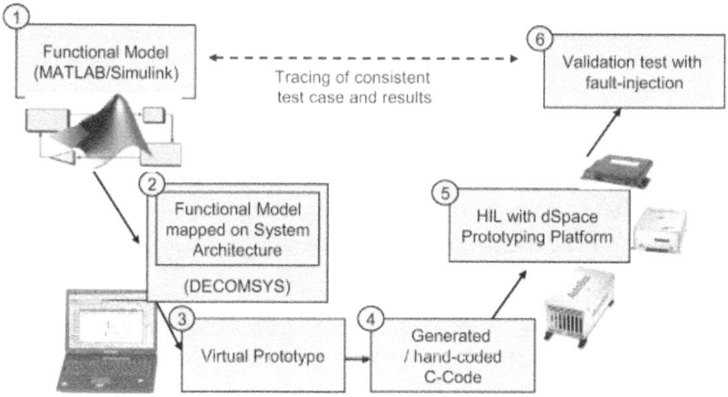

Figure 10-1: Tool chain and development process

As shown in Figure 10-1, after requirement analysis of the safety applications,

- The first step the system functional design is initiated by building and modeling the whole system including the simulation environment with MATLAB/Simulink.
- The dependability software services are prototyped and simulated on a PC as a virtual prototype in the second step.
- Based on the principle of fault injection, the virtual prototype is evaluated as a front-loading in the third step.
- In the fourth step, based on the knowledge gained from the virtual prototype and other automotive constraints such as resource and timing requirements, the virtual prototype of the dependability software services and the modeled safety application will be mapped onto tasks and scheduled on the system architecture. For the rapid-prototyping platform C-code can be generated automatically. For the evaluation-board source-code can be manually implemented with the development environment of the certain microcontroller.
- In the fifth step, source codes will then be compiled and loaded onto the rapid prototyping platform using AutoBox or MicroAutoBox [Eck06] from dSPACE or evaluation-board. Dependability software services and concepts are evaluated in the real hardware environment and tested against the requirement specification, by injecting the faults which are supposed to be tolerated by the dependable software platform. The consistent evaluation cases and validation concept will be applied through the virtual prototype, rapid-prototyping and on evaluation-board, thus the traceability of the testing results to the design specification through different platforms and environments can be guaranteed.

10.1.2 Architecture design of the validator

As mentioned in Chapter 5, the validation of concepts in this dissertation took place in the EASIS WT1.3 Prototypical Implementation [VAL06] and WP 5 – WT5.1 Architecture Validator [VAL06] and WT 5.2 Process Validator (Commercial Vehicle validator). The EASIS architecture validator [VAL06] is focused on the prototyping and validation of some of the most relevant properties of the EASIS architecture both in hardware and software, including fault tolerant hardware, dependability software services and telematics services etc.

The implemented EASIS architecture validator (as depicted in Figure 10-2) incorporates the following components:

- Fault tolerant communication with dual-channel FlexRay network
- Fault tolerant ECUs with dual duplex architecture
- Dependability software services providing dependability software services, upon which further applications can be built, including:
    - Agreement Protocol
    - Software Watchdog
    - Fault Management Framework
    - Dual-duplex fault tolerant signal processing

- Telematics gateway between internal buses (dual channel FlexRay inter-connected with two active stars and CAN) and external Telematics network (TCP/IP) with protocol conversion and security services.

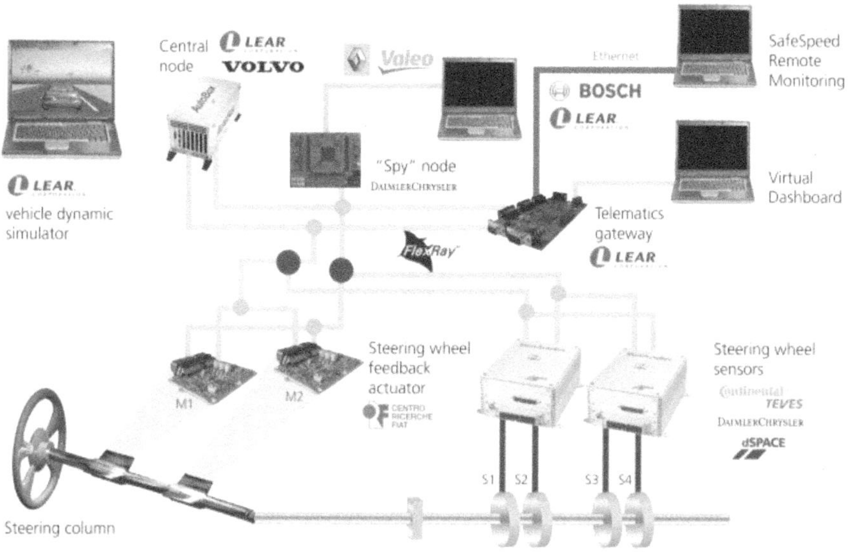

Figure 10-2: System topology of EASIS architecture validator [AfS07]

The most important hardware nodes are the Sensor Node (steering wheel sensors), Actuator Node (steering wheel force feedback actuator), Central Node, Spy Node and Telematics Gateway.

In order to show the behavior of EASIS architecture from the application's point-of view, the EASIS validator includes a safety-critical function (steer-by-wire) and three demonstrative integrated safety applications:

- SafeLane – a lane departure warning system
- SafeSpeed – limitation of vehicle speed by an external commanded value and remote monitoring using Internet browser.
- SafeLight – an enhanced adaptive lighting system with help of vehicle dynamic and environment information.

The EASIS architecture validator also includes several interactive graphical interfaces to show, in real time, the behavior of the EASIS architecture and its resilience in front of faults. It is possible to inject faults (both hardware and software) on the system and monitor the system behavior including a 3D vehicle dynamics simulation.

10.2 Validation of hardware architectures for ISS

In the following, the validation process and results of the most important safety concepts of hardware architecture in Chapter 8 will be described.

System topology with backbone concept

In the three system topologies discussed in Subsection 8.1, we chose the backbone concept for the validator, since the integrated safety applications based on the steer-by-wire technologies require fault tolerant communication with high-bandwidth. Compared with the Figure 8-14, the system topology of the validator is almost the right part with the dual-channel chassis FlexRay and telematics gateway. As mentioned in Subsection 3.3.4.2, the active stars provide better fault tolerance than the passive star topology of CAN due to the detection of short cuts and deactivation of branches in the active star.

Result: The backbone concept with FlexRay enables integration of ISS-communication with large mount of data transfer. The safety integrity requirement of communication system can be reached without any additional safety concept.

Distribution of ISS-application to the HW-architecture

The three high level ISS-application as well as two basis functions of Steering Wheel Sensing and Steering Wheel Feedback were mapped to the HW-architecture. The guidelines about portioning and mapping of ISS-applications in Subsection 8.4.1 are considered here.

1. According to rule 1, the actuator control with steering wheel feedback is mapped to dual-duplex ECUs directly connected with the electronic motor.
2. Since the ECUs for actuator control have also FlexRay interface, the sensor for vehicle speed and steering wheel angle can be placed freely.
3. FlexRay transfer rate was configured to 5 Mbit/s, no limitation of bandwidth is needed to be considered.
4. In the validator, the signals of speed and steering wheel angle are distributed with FlexRay. Steering wheel angle, for example, is sensed redundantly in a dual-duplex manner. The redundant signals are checked and voted for its plausibility in the value and time domain before they are sent to the bus.
5. The high level application SafeLane and SafeSpeed are integrated together on the Central Node, since they are of the same safety integrity level and the hardware platform AutoBox provides enough resource for them. Dependability software services facing the challenges of application integration were also mapped to the Central Node.

Design concept of communication networks

There are mainly 3 communication networks integrated in the validator: CAN and FlexRay are mainly used for the vehicle dynamic control while Ethernet is applied to represent the telematics communication from extern. CAN is applied for the coupling between physical sensors and sensor node, and also monitoring/fault injection (see Subsection 10.3.3). While the latter use-case is not safety relevant, safety concept mentioned in Subsection 8.2 is implemented to tolerate the timing drift of sensors.

Result: Switches for fault injection were built into the validator to emulate DC fault model of the network. FlexRay Scheduling requires a high amount of front-loaded design work, since all the signals have their fixed place in the communication schedule. OEM, together with Teer-1 suppliers should spend more effort in the system requirement engineering and design phase to fix the communication matrix. This effort, however, is paid back upon the system integration, the FlexRay communication was integrated almost like plug-in & play and works very reliable and.

Dual-duplex architecture

Dual-duplex architecture can be found both in the safety concepts of the actuator and sensor-coupling in the validator as shown in Figure 10-3. The steering wheel angle is the one of the most important inputs of steer-by-wire function. The sensing function, with the safety integrity of ASIL-D, should be redundant and independent in the path of signal processing. Thus the two sensor node, each connected with two physical steering angle sensors, can work completely independently in a fail-operational mode. The exchange and voting of the angle is carried out with FlexRay.

The dual-duplex actuator has architecture of 1oo2D-system, with independent power supply and communication path to both FlexRay channel A and channel B. The application software is partitioned between the microcontroller and the DSC in such a way that the DSC practically acts as an actuation co-processor, providing the management of the electric motor at low level, while the main microcontroller handles the actual application software and provides interface to the rest of the system and to the end user for calibration and diagnosis (for more details see Appendix 1).

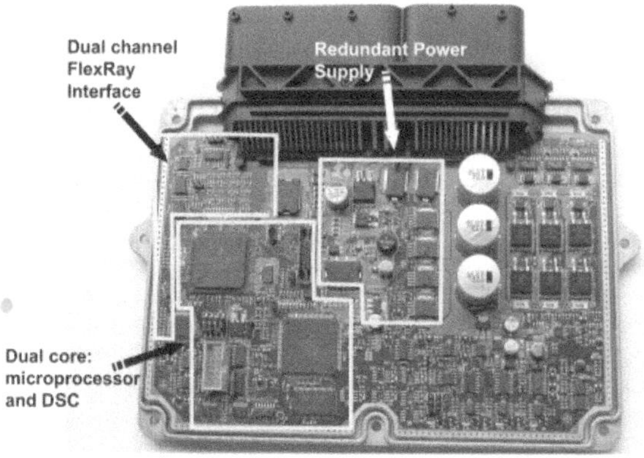

Figure 10-3: Fail-safe unit of dual-duplex actuator ECU

Result: The dual-duplex architecture guarantees enough safety integrity, which is required for the fail-operational behavior.

10.3 Validation of dependability software platform and services

The validation of the dependability software platform as shown in Subsection 9.2 is performed during the design and integration of ISS-application with the underlying software platform. Design of the software architecture on the Central Node, Sensor Node and Spy Node followed strictly the layered architecture. The smooth integration of applications and dependability software services from different consortium partners in very short period demonstrates the effectiveness of the approach.

Based on the requirements and fault hypothesis in Chapter 6, fault injection is applied to validate the dependability software services along with exemplary fault handling strategies from the praxis. Before we will go into the details about the validation of each software services, the principle of fault injection is given as following:

Typically fault injection can be divided into 3 main types ([Hed01] and [Ehr03]), physical fault injection, software based fault injection and fault injection in simulation model. For the validation of dependability software services, because physical fault injection (EMC, circuits stuck-open, open or high impedance outputs) are inflexible and not versatile enough in cope of different fault types, it was only applied in the validation phase for the validation on the communication and hardware architecture. For the validation of dependability software services, software based fault injection and fault injection in simulation model are intensively applied.

Following the basic theory of fault tolerance as shown in Subsection 2.2, there exist two basic approaches: fault injection and error injection. Fault injection simulates software design faults by targeting the code. Here the injection considers the syntax of the software to modify it in various ways with the goal of replacing existing code with new code that is semantically different [Voa98]. This "code mutation" can be performed at the source code level before compilation if the source code is available. Error injection, called "data-state mutation" in [Voa98], targets the state of the program to simulate fault manifestations. Actual state injection can be performed by modifying the data of a program using any of various available mechanisms.

The relationship between faults and errors is depicted in Figure 10-4. We can see that one fault can result in several errors (common mode) and on the other hand several faults can cause the same error (same syndrome). Generally speaking there exist much more faults than errors, e.g. a time violation of a task may have many different reasons (e.g. corrupted scheduling table, corrupted memory or bugs in the tasks).

As long as all the errors can be tolerated, all the faults in the fault hypothesis can be tolerated with the dependability services as well. Here it is to mention that no 100% complete mapping between all the possible faults and errors are needed to be identified. Another experience is that errors are generally easier to be injected than faults, e.g. emulation of memory faults to influence the control flow of runnables is more difficult to implement than the direct error injects in the control flow with manipulation of program counter. Because of the reasons mentioned above, the validation of the dependability software services in the following subsections is based mainly on the "error injection", strictly speaking.

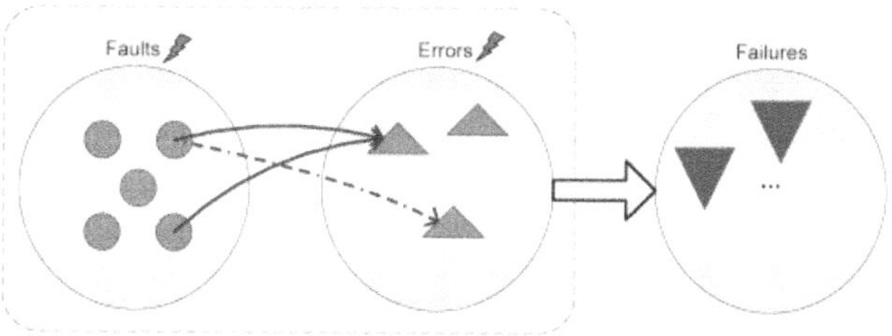

Figure 10-4: Mapping between faults and errors

10.3.1 Prototyping and validation of Agreement Protocol

As discussed in Subsection 9.3.1.4, Signed Messages Protocol was chosen as a basic concept, upon which an Agreement Protocol Service was designed and prototyped, while the following requirements and constraints are considered:

Firstly, the Agreement Protocol Service executed on a node must deliver not only an agreed value, but also a state vector, which includes the state of each participating node. This state is gained from last agreement process from the point of view of the node that generated it. The state vector could contain useful information in case of a failure to identify the faulty nodes in a system.

Moreover, the signature mechanism should not be as complex as for IT applications, since efficiency and short execution times are required. Therefore a simpler and thus more efficient signature mechanism was designed and integrated in the Agreement Protocol Service.

Usually, fault-tolerant applications have backup system states, which are applied when faulty components or other abnormal states are detected. In this case, all the distributed safety components of the application must be consistently informed of this change in order to take synchronized and coordinated fault treatment measures. Therefore, the SafeLane application in the EASIS validator has, besides the "operational mode", a "faulty sensor mode" and a "failure mode". Since the application includes the central node and the two fail-safe sensor nodes. Each fail-safe node is equipped with two physical steering wheel angle sensors [SWS03], thus they implemented a fail-operational sensor node in a dual-duplex structure. The global application state is computed from the single states of these nodes.

As shown in Figure 10-5, the Agreement Protocol Service was embedded and validated on the three nodes (Central Node and two steering wheel sensor nodes) and is cyclically executed in order to maintain consistency among them with respect to the global state information of SafeLane. The private value of each node is its own state. The state of each node is determined, as depicted in Figure 10-5, by a state machine as demonstrated in Figure 10-6, whose transitions depend on local node information, as well as the global application state as an output of the Agreement Protocol Service. The state information is exchanged between the nodes over FlexRay which is not only reliable and deterministic, but also exhibits the required cyclic behavior. Upon the execution of the agreement protocol, the non-faulty nodes consistently build a vector consisting of the different nodes states. Based on this vector, each node consistently computes the global application state by applying a specific decision function designed for the SafeLane application.

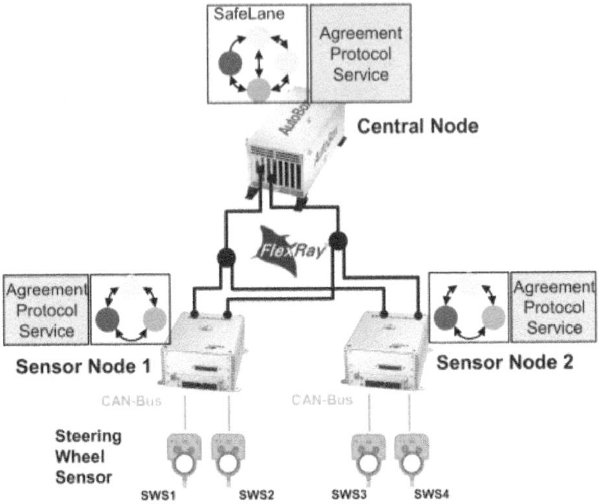

Figure 10-5: Outline of system architecture for the validation of Agreement Protocol Service

Hence, the use of the Agreement Protocol Service guarantees a reliable and deterministic global state transition among the three nodes so that an agreed and safe state can be maintained for the SafeLane application.

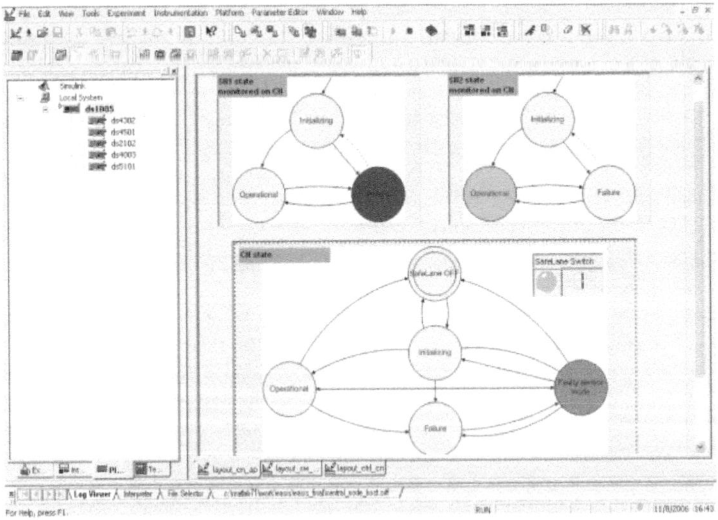

Figure 10-6: User interface for agreement protocol validation

The validation of Agreement Protocol follows the principle of fault injection according to the fault types defined in the design phase. Different faults like FlexRay communication fault, Byzantine Fault whereby inconsistent data is exchanged between the nodes and memory fault of inconsistent data read, etc. are injected to the validator. The faults can be injected with hardware failure or emulated with software using dSPACE ControlDesk. Figure 10-6 shows the user interface with the state machines of each node implementing the agreement protocol changing in real time.

10.3.2 Prototyping and validation of Software Watchdog

The prototyping and validation of Software Watchdog follows the complete steps as shown in Figure 10-1 from virtual prototype to evaluation-board, which is explained in depth as following:

Modeling, simulation and prototyping of the Software Watchdog

For the modeling of task dispatching and program flow of runnables in OSEK, Stateflow in Matlab/Simulink was applied. Stateflow is a design and development tool used for modeling complex system behavior based on finite state machines. Runnables are modeled with function-call subsystems and triggered by events sent by Stateflow in a defined execution sequence. A function-call subsystem is a block in Matlab/Simulink which can be invoked as a function by another block. For instance, as illustrated in Figure 10-7, the application SafeSpeed can be divided into three runnables: sensor value reading in *GetSensorValue*, the control algorithm in *SAFE_CC_process* and setting of the actuator in *Speed_process*. These are triggered as function-call subsystems by the Stateflow chart SafeSpeed, in which the execution sequence of runnables is implemented. To indicate the aliveness of the runnables, further function-call subsystems to simulate the glue code are also implemented, which report the execution of the runnables.

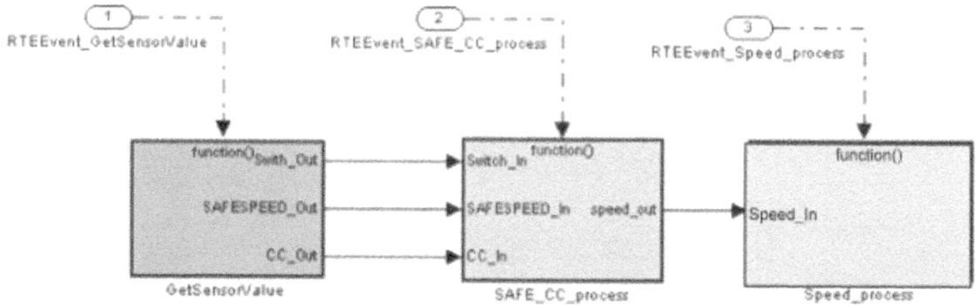

Figure 10-7: Modeling of runnables and program flow in SafeSpeed

The time-triggered behavior of the heartbeat monitoring unit and task state indication unit was modeled with time counters. Thus, in order to simulate the mechanism of task scheduling with different periods in the operating system, different time counters can be assigned to the Stateflow charts. On the other hand, the program flow checking unit was modeled using an event-triggered Stateflow chart.

Evaluation of the Software Watchdog in EASIS validator

The evaluation of the Software Watchdog is performed based on the fault/error definition in the design phase. Since different faults can result in the same error, error injection is applied for the evaluation of the design and prototyping of the Software Watchdog. Such an approach has the advantage that the dependability requirements can be tested in a front-loading manner of system development. The concept can be validated independently from the specific fault-types. Faults, which are difficult to inject into the test bench or on-road test, can be relatively easily emulated with errors.

Here again Stateflow is used to manipulate the execution frequency and sequence of runnables by changing the timing parameter of runnables, manipulation of loop counters and building invalid execution branches, etc. The experiment environment ControlDesk provides the possibility to manipulate the data assigned to the timing parameter of runnables to the condition that determine the invalid execution branches in the runtime. Therefore, it is used to trigger the error injection during the execution of the applications and visualize the results as well.

By building different evaluation cases, the three chief functionalities of the Software Watchdog, i.e. the detection of the aliveness error, the arrival rate error and the program flow error, are successfully validated. The following screenshots demonstrate some of the evaluation cases generated by injecting heartbeat or program flow errors. The x-axes of each plot in the diagram indicate the time lapse, which has a scalar of 10ms. The y-axes indicate the value of the counter and number of detected error. In order to inject heartbeat errors, a time scalar is connected to a slider instrument to change the execution frequency, For example, Figure 10-8 shows the test with an injected aliveness error. Similar test with arrival rate error and control flow error were performed as well. The increase in the y-value in the last plot "AM Result" (Aliveness Monitoring Result) indicates the detection of the errors.

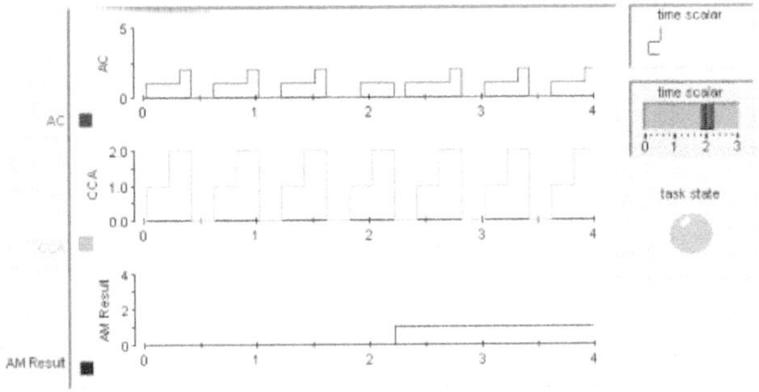

Figure 10-8: Test with injected aliveness error

Figure 10-9 shows the case in which the real cause of the erroneous state is identified through the collaboration of the units of the Software Watchdog. Here, the aliveness errors detected by the heartbeat monitoring unit are actually caused by program flow errors, which are reported with the plot "PFC Result" (Program Flow Checking Result). After the detection of three program flow errors (which here is set as the threshold), the task state is set to "faulty". Only one accumulated aliveness error is reported.

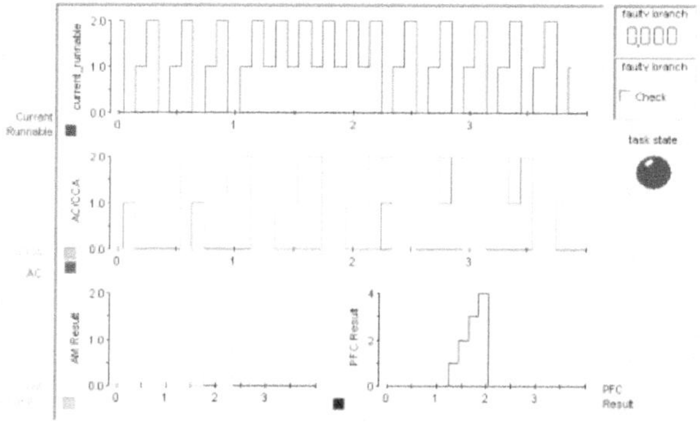

Figure 10-9: Collaboration of fault detection units

10.3.3 Prototyping and validation of Fault Management Framework

As well as other dependability software services, in the first step the system functional design of the FMF was initiated by building and modeling the whole system with Matlab/Simulink. The Fault Management Framework was prototyped and simulated on a PC as a virtual prototype in the second and third step. In the fourth step, based on the knowledge gained from the virtual prototype and other automotive constraints such as memory and timing requirements, an actual microcontroller was chosen here for the implementation and system validation. Individual hardware specific C-codes were generated or hand-coded. This version was integrated and compiled with the safety applications light control and other basic software modules, such as the operating system OSEK OS with an extended task scheduler. In the fifth step, the FMF along with the applications and environment were loaded onto the target platform and tested in a HIL. The development environment CodeWarrior from Freescale [CoW05] was applied for the debugging, compiling and linking. Thus, the virtual prototyping in MATLAB/Simulink eased the development of the FMF functionalities on the target hardware. More details about the prototyping of the FMF can be found in [FMF06] and [Spi07].

Validation with Fault Injection and Rest-bus-Simulation

The validation of the FMF followed the principle of fault injection. Generally speaking, fault injection can be possible at different system components and it depends on the test object. In this case we want to examine the designed functionality of the FMF. Therefore, one possibility would be to insert faults right into the FMF, that is to say, fictive faults can be injected and reported to the FMF to test the designed fault-treatment. On the other hand, the FMF works with application software components and other dependability software services like the Software Watchdog mentioned above, so we can inject the faults by real faults as well. That is to say, the faults are injected on side of the application, which are detected by the Software Watchdog and then reported to the FMF. This second approach is more appropriate because it takes also the interaction between the FMF and other components into consideration.

In order to evaluate the fault management framework service, different evaluation cases must be considered. The main cases of fault injection are depicted in Figure 10-10.

1. In the first case, the fault injection will take place at application level (as demonstrated in Figure 10-10: faults are injected in light/rain sensors to result in the faults of light control).
2. In the second case, the Software Watchdog was applied to report errors to the FMF. Thus, we manipulate the scheduling table (as an extension to the OS) in order to trigger errors concerning runnables, which should then be detected by software watchdog and then reported to the FMF.

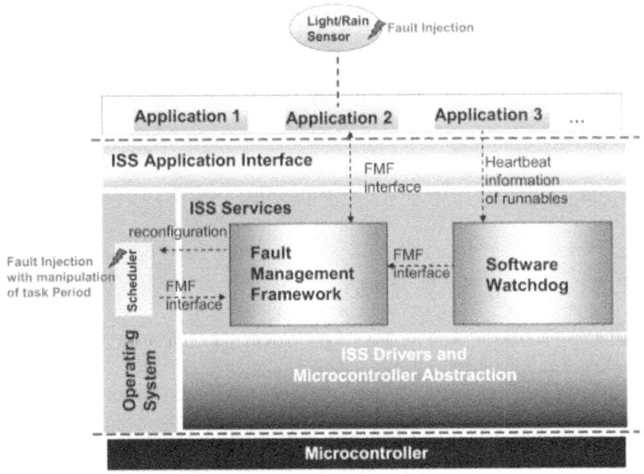

Figure 10-10: Evaluation cases with fault injection for FMF

In order to trigger fault injection and to display the evaluation results, a simulation and diagnosis tool (CANoe) was used, which was connected via a CAN card to the CAN interface of the evaluation board. The structure is depicted in the Figure 10-11.

The evaluation microcontroller S12XF was connected to the FlexRay network to collect the vehicle status information (speed, park, start...). A computer, running the tool CANoe, was linked to the ECU via CAN bus for simulation and demonstration effects, but also for rest-bus simulation. The graphical interface implemented with the CANoe panel editor was used to trigger fault injection on the one hand and to display the actual status of the ISS services, e.g. faulty state of a runnable, state of FMF and application sensors, etc.

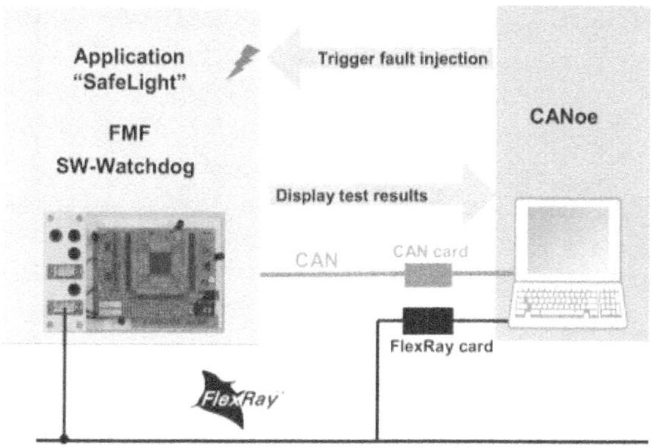

Figure 10-11: Fault injection triggered with CANoe to FMF

The rest-bus simulation here enabled a smooth integration in the EASIS overall architecture validator. The same test cases could be reused and traced in the fully networked Hardware-in-the-Loop test bench.

10.4 Validation of the ISS Engineering Process

As stated in Subsection 7.1 the two most important concepts of EASIS Engineering Process are

- Virtual Frontloading (virtual integration with structured test) and
- Correct by construction approach

These two principles are validated during the participation in EASIS work-package of the validator. Following the steps specified in the ISS Engineering Process, some of the most important validation results are summarized as following:

Development of Functional Analysis Architecture

The FAA-model, using model-based specification is depicted in Figure 10-12. As a hardware mapping and platform independent model, the FAA is validated with "model-in-the-loop" approach. The hardware independent FAA-model enables a clear specification of interface and data model, which can be easily transferred and further developed by the OEM and suppliers in a decentralized manner. Based on the safety analysis (e.g. preliminary hazard analysis) and fault types, the potential hazards of the system are simulated and injected into the simulation environment to the FAA-model with virtual front-loading.

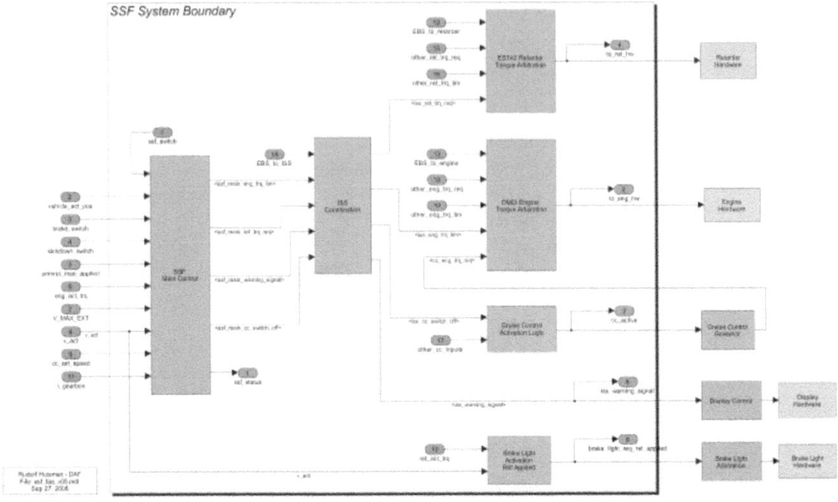

Figure 10-12: FAA-model of SafeSpeed

Development of basic Functional Design Architecture

By mapping the FAA-model to the HW-architecture and split the whole model to the subsystems, we can get the FDA-model as shown in Figure 10-13. The FDA-model here can be further supplement to the interfaces of basic software services. With the virtual integration of FDA with these HW-platform independent basic software services, the FDA-model can be validated with the "Software-in-the-loop". The same fault injections from the last step are recycled. With the suggested unified simulation environment and tool chains, the FDA-model can be exchanged simultaneously to any time point of the development between the development partners without any overhead customization. Moreover, software bugs and test scenario can be distributed and reproduced easily. The ISS EP enables a significant improvement of the interaction of partners to manage the complexity of ISS.

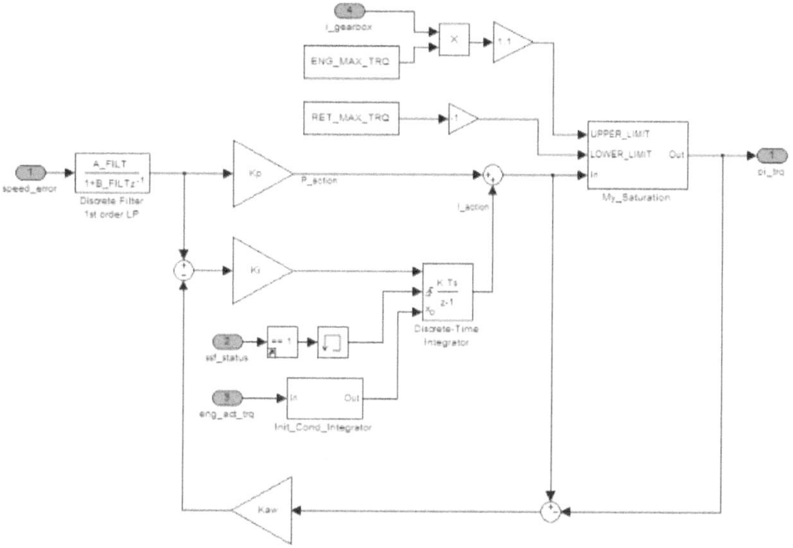

Figure 10-13: FDA-model of the SpeedControl in SafeSpeed

Refinement of and validation of FDA model with SiL-test

With the application of dependability architecture framework, the FDA can be integrated to the layer software topology with the dependability software services. With the standardization of dependability software services, distributed test of safety relevant requirements (to which, the dependability services are designed for) can be carried out without costly shipment of special equipment. In the following Figure 10-14, we can see the FlexRay scheduling requires a global communication road-map at early development phase. Any changes of the scheduling will result in a chained reaction to the scheduling and resynchronization between application tasks and communication services on the other nodes. The consistent tool chain and standardized FT-Com services enables an uncomplicated exchange and update between the partners.

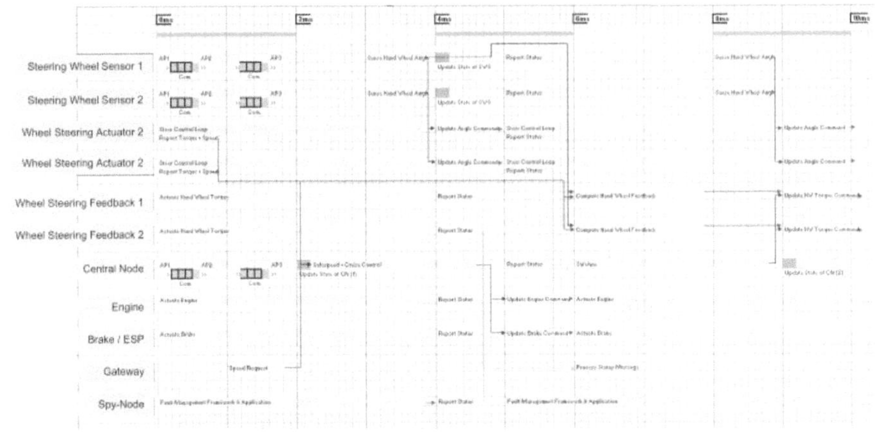

Figure 10-14: FlexRay communication scheduling in the validator

Experience in setting up of the validator, using a hybrid system, made up of the nodes and communication systems listed in ISS EP step 5.3, shows the flexibility and effectiveness the approach. The counter-part (remote nodes in the ISS) can be simulated and virtual integrated in the most cases. The definition of dependability architecture frameworks shortens the transition time between different development phases and prototyping platforms, e.g. PC-based simulation environment, RP to evaluation-board.

In all the integration and validation of total EASIS-validator was quite efficient. The physical assembling and system test of the different nodes is almost plug-in and play, no iterative design loop was required.

10.5 Evaluation and optimization of the concepts

Safety concepts proposed here are implemented with rapid-prototyping and serial near equipment, while the validation follows the principle of fault injection. During this validation phase, faults were injected into the system on hardware, application level as well as directly in the dependability software services. All injected faults are proved to be detected by the system correctly and the appropriated fault treatment strategies have been executed. However, it has to be noted, that only fault scenarios identified and defined in advance by the developers could be handled. As the aim of the work here is to validate the concepts, fault injection does not provide results about quantitative fault detection coverage and test coverage. Since the ISS-applications here are not directly from the serial projects, no confident performance benchmarking can be carried out.

As a summary, the results here with regard to hardware/software architecture and platform are here shown to be functioning properly, and they are also identified as being pertinent for upcoming Integrated Safety Systems. Therefore, services like those defined in the dependable platform have to find their way into real-world platform. A step towards bringing the results closer to being deployed in products will be to advocate the use of the results as input and inspiration for on-going projects concerning automotive software platforms, such as AUTOSAR.

10.6 Experience and findings from the prototyping and validation

In all, most of the concepts suggested in this thesis, including ISS Engineering Process, fault-tolerant hardware architecture and dependable software platform was prototyped and validated during the EASIS work-packages of validator. The prototyping and validation process here is not characterized with the traditional OEM/supplier (customer/vender) relationship, but with a more decentralized peer-to-peer relationship as project partners. It was quite challenging to make a master plan for the distributed development, such as to consolidate the design and prototyping approach and to control and synchronized the distributed activities. The common prototyping platforms as well as development environments despite of a few features (e.g. using interchange format of communication scheduling to synchronize the changes, migration from virtual prototype to the hardware platform) have contribute to the successful cooperation significantly.

11 Conclusion of the results, discussion and outlook

Future Integrated Safety System exhibits an ever higher system complexity with the integration of safety functions with different integrity levels. The dependability requirements as to the fault treat behavior, triggered by the industry norms, increase from fail-safe/fail-silent to fail-operational.

In the dissertation presented here, with the analysis of the key aspects of ISS, the requirements of future ISS are derived. With the assessment of these requirements with state of the art safety electronics as well as the current trends of the in-vehicle electronics and safety norms, a delta analysis for the concept development was carried out. To address these challenges and close the gap, an integrated dependability architecture framework for the ISS was specified in the three main levers: hardware, software and development process. As a conclusion the most important results and their implication are concluded as following.

11.1 Conclusion and implication of dependability architecture framework

Development process

ISS-applications are characterized with intensified interaction between applications of different safety integrity levels cross domain borders in a deterministic real-time manner. Thus it requires by its nature a decentralized engineering process. In order to manage the complexity, the correctness by construction approach is ensured with the fine specified ISS engineering steps. For a better cooperation between the development partners towards higher system dependability, approach of virtual front-loading is suggested. The engineering steps, virtual integration of FAA-model, derivation of FDA-model with the mapping procedure, prototyping on different platforms are supplemented with a systematic test methodic with consistent test cases through out the engineering process. The dependability E/E-architecture framework, common simulation, development and prototyping environment provide one of the most essential fundaments for the safety justified engineering process between the involved partners.

Hardware

The hardware architecture framework is addressed with the three aspects system topology, communication systems and ECU architecture. We tried to give a few vertexes and Lego-modules to provide a framework with possible configurations to dependable hardware architecture. A few use-cases are listed here to demonstrate the application of the hardware architecture framework.

The traditional work distribution is that OEM defines the system topology, communication systems and its interfaces to ECU, while the suppliers define and implement the ECU HW-architecture. With ISS, however, this distribution border is more and more unclear between the participants. Triggered by the aimed safety integrity level as to the safety norms, taking the constraints as development complexity into consideration, the hardware architecture framework should be worked out and configured with a much intensified interaction and cooperation between the development partners.

Software

The dependability software platform was specified in a layered reference model after a short benchmarking of automotive software with the evolution of IT. In order to design a hardware and application transparent fault tolerance, dependability software services such as ISS-communication service with redundancy, end-to-end CRC, Agreement Protocol, time partitioning and space partitioning for the integration of applications, fault management, fault treatment of dynamic reconfiguration and gateway services are introduced as an integrated manner and merged in the dependability software platform as mentioned above. The explanation about interaction and configuration of the dependability software services as a library concluded the software architecture framework.

Implication on the business model

The reaction of technologies and business process is always in mutual directions. The way how OEM and supplier should interact in the development of ISS must change in response to the E/E-architecture framework defined here. Again a benchmarking with the development of IT-industry shows a decentralized network in the development of ISS-application, with OEMs, suppliers and the third-party companies (e.g. software vendors and companies specialized for integration and test) can grant a more effective cooperation towards system with higher dependability. It brings also a few chances and challenges to the business model as well:

- Decentralized structure requires more effort to manage
- New market opportunities for suppliers esp. software supplier and system integrator
- Responsibility and warranty by the integration of safety components from different partners
- Systems integration and test can be offered as an independent service

11.2 Outlook for the future work

The concepts discussed in this dissertation provide potential for the further research, esp. taking the fact into the consideration that automotive standards and architecture usually takes a long time to establish, e.g. it took CAN 10 years to be introduced into the market and another 10 years to be widely used. The dependable E/E-architecture and engineering process will take a considerable time and adaptations to find their way in the real life.

The presented work in this thesis raises a few interesting topics for future research activities.

- Setting up a library of dependability services for the ISS-application from practice
- Further quantitative analysis of real-time performance of the dependability software services
- As mentioned in the Subsection 10.6, based on the experiences and finding during the validation, ISS Engineering Process and tool chain should be extended from the rapid prototyping to the later product phase of the serial development. One interesting topic is the safety analysis according to the special requirements of ISS.
- Further development of dynamic reconfiguration in the direction of service displacement
    - Concept of dynamic reconfiguration of communication paths for service reallocation
    - Resource reconfiguration, management and optimization for the service displacement

In all, the dependable E/E-architecture framework combined with the ISS Engineering Process is a promising approach for the development of future Integrated Safety Systems. Therefore, it has to find the way into real-world platform since an architecture framework provides only a reference model and guidelines for the practical solution. A step towards bringing the results closer to being deployed in products will be to advocate the use of the results as input and inspiration for on-going projects concerning automotive software platforms, such as the safety group in AUTOSAR [AUT07].

Appendix 1: Implementation details of the validator

In this chapter more details about the prototyping and validation of the safety concepts in this dissertation will be given. In order to keep an overview of the work in the scope the chapter, more details as to software implementation, simulation modes can be found the appropriate references and EASIS deliverables.

In Figure AI-1, the EASIS validator presented at the end of 2006 on the conference "Intelligent Transport System" in London with the most important nodes are shown.

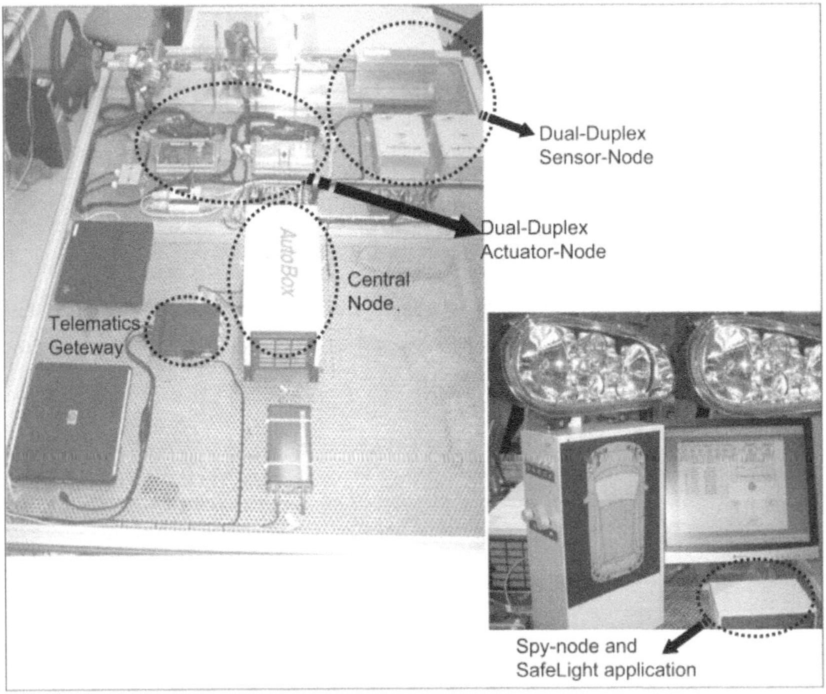

Figure AI-1: EASIS validator demonstrated on ITS 2006 in London

Appendix 1.1 Validation of hardware architectures for ISS

The dual-duplex 1oo2D ECU hardware architecture depicted in Subsection 8.3.1 Figure 8-8 is made up of 2 FSU. The hardware architecture of each FSU is demonstrated in Figure AI-2.

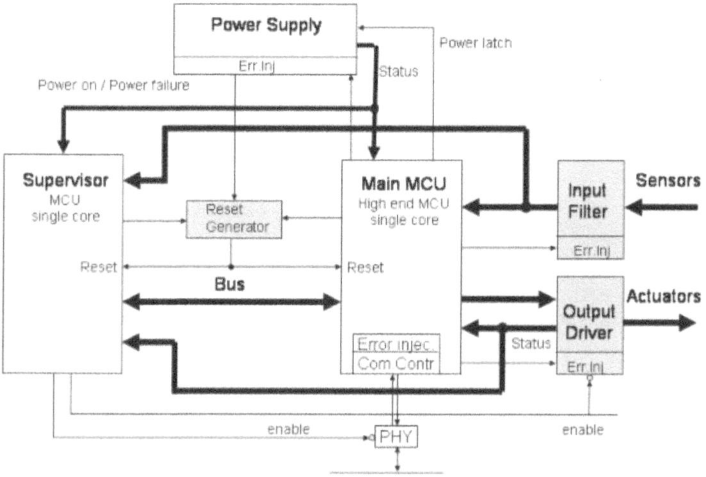

Figure AI-2: FSU processor block diagram [HWV06]

This power supply of FSU has two independent battery inputs (V_{b1} and V_{b2}) and generates the different supply voltages for the FSU internal circuitry. This block also generates the reset signals, which activates the logic devices once their supply voltages are stable, after the FSU has been switched on. It also generates the sensor power supply, both for the synchronous motor Hall Effect sensors and for the torque sensors. Feed-back on the status of these external supply lines is provided to the logic core of the FSU for fault handling. Finally, this block also provides power to the output stage, deriving it from one of the battery connections; should a failure occur on one of the FSU power supply lines, a switch over to the other power supply can takes place.

The FSU core is made up of two units, namely a microcontroller managing the actual application software, the interface to the rest of the system, whilst a second device, a digital signal processor, acts as a slave to the former device, managing the interface to the output stage. The crosscheck mechanism let each unit monitor the operation of the other and can make the FSU enter a fail-safe state in case of fault. In this way, the processing core architecture of the FSU reflects the architecture depicted in Figure 8-8 of Subsection 8.3.1.

The validation of dual-duplex hardware architecture was carried out also according to the principle of fault injection. In the following a list of fault injection is given:

Test number	Test case	Description	Expected results
1.	Fail-silent and fail-operational by internal FSU fault	An internal simulated fault is injected in one of the two FSUs that belong to the fail-operational unit. The fault can be injected through the activation of an internal switch.	Fail-silent characteristics: The faulty FSU detects the fault, sends a diagnostic message on the FlexRay network and switches itself off. Fail-operational characteristics: The second FSU receives the diagnostic message, detects the fault of the other ECU and keeps the system fully operative.
2.	Fail-silent and fail-operational by communication fault	Both FlexRay channels of one fail silent node are disconnected.	Fail-silent characteristics: The faulty FSU detects the loss of communication with all the validator nodes and switches itself off. Fail-operational characteristics: The second FSU does not receive any diagnostic message, but detects the communication loss with the other FSU and keeps the system fully operative.
3.	Redundant communication with FlexRay	One of the two FlexRay channels of both fail silent nodes is disconnected.	The two fail-silent nodes detect the line disconnection and send a diagnostic message. Both the FSUs are fully operative.
4.	Redundant power supply by automatic switch	The system is supplied by line A. A fault is injected by disconnecting line A on one of the two fail silent FSUs.	The fail-silent FSU detects the fault and sends a diagnostic message. The switch is triggered. The two fail-silent nodes are fully operative.

Table AI-1: Fault injection for the validation of the HW-Architecture

Appendix 1.2 Validation of dependability software platform and services

As discussed in Subsection 10.3, the validation of dependability software platform and services are carried out based on the fault/error injection in the ISS-Applications SafeLane, SafeSpeed, SafeLight and the prototyping environment with a mixture of three fault injection methods as introduced in Subsection 10.3 [TRV06].

Appendix 1.2.1 Prototyping and validation of Agreement Protocol

As discussed in Subsection 10.3.1, the validation of Agreement Protocol is based on the coordinated state reconfiguration on the Central Node and the two Steering Wheel Sensor Nodes. The validation is the following three scenarios:

- All nodes agrees on consistent state information
- One node doesn't send any information / sends inconsistent information. The Agreement Protocol has to guarantee the toleration of one fault.
- More than one node doesn't send any information / sends inconsistent information. The Agreement Protocol helps to switch to a safe state by delivering status feedback of the agreement process.

The list of mechanisms of the Agreement Protocol service in the validator is given in the following as sensor and actuator control.

Test number	Test case	Description	Expected results
1.	Single Node Fault	The Central Node fails to receive one sensor signal	System detects single faulty sensor, fully operative
2.	Double Node Fault	The Central Node fails to receive both sensor signals	System detects faulty sensor input, fail-safe mode
3.	Single Byzantine Fault	One Sensor Node sends inconsistent information	System detects faulty sensor, fully operative
4.	Double Byzantine Fault	Two Sensor Nodes send inconsistent information	System detects faulty sensor, fail-safe mode

Table AI-2: Test cases for the validation of Agreement Protocol

In the test cases 1 and 2, the communication between Central Node and Sensor Nodes are carried out with Agreement Protocol, but they mainly test the system wide coordinated reconfiguration. In the following, test cases 3 and 4, specially designed for Byzantine Fault are specified in details.

Test Case 3 – Single Byzantine Fault

Figure AI-3 shows the System State when a Byzantine fault is injected in one Steering Wheel Sensor node (the node sends different information to the other Steering Wheel Sensor node and to the Central Node). The node tries to reinitialize (five attempts) and then fixes its state on faulty. Central Node recognizes the situation, sets the flag of "Faulty Sensor Node" but since the other Steering Wheel Sensor node is still working, complete steering function and SafeLane functions are working.

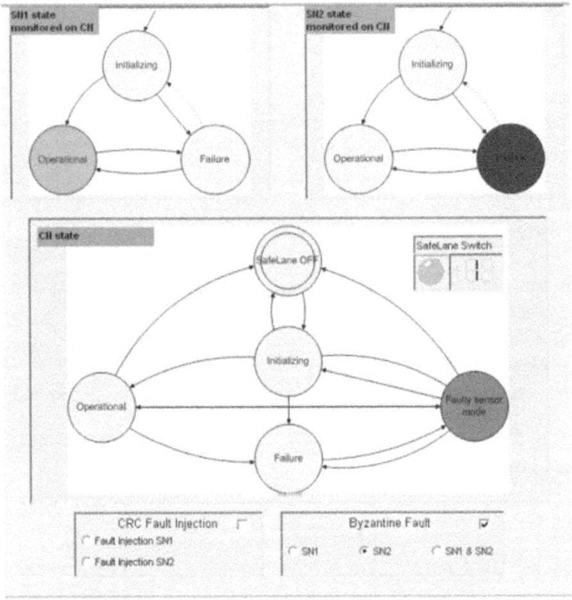

Figure AI-3: Single Byzantine Fault

Test Case 4 – Double Byzantine Fault

Figure AI-4 demonstrates the System State when Byzantine faults are injected in both Steering Wheel Sensor nodes (both nodes send different information to the other Steering Wheel Sensor node and to the Central Node). Both Steering Wheel Sensor nodes try to reinitialize (five attempts) and then fix their states on faulty. Central Node recognizes the situation, sets the flag of "Faulty". Therefore, SafeLane functions are disabled and steering function goes to the fail-safe mode.

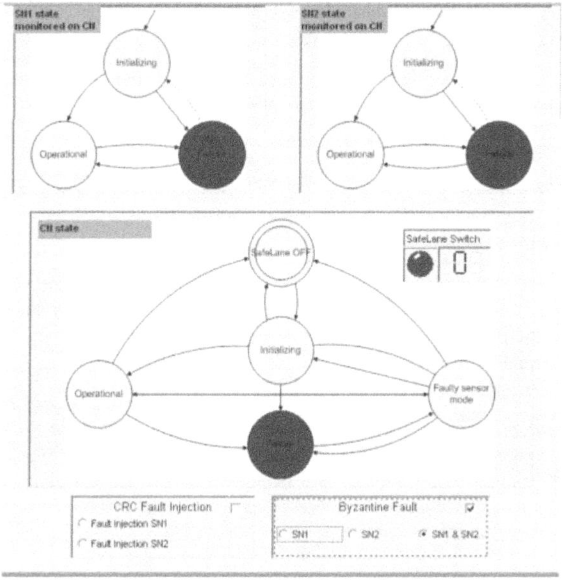

Figure AI-4: Double Byzantine Fault

Appendix 1.2.2 Prototyping and validation of Software Watchdog

As specified in Subsection 10.3.2, the SW-Watchdog has been tested on the Spy-Node in the context of a simplified SafeLight application. The SW-Watchdog is integrated with the Fault Management Framework. Complete test results on FMF validation are detailed in Appendix 1.2.3. The SW-Watchdog is configured to monitor all the Runnables: (Aliveness Monitoring, Arrival Rate Monitoring and Control Flow Checking).

The demonstrative application is composed of three tasks that manage nine runnables:

- Task T1 (execution period = 0200 ms) for Runnables R1, R2, R3, R4, R5, R6, R7
- Task T2 (execution period = 1000 ms) for Runnable R8
- Task T3 (execution period = 2000 ms) for Runnable R9

For each runnable of the demonstration application, three errors are defined: Aliveness Error, Arrival Rate Error and Control Flow Error (e.g. Runnable R1 is associated with R1_Aliveness, R1_Arrival_Rate and R1_Control_Flow errors). Using the HMI of the Control PC we can:

- Display in real time the current internal values of the heartbeat monitoring counters (AC, CCA, ARC, CCAR) and the configured parameters (m, n, p, q)
- Display the last executed runnable of each task

- Display in real time the current internal state of each error handled by the Fault State Manager associated with the SW Watchdog
- Inject a set a predefined faults : deactivation of runnables, inversion of order of execution of runnables, modification of the execution period of runnables
- Execute the scheduler cycle by cycle in order to see the evolution of the counters step by step

Test Case 1 – Aliveness Monitoring

The SW-Watchdog monitors the aliveness of runnables. The execution period of the SW-Watchdog defines the monitoring cycle. If the number of executions of a runnable (counted by the Aliveness Counter AC) is lower than the configured limit (m), which is called by this runnable within a certain monitoring period, an error will be detected. The monitoring period is defined in number of monitoring cycles (counted with the Cycle Counter for Aliveness CCA).

In the configuration of test environment, the execution period of Runnable R8, which belongs to Task T2, was set to1000ms. The monitoring cycle is 100ms. The fault assumption is "less than 10 executions of R8 during 10,000 ms". So the minimum number of executions m is 10 and the monitoring period for R8 is 10,000 ms = n*monitoring cycle = 100 * 100 ms. As depicted in Figure AI-5, for the Aliveness checking of R8, the number of monitoring cycles n is chosen as 100.

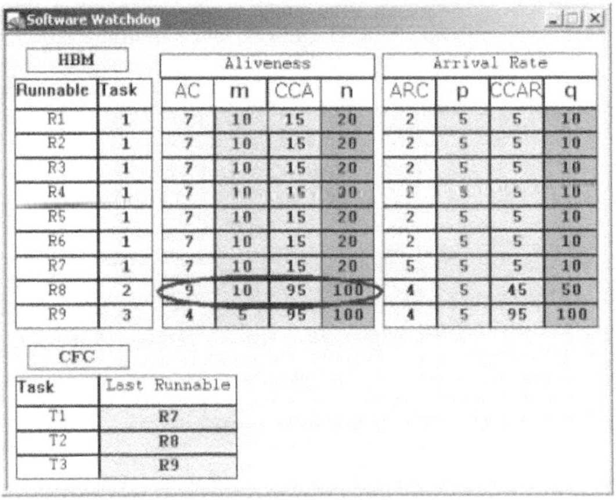

Figure AI-5: Aliveness Monitoring without error

The fault Period T2 * 2 is injected. Now the execution period of R8 is 2000 ms instead of 1000 ms, so R8 will be executed two times less than expected. After the monitoring period of 10,000 ms, R8 has been executed only 5 times (less than the low limit m = 10) and the error R8_Aliveness is detected.

Test Case 2 – Arrival Rate Monitoring

The SW-Watchdog monitors the arrival rate of runnables. If the number of executions of a runnable (counted by the Arrival Rate Counter ARC) is higher than the configured limit (p), an error is detected. The number of monitoring cycles is registered with the Cycle Counter for Arrival Rate CCAR. Here, the monitoring period equals the monitoring cycle, which means that ARC is checked at each monitoring cycle.

In the configuration of test environment, the execution period of runnable R8 in task T2 is 1000 ms, while the monitoring cycle is 100 ms. The fault assumption is "more than 5 executions of R8 during 5000 ms". So the maximum number of executions p is 5 during 5000 ms = q*monitoring cycle = 50 * 100 ms, so q is 50.

By the fault injection period T2 / 2 is set. Now the execution period of R8 is 500 ms instead of 1000 ms, thus R8 will be executed twice as often as expected. As the arrival rate checking is performed at each monitoring cycle, after 2600 ms R8 has been executed 5 times and the error R8_Arrival_Rate is detected.

Test Case 3 – Control Flow Checking

The SW-Watchdog with the function Control Flow Checking monitors the execution sequence of the runnables inside a task. For task T1 the expected execution order is R1 R2 R3 R4 R5 R6 R7 R1 R2 R3 etc.

- Fault injection test case 1

The fault *R3 disabled* is injected; the following errors are detected by the SW-Watchdog:

  - R4_Control_Flow (because R4 is not the successor of R2): after the monitoring period for R3 (where the Aliveness checking is performed for R3), another error is detected
  - R3_Aliveness (because R3 has not been executed)

- Fault injection test case 2

The fault "inverse R1 / R2" is injected; the following errors are detected by the SW-Watchdog:

  - R1_Control_Flow (because R1 is not the successor of R2)
  - R2_Control_Flow (because R2 is not the successor of R7)
  - R3_Control_Flow (because R3 is not the successor of R1)

Appendix 1.2.3 Prototyping and validation of Fault Management Framework

The Fault Management Framework is tested on the Spy-Node in the prototyping environment as for the SW-Watchdog. The test cases are organized in three scenarios.

1. Test of the Fault State Manager part
2. Test of the Supervision – Reconfiguration part

A synthesis test demonstrates an integration example of the Fault Management Framework with the demonstration application:

3. Test of the Fault Management Framework

The list of characteristics of the Fault Management Framework tested in the validator is given in the following tables.

Test number	Test Case	Description	Expected results
1	Off line configuration	Demonstration of the Python tool to generate configuration source files for Fault State Manager software component	Functional code generated
2	Filtering mechanism and error state management	Demonstration of the filter counter, demonstration of "No Error, Present, Present & Confirmed, and Confirmed states management"	Functional filter counter
3	Combined errors	Demonstration of resilience to logical errors combination	Functional code generated

Table AI-3: Test cases for the Fault State Manager in FMF

Test number	Test Case	Description	Expected results
4	Mode management by dynamic reconfiguration of fail-degradation	Demonstration of call-back function mechanism for supervision integration, demonstration of degraded mode	Functional call back mechanism working
5	Dynamic reconfiguration on the task layer	Demonstration of disabling tasks and scheduler reconfiguration, demonstration of SW-Watchdog reconfiguration	Functional reconfigurable scheduler working
6	Dynamic reconfiguration with ECU reset	Demonstration of reset capability	Reset capability with context-awareness

Table AI-4: Test cases Supervision – Reconfiguration in FMF

In the following the test cases 4, 5 and 6 will be specified in details:

Test Case 4 – Mode management by dynamic reconfiguration of fail-degradation

By association an error with a call back function, the Fault State Manager calls this function when the error passes to the state "Confirmed".

```
conf.SetEventCallBack( 'SWCIOIN_ERR_STS_LAMP_INDIC_FL', ('SUP_vidErrorStsLampIndicFL',) )
conf.SetEventCallBack( 'SWCIOIN_ERR_STS_LAMP_SIDE_FL',  ('SUP_vidErrorStsLampSideFL',) )
conf.SetEventCallBack( 'SWCIOIN_ERR_STS_LAMP_LBEAM_FL', ('SUP_vidErrorStsLampLBeamFL',) )
conf.SetEventCallBack( 'SWCIOIN_ERR_STS_LAMP_HBEAM_FL', ('SUP_vidErrorStsLampHBeamFL',) )
conf.SetEventCallBack( 'SWCIOIN_ERR_STS_LAMP_FOG_FL',   ('SUP_vidErrorStsLampFogFL',) )
```

This design mechanism is used to integrate the Fault State Manager and the Supervision and Reconfiguration sub-components. The management of the demonstration application modes (Normal mode, Degraded mode, Restarted mode, Fail safe mode) is realized in the Supervision sub-component. For example as depicted in Figure AI-6, when the error STS_LAM_INDIC_FL (open load detection on front left indicator lamp) is detected, the application passes in a 'Degraded mode' (front left position lamp replaces front left indicator lamp for left turning indication).

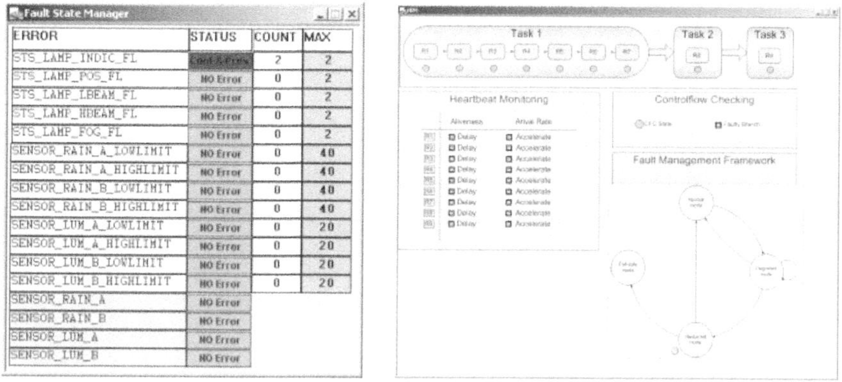

Figure AI-6: Supervision Reconfiguration – Mode management

Test Case 5 – Dynamic reconfiguration on the task layer

In this SafeLight application, runnables correspond to elementary atomic treatments that are executed periodically (e.g. Runnable R1 reads inputs, Runnable R2 manages blinking, …). Runnables of the same period of the same safety integrity are grouped in a task (e.g. Task T1 200ms, Task T2 1000 ms, etc.) The tasks are organized functionally in configurations that correspond to a set of tasks executed in a major stable state (e.g. Normal Configuration, Fail Safe Configuration, etc.). A synchronous real-time scheduler manages the execution of all tasks of a configuration with an elementary cycle of 100 ms. The current configuration can be switched. A task of a configuration can be activated / deactivated.

Two types of reconfiguration of the execution flow have been tested:

- Deactivation of tasks. The tasks associated with the automatic lighting functions are deactivated in case of sensor fault detection. As depicted in Figure AI-7, the deactivation of task1 leads to the state transition of SafeLight to the fail-safe mode.
- Switch of configuration. The application switches in a "Fail Safe Configuration" in case of non recoverable software fault.

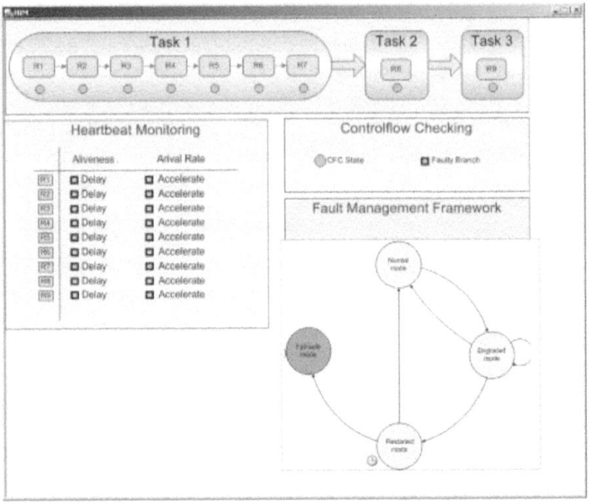

Figure AI-7: Supervision Reconfiguration – Scheduler

Test Case 5 – Dynamic reconfiguration with ECU reset

The supervision and reconfiguration unit in FMF can trigger an ECU reset. In this implementation, the design solution was to use the internal hardware watchdog of the micro-controller of the ECU. In the demonstration application, the ECU reset is used as a recovering strategy for some critical software failures, e.g. a critical control flow error as depicted in Figure AI-8.

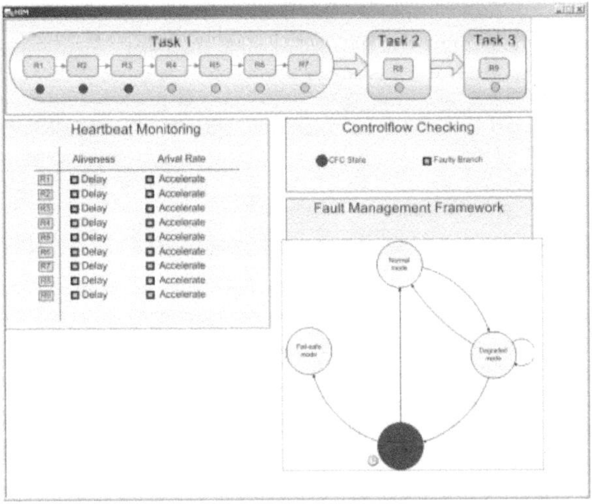

Figure AI-8: Supervision Reconfiguration – ECU Reset

Appendix 2: Mathematic derivation of the communication overhead of agreement protocol

$$\underbrace{n\cdot(n-1)}_{\text{1st Round}}+\underbrace{n\cdot(n-1)^2}_{\text{2nd Round}}+\ldots+\underbrace{n.(n-1)^{m+1}}_{\text{(m+1)-th Round}}=$$

$$n\cdot((n-1)+(n-1)^2+\ldots+(n-1)^{m+1})=$$

$$\frac{n\cdot(n-1)((n-1)^{m+1}-1)}{n-2}$$

Figure AII-1: Derivation of the communication overhead for the OM

$$\underbrace{n\cdot(n-1)}_{\text{1st Round}}+\underbrace{n\cdot(n-1)\cdot(n-2)}_{\text{2nd Round}}+\ldots+\underbrace{n.(n-1)\cdot\ldots\cdot(n-m-1)}_{\text{(m+1)-th Round}}=$$

$$\frac{n!}{(n-2)!}+\frac{n!}{(n-3)!}+\ldots+\frac{n!}{(n-m-2)!}$$

Figure AII-2: Derivation of the communication overhead for the SM

Appendix 3: Glossary

A few terms used in this dissertation are explained here as a short summary. For more complete and detailed glossary please refer to EASIS glossary [Glo07] and AUTOSAR glossary [Glo03].

Application Software Component

An Application Software Component is a specific Software Component realizing a defined functionality of a set (one or more) of features, which consist out of 1..n (n>=1) functions (in terms of C-Code functions) and 1..m (m>=1) executable runnables. In EASIS, Application Software Components are resident in the most upper layer of EASIS software topology.

Basic Software Module

Basic Software Module provides the infrastructural (schematic dependent and schematic independent) functionalities of an ECU. It consists of ECU Firmware and Standard Software.

Compositionality

Compositionality is given when the behavior of a software component or subsystem of a system has to be independent of the overall system load and configuration. Notes: Compositionality is an important property of deterministic systems. This property leads to a complete decoupling of systems. Smooth subsystem integration without backlashes is then easily achievable.

Software Configuration

The configuration of software elements in a SW system. Notes: A software element is a clearly definable software part. A software configuration is a selected version of software modules, software components, parameters and generator configurations. Calibration and variant coding can be regarded as subset of software configuration.

Control Flow

The directed transmission of information between multiple entities, directly resulting in a state change of the receiving entity. Notes: A state change could result in an activation of a schedulable entity.

Event

The occurrence of a state change of a hardware or software entity or system or its environment. Notes: (1) Examples of different types of events: Failure, exception, interrupt, hazardous event mishap, loss event (2) Event is a specialization of Signal.

Fail-Degraded

Property of a system or functional unit, describes the ability of a system to continue with intended degraded operation at its output interfaces despite the presence of hardware or software faults. Notes: e.g. "Limp home" functionality for ECU (reduce torque to assure an arrival at home or service station).

Fail-Operational

Property of a system or functional unit, describes the ability of a system or functional unit to continue normal operation at its output interfaces despite the presence of hardware or software faults. This is a characteristic of a unit that means that the system is designed in such a way, that it can (without repair) fail once and remain fully operational without any limit of time, e.g. braking system

Fail-Safe

Property of a system or functional unit. In case of a fault the system or functional unit transits to a safe state. This is the behavior of a unit such that the outputs of the unit must be correct or otherwise must be in any electrical or logical state that is supposed to lead the vehicle in an intrinsic safe state.

Fail-Silent

Property of a system or functional unit. In case of a fault the output interfaces are disabled in a way that no further outputs are made. This is the behavior of a unit such that the outputs of the unit must be correct or otherwise must be "silent" that is null or in any other electrical or logical state that does not lead to any kind of actuation or interference with other parts of the system. This applies either to the communication bus or to sensors and actuators directly connected to the unit. This is a special case of fail safe. Fail-silent is a special case of the fail-safe property.

Fault Containment

As an arbitrary propagation of errors cannot be tolerated, fault containment regions, which restrict errors to certain parts of the system, are needed. Methods for error containment are access-restriction, consistency checking, physical isolation, and others.

Front Loading

During a standard product development process development and validation/verification workload is steadily increasing over time until it reaches its maximum shortly before start-of-production. This may lead to "fire-fighting" in some projects, where unexpected trouble occurs during validation. To solve this issue, a maximum of validation and verification should be done in early phases of the development process as early as possible

Redundancy

Existence of means, in addition to the means which would be sufficient, for a system or functional unit to perform a required function.

Appendix 4: List of figures

Appendix 5: Literature index

[EUC01] EU, "White Paper - European transport policy for 2010, time to decide", European Commission, Office for Official Publications of the European Communities, Luxemburg, 2001, http://europa.eu.int

[ADS07] M. Lannoije, J. Dr. Schuller, et. al., "Entwurf und Realisierung des Funktions- und Sicherheitskonzepts der Audi Dynamiklenkung", presented at Haus der Technik, 27. Conference "Elektronik im Kraftfahrzeug" (In-vehicle Electronics), Dresden, Germany, 2007

[ADL05] W. Bernhardt, H. Erl, "Markt- und Technologiestudie Leistungselektronik Automotive 2015", Arthur D Little GmbH - Consulting, 26. Sep. 2005

[AfS07] J. Stroop, X. Chen, et. al., "Architectures for Safety", January, 2007, http://www.easis-online.org/wEnglish/img/pdf-files/dSPACENEWS2007-1_Architectures_for_Safety_en_504

[AOS06] AUTOSAR, "AUTOSAR Specification of Operating System", 2006 http://www.autosar.org/download/AUTOSAR_SWS_OS.pdf

[ASM06] ASAM, "Association for Standardization of Automation and Measuring Systems", 2006, http://www.asam.net

[ASR04] ASRB, Automotive Safety Restraints Bus, 2004, http://www.electronic-data.com/unternehmen/4257.asp

[ATZ05] ATZ/MTZ, "Aktive und passive Sicherheit," ATZ/MTZ extra S-Klasse BR221, pp. 118-125, 2005

[AUT07] AUTOSAR, "AUTOSAR - Automotive Open System Architecture", 2007, www.autosar.org

[ATO07] AUTOSAR, "AUTOSAR – Technical Overview", 2007, http://www.autosar.org/download/AUTOSAR_TechnicalOverview.pdf

[Bel01] R. Belschner, et al., "FlexRay - The Communication System for Advanced Automotive Control Systems", presented at SAE Technical Paper Series Nr. 2001-01-0679, Detroit, 2001

[Ben04] S. Benz, "Eine Entwicklungsmethodik für sicherheitsrelevante Elektroniksysteme im Automobil", Dissertation in Institut für Technik der Informationsverarbeitung: Universität Karlsruhe, 2004

[BER04] J. Ferreira, "An Experiment to Assess Bit Error Rate in CAN", presented at RTN 2004 - 3rd Int. Workshop on Real-Time Networks, 2004

[Bes02] A. Best, Echtle, K., "Standardverfahren und spezielle Ansätze zur Fehlertoleranz von Steuergeräten in Automobilen", Projekt FeSTA, Daimler AG, Uni Essen, 2002

[Bro03] M. Broy, "Automotive Software and Systems Engineering", presented at 25th International Conference on Software Engineering, 2003, pp. 719 - 720

[BSW06] EASIS, "Deliverable D1.2-2 - Basic services", EASIS Partnership, 2006 www.easis.org

[Gri00] R. Grießbacher, "Byteflight - neues Datenbussystem für sicherheitsrelevante Anwendungen", Automotive Electronics of ATZ/MTZ, 01/2000, http://www.byteflight.com/presentations/atz_sonderausgabe.pdf

[CHS06] X. Chen, M. Limam, M. Wedel, et. al., "Concept and Integration of an Agreement Protocol in a Dependable Software Platform for Automotive Integrated Safety Systems", presented at Automotive - Safety & Security 2006, Stuttgart, Germany, 2006, http://www.easis-online.org/wEnglish/img/pdf-files/Paper_Automotive-Safety&Security2006.pdf

[CoW05] Freescale, "CodeWarrior™ Development Studio for Freescale™ HCS12(X) Microcontrollers, product information", 2005

[Con06] Conti Temic microelectronic GmbH 2006, Electronic Brake Systems MK25, http://www.conti-online.com/generator/www/de/en/cas/cas/themes/products/electronic_brake_and_safety_systems/electronic_brake_systems/abs_tcs_esc/ebs_1003_en.html

[Cri91] Cristian, F. „Understanding Fault Tolerant Distributed Systems", Communications of the ACM, Ausgabe 34, Heft 2, 1991, pp. 56-78

[DFP06] EASIS, "Deliverable D1.2-13 - Description of fault types", EASIS Partnership, 2006 http://www.easis.org

[DIN93] DIN60870-5-1, "Norm Fernwirkeinrichtungen und -Systeme Teil 5: Übertragungsprotokoll", 1993

[DSC05] Decomsys, "Simcom User Manual Version 2.2, „DECOMSYS::SIMCOM<FlexRay> Matlab/Simulink blockset for FlexRay, Generator User Manual Version 2.2.8, Designer User Manual Version 2.0.6, Simsytem User Manual Version 2.2", Dependable Computer Systems GmbH, Vienna 2005

[DyR96] D. B. Stewart, G. Arora, "Dynamically Reconfigurable Embedded Software - Does It Make Sense?" presented at IEEE Intl. Conf. on Engineering of Complex Computer Systems and Real Time Application Workshop, Montreal, Canada, 1996, pp. 217-220,

[EAD42] EASIS Deliverable D4.2 A Prototype Tool Interaction Software Layer

[EAD41] EASIS Deliverable D4.1 Appendix 1: EASIS Engineering Process

[EAST03] EAST-EEA Embedded Electronic Architecture "Glossary", Version 6.1, ITEA EAST-EEA Project, www.east-eea.net, Version 7.2, 2003

[Ech90] Echtle, K. „Fehlertoleranzverfahren", Springer Verlag, Berlin [u.a.], 1990, ISBN 3-540-52680-3

[Eck06] M. Eckmann, F. Mertens, "Close-to-Production Prototyping, Flexible and Cost-efficient", in ATZ Elektronik, vol. 01/2006, 2006, pp. 22-27

[Ehr03] J. Ehret, "Validation of Safety-Critical Distributed Real-Time Systems", Dissertation in Electrical and Computer Engineering. Munich: Technische Universität München, 2003

[EiK06] X. Chen, et. al., "A Dependable Software Platform for Automotive Integrated Safety Systems", presented at Haus der Technik, 26. Conference "Elektronik im Kraftfahrzeug" (In-vehicle Electronics), Dresden, Germany, 2006, http://www.easis-online.org/wEnglish/img/pdf-files/Paper_Elektronik_im_Fahrzeug.pdf

[EiK07] X. Chen, F. Salewski, et. al., "Concept and Prototyping of a Fault Management Framework for Automotive Safety Relevant Systems", presented at Haus der Technik, 27. Conference "Elektronik im Kraftfahrzeug" (In-vehicle Electronics), Dresden, Germany, 2007, http://www.easis-online.org/wEnglish/img/pdf-files/Paper_Elektronik_im_Fahrzeug_2007.pdf

[ElA04] Elektronik automotive, "Systemvernetzung: Künftig nur noch 20 Steuergeräte", 3/2004, vol. Heft 3, pp. 21, 2004

[Ele04] B. Elend, "FlexRay Netzwerk Topologie für sicherheitsrelevante Applikationen", auto&elektronik, vol. 2/2004, pp. 44-46, 2004, http://www.flexray.com/publications/Philips_FR_auto_elektronik02_2004.pdf

[FMF06] EASIS, "Deliverable D1.2-8 - Fault management framework", EASIS Partnership, 2006, http://www.easis-online.org/wEnglish/download/Deliverables/EASIS_Deliverable_D1.2-8_V1.0.pdf

[FTD06] EASIS, "Deliverable D1.2-5 - Discussions and findings on Fault Tolerance", EASIS Partnership, 2006, www.easis.org

[FTS06] EASIS, "Deliverable D1.2-3 - Fault Tolerance Services", EASIS Partnership, 2006, www.easis.org

[FxR02] H. Heinecke, A. Schedl, B. Hedenetz, "FlexRay - ein Kommunikationssystem für das Automobil der Zukunft", Elektronik Automotive, pp. 36-45, Sep-2002

[GAN03] J. Ganssle, "Watching the Watchdog," embedded world, 2003

[GDi02] G. Leen, D. Heffernan: "Expanding Automotive Electronic Systems", IEEE Computer, pp. 5-6, January 2002

[Glo03] AUTOSAR, "AUTOSAR Glossary", 2003, www.autosar.org

[Glo07] EASIS, "EASIS Glossary", 2007 http://www.easis-online.org/wEnglish/download/deliverables_WP0.shtml?navid=15

[GTD06] EASIS, "Deliverable D1.2-10 - Data exchange concepts for gateways", EASIS Partnership, 2006, www.easis.org

[GTS06] EASIS, "Deliverable D1.2-12 - Security and firewall concepts for gateways", EASIS Partnership, 2006, www.easis.org

[Har02] F. Hartwich, T. Führer, B. Müller and R. Hugel, Robert Bosch GmbH, "Integration of Time Triggered CAN (TTCAN_TC)," SAE, 2002

[Hed01] Hedenetz, B. "Entwurf von verteilten fehlertoleranten Elektronikarchitekturen in Kraftfahrzeugen", Dissertation, Eberhard-Karls-Universität Tübingen, 2001

[HIL06] M. Hiller, X. Chen, et. al., "Dependability Services in the EASIS Software Platform", presented at DSN 2006 Workshop on Architecting Dependable Systems, 2006, http://www.easis-online.org/wEnglish/img/pdf-files/wads_2006_easis.pdf

[HOS07] Wikipedia, "History of Operating System", 2007, http://en.wikipedia.org/wiki/History_of_operating_systems

[HRK00] B. Hedenetz, J. Ruh, M. Kühlerwein, Th. Ringler, et. al., "OSEKtime - the new fifth OSEK/VDX group: A Dependable Fault-Tolerant Real-Time Operating System and Communication Layer for By-Wire Applications", presented at VDI Baden-Baden, 2000, pp. 59

[HWA06] EASIS, "Deliverable D2.2 - Conceptual Hardware Architecture Specification", EASIS Partnership, 2006, www.easis.org

[HWAA06] EASIS, "Appendix to Deliverable D2.2 - Conceptual Hardware Architecture Specification", EASIS Partnership, 2006, www.easis.org

[HWV06] EASIS, "Deliverable D2.4 An ECU prototype with fault tolerance characteristics", EASIS Partnership, 2006, www.easis.org

[IEC01] IEC61508, "Functional safety of electronical/electronic/programmable electronic safety-related systems - Part 1: General requirements", 1998

[IEC02] IEC61508, "Functional safety of electronical/electronic/programmable electronic safety-related systems - Part 2: Requirements for electrical/electronic/programmable electronic safety-related systems", 1998

[IEC03] IEC61508, "Functional safety of electronical/electronic/programmable electronic safety-related systems - Part 3: Software requirements", 1998

[IEC04] IEC61508, "Functional safety of electronical/electronic/programmable electronic safety-related systems - Part 4: Definitions and abbreviations", 1998

[IEC05] IEC61508, "Functional safety of electronical/electronic/programmable electronic safety-related systems - Part 5: Examples of methods for the determination of safety integrity levels", 1998

[IEC06] IEC61508, "Functional safety of electronical/electronic/programmable electronic safety-related systems - Part 6: Guidelines on the application of IEC 61508-2 and IEC 61508-3", 1998

[IEC07] IEC61508, "Functional safety of electronical/electronic/programmable electronic safety-related systems - Part 7: Overview of techniques and measures", 1998

[IECFAQ] IEC61508, "IEC 61508 FAQs", http://www.iec.ch/zone/fsafety/questions.htm

[ISO11898] International Organization for Standardization "ISO 11898: Road vehicles — Controller area network (CAN)", Geneva, 2003

[ISO15] CMMI/ISO 15504: Information technology - (Software) Process assessment - Parts 1 to 5

[ISO21] ISO/WD 26262-1, TC 22/SC3 N 035, Road vehicles – functional safety – part 1: Glossary, 2005-12-22.

[ISO22] ISO/WD 26262-2, TC 22/SC3 N 007, Road vehicles – functional safety – part 2: Management of functional safety, 2005-09-26

[ISO23] ISO/WD 26262-3, TC 22/SC3 N 008, Road vehicles – functional safety – part 3: Concept phase, 2005-09-26

[ISO24] ISO/WD 26262-4, TC 22/SC3 N 009, Road vehicles – functional safety – part-4: Product development system, 2005-09-25

[ISO25] ISO/WD 26262-5, TC 22/SC3 N 010, Road vehicles – functional safety – part 5: Product development hardware, 2005-10-7

[ISO26] ISO/WD 26262-6, TC 22/SC3 N 011, Road vehicles – functional safety – part 6: Product development software, 2005-09-19

[ISO27] ISO/WD 26262-7, TC 22/SC3 N 012, Road vehicles – functional safety – part 7: Production and operation, 2005-09-16

[ISO28] ISO/WD 26262-8, TC 22/SC3 N 013, Road vehicles – functional safety – part 8: Supporting processes, 2005-09-26

[ISO7498-1] International Organization for Standardization „ISO 7498-1: Information Technology – Open System Interconnection – Basic Reference Model: The Basic Model“, Standard, Geneva, 1994

[Kop02] P. Koopman, "Dependability for Embedded Systems", Lecture, Carnegie Mellon, 2002

[Kop04] P. Koopman, T. Chakravarty, "Cyclic Redundancy Code (CRC) Polynomial Selection For Embedded Networks", presented at The International Conference on Dependable Systems and Networks, DSN-2004, 2004

[Kru98] M. Krug, "Concept and Implementation of a Dependable Automotive Operating System", Dissertation in Computer of Science: University of Tübingen, 1998, pp. 57, Chapter 5: Defining a Fault Hypothesis

[Lam80] L. Lamport; R.Shostak; M.Pease, "Reaching Agreement in the Presence of Faults," Journal of the ACM, vol. 27, no. 2, pp. 228..234, Apr-1980.

[Lam82] L. Lamport, M. Pease, R. Shostak, "The Byzantine Generals Problem," ACM Transactions on Programming Languages and Systems, vol. 4, pp. 382-401, 1982

[LAN97] D. Lantrip, "General Purpose Watchdog Timer Component for a Multitasking System": embedded world, 1997, http://www.embedded.com/97/feat29704.htm

[Lap95] J.-C. Laprie, Dependability - its attributes, impairments and means, Predictably Dependable Computing Systems, pp. 3-24, Springer-Verlag, 1995

[Lim05] M. Limam, " Conception and Implementation of an Agreement Protocol for Fault-Tolerant Automotive Embedded Systems", Diploma thesis in Institut für Automatisierungs- und Softwaretechnik: University of Stuttgart, 2005

[LIN2.0] LIN Consortium „LIN Specification Package", Revision 2.0, Motorola, Munich, 2003

[LSP80] Lamport L.; Shostak R.; Pease M.: Reaching Agreement in the Presence of Faults. In: Journal of the ACM, vol. 27, no. 2, pp. 228..234, April 1980

[LSP82] Lamport L.; Shostak R.; Pease M.: The Byzantine Generals Problem, In: ACM Transactions on Programming Languages and Systems, vol. 4, no. 3, pp. 382..401, July 1982

[McK05] McKinsey&Company, "Automotive Electronics", McKinsey Automotive Electronics Initiative, 2005

[McK06] Hoch, et. al., "The Race to Master Automotive Embedded Systems Development", McKinsey&Company, 2006

[Nah02] O. Nahmsuk, P. Shirvani, E. McCluskey, "Control-Flow Checking by Software Signatures," IEEE Transaction on Reliability, vol. 51, pp. 111-121, Mar-2002

[OSK01] OSEK, "OSEK/VDX Time-Triggered Operating System Specification 1.0", 2001 www.osek-vdx.org

[OSK05] OSEK, "OSEK/VDX Operating System Specification 2.2.3", ISO 17356-3:2005, Road vehicles -- Open interface for embedded automotive applications -- Part 3: OSEK/VDX Operating System (OS), 2005, www.osek-vdx.org

[Rei05] W. Reinelt, W. Klier, G. Reimann., "Systemsicherheit des Active Front Steering", in Automatisierungstechnik, 53(1):36-43, 2005

[Ruh04] J. Ruh, "Entwurf von fehlertoleranten Steuergeräteapplikationen in Kraftfahrzeugen unter Berücksichtigung moderner Entwicklungsmethodiken", in Fakultät für Informatik der Eberhard-Karls-Universität Tübingen, 2004

[RWT02] RWTH Aachen, Lecture Notes "Fault Tolerant Computer Systems" Summer semester 02, Lectures 9: Byzantine agreement, http://www-i4.informatik.rwth-aachen.de/lufg/lvs/teaching/ss02/ftcs/index.html

[SCK04] C. Shelton, P. Charles, P. Koopman, "Improving System Dependability with Functional Alternatives", presented at The International Conference on Dependable Systems and Networks, DSN, 2004

[SIV04] H. Sivencrona, "On the Design and Validation of Fault Containment Regions in Distributed Communication Systems", Dissertation in Department of Computer Engineering. Göteborg, Sweden: Charmers University of Technology, 2004, pp. 165-185

[SMT06] M. Jörg, "Dynamic Software Module Tests - A Process Component for Better Software Quality", in ATZ Elektronik, 2006, pp. 36

[Spi07] K. Spisic, "Concept and Implementation of a Fault Management Framework for a dependable in-vehicle E/E-architecture", Diploma thesis in Lehrstuhl Informatik XI Software für eingebettete Systeme: RWTH Aachen, 2007

[SWA06] AUTOSAR, "AUTOSAR Layered Software Architecture, v2.0", 2006 http://www.autosar.org/download/AUTOSAR_LayeredSoftwareArchitecture.pdf

[SWI06] EASIS, "Deliverable D1.3 A prototypical implementation of the common software platform", EASIS Partnership, 2006 www.easis.org

[SWP06] EASIS, "Deliverable D1.2-0 - Overall Description of the EASIS Software Platform", EASIS Partnership, 2006, www.easis.org

[SWR07] EASIS, "Requirements Specification WP1 Software Architecture", EASIS Consortium, 2007, www.easis.org

[SWS03] Bosch, "Steering Wheel Angle Sensor LWS3.6 Technical Customer Documentations", 2003

[SWT06] EASIS, "Deliverable D1.2-1 - Software topology", EASIS Partnership, 2006, www.easis.org

[SML06] M. Hause, A. Korff, "An Overview of SysML for Automotive Systems Engineers", ATZ-Elektronik, Mar-2007, pp. 22-29, 2007

[Tin03] K. Tindell, F. Wolf, R. Ernst, "Safe Automotive Software Development, Chapter 2: The need for a protected OS in high integrity automotive systems", presented at Proceedings of the Design, Automation and Test in Europe Conference and Exhibition (DATE'03), IEEE Computer Society, Munich, Germany, Mar-2003

[TMg02] T. Bertram, M. Torlo: Globale dynamische Fehlertoleranz – Systemverband als Basis für sichere X-by-Wire-Systeme, ATZ Automobiltechnische Zeitschrift 104, Nr. 11, 2002, 1008-1014

[TRA06] V. Lauer, et. al.,, "EASIS – Electronic Architecture and System Engineering for Integrated Safety Systems", presented at Transport Research Arena Europe, 2006, http://www.easis-online.org/wEnglish/img/pdf-files/tra_2006_easis.pdf

[TRV06] EASIS, "Deliverable D5.3 - Functional Test Report of EASIS Architecture Validator", EASIS Partnership, 2006, www.easis.org

[µCI06] P. Sundaram, J. G. D'Ambrosio, "Controller Integrity in Automotive Failsafe System Architectures", presented at 2006 SAE World Congress, Detroit, Michigan, Apr-2006

[VAL06] EASIS, "Specification of EASIS Validator with Telematics Gateway, WT5.1 Deliverable", EASIS Consortium, 2006, www.easis.org

[Voa98] J. M. Voas, G. McGraw, Software Fault Injection: Inoculating Programs Against Errors: John Wiley & Sons, 1998

Appendix 6: Author's biography

Xi Chen was born in Wuhan, Hubei Province, China, on March 11, 1978. After two years of study in Applied Physics he came to Germany for his graduate study in 1998. He received his Diploma of Electrical and Computer Engineering and Master of Science degree in Information Technology at the University of Stuttgart in December 2003. In February 2004 he joined Daimler AG Group Research and Advanced Engineering as a research associate in the department of Electric/Electronic Architecture. In Spring 2004, he started pursuing his Ph.D. in the Faculty of Electrical and Computer Engineering in the Institute für Technik der Informationsverarbeitung at the University of Karlsruhe (TH). In February 2007 he joined ZF-Lenksysteme in Schwäbisch Gmünd as a Safety Manager in the Central Development of Functional Safety.

Appendix 7: Lebenslauf

Persönliche Daten

Familienname:	Chen
Vorname:	Xi
Geburtsdatum:	11. März 1978
Geburtsort:	Wuhan, China
Familienstand:	ledig
E-Mail:	xi.chen@gmx.de

Akademische Bildung

02/2004– 02/2008	Promotion am Institut für Technik der Informationsverarbeitung der Universität Karlsruhe in der Zusammenarbeit mit der Zentralen Forschung der Daimler AG
10/1999 – 12/2003	Doppelstudium in Elektrotechnik und Informationstechnologie an der Universität Stuttgart mit den Abschlüssen Dipl.-Ing. und Master of Science
10/1998 – 10/1999	Intensiver Deutsch-Sprachkurs an der Universität Stuttgart
10/1996 – 10/1998	Vier Semester Physikstudium an der Central China Normal University
09/1984 – 07/1996	Schulbesuch in Wuhan, Erwerb der allgemeinen chinesischen Hochschulreife

Berufserfahrung

Seit 02/2007	Teilprojektleiter für die Entwicklung der Automobilsicherheit in den Kundenprojekten bei der ZF-Lenksysteme GmbH in Schwäbisch Gmünd
02/2004 – 01/2007	Projektarbeit während der Industriepromotion bei der Daimler AG ▪ Ingenieur in der Zentralen Forschung der Daimler AG - Fahrzeugkomfortelektronik ▪ Teilprojektleiter im EU-Projekt EASIS - Elektronikarchitektur and Entwicklungsprozess ▪ Mitarbeiter in der AUTOSAR-Konsortium - Arbeitsgruppe „Sicherheit"
05/2003 – 12/2003	Industriediplomarbeit bei der Daimler AG in Esslingen zum Thema „Testautomatisierung der Steuergerätsoftware im Fahrzeug"
11/2002 – 03/2003	Industriepraktikum und Werkstudententätigkeit bei Sony International, Mobile Multimedia Lab
03/1999 – 12/2003	Wissenschaftliche Hilfskraft beim ▪ Institut für elektrische und optische Nachrichtentechnik (INT) ▪ Institut für Plasmaforschung (IPF) ▪ Fraunhofer Institut für Arbeitswirtschaft und Organisation (IAO)
08/1999 – 04/2000	Werkstudenten- und Ferientätigkeit bei der Robert Bosch GmbH

Soziales Engagement

12/2005 – 01/2007	Doktorandensprecher der Daimler AG
Seit 09/2001	VDE-Mitglied der Hochschulgruppe und Young-Professional

College Mathematics: Series 1

HACK ALGEBRA

THE EASY WAY

HACK ALGEBRA

THE EASY WAY

EZEOGU CHINONSO APOLLOS

KINGS BUSINESS INSTITUTE OF INFORMATION TECHNOLOGY LTD.
KBIIT
...Where scholars are made

ISBN: 978-978-767-022-4

Published by:
Kings Business Institute of Information Technology, Limited.
Plot 501 Maxcare Mart Plaza, Gaduwa District, Off Apo
Lokogoma Expressway, FCT-Abuja, Nigeria.

E-mail: kbittraining@gmail.com
E-mail: ae0600@my.bristol.ac.uk
Telephone: +2348062099567
WhatsApp: +2348062099567
Author's research works: https://www.researchgate.net/profile/Apollos-Ezeogu

College Mathematics Series:
#1: Hack Algebra the Easy Way
#2: Unlock Geometry: Become A Genius
#3: Calculus: Easy Guide

CONTENTS

Contents

UNIT II: ALGEBRA

Chapter 9 Algebraic Process I

Chapter 10 Inequality

Chapter 11 Polynomial and Partial Fraction

Contents

UNIT III APPENDIXES AND INDEX

Preface

This book is a compilation work on mathematical concepts of Algebra, with the intention to elucidate the difficulties faced by students taking courses in Algebra in high school and junior undergraduate classes. In addition, those preparing for the International General Certificate of Secondary Education (IGCSE), Scholastic Aptitude Test (SAT), West Africa Senior Secondary Schools Examinations (WASSCE), Unified Tertiary Matriculation Examination (UTME), Graduate Record Exam (GRE), General Management Aptitude Test (GMAT) and Olympiads. This book helps you ace the topics in Algebra with so much ease. This book authoritatively aver that Algebra can be hacked when studied with diligence by every student regardless of the student's background in mathematics. It is concise and straight to the point book with average book length of 400 pages. It contains eighteen chapters and exhaustively treats each chapter using numerous examples and annotations where necessary then with objective tests and exercises at the end of each chapter including the solution appended. It is a self-teaching book on Algebra for anyone. The Author's idea is to keep helping students succeed effortlessly in mathematics, therefore, he dedicated years of hard work to bring the book to perfection and fruition. "What is so good must talk-less of itself and be concise". So, when the student is ready then investigate the truth; and the truth shall make you whole only through self-evident conviction.

Ezeogu C. Apollos

Dedication

This work is dedicated to the Author of the universe whose language is mathematics.

Acknowledgements

I am grateful to Mrs Halima Buba, the CEO/Managing Director of Sun Trust Bank, for her encouragement and financial support for this book project; Architect Akahara Anthony, for his insights, guidance, and ideas on this book including financial support; Mr and Mrs Haliru Salihu who gave me parental guidance over a decade and took me like a son; Dr Suleiman Abdul-Malik for unexpected productivity allowance for my teaching skills, quality and abilities; Dr Muazu Omeji, Nigeria Tulip International Colleges, for the continuous advice, ideas and contributions for over a decade on the College Mathematics book series; Engr Ndubuisi Ishiwu for insights and many criticisms, he plays the role of the Devil's Advocate; Ndujiuba Leonard for managing the entire process of this book; Godwin Ezeogu, my uncle who always shows continuous concern for my welfare and education; Barister Anthony Ezeogu and Chief Abuchi Ezeogu for their financial support in this book project; Prof. Nwojo Nnanna and Prof. Steve Adeshina of Nile University of Nigeria, for their academic input in my life especially in my PhD program and my other academics research works; Dr Adeyemo Adedotun, for the regular checkup, calls, and ideas on many of my research projects; Dr Amar Musa, for his friendly support and constructive criticism; Mr Chukwuebuka Kingsley and Mr Sheriff Tahir, my co-directors at KBIIT Ltd. for their input and contributions to this book becoming a finished product; Barrister Ifeoma Onuike, who have contributed immensely to the growth of KBIIT Ltd; Miss Alonge Mercy, for taking time to also proof read the book; not forgetting my siblings and those who contributed immensely in the completion of this book series by investing their intellectual advice, time, money and energy to bring this work to completion. In addition, I wish to thank those that will purchase this book for themselves, loved ones, as a gift, or whatever reason including for the love and appreciation of mathematics.

UNIT I NUMBER AND NUMERATION

Specific Objectives:

At the end of this unit, the student should be able to:

(a) Perform simple arithmetic operations such as: addition, subtraction, multiplication and division;

(b) Find the LCM and HCF of numbers;

(c) Approximate and express numbers in standard form;

(d) Solve problems involving interest rates, profit and loss;

(e) Evaluate problems on standard base 10 and other bases like binary, octenary and hexadecimal;

(f) Simplify surd expressions and rationalise the denominator of a surd;

(g) Solve problems on variation: direct, inverse, joint and partial variation;

(h) Solve problems on sequence: Arithmetic progression and geometric progression;

(i) Understand the basic concept of set theory: Solve problems on set and use Venn diagram to represent information.

1 REAL NUMBER SYSTEM, FACTORS, MULTIPLES, AND APPROXIMATION

1.1 REAL NUMBER SYSTEM

The real number system is the set of all numbers, including the terminating and non-terminating decimals. Thus, the real number system is classified into rational numbers and irrational numbers.

(a) **Rational Numbers**: These are set of numbers that can be expressed as fraction.

That is, any number of the form $\frac{p}{q}$ where $p, q \in \mathbb{Z}$; such that $q \neq 0$.

Rational numbers are made up of *integers* and *non-integers.*

(i) **Integers**: These are made up of zero, positive and negative whole numbers.
The set of all integers is denoted by $\mathbb{Z}$.

- **zero** is denoted as 0
- **positive integers** are natural (counting) numbers which are 1, 2, 3, 4, 5 and so on. Positive integers are sub-divided into *odd, even, prime* and *composite* numbers.
  - *Odd numbers* are the natural numbers that leave a remainder when divided by 2, e.g., 1, 3, 5, 7, 9, 11, 13, 15, ...
  - *Even numbers* are natural numbers that do not leave a remainder when divided by 2, e.g 2, 4, 6, 8, 10, 12, ...
  - *Prime numbers* are numbers divisible by just itself and 1 only. That means it must have only two factors, i.e., itself and 1 only. E.g. 2, 3, 5, 7, 11, 13, 17,19, 23, 29, ...

N/B : 1 is not a prime number because it has just one factor; that is "itself only."

  - *composite numbers* are not prime numbers. However, they are numbers that can be divided by other numbers besides itself, e.g. 4, 6, 8, 9, 10, 12, 14, ...

N/B : 1 is not a composite number because it cannot be divided by other numbers.

- **Negative integers** are whole numbers having negative values, e.g. $-1, -2, -3, -4, -5, -6, -7, -8, \ldots$

(b) **Irrational Numbers:** These are numbers that cannot be rationalised or expressed as quotient of two integers. Thus, they have non-terminating decimals that do not repeat.
E.g. $\sqrt{2} = 1.4142135623730...$, $\sqrt{3} = 1.73205080756887...$, and π.

1.2 FACTORS

A factor is a number that can divide another number exactly without a remainder.
e.g., $24 = 1 \times 24 = 2 \times 12 = 3 \times 8 = 4 \times 6$
The numbers 1, 2, 3, 4, 6, 8, 12 and 24 are factors of 24. Thus, any number can be expressed as a product of its factors.

1.2.1 PRIME FACTORS: A prime factor is a factor which is a prime number.

Example 1

Find the prime factors of 63

Solution

Factors of $63 = 1 \times 63 = 3 \times 21 = 7 \times 9$
$\therefore$ prime factors of 63 are 3 and 7.
N/B : 9, 21 and 63 are factors, but they are not prime numbers. Therefore, they are not prime factors.

Example 2

Express 210 as a product of prime factors.

Solution

$210 = 2 \times 105 = 2 \times 5 \times 21 = 2 \times 5 \times 3 \times 7$
$\Rightarrow 210 = 2 \times 5 \times 3 \times 7$

1.2.2 COMMON FACTOR: It is the set of all the possible numbers that are factors of given numbers.

Example 3

Find the Common Factors of 16, 32 and 40.

Solution

Factors of 16 = 1, 2, 4, 8 and 16
Factors of 32 = 1, 2, 4, 8, 16 and 32
Factors of 40 = 1, 2, 4, 5, 8, 10, 20 and 40
$\therefore$ Common Factor (C.F) =1, 2, 4 and 8

Example 4

Find the Common Factors of $2^2 \times 3^3 \times 5$ and $2^2 \times 3^2 \times 7$

Solution

Factors of $2^2 \times 3^3 \times 5 = 2 \times 2 \times 3 \times 3 \times 3 \times 5$

Factors of $2^2 \times 3^2 \times 7 = 2 \times 2 \times 3 \times 3 \times 7$

$\therefore$ Common Factor $= 2 \times 2 \times 3 \times 3$

1.2.3 HIGHEST COMMON FACTOR (HCF): It is the largest factor common to given numbers. It is found by "multiplying the common prime factors" of given numbers, or by picking out the "largest value of the common factor" of given factors of numbers.

Example 5

Find the Highest Common Factor of (a) 16, 32 and 40 (b) 208, 302 and 416

Solution

(a) Using the method of picking out "largest value of the common factor," we have that:

Factors of 16 = 1, 2, 4, 8 and 16

Factors of 32 = 1, 2, 4, 8, 16 and 32

Factors of 40 = 1, 2, 4, 5, 8, 10, 20 and 40

$\therefore$ Highest Common Factor (HCF) is **8**

(b) Using the method of "multiplying the common prime factors," we have that:

$208 = 1 \times 2 \times 2 \times 2 \times 2 \times 13$

$312 = 1 \times 2 \times 2 \times 2 \times 3 \times 13$

$416 = 1 \times 2 \times 2 \times 2 \times 2 \times 2 \times 13$

$\therefore$ Highest Common Factor (HCF) is $1 \times 2 \times 2 \times 2 \times 13 =$ **104**

Example 6

Find the Highest Common Factors of $2^2 \times 3^3 \times 5$ and $2^2 \times 3^2 \times 7$

Solution

Factors of $2^2 \times 3^3 \times 5 = 2 \times 2 \times 3 \times 3 \times 3 \times 5$

Factors of $2^2 \times 3^2 \times 7 = 2 \times 2 \times 3 \times 3 \times 7$

$\therefore$ Highest Common Factor (HCF) is $2 \times 2 \times 3 \times 3 =$ **36**

1.3 MULTIPLES

This is obtained when certain whole numbers are multiplied by a constant integer, e.g., when the numbers 0, 5, 10, 15, 20, 25, 30, ... are multiplied by 6, then the results are 0, 30, 60, 90, 120, 150, ... Therefore, the numbers 0, 30, 60, 90, 120, 150, ... are called multiples of 6.

LOWEST COMMON MULTIPLE (LCM): It is also called least common multiple. It is defined as the smallest multiple that is exactly divisible by every member of a set of numbers.

Example 7

Find the LCM of the following:
(a) 6 and 8 (b) 18, 30 and 48 (c) $2^4 \times 3 \times 5$ and $2^2 \times 3^2 \times 5$

Solution

(a) 6 and 8

$6 = 2 \times 3$
$8 = 2 \times 2 \times 2$

$\therefore$ LCM $= 2 \times 2 \times 2 \times 3 = 24$

In each case, find the factors of each given number, and pick out the "common factors" if any, then multiply the common factors by the other factors of all the given numbers.

(b) 18, 30 and 48

$18 = 2 \times 3 \times 3$
$30 - 2 \times 3 \times 5$
$48 = 2 \times 2 \times 2 \times 2 \times 3$
$\therefore$ LCM $= 2 \times 3 \times 2 \times 2 \times 2 \times 3 \times 5 = 720$

(c) $2^4 \times 3 \times 5$ and $2^2 \times 3^2 \times 5$
$2^4 \times 3 \times 5 = 2 \times 2 \times 2 \times 2 \times 3 \times 5$
$2^2 \times 3^2 \times 5 = 2 \times 2 \times 3 \times 3 \times 5$

$\therefore$ LCM $= 2 \times 2 \times 2 \times 2 \times 3 \times 3 \times 5 = 2^4 \times 3^2 \times 5$

1.4 APPROXIMATION AND STANDARD FORM

Approximation is the process of estimating figures to a certain degree of accuracy – this is achieved by rounding off numbers. Thus, in order to round off numbers, the digits 1, 2, 3, 4 are rounded down to zero while the digits 5, 6, 7, 8, 9 are rounded up to one.

1.4.1 DECIMAL PLACES (d.p.): This is an approximation technique that allows certain number of digits to the right of the decimal point.

Example 8

Approximate 13.3487 to (a) 3 d.p. (b) 2 d.p. (c) 1 d.p.

Solution

(a) 13.349 which is rounded-off to 3 d.p.
(b) 13.35 which is 2 d.p.
(c) 13.3 which is 1 d.p.

Example 9

Approximate the following to 1 d.p.:

(a) 10.834 (b) 1.573 (c) 3.142

Solution

(a) $10.834 \doteqdot 10.8$

(b) $1.573 \doteqdot 1.6$

(c) $3.142 \doteqdot 3.1$

1.4.2 **SIGNIFICANT FIGURE (s.f.)**: This is any non-zero digit of a number beginning from the left and ending with non-zero digit to the right. Thus, it is used for expressing accuracy when we take into consideration the necessary(significant) non- zero digits as required.

Example 10

(a) 2052.37 has 6 significant figures (b) 0.00206153 has 6 s.f.
Approximate (a) and (b) respectively to:
(i) 5 s.f. (ii) 4 s.f. (iii) 3 s.f. (iv) 2 s.f. (v) 1 s.f.

Solution

(a) (i) $2052.37 \Rightarrow 2052.4$ is now to 5 s.f.

(ii) $2052.37 \Rightarrow 2052$ is now to 4 s.f.
(iii) $2052.37 \Rightarrow 2050$ is now to 3 s.f.

(iv) $2052.37 \Rightarrow 2100$ is now to 2 s.f.
(v) $2052.37 \Rightarrow 2000$ is now to 1 s.f.

the zeroes between two non-zero digits are regarded as significant figures.

The zeroes after a non-zero digit, and preceding no other non-zero digit are not considered as significant figure.

(b) (i) $0.00206153 \Rightarrow 0.0020615$ which is now to 5 s.f.
(ii) $0.00206153 \Rightarrow 0.002062$ which now to 4 s.f.
(iii) $0.00206153 \Rightarrow 0.00206$ which is now to 3 s.f.
(iv) $0.00206153 \Rightarrow 0.0021$ which is now to 2 s.f.
(v) $0.00206153 \Rightarrow 0.002$ which is now 1 s.f.

When zeroes come after a decimal point, and precedes a non-zero digit; then it is not a significant figure.

1.4.3 **NEAREST WHOLE NUMBER**: This is an approximation technique that round - off the digits to become a whole number. Therefore, it leaves no decimal point in the estimated figure.

Example 11

Approximate the following figures to the nearest whole number:
(a) 13.3487 (b) 17.892 (c) 0.85 (d) 0.43

Solution

(a) $13.3487 \triangleq 13$ (b) $17.892 \triangleq 18$ (c) $0.85 \triangleq 1$ (d) $0.43 \triangleq 0$

Example 12

Approximate the following to the nearest whole number:
(a) 5.835 (b) 3.142 (c) 7.259

Solution

(a) $5.835 \simeq 6$ (b) $3.142 \simeq 3$ (c) $7.259 \simeq 7$

1.4.4 STANDARD FORM: When a number is written in the form $K \times 10^n$, where K is a real number such that $1 \leq K < 10$ and n is an integer (i.e., a positive or negative whole number) then we say the expression is in standard form.

Example 13

Express the following in standard form:
(a) 0.0365 (b) 315.6 (c) 61.208 (d) 0.0000082

Solution

(a) $0.0365 = 3.65 \times 10^{-2}$ (b) $315.6 = 3.156 \times 10^2$
(c) $61.208 = 6.1208 \times 10^1$ (d) $0.0000082 = 8.2 \times 10^{-6}$

Example 14

Divide the following numbers and leave the answers in standard form.
(a) $4.8 \times 10^{-2} \div 2.4 \times 10^{-4}$ (b) $10.34 \div 0.05$ (c) $2.842 \div 0.25$

Solution

(a) $4.8 \times 10^{-2} \div 2.4 \times 10^{-4} = 2 \times 10^{(-2)-(-4)} = 2 \times 10^{-2+4} = 2 \times 10^2$
(b) $10.34 \div 0.05 = 206.8 = 2.068 \times 10^2$
(c) $2.842 \div 0.25 = 11.368 = 1.1368 \times 10^1$

OBJECTIVE TEST 1

1. Which of these is an irrational number?
   (A) $\frac{1}{2}$ (B) π (C) $\frac{2}{5}$ (D) None
2. Which of these contains only prime numbers?
   (A) 1, 2, 4 (B) 2, 3, 5 (C) 2, 3, 5, 9 (D) 1, 7, 9
3. What is the HCF of 45, 90 and 150?
   (A) 45 (B) 15 (C) 35 (D) 90
4. The prime factors of 2520 are:
   (A) 2, 9, 5 (B) 2, 9, 7 (C) 2, 3, 5, 7 (D) 2, 3, 7, 9
5. Find the LCM of $3 \times 5^2 \times 7$ and $2^2 \times 3^2 \times 5$
   (A) $2^2 \times 3^2 \times 5^2 \times 7$ (B) $2^2 \times 3^2 \times 5$ (C) $2^2 \times 3 \times 5^2 \times 7$ (D) $2 \times 3^2 \times 5^2 \times 7$
6. Express the product of 0.0014 and 0.011 in standard form.
   (A) 1.54×10^{-2} (B) 1.54×10^{-3} (C) 1.54×10^{-4} (D) 1.54×10^{-5}
7. Simplify $4.8 \times 10^{-2} \div 1.2 \times 10^{-4}$ and leave your answer in standard form.
   (A) 4×10^3 (B) 4×10^2 (C) 4×10^{-2} (D) 4×10^{-3}
8. Evaluate $\frac{2.8 \times 10^{-3} \times 1.1}{5.6 \times 10^{-2}}$ and then leave your answer in 1 s.f.
   (A) 0.055 (B) 0.06 (C) 0.054 (D) 0.6
9. Approximate 0.0585 to 1 d.p.
   (A) 0.06 (B) 0.01 (C) 0.1 (D) 1.0
10. Approximate 24.639 to the nearest whole number.
    (A) 25 (B) 24.6 (C) 25.60 (D) 2.5

Exercise 1

1. Find the common factors (CF) and Highest Common Factor (HCF) of the following:

   (a) 9, 15 and 21 (b) 18, 24 and 56 (c) 12, 18 and 96
   (d) 30, 45 and 48 (e) 121, 132 and 156 (f) 45, 90 and 150
   (g) $2^3 \times 3^2$ and $2^2 \times 3$ (h) $7^2 \times 3^3$ and $7^3 \times 3^2$ (i) $4 \times 5 \times 3^2$ and $4 \times 7 \times 5^2$

2. (a) Find the prime factors of 216, 288 and 360.
   (b) Find also the C.F and HCF of question 2a.
3. Find the LCM of each of the following:
   (a) 2 and 3 (b) 5, 7 and 9 (c) 6, 12 and 18 (d) $x, x^2, 5x$
   (e) $(x-1)$ and $(x+1)^2$ (f) $(x-1), (x-1)^2$ and $(x-1)^3$
   (g) 2×3^2 and $2^2 \times 3$ (h) $5^2 \times 7$ and $5^4 \times 7^3$

4. Write the value of the following numbers to 3 d.p.:
   (a) 4.89247 (b) 0.0766 (c) 15.25624 (d) 456.8834765
5. Evaluate and approximate the answers of the following to 2 d.p.:
   (a) 50.56×2.3 (b) $0.3853 \div 0.19$ (c) 827.51×0.015
   (d) $100 + \frac{1}{100} + \frac{3}{1000} + \frac{27}{10000}$
6. Express the following in 2 s.f:
   (a) 208.6 (b) 0.2053 (c) 4081.62 (d) 3065
7. Approximate the following to the nearest whole number:
   (a) 0.8899 (b) 287.467 (c) 604.08 (d) 0.09
8. Express in standard form:
   (a) 0.0318×10^{-3} (b) 131.5×10^2 (c) 0.0000863 (d) 456756
9. Approximate to 2 s.f, then express in standard form:
   (a) 285.725 (b) 0.00829×10^{-5} (c) 7058 (d) 4235
10. Simplify and leave your answer in standard form:

    (a) $\frac{2.5\times10^8}{0.05\times10^6}$ (b) $\frac{0.007 \times 0.081}{0.042}$ (c) $\sqrt{\frac{1.21\times5}{0.01\times10^{-3}}}$ (d) $\sqrt{\frac{0.9\times10^{-5}}{2.25\times10^{-4}}}$
11. Express in standard form, and correct to 2 s.f:
    (a) $\frac{3}{8}$ (b) $18\frac{1}{4}$ (c) $652\frac{2}{5}$ (d) $\frac{10}{4}$ (e) $\frac{428}{8}$
12. Find the value of:
    (a) $8.3 \times 10^2 + 5.2 \times 10^2$ (b) $6.3 \times 10^{-2} - 4.3 \times 10^{-2}$ (c) $20.32 \times 10^3 + 5.8 \times 10^2$
    leaving the answers in standard form.
13. Round off each of the following to 2 significant figures:
    (a) 23.56×10^{-1} (b) 894×10^2 (c) $\frac{4}{9}$ (d) $\frac{1}{3}$

2 ARITHMETICS

2.1 FRACTIONS

A fraction is a number that is not a whole number. It has both numerator and denominator, and it is of the form $\frac{a}{b}$; where "a" is the numerator and "b" is the denominator (this form of fraction is said to be in *vulgar form*). A fraction in which $a < b$ is called a proper fraction. While, a fraction in which $a > b$ is called an improper fraction; and when an improper fraction is divided to give a mixture of a whole number and a proper fraction, then it is called a mixed fraction.
(N/B: A fraction can also be expressed in decimal form, e.g. 0.005)

For instance:

$\frac{2}{7}$ is a proper fraction; and 2 is the numerator, while 7 is the denominator.

The fraction is a proper fraction because the numerator (i.e., 2) is smaller than the denominator (i.e., 7).

$\frac{5}{3}$ is an improper fraction; and 5 is the numerator, while 3 is the denominator

The fraction is an improper fraction because the numerator (i.e. 5) is larger than the denominator (i.e. 3).

$\frac{5}{3} = 1\frac{2}{3}$ is a mixed fraction because it contains a whole number and a proper fraction

The fraction is $\frac{2}{3}$ and it is a proper fraction while the whole number is 1.

Example 1

Which of the fractions is smaller $\frac{1}{2}$ or $\frac{3}{5}$?

Solution

Rename each of these fractions so that they can have the same denominator. This is done by finding the LCM of the denominators of the fractions.

$$\frac{1}{2} = \frac{1 \times 5}{2 \times 5} = \frac{5}{10}$$

$$\frac{3}{5} = \frac{3 \times 2}{5 \times 2} = \frac{6}{10}$$

$\frac{5}{10}$ is smaller than $\frac{6}{10}$ when we compare the numerators.

$\therefore$ $\frac{1}{2}$ is smaller than $\frac{3}{5}$

Example 2

Arrange $\frac{7}{12}$, $\frac{3}{10}$, $\frac{3}{4}$ and $\frac{2}{5}$ in ascending order (i.e., starting from the smallest to the largest).

Solution

$$\frac{7}{12} = \frac{7 \times 5}{12 \times 5} = \frac{35}{60}$$

$$\frac{3}{10} = \frac{3 \times 6}{10 \times 6} = \frac{18}{60}$$

$$\frac{3}{4} = \frac{3 \times 15}{4 \times 15} = \frac{45}{60}$$

$$\frac{2}{5} = \frac{2 \times 12}{5 \times 12} = \frac{24}{60}$$

Find the **LCM** of **12**, **10**, **4** and **5**; which is **60**. Then divide **60** by each denominator e.g., $\frac{60}{12} = 5$, $\frac{60}{10} = 6$; and so on. Now, multiply these values to both numerator and denominator of its fraction. Finally, compare the sizes of the numerators of all the fractions to find the smallest up to the largest (i.e., in ascending order).

When arranged in ascending order, we have $\frac{18}{60}$, $\frac{24}{60}$, $\frac{35}{60}$ and $\frac{45}{60}$

thus $\frac{3}{10}$, $\frac{2}{5}$, $\frac{7}{12}$ and $\frac{3}{4}$ is the ascending order of magnitude.

Example 3

Simplify

(a) $2\frac{2}{5} + 6\frac{1}{5} + 3\frac{3}{4}$ (b) $6\frac{11}{12} - 2\frac{5}{6} - 1\frac{1}{4}$

(c) $\dfrac{4\frac{3}{5} - 6\frac{1}{4}}{4\frac{1}{5}\ of\ 1\frac{1}{4}}$ (d) $\frac{10}{9} + \frac{5}{7} + 3\frac{2}{5} + \frac{5}{3}$

Solution

(a) $2\frac{2}{5} + 6\frac{1}{5} + 3\frac{3}{4} = 2 + 6 + 3 + \frac{2}{5} + \frac{1}{5} + \frac{3}{4}$

Add the whole numbers.

Find the **LCM** of the denominators of all the fractions; and that is **20** in this case.

$$\Rightarrow 11 + \frac{\frac{2}{5}(20) + \frac{1}{5}(20) + \frac{3}{4}(20)}{20} = 11 + \frac{8 + 4 + 15}{20}$$

$$= 11\frac{27}{20} = 11 + \left(1 + \frac{7}{20}\right) = 12\frac{7}{20}$$

> Now, multiply 20 by each fraction (e.g., $\frac{2}{5}(20) = 8$). Repeat this approach for all the other fractions, and finally add them as required.

(b) $6\frac{11}{12} - 2\frac{5}{6} - 1\frac{1}{4} = (6 - 2 - 1) + \left(\frac{11}{12} - \frac{5}{6} - \frac{1}{4}\right)$

$$= 3 + \frac{\frac{11}{12}(12) - \frac{5}{6}(12) - \frac{1}{4}(12)}{12}$$

$$= 3 + \frac{11 - 10 - 3}{12}$$

$$\therefore 3 + \frac{11 - 10 - 3}{12} = 3 + \frac{-2}{12}$$

$$= 2 + \frac{12}{12} + \frac{-2}{12}$$

$$= 2 + \frac{12 - 2}{12} = 2 + \frac{10}{12}$$

$$\Rightarrow 2 + \frac{5}{6} = 2\frac{5}{6}$$

> Find the **LCM** of the denominators of the fractions; and that is 12 in this case.

> Then multiply 12 by each fraction (e.g., $\frac{11}{12}(12) = 11$). Repeat this approach for all other fractions, and finally add them as required.

Note: the manipulation of $3 = 2 + \frac{12}{12}$ was to help us easily evaluate the negative fraction i.e., $\frac{-2}{12}$.

(c) $\dfrac{4\frac{3}{5} - 6\frac{1}{4}}{4\frac{1}{5} \; of \; 1\frac{1}{4}} \Rightarrow \dfrac{4 - 6 + \left(\frac{3}{5} - \frac{1}{4}\right)}{\frac{21}{5} \times \frac{5}{4}} = \dfrac{-2 + \frac{12 - 5}{20}}{\frac{21}{4}}$

$$= \frac{-2 + \frac{7}{20}}{\frac{21}{4}} = \frac{-\frac{33}{20}}{\frac{21}{4}} = -\frac{33}{20} \times \frac{4}{21}$$

$$= -\frac{33}{105}$$

(d) $\frac{10}{9} + \frac{5}{7} + 3\frac{2}{5} + \frac{5}{3} = 1\frac{1}{9} + \frac{5}{7} + 3\frac{2}{5} + 1\frac{2}{3}$

$$= (1 + 3 + 1) + \left(\frac{1}{9} + \frac{5}{7} + \frac{2}{5} + \frac{2}{3}\right) = 5 + \frac{35 + 225 + 126 + 206}{315}$$

$$= 5 + \frac{592}{315} = 5 + 1\frac{277}{315}$$

$$\therefore 5 + 1\frac{277}{315} = 6\frac{277}{315}$$

Example 4

A Fulani man shared 17 cows to three of his sons. If the first son received $\frac{1}{2}$ of the cows, the second received $\frac{1}{3}$ of the cows, and the last son received $\frac{1}{9}$ of the cows. How many cows wholly will each of them get?

Solution

Let: the first son's share be $= F$
second son's share $= S$
and last son's share $= L$

Note that the number of cows is an odd number (i.e., 17). It will be difficult to share them wholly according to the fractions given. So, we have to introduce 1 imaginary cow to make it a total of 18 cows (i.e., even number of cows)

$\therefore \quad F = \frac{1}{2}$ of (17 cows + 1 imaginary cow)

$\Rightarrow F = \frac{1}{2} \times 18 = 9$ cows

$S = \frac{1}{3}$ of (17 cows + 1 imaginary cow) $= \frac{1}{3} \times 18 = 6$ cows

$L = \frac{1}{9}$ of (17 cows + 1 imaginary cow) $= \frac{1}{9} \times 18 = 2$ cows

Therefore, the first son gets 9 cows; the second gets 6 cows; and the last son gets 2 cows, thus making a total of 17 cows.

Despite the "imaginary" cow added we still have the total number of cows shared to be 17 (i.e., 9+6+ 2). Since it is only abstract, it will not appear in the "actual number of cows."

Example 5

Convert the following decimals to their equivalent fraction:

(a) $0.555\dot{5}$ (b) $0.8\frac{3}{15}$ (c) $2.4\frac{2}{5}$

Solution

Notice the dot on top of the last digit, which means it is a recurring number. $0.555\dot{5} = 0.5555\dot{5} = 0.55555\dot{5}$ and so on.

(a) $0.555\dot{5}$

Let $R = 0.555\dot{5}$ ....(i)

multiply equation (i) by 10

$\therefore 10R = 5.55\dot{5}$ ....(ii)

Subtract (i) from (ii)

$\Rightarrow 10R - R = 5.555\dot{5} - 0.55\dot{5} = 5.555\dot{5} - 0.555\dot{5}$

$9R = 5$

$\therefore R = \frac{5}{9}$

Notice the change from: $0.55\dot{5} = 0.555\dot{5}$ to keep the recurring decimal the same for easy subtraction.

(b) $0.8\frac{3}{15} \Rightarrow \frac{8\frac{3}{15}}{10}$

$\therefore \frac{8\frac{3}{15}}{10} = \frac{\frac{123}{15}}{10} = \frac{\frac{123}{15} \times \frac{15}{1}}{10 \times \frac{15}{1}} \Rightarrow \frac{\frac{123}{15} \times \frac{15}{1}}{10 \times \frac{15}{1}} = \frac{123}{150} = \frac{41}{50}$

$\therefore 0.8\frac{3}{15} = \frac{41}{50}$

(c) $2.4\frac{2}{5} = 2 + 0.4\frac{2}{5} \Rightarrow 2 + \frac{4\frac{2}{5}}{10} = 2 + \frac{\frac{22}{5}}{10}$

$\therefore 2 + \frac{\frac{22}{5}}{10} = 2 + \frac{\frac{22}{5} \times \frac{5}{1}}{10 \times \frac{5}{1}} = 2 + \frac{22}{50} = 2 + \frac{11}{25}$

$\therefore 2 + \frac{11}{25} = 2\frac{11}{25}$

Example 6

Find the value of x for which the fraction is undefined:

(a) $\frac{x+1}{x-3}$ (b) $\frac{1}{x^2-9}$ (c) $\frac{x^2+3x+21}{(x+6)(2+x)}$ (d) $\frac{1-x}{\frac{3}{2}x-8}$

A fraction is undefined when the denominator is equated to zero.

Solution

(a) $\frac{x+1}{x-3} \Rightarrow x - 3 = 0$

$\therefore\ x = 3$

(b) $\frac{1}{x^2-9} \Rightarrow x^2 - 9 = 0$

$\therefore\ x = \sqrt{9} = \pm 3$

$\therefore\ x = +3 \text{ or} - 3$

For each of the fraction here, equate the denominator to be equal to zero and then make x the subject of the formula

(c) $\frac{x^2+3x+21}{(x+6)(2+x)} \Rightarrow (x+6)(2+x) = 0$

$\therefore\ (x+6) = 0$ and $(2+x) = 0$

$\therefore\ x = -6$ and -2

(d) $\frac{1-x}{\frac{3}{2}x-8} \Rightarrow \frac{3}{2}x - 8 = 0$

$\therefore\ \frac{3}{2}x = 8 \Rightarrow x = \frac{8 \times 2}{3}$

$\therefore\ x = \frac{16}{3} = 5\frac{1}{3}$

Example 7

For what values of x will the following fractions be equal to zero?

(a) $\frac{x^2-36}{x^2-5x-6}$ (b) $\frac{7x}{x^2+6x+8}$ (c) $\frac{5x^2+7x-6}{x^2+3x-8}$ (d) $\frac{1}{x+1} - \frac{2}{3x+5}$

Solution

(a) $\frac{x^2-36}{x^2-5x-6} \Rightarrow \frac{x^2-36}{x^2-5x-6} = 0$

$\therefore x^2 - 36 = (x^2 - 5x - 6) \times 0 = 0$

$x = \sqrt{36} = \pm 6$

$\therefore x = 6 \ or - 6$

(b) $\frac{10x}{x^2+6x+8} \Rightarrow \frac{10x}{x^2+6x+8} = 0$

In each case, equate the entire fraction to be zero and then simplify to solve for x.

$10x = 0 \times (x^2 + 6x + 8)$

then, $10x = 0$

$\therefore \ x = 0$

(c) $\frac{5x^2+7x-6}{x^2+3x-8} \Rightarrow \frac{5x^2+7x-6}{x^2+3x-8} = 0$

$\therefore \ 5x^2 + 7x - 6 = 0 \times (x^2 + 3x - 8) = 0$

$5x^2 + 7x - 6 = 0$

$\Rightarrow \ (x + 2)(5x - 3) = 0$

$\therefore (x + 2) = 0$ and $(5x - 3) = 0$

Thus, $x = -2$ and $x = \frac{3}{5}$

(d) $\frac{1}{x+1} - \frac{2}{3x+5} \Rightarrow \frac{1}{x+1} - \frac{2}{3x+5} = 0$

$\frac{1}{x+1} = \frac{2}{3x+5} \Rightarrow 3x + 5 = 2(x + 1)$

$3x + 5 = 2x + 2$

$3x - 2x = 2 - 5$

$\therefore \ x = -3$

Exercise 2a

1. Simplify $5 + 6\frac{1}{2} + 2\frac{7}{10}$
2. Evaluate $2\frac{4}{21} - \frac{3}{4}\left(\frac{1}{6} - 2\frac{2}{3}\right)$
3. An N.Y.S.C member spends $\frac{3}{7}$ of his allowance on drinks, and $\frac{7}{8}$ of the remainder on the buying of books. If he has ₦2,610 left, how much is his allowance?

4. Change the following decimals into vulgar fractions:
   (a) 0.0125 (b) 0.25 (c) 0.002 (d) 0.15 (e) 0.85 (f) 0.375
5. Change the following vulgar fractions into decimals:
   (a) $\frac{1}{2}$ (b) $\frac{1}{20}$ (c) $\frac{3}{8}$ (d) $\frac{2}{3}$ (e) $\frac{3}{5}$
6. Arrange $\frac{7}{8}$, $\frac{3}{4}$, $\frac{8}{9}$ and $\frac{4}{7}$ in ascending order.
7. Add the fractions: (a) $7\frac{1}{3} - 2\frac{5}{6} - 1\frac{1}{4}$ (b) $5\frac{3}{8} - 2\frac{1}{5}$
8. Simplify (a) $\frac{1}{5} \div 2\frac{1}{3} + \frac{1}{4} \div \frac{1}{5}$ (b) $\frac{1}{3} - 5\frac{2}{3} \div 8\ of\ \frac{1}{5}$
9. Add $\frac{1}{3}$ to the difference of 10 and $3\frac{1}{5}$
10. $\frac{2}{5}$ of a class are girls, what fraction of the class consist of boys? And if there are 60 boys, how many girls are in the class?
11. A drum is filled with water, after $\frac{2}{3}$ of the water had been poured out, 80 litres of water was still left. What is the capacity of the container?
12. Find the equivalent fraction of the following decimals:
   (a) $0.833\dot{3}$ (b) 0.6666 (c) 61.25 (d) $0.5\frac{1}{2}$ (e) 17.0124
13. Find the value(s) of x for which each of the given fractions is undefined:
   (a) $\frac{1}{x-3}$ (b) $\frac{2x-1}{x^2-3x-4}$ (c) $\frac{x+8}{x^2-4}$ (d) $\frac{5}{x^2-9}$ (e) $\frac{3x+8}{(x-1)(2x-5)(x+7)}$
14. Find the value(s) of x for which the fraction is equal to zero:
   (a) $\frac{x+1}{10x-1}$ (b) $\frac{2x-1}{x^2-4}$ (c) $\frac{x+1}{x+3} + \frac{x-5}{x-4}$ (d) $\frac{1}{x-25} - \frac{2-x}{x+19}$

2.2 RATIO

Ratio is defined as the relative magnitudes of two quantities, whereby one of the quantities is assigned the numerator, while the other is assigned the denominator.

$$\frac{a}{b} = a : b$$

If a set is divided into quantities in the ratio $a : b$, then the first quantity contains $\frac{a}{a+b}$ of the total quantity; while the second quantity contains $\frac{b}{a+b}$ of the total quantity.

Similarly, ratio can be extended into 3, 4, 5, ... terms when comparing quantities. And each term is evaluated using the same principle as explained above.

Example 8

25% of the students in a class are female, find the ratio of female students to male students.

Solution

Let female = $\boldsymbol{a}$ and male = $\boldsymbol{b}$
But percentage of females = 25%
This implies that percentage of male is 100% −25% = 75%

$\therefore \quad \frac{a}{b} = \boldsymbol{a}{:}\boldsymbol{b} = \frac{25\%}{75\%} = \frac{1}{3} \quad \Rightarrow 1:3$

$\therefore$ the ratio of female to male is 1: 3

Example 9

The ratio of economics students to biology students is 4:3, how many students study economics if the total number of students is 70?

Solution

Let economics = $\boldsymbol{a}$; biology = $\boldsymbol{b}$

then, number for economics = $\frac{a}{a+b} \times$ Total number of students

$= \frac{4}{4+3} \times 70 = \frac{4}{7} \times 70 = 40$

$\therefore$ the number of students that study economics is 40 students.

Example 10

The ratio of the age of a man to his sister's is now 8:5. If the sister is 20 years old now, find the ratio of the age of the man to that of his sister two years ago.

Solution

Let: the man's age = y, and the sister's age = x

$\therefore \frac{y}{x} = \frac{8}{5} = 8{:}5$

If x is 20 years now, then we have $\frac{y}{20} = \frac{8}{5}$

$\therefore y = \frac{20 \times 8}{5} = \frac{160}{5} = 32$ years

Make $\boldsymbol{y}$ the subject formula

$\therefore$ the man is 32 years now.
But, two years ago he was, $y - 2 = 32 - 2 = 30$ years
Also, the sister's age two years ago was, $x - 2 = 20 - 2 = 18$ years

$\Rightarrow$ ratio of their ages two years ago was, $\frac{y-2}{x-2} = \frac{30}{18} = \frac{5}{3} = 5{:}3$

Example 11

Three boys share 80 oranges in the ratio of 2:3:5, how many oranges will each get?

Solution

Let the boys be represented as p, q and r;
then, total oranges, $T = p + q + r$
Also, let their ratio be $x : y : z = 2 : 3 : 5$

$\Rightarrow$ share of $p = \frac{x}{x+y+z} \times T = \frac{2}{2+3+5} \times 80$

$= \frac{2}{10} \times 80 = 16$

Thus, share of $q = \frac{y}{x+y+z} \times T = \frac{3}{2+3+5} \times 80$

$= \frac{3}{10} \times 80 = 24$

And, share of $r = \frac{z}{x+y+z} \times T = \frac{5}{2+3+5} \times 80$

$= \frac{5}{10} \times 80 = 40$

$\therefore p = 16$ oranges, $q = 24$ oranges and $r = 40$ oranges

2.3 PERCENTAGES

This is the proportion of a set of quantities. It expresses the ratio of each quantity in the set by multiplying by 100%.

Example 12

Express the following as a percentage:

(a) $\frac{1}{2}$ (b) $\frac{1}{3}$ (c) $\frac{2}{5}$

Solution

(a) $\frac{1}{2} = \frac{1}{2} \times 100\% = 50\%$

(b) $\frac{1}{3} = \frac{1}{3} \times 100\% = 33.\dot{3}\%$

(c) $\frac{2}{5} = \frac{2}{5} \times 100\% = 40\%$

Example 13

A sales man sold goods worth ₦500,000 and was paid a commission of ₦31,750. Find the percentage of his commission.

Solution

$\%\ \text{commission} = \frac{\textit{commission}}{\textit{price of goods}} \times 100\% = \frac{₦31{,}750}{₦500{,}000} \times 100\% = 6.35\%$

Example 14

The price of a pair of shoes rises from ₦5,000 to ₦6,000, what is the percentage increase?

Solution

Price increase = New price − Original price
= ₦6,000 − ₦5,000 = ₦1,000

$\therefore$ % increase $= \frac{\textit{price increase}}{\textit{original price}} \times 100\%$

$= \frac{₦1{,}000}{₦5{,}000} \times 100\% = 20\%$

Example 15

Assuming from year 2019 to 2020, the number of students applying for the SAT examination in Nigeria increased by 10% to a number of 2,200,000. How many students applied in 2019?

Solution

Let old value = number of candidates in 2019 = x
and new value = number of candidates in 2020 = y
new value = old value (1 + % increase)
$y = x(1 + 10\%) = x(1 + 0.1)$
$\therefore\ y = 1.1x$
But, $y = 2{,}200{,}000 \Rightarrow 2{,}200{,}000 = 1.1x$

$x = \frac{2{,}200{,}000}{1.1} = 2{,}000{,}000$

$\therefore$ the number of candidates in 2019 was 2,000,000

Example 16

The length of a square 12cm was measured as 12.2cm. Calculate and correct to two decimal places:
(a) the percentage error in length (b) the percentage error in area.

Solution

(a) % Error $= \frac{\textit{Actual error in length}}{\textit{Original length}} \times 100\% = \frac{l_1 - l_0}{l_0} \times 100\%$

Now let $l_0 = 12\text{cm},\ l_1 = 12.2\text{cm}$

$\Rightarrow$ % Error $= \frac{12.2 - 12}{12} \times 100\% = \frac{0.2}{12} \times 100\% = 1.67\%$

$\therefore$ the percentage error in length is 1.67%

(b) % Error $= \frac{\textit{Actual error in area}}{\textit{Original area}} \times 100\% = \frac{A_1 - A_0}{A_0} \times 100\%$

Since the figure is a square, then we have:

$A_0 = 12 \times 12 = 144\text{cm}^2$ and $A_1 = 12.2 \times 12.2 = 148.84\text{cm}^2$

$\Rightarrow$ % Error $= \frac{148.84 - 144}{144} \times 100\% = \frac{4.84}{144} \times 100\% = 3.36\%$

$\therefore$ the percentage error in area is 3.36%

2.4 PROPORTION AND RATES

2.4.1 PROPORTION: This is when a change in a quantity causes a change in another related quantity. Thus, the quantities are said to be in proportion to each other.

Example 17

If 5 mangoes cost ₦120, find the cost of 15 mangoes.

Solution

If 5 mangoes cost ₦120

then, 1 mango will cost $\frac{₦120}{5} = ₦24$

$\therefore$ 15 mangoes will cost, $15\times ₦24 = ₦360$

Example 18

If it takes 3 girls 30 minutes to clean a house, how long will it take 5 girls working at the same rate to do it?

Solution

If 3 girls clean a house in 30 minutes, then 1 girl will do it in $3 \times 30 = 90$ minutes

Thus, 5 girls can do the same in $\frac{90}{5} = 18$ minutes

2.4.2 RATES: This is the measure/amount of a quantity relative to a time unit. Examples are measurement of the rate of distance moved by a car with time; the rate of people doing something between a time interval; the number of phone calls made between a time interval.

Example 19

If a car covered a distance of 1000km in 2hrs. What is the rate at which the car travelled?

Solution

Let $x =$ distance travelled $= 1000km$

and $t =$ time of travel $= 2hrs$

Then, rate $= \frac{x}{t} = \frac{1000}{2} = 500$km/hr

Example 20

Given that two boys and five girls can cut a field in 4 hours. If the boys work at $\frac{5}{4}$th the rate which the girls work, how many boys only will be needed to cut the field in:

(a) 3 hours (b) 6 hours

Let x boys work at the $\frac{5}{4}$th rate of the five (5) girls

$$\therefore\ x = \frac{1}{5/4} \times 5 = 4$$

$\Rightarrow$ 4 boys are needed to do the same work that 5 girls would have done.
Thus, it will then take a total of two (2) boys + four(4) boys = Six (6) boys to cut the field in 4 hours.

(a) $\Rightarrow$ 6 boys can do the job in 4 hours
Then, 1 boy will do it in $6 \times 4 = 24hrs$
Now, let N_{boys} do it in 3 hrs, $= \frac{24}{3} = 8$ boys

(b) Similarly, for 6hrs, we have:
$N_{boys} = \frac{24}{6} = 4$ boys

2.5 FINANCIAL/COMMERCIAL ARITHMETIC

Financial or/and commercial mathematics simply means how we use rate, ratio, percentages and proportion in daily business transactions.

Terms used:

Cost Price: This is the price paid on a commodity by a person who wishes to sell it again if possible.
Selling Price: This is the price at which the article/commodity is finally sold.
Discount: This is the amount removed/subtracted from the price list of a commodity.
Allowance: This is the additional money paid to the salary of an individual.
Commission: This is the extra money paid for doing extra/overtime work.
Profit: This is when the selling price is greater than the cost price
Loss: This is when the selling price is less than the cost price; in other words, the cost price is greater than the selling price.
Overhead: This is the expenses added to the cost price in order to determine the total cost when calculating profit or assigning selling price.
Tax: This is an amount of money or charge against a citizen, person, property, company or activity used for the support of government.
Stock: This is the capital raised by a firm or corporation through the issue of shares entitling holders to an ownership interest called "equity."
Shares: This is defined as assets belonging to or contributed by an individual or groups or corporation.
Salary: This is the earning or payment given to somebody for his/her effort.

2.5.1 PROFIT AND LOSS: **Profit** simply means gain resulting from a transaction. While, **loss** means a depreciation in the value of a given quantity in a transaction.

Example 21

The annual profit of a transport business was divided between two partners X and Y in the ratio 3:5. If Y received ₦30,000 more than X, what is the total profit?

Solution

Let T represent the total profit

$\therefore$ X receives $= \frac{3}{3+5} \times T = \frac{3T}{8}$

and Y receives $= \frac{5}{3+5} \times T = \frac{5T}{8}$

But, $Y = X + 30,000$

$Y = X + 30,000$ because from the question Y receives 30,000 more than X.

$\Rightarrow \; Y = X + 30,000 = \frac{5T}{8}$

and X $+Y = T =$ total profit made.

$\therefore X + (X + 30,000) = T$

$\Rightarrow \; 2X + 30,000 = T$

$\therefore \; 2\left(\frac{3T}{8}\right) + 30,000 = T$

$30,000 = T - \frac{6T}{8} = \frac{(8-6)T}{8} = \frac{2T}{8}$

$T = \frac{30,000 \times 8}{2} = \frac{240,000}{2} =$ ₦120,000

Therefore, the total profit shared by X and Y is ₦120,000

Example 22

By selling 20 oranges for ₦297 a trader makes a profit of 10%. What is his percentage gain or loss if he sells the same oranges for ₦250?

Solution

10% profit $= \frac{Selling\ price - Cost\ price}{Cost\ price} = \frac{297 - Cost\ price}{Cost\ price}$

$\therefore \; \frac{10}{100} = \frac{297-C}{C} \quad \Rightarrow 10C + 100C = 29700$

Thus, solving for the cost price, C, we have $C =$ ₦270

Now, if he sells the 20 oranges for ₦ 250 that means he made a loss because it was sold below the cost price (i.e. ₦250 < ₦270)

$\therefore$ % Loss $= \frac{Cost\ price - Selling\ price}{Cost\ Price} \times 100\% = \frac{₦270 - ₦250}{₦270}$

$= \frac{₦20}{₦270} \times 100\% = 7.4\%$

2.5.2 **SIMPLE INTEREST**: This is the amount accumulated for a period of time (T) on an invested capital called the principal (P).

Mathematically,

$$\text{Simple interest}, I = \frac{Principal \times Rate \times Time}{100} = \frac{PRT}{100}$$

∴ the total Accumulated Amount, $A = I + P$

Where: I is the Interest amount; P is the Principal Amount (i.e., invested capital);
R is the Rate in percentage (%); T is the total Time of investment;
A is the total (overall Accumulated) amount.

Example 23

A man deposited ₦ 10,000 in a bank with a savings account of 5% interest rate per annum. Find: (a) the interest yielded after 3 years (b) the total amount accumulated over the years.

Solution

Principal = P = ₦10,000, Rate = R = 5% and Time = $T = 3yrs$

(a) $I = \frac{PRT}{100} = \frac{₦10,000 \times 5 \times 3}{100} = \frac{₦150,000}{100} = ₦1,500$

∴ the interest after 3 years is ₦1,500

(b) $A = I + P =$ ₦1,500 + ₦ 10,000 = ₦11,500

∴ this is the total amount accumulated over the time period.

Example 24

The rate of interest in a bank is increased from 10% to 12% per annum. How much extra interest will be paid on a deposit of ₦250,000 after 4 years?

Solution

% increase in rate = 12% − 10% = 2%

∴ $I = \frac{PRT}{100} = \frac{₦250,000 \times 2 \times 4}{100} = \frac{₦2,000,000}{100} = ₦20,000$

⇒ extra interest of ₦ 20,000 is paid on the deposit.

Example 25

Find the simple interest rate per annum at which ₦ 10,000 accumulates to ₦12,400 in 3 years.

Solution

P = ₦10,000; A = ₦12,400; and $T = 3yrs$

∴ $I = A - P$ = ₦12,400 − ₦10,000 = ₦2,400

Hence, $R = \frac{100\,I}{P\,T} = \frac{100 \times ₦2,400}{₦10,000 \times 3} = 8\%$

2.5.3 COMPOUND INTEREST: In compound interest, the interest earned in the first year of investment is added to the principal amount; then this new amount becomes the new principal amount used for finding the interest of the next year. This is repeated year after year till the end of the last year. Hence, the total interest is now called the "compound interest" for the given period of time. So, the process whereby the future value of a payment or series of payments is calculated when interest is earned on the original principal and the interest accumulated from prior periods is called "*compound interest.*"

We can do calculations on compound interest either by analytical method or by using the compound interest formula method.

Compound interest Formula method is given by:

$C = A - P \quad \therefore \quad C = P\left(1 + \frac{R}{100}\right)^t - P$ and $A = P\left(1 + \frac{R}{100}\right)^t$

Where: $A =$ *Total compounded amount*
$P =$ *Principal amount*
$C =$ *Compound interest*
$R =$ *Rate of interest*
$t =$ *Time*

Example 26

Using the (a) analytical method (b) formula method, find the compound interest on ₦5,000 for 3 years at 5% per annum. Also, find the amount at the end of the 3 years.

Solution

(a) Analytical method:

For the first year:

$Principal\ amount = P = ₦5{,}000$
$Rate = R = 5\%$

$$\therefore Interest = \frac{PRT}{100} = \frac{₦5{,}000 \times 5 \times 1}{100} = \frac{₦25{,}000}{100} = ₦250$$

$\Rightarrow I = ₦250$

For the second year:

$Principal\ amount = ₦5{,}000 + ₦250 = ₦5{,}250$

$$Interest = \frac{PRT}{100} = \frac{₦5250 \times 5 \times 1}{100} = \frac{₦26{,}250}{100} = ₦262.5$$

For the third year:

$Principal\ amount = ₦5{,}250 + ₦262.5 = ₦5{,}512.5$

$$Interest = \frac{PRT}{100} = \frac{₦5{,}512.5 \times 5 \times 1}{100} = \frac{₦27{,}562.5}{100} = ₦275.63$$

$\therefore$ The compound interest = C = ₦250 + ₦262.5 + ₦275.63

$\Rightarrow$ C = ₦788.13

And the total amount (A) at the end of the 3 years = $P + C$

$\Rightarrow A$ = ₦5,000 + ₦788.13 = ₦5,788.13

The compound interest here is the total calculated interest for the first, second and third year.

(b) Formula method:

$$C = A - P = P\left(1 + \frac{R}{100}\right)^t - P$$

where: $A = P\left(1 + \frac{R}{100}\right)^t = \textit{Total compounded amount}$

$P = \textit{principal amount}$ = ₦5,000

$R = 5\%$

$t = \textit{Time} = 3yrs$

$C = \textit{Compound Interest}$

$C = P\left(1 + \frac{R}{100}\right)^t - P$ = ₦5,000$\left(1 + \frac{5}{100}\right)^3$ − ₦5,000

= ₦5,000$(1 + 0.05)^3$ − ₦5,000

$\therefore C$ = ₦5,000$(1.05)^3$ − ₦5,000

$\therefore C$ = ₦5,000 $\times$ 1.157625 − ₦5,000

= ₦5,788.13 − ₦5,000 = ₦788.13

$\Rightarrow$ Total Amount after 3 years = $A = P\left(1 + \frac{R}{100}\right)^t$ = ₦5,000$\left(1 + \frac{5}{100}\right)^3$

= ₦5,000$(1.05)^3$ = ₦5,000 $\times$ 1.157625 = ₦5,788.13

Example 27

John wishes to keep some money in a savings deposit at 25% compound interest so that after 2 years he can buy a car for ₦250,000. How much does he need to deposit now?

Solution

Make P the subject formula in the equation, then substitute the respective values of A, R and t to find P.

$A = P\left(1 + \frac{R}{100}\right)^t$

where $A = \textit{Accumulated amount}$ = ₦250,000

$R = \textit{Rate} = 25\%$

$t = \textit{Time} = 2yrs$

$$\therefore P = \frac{A}{\left(1 + \frac{R}{100}\right)^t} = \frac{₦250{,}000}{\left(1 + \frac{25}{100}\right)^2} = \frac{₦250{,}000}{(1 + 0.25)^2}$$

$$\Rightarrow P = \frac{₦250{,}000}{(1.25) \times (1.25)} = \frac{₦250{,}000}{1.5625} = ₦160{,}000$$

$\therefore$ John needs to deposit ₦160,000 so that he can buy the car after 2 years.

Example 28

If Adamu deposits ₦15,000 in a bank which pays 2.4% interest compounded annually on a savings account. What will be his amount in the bank at the end of 1 year? 3 years? and 10 years?

Solution

Using the compound interest formula, we have: $A = P\left(1 + \frac{R}{100}\right)^t$

Thus, for the 1st year:

$P =$ ₦15,000, $R = 2.4\%$ and $t = 1$ year

$A = P\left(1 + \frac{R}{100}\right)^t =$ ₦15,000 $\left(1 + \frac{2.4}{100}\right)^1 =$ ₦15,000 $(1.24)^1$

$=$ ₦18,600

Thus, for the 3rd year:

$P =$ ₦15,000, $R = 2.4\%$, $t = 3$ years

$A = P\left(1 + \frac{R}{100}\right)^t =$ ₦15,000 $\left(1 + \frac{2.4}{100}\right)^3 =$ ₦15,000 $(1.24)^3$

$=$ ₦15,000 $\times 1.906624 =$ ₦28,599

Thus, for the 10th year:

$P =$ ₦15,000, $R = 2.4\%$, $t = 10$ years

$A = P\left(1 + \frac{R}{100}\right)^t =$ ₦15,000 $\left(1 + \frac{2.4}{100}\right)^{10} =$ ₦15,000 $(1.24)^{10}$

$=$ ₦15,000 $\times 8.59443 =$ ₦128,916

Example 29

Assuming that United Bank of Africa (UBA) pays interest of 6% on savings account which is compounded quarterly. If ₦50,000 is deposited today, how much will this yield at the end of 5 years?

Solution

The compound interest formula can be modified from $A = P\left(1 + \frac{R}{100}\right)^t$ to $A = P\left(1 + \frac{R}{100m}\right)^{tm}$ when we are dealing with interest that are not compounded annually.

Thus, $P =$ ₦50,000; $R = 6\%$; $t = 5$; and $m = 4$ (i.e., quarterly)

$A =$ ₦50,000 $\left(1 + \frac{6}{100\times4}\right)^{5\times4} =$ ₦50,000 $\left(1 + \frac{6}{400}\right)^{20} =$ ₦50,000 $(1.015)^{20}$

$=$ ₦50,000 $\times 1.3469 =$ ₦67,345

2.5.4 DISCOUNT, COMMISSION AND TAX

Example 30

A buyer paid ₦2,760 for an article on which he was given 8% discount, how much was the marked price on the article?

Solution

Discount = 8% and Amount paid = ₦2,760
Thus, let x be the amount tagged on the article.

$\therefore$ % paid = 100 − 8 = 92%

$$\Rightarrow \frac{92}{100} of\ x = ₦2{,}760 \quad \therefore\ x = \frac{100}{92} \times ₦2{,}760 = ₦3{,}000$$

Example 31

A land agent receives 5% commission for every buyer who buys a land from the land owner, calculate the commission for a land sold for ₦500,000.

Solution

% commission = 5%, Amount paid for land = ₦500,000

$$\Rightarrow \text{Amount of commission} = \frac{5}{100} of\ ₦500{,}000 = \frac{5}{100} \times ₦500{,}000 = ₦25{,}000$$

$\therefore$ Commission = ₦25,000

Example 32

A company gives 2% commission to its agent for a business transaction; assuming an agent receives ₦18,500, find:
(a)The total amount involved in the transaction (b)The net payment of the company.

Solution

% commission = 2%;
Amount of commission=₦18,500;
Net % to company = 100 −2 = 98%

(a) Let the total amount in the transaction be x;

$$\therefore 2\%\ of\ x = ₦18{,}500 \Rightarrow \frac{2}{100} \times x = ₦18{,}500$$

$$\therefore x = \frac{100}{2} \times ₦18{,}500 = ₦925{,}000$$

(b) Net payment to Company $= \frac{98}{2} \times ₦18{,}500$

$$= \frac{1{,}813{,}000}{2} = ₦906{,}500$$

OR

Net payment to company = total amount of transaction − commission
= ₦925,000 − ₦18,500 = ₦906,500

Example 33

A man bought a car for ₦2,000,000, and pays 5% tax rate every month for his car, how much does he pay?

Solution

Cost of car = ₦2,000,000; tax = 5%

Tax = ₦2,000,000 $\times \frac{5}{100}$ = ₦100,000

Hence, tax paid is ₦100,000 per month.

Example 34

A house is to be sold for thirty-five million Naira with payable tax of 7%, what is the actual cost of the house?

Solution

price of house = ₦35,000,000; tax = 7%

Tax = ₦35,000,000 $\times \frac{7}{100}$ = ₦2,450,000

Actual cost of the house = ₦35,000,000 + ₦2,450,000 = ₦37,450,000.

Example 35

The list price of a refrigerator is ₦23,000. If buyer is given a discount of 8% and pays a tax of 3%, find the actual cost of the refrigerator.

Solution

price of refrigerator = ₦23,000; discount = 8%; tax = 3%

Discount = ₦23,000 $\times \frac{8}{100}$ = ₦1,840

Sales price = list price − discount = ₦23,000 − ₦1,840 = ₦21,160

then, tax paid = ₦21,160 $\times \frac{3}{100}$ = ₦634.80 $\triangleq$ ₦635

therefore, actual cost of refrigerator = Sales price + tax
= ₦21,160 + ₦635 = ₦21,795

OBJECTIVE TEST 2

1. If $a : x = 1 : 2$ and $a^2 - x^2 = a + x$, find a.
   (A) −2 (B) −1 (C) 1 (D) 2

2. If ₦560.7 is shared in the ratio 7 : 2 : 1 what is the smallest share?
   (A) ₦392.49 (B) ₦56.7 (C) ₦118.4 (D) ₦56.07

3. If $a : b = 5 : 8$ and $x : y = 25 : 16$, evaluate $\frac{a}{x} : \frac{b}{y}$
   (A) 125 : 128 (B) 3 : 5 (C) 3 : 4 (D) 2 : 5

4. Silas deposited ₦8,000 in the bank at the rate of $12\frac{1}{2}\%$ simple interest. After some time, the total amount was one and half times the principal. For how many years was the money left in the bank?
   (A) 2 (B) 4 (C) 5 (D) 8

5. By selling 20 oranges for ₦135 a trader makes a profit of 8%. What is his percentage gain or loss if he sells the same 20 oranges for ₦110?
   (A) 12% gain (B) 10% gain (C) 12% loss (D) 15% gain

6. Chidinma deposited ₦15,000 in the bank. At the end of 5 years the simple interest on the principal was ₦5,500. At what rate per annum was the interest paid?
   (A) 11% (B) $7\frac{1}{3}\%$ (C) 5% (D) $3\frac{1}{3}\%$

7. What is the cost price of an article selling for ₦200 on which there has been a loss of 6% of the selling price?
   (A) ₦112 (B) ₦215 (C) ₦212 (D) ₦312

8. If the interest rate of a bank increases from 5.25% to 5.5% per annum. How much extra interest is paid on a deposit of ₦5,000 ?
   (A) ₦275 (B) ₦15 (C) ₦12.5 (D) ₦262.5

9. Find the compound interest on ₦3000 for a period of 3 years at the rate of 5% per annum.
   (A) ₦3,473 (B) ₦472.88 (C) ₦450 (D) ₦3,450

10. Calculate the total amount received after 5 years on a deposit of ₦120,000 if the interest rate is 10% compounded.
    (A) ₦ 1,932 (B) ₦19,326 (C) ₦73,261 (D) ₦193,261

Exercise 2b

1. A worker's present salary is ₦24,000 per annum. His annual increment is 10% of his basic salary. What would be his annual salary at the beginning of the third year?
2. Find the difference between the simple interest and compound interest on ₦10,000 for 3 years at 6% per annum.
3. A sum of money invested at 5% per annum simple interest amounts to ₦4,500 after 3 years. How long will it take the same sum to amount to ₦8,300 at 12.5% per annum simple interest?
4. By selling an article for ₦540 a man makes a profit of 8%, for how much should he have sold it in order to make a profit of 25%?
5. At what rate would a sum of ₦1,000 deposited for 5 years raise an interest of ₦75?
6. John bought an article for ₦50,000 and sold it to Kelechi at a loss of x %. Kelechi later sold the article to Alex at a profit of 40%. If Kelechi made a profit of ₦10,000, find the value of x.
7. A trader bought 12 Mangoes for a price of ₦50. She sold 4 of the mangoes at ₦7 each, then another 3 at ₦5 each and the remainder at ₦2 each; find her percentage gain or loss.
8. A book of dimension 20cm × 24cm was wrongly estimated by a student as 16cm× 30cm. What is the percentage error in the estimate of the perimeter of the book?
9. A circular disc of radius 4.1cm was measured as 4cm. What is the percentage error in the measurement of the area of the disc? Give your answer correct to one decimal place.
10. The correct length of a line segment is 1.25cm. If Femi incorrectly measured it as 1.15cm, find the percentage error in the measurement, correct to one decimal place.
11. A man is 5 years older than his wife. Four years ago, the ratio of their ages was 7: 6, find their present ages.

3 NUMBER BASES

A "base" is simply the fundamental standard that we use when performing operations on numbers. For instance, the human brain can simply solve problems easily using the base 10 standard while in electronics, digital and computer systems base like binary (base 2), Octenary (base 8) and Hexadecimal (base 16) are used when processing information. Therefore, it is necessary to understand these bases, their operations and how to interconvert from one base to another.

3.1 OPERATIONS ON BASES

3.1.1 **ADDITION OF BASES**: Addition in any base other than base 10 is the same technique that we use in base 10. It will be necessary to note that a base, for example, base 2 can only have digits 0 and 1; also base 3 can have digits 0, 1 and 2; and base 8 will have digits 0, 1, 2, 3, 4, 5, 6 and 7. Thus, if you count the number of digits in each case you will observe that it corresponds to the given base in view.

Example 1

(a) Add $75_8 + 523_8$ (b) Add $211_3 + 112_3$

Solution

(a) $75_8 + 523_8 \Rightarrow$

$$\begin{array}{r} 0\,7\,5 \\ +\,\underline{5\,2\,3} \\ 6\,2\,0_8 \end{array}$$

Add 5 + 3 = 10 (because we are in base 8, so, whenever we get to sum of 8 it is equivalent to 10); then write 0 and carry 1. Next, add 7 + 2 = 7 + 1+ 1 = 10 + 1 = 11, now add the 1 we are carrying to 11, i.e., 11 + 1= 12. So, we write 2 and carry 1. Finally, add the last digits, 5+0 =5; then add the carry, 1, to give 5+1 = 6.

(b) $211_3 + 112_3 \Rightarrow$

$$\begin{array}{r} 2\,1\,1 \\ +\,\underline{1\,1\,2} \\ 1\,1\,0\,0_3 \end{array}$$

Add 1 + 2 = 10 (because we are in base 3, so, whenever we get to sum of 3 it is equivalent to 10); then write 0 and carry 1. Next, add 1 +1 = 2, then add the 1 we are carrying to 2, i.e. 2 + 1= 10; then write 0 and carry 1. Next and finally, add the last digits, 2 + 1 = 10; then plus the carry, 1, to give 10 + 1 = 11

Example 2

(a) Add $1011_2 + 1101_2$ (b) Add $11101_2 + 111011_2$

Solution

(a) $1011_2 + 1101_2$

$$\Rightarrow \begin{array}{r} 1011 \\ +\,1101 \\ \hline 11000_2 \end{array}$$

Add 1 + 1= 10 (because we are in base 2, so, whenever we get to sum of 2 it is equivalent to 10); then write 0 and carry 1. Then add 1 + 0 = 1, now add the 1 we are carrying to 1, i.e., 1 + 1= 10; so we write 0 and carry 1. Now, add 0 + 1 = 1 then plus the carry, 1, i.e., 1 + 1 =10 then write 0 and carry 1. Finally, add the last digits 1 + 1 = 10; then plus the carry, 1, to give 10 + 1 = 11.

(b) $11101_2 + 111011_2$

$$\Rightarrow \begin{array}{r} 011101 \\ +111011 \\ \hline 1011000_2 \end{array}$$

Notice that 11101_2 is five digits while 111011_2 is six digits, so we put 0 preceding 11101_2, i.e. 011101_2 so as to make it six digits. Then follow and use the same concept as in (a) above for the addition.

3.1.2 SUBTRACTION OF BASES

The same approach as in addition is also applied in subtraction.

Example 3

(a) $2312_4 - 112_4$ (b) $826_9 - 685_9$

Solution

(a) $2312_4 - 112_4 \Rightarrow \begin{array}{r} 2312 \\ -\,0112 \\ \hline 2200_4 \end{array}$

2−2=0, 1−1=0, 3−1=2, and 2−0 = 2. Final answer is 2200_4

(b) $826_9 - 685_9 \Rightarrow \begin{array}{r} 826 \\ -\,685 \\ \hline 131_9 \end{array}$

6−5=1, then 2−8 = is illogical here, so we borrow 1 from the next digit (this is because we are in base 9, thus borrowing 1 from the next digit means taking 9). Now, 9+2 =11. Then 11−8 = 3. Finally, we have the last digit 8 reduced to 7 because we borrowed a digit from it. So, 7−6 =1. Therefore, the final answer = 131_9

Example 4

(a) $1111_2 - 1011_2$ (b) $11011_2 - 10100_2$

Solution

(a) $1111_2 - 1011_2$

$$\begin{array}{r} \Rightarrow 1111 \\ -1011 \\ \hline 0100_2 \end{array}$$

(b) $11011_2 - 10100_2$

$$\begin{array}{r} \Rightarrow 11011 \\ -10100 \\ \hline 00111_2 \end{array}$$

In solution (b) we have: 1−0 =1, 1−0=1, 0−1= illogical here, so we borrow 1 from the next digit (since we are in base 2, borrowing 1 means taking 2) and add to digit 0, i.e., 2+0 =2, then 2−1=1. Now, the digit we borrowed 1 is left with nothing. So, 0−0=0. Finally, 1−1= 0. Thus, the answer becomes 00111_2

3.1.3 MULTIPLICATION

Example 5

Simplify $614_8 \times 45_8$

Solution

$$\begin{array}{r} 614 \\ \times\ 45 \\ \hline 3674 \\ 3060 \\ \hline 34474_8 \\ \hline \end{array}$$

<u>Step 1</u>: Note that in base 10 we have 5× 4 = 20 = 8+8+4. So, changing to base 8 we represent each 8 as "10"
⇒ 8+8+4 =10+10+4=24; write 4 and carry 2. Then 5 × 1 = 5 plus the carry, 2, ⇒ 5 + 2 =7. Next, 5× 6 = 30 = 8 + 8 + 8 +6 then in base 8 it is 10+10+10+6 = 36.

<u>Step 2</u>: In base 10 we have 4 × 4 =16 = 8+8. So, in base 8 it is equivalent to 10+10 =20. Write 0 and carry 2; then 4×1=4 plus the carry, 2, ⇒ 4 + 2 = 6. Next, 4×6=24=8+8+8. Its equivalent in base 8 is 10+10+10 =30

<u>Step 3</u>: Add step 1 and step 2 in base 8.

Example 6

Simplify (a) $111_2 \times 11_2$ (b) $1011_2 \times 1010_2$ (c) $101.11_2 \times 11.01_2$

Solution

(a)

$$\begin{array}{r} 111 \\ \times\ 11 \\ \hline 111 \\ +111 \\ \hline 10101_2 \end{array}$$

(b)

$$\begin{array}{r} 1011 \\ \times\ 1010 \\ \hline 0000 \\ +1011 \\ +0000 \\ +1011 \\ \hline 1101110_2 \end{array}$$

(c)

$$\begin{array}{r} 101.11 \\ \times\ 11.01 \\ \hline 10111 \\ +00000 \\ +10111 \\ +10111 \\ \hline 10010.1011_2 \end{array}$$

In solution (c):

<u>Step 1</u>: Multiply 101.11_2 and 11.01_2 without considering the decimal points.

<u>Step 2</u>: After arriving at the final answer, 10010.1011_2 count back four digits from the left and then put the decimal point; this is because the total number of digits after the decimal point is four, i.e., .11 and .01

Example 7

(a) $302_4 \times 21_4$ (b) $536.12_7 \times 32.45_7$

Solution

(a)

$$\begin{array}{r} 302 \\ \times\ 21 \\ \hline 302 \\ +1110 \\ \hline 12002_4 \end{array}$$

<u>Step 1</u>: 2×1=2, 0 × 1= 0, 3 × 1 =3

<u>Step 2</u>: 2 ×2 =10 (because we are in base 4, so 4≡ 10)

<u>Step 3</u>: Add step 1 and Step 2.

(b)
```
     5 3 6 . 1 2₇
   × 3 2 . 4 5₇
   ------------
     3 6 5 2 6 3
   3 1 1 3 5 1
 1 4 0 5 2 4
2 2 4 4 3 6
--------------
2 4 5 3 3 . 3 5 0 3₇
```

Step 1: Multiply 536.12_7 and 32.45_7 without considering the decimal points.
Step 2: After arriving at the final answer, 245333503_7, count back four digits from the right and then put the decimal point; this is because the total number of digits after the decimal point is four, i.e., .1 2 and .4 5

3.1.4 DIVISION

Division in other bases except in our familiar base 10 seems more technical, though it uses the same procedure of "*long division*" as in base 10.

Example 8

Evaluate the following:
(a) $10111_2 \div 110_2$ (b) $11011_2 \div 101_2$ (c) $11111_2 \div 11_2$

Solution

(a)
```
          0 1 1
     ----------
110 |  1 0 1 1 1
      -1 1 0
      ---------
       1 0 1 1
      - 1 1 0
      ---------
       1 0 1 1
      -  1 1 0
      ---------
         1 0 1
```

The dividend is 1 0 1 1 1 while the divisor is 1 1 0
Step 1: Subtract the divisor (1 1 0) from the leftmost three digits (1 0 1) of the dividend; if it won't go, then write 0 on the top bar.
Step 2: Introduce down the fourth digit of the dividend to become 1 0 1 1.
Step 3: Now, shift the divisor just one step to the right; then subtract the divisor from 1 0 1 1 and write 1 on the top bar.
Step 4: Continue this process until the digits of the dividend are exhausted.

∴ the quotient = 011_2, with a remainder of 101_2

(b)
```
          1 0 1
     ----------
101 |  1 1 0 1 1
      -1 0 1
      ---------
         0 1 1
      -  1 0 1
      ---------
         0 1 1 1
      -    1 0 1
      ---------
             1 0
```

∴ the quotient = 101_2, with a remainder of 10_2

(c)

```
        1010
    ┌───────
    │ 11111
    │-11↓||
    │  01 ||
    │- 11↓|
 11 │  011|
    │-  11↓
    │   001
    │-   11
    │    01
```

∴ the quotient $= 1010_2$, with a remainder of 01_2

Example 9

Divide (a) 1452_7 by 23_7 (b) 2134_5 by 12_5

Solution

(a) 1452_7 by 23_7

```
        45
    ┌──────
 23 │ 1452
    │-125↓
    │  202
    │ -151
    │   21
```

The dividend is 1 4 5 2 while the divisor is 23 in base 7

Step 1: By inspection, multiply the divisor by a certain number to get a value closer to the first three digits of the dividend; this number is 4 , write it on the top bar; this is because $23 * 4 = 125$ in base 7. Note that if you had used 5 then you would have gotten 151 in base 7 which exceeds 1 4 5

Step 2: Subtract 125 from 1 4 5 to get 20.

Step 3: introduce down the fourth digit of the dividend so that it becomes 2 0 2.

Step 4: Now repeat step 1, this time by inspection $23 * 5 = 151$ in base 7 write 5 on the top bar. Then shift 151 just one step to the right; then subtract it from 2 0 2 to get 0 2 1

Step 5: Since 0 2 1 is less than the divisor, 2 3 , then stop.

∴ the quotient $= 45_7$, with a remainder of 21_7

Alternative method:

1452_7 by 23_7

$\Rightarrow 1452_7 = 576_{10}$

and $23_7 = 17_{10}$

$$\frac{576}{17} = 33\frac{15}{17}$$

Alternative method used in evaluating base division is to first convert the dividend and divisor from their respective bases to base 10, after which they are then divided in base 10. Finally, the quotient is translated back to the required base (i.e., base 7 in this case).

```
33   ⇒  7 | 33 |
        7 |  4 | R 5 ↑
          |  0 | R 4 |    = 45₇
```

and $\left(\frac{15}{17}\right)_{10} = \left(\frac{21}{23}\right)_7$

$\therefore$ quotient $= 45_7$

and remainder $= 21_7$

Convert the numerator, i.e., 15_{10} to base 7

$$\begin{array}{r|l} 7 & 15 \\ & 2 \text{ R } 1 \uparrow \\ & 0 \text{ R } 2 \end{array} \qquad 15_{10} = 21_7$$

Similarly, convert the denominator, i.e., 17_{10} to base 7

$$\begin{array}{r|l} 7 & 17 \\ & 2 \text{ R } 3 \uparrow \\ & 0 \text{ R } 2 \end{array} \qquad 17_{10} = 23_7$$

(b)

$$\begin{array}{r|l} & \;\;132 \\ \hline 12 & 2134 \\ & -12\downarrow \\ & \;\;\;43 \\ & -41\downarrow \\ & \;\;\;\;24 \\ & -24 \\ & \;\;\;\;00 \end{array}$$

$\therefore$ the quotient $= 132_5$

3.2 CONVERSION FROM OTHER BASES TO DENARY

3.2.1 CONVERSION FROM BASE 2 (BINARY) TO BASE 10

Example 10

Convert 10111_2 to base 10

Solution

$1^4 0^3 1^2 1^1 1^0$

$\Rightarrow 1\times 2^4 + 0\times 2^3 + 1\times 2^2 + 1\times 2^1 + 1\times 2^0 = 1\times 16 + 0\times 8 + 1\times 4 + 1\times 2 + 1\times 1$

$\therefore\ 16 + 0 + 4 + 2 + 1 = 23_{10}$

Example 11

Convert 101101_2 to base 10

Solution

$1^5 0^4 1^3 1^2 0^1 1^0$

$1\times 2^5 + 0\times 2^4 + 1\times 2^3 + 1\times 2^2 + 0\times 2^1 + 1\times 2^0$

$= (1\times 32) + (0\times 16) + (1\times 8) + (1\times 4) + (0\times 2) + (1\times 1)$

$\Rightarrow\ 32 + 0 + 8 + 4 + 0 + 1 = 45_{10}$

3.2.1 CONVERSION FROM OTHER BASES TO BASE 10

Example 12

Convert the following bases to base 10

(a) 132_4 (b) 156_{16}

Solution

(a) 132_4

$1^2 3^1 2^0$

$\Rightarrow 1 \times 4^2 + 3 \times 4^1 + 2 \times 4^0 = 1 \times 16 + 3 \times 4 + 2 \times 1 = 16 + 12 + 2$

$\therefore 16 + 12 + 2 = 30$

(b) 156_{16}

$1^2 5^1 6^0$

$\Rightarrow 1 \times 16^2 + 5 \times 16^1 + 6 \times 16^0 = 1 \times 256 + 5 \times 16 + 6 \times 1 = 256 + 80 + 6$

$\therefore 256 + 80 + 6 = 342$

Example 13

Convert each base to base 10

(a) 1256_8 (b) 378_{12} (c) 101.11_2 (d) 321.23_4

Solution

(a) 1256_8

$1^3 2^2 5^1 6^0$

$\Rightarrow 1 \times 8^3 + 2 \times 8^2 + 5 \times 8^1 + 6 \times 8^0$

$\therefore 1 \times 512 + 2 \times 64 + 5 \times 8 + 6 \times 1 = 512 + 128 + 40 + 6 = 686$

(b) 378_{12}

$3^2 7^1 8^0$

$\Rightarrow 3 \times 12^2 + 7 \times 12^1 + 8 \times 12^0 = 3 \times 144 + 7 \times 12 + 8 \times 1$

$\therefore 3 \times 144 + 7 \times 12 + 8 \times 1 = 432 + 84 + 8 = 524$

(c) 101.11_2

$1^2 0^1 1^0 . 1^{-1} 1^{-2}$

$\therefore 1 \times 2^2 + 0 \times 2^1 + 1 \times 2^0 + 1 \times 2^{-1} + 1 \times 2^{-2}$

$= 1 \times 4 + 0 \times 2 + 1 \times 1 + 1 \times \frac{1}{2} + 1 \times \frac{1}{4} = 4 + 0 + 1 + \frac{1}{2} + \frac{1}{4}$

$= 5 + \frac{3}{4} = 5\frac{3}{4}$

(d) 321.23_4

$3^2 2^1 1^0 . 2^{-1} 3^{-2}$

$\Rightarrow 3 \times 4^2 + 2 \times 4^1 + 1 \times 4^0 + 2 \times 4^{-1} + 3 \times 4^{-2}$

$= 3 \times 16 + 2 \times 4 + 1 \times 1 + 2 \times \frac{1}{4} + 3 \times \frac{1}{16}$

$= 48 + 8 + 1 + \frac{1}{2} + \frac{3}{16} = 57 + \frac{11}{16} = 57\frac{11}{16}$

Example 14

Convert the following to base 10:

(a) $5AC_{16}$ (b) $8AEC_{16}$ (c) 615.432_8 (d) $7tet_{12}$ (e) $3AD.AC_{14}$

Solution

(a) $5AC_{16}$

$5^2A^1C^0$

$\Rightarrow 5 \times 16^2 + 10 \times 16^1 + 12 \times 16^0 = 5 \times 256 + 10 \times 16 + 12 \times 1$

$= 1280 + 160 + 12 = 1452$

(b) $8AEC_{16}$

$8^3A^2E^1C^0$

$\Rightarrow 8 \times 16^3 + 10 \times 16^2 + 14 \times 16^1 + 12 \times 16^0$

$= 8 \times 4096 + 10 \times 256 + 14 \times 16 + 12 \times 1$

$= 32768 + 2560 + 224 + 12 = 35564$

(c) 615.432_8

$6^2 1^1 5^0 . 4^{-1} 3^{-2} 2^{-3}$

$\Rightarrow 6 \times 8^2 + 1 \times 8^1 + 5 \times 8^0 + 4 \times 8^{-1} + 3 \times 8^{-2} + 2 \times 8^{-3}$

$= 6 \times 64 + 1 \times 8 + 5 \times 1 + 4 \times \frac{1}{8} + 3 \times \frac{1}{64} + 2 \times \frac{1}{512}$

$= 384 + 8 + 5 + \frac{1}{2} + \frac{3}{64} + \frac{1}{256}$

$= 397 + \frac{141}{256} = 397\frac{141}{256}$

(d) $7tet_{12}$

$7^3t^2e^1t^0$

$\Rightarrow 7 \times 12^3 + 10 \times 12^2 + 11 \times 12^1 + 10 \times 12^0$

$= 7 \times 1728 + 10 \times 144 + 11 \times 12 + 10 \times 1$

$= 12096 + 1440 + 132 + 10 = 13678$

(e) $3AD.AC_{14}$

$\Rightarrow 3 \times 14^2 + 10 \times 14^1 + 13 \times 14^0 + 10 \times 14^{-1} + 12 \times 14^{-2}$

$= 3 \times 196 + 10 \times 14 + 13 \times 1 + 10 \times \frac{1}{14} + 12 \times \frac{1}{196}$

$= 588 + 140 + 13 + \frac{5}{7} + \frac{3}{49}$

$= 741 + \frac{38}{49} = 741\frac{38}{49}$

3.3 CONVERSION FROM BASE TEN TO OTHER BASES

Example 15

Convert the following from base 10 to binary:

(a) 53 (b) 42 (c) 32 (d) 65

Solution

(a) 53 ⇒

2	53	
2	26	R 1 ↑
2	13	R 0
2	6	R 1
2	3	R 0
2	1	R 1
	-	R 1

⇒ 110101_2

Divide the decimal number by the base we are converting it to. Repeat this division and keep writing down the remainder (i.e., **R**) for every division until the number becomes less than the base. Finally, follow the upward arrow direction and write down the entire remainder in that order.

(b) 42 ⇒

2	42	
2	21	R 0 ↑
2	10	R 1
2	5	R 0
2	2	R 1
2	1	R 0
	-	R 1

⇒ 101010_2

(c) 32 ⇒

2	32	
2	16	R 0 ↑
2	8	R 0
2	4	R 0
2	2	R 0
2	1	R 0
	-	R 1

⇒ 100000_2

(d) 65 ⇒

2	65	
2	32	R 1 ↑
2	16	R 0
2	8	R 0
2	4	R 0
2	2	R 0
2	1	R 0
	-	R 1

⇒ 1000001_2

Example 16

Convert the denary figures to their respective bases as stated:
(a) 156 to base 3 (b) 74 to base 5 (c) 67 to base 8 (d) 100 to base 9

Solution

(a) 156 ⇒

3	156	
3	52	R 0 ↑
3	17	R 1
3	5	R 2
3	1	R 2
	-	R 1

$\Rightarrow 12210_3$

(b) 74 ⇒

5	74	
5	14	R 4 ↑
5	2	R 4
	-	R 2

$\Rightarrow 244_5$

(c) 67 ⇒

8	67	
8	8	R 3 ↑
8	1	R 0
	-	R 1

$\Rightarrow 103_8$

(d) 100 ⇒

9	100	
9	10	R 1 ↑
9	1	R 1
	-	R 1

$\Rightarrow 111_9$

Example 17

Convert the denary numbers to the following bases:
(a) 12.5 to binary (b) 234.75 to octal (c) 94.2 to base 12 (d) 45.725 to base 3

Solution

(a) $12.5 \Rightarrow 12 + 0.5$

step 1:

2	12	
2	6	R 0 ↑
2	3	R 0
2	1	R 1
	-	R 1

$\Rightarrow 1100$

> The <u>Step 1</u> value is for the whole number part (i.e., **12**). Whereas, <u>step 2</u> is the fraction or decimal part (i.e., **0.5**) of the number. Then add step **1** and **2** to get the final answer.

step 2: $0.5 \Rightarrow 0.5 \times 2 = 1.0 \Rightarrow 1$

Final answer $= 1100 + 0.1 = 1101.10_2$

(b) $234.75 \Rightarrow 234 + 0.75$

step 1:

8	234	
8	29	R 2 ↑
8	3	R 5
	-	R 3

step 2: $0.75 \Rightarrow 0.75 \times 8 = 6.00 \Rightarrow 6$

Final answer $= 352 + 0.6 = 352.6_8$

(c) $94.2 \Rightarrow 94 + 0.20$

step 1:

12	94	
12	7	R 10 = t
	-	R 7 ↗

$\Rightarrow 7t$

step 2: $0.20 \Rightarrow 0.2\,0$

$$\begin{array}{r} \times\ 12 \\ \hline 2.40 \end{array} \Rightarrow 2$$

next: $$\begin{array}{r} 0.40 \\ \times\ 12 \\ \hline 4.80 \end{array} \Rightarrow 4$$

next: $$\begin{array}{r} 0.80 \\ \times\ 12 \\ \hline 9.60 \end{array} \Rightarrow 9$$

next: $$\begin{array}{r} 0.60 \\ \times\ 12 \\ \hline 7.20 \end{array} \Rightarrow 7$$

From **step 2** in the last value for the fraction part (i.e., **7.2=7+0.2**) it can be seen that **0.2** is a repetition of the original fraction part of the question. If this recurs, then stop because the values will keep recurring as **249724972497...** and this recurrence can be indicated by the "**bar**" notation on top of the digits.

Final answer $= 7t + 0.2497 = 7t.\overline{2497}_{12}$

(d) $45.725 \Rightarrow 45 + 0.725$

step 1:

$$\begin{array}{r|r|l} 3 & 45 & \\ 3 & 15 & R\,0 \\ 3 & 5 & R\,0 \\ 3 & 1 & R\,2 \\ & - & R\,1 \end{array} \Rightarrow 1200_3$$

step 2: $0.725 \Rightarrow$ $$\begin{array}{r} 0.725 \\ \times\ 3 \\ \hline 2.275 \end{array} \Rightarrow 2$$

next: $$\begin{array}{r} 0.275 \\ \times\ 3 \\ \hline 0.825 \end{array} \Rightarrow 0$$

next: $$\begin{array}{r} 0.825 \\ \times\ 3 \\ \hline 2.475 \end{array} \Rightarrow 2$$

Multiply the fraction by the base **3**, then write out the whole number. Then use the next fraction (i.e. the decimal part) and multiply by the base; then write out the whole number part. Repeat until all values become **0.000**. If peradventure we did not get **0.000** and the original fraction repeats itself again then stop, because the values will keep recurring till infinity.

$$\begin{array}{r} \text{next: } 0.475 \\ \times \quad 3 \\ \hline 1.425 \end{array} \quad \Rightarrow 1$$

$$\begin{array}{r} \text{next: } 0.425 \\ \times \quad 3 \\ \hline 1.275 \end{array} \quad \Rightarrow 1$$

$$\begin{array}{r} \text{next: } 0.275 \\ \times \quad 3 \\ \hline 0.725 \end{array} \quad \Rightarrow 0$$

$\therefore$ Final answer = $1200 + 0.202110 = 1200.202110_3$

3.4 CONVERSION FROM ONE BASE TO ANOTHER BASE

Converting a base that is not in base 10 to another base requires that we first change the given base to base 10, then convert it from base 10 to the specified base.

Example 18

Change the following bases to base to base 7.

(a) 432_5 (b) 2015_6

Solution

(a) 432_5 is first changed to base 10

$$\Rightarrow 4 \times 5^2 + 3 \times 5^1 + 2 \times 5^0 = 4 \times 25 + 3 \times 5 + 2 \times 1$$

$$= 100 + 15 + 2 = 117_{10}$$

Now change to base 7

$$\begin{array}{r|r|l} 7 & 117 & \\ 7 & 16 & R\,5 \uparrow \\ 7 & 2 & R\,2 \\ & 0 & R\,2 \end{array} \Rightarrow 225_7$$

(b) 2015_6 is first changed to base 10

$$\Rightarrow 2 \times 6^3 + 0 \times 6^2 + 1 \times 6^1 + 5 \times 6^0 = 2 \times 216 + 0 \times 36 + 1 \times 6 + 5 \times 1$$
$$= 432 + 0 + 6 + 5 = 443_{10}$$

Now change to 443_{10} base 7

$$\begin{array}{r|r|l} 7 & 443 & \\ 7 & 63 & R\,2 \\ 7 & 9 & R\,0 \\ 7 & 1 & R\,2 \\ & 0 & R\,1 \end{array} \uparrow \Rightarrow 1202_7$$

Example 19

Convert 1101_2 to base 3

Solution

$1101_2 \Rightarrow 1\times2^3+1\times2^2+0\times2^1+1\times2^0 = 1\times8+1\times4+1\times1$

$= 8+4+1 = 13_{10}$

$$\begin{array}{r|r|l} 3 & 13 & \\ 3 & 4 & R\,1 \\ 3 & 1 & R\,1 \\ & 0 & R\,1 \end{array} \uparrow \Rightarrow 111_3$$

Example 20

Express 201_3 to octal.

Solution

$201_3 = (2 \times 3^2) + (0\times3^1) + (1\times3^0) = 2\times9 + 0\times3 + 1\times1$

$= 18+0+1 = 19_{10}$

$$\begin{array}{r|r|l} 8 & 19 & \\ 8 & 2 & R\,3 \\ & 0 & R\,2 \end{array} \uparrow \Rightarrow 23_8$$

Example 21

Express the following to base 5 (a) $F3FB_{hexadecimal}$ (b) 476_8

Solution

(a) $F3FB_{hexadecimal} = 15 \times 16^3 + 3\times 16^2 + 15\times 16^1 + 11\times 16^0$

$= 15 \times 4096 + 3 \times 256 + 15\times 16 + 11\times 1$

$= 61440 + 768 + 240 + 11 = 62459_{10}$

Then,

5	62459	
5	12491	R 4
5	2498	R 1
5	499	R 3
5	99	R 3
5	19	R 4
5	3	R 4
	0	R 3

$\Rightarrow 3443314_5$

(b) $476_8 \Rightarrow (4 \times 8^2) + (7 \times 8^1) + (6 \times 8^0) = (4 \times 64) + (7 \times 8) + (6\times 1)$

$= 256 + 56 + 6 = 318_{10}$

5	318	
5	63	R 3
5	12	R 3
5	2	R 2
	0	R 2

$\Rightarrow 2233_5$

3.5 OTHER WORKED EXAMPLES ON BASES

Example 22

Solve for X if $1011_2 + X_7 = 25_{10}$

Solution

$1011_2 = (1\times 2^3) + (0\times 2^2) + (1\times 2^1) + (1\times 2^0) = 8+0+2+1 = 11_{10}$

$\Rightarrow 11 + X_7 = 25$

$X_7 = 25 - 11 = 14_{10}$

Now convert from $X_7 = 14_{10}$ from base 10 to base 7

7	14		
7	2	R 0 ↑	
7	0	R 2	$\Rightarrow 20_7$

$\therefore X = 20$

Example 23

Find n if $34_n = 10011_2$

Solution

$10011_2 = (1\times 2^4) + (0\times 2^3) + (0\times 2^2) + (1\times 2^1) + (1\times 2^0)$
$= 16 + 0 + 0 + 2 + 1 = 19_{10}$
$\Rightarrow 34_n = 19$
But, $34_n = (3\times n^1) + (4\times n^0) = 3n + 4$
$\therefore 34_n = 19 \quad \Rightarrow 3n + 4 = 19$
$3n = 19 - 4 = 15$
$n = \frac{15}{3} = 5$
$\therefore \quad n = 5$

Example 24

$(X_5)^2 = 100100_2$, find x.

Solution

$(X_5)^2 = 100100_2 = 1\times2^5+0\times2^4+0\times2^3+1\times2^2+0\times2^1+0\times2^0$
$= 32 + 0 + 0 + 4 + 0 + 0 = 36_{10}$
$\Rightarrow (X_5)^2 = 36$
$\therefore X_5 = \sqrt{36} = 6$
$X_5 = 6$

Now converting 6 from base 10 to base 5, we have:

5	6	
5	1	R 1 ↑
	0	R 1

$\Rightarrow X = 11$

OBJECTIVE TEST 3

1. Express 29 from base 10 to base 2
   (A) 11101_2 (B) 10111_2 (C) 11011_2 (D) 11001_2

2. Convert $4AD_{16}$ to base10
   (A) 9711 (B) 1791 (C) 1197 (D) 1198

3. Convert 110.11_2 to base 10
   (A) $6\frac{1}{2}$ (B) $6\frac{3}{4}$ (C) $6\frac{2}{5}$ (D) $6\frac{3}{5}$

4. $y_3 = 100100_2$, find y
   (A) 0101 (B) 1100 (C) 1011 (D)1001

5. Convert 268_{10} to base 16
   (A) 1012_{16} (B) $10C_{16}$ (C) $C01_{16}$ (D) 1201_{16}

6. Evaluate $225_6 + 234_6$
   (A) 305_6 (B) 350_6 (C) 503_6 (D) 530_6

7. Simplify $163_7 \times 25_7$
   (A) 5131_7 (B) 5064_7 (C) 5104_7 (D) 1315_7

8. Evaluate $212_3 - 121_3$
   (A) 021_3 (B) 210_3 (C) 102_3 (D) 120_3

9. If $26_x = 103_5$, find x
   (A) 6 (B) 11 (C) 8 (D) 12

10. Find the binary equivalent of the hexadecimal number 2ED3
    (A) 10111011010011_2 (B) 10111000010011_2
    (C) 10011011010111_2 (D) 11001011011101_2

Exercise 3

1. Add each of the following bases:
(a) $75_9 + 346_9$ (b) $231_5 + 103_5$ (c) $1011_2 + 1111_2$
(d) $1101101_2 + 1110011_2$ (e) $4052_8 + 2106_8$ (f) $131_5 + 102_5$

2. Subtract each of the following bases:
(a) $11_2 - 10_2$ (b) $111_2 - 110_2$ (c) $11011_2 - 10011_2$ (d) $331_4 - 301_4$
(e) $560_8 - 430_8$ (f) $21345_7 - 11423_7$ (g) $437_8 - 326_8$ (h) $ABC_{16} - 626_{16}$
(i) $508AB_{16} - 450DE$ (j) $FAB2B_{16} - CADA5_{16}$

3. Multiply each of the following bases:
(a) $101_2 \times 11_2$ (b) $1011_2 \times 1011_2$ (c) $11101_2 \times 1011_2$
(d) $1011_2 \times 10.11_2$ (e) $11.101_2 \times 10_2$ (f) $11.01_2 \times 11.1_2$
(g) $1010.11_2 \times 10.1_2$ (h) $23.12_4 \times 1.3_4$
(i) $534.2_6 \times 25.3_6$ (j) $63.25_8 \times 15.4_8$

4. Divide each of the following bases:
(a) $1010_2 \div 10_2$ (b) $1001_2 \div 11_2$ (c) $110111_2 \div 101_2$
(d) $4236_7 \div 25_7$ (e) $2145_8 \div 12_5$

5. Convert each of the following binary digits to denary:
(a) 101_2 (b) 1101_2 (c) 111011_2 (d) 1110_2 (e) 11100101101_2
(f) 11.011_2 (g) 10.11_2 (h) 100.111_2 (i) 1101.11_2

6. Convert the octenary bases to denary:
(a) 54_8 (b) 723_8 (c) 604526_8 (d) 35.27_8 (e) 103.26_8 (f) 4706.25_8

7. Convert each of the following bases to denary:
(a) $AB54_{Hex}$ (b) 2012_3 (c) 4563_8 (d) 54063_9 (e) 2301_5

8. Change the following denary bases to binary digits:
(a)15 (b) 0.52 (c) 72 (d) 105 (e) 729 (f) 512 (g) 57
(h) 106 (i) 0.4 (j) $16\frac{2}{3}$

9. Convert the hexadecimal digits to denary:
(a) 548_{16} (b) ABA_{16} (c) $758E_{16}$ (d) $EF3_{16}$

10. Convert each base to the required base as stated:
(a) 1011_2 to base 3 (b) 305_7 to base 2 (c) ACF_{16} to octenary
(d) 10111011_2 to hexadecimal (e) 56.32_{10} to base 6 (f) 3201_5 to base 9
(g) 10231_7 to base 4 (h) $tete_{12}$ to base 10 (i) 849_{10} to base 12

11. Solve for X in each of the following:
(a) $203_5 + X_3 = 58_{10}$ (b) $X_2 - 24_5 = 13_7$ (c) $(X_3)^2 + 101_2 = 25_8$ (d) $\frac{X_{16}}{32_5} = 1101_2$

4 INDICES AND LOGARITHM

4.1 LAWS OF INDICES

1. $0^n = 0$ (for $n \neq 0$)
2. $a^0 = 1$
3. $a^m \times a^n = a^{m+n}$ (Equal Base law)
4. $a^m \div a^n = a^{m-n}$ (Equal Base law)
5. $a^m \times b^m = (ab)^m$ (Equal indices law)
6. $\frac{a^m}{b^m} = \left(\frac{a}{b}\right)^m$ (Equal Indices law)
7. $a^{\frac{1}{m}} = \sqrt[m]{a}$ and $a^{n/m} = \sqrt[m]{a^n}$
8. $(a^m)^n = a^{m \times n} = a^{mn}$ (Power Law)
9. $a^{-m} = \frac{1}{a^m}$

Example 1

Evaluate: (a) $2^3 \times 4^2$ (b) $125^{-\frac{1}{3}}$ (c) $16^{\frac{1}{2}} \times 25^{-\frac{1}{2}}$

Solution

(a) $2^3 \times 4^2 = 2^3 \times (2^2)^2 = 2^3 \times 2^4$ from law $a^m \times a^n = a^{m+n}$

$\Rightarrow 2^3 \times 2^4 = 2^{3+4} = 2^7 = 2 \times 2 \times 2 \times 2 \times 2 \times 2 \times 2 = 128$

(b) $125^{-\frac{1}{3}} = \frac{1}{125^{1/3}}$ from law $a^{-m} = \frac{1}{a^m}$

$\Rightarrow \frac{1}{125^{1/3}} = \frac{1}{\sqrt[3]{125}} = \frac{1}{5}$

(c) $16^{\frac{1}{2}} \times 25^{-\frac{1}{2}} = \sqrt{16} \times \frac{1}{25^{\frac{1}{2}}} = 4 \times \frac{1}{\sqrt{25}}$

$= 4 \times \frac{1}{5} = \frac{4}{5}$

Example 2

Simplify:

(a) $y^3 \times x^3 \div y^2 \times x^2$ (b) $7^{3n} \div 7^{2n}$

Solution

(a) $y^3 \times x^3 \div y^2 \times x^2 = \frac{y^3 \times x^3}{y^2 \times x^2} = y^{3-2} \times x^{3-2}$

$= y^1 \times x^1 = xy$

(b) $7^{3n} \div 7^{2n} = 7^{3n-2n} = 7^n$

Example 3

Solve the following equations:

(a) $2^x = 4$ (b) $3^{2x} = 27$ (c) $3^{2x} = 81^{\frac{1}{4}}$

Solution

(a) $2^x = 4$

$\Rightarrow 2^x = 2^2$

$\therefore x = 2$

(b) $3^{2x} = 27$

$\Rightarrow 3^{2x} = 3^3$

$2x = 3$

$\therefore x = \frac{3}{2}$

Then compare the index powers of the RHS and the LHS.

(c) $3^{2x} = 81^{\frac{1}{4}}$

$\Rightarrow 3^{2x} = 81^{\frac{1}{4}} = 3^{4(\frac{1}{4})}$

$3^{2x} = 3^{\frac{4}{4}} = 3^1$

then $3^{2x} = 3^1$

Now, compare the powers, since the RHS and LHS have the same base:

so, $2x = 1$

$\therefore x = \frac{1}{2}$

Example 4

Simplify $\frac{2(2^{x+1}) - 5(2^{x-1})}{2^{x-1} - 2^x}$

Solution

$$\frac{2(2^{x+1}) - 5(2^{x-1})}{2^{x-1} - 2^x} = \frac{2(2^x.2^1) - 5(\frac{2^x}{2^1})}{\frac{2^x}{2^1} - 2^x}$$

Let $2^x = y$

$$\Rightarrow \frac{2(2^x.2^1) - 5(\frac{2^x}{2^1})}{\frac{2^x}{2^1} - 2^x} = \frac{2(y.2^1)\quad 5(\frac{y}{2^1})}{\frac{y}{2^1} - y} = \frac{4y - \frac{5y}{2^1}}{\frac{y}{2^1} - y}$$

$$\Rightarrow \frac{4y - \frac{5y}{2^1}}{\frac{y}{2^1} - y} = \frac{8y - 5y}{y - 2y} = \frac{3y}{-y} = -3$$

Example 5

Solve for x in each of the following:

(a) $2^{2x} - 5(2^x) + 4 = 0$ (b) $3(5^{2x}) - 12(5^x) + 9 = 0$

Solution

(a) $2^{2x} - 5(2^x) + 4 = 0$

Let $2^x = y$

$\Rightarrow 2^{2x} - 5(2^x) + 4 = 0$; this can be rewritten as $(2^x)^2 - 5(2^x) + 4 = 0$

$y^2 - 5y + 4 = 0$ factorising this expression, we have:

$(y - 4)(y - 1) = 0$

$\therefore\ y = 4$ or 1

But, $2^x = y \quad \Rightarrow 2^x = 4$ or $2^x = 1$

for $2^x = 4 \quad \Rightarrow 2^x = 2^2$

$\therefore\ x = 2$

for $2^x = 1 \quad \Rightarrow 2^x = 2^0$

$\therefore\ x = 0$

Hence, the values of x are: 2 or 0

(b) $3(5^{2x}) - 12(5^x) + 9 = 0$

Let $5^x = y$

$\Rightarrow 3(5^{2x}) - 12(5^x) + 9 = 0$

this can be re-written as $3(5^x)^2 - 12(5^x) + 9 = 0$

$3y^2 - 12y + 9 = 0$ Factorise this expression

$3(y - 1)(y - 3) = 0$

$\therefore y = 1$ or 3

But, $5^x = y \quad \Rightarrow 5^x = 1$ or $5^x = 3$

for $5^x = 1 \quad \Rightarrow 5^x = 5^0$

$\therefore x = 0$

for $5^x = 3 \quad \Rightarrow \log 5^x = \log 3$

$$x = \frac{\log 3}{\log 5} = \frac{0.4771}{0.6990} = 0.6825$$

Hence, the values of x are: 0 or 0.6825

Exercise 4a

1. Simplify the following:

   (a) $8^{1/3}$ (b) 13^{-2} (c) $32^{-\frac{2}{5}}$ (d) $\left(\frac{25}{16}\right)^{-\frac{3}{2}}$ (e) $\left(\frac{36}{81}\right)^{-\frac{1}{2}}$

2. Evaluate the following:

   (a) $2a \times (5a)^{-3}$ (b) $(9a^2)^{-\frac{1}{2}} \times 6a^3$

3. Find the value of x in each of the following expressions:

   (a) $2^x = 4$ (b) $4^{x-3} = \frac{32^{x+2}}{2^{-x}}$ (c) $5^{x+2} = 625$

   (d) $8^x = \frac{1}{1024}$ (e) $(0.25)^{x+1} = 16$

   (f) $8^{x/2} = 2^{3/8} \times 4^{1/4}$ (g) $3^{2x} = 81$

4. Simplify each of the following:

   (a) $125^{-1/3} \times 25^{-1/2} \times 10^0$ (b) $16^{-1/4} \times 25^{-1/2} \times 10^3$

5. Solve for x:

   (a) $9 = \frac{1}{27^x}$ (b) $\frac{2^x}{3^x} = \frac{27}{8}$ (c) $3^x + 3^{x+2} = 810$

   (d) $2^{x+2} = \frac{1}{2^x}$ (e) $5^{2x-1} = 625^x$ (f) $\sqrt{4^{2x+1}} = 2^{x-3}$

6. Solve for x:

   (a) $2^{2x} - 3(2^x) + 2 = 0$ (b) $3^{3x} - 237(3^{2x}) + 279(3^x) = 243$

4.2 LAWS OF LOGARITHMS

1. $\log_u 1 = 0$
2. $\log_a a = 1$
3. $\log_a x + \log_a y = \log_a xy$
4. $\log_a x - \log_a y = \log_a \frac{x}{y}$
5. $\log_a x^n = n \log_a x$
6. $\log_a x = \frac{\log_{10} x}{\log_{10} a} = \frac{1}{\log_x a}$
7. $\log_{a^n} x = \frac{1}{n} \log_a x$
8. $a^{\log_a x} = x$

Example 6

Simplify: (a) $\log_{10} 1$ (b) $\log_5 5$ (c) $\log_2 8$ (d) $\log_{3^2} 5$

Solution

(a) $\log_{10} 1 = 0$
(b) $\log_5 5 = 1$
(c) $\log_2 8 = \log_2 2^3$
$\Rightarrow \log_2 2^3 = 3 \log_2 2 = 3 \times 1 = 3$
(d) $\log_{3^2} 5 = \frac{1}{2} \log_3 5$

Example 7

Evaluate (a) $\log_{10} 4 + \log_{10} 25$ (b) $\log_8 2 - \log_8 0.25 + \log_8 32$

Solution

(a) $\log_{10} 4 + \log_{10} 25 = \log_{10} 4 \times 25 = \log_{10} 100$
then, $\log_{10} 100 = \log_{10} 10^2 = 2 \log_{10} 10$
$\Rightarrow 2 \log_{10} 10 = 2 \times 1 = 2$

(b) $\log_8 2 - \log_8 0.25 + \log_8 3 2 = \log_8 \frac{2}{0.25} + \log_8 32$
$= \log_8 8 + \log_8 32 = \log_8 (8 \times 32) = \log_8 2^3 \times 2^5$
$= \log_8 2^{(3+5)} = \log_8 2^8$
$\Rightarrow \log_8 2^8 = 8 \times \log_8 2 = 8 \times \log_{2^3} 2 = 8 \times \frac{1}{3} \log_2 2$
$\therefore\ 8 \times \frac{1}{3} \log_2 2 = 8 \times \frac{1}{3} \times 1 = \frac{8}{3} = 2\frac{2}{3}$

Example 8

Evaluate (a) $\log_{10}\sqrt{35}-\log_{10}\sqrt{7}+\log_{10}\sqrt{2}$ (b) $\log_9 3-2\log_9 3+\log_9 243$

Solution

(a) $\log_{10}\sqrt{35}-\log_{10}\sqrt{7}+\log_{10}\sqrt{2}=\log_{10}\frac{\sqrt{35\times 2}}{\sqrt{7}}=\log_{10}\frac{\sqrt{70}}{\sqrt{7}}$

$\Rightarrow \log_{10}\frac{\sqrt{70}}{\sqrt{7}}=\log_{10}\sqrt{\frac{70}{7}}=\log_{10}\sqrt{10}=\log_{10}10^{\frac{1}{2}}$

$=\frac{1}{2}\log_{10}10=\frac{1}{2}\times 1=\frac{1}{2}$

(b) $\log_9 3-2\log_9 3+\log_9 243=\log_9 3-\log_9 3^2+\log_9 243$

$=\log_9 3-\log_9 9+\log_9 243=\log_9\frac{3\times 243}{9}=\log_9 81$

$=\log_9 9^2 \;=2\times\log_9 9=2\times 1=2$

Example 9

Simplify

(a) $\log_{10}0.001$ (b) $\frac{\log 32}{\log 8}$ (c) $\frac{\log 125}{\log 5}$

Solution

(a) $\log_{10}0.001=\log_{10}10^{-3}$

$=-3\log_{10}10=-3\times 1=-3$

(b) $\frac{\log 32}{\log 8}=\frac{\log 2^5}{\log 2^3}=\frac{5\log 2}{3\log 2}$

$=\frac{5}{3}\times 1=\frac{5}{3}=1\frac{2}{3}$

Note that when a logarithm is written without the base number, that means it is logarithm to base 10, e.g., $\log 32=\log_{10}32$

(c) $\frac{\log 125}{\log 5}=\frac{\log 5^3}{\log 5}=\frac{3\log 5}{\log 5}=3\times 1=3$

Example 10

Evaluate (a) $\frac{\log_2 6-\log_2 3}{\log_2 8-2\log_2\frac{1}{2}}$ (b) $2\log_2 16+4\log_2 8-3\log_2 64$

Solution

(a) $\frac{\log_2 6-\log_2 3}{\log_2 8-2\log_2\frac{1}{2}}=\frac{\log_2\frac{6}{3}}{\log_2 8-\log_2\left(\frac{1}{2}\right)^2}=\frac{\log_2 2}{\log_2 8-\log_2\left(\frac{1}{4}\right)}$

$$= \frac{\log_2 2}{\log_2 \frac{8}{2^{1/4}}} = \frac{\log_2 2}{\log_2 8 \times \frac{4}{1}} = \frac{\log_2 2}{\log_2 32}$$

$$\Rightarrow \frac{\log_2 2}{\log_2 2^5} = \frac{\log_2 2}{5 \log_2 2} = \frac{1}{5} \times 1 = \frac{1}{5}$$

(b) $2\log_2 16 + 4\log_2 8 - 3\log_2 64 = 2\log_2 2^4 + 4\log_2 2^3 - 3\log_2 2^6$
$= 2 \times 4\log_2 2 + 4 \times 3\log_2 2 - 3 \times 6\log_2 2 = 8 \times 1 + 12 \times 1 - 18 \times 1$
$= 8 + 12 - 18 = 2$

Example 11

Find the value of x if (a) $2\log_{10} x - \log_{10} 2 = 2$ (b) $\log_8 x - 4\log_8 x = 2$

Solution

(a) $2\log_{10} x - \log_{10} 2 = 2$

$\Rightarrow \log_{10} x^2 - \log_{10} 2 = 2$

$\Rightarrow \log_{10} x^2 - \log_{10} 2 = \log_{10} \frac{x^2}{2} = 2$

$\therefore \frac{x^2}{2} = 10^2 = 100$

$x^2 = 2 \times 100 = 200$

$x = \sqrt{200} = 10\sqrt{2}$

$\therefore x = 10\sqrt{2} = 14.14$

(c) $\log_8 x - 4\log_8 x = \log_8 x - \log_8 x^4 = 2$

$\Rightarrow \log_8 \frac{x}{x^4} = 2$ then, $\frac{x}{x^4} = 8^2 = 64$

$x^{1-4} = 64$

$x^{-3} = 64$

$x = 64^{-3} = \frac{1}{\sqrt[3]{64}} = \frac{1}{4}$

$\Rightarrow x = \frac{1}{4}$

Example 12

Given that $6\log n + 4 = 2\log 8$, find the possible values of n.

Solution

$6\log n + 4 = 2\log 8$

$6\log n + 4 = \log(n+4)^6 = \log 8^2$

$\Rightarrow \log(n+4)^6 = \log 8^2$

$\log(n+4)^6 = \log 64$

$\therefore (n+4)^6 = 64$

$n + 4 = \sqrt[6]{64} = \pm 2$

$\Rightarrow n = 2 - 4, \text{or} -2 - 4$

$n = -2 \text{ or } -6$

Comparing the logarithms of both sides (i.e. **LHS** and **RHS**) of the equation, we obtain the functions $(n+4)^6$ and **64** to be equal.

Example 13

Given that $\log_{10} 2 = 0.3010$, $\log_{10} 3 = 0.4771$ and $\log_{10} 7 = 0.8451$, evaluate:
(a) $\log_{10} 9$ (b) $\log_{10} 14$ (c) $\log_{10} 21$ (d) $\log_{10} 42$

Solution

(a) $\log_{10} 9 = \log_{10} 3^2 = 2\log_{10} 3$
$= 2 \times 0.4771 = 0.9542$

(b) $\log_{10} 14 = \log_{10} 2 \times 7 = \log_{10} 2 + \log_{10} 7$
$= 0.3010 + 0.8451 = 1.1461$

(c) $\log_{10} 21 = \log_{10} 3 \times 7 = \log_{10} 3 + \log_{10} 7$
$= 0.4771 + 0.8451 = 1.3322$

(d) $\log_{10} 42 = \log_{10} 2 \times 3 \times 7 = \log_{10} 2 + \log_{10} 3 + \log_{10} 7$
$= 0.3010 + 0.4771 + 0.8451 = 1.6232$

Example 14

Given that $\log_{10} 2 = 0.3010$, $\log_{10} 3 = 0.4771$, and $\log_{10} 15 = 1.1761$.
Evaluate: (a) $\log_{10} 22.5$ (b) $\log_{10} 1.5$ (c) $\log_{10} 5$

Solution

(a) $\log_{10} 22.5 = \log_{10} \frac{15\times3}{2} = \log_{10} 15 + \log_{10} 3 - \log_{10} 2$
$= 1.1761 + 0.4771 - 0.3010 = 1.3522$

(b) $\log_{10} 1.5 = \log_{10} \frac{3}{2} = \log_{10} 3 - \log_{10} 2$
$= 0.4771 - 0.3010 = 0.1761$

(c) $\text{Log}_{10} 5 = \log_{10} \frac{15}{3} = \log_{10} 15 - \log_{10} 3$
$= 1.1761 - 0.4771 = 0.6990$

Example 15

Simplify (a) $7^{-2\log_7 3} \times 5^{3\log_5 2}$ (b) $\frac{3^{-2\log_3 5} \times 6^{-\log_6 2}}{2^{3\log_2 5}}$

Solution

(a) $7^{-2\log_7 3} \times 5^{3\log_5 2}$
Let $\log_7 3 = p \quad \Rightarrow 7^p = 3$
Also, let $\log_5 2 = q \quad \Rightarrow 5^q = 2$
$\therefore\ 7^{-2\log_7 3} \times 5^{3\log_5 2} = 7^{-2p} \times 5^{3q} = 7^{p(-2)} \times 5^{q(3)}$
But, $7^p = 3$ and $5^q = 2$
then, $7^{p(-2)} \times 5^{q(3)} = 3^{(-2)} \times 2^{(3)}$
$= \frac{1}{9} \times 8 = \frac{8}{9}$

(b) $\dfrac{3^{2\log_3 5} \times 6^{-\log_6 2}}{2^{3\log_2 5}}$

Let: $\log_3 5 = p \quad \Rightarrow 3^p = 5$

$\log_6 2 = q \quad \Rightarrow 6^q = 2$

$\log_2 5 = r \quad \Rightarrow 2^r = 5$

$$\therefore \frac{3^{2\log_3 5} \times 6^{-\log_6 2}}{2^{3\log_2 5}} = \frac{3^{2p} \times 6^{-q}}{2^{3r}} = \frac{3^{p(2)} \times 6^{q(-1)}}{2^{r(3)}}$$

But, $3^p = 5$, $6^q = 2$, and $2^r = 5$

then,

$$\frac{3^{p(2)} \times 6^{q(-1)}}{2^{r(3)}} = \frac{5^2 \times 2^{(-1)}}{5^3} = \frac{25 \times \frac{1}{2}}{125}$$

$$= \frac{1}{10}$$

4.3 EVALUATING LOGARITHM EXPRESSIONS USING LOGARITHM TABLE

There are two major logarithm bases: The common logarithm base known as logarithm to base 10 (**$\log_{10}$** sometimes written as **log**), and the natural logarithm base e (denoted as **$\log_e$** sometimes written as ***ln***) where $e \doteq 2.718$. The logarithm to base 10 is chosen as the standard for the evaluation of logarithm expressions using the logarithm table. The numbers written for a logarithm expression consist of two parts namely:

- *Characteristics*
- *Mantissa*

The *Characteristic* is an integer which can be positive, negative or zero. Thus, if a number is expressed in standard form as, $m \times 10^n$; then n is the "characteristics". While the "*mantissa*" is the decimal part whose logarithm is found.

Example 16

Find the characteristics of the following: (a) 4682 (b) 1.545 (c) 65.54 (d) 43568

Solution

(a) $4682 = 4.682 \times 10^3$

$\therefore$ characteristics = 3

(b) $1.554 = 1.554 \times 10^0$

$\therefore$ characteristics = 0

(c) $65.54 = 6.554 \times 10^1$

$\therefore$ characteristics = 1

(d) $43568 = 4.3568 \times 10^4$

$\therefore$ characteristics = 4

Example 17

Find the characteristics of the numbers:

(a) 0.0000333 (b) 0.08922 (c) 0.000000006 (d) 0.112

Solution

(a) $0.0000333 = 3.33 \times 10^{-5}$

$\therefore$ characteristics $= -5 = \bar{5}$

(b) $0.08922 = 8.922 \times 10^{-2}$

$\therefore$ characteristics $= -2 = \bar{2}$

(c) $0.000000006 = 6.0 \times 10^{-9}$

$\therefore$ characteristics $= -9 = \bar{9}$

(d) $0.112 = 1.12 \times 10^{-1}$

$\therefore$ characteristics $= -1 = \bar{1}$

> For a negative characteristic we denote a "bar" on the number rather than minus, this is just the rule.
> $\therefore$ characteristics $= \bar{5}$

Example 18

Find the mantissa of the following: (a) 4682 (b) 1.545 (c) 65.54 (d) 43568

Solution

(a) 4682 = 0.6704

(b) 1.545 = 0.1889

(c) 65.54 = 0.8165

(d) 43568 = 0.6391

> The mantissa of this expression is read from the logarithm table by looking up for the first two digits **46** in the left hand column of the log table, then follow the line across till you arrive at the corresponding digit **8**; then add the difference value of **2**, $\Rightarrow$ **0.6704** . Do the same for (b), (c) and (d).

Example 19

Find the logarithm expression of the following numbers:

(a) 4682 (b) 1.545 (c) 65.54 (d) 43568

Solution

(a) 4682 = 3.6704

(b) 1.545 = 0.1889

(c) 65.54 = 1.8165

(d) 43568 = 4.6392

Example 20

Find the logarithm expression of the following numbers:

(a) 0.0000333 (b) 0.08922 (c) 0.000000006 (d) 0.112

Solution

(a) $0.0000333 = \bar{5}.5224$

(b) $0.08922 = \bar{2}.9505$

(c) $0.000000006 = \bar{9}.7782$

(d) $0.112 = \bar{1}.0492$

Example 21

Find the anti-logarithm of the following expressions:

(a) 1.9101 (b) 2.8647 (c) 0.1889 (d) 4.6391

Solution

(a) $1.9101 = 81.3$
(b) $2.8647 = 732.3$
(c) $0.1889 = 1.545$
(d) $4.6391 = 4356$

The anti-logarithm (Anti-log) of a number is the reverse of the logarithm, e.g., **1.9101** has its characteristics as **1** and the mantissa is **.9101**. Then go to the anti-logarithm table and check for **91** under **0** difference **1**, = **813**. Now, because the characteristics is **1**, then put the decimal point after "two" digits (i.e., **81.3**).

Example 22

Find the anti-logarithm of (a) $\bar{1}.6602$ (b) $\bar{2}.7308$ (c) $\bar{3}.3962$ (d) $\bar{4}.9154$

Solution

(a) $\bar{1}.6602 = 0.4573$
(b) $\bar{2}.7308 = 0.0538$
(c) $\bar{3}.3962 = 0.00249$
(d) $\bar{4}.9154 = 0.000823$

Example 23

Simplify (a) $\bar{1}.8650 \div 2$ (b) $\bar{2}.4126 \div 3$ (c) $\bar{5}.7864 \div 4$ (d) $\bar{8}.4654 \div 5$

Solution

(a) $\bar{1}.8650 \div 2$
$\Rightarrow \bar{2} + 1.8650 \div 2$
$= \bar{1}.9325$

(b) $\bar{2}.4126 \div 3$
$\Rightarrow \bar{3} + 1.4126 \div 3$
$= \bar{1}.4709$

(c) $\bar{5}.7864 \div 4$
$\Rightarrow \bar{8} + 3.7864 \div 4$
$= \bar{2}.9466$

(d) $\bar{8}.4654 \div 5$
$\Rightarrow \overline{10} + 2.4654 \div 5$
$= \bar{2}.4931$

Example 24

Evaluate the following using logarithm table

(a) $\sqrt{25.38}$ (b) $22.4^{1/3}$ (c) 27.74^3 (d) 253.8×0.4573

Solution

(a) $\sqrt{25.38}$

No	Log
$\sqrt{25.38}$	$1.4045 \div 2$
	0.7023 Anti-log
$= 5.039$	

Square root, $\sqrt{\square}$, implies division by **2**.

Anti-log is to translate the logarithm value back to number. The value is then written on the number column.

(b) $22.4^{1/3}$

No	Log
$22.4^{1/3}$	$1.3502 \div 3$
	0.4501
$= 2.819$	

Index power of $\frac{1}{3}$ is same as cube root, $\sqrt[3]{\square}$, which implies division by 3.

(c) 27.74^3

No	Log
27.74^3	1.4431×3
	4.3293 Anti-log
$= 21340$	

Index power of 3 is same as cube, which implies multiplication by 3.

(d) 253.8×0.4573

No	Log
253.8	2.4045 (Add)
0.4523	$\bar{1}.6602$ (Add)
	2.0647 Anti-log
$= 116.1$	

Example 25

Evaluate $732.4 \times \sqrt{32}$

Solution

No	Log
732.4	2.8647 (Add)
$\sqrt{32}$	$1.5051 \div 2$
	0.7526 (Add)
	3.6173 Anti-log.
$= 4143$	

Example 26

Evaluate (a) $87.32 \div 25.78$ (b) $0.006924 \div 0.004523$

Solution

(a)

No	Log	
87.32	1.9411	Subtract
25.78	1.4114	
	0.5297	Anti- log.
= 3.386		

(b)

No	Log	
0.006924	$\bar{3}.8403$	Subtract
0.004523	$\bar{3}.6554$	
	0.1849	Anti – log.
= 1.531		

Example 27

Evaluate 0.0538×1.325^2

Solution

No	Log	
0.0538	$\bar{2}.7308$	Add
1.325^2	0.1222×2	
	0.2444	
	$\bar{2}.9752$	Anti-log.
= 0.09445		

Example 28

Evaluate $\dfrac{\sqrt[3]{0.0072} \times 81.3^2}{\sqrt{23140}}$

Solution

No	Log	
$\sqrt[3]{0.0072}$	$\bar{3}.8573 \div 3$	
	$\bar{1}.2858$	Add
81.3^2	1.9101×2	
	3.8202	
	3.1060 Numerator	Subtract
$\sqrt{23140}$	$4.3643 \div 2$	
	2.1822 Denominator	
	0.9238 Anti-log.	
$= 8.390$		

Example 29

Evaluate $\dfrac{\sqrt{0.006942} \times \sqrt[3]{0.6418}}{0.002543^3}$

Solution

No	Log	
$\sqrt{0.006942}$	$\bar{3}.8415 \div 2$	
	$\bar{4} + 1.8415 \div 2$	
	$\bar{2}.9208$	Add
$\sqrt[3]{0.6418}$	$\bar{1}.8074 \div 3$	
	$\bar{3} + 2.8074 \div 3$	
	$\bar{1}.9358$	
	$\bar{2}.8566$ Numerator	Subtract
0.002543^3	$\bar{3}.4053 \times 3$	
	$\bar{8}.2159$ Denominator	
	6.6407 Anti-log.	
$= 4372000$		

$\bar{3}$ cannot give a whole number when divided by **2**. In this case, we replace $\bar{3}$ with $\bar{4}$ +1 which is still equivalent to $\bar{3}$. Now, $\bar{4}$ can be divisible by **2** without a remainder to get the characteristics, while **+1.8415** is divided by **2** to get the mantissa.

Similarly, $\bar{1}$ is replaced with $\bar{3} + 2$; then $\bar{3}$ is divided by **3** to get the characteristics, while **2.8074** is also divided by **3** to get the mantissa.

Example 30

Evaluate $\frac{16.73^3 \times \sqrt[4]{0.04716}}{13.25 \times \sqrt{22.41}}$

Solution

No	Log	
16.73^3	1.2235×3	
	3.6705	Add
$\sqrt[4]{0.04716}$	$\bar{2}.6735 \div 4$	
	$\bar{4} + 2.6735 \div 4$	
	$\bar{1}.6684$	
	3.3389 Numerator	Subtract
13.25	1.1222	Add
$\sqrt{22.41}$	$1.3504 \div 2$	
	0.6752	
	1.7974 Denominator	
	1.5415 Anti-log.	
$= 34.79$		

Example 31

Evaluate $\sqrt[3]{\frac{0.2569^2 \times \sqrt{25.69}}{0.007432^2}}$

Solution

No	Log	
0.2569^2	$\bar{1}.4097 \times 2$	
	$\bar{2}.8194$	Add
$\sqrt{25.69}$	$1.4097 \div 2$	
	0.7049	
	$\bar{1}.5243$ Numerator	Subtract
0.007432^2	$\bar{3}.8711 \times 2$	
	$\bar{5}.7422$ Denominator	
	3.7821	
$\sqrt[3]{3.7821}$	$3.7821 \div 3$	
	1.2607 Anti-log.	
$= 18.23$		

Example 32

Evaluate:

(a) $\dfrac{1+\sqrt[3]{0.00625}}{1-\sqrt[3]{0.00625}}$ (b) $\dfrac{1.5320+(3.581)^2}{1.9840-\sqrt{0.0368}}$

Solution

(a) $\dfrac{1+\sqrt[3]{0.00625}}{1-\sqrt[3]{0.00625}}$

No	Log	
$\sqrt[3]{0.00625}$	$\bar{3}.7959 \div 3$	
	$\bar{1}.2653$	Anti-log.
$= 0.1842$		
$1+\sqrt[3]{0.00625}$		
$\Rightarrow 1 + 0.1842 = 1.1842$		
1.1842	0.7340	Numerator (Subtract)
$1-\sqrt[3]{0.00625}$		
$\Rightarrow 1 - 0.1842 = 0.8158$		
0.8158	0.9116	Denominator
	$\bar{1}.8224$	Anti-log.
$= 0.6643$		

(b) $\frac{1.5320 + (3.581)^2}{1.9840 - \sqrt{0.0368}}$

solution

No	Log
3.581	0.5540
3.581^2	0.5540×2
	1.1082 Anti- log.
= 12.83	
$1.5320 + (3.581)^2$	
$\Rightarrow 1.5320 + 12.83 = 14.362$	
14.362	1.1571 Numerator
$\sqrt{0.0368}$	$\bar{2}.5658 \div 2$
	$\bar{1}.2829$ Anti-log
= 0.1918	
$1.9840 - \sqrt{0.0368}$	
$\Rightarrow 1.9840 - 0.1918 = 1.7922$	
1.7922	0.2534 Denominator
	0.9037 Anti-log.
= 8.011	

Subtract (Numerator − Denominator)

OBJECTIVE TEST 4

1. Solve for x and y if $(x+y)\log_4 2 = \log_4 32$ and $(x-y)\log_3 27 = \log_2 8$
   (A) $x = 2,\ y = 3$ (B) $x = -3,\ y = -2$ (C) $x = 3,\ y = 2$ (D) $x = -2,\ y = 3$

2. Solve $3^{\sqrt{x}} = 27^x$
   (A) $0, \frac{1}{9}$ (B) $\frac{1}{3}, \frac{1}{9}$ (C) $0, 9$ (D) $\frac{1}{9}, \frac{1}{3}$

3. If $\log_x 243 = 5$, then x is ?
   (A) 2 (B) 3 (C) 4 (D) 5

4. Solve for x if $2^{2x+1} - 10 \times 2^x + 8 = 0$
   (A) $0, 2$ (B) $1, 4$ (C) $1, -4$ (D) $0, -2$

5. Solve for y if $4^{\log_2 8} = 8^y$
   (A) 0 (B) 1 (C) 2 (D) 3

6. Given that $\log_{10} 3 = 0.4770$ and $\log_{10} 2 = 0.3010$, find without using tables $\log_{10}\left(\frac{3}{20}\right)$
   (A) $\bar{1}.1760$ (B) 0.1760 (C) $\bar{1}.3824$ (D) $\bar{1}.8802$

7. Simplify $\dfrac{\log 25}{\log\frac{1}{5}}$
   (A) 5 (B) 125 (C) 2 (D) -2

Use mathematical table to evaluate question 8 -10.

8. 23.85^2
   (A) 586.9 (B) 869.5 (C) 568.9 (D) 47.70

9. 0.03872×0.005681
   (A) 0.002199 (B) 0.0002199 (C) 0.00002199 (D) 0.000002199

10. $\sqrt[3]{35.83} \times 0.082$
   (A) 0.02704 (B) 0.2704 (C) 2.704 (D) 0.002704

Exercise 4b

1. Simplify (a) $3^{-2\log_3 5} \times 2^{2\log_2 3}$ (b) $\dfrac{7^{-3\log_7 2} \times 2^{2\log_2 3}}{3^{-2\log_3 10}}$

2. $\frac{1}{2}\log_{10} q = 1$, find q.

3. Simplify (a) $\dfrac{\log\sqrt{27}}{\log 243}$ (b) $\dfrac{2\log_5 125 \times \log_3\sqrt{27}}{\log_3 81 \times \log_5 625}$

4. Given that $2\log x - 3\log x + 5\log x = 8$, find x.

5. Evaluate the following:
   (a) $\log_{\sqrt{3}} 9 + \log_{\frac{1}{3}} 27 + \log_9 243$ (b) $\log_7 49 + \log_7 \frac{1}{7} + \log_{\sqrt{7}} 343$
   (c) $\dfrac{\log_{10} 8 + \log_{10} 2}{\log_{10} 4 + \log_{10} 16}$ (d) $\dfrac{\log_{10} 9 - \log_{10} 3}{\log_{10} 27 - \log_{10} 81}$

6. Find the value of the following:
   (a) $\log_4 16$ (b) $\log_{\frac{1}{2}} 8$ (c) $\log_2 128$ (d) $\log_5 25$ (e) $\log_5 625$
   (f) $\log_4 0.25$ (g) $\log_{\sqrt{7}} 7$ (h) $\log_3 243$ (i) $\log_{10} 0.0001$
   (j) $\log_{\sqrt{5}} 125$ (k) $\log_2\sqrt{32}$ (l) $(\log_8 64)^2$ (m) $\sqrt{\log_6 36}$

7. Evaluate the following logarithmic expressions:
   (a) $\dfrac{\log 16}{\log 4}$ (b) $\dfrac{\log 32}{\log 8}$ (c) $\dfrac{\log 5}{\log 125}$ (d) $\dfrac{\log 9}{\log 81}$ (e) $\dfrac{\log 36}{\log\sqrt{6}}$

8. Evaluate the following:
   (a) $\log_2 4 + \log_2 8$ (b) $\log_5 25 - \log_5 625$ (c) $\log_{\sqrt{3}} 9 + \log_{\frac{1}{3}} 27 - \log_3 243$

9. Use mathematical table to find the anti- logarithm of the following:
   (a) 0.6324 (b) $\bar{3}.8426$ (c) $\bar{1}.4367$ (d) 2.5430 (e) 0.02986 (f) $\bar{2}.8249$

10. Use mathematical (logarithm) table to evaluate the following:
   (a) $\dfrac{2.86^2 \times 6705}{0.058}$ (b) $\dfrac{\sqrt[3]{1.382}}{\sqrt[4]{0.056}}$ (c) $\dfrac{\sqrt{0.836 \times 3.572}}{0.00425}$ (d) $\dfrac{\sqrt{10.63} \times \sqrt[3]{0.6540}}{(0.000823)^3}$
   (e) $\dfrac{(15.47)^4 \times \sqrt[3]{0.08340}}{(12.82)^3 \times \sqrt{11.45}}$ (f) $\sqrt[3]{\dfrac{(0.8232)^2 \times \sqrt{457.3}}{(0.00637)^3}}$ (g) $\dfrac{1 + \sqrt[3]{0.0081}}{1 - \sqrt[3]{0.0081}}$
   (h) $\dfrac{\sqrt[4]{1984} - 1}{1 - \sqrt[3]{0.00625}}$ (i) $\dfrac{1 + \sqrt{0.0283}}{1 - \sqrt{0.0283}}$

11. Find the value of x in each of the following:
   (a) $\log_2 8 + \log_2 5 - \log_2 x = 3$ (b) $3\log_2 x - \log_2 8 + \log_2 2048 = 0$

12. If $2\log_3 n - 1 = 4$, find n.

13. Find x in each of the following expressions:
   (a) $5^{3\log x} = 125$ (b) $3^{2\log_5 x} = 729$

5 SURD

A surd is any function with radicals; examples of radicals are $\sqrt{2}, \sqrt[3]{5}, \sqrt[4]{17}$. While 2, 5 and 17 are called the radicands. Surds can also be expressed in the form of ratios, i.e. $\frac{1}{\sqrt{2}}, \frac{3}{1+\sqrt{2}}, \frac{\sqrt{3}}{\sqrt{5}}$ etc.

5.1 RULES OF SURD

1. $\sqrt{a \times b} = \sqrt{a} \times \sqrt{b}$
2. $\sqrt{\frac{a}{b}} = \frac{\sqrt{a}}{\sqrt{b}}$
3. $\sqrt{a} + \sqrt{b} \neq \sqrt{a+b}$
4. $(\sqrt{a})^2 = a$

5.2 BASIC FORM OF SURD

This is a reduced form of surd which expresses the surd in its simplest form.

Example 1

(a) $\sqrt{27}$ (b) $\sqrt{128}$ (c) $\sqrt{108}$

Solution

(a) $\sqrt{27} = \sqrt{9 \times 3} = \sqrt{9} \times \sqrt{3}$
$= 3 \times \sqrt{3} = 3\sqrt{3}$

(b) $\sqrt{128} = \sqrt{64 \times 2} = \sqrt{64} \times \sqrt{2}$
$= 8 \times \sqrt{2} = 8\sqrt{2}$

(c) $\sqrt{108} = \sqrt{4 \times 27} = \sqrt{4} \times \sqrt{27}$
$= \sqrt{4} \times \sqrt{9 \times 3} = \sqrt{4} \times \sqrt{9 \times 3} = \sqrt{4} \times \sqrt{9} \times \sqrt{3}$
$= 2 \times 3 \times \sqrt{3} = 6\sqrt{3}$

Example 2

Express the basic form of each surd to a single surd form:

(a) $2\sqrt{3}$ (b) $3\sqrt{5}$ (c) $2\sqrt{7}$ (d) $10\sqrt{2}$

Solution

(a) $2\sqrt{3} = 2 \times \sqrt{3} = \sqrt{2 \times 2} \times \sqrt{3} = \sqrt{4} \times \sqrt{3}$
$= \sqrt{4 \times 3} = \sqrt{12}$

(b) $3\sqrt{5} = 3 \times \sqrt{5} = \sqrt{3 \times 3} \times \sqrt{5} = \sqrt{9} \times \sqrt{5}$
$\sqrt{9 \times 5} = \sqrt{45}$

(c) $2\sqrt{7} = 2 \times \sqrt{7} = \sqrt{2 \times 2} \times \sqrt{7} = \sqrt{4} \times \sqrt{7}$
$= \sqrt{4 \times 7} = \sqrt{28}$

(d) $10\sqrt{2} = \sqrt{10 \times 10} \times \sqrt{2} = \sqrt{100} \times \sqrt{2}$
$= \sqrt{100 \times 2} = \sqrt{200}$

5.3 SIMILAR SURDS

These are surds which are multiples of the same surd, e.g. $3\sqrt{2}, 7\sqrt{2}, 9\sqrt{2}$, and $15\sqrt{2}$. Each of these surds is a multiple of $\sqrt{2}$, and so they are similar surds.

Example 3

Simplify the following:

(a) $\sqrt{27} + \sqrt{75} + \sqrt{192}$ (b) $\sqrt{8} + \sqrt{18} - \sqrt{50} - \sqrt{72}$

Solution

(a) $\sqrt{27} + \sqrt{75} + \sqrt{192}$
$= \sqrt{9 \times 3} + \sqrt{25 \times 3} + \sqrt{64 \times 3}$
$= 3\sqrt{3} + 5\sqrt{3} + 8\sqrt{3}$
$= (3 + 5 + 8)\sqrt{3} = 16\sqrt{3}$

Reduce each single surd to its basic form; then factorise the multiple terms of $\sqrt{3}$. Now, add **3**, **5** and **8**.

(b) $\sqrt{8} + \sqrt{18} - \sqrt{50} - \sqrt{72}$
$= \sqrt{4 \times 2} + \sqrt{9 \times 2} - \sqrt{25 \times 2} - \sqrt{36 \times 2}$
$= 2\sqrt{2} + 3\sqrt{2} - 5\sqrt{2} - 6\sqrt{2}$
$= (2 + 3 - 5 - 6)\sqrt{2} = -6\sqrt{2}$

Use the same method as in (a) above

5.4 CONJUGATE SURDS

Two surds are conjugates of each other if the product of the two surds gives rise to a rational number. Similarly, $(\sqrt{a} + \sqrt{b})(\sqrt{a} - \sqrt{b}) = (\sqrt{a})^2 + \sqrt{ab} - \sqrt{ab} - (\sqrt{b})^2$
$= (\sqrt{a})^2 - (\sqrt{b})^2 = a - b$
where $a - b$ is a rational number gotten from the product of the two surds which are conjugates of each other. Hence, $\sqrt{a} + \sqrt{b}$ and $\sqrt{a} - \sqrt{b}$ are conjugate surds to each other.

Example 4

Write the conjugate of the following surd:

(a) $\sqrt{2} + \sqrt{5}$ (b) $7 + \sqrt{3}$ (c) $\sqrt{5} - 12$ (d) $1 - \sqrt{6}$ (e) $2\sqrt{3} + 9\sqrt{5}$

Solution

(a) $\sqrt{2} + \sqrt{5}$ the conjugate is $\sqrt{2} - \sqrt{5}$
(b) $7 + \sqrt{3}$ the conjugate is $7 - \sqrt{3}$
(c) $\sqrt{5} - 12$ the conjugate is $\sqrt{5} + 12$
(d) $1 - \sqrt{6}$ the conjugate is $1 + \sqrt{6}$
(e) $2\sqrt{3} + 9\sqrt{5}$ the conjugate is $2\sqrt{3} - 9\sqrt{5}$

Example 5

Simplify each of the following product of surd:

(a) $(\sqrt{2}+\sqrt{5})(\sqrt{2}-\sqrt{5})$ (b) $(\sqrt{3}+2\sqrt{5})(\sqrt{3}-2\sqrt{5})$ (c) $(3-\sqrt{2})(3+\sqrt{2})$

Solution

(a) $(\sqrt{2}+\sqrt{5})(\sqrt{2}-\sqrt{5}) = (\sqrt{2})^2 - (\sqrt{5})^2$
$= 2 - 5 = -3$

(b) $(\sqrt{3}+2\sqrt{5})(\sqrt{3}-2\sqrt{5}) = (\sqrt{3})^2 - (2\sqrt{5})^2$
$= 3 - (4 \times 5) = 3 - 20 = -17$

Since each surd here is a conjugate surd we can evaluate their products as a difference of two square.

(c) $(3-\sqrt{2})(3+\sqrt{2}) = (3)^2 - (\sqrt{2})^2$
$= 9 - 2 = 7$

5.5 RATIONALIZING SURDS

We rationalise a surd when the denominator has any surd that looks like:

$\frac{a}{\sqrt{b}}, \frac{a}{m+\sqrt{b}}, \frac{a}{m-\sqrt{b}}, \frac{a}{\sqrt{m}-\sqrt{b}}, \frac{a}{\sqrt{m}+\sqrt{b}}, \frac{a+\sqrt{n}}{\sqrt{m}+\sqrt{b}}, \frac{a-\sqrt{n}}{\sqrt{m}+\sqrt{b}}, \frac{\sqrt{a}+\sqrt{n}}{\sqrt{m}+\sqrt{b}}, \frac{\sqrt{a}-\sqrt{n}}{\sqrt{m}+\sqrt{b}}, \frac{\sqrt{a}+\sqrt{n}}{\sqrt{m}\ \ \sqrt{b}}$, or $\frac{\sqrt{a}-\sqrt{n}}{\sqrt{m}-\sqrt{b}}$;

where a, b, m and n are whole numbers (i.e. constants).

As for the expression, $\frac{a}{\sqrt{b}}$, we can rationalise it by multiplying the numerator and denominator by $\sqrt{b}$. $\therefore\ \frac{a}{\sqrt{b}} = \frac{a}{\sqrt{b}} \times \frac{\sqrt{b}}{\sqrt{b}} = \frac{a\sqrt{b}}{b}$

Whereas, the expressions $\frac{a}{m+\sqrt{b}}, \frac{a}{m-\sqrt{b}}, \frac{a}{\sqrt{m}-\sqrt{b}}, \frac{a}{\sqrt{m}+\sqrt{b}}, \frac{a+\sqrt{n}}{\sqrt{m}+\sqrt{b}}, \frac{a-\sqrt{n}}{\sqrt{m}+\sqrt{b}}, \frac{\sqrt{a}+\sqrt{n}}{\sqrt{m}+\sqrt{b}}, \frac{\sqrt{a}-\sqrt{n}}{\sqrt{m}+\sqrt{b}}, \frac{\sqrt{a}+\sqrt{n}}{\sqrt{m}-\sqrt{b}}$ or $\frac{\sqrt{a}-\sqrt{n}}{\sqrt{m}-\sqrt{b}}$ can be rationalised by multiplying the "numerator and denominator" by a conjugate surd, then simplifying to the lowest possible expression.

Thus,

$$\frac{a}{m+\sqrt{b}} = \frac{a}{m+\sqrt{b}} \times \frac{m-\sqrt{b}}{m-\sqrt{b}}$$

$$\frac{a}{m-\sqrt{b}} = \frac{a}{m-\sqrt{b}} \times \frac{m+\sqrt{b}}{m+\sqrt{b}}$$

$$\frac{a}{\sqrt{m}-\sqrt{b}} = \frac{a}{\sqrt{m}-\sqrt{b}} \times \frac{\sqrt{m}+\sqrt{b}}{\sqrt{m}+\sqrt{b}}$$

$$\frac{a}{\sqrt{m}+\sqrt{b}} = \frac{a}{\sqrt{m}+\sqrt{b}} \times \frac{\sqrt{m}-\sqrt{b}}{\sqrt{m}-\sqrt{b}}$$

$$\frac{a+\sqrt{n}}{\sqrt{m}+\sqrt{b}} = \frac{a+\sqrt{n}}{\sqrt{m}+\sqrt{b}} \times \frac{\sqrt{m}-\sqrt{b}}{\sqrt{m}-\sqrt{b}}$$

Example 6

Rationalise the following expression:

(a) $\frac{3}{\sqrt{2}}$ (b) $\frac{1}{\sqrt{3}}$ (c) $\frac{5\sqrt{3}}{2\sqrt{2}}$ (d) $\frac{3+\sqrt{2}}{\sqrt{5}}$

Solution

(a) $\frac{3}{\sqrt{2}} = \frac{3}{\sqrt{2}} \times \frac{\sqrt{2}}{\sqrt{2}} = \frac{3\sqrt{2}}{2}$

(b) $\frac{1}{\sqrt{3}} = \frac{1}{\sqrt{3}} \times \frac{\sqrt{3}}{\sqrt{3}} = \frac{\sqrt{3}}{3}$

(c) $\frac{5\sqrt{3}}{2\sqrt{2}} = \frac{5\sqrt{3}}{2\sqrt{2}} \times \frac{2\sqrt{2}}{2\sqrt{2}} = \frac{10\sqrt{6}}{4\times\sqrt{4}} = \frac{10\sqrt{6}}{4\times2} = \frac{5\sqrt{6}}{4}$

(d) $\frac{3+\sqrt{2}}{\sqrt{5}} = \frac{3+\sqrt{2}}{\sqrt{5}} \times \frac{\sqrt{5}}{\sqrt{5}} = \frac{\sqrt{5}(3+\sqrt{2})}{5} = \frac{3\sqrt{5}+\sqrt{10}}{5}$

Example 7

Given that $\sqrt{2} = 1.414$, and $\sqrt{3} = 1.732$

Simplify: (a) $\frac{3}{\sqrt{2}}$ (b) $\frac{6}{\sqrt{3}}$ (c) $\frac{2+\sqrt{2}}{\sqrt{2}}$ (d) $\frac{5}{\sqrt{2}+\sqrt{3}}$

Solution

(a) $\frac{3}{\sqrt{2}} = \frac{3}{\sqrt{2}} \times \frac{\sqrt{2}}{\sqrt{2}} = \frac{3\sqrt{2}}{2}$

$= \frac{3\times1.414}{2} = \frac{4.242}{2}$

$= 2.121$

(b) $\frac{6}{\sqrt{3}} = \frac{6}{\sqrt{3}} \times \frac{\sqrt{3}}{\sqrt{3}} = \frac{6\sqrt{3}}{3}$

$= 2\sqrt{3} = 2 \times 1.732$

$= 3.464$

(c) $\frac{2+\sqrt{2}}{\sqrt{2}} = \frac{2+\sqrt{2}}{\sqrt{2}} \times \frac{\sqrt{2}}{\sqrt{2}} = \frac{\sqrt{2}(2+\sqrt{2})}{2}$

$= \frac{2\sqrt{2}+2}{2} = \frac{2(\sqrt{2}+1)}{2}$

$= \sqrt{2} + 1 = 1.414 + 1$

$= 2.414$

(d) $\frac{5}{\sqrt{2}+\sqrt{3}} = \frac{5}{\sqrt{2}+\sqrt{3}} \times \frac{\sqrt{2}-\sqrt{3}}{\sqrt{2}-\sqrt{3}} = \frac{5(\sqrt{2}-\sqrt{3})}{2-3}$

$= \frac{5(\sqrt{2}-\sqrt{3})}{-1} = -5(\sqrt{2} - \sqrt{3})$

$= -5(1.414 - 1.732) = -5 \times (-0.318)$

$= 1.59$

Example 8

Rationalise each of the following expression:

(a) $\frac{5}{1+\sqrt{3}}$ (b) $\frac{1}{\sqrt{2}+\sqrt{5}}$ (c) $\frac{1}{3\sqrt{2}+2\sqrt{3}}$

Solution

(a) $\frac{5}{1+\sqrt{3}} = \frac{5}{1+\sqrt{3}} \times \frac{1-\sqrt{3}}{1-\sqrt{3}} = \frac{5(1-\sqrt{3})}{(1+\sqrt{3})(1-\sqrt{3})} = \frac{5(1-\sqrt{3})}{(1)^2-(\sqrt{3})^2}$

$= \frac{5(1-\sqrt{3})}{1-3} = \frac{5(1-\sqrt{3})}{-2} = \frac{5(\sqrt{3}-1)}{2}$

(b) $\frac{1}{\sqrt{2}+\sqrt{5}} = \frac{1}{\sqrt{2}+\sqrt{5}} \times \frac{\sqrt{2}-\sqrt{5}}{\sqrt{2}-\sqrt{5}} = \frac{\sqrt{2}-\sqrt{5}}{(\sqrt{2}+\sqrt{5})(\sqrt{2}-\sqrt{5})}$

$= \frac{\sqrt{2}-\sqrt{5}}{(\sqrt{2})^2-(\sqrt{5})^2} = \frac{\sqrt{2}-\sqrt{5}}{2-5} = \frac{\sqrt{2}-\sqrt{5}}{-3} = \frac{\sqrt{5}-\sqrt{2}}{3}$

(c) $\frac{1}{3\sqrt{2}+2\sqrt{3}} = \frac{1}{3\sqrt{2}+2\sqrt{3}} \times \frac{3\sqrt{2}-2\sqrt{3}}{3\sqrt{2}-2\sqrt{3}} = \frac{3\sqrt{2}-2\sqrt{3}}{(3\sqrt{2}+2\sqrt{3})(3\sqrt{2}-2\sqrt{3})}$

$\frac{3\sqrt{2}-2\sqrt{3}}{(3\sqrt{2})^2-(2\sqrt{3})^2} = \frac{3\sqrt{2}-2\sqrt{3}}{18-12} = \frac{3\sqrt{2}-2\sqrt{3}}{6}$

$= \frac{\sqrt{2}}{2} - \frac{\sqrt{3}}{3}$

Example 9

Simplify (a) $\frac{1-\sqrt{5}}{1+\sqrt{5}}$ (b) $\frac{2\sqrt{5}+3\sqrt{2}}{7\sqrt{2}+4\sqrt{5}}$

Solution

(a) $\frac{1-\sqrt{5}}{1+\sqrt{5}} = \frac{1-\sqrt{5}}{1+\sqrt{5}} \times \frac{1-\sqrt{5}}{1-\sqrt{5}} = \frac{(1-\sqrt{5})(1-\sqrt{5})}{(1+\sqrt{5})(1-\sqrt{5})}$

$= \frac{1-\sqrt{5}-\sqrt{5}+5}{(1)^2-(\sqrt{5})^2} = \frac{6-2\sqrt{5}}{1-5} = \frac{6-2\sqrt{5}}{-4}$

$\Rightarrow \frac{2\sqrt{5}-6}{4} = \frac{2(\sqrt{5}-3)}{4} = \frac{\sqrt{5}-3}{2}$

(b) $\frac{2\sqrt{5}+3\sqrt{2}}{7\sqrt{2}+4\sqrt{5}} = \frac{2\sqrt{5}+3\sqrt{2}}{7\sqrt{2}+4\sqrt{5}} \times \frac{7\sqrt{2}-4\sqrt{5}}{7\sqrt{2}-4\sqrt{5}} = \frac{(2\sqrt{5}+3\sqrt{2})(7\sqrt{2}-4\sqrt{5})}{(7\sqrt{2})^2-(4\sqrt{5})^2}$

$= \frac{14\sqrt{10}-40+42-12\sqrt{10})}{98-80} = \frac{2-2\sqrt{10}}{18}$

$= \frac{2(1-\sqrt{10})}{18} = \frac{1-\sqrt{10}}{9}$

Example 10

Given $\frac{3}{1+\sqrt{2}} - \frac{2}{1-\sqrt{2}} = a + b\sqrt{n}$; where a, b and n are constants. Find a, b and n

Solution

$$\frac{3}{1+\sqrt{2}} - \frac{2}{1-\sqrt{2}} = \frac{3(1-\sqrt{2})-2(1+\sqrt{2})}{(1+\sqrt{2})(1-\sqrt{2})} = \frac{3-3\sqrt{2}-2-2\sqrt{2}}{(1)^2-(\sqrt{2})^2} = \frac{1-5\sqrt{2}}{-1}$$

$= -1 + 5\sqrt{2} = a + b\sqrt{n}$

$\therefore\ a = -1,\ b = 5,\ n = 2$

Example 11

Simplify $\frac{\sqrt{1+x}-\sqrt{x}}{\sqrt{1+x}+\sqrt{x}}$

Solution

$$\frac{\sqrt{1+x}-\sqrt{x}}{\sqrt{1+x}+\sqrt{x}} = \frac{\sqrt{1+x}-\sqrt{x}}{\sqrt{1+x}+\sqrt{x}} \times \frac{\sqrt{1+x}-\sqrt{x}}{\sqrt{1+x}-\sqrt{x}} = \frac{1+x-\sqrt{x(x+1)}-\sqrt{x(x+1)}+x}{\left(\sqrt{1+x}\right)^2-\left(\sqrt{x}\right)^2}$$

$$= \frac{1+2x-2\sqrt{x(x+1)}}{1+x-x} = 1 + 2x - 2\sqrt{x(x+1)}$$

Example 12

Find the value of P if $\frac{P}{\sqrt{2}+5} = P\sqrt{2} - 5$

Solution

$P = \left(P\sqrt{2} - 5\right)\left(\sqrt{2} + 5\right) = 2P + 5P\sqrt{2} - 5\sqrt{2} - 25$

$2P - P + 5P\sqrt{2} = 5\sqrt{2} + 25$

$P\left(1 + 5\sqrt{2}\right) = 5\sqrt{2} + 25$

$\therefore\ P = \frac{5\sqrt{2}+25}{1+5\sqrt{2}}$

Rationalising the denominator, we have:

$$P = \frac{5\sqrt{2} + 25}{1 + 5\sqrt{2}} = \frac{5\sqrt{2} + 25}{1 + 5\sqrt{2}} \times \frac{1 - 5\sqrt{2}}{1 - 5\sqrt{2}} = \frac{5\sqrt{2} - 50 + 25 - 125\sqrt{2}}{(1)^2 - (5\sqrt{2})^2}$$

$$= \frac{-25 - 120\sqrt{2}}{1 - 50} = \frac{-(25 + 120\sqrt{2})}{-49} = \frac{25 + 120\sqrt{2}}{49}$$

$\therefore\quad P = \frac{25+120\sqrt{2}}{49}$

5.6 EQUALITY OF SURDS

Given two surds $a + \sqrt{m}$ and $b + \sqrt{n}$, then the surds are equal if $a + \sqrt{m} = b + \sqrt{n}$

$\Rightarrow a - b = \sqrt{n} - \sqrt{m}$

Where $a - b = 0$

$\therefore a = b$

Similarly, $\sqrt{n} - \sqrt{m} = 0$

$\therefore \sqrt{n} = \sqrt{m}$

$\therefore n = m$

The equality of two surds is used for determining the square roots of a number.

Example 13

Find the square root of $8 + 2\sqrt{15}$

Solution

Let the square root of $8 + 2\sqrt{15}$ be $\sqrt{m} + \sqrt{n}$

$\Rightarrow \sqrt{8 + 2\sqrt{15}} = \sqrt{m} + \sqrt{n}$

$\therefore 8 + 2\sqrt{15} = (\sqrt{m} + \sqrt{n})^2 = m + n + 2\sqrt{mn}$

From equality of two surds, we have:

$m + n = 8$ ....(i)

Also, $2\sqrt{mn} = 2\sqrt{15}$

$\therefore mn = 15$ ....(ii)

From (i) make n subject of the formula

$\Rightarrow n = 8 - m$, then put into equation (ii)

$m(8 - m) = 15 \quad \Rightarrow 8m - m^2 = 15$

Rearrange this equation and then factorise.

$\therefore m^2 - 8m + 15 = 0$

$(m - 5)(m - 3) = 0$

$\therefore m = 5$ or 3

when $m = 5$, then $n = 8 - m = 8 - 5 = 3$

when $m = 3$, then $n = 8 - m = 8 - 3 = 5$

Hence, the square root of $8 + 2\sqrt{15}$ will be: $\sqrt{m} + \sqrt{n} = \sqrt{5} + \sqrt{3}$ or $\sqrt{3} + \sqrt{5}$

$\Rightarrow \sqrt{5} + \sqrt{3}$ twice

Example 14

Find the square root of $7 - 2\sqrt{6}$

Solution

Let the square root of $7 - 2\sqrt{6}$ be $\sqrt{m} - \sqrt{n}$

$\Rightarrow \sqrt{7 - 2\sqrt{6}} = \sqrt{m} - \sqrt{n}$

$\therefore 7 - 2\sqrt{6} = (\sqrt{m} - \sqrt{n})^2 = m + n - 2\sqrt{mn}$

From equality of two surds, we have:

$m + n = 7$ ....(i)

And, $-2\sqrt{mn} = -2\sqrt{6}$

$\therefore mn = 6$ ....(ii)

From (i) make n subject of the formula:

Hence $n = 7 - m$, then put into equation (ii) to have:

$m(7 - m) = 6$

$\therefore 7m - m^2 = 6$

Rearranging the equation, we have:

$m^2 - 7m + 6 = 0$

Now, factorise the expression to have:

$(m - 6)(m - 1) = 0$

$\therefore m = 6$ or 1

$\Rightarrow$ when $m = 6$, then $n = 7 - m = 7 - 6 = 1$

when $m = 1$, then $n = 7 - m = 7 - 1 = 6$

Hence, the square root of $7 - 2\sqrt{6}$ will be:

$\sqrt{m} - \sqrt{n} = \sqrt{6} - \sqrt{1} = \sqrt{6} - 1$

or $\sqrt{1} - \sqrt{6} = 1 - \sqrt{6}$

$\Rightarrow \sqrt{6} - 1$ or $1 - \sqrt{6}$

5.7 SOLVING RADICAL EQUATIONS OF SURD

A radical equation is any equation of a surd in the form $\sqrt{x + m} = c$. Where the expression $\sqrt{x + m}$ is called the radical; then $x + m$ is the radicand; and the value c is a constant.

Example 15

Solve for x if $\sqrt{x^2 - 21} = 10$

Solution

$\sqrt{x^2 - 21} = 10$

squaring both sides of the equation, we have:

$(\sqrt{x^2 - 21})^2 = 10^2$

Thus, $x^2 - 21 = 100$

$x^2 = 100 + 21 = 121$

$x = \sqrt{121} = 11$

$\Rightarrow x = 11$

Example 15

$3\sqrt{x+3} + x = 25$, find x.

Solution

$3\sqrt{x+3} + x = 25$

$\therefore 3\sqrt{x+3} = 25 - x$

$(3\sqrt{x+3})^2 = (25 - x)^2$

$9(x+3) = 625 - 50x + x^2$

$9x + 27 = 625 - 50x + x^2$

Collecting like terms we have:

$x^2 - 50x - 9x + 625 - 27 = 0$

$\therefore x^2 - 59x + 598 = 0$

Factorising the expression, we have:

$(x - 46)(x - 13) = 0$

$x = 46$ or 13

Now, check for the true solution of the equation by putting the values of x into the equation, $3\sqrt{x+3} + x = 25$.

$\Rightarrow 3\sqrt{x+3} + x = 25$

$\therefore$ when $x = 13$

$\Rightarrow 3\sqrt{13+3} + 13 = 3\sqrt{16} + 13 = 3 \times 4 + 13$

$= 12 + 13 = 25$

It can be seen that $x = 13$ satisfies the condition of the equation.

Hence, $x = 13$ is a correct solution.

Now using $x = 46$

$\Rightarrow 3\sqrt{46+3} + 46 = 3\sqrt{49} + 46 = 3 \times 7 + 46$

$= 21 + 46 = 67 \neq 25$

When $x = 46$ it does not satisfy the condition of the equation; therefore, $x = 46$ is *extraneous.*

$\Rightarrow x = 13$ is the only solution for the radical equation.

OBJECTIVE TEST 5

1. Write the surd $\sqrt{162}$ in its basic form
   (A) $2\sqrt{3}$ (B) $9\sqrt{2}$ (C) $3\sqrt{2}$ (D) $5\sqrt{3}$
2. Express $3\sqrt{5}$ as a single surd
   (A) $\sqrt{45}$ (B) $\sqrt{50}$ (C) $\sqrt{90}$ (D) $\sqrt{15}$
3. The conjugate of $5 - \sqrt{2}$ is
   (A) $5 - \sqrt{2}$ (B) $2 - \sqrt{5}$ (C) $5 + \sqrt{2}$ (D) $2 + \sqrt{5}$
4. Rationalize $\frac{3}{\sqrt{2}}$
   (A) $\frac{3}{2}\sqrt{2}$ (B) $\frac{2}{3}\sqrt{2}$ (C) $\frac{3}{2}$ (D) $\frac{\sqrt{3}}{2}$
5. Simplify $(2\sqrt{3} - 1)(2\sqrt{3} + 1)$
   (A) 13 (B) 12 (C) 11 (D) 10
6. Rationalize $\frac{1+\sqrt{5}}{1-\sqrt{5}}$
   (A) $\frac{-3-\sqrt{5}}{2}$ (B) $\frac{\sqrt{5}-3}{2}$ (C) $\frac{3-\sqrt{5}}{2}$ (D) $\frac{5-\sqrt{3}}{2}$
7. Find the square root of $5 + 2\sqrt{6}$
   (A) $(\sqrt{3} + \sqrt{2})$ twice (B) $(\sqrt{3} - \sqrt{2})$ twice
   (C) $(\sqrt{3} + \sqrt{2})$ and $(\sqrt{3} - \sqrt{2})$ (D) $(3 + \sqrt{2})$twice
8. Solve the equation $3\sqrt{x - 1} + x = 5$
   (A) $x = 2$ or 17 (B) $x = -2$ or -17 (C) $x = -2$ or 17 (D) $x = 2$ or -17
9. Given that $\sqrt{3} = 1.732$, find $\frac{1}{\sqrt{3}}$
   (A) 0.5737 (B) 0.5377 (C) 0.5773 (D) 5.773
10. Find the value of P if $\frac{P}{2+\sqrt{3}} = P\sqrt{3} - 1$
    (A) $\frac{\sqrt{3}+1}{2}$ (B) $\frac{\sqrt{3}-1}{2}$ (C) $\frac{\sqrt{3}+1}{4}$ (D) $\frac{\sqrt{3}-1}{2}$

1. Write each of the following as a single square root:
   (a) $2\sqrt{5}$ (b) $3\sqrt{2}$ (c) $5\sqrt{2}$ (d) $7\sqrt{3}$ (e) $10\sqrt{2}$ (f) $2\sqrt{11}$ (g) $9\sqrt{3}$
2. Express each of the following in basic form:
   (a) $\sqrt{8}$ (b) $\sqrt{18}$ (c) $\sqrt{50}$ (d) $\sqrt{75}$ (e) $\sqrt{40}$ (f) $\sqrt{512}$ (g) $\sqrt{600}$ (h) $\sqrt{48}$
3. Simplify the following:
   (a) $(\sqrt{2}+1)^2$ (b) $(\sqrt{3}+1)^2$ (c) $(2\sqrt{3}-1)^2$ (d) $(2\sqrt{5}+3\sqrt{2})^2$ (e) $(\sqrt{3}-\sqrt{2})^2$
4. Simplify each of the following:
   (a) $2\sqrt{12}+3\sqrt{48}-\sqrt{75}$ (b) $3\sqrt{6}-2\sqrt{24}+\sqrt{150}$
   (c) $5\sqrt{7}+\sqrt{28}+\sqrt{63}$ (d) $4\sqrt{18}+3\sqrt{72}-7\sqrt{50}$
5. Given that $\sqrt{2}=1.414$ and $\sqrt{3}=1.732$. Evaluate each of the following given your answers correct to four significant figure:
   (a) $\sqrt{75}$ (b) $\sqrt{8}$ (c) $1+\sqrt{128}$ (d) $\sqrt{12}-1$
6. Given that $\sqrt{2}=1.414$, and $\sqrt{7}=2.656$. Evaluate each of the following leaving your answer in three significant figures:
   (a) $\frac{5}{\sqrt{2}}$ (b) $\frac{1}{1+\sqrt{7}}$ (c) $\frac{1}{2\sqrt{2}-\sqrt{7}}$ (d) $\frac{10}{\sqrt{2}-\sqrt{7}}$
7. Rationalize the denominator of each of the following:
   (a) $\frac{10}{\sqrt{5}}$ (b) $\frac{1+\sqrt{3}}{\sqrt{2}}$ (c) $\frac{3\sqrt{2}+2\sqrt{3}}{5\sqrt{2}}$ (d) $\frac{\sqrt{5}}{\sqrt{3}}$
8. Rationalise the denominator of each of the following:
   (a) $\frac{2}{1+\sqrt{7}}$ (b) $\frac{1-\sqrt{2}}{1+\sqrt{2}}$ (c) $\frac{1-\sqrt{5}}{1+\sqrt{5}}$ (d) $\frac{1}{8-\sqrt{5}}$ (e) $\frac{2\sqrt{3}+3\sqrt{2}}{5\sqrt{2}-2\sqrt{3}}$
9. Given $\frac{1}{2\sqrt{3}-5}=a+b\sqrt{n}$; where a, b and n are constants. Find a, b and n.
10. Express $\frac{2\sqrt{3}+3\sqrt{2}}{2\sqrt{3}-3\sqrt{2}}$ in the form of $p+q\sqrt{r}$ where p, q and r are rational numbers.
11. Express $\frac{\sqrt{2}+\sqrt{7}}{\sqrt{14}}$ in the form $a\sqrt{2}+b\sqrt{7}$
12. Express $\sqrt{32}+\frac{6}{\sqrt{2}}$ as a single surd.
13. Find the square root of each of the following:
    (a) $6+\sqrt{35}$ (b) $21-4\sqrt{5}$ (c) $14-4\sqrt{6}$ (d) $7+2\sqrt{10}$
14. Rationalise the denominator of the expression $\frac{\sqrt{x+1}+\sqrt{x}}{\sqrt{x+1}-\sqrt{x}}$
15. Simplify$(\sqrt{2}+1)^4-(\sqrt{2}-1)^4$
16. Find the expression for P, given that $\frac{P}{\sqrt{2}+\sqrt{5}}=P\sqrt{2}-\sqrt{5}$

6 VARIATION

Variation is the relationship between two or more quantities such that any change in one of the quantities affects the other. Thus, if there is any factor in such a relationship which does not change or not affected in the relationship then it is called a "constant." Variation can be classified as direct, inverse, joint, and partial variation.

6.1 DIRECT VARIATION

In direct variation, an increase in a variable (e.g., x) will cause an increase in another variable (e.g. y) vice – versa.
Then, $y \propto x$ where $\propto$ is the sign of proportionality or variation of the quantity.

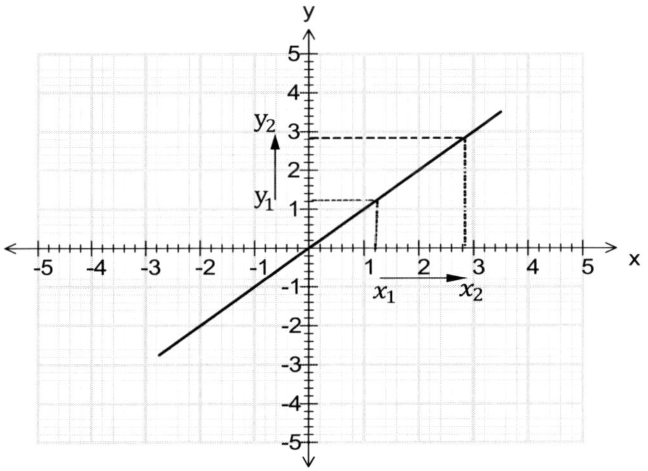

Fig 6.1 Direct Variation

Example 1

A quantity y varies directly as another quantity x if $y = 10$ when $x = 5$, find:
(a) the relationship between y and x (b) the value of y whe $x = 20$

Solution

(a) $y \propto x \Rightarrow y = kx$ where k is the relationship constant between y and x.
when $y = 10$, and $x = 5$

$k = \frac{y}{x} = \frac{10}{5} = 2 \quad \therefore y = 2x$ This is the equation relating y and x

(b) When $x = 20, y = ?$
Using $y = 2x \Rightarrow y = 2 \times 20 = 40$
$\therefore y = 40$

Example 2

A quantity p varies directly as the square of q. When $p = 27, q = 6$, find:
(a) the connecting factor between p and q (b) the value of p when $q = 10$
(c) value of q when $p = 48$.

Solution

$p \propto q^2 \Rightarrow p = kq^2$ where k is the connecting factor between p and q.

(a) When $p = 27$, and $q = 6$

Then, $k = \frac{p}{q^2} = \frac{27}{6^2} = \frac{27}{36} = 0.75$

$p = 0.75q^2$

(b) When $q = 10$

Then, $p = 0.75q^2 = 0.75 \times 10^2 = 0.75 \times 100 = 75$

$\therefore p = 75$

(c) When $p = 48$

Then, $p = 0.75q^2 \Rightarrow 48 = 0.75q^2$

$q^2 = \frac{48}{0.75} = 64$

$q = \sqrt{64} - 8 \quad \therefore q - 8$

Example 3

The quantity of a commodity M varies directly as the square root of N. When $M = 3$, $N = 4$, find:
(a) the equation relating M and N
(b) the value of N when $M = 15$
(c) the value of M when $N = 36$

Solution

$M \propto \sqrt{N} \Rightarrow M = k\sqrt{N}$ where k is the relating factor between M and N

(a) When $M = 3$, $N = 4$; then $k = \frac{M}{\sqrt{N}} = \frac{3}{\sqrt{4}} = \frac{3}{2}$

$\therefore M = \frac{3}{2}\sqrt{N}$

(b) When $M = 15 \Rightarrow 15 = \frac{3}{2}\sqrt{N}$

$\sqrt{N} = \frac{15 \times 2}{3} = \frac{30}{3} = 10$

$N = 10^2 = 100$

$\therefore N = 100$

(c) When $N = 36$

$\Rightarrow M = \frac{3}{2}\sqrt{N} = \frac{3}{2}\sqrt{36}$

$= \frac{3}{2} \times 6 = 9 \quad \therefore M = 9$

Exercise 6a

1. If $y \propto x$, and $y = 10$ when $x = 2$. Find the relationship between y and x. Hence, find x when $y = 30$.
2. P varies directly as the square of Q and when $P = 18$, $Q = 3$; find:
   (a) the equation relating P and Q (b) P when $Q = 5$ and Q when $P = 32$
3. $R \propto \sqrt{M}$ and $R = 2$ when $M = 9$; find:
   (a) Law connecting R and M (b) R when $M = 36$,and M when $R = 12$.
4. The volume of a sphere is directly proportional to the cube of the radius. If the volume is 34cm^3 when the radius is 2cm. Find (a) equation relating V and R (b) the volume when $R = 3\text{cm}$ (c) the radius when $V = 532\text{cm}^3$.
5. $V \propto d$, and $V = 120$ when $d = 30$. Find the formula connecting V and d.
6. If $P \propto Q^2$ and $P = 16$ when $Q = 2$. Find P when $Q = 5$ and Q when $P = 128$.
7. The number of students enrolling to the university is directly proportional with time. If 200,000 students enrolled in the year 2009, find the number of students that will enrolled in the year 2014.
8. Given that $A \propto N$, and $A = 8$ when $N = 10$. Find:
   (a) The law of the variation (b) A when $N = 15$ and N when $A = 20$.
9. If Dollar($) and Naira (₦) are related by $\$ \propto ₦$, and if 1$ is equivalent to ₦25. Find the equivalent of $80.
10. Given that $y \propto x^{\frac{3}{2}}$, and $y = 27$, when $x = 4$. Find:
    (a) the law relating y and x (b) y when $x = 25$.

6.2 INVERSE VARIATION

Inverse variation is a type of variation such that the increase in a given variable (e.g. y) causes a decrease in another variable (e.g. x) vice – versa. Thus, $y \propto \frac{1}{x}$ where $\propto$ is the sign of proportionality connecting y and x.

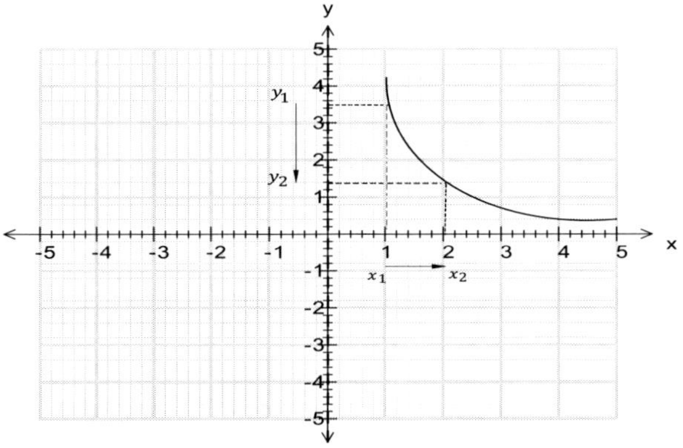

Fig 6.2 Inverse Variation

Example 4

Given that y varies inversely as x, if $y = 8$ when $x = 2$. Find:
(a) The relationship between x and y;
(b) The value of y when $x = 12$ and x when $y = \frac{1}{2}$.

Solution

$y \propto \frac{1}{x} \quad \Rightarrow y = k\frac{1}{x}$

(a) When $y = 8, x = 2 \quad \therefore k = yx = 8 \times 2 = 16$

$\therefore y = \frac{16}{x}$ This is the relationship between y and x.

(b) When $x = 12 \quad \Rightarrow y = \frac{16}{x} = \frac{16}{12} = \frac{4}{3}$

$\therefore \ y = \frac{4}{3}$

Similarly, when $y = \frac{1}{2}$

$y = \frac{16}{x} \quad \Rightarrow \frac{1}{2} = \frac{16}{x}$

$x = 2 \times 16 = 32$

$\therefore \ x = 32$

Example 5

The power P of received signal from a radio transmitter is inversely proportional to the square of the signal coverage distance, D. If $P = 50$Watt when $D = 2$km, find:
(a) the relationship connecting the power and signal coverage distance.
(b) the power needed to cover a signal of distance (i) 5km (ii) 0.3km.

Solution

$P \propto \frac{1}{D^2} \quad \Rightarrow P = k\frac{1}{D^2}$ where k is the proportionality constant of P and D.

(a) When $P = 50$ Watt, $D = 2$km

$\therefore \ P = k\frac{1}{D^2} \quad \Rightarrow 50 = k\frac{1}{2^2}$

$k = 50 \times 2^2 = 50 \times 4 = 200$

$\therefore P = \frac{200}{D^2}$ This is the relationship between P and D.

(b) (i) When $D = 5$km

$P = \frac{200}{D^2} \quad \Rightarrow P = \frac{200}{5^2} = \frac{200}{25} = 8$ Watt

(ii) When $D = 0.3$ km

$P = \frac{200}{D^2} = \frac{200}{0.3^2}$

$= \frac{200}{0.09} = 2222.2$ Watt

$\therefore P = 2222.2$ Watt

Example 6

A quantity M varies inversely as the cube of Q, and $M = 180$ when $Q = 3$. Find M when $Q = 2$.

Solution

$M \propto \frac{1}{Q^3} \quad \Rightarrow M = k\frac{1}{Q^3}$

When $M = 180, Q = 3$

$180 = k\frac{1}{3^3} \quad \Rightarrow k = 180 \times 3^3 = 180 \times 27 = 4860$

$\therefore M = \frac{4860}{Q^3}$

when $Q = 2$

$M = \frac{4860}{Q^3} = \frac{4860}{2^3} = \frac{4860}{8} = 607.5$

$\therefore M = 607.5$

Example 7

Graham's law of diffusion of a gas states that the rate of diffusion of a gas is inversely proportional to the square root of its vapour density. At a rate of 10cm³/s the vapour density of a gas was 9, find:

(a) the relationship between the rate of diffusion of gas and vapour density.

(b) the vapour density if the rate is increased to 25cm³/s.

Solution

From Graham's law $R \propto \frac{1}{\sqrt{d}} \quad \Rightarrow R = k\frac{1}{\sqrt{d}}$

(a) When $R = 10\text{cm}^3/\text{s},\ d = 9$

$R = k\frac{1}{\sqrt{d}} \quad \Rightarrow 10 = k\frac{1}{\sqrt{9}}$

$k = 10 \times \sqrt{9} = 10 \times 3 = 30$

$\therefore R = \frac{30}{\sqrt{d}}$

(b) When $R = 25\ \text{cm}^3/\text{s}$

$R = \frac{30}{\sqrt{d}} \quad \Rightarrow 25 = \frac{30}{\sqrt{d}}$

$\sqrt{d} = \frac{30}{25} = 1.2 \quad \Rightarrow d = 1.2^2 = 1.44$

$\therefore d = 1.44$

Exercise 6b

1. Given that $y \propto \frac{1}{x}$ and $y = 18$ when $x = 2$. Find the formula connecting x and y.
2. y is inversely proportional to x. If $y = 6$ when $x = 10$
   (a) what is the value of y when $x = 12$? (b) the value of x when $y = 8$
3. $P \propto \frac{1}{Q^2}$ and $P = 25$ when $Q = 3$, find:
   (a) the formula connecting P and Q
   (b) the value of P when $Q = 2$
   (c) value of Q when $P = 56\frac{1}{4}$
4. The time(t) taken to make a journey is inversely proportional to the velocity (v) of the car. When $t = 5\text{s}, v = 30\text{m/s}$, determine:
   (a) the law connecting v and t (b) v when $t = 20\text{s}$.
5. The cost of a shirt is inversely proportional to the number sold. If the cost is ₦25,000 for 3 shirts sold, find:
   (a) the law connecting the cost and number of shirts sold
   (b) the cost if 5 shirts were sold
6. $R \propto \frac{1}{\sqrt{M}}$, and $R = 25$ when $M = 9$.
   (a) find the relationship between R and M.
   (b) the value of R when $M = 64$.

6.3 JOINT VARIATION

Joint variation combines three or more quantities which can be related to one another through " Direct–Direct", " Direct–Inverse," or "Inverse –Inverse."

(i) If $y \propto x$, and $y \propto m$

$\therefore y \propto xm$

This is a direct- direct relationship between y, x and m.

(ii) If $y \propto x$, and $y \propto \frac{1}{m}$

$\therefore y \propto \frac{x}{m}$

This is a direct- inverse relationship between y, x and m.

(iii) If $y \propto \frac{1}{x}$, and $y \propto \frac{1}{m}$

$\therefore y \propto \frac{1}{xm}$

This is an inverse- inverse relationship between y, x and m.

Example 8

If y varies jointly as x and m; and $y = 50, m = 8$ when $x = 5$. Find:
(a) the equation relating y, x and m (b) y when $x = 10$ and $m = 16$

Solution

$y \propto xm \qquad \Rightarrow y = kxm$

(a) $y = 50, m = 8$ when $x = 5$

$50 = k \times 5 \times 8 \qquad \Rightarrow k = \frac{50}{5\times 8} = \frac{50}{40} = 1.25$

$\therefore\ y = 1.25xm$

(b) To find y when $x = 10$ and $m = 16$

$y = 1.25xm \qquad \Rightarrow y = 1.25xm = 1.25 \times 10 \times 16 = 200$

$\therefore\ y = 200$

Example 9

If $A \propto \frac{B}{C^2}$ where $A = 25, B = 20,$ and $C = 2$

(a) Express A in terms of B and C (b) Find A when $B = 15,$ and $C = 3$

(b) Find C when $A = 120,$ and $B = 6$

Solution

$A \propto \frac{B}{C^2} \qquad \Rightarrow A = k\frac{B}{C^2}$

(a) When $A = 25, B = 20,$ and $C = 2$

$A = k\frac{B}{C^2} \qquad \Rightarrow 25 = k\frac{20}{2^2}$

$k = \frac{25\times 2^2}{20} = \frac{25\times 4}{20} = \frac{100}{20} = 5$

$\therefore A = \frac{5B}{C^2}$

This expresses A in terms of B and C, which is also the same as the relationship between A, B and C.

(b) To Find A when $B = 15$ and $C = 3$

$A = \frac{5B}{C^2} \qquad \Rightarrow A = \frac{5\times 15}{3^2} = \frac{75}{9} = 8\frac{1}{3}$

(c) Find C when $A = 120$ and $B = 6$

$A = \frac{5B}{C^2} \qquad \Rightarrow 120 = \frac{5\times 6}{C^2}$

$C^2 = \frac{5\times 6}{120} = \frac{30}{120} = \frac{1}{4}$

$C = \sqrt{\frac{1}{4}} = \frac{1}{2}$

$\therefore\ C = \frac{1}{2}$

Example 10

P varies inversely as the cube of Q and square root of R. When $P = 16, Q = 2$ and $R = 9$, Find:

(a) the connecting equation between P, Q and R

(b) the value of P when $Q = 5,\ R = 36$.

Solution

$P \propto \frac{1}{Q^3\sqrt{R}} \quad \Rightarrow P = k\frac{1}{Q^3\sqrt{R}}$

(a) When $P = 16, Q = 2$, and $R = 9$

$P = k\frac{1}{Q^3\sqrt{R}} \quad \Rightarrow 16 = k\frac{1}{2^3\sqrt{9}} = k\frac{1}{8\times3}$

$k = 16 \times 8 \times 3 = 384$

$\therefore P = \frac{384}{Q^3\sqrt{R}}$

(b) Value of P when $Q = 5, R = 36$

$P = \frac{384}{Q^3\sqrt{R}} = \frac{384}{5^3\times\sqrt{36}} = \frac{384}{125\times6}$

$P = \frac{384}{750} = 0.512$

$\therefore P = 0.512$

Exercise 6e

1. $y \propto x$ and $y \propto z$ when $y = 6, x = 5, z = 2$. Determine:
(a) the relationship between y, x and z (b) the value of x when $y = 18,\ z = 3$.
2. $y \propto \frac{x}{z}$, $y = 12$ when $x = 6,\ z = 2$.
(a) find the law of variation (b) find y when $x = 15,\ z = 6$.
3. $P \propto \frac{Q^2}{R}$, and $Q = 3, R = 2$ when $P = 12$.
(a) find the equation relating P, Q and R (b) find P when $Q = 5,\ R = 4$.
4. The general gas law states that the volume(V) of a gas is directly proportional to its temperature(T) and inversely proportional to the pressure(P). When $V = 200\text{cm}^3$, $T = 240\text{k}$, and $P = 120\text{mmHg}$.
(a) Find the relationship between V, P and T
(b) Find the value of V when $P = 180\text{mmHg}$ and $T = 450\text{k}$
(c) Find P when $V = 150\text{cm}^3$ and $T = 600\text{k}$.
5. The capacitance of a capacitor(C) varies directly with the area (A) of the parallel plates and inversely with their distance (L) apart. When $C = 120\text{F}$, $A = 0.5\text{mm}^2$ and $L = 0.1\text{mm}$.
(a) Find the connecting equation between C, A and L.
(b) What is the value of the capacitance, C, when $A = 0.2\text{mm}^2$ and $L = 1.2\text{mm}$?
6. The signal power (P) of an antenna is directly proportional to the height (H) above the ground, and inversely to the square root of the radius (R) of the dish. Determine:
(a) the connecting factor between P, H and R when $P = 75\text{Watt}$, $H = 10\text{m}$, $R = 100\text{cm}$
(b) the height (H) required to produce a signal power of 120Watt with a dish of given radius, 25cm.

7. The kinetic energy (E) of a car is jointly dependent on the mass (m) of the car and square of its velocity (v). When the mass was 500kg, the velocity is 20m/s and kinetic energy is 1×10^5J.
(a) What is the connecting formula between the kinetic energy, velocity, and mass of the car?
(b) Find the kinetic energy of a car of mass 1000Kg and velocity 30m/s.

8. x is directly proportional to y and inversely proportional to z. If $x = 18$ when $y = 36$, $z = 16$. What is the value of x when $y = 15,\ z - 12$?

6.4 PARTIAL VARIATION

This is sometimes called "part variation". It is a variation that has at least two constants; whereby, the variables depend partly on one of the constants.
Partial variation can be of the following forms:
(i) A variable (e.g. y) is partly constant and partly varies directly as another variable (e.g. x)

$y \propto k + x$

$\therefore y = k + cx$ where: k is the part constant that y depends upon;
c is the constant of variation;
x is the variable that y depends upon.

(ii) A variable (e.g. y) is partly constant and partly varies inversely as another variable (e.g. x)

$$y \propto k + \frac{1}{x}$$

$$\therefore y = k + c\frac{1}{x}$$

(iii) A variable (e.g. y) is partly constant and partly varies jointly with other variables (e.g. x and z).

$y \propto k + xz$ or $y \propto k + \frac{z}{x}$ or $y \propto k + \frac{1}{xz}$

$\therefore y = k + cxz$ or $y = k + c\frac{z}{x}$ or $y = k + c\frac{1}{xz}$

Example 11

The cost (C) of a telephone call is partly constant and partly varies with the time(T) it takes to make the call. If the cost for 2min is ₦20, and 5min is ₦ 41. Find:
(a) the formula connecting the cost, C, and time T;
(b) The cost per minute;
(c) the cost of 60-minute call.

Solution

$C \propto k + T \quad \Rightarrow C = k + bT$ where k is the part constant and b is the constant of variation

(a) When $C =$ ₦20, $T = 2\text{min}$

$C = k + bT \quad \Rightarrow 20 = k + 2b$ ....(i)

When $C =$ ₦41, $T = 5\text{min}$

$C = k + bT \quad \Rightarrow 41 = k + 5b$ ....(ii)

Solve (i) and (ii) simultaneously using elimination method

$20 = k + 2b$

$41 = k + 5b$

$21 = 0 + 3b$

$\therefore b = \frac{21}{3} = 7$

Put b into equation (i) (i.e. $20 = k + 2b$)

$20 = k + 2(7) \quad \Rightarrow k = 20 - 14 = 6$

But, $C = k + bT \quad \Rightarrow C = 6 + 7T$

$\therefore$ the connecting factor between C and T is given by: $C = 6 + 7T$

(b) Cost per minute (i.e. when $T - 1$minute)

$C = 6 + 7T = 6 + 7(1) = 13$

$\therefore C =$ ₦13 per minute

(c) Cost for 60 minute call (i.e. when $T = 60$ minute)

$C = 6 + 7T = 6 + 7(60) = 6 + 420 = 426$

$\therefore C =$ ₦426

Example 12

P is partly constant with a and partly varies inversely with square root of a variable y. When $P = 20, y = 4$ and $P = 50, y = 9$. Find (a) The relationship between P and y

(b) value of P when $y = 100$.

Solution

$P \propto a + \frac{1}{\sqrt{y}} \Rightarrow P = a + c\frac{1}{\sqrt{y}}$ where: a is the part constant;

c is the constant of variation.

(a) When $P = 20, y = 4$

$20 = a + c\frac{1}{\sqrt{4}}$ ....(i)

When $P = 50, y = 9$

$50 = a + c\frac{1}{\sqrt{9}}$ ....(ii)

Solve (i) and (ii) simultaneously using elimination method (i.e. subtract equation (i) from (ii))

$$20 = a + c\frac{1}{\sqrt{4}}$$
$$50 = a + c\frac{1}{\sqrt{9}}$$
$$30 = c\left(\frac{1}{\sqrt{9}} - \frac{1}{\sqrt{4}}\right)$$

Thus, $c\left(\frac{1}{3} - \frac{1}{2}\right) = 30 \quad \Rightarrow -\frac{1}{6}c = 30$

$\therefore c = 30 \times (-6) = -180$

Put c into equation (i) to find a:

$20 = a + c\frac{1}{\sqrt{4}} \quad \Rightarrow 20 = a + (-180)\frac{1}{\sqrt{4}} = a - \frac{180}{2}$

$a - \frac{180}{2} = 20$

$a - 90 = 20 \quad \Rightarrow a = 20 + 90 = 110$

$\therefore a = 110$

But, $P = a + c\frac{1}{\sqrt{y}} \quad \Rightarrow P = 110 - 180\frac{1}{\sqrt{y}}$

$\therefore P = 110 - 180\frac{1}{\sqrt{y}}$ This is the relationship between P and y.

(b) Value of P when $y = 100$

$P = 110 - 180\frac{1}{\sqrt{y}} = 110 - 180\frac{1}{\sqrt{100}} = 110 - \frac{180}{10}$

$P = 110 - 18 = 92$

$\therefore P = 92$

Example 13

Two variables are connected by the formula $P = k + cq^2$, where k and c are constants. When $P = 200,\ q = 5$, and when $P = 167, q = 6$. Find:
(a) the value of k and c (b) value of P when $q = 10$.

Solution

$P = k + cq^2$

(a) When $P = 200,\ q = 5$

$200 = k + c(5)^2 = k + 25c$ ....(i)

when $P = 167,\ q = 6$

$167 = k + c(6)^2 = k + 36c$ ....(ii)

Solve equation (i) and (ii) simultaneously using elimination method (i.e. subtract equation (ii) from (i))

Thus,

$$200 = k + 25c$$
$$- \quad 167 = k + 36c$$
$$33 = 0 - 11c$$

$\therefore\ c = \dfrac{33}{-11} = -3$

Put this value into equation (i)

$200 = k + 25c \quad \Rightarrow \ 200 = k + 25(-3) = k - 75$

$\therefore\ k - 200 + 75 - 275$

$\therefore c = -3, \qquad k = 275$

(b) value of P when $q = 10$

$P = k + cq^2 = 275 + (-3)q^2$

$P = 275 - 3(10)^2 = 275 - 3 \times 100 = 275 - 300$

$\therefore P = -25$

Exercise 6d

1. y is partly constant and partly varies as x. If $y = 30$, when $x = 2$; and $y = 45$ when $x = 5$.
   (a) Find the relationship between x and y.
   (b) What is the value of y when $x = 10$?
2. Two variables are connected by the formula $P = k + \frac{c}{q}$, where k and c are constants. When $P = 250$, $q = 10$, and when $P = 100, q = 5$. Find k and c.
3. M is partly constant and partly varies as the square of N. When $M = 40, N = 5$ and when $M = 90, N = 4$.
   (a) Find the equation relating M and N
   (b) Find the value of M when $N = 3$ and the value of N when $M = 120$.
4. The escape velocity (V) of a spaceship is partly constant and jointly depends on the acceleration due to gravity(g) and radius(R) of a given planet.
   When $V = 2$ km/s, g=10m/s^2 and $R = 6400$km, while g =15m/s^2 and $R = 5000$km when $V = 5$ km/s. Find:
   (a) the formula connecting V, R and g.
   (b) the escape velocity for a planet with $R = 7000$km and g = 12m/s^2?

OBJECTIVE TEST 6

1. If y varies directly as $\sqrt{x}$ and $y = 10$ when $x = 16$. Find the value of x for which $y = 20$.
   (A) 4 (B) 8 (C) 16 (D) 64
2. If $P \propto R$; and $P = 10$ when $R = 6$. Find the relationship between P and R. (A) $P = \frac{5}{3}R$
   (B) $P = \frac{4}{3}R$ (C) $P = \frac{3}{5}R$ (D) $P = \frac{2}{3}R$
3. y varies directly as the cube root of x, and $y = 3$ when $x = 64$. Find x when $y = \frac{15}{4}$
   (A) 125 (B) 25 (C) 625 (D) 5
4. If y varies inversely as x^3 and $y = 4$ when $x = 2$. Find the value of y when $x = 1$.
   (A) 8 (B) 24 (C) 32 (D) 48
5. The number of reams of paper to be produced varies inversely as the cost; at a volume of 50 reams the cost is ₦35. Find the cost of 700 reams.
   (A) ₦3.5 (B) ₦2.5 (C) ₦1.5 (D) ₦1
6. The power (P) generated by an electrical machine is inversely proportional to the period (T) of rotation of the coil. When $P = 1000$ Watt, $T = 0.2$s. Find the period with a power of 400 Watt.
   (A) 0.5s (B) 0.4s (C) 0.3s (D) 2s
7. If y varies directly as d and inversely as the square of x. When $y = \frac{1}{18}, x = 24$ and $d = 30$. Find the law connecting the variables.
   (A) $y = \frac{3d}{5x^2}$ (B) $y = \frac{12d}{13x^2}$ (C) $y = \frac{15d}{16x^2}$ (D) $y = \frac{16d}{15x^2}$
8. A variable w varies jointly as h and the square of r. If $w = 15$ when $h = 3,\ r = \frac{5}{2}$, find the value of w when $h = 1,\ r = 10$.
   (A) 8 (B) 80 (C) 125 (D) 40
9. z varies partly as x^2 and partly as $\frac{y}{x}$. When $x = 1, y = 1$, then $z = 5$; and when $x = -1$, $y = 2,\ z = 20$. Find the value of z when $x = 2$ and $y = 4$.
   (A) -30 (B) 30 (C) 10 (D) -50
10. M is partly constant and partly varies as P. When $M = 10,\ P = 3$; and when $M = 25$, $P = 8$. Find the relationship between M and P.
   (A) $M = 3P - 1$ (B) $M = 2P + 1$ (C) $M = 3P + 1$ (D) $P = 3M + 1$

7 SEQUENCE AND SERIES

A sequence is a set of numbers arranged consecutively following one another in a defined order, whereby, the members of the sequence are called the "terms" of the sequence. The common types of sequence are Arithmetic progression (A.P.) and Geometric progression (G.P.).

7.1 ARITHMETIC PROGRESSION (A.P)

Arithmetic progression or sequence is a sequence in which the difference between a term and the immediate preceding term of the sequence is constant. Hence, we say that the arithmetic progression has a common difference - usually denoted as d.
Thus, $d = U_2 - U_1 = U_3 - U_2 = \cdots = U_n - U_{n-1}$
And an A.P is of the form $a,\ a+d, a+2d, a+3d, a+4d, \ldots$

7.1.1 Nth TERM OF AN ARITHMETIC PROGRESSION

The nth term of the A.P represented as U_n.

$$\therefore\ U_n = a + (n-1)d$$

where: a is the first term of the sequence;
n is the number of terms of the sequence; d is the common difference.

7.1.2 SUM OF ARITHMETIC PROGRESSION

Let S_n denote the sum of an A.P where n is the number of terms in the series :
$a, \quad a+d, \quad a+2d, \quad a+3d, \quad a+4d, \ldots, U_n$
thus, $S_n = 1^{st} term + 2^{nd} term + 3^{rd} term + 4^{th} term + \cdots + n^{th}\ term$
$S_n = a + (a+d) + (a+2d) + \cdots + U_{n-1} + U_n \qquad$ where $U_n = a + (n-1)d$
$S_n = a + (a+d) + (a+2d) + \cdots + \{a + (n-2)d\} + \{a + (n-1)d \qquad$ ....(i)
Now, add this terms from the last term (i.e., in reverse order)
$S_n = \{a + (n-1)d\} + \{a + (n-2)d\} + \{a + (n-3)d + \cdots + (a+d) + a$ ....(ii)
Combine each term in column of equation (i) to that in column of equation (ii)
$\therefore 2S_n = \{2a + (n-1)d\} + \{2a + (n-1)d\} + \{2a + (n-1)d\} + \cdots + \{2a + (n-1)d\}$
That means all the terms on the right-hand side equation are the same. Hence, we represent their summation for n terms as:
$2S_n = n \times \{2a + (n-1)d\}$

$$\therefore\ S_n = \frac{n}{2}\{2a + (n-1)d\}$$

Alternatively, the sum of n terms of an AP can be proved using:
$S_n = \frac{1}{2}n(1^{st} term + last\ term)$
where the last term is given as $U_n = a + (n-1)d$
then, $S_n = \frac{1}{2}n(a + a + (n-1)d) = \frac{n}{2}(2a + (n-1)d)$

$$\therefore\ S_n = \frac{n}{2}\{2a + (n-1)d\}$$

7.1.3 ARITHMETIC MEAN

The arithmetic mean of two numbers x and y, is a number z between x and y whereby x, y and z are terms in the arithmetic progression.

Thus, $z = \frac{x+y}{2}$

where z is the arithmetic mean of two numbers x and y in an Arithmetic Progression.

Example 1

Find the common difference in each of the following sequence:
(a) $5, 9, 13, 17, \ldots$ (b) $100, 92, 84, 76, \ldots$
(c) $-\frac{1}{2}, \frac{1}{2}, \frac{3}{2}, \frac{5}{2}, \ldots$ (d) $\frac{2}{3}, -\frac{1}{3}, -\frac{4}{3}, -\frac{7}{2}, \ldots$

Solution

(a) $5, 9, 13, 17, \ldots$
$$d = U_2 - U_1 = 9 - 5 = 4$$
$$\therefore\ d = 4$$
(b) $100, 92, 84, 76, \ldots$
$$d = U_2 - U_1 = 92 - 100 = -8$$
$$\therefore d = -8$$
(c) $-\frac{1}{2}, \frac{1}{2}, \frac{3}{2}, \frac{5}{2}, \ldots$
$$d = U_2 - U_1 = \frac{1}{2} - \frac{-1}{2} = \frac{1}{2} + \frac{1}{2} = 1$$
$$\therefore\ d = 1$$
(d) $\frac{2}{3}, -\frac{1}{3}, -\frac{4}{3}, -\frac{7}{2}, \ldots$
$$d = U_2 - U_1 = \frac{-1}{3} - \frac{2}{3} = \frac{-3}{3} = -1$$
$$\therefore\ d = -1$$

Example 2

Find the 8th term and sum of the first ten terms of the sequence
5, 8, 11, 14, 17,...

Solution

5, 8, 11, 14, 17,...

$a = 5, d = U_2 - U_1 = 8 - 5 = 3$

The 8th term $= U_8 = a + (n-1)d$ where $n = 8$

$U_8 = 5 + (8-1) \times 3 = 5 + 7 \times 3 = 5 + 21 = 26$

$\therefore U_8 = 26$

Now, the sum of ten terms is given by:

$S_n = \frac{n}{2}\{2a + (n-1)d\}$ where $n = 10$

$S_{10} = \frac{10}{2}\{2 \times 5 + (10-1) \times 3\} = 5\{10 + 9 \times 3\} = 5\{10 + 27\}$

$= 5 \times 37 = 185$

$\therefore S_{10} = 185$

Example 3

The 4th term of an A.P is 10, and the 6th term is 16. Find the 9th term of the A.P.

Solution

$a + 3d = 10$ ....(i)

$-\ a + 5d = 16$ ....(ii)

$0 + 2d = 6$

$\therefore d = \frac{6}{2} = 3$

using equation (i), we have:

$a + 3(3) = 10$

$a + 9 = 10 \Rightarrow a = 10 - 9 = 1$

$\therefore a = 1, \quad d = 3$

The 9th term $= U_9 = a + (n-1)d$

$U_9 = 1 + (9-1) \times 3 = 1 + 8 \times 3$

$= 1 + 24 = 25$

$\therefore U_9 = 25$

Subtract equation (i) from (ii) and then solve for d. After which d is put into either equation (i) or (ii) to find a. Then, both a and d are now substituted into the equation of the nth term of an A.P in order to find the 9th term.

Example 4

5, x, y, z, 25 is an Arithmetic Progression. Find x, y and z.

Solution

$a = 5, U_2 = x, U_3 = y, U_4 = z, U_5 = 25$

But, $U_5 = a + (n-1)d = 25 \quad \Rightarrow U_5 = 5 + (5-1)d = 25$

$4d = 25 - 5 = 20 \quad \Rightarrow d = \frac{20}{4} = 5$

$U_2 = a + (n-1)d = a + d = 5 + 5 = 10$

$\therefore x = 10$

$U_3 = a + (n-1)d = a + 2d = 5 + 2 \times 5 = 5 + 10 = 15$

$\therefore y = 15$

$U_4 = a + (n-1)d = a + 3d = 5 + 3 \times 5 = 5 + 15 = 20$

$\therefore z = 20$

Example 5

A sequence is given by $\frac{(n+1)}{2}$ as the nth term, find the:

(a) sequence (b) 10th term of the progression.

Solution

(a) $U_n = \frac{n+1}{2}$

$U_1 = \frac{(1+1)}{2} = \frac{2}{2} = 1$

$\therefore\ U_1 = 1$

$U_2 = \frac{(2+1)}{2} = \frac{3}{2}$

$\therefore\ U_2 = \frac{3}{2}$

$U_3 = \frac{(3+1)}{2} = \frac{4}{2} = 2$

$\therefore U_3 = 2$

$U_4 = \frac{(4+1)}{2} = \frac{5}{2}$

$\therefore\ U_4 = \frac{5}{2}$

$U_5 = \frac{(5+1)}{2} = \frac{6}{2} = 3$

$\therefore U_5 = 3$

The sequence is $1,\ \frac{3}{2},\ 2,\ \frac{5}{2}, 3, \ldots$

(b) The 10th term will become, $U_{10} = \frac{(10+1)}{2} = \frac{11}{2}$

$\therefore\ U_{10} = \frac{11}{2}$

Example 6

A sequence is defined by $15,\ 10,\ 5,\ 0, -5, \ldots$

find: (a) the common difference (b) the 8th term (c) sum of 20 terms

Solution

$15, 10, 5, 0, -5, \ldots$

(a) $d = U_2 - U_1 = 10 - 15 = -5$

$\therefore$ the common difference, $d = -5$

(b) 8th term:

$a = 15, d = -5$

$U_8 = a + (n-1)d = a + (8-1)d = a + 7d = 15 + 7(-5)$

$= 15 - 35 = -20$

$\therefore U_8 = -20$

(c) Sum of 20 terms :

$S_n = \frac{n}{2}\{2a + (n-1)d\}$ where $n = 20$

$S_n = \frac{20}{2}\{2 \times 15 + (20-1) \times -5\} = 10\{30 + 19 \times (-5)\}$

$= 10(30 \quad 95) = 10 \times (-65) = -650$

$\therefore S_n = -650$

Example 7

Find the arithmetic mean of a sequence given that the third term is 5 and seventh term is 17.

Solution

$U_3 = a + 2d = 5$ ....(i)

$U_7 = a + 6d = 17$ ....(ii)

Subtracting equation (i) from (ii), we have:

$4d = 12 \quad \Rightarrow d = \frac{12}{4} = 3$

put d into equation (i) so as to find the value of a.

$a + 2d = 5 \quad \Rightarrow a + 2(3) = 5$

$a = 5 - 6 = -1$

$\therefore a = -1, d = 3$

the sequence is $-1, 2, 5, 8, 11, 14, 17$.

Arithmetic mean $= \frac{first\ term + last\ term}{2} = \frac{a + U_7}{2}$

$= \frac{-1+17}{2} = \frac{16}{2}$

Arithmetic mean $= 8$

Example 8

Evaluate:

(a) $\sum_{n=1}^{20} 5n$ (b) $\sum_{n=1}^{15} 3n - 2$

Solution

(a) $\sum_{n=1}^{20} 5n$ for $n = 1$ to $n = 20$

when $n = 1$

$U_1 = 5n = 5(1) = 5$

$\therefore$ first term $= a = U_1 = 5$

when $n = 2$

$U_2 = 5n = 5(2) = 10$

The sigma notation $\sum$ means the summation of the given sequence. Thus, it requires that we find the sum of the given A.P.

$d = U_2 - U_1 = 10 - 5 = 5$

$\therefore$ common difference $d = 5$

$$\Rightarrow S_n = \frac{20}{2}\{2a + (n-1)d\} = \frac{20}{2}\{2 \times 5 + (20-1) \times 5\} = 10 \times \{10 + 19 \times 5\}$$

$$= 10 \times \{10 + 95\} = 10 \times 105 = 1050$$

$\therefore S_{20} = 1050$

(b) $\sum_{n=1}^{15} 3n - 2$ from $n = 1$, to $n = 15$

when $n = 1$

$U_1 = 3n - 2 = 3(1) - 2 = 3 - 2 = 1$

$\therefore$ first term $= a = U_1 = 1$

when $n = 2$

$U_2 = 3n - 2 = 3(2) - 2 = 6 - 2 = 4$

$d = U_2 - U_1 = 4 - 1 = 3$

$\therefore$ common difference $d = 3$

$$\Rightarrow S_{15} = \frac{15}{2}\{2a + (n-1)d\} = \frac{15}{2}\{2 \times 1 + (15-1) \times 3\} = \frac{15}{2}\{2 + 14 \times 3\}$$

$$= \frac{15}{2}\{2 + 42\} = \frac{15}{2} \times 44 = \frac{660}{2} = 330 \quad \therefore \quad S_{15} = 330$$

Example 9

The n terms of an A.P is given by $5n^2 + 3n$, find:

(a) the first term and common difference of the sequence

(b) the fifth term of the sequence.

Solution

(a) The n terms of the sequence is defined by, $U_n = 5n^2 + 3n$.

when $n = 1$

$U_1 = 5n^2 + 3n = 5(1)^2 + 3(1) = 5 + 3 = 8$

But, first term $= a = S_1 = 8$

$\therefore a = 8$

when $n = 2$, we have that:

$U_2 = 5n^2 + 3n = 5(2)^2 + 3(2) = 5 \times 4 + 3 \times 2$

$= 20 + 6 = 26$

$d = S_2 - S_1 = 26 - 8 = 18$

$\therefore$ common difference $= d = 18$

(b) The fifth term(i.e. when $n = 5$) becomes:

$U_5 = 5n^2 + 3n = 5(5)^2 + 3(5) = 5 \times 25 + 15$

$= 125 + 15 = 140$

$\therefore U_5 = 140$

The sum of three consecutive numbers in an A.P is 12, and the sum of their squares is 66, find the numbers.

Solution

Let the numbers be x, y and z; then their sequence is given by x, y, z.

$\therefore$ first term $= U_1 = x$

second term $= U_2 = y = x + d$

where d is the common difference $= y - x$

and, third term$= U_3 = z = y + d = x + 2d$

$x + y + z = 12$ ....(i)

$x^2 + y^2 + z^2 = 66$ ....(ii)

Put $y = x + d$ and $z = x + 2d$ into equation (i)

$\Rightarrow x + (x + d) + (x + 2d) = 12$

$3x + 3d = 12$

$\therefore x + d = 4$ ....(iii)

Similarly, put $y = x + d$ and $z = x + 2d$ into equation (ii)

$\Rightarrow x^2 + (x + d)^2 + (x + 2d)^2 = 66$

$x^2 + (x + d)^2 + (x + d + d)^2 = 66$

But from equation (iii), $x + d = 4$ or $x = 4 - d$

$\therefore x^2 + (x + d)^2 + (x + d + d)^2 = (4 - d)^2 + 4^2 + (4 + d)^2 = 66$

$(16 - 8d + d^2) + 16 + (16 + 8d + d^2) = 66$

$2d^2 + 48 = 66 \qquad \Rightarrow 2d^2 = 66 - 48 = 18$

$d^2 = \frac{18}{2} = 9$

$d = \sqrt{9} = \pm 3 \qquad \Rightarrow d = 3 \text{ or } -3$

when $d = 3$

then, $x = 4 - d = 4 - 3 = 1$

and $\quad y = x + d = 1 + 3 = 4$

$z = x + 2d = 1 + 2(3) = 1 + 6 = 7$

when $d = -3$

$x = 4 - d = 4 - (-3) = 4 + 3 = 7$

$y = x + d = 7 + (-3) = 7 - 3 = 4$

$z = x + 2d = 7 + 2(-3) = 7 - 6 = 1$

The sequence is x, y, z

$\Rightarrow 1, 4, 7 \quad$ OR $\quad 7, 4, 1.$

Exercise 7a

1. Find the common difference of each of the following sequence:
   (a) $3, 5, 7, 9, \dots$
   (b) $256, 248, 240, \dots$
   (c) $-15, 10, 35, 60, \dots$
   (d) $\frac{1}{2}, \frac{3}{2}, \frac{5}{2}, \dots$
   (e) $-\frac{1}{5}, \frac{3}{10}, \frac{4}{5}, \dots$
2. The fourth term of an A.P is 10, and the seventh term is 19. Find the tenth term of the A.P.
3. A sequence is given by $5, 9, 13, \dots$
   (a) Determine the tenth term
   (b) What term of the sequence has a value of 101?
4. What are the next three terms of the sequence $\frac{1}{3}, \frac{3}{5}, \frac{13}{15}, \dots$?
5. Find the common difference of the nth term of the sequence, $12 - 5n$
6. Write out the nth term of the following sequence
   (a) $20, 24, 28, \dots$
   (b) $5, 2, -1, \dots$
   (c) $-10, -7, -4, \dots$
7. If $5, 10, 15, \dots$ is an Arithmetic Progression, find:
   (a) the seventh term
   (b) twelfth term of the sequence
   (c) the sum of the first ten terms
8. Find the three-arithmetic means between 23 and 35
9. An A.P has a sum given as $S_n = 7n^2 - 3n$. Find:
   (a) the first term of the sequence
   (b) the third term of the sequence
10. The last term in a sequence of seven terms is 25. If the common difference is 2, find the arithmetic mean.
11. Find the value of the arithmetic series $2 + 5 + 8 + 11 + 14 + \cdots + 50$.
12. Evaluate:
   (a) $\sum_{n=1}^{15} 3n$ (b) $\sum_{n=1}^{5} 2n - 1$
13. Prove that the sum of an A.P is $S_n = \frac{n}{2}\{2a + (n-1)d\}$
14. The first term of an A.P is 5 and the nineth term is 15, find the number of terms in the progression if the sum is 75.

7.2 GEOMETRIC PROGRESSION (G.P)

This is also known as exponential sequence. It is a sequence in which the ratio of one term and that immediately preceding it is constant. This constant is called the common ratio, and it is denoted as r.

Thus, $r = \frac{U_2}{U_1} = \frac{U_3}{U_2} = \frac{U_4}{U_3} = \cdots = \frac{U_n}{U_{n-1}}$

7.2.1 N^th^ TERM OF A GEOMETRIC PROGRESSION

The nth term of a geometric progression is given as:

$$U_n = ar^{n-1}$$

where: $n = 1, 2, 3, 4, 5, \ldots$

a = first term of the geometric progression; U_n = the nth term.

7.2.2 SUM OF GEOMETRIC PROGRESSION

Let the sum of a G.P be S_n

$\therefore S_n = a + ar + ar^2 + ar^3 + ar^4 + \cdots + ar^{n-1}$ ....(i)

multiply equation (i) by the common ratio r

$rS_n = ar + ar^2 + ar^3 + ar^4 + ar^5 \ldots + ar^n$ ....(ii)

Subtract equation (ii) from (i)

$S_n - rS_n = a - ar^n$

$S_n(1 - r) = a(1 - r^n)$

$$\therefore S_n = \frac{a(1-r^n)}{1-r} \quad \text{for } r < 1$$

$$\therefore S_n = \frac{a(r^n-1)}{r-1} \quad \text{for } r > 1$$

7.2.3 SUM TO INFINITY

The sum to infinity of a geometric progression is given as:

$$S_\infty = \frac{a}{1-r}$$

where a is the first term, and r is the common ratio.

7.2.4 GEOMETRIC MEAN

Given three consecutive terms x, y, z of a G.P; then the common ratio is given by, $r = \frac{y}{x} = \frac{z}{y}$

Thus, $\frac{y}{x} = \frac{z}{y} \Rightarrow y^2 = xz$

$\therefore \quad y = \sqrt{xz}$ where y is the geometric mean of x and z.

Example 11

Find the common ratio of each of the following series:

(a) $2, 6, 18, 54, \ldots$ (b) $3, 9, 27, 81, \ldots$ (c) $-2, 1, -\frac{1}{2}, \frac{1}{4}, \ldots$

(d) $24, 48, 96, \ldots$ (e) $x^2, x, 1, \frac{1}{x}, \frac{1}{x^2}, \ldots$ (f) $\log a, \log a^2, \log a^4, \ldots$

Solution

(a) $2, 6, 18, 54, \ldots$

$r = \frac{U_2}{U_1} = \frac{ar}{a}$

$U_2 = 6, U_1 = 2$

$\therefore r = \frac{6}{2} = 3$

(b) $3, 9, 27, 81, \ldots$

$r = \frac{U_2}{U_1} = \frac{ar}{a}$

$U_2 = 3, U_1 = 9$

$\therefore r = \frac{9}{3} = 3$

(c) $-2, 1, -\frac{1}{2}, \frac{1}{4}, \ldots$

$r = \frac{U_2}{U_1} = \frac{ar}{a}$

$U_2 = 1, U_1 = -2$

$\therefore r = \frac{1}{-2}$

(d) $24, 48, 96, \ldots$

$r = \frac{U_2}{U_1} = \frac{ar}{a}$

$U_2 = 48, U_1 = 24$

$\therefore r = \frac{48}{24} = 2$

(e) $x^2, x, 1, \frac{1}{x}, \frac{1}{x^2}, \ldots$

$r = \frac{U_2}{U_1} = \frac{ar}{a}$

$U_2 = x, U_1 = x^2$

$\therefore r = \frac{x}{x^2} = \frac{1}{x}$

(f) $\log a, \log a^2, \log a^4, \ldots$

$r = \frac{U_2}{U_1} = \frac{ar}{a}$

$U_2 = \log a^2, U_1 = \log a$

$\therefore r = \frac{\log a^2}{\log a} = \frac{2\log a}{\log a} = 2$

Example 12

The second term and fifth term of a G.P are 12 and 96 respectively, find:
(a) the common ratio and first term of the progression
(b) the eight term
(c) the sum of the first five terms.

Solution

(a) $U_2 = ar = 12$ ....(i)

$U_5 = ar^4 = 96$ ....(ii)

Divide equation (ii) by (i)

$\frac{ar^4}{ar} = \frac{96}{12} \Rightarrow r^3 = 8$

$r = \sqrt[3]{8} = 2$ using the positive root of r

$\therefore r = 2$

Put r into equation (i) to find the value of a.
Thus, $ar = 12 \quad \Rightarrow a(2) = 12$
$a = \frac{12}{2} = 6 \quad \therefore a = 6$

(b) The eighth term (i.e., when $n = 8$):
$U_8 = ar^{n-1} = ar^{8-1} = ar^7$
$= 6 \times 2^7 = 6 \times 128 = 768$
$\therefore U_8 = 768$

(c) Sum of the first five terms (i.e., S_n for $n = 5$)
$S_n = \frac{a(r^n-1)}{r-1}$ where: $a = 6, r = 2$ and $n = 5$
$S_8 = \frac{6 \times (2^5 - 1)}{2 - 1} = \frac{6 \times (32 - 1)}{1} = 6 \times 31 = 186$
$\therefore S_8 = 186$

Example 13

Given the sequence $1, x, x^2, x^3, \ldots$
(a) Find the common ratio (b) write the sum of the n terms.

Solution

(a) $r = \frac{U_2}{U_1} = \frac{ar}{a}$ where $U_2 = x, U_1 = 1$
$r = \frac{x}{1} = x \quad \therefore r = x$
(b) $S_n = \frac{a(r^n-1)}{r-1}$ where $a = 1, r = x$
Thus, $S_n = \frac{1(x^n-1)}{x-1}$
$\therefore S_n = \frac{(x^n-1)}{x-1}$

Example 14

The sum of the first n terms of a G.P is given as $S_n = 2^n - 1$, find:
(a) the first term of the G.P
(b) the common ratio
(c) the sum of ten terms of the progression.

Solution

(a) $S_n = 2^n - 1$
when $n = 1$
$S_1 = 2^1 - 1 = 2 - 1 = 1$
the first term $a = S_1 = U_1$
$\therefore a = 1$

(b) Common ratio:

$$r = \frac{U_2}{U_1} = \frac{ar}{a}$$

$S_2 = U_1 + U_2 \qquad \Rightarrow U_2 = S_2 - U_1$

Using $S_n = 2^n - 1$

$S_2 = 2^2 - 1 = 4 - 1 = 3$

$U_2 = S_2 - U_1 = 3 - 1 = 2$

$\therefore r = \frac{U_2}{U_1} = \frac{2}{1} = 2$

(c) The sum of ten terms of the progression (i.e., S_n for $n = 10$)

$S_n = 2^n - 1 \quad \Rightarrow S_{10} = 2^{10} - 1 = 1024 - 1 = 1023$

$\therefore S_{10} = 1023$

Example 15

Find the common ratio and sum to infinity of the following sequence:

(a) $-1 + \frac{2}{3} - \frac{4}{9} + \cdots$ (b) $\frac{1}{2} - \frac{1}{4} + \frac{1}{8} - \frac{1}{16} + \cdots$ (c) $1 + \frac{3}{10} + \left(\frac{3}{10}\right)^2 + \left(\frac{3}{10}\right)^3 + \cdots$

Solution

(a) $-1 + \frac{2}{3} - \frac{4}{9} + \cdots$

$U_1 = -1, U_2 = \frac{2}{3} \qquad \Rightarrow r = \frac{U_2}{U_1} = \frac{\frac{2}{3}}{-1} = -\frac{2}{3}$

$\therefore$ Common ratio, $r = -\frac{2}{3}$

Sum to infinity, $S_\infty = \frac{a}{1-r}$ where: a = first term $= -1$

r = common ratio $= -\frac{2}{3}$

$$\therefore S_\infty = \frac{a}{1-r} = \frac{-1}{1-(-\frac{2}{3})} = \frac{-1}{1+\frac{2}{3}} = \frac{-1}{5/3}$$

$\therefore S_\infty = -\frac{3}{5}$

(b) $\frac{1}{2} - \frac{1}{4} + \frac{1}{8} - \frac{1}{16} + \cdots$

$U_1 = \frac{1}{2}, U_2 = -\frac{1}{4} \qquad \Rightarrow r = \frac{U_2}{U_1} = \frac{-1/4}{1/2} = -\frac{1}{4} \times \frac{2}{1} = -\frac{1}{2}$

$\therefore$ Common ratio, $r = -\frac{1}{2}$

Sum to infinity, $S_\infty = \frac{a}{1-r}$ where: a = first term $= \frac{1}{2}$

r = common ratio $= -\frac{1}{2}$

$\therefore\ S_\infty = \frac{a}{1-r} = \frac{\frac{1}{2}}{1-(-\frac{1}{2})} = \frac{\frac{1}{2}}{1+\frac{1}{2}} = \frac{1/2}{3/2}$

$= \frac{1}{2} \times \frac{2}{3} = \frac{1}{3}$

$\therefore\ S_\infty = \frac{1}{3}$

(c) $1+\frac{3}{10} + \left(\frac{3}{10}\right)^2 + \left(\frac{3}{10}\right)^3 + \ldots$

$U_1 = 1,\ U_2 = \frac{3}{10} \qquad \Rightarrow r = \frac{U_2}{U_1} = \frac{3/10}{1} = \frac{3}{10}$

$\therefore$ Common ratio, $r = \frac{3}{10}$

Sum to infinity, $S_\infty = \frac{a}{1-r}$ where: a = first term = 1, r = common ratio $= \frac{3}{10}$

$\therefore\ S_\infty = \frac{a}{1-r} = \frac{1}{1-\left(\frac{3}{10}\right)} = \frac{1}{1-\frac{3}{10}}$

$= \frac{1}{7/10} = 1 \times \frac{10}{7} = \frac{10}{7}$

$\therefore\ S_\infty = \frac{10}{7}$

Example 16

Evaluate the following:

(a) $\left(\frac{1}{2} + \frac{1}{4} + \frac{1}{8} + \cdots\right) - 2$ (b) $0.\dot{3}\dot{2}$ (c) $0.\dot{8}5\dot{4}$

Solution

(a) $\left(\frac{1}{2} + \frac{1}{4} + \frac{1}{8} + \cdots\right) - 2$

A careful look at this series shows that the final sum of the series is the subtraction of 2 from the sum to infinity of the G.P. (i.e. $S = S_\infty - 2$).

$U_1 = \frac{1}{2}, U_2 = \frac{1}{4}$

$\Rightarrow r = \frac{U_2}{U_1} = \frac{1/4}{1/2} = \frac{1}{4} \times \frac{2}{1} = \frac{1}{2}$

$\therefore$ Common ratio, $r = \frac{1}{2}$

Then, $S = S_\infty - 2 = \frac{a}{1-r} - 2$

where: a = first term $= \frac{1}{2}$ and r = common ratio $= \frac{1}{2}$

$\therefore\ S = \frac{a}{1-r} - 2 = \frac{\frac{1}{2}}{1-\left(\frac{1}{2}\right)} - 2 = \frac{\frac{1}{2}}{1-\frac{1}{2}} - 2$

$= \frac{1/2}{1/2} - 2 = 1 - 2 = -1$

$\therefore\ S = -1$

(b) $0.\dot{3}\dot{2} = 0.32323232\ldots = \frac{32}{100} + \frac{32}{1000} + \frac{32}{10000} + \cdots$

$U_1 = \frac{32}{100},\ U_2 = \frac{32}{1000} \quad \Rightarrow r = \frac{U_2}{U_1} = \frac{32/1000}{32/100} = \frac{1}{10}$

$\therefore$ Common ratio, $r = \frac{1}{10}$

Sum to infinity, $S_\infty = \frac{a}{1-r}$ where a = first term $= \frac{32}{100}$, r = common ratio $= \frac{1}{10}$

$\therefore\ S_\infty = \frac{a}{1-r} = \frac{32/100}{1-\left(\frac{1}{10}\right)} = \frac{32/100}{1-\frac{1}{10}}$

$= \frac{32/100}{9/10} = \frac{32}{100} \times \frac{10}{9} = \frac{16}{9}$

$\therefore\ S_\infty = \frac{16}{45}$

(c) $0.\dot{8}5\dot{4} = 0.854854854\ldots = \frac{854}{1000} + \frac{854}{10000} + \frac{854}{100000} + \cdots$

$U_1 = \frac{854}{1000},\ U_2 = \frac{854}{10000} \quad \Rightarrow r = \frac{U_2}{U_1} = \frac{854/1000}{854/10000} = \frac{1}{10}$

$\therefore$ Common ratio, $r = \frac{1}{10}$

Sum to infinity, $S_\infty = \frac{a}{1-r}$ where: a = first term $= \frac{854}{1000}$, r = common ratio $= \frac{1}{10}$

$\therefore S_\infty = \frac{a}{1-r} = \frac{854/1000}{1-\left(\frac{1}{10}\right)} = \frac{854/1000}{1-\frac{1}{10}}$

$= \frac{854/1000}{9/10} = \frac{854}{1000} \times \frac{10}{9} = \frac{854}{900}$

$\therefore S_\infty = \frac{854}{900}$

Example 17

Find the geometric mean between:

(a) 2 and 8 (b) 6 and 24

Solution

(a) 2 and 8

Let the geometric mean be x

$\therefore \frac{2}{x} = \frac{x}{8}$

$x^2 = 2 \times 8 = 16$

$x = \sqrt{16} = 4$

$\therefore$ the geometric mean $= 4$

(b) 6 and 24

$\therefore \frac{6}{x} = \frac{x}{24}$

$x^2 = 6 \times 24 = 144$

$x = \sqrt{144} = 12$

$\therefore$ the geometric mean $= 12$

Example 18

Insert three geometric mean between (a) 2 and 162 (b) 4 and 2500

Solution

(a) In order to insert three geometric mean between 2 and 162 let the G. P be $2, x, y, z, 162$

$\Rightarrow U_1 = a = 2$ ....(i)

$U_5 = ar^4 = 162$ ....(ii)

Dividing equation (ii) by (i)

$\frac{ar^4}{a} = \frac{162}{2}$

$r^4 = 81 \quad \Rightarrow r = \sqrt[4]{81} = 3$

Now, $x = U_2 = ar = 2 \times 3 = 6$

$y = U_3 = ar^2 = 2 \times 3^2 = 2 \times 9 = 18$

$z = U_4 = ar^3 = 2 \times 3^3 = 2 \times 27 = 54$

$\therefore$ the three geometric mean are: $x = 6, y = 18$ and $z = 54$

Thus, the sequence is $2, 6, 18, 54, 162$.

(b) 4 and 2500

Let the three geometric mean between 4 and 2500 be p, q, s.

$\therefore$ the sequence is $4, p, q, s, 2500$.

$\Rightarrow U_1 = a = 4$ ....(i)

$U_5 = ar^4 = 2500$ ....(ii)

Dividing equation (ii) by (i)

$\frac{ar^4}{a} = \frac{2500}{4}$

$r^4 = 625 \quad \Rightarrow r = \sqrt[4]{625} = 5$

Now, $p = U_2 = ar = 4 \times 5 = 20$

$q = U_3 = ar^2 = 4 \times 5^2 = 4 \times 25 = 100$

$s = U_4 = ar^3 = 4 \times 5^3 = 4 \times 125 = 500$

$\therefore$ the three geometric mean are: $p = 20, \quad q = 100$ and $s = 500$

Thus, the sequence is 4, 20, 100, 500, 2500.

7.3 OTHER EXAMPLES ON SEQUENCE

Example 19

The nth term of a sequence is given by 2^{n-3}, find the sum of the first four terms of the sequence.

Solution

$U_n = 2^{n-3}$ where $n = 1, 2, 3, 4, \ldots$

When $n = 1$

$U_1 = 2^{1-3} = 2^{-2} = \frac{1}{4}$

When $n = 2$

$U_2 = 2^{2-3} = 2^{-1} = \frac{1}{2}$

When $n = 3$

$U_3 = 2^{3-3} = 2^0 = 1$

When $n = 4$

$U_4 = 2^{4-3} = 2^1 = 2$

Thus, the sequence is $\frac{1}{4}, \frac{1}{2}, 1, 2, \ldots$

But, $U_1 = a = \frac{1}{4}$ and $U_2 = ar = \frac{1}{4}$

Then, $r = \frac{U_2}{U_1} = \frac{1/2}{1/4} = 2$

$S_n = \frac{a(r^n - 1)}{r - 1}$ where: $a = \frac{1}{4}, r = 2$, and $n = 4$.

$S_4 = \frac{\frac{1}{4}(2^4 - 1)}{2 - 1} = \frac{\frac{1}{4}(16-1)}{1} = \frac{15}{4} = 3\frac{3}{4}$

$\therefore S_4 = 3\frac{3}{4}$

> The sequence is a geometric progression. So, find the common ratio and then use the formula for the sum of n terms to find the sum of the first **four** terms (i.e. when $n = 4$) as required.

Example 20

A sequence is given as $1, 3, \frac{1}{3}, 5, 7, \frac{1}{7}, 9, 11, \frac{1}{11}, \ldots$

What is the value of the next three terms?

> The first two sequence increases with a common difference of 2, then the next term becomes an inverse of the previous term.

Solution

The next three sequence will be $13, 15, \frac{1}{15}$.

OBJECTIVE TEST 7

1. Find the common difference of the sequence $\frac{1}{3}, \frac{5}{6}, \frac{4}{3}, \ldots$

   (A) $\frac{1}{3}$ (B) $\frac{2}{3}$ (C) $\frac{1}{2}$ (D) $-\frac{1}{2}$

2. Find the 30th term of the linear sequence $5, 11, 17, 23, \ldots$

   (A) 174 (B) 179 (C) 180 (D) 186

3. How many terms are in the A.P. $3, 4\frac{1}{2}, 6, 7\frac{1}{2}, \ldots$ if the sum is 156.

   (A) 10 (B) 11 (C) 12 (D) 13

4. Find the sum of the first 10 terms of the series $1, 8, 15, \ldots$

   (A) 320 (B) 325 (C) 350 (D) 360

5. What is the common ratio of the sequence $\log x,\ \log x^2, \log x^4, \ldots$

   (A) 2 (B) -2 (C) $\frac{1}{2}$ (D) $-\frac{1}{2}$

6. The second and fifth term of a G.P are 12 and -96 respectively. Find the sum of the first six terms of the progression.

   (A) 126 (B) -126 (C) 0 (D) 62

7. What is the sum to infinity of the series $\frac{1}{5}, \left(\frac{1}{5}\right)^2, \left(\frac{1}{5}\right)^3, \ldots$

   (A) 4 (B) $\frac{1}{4}$ (C) -4 (D) $-\frac{1}{4}$

8. Find the sum to infinity of the series $0.\dot{3}$

   (A) $\frac{1}{9}$ (B) -9 (C) $\frac{1}{3}$ (D) -3

9. Evaluate $\left(\frac{1}{3} - \frac{1}{6} + \frac{1}{12} - \frac{1}{24} + \cdots\right) - 1$

   (A) $\frac{2}{9}$ (B) 0 (C) $\frac{7}{9}$ (D) $-\frac{7}{9}$

10. The nth term of a sequence is given by 2^{n-1}, find the sum of the first five terms.

    (A) 31 (B) 30 (C) 29 (D) 28

Exercise 7b

1. Find the common ratio of each of the following geometric sequence:

   (a) $5, 10, 20, 40, \ldots$ (b) $\frac{1}{2}, 1, 2, 4, \ldots$

   (c) $-\frac{1}{3}, 2, -12, 72, \ldots$ (d) $64, 8, 1, \frac{1}{8}, \ldots$
2. The second and fifth term of a geometric progression are 6 and -48 respectively. Find:

   (a) the common ratio (b) the first term

   (c) the sum of the first six terms of the sequence.
3. The sum of the first n terms of a G.P is given by $S_n = 3^n - 1$, find:

   (a) the first term of the G.P (b) the common ratio.
4. If $3, x, y, 24$ are consecutive terms of a geometric progression; find the values of x and y.
5. Show that the sum of the n terms of a G.P is given as,

$$S_n = \frac{a(r^n - 1)}{r - 1} \quad \text{for } r \geq 1$$

6. The fourth term of a G.P is greater than the third term by 8, and the third term is greater than the second term by 2. Find the common ratio and first term of the sequence.
7. The sum of the first two terms of a G.P is 36, and the sum of the third and fourth term is 100. Given that the common ratio is positive, calculate:

   (a) the common ratio (b) first term of the series.
8. Write the next four terms of the sequence $\frac{1}{3}, \frac{1}{6}, \frac{1}{12}, \ldots$
9. If the first term of a G.P is 36, and the common ratio is $\frac{1}{2}$; how many terms of the G.P are greater than 1.
10. How many terms of the G.P $3, 6, 12, 24, \ldots$ must be taken to give a sum greater than 300?
11. Insert two geometric means between 9 and $\frac{125}{3}$.
12. U_n is the nth term of a G.P, show that $U_1 U_n = U_2 U_{n-1}$.
13. Find the sum to infinity of the sequence $1 + \frac{1}{8} + \frac{1}{64} + \cdots$
14. Find the first term, common ratio, and sum to infinity of the following series:

    (a) $0.\dot{0}\dot{6}$ (b) $0.\dot{4}\dot{3}$
15. Find the sum to infinity of the series:

    (a) $1 + \frac{1}{3} + \frac{1}{9} + \frac{1}{27} + \cdots$ (b) $2 + \frac{1}{4} + \frac{1}{32} + \cdots$
16. Evaluate:

    (a) $1 + \frac{1}{6} + \left(\frac{1}{6}\right)^2 + \left(\frac{1}{6}\right)^3 + \cdots$ (b) $\left(1 - \frac{1}{2} + \frac{1}{4} - \cdots\right) + 1$

8 SET THEORY

A set is a collection of objects/data which are grouped according to their property or common elements. The grouping helps us to make easy identification, analysis and interpretation.

8.1 DEFINITIONS AND NOTATIONS OF SET

A set is defined using a capital letter (e.g., A), whereby, the elements of the set are enclosed with a curly brace, { }; and each of the element is separated from each other using a comma. Therefore, if the elements are so many that we cannot easily list them out, then few are listed followed by a comma, then three dots, and followed by a comma again; then the final element is listed (e.g., $A = \{1, 2, 3, ..., 10\}$). Furthermore, if x is an element of a set A, then it is written as $x \in A$, otherwise it is $x \notin A$. The following definitions and notations will further help us to understand more about set:

PROPERTY FORM: This is a way of representing a set using their common property to describe the elements of the set.
e.g., $P = \{x: x \text{ are odd numbers } 0 < x < 20\}$
$Q = \{x: x \text{ are months of the year}\}$
The set P is read as: the set of all the odd numbers defined by x which is greater than zero, but less than twenty. Thus, $P = \{1, 3, 5, 7, 9, \dots, 19\}$.
Similarly, the set Q is read as the set of all the months in a year defined by the variable x. Thus, $Q = \{\text{January, February, March, \dots, December}\}$.

LISTING METHOD: This is a method of representing a set by listing all the members of the set.
e.g. $P = \{\text{Ade, Abiola, Tony, Abdulazeez, Chinwe}\}$
$Q = \{1, 2, 3, 4, 5, 6, 7, 8, 9\}$

FINITE SET: This is a set with definite number of elements.
e.g., $P = \{\text{Ade, Abiola, Tony, Abdulazeez, Chinwe}\}$
$Q = \{1, 2, 3, 4, 5, 6, 7, 8, 9\}$

INFINITE SET: This is a set without definite number of elements, such as the set of natural numbers or counting numbers.
e.g, $X = \{1, 2, 3, 4, 5, 6, 7, 8, 9, \dots\}$

EQUIVALENCE OF SETS: Two sets are equivalent if the sets have equal members or elements.
e.g., If $X = \{a, e, c, d, e\}$ and $Y = \{5, 10, 15, 20, 25\}$

Then, $X \equiv Y$ this is because they have equal number of elements, $n(X) = n(Y)$.
Thus, $\{a, e, c, d, e\} \equiv \{5, 10, 15, 20, 25\}$
where the symbol $\equiv$ means equivalence

EQUALITY OF SETS: If two sets are equal, then their elements are the same irrespective of the order of arrangement.
Thus, if, $X = \{5, 10, 15, 20, 25\}$ and $Y = \{25, 20, 15, 10, 5\}$
Then, $X = Y$ because they have the same elements or members.

COMPARABILITY OF SETS: Two sets are said to be comparable if $X \subset Y$ or $Y \subset X$. Similarly, if $X \not\subset Y$ or $Y \not\subset X$, then the two sets are not comparable.

Example 1

Given the following sets, state with reason if they are comparable.
(a) $X = \{2, 4, 5, 7, 9\}$ and $Y = \{5, 4, 7\}$
(b) $P = \{1, 2, 3\}$ and $Q = \{3, 5\}$

Solution

(a) $X = \{2, 4, 5, 7, 9\}$ and $Y = \{5, 4, 7\}$
Since Y is a subset of X (i.e., $Y \subset X$), then X is comparable to Y.

(b) $P = \{1, 2, 3\}$ and $Q = \{3, 5\}$
Since Q is not a subset of P (i.e., $Q \not\subset P$), nor P a subset of Q (i.e., $P \not\subset Q$), then P and Q are not comparable.

UNIVERSAL SET: This is the set of all objects/elements being considered in a given problem and may also contain elements not considered in that problem. Thus, a universal set is denoted as U or E.

Example 2

Write out the members of the set of the English alphabet.

Solution

Let the universal set be U, which contains 26 letters.
$U = \{a, b, c, d, e, f, g, h, i, j, k, l, m, n, o, p, q, r, s, t, u, v, w, x, y, z\}$
This could be written easily as:
$U = \{a, b, c, d, e, f, \ldots, x, y, z\}$

NUMBER OF ELEMENTS OF A SET: This gives the total number of elements/members in a set. Sometimes, it is called "cardinality of a set". It is denoted as $n(A)$ where the symbol n stands for number, and A is the given set. Thus, using example 2, the universal set U of the English alphabet contains 26 letters, that means $n(U) = 26$.

NULL/EMPTY SET: This is a set that contains no element. Thus, it is denoted as { } or $\emptyset$.

SINGLETON: This is any set that has only one element, e.g., $A = \{a\}$.

SUBSET AND SUPERSET: These are sets that are present in another set. Thus, given two sets X and Y, then set X is a subset of Y if all the elements of set X are present in set Y; and it is represented as $X \subset Y$. Whereas, Y is a superset of X, and it is represented as $Y \supset X$.

For instance, if $X = \{a, d, e\}$ and $Y = \{a, b, c, d, e\}$; then X is a subset of Y because all the elements in X are also found in Y. While Y is a superset of X because it contains all the elements in X and has additional elements (i. e., b and c).

POWER SET: The power set of a given set, M, is denoted by $P(M)$; and mathematically defined by 2^m, which is the family of all the subsets of set M. Thus, if $M = \{1, 2, 5\}$, then the power set is the set $2^m = 2^3 = 8$. Therefore, the number of all possible subsets of M are: $\emptyset$, $\{1\}$, $\{2\}$, $\{5\}$, $\{1, 2\}$, $\{1, 5\}$, $\{2, 5\}$, $\{1, 2, 5\}$.

VENN DIAGRAM: The Venn diagram makes use of circle and rectangle to represent a set of information. This was first proposed by the British mathematician, "John Venn." Hence, the word "VENN" diagram was given in honour of John Venn. Thus, Venn diagram enables the user (reader) to understand the relationship between sets, subsets, and elements.

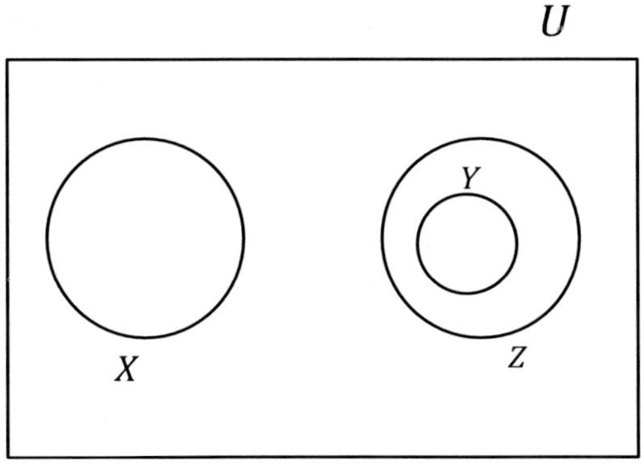

Fig. 8.1 Venn diagram

Where U is the universal set represented by a rectangle.
X, Y and Z are sets represented by circles; and they are subsets of the universal set.

JOINT SET: A set is said to be joint, if two subsets of the set have common elements. See fig. 8.1, in the Venn diagram the subsets Y and Z are joint sets. Thus, for joint sets we have:

$$n(A \cup B) = n(A) + n(B)$$

DISJOINT SET: A set is said to be disjoint, if two subsets of the set do not have any element in common; that means there is no union between the two subsets. In fig. 8.1, the subsets, X and Y, or X and Z are disjoint sets. Thus, for disjoint sets, $n(A \cup B) = n(A) + n(B) - n(A \cap B)$.

UNION OF SET: The union of two sets X and Y, is the set that contains all the elements of X and all the elements of Y. It is denoted as $X \cup Y$ which reads X union Y, and it is represented by the shaded portion of the Venn diagram in fig. 8.2.

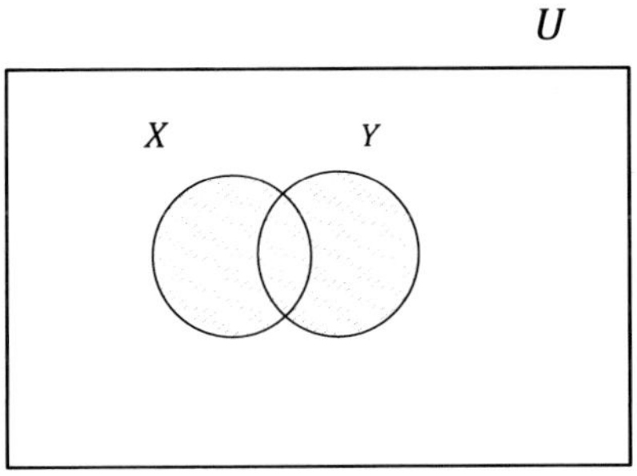

Fig. 8.2 Union of Sets

Example 3

Given the sets: $X = \{2, 3, 4, 8, 11\}$ and $Y = \{1, 3, 4, 5\}$. Find $X \cup Y$, and represent the information using a Venn diagram.

Solution

$X \cup Y = \{1, 2, 3, 4, 5, 8, 11\}$

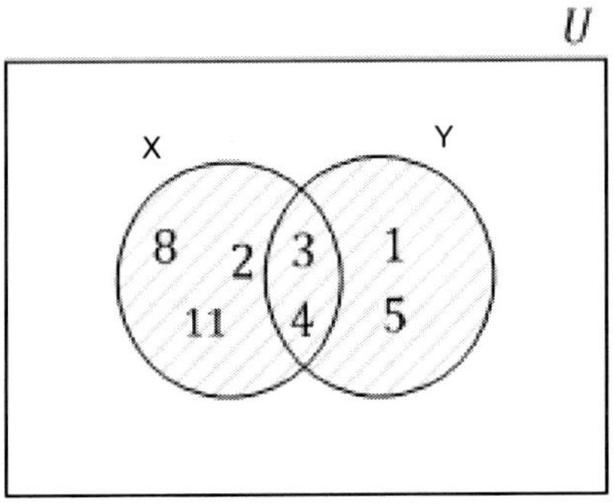

Fig. 8.3 X U Y shaded

Example 4

Given the following sets: $A = \{2, 4, 6, 8\}$, $B = \{12, 14, 16, 18\}$, $C = \{a, b, c\}$ and $D = \{a, b, c, d, e\}$. Find: (a) $A \cup B$ (b) $C \cup D$

Solution

(a) $A \cup B = \{2, 4, 6, 8, 12, 14, 16, 18\}$
(b) $C \cup D = \{a, b, c, d, e\}$

See fig. 8.4 (a) and (b) on next page for the Venn diagram of example 4.

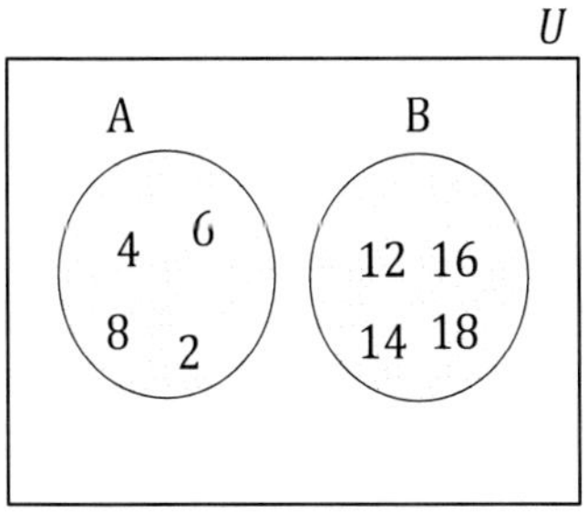

Fig. 8.4(a) A U B

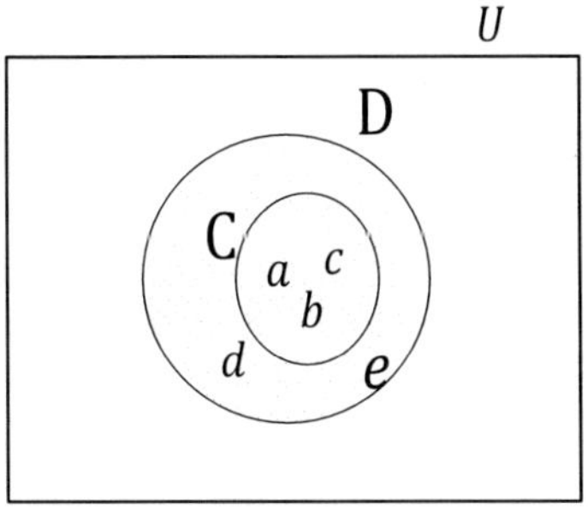

Fig. 8.4(b) CU D

INTERSECTION OF SETS: The intersection of two sets X and Y, is the set containing all the elements that are common in both sets X and Y. This is denoted as $X \cap Y$ which reads "X intersection Y."

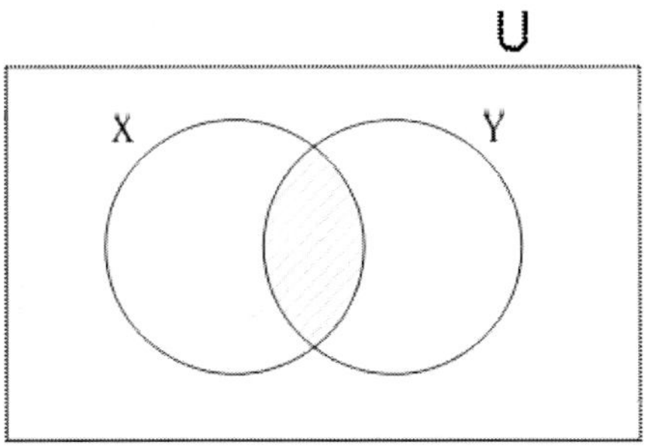

Fig. 8.5 $X \cap Y$ shaded

Example 5

$X = \{a, e, i, o, u\}, Y = \{a, b, c, o, u\},\ P = \{1, 2, 3, 4, 5\}$ and $Q = \{3, 4, 5\}$. Find:
(a) $X \cap Y$ (b) $P \cap Q$

Solution

(a) $X \cap Y = \{a, o, u\}$

(b) $P \cap Q = \{3, 4, 5\}$

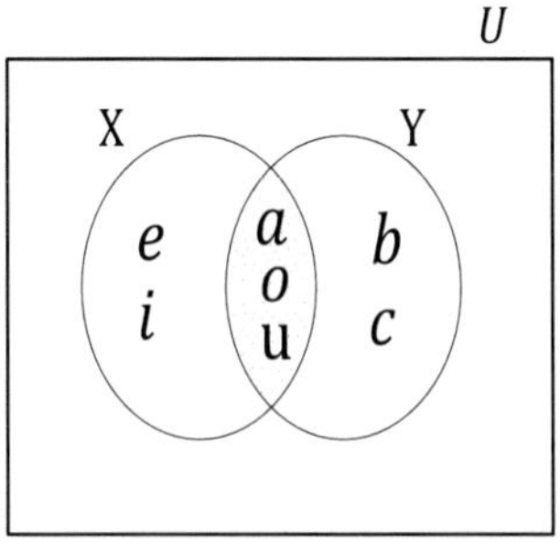

Fig. 8.6(a) X∩Y shaded

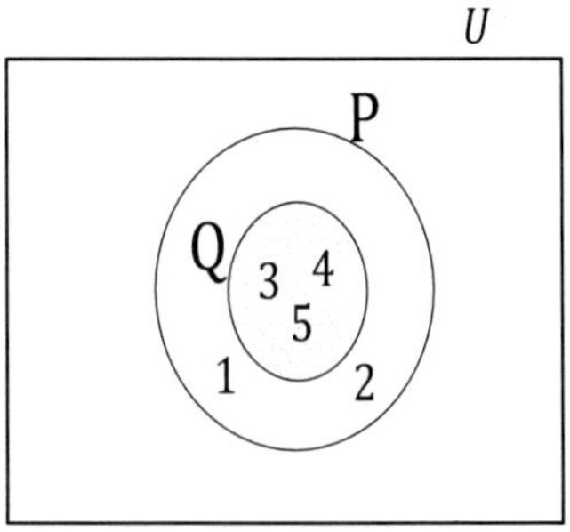

Fig. 8.6(b) P∩Q

Example 6

Given $A = \{1,2,3,4,5\}$, $B = \{6,7,8,9,10\}$. Find $A \cap B$

Solution

$$A \cap B = \emptyset$$

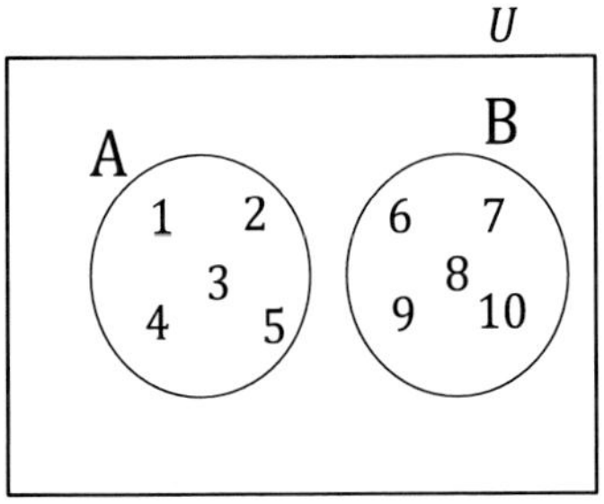

Fig. 8.7 A∩B Null set

COMPLEMENT OF SETS: The complement of a set X, is the set of all the elements that do not belong to X, but belongs to the universal set, U. The complement of a set X is thus denoted as X'or X^c.

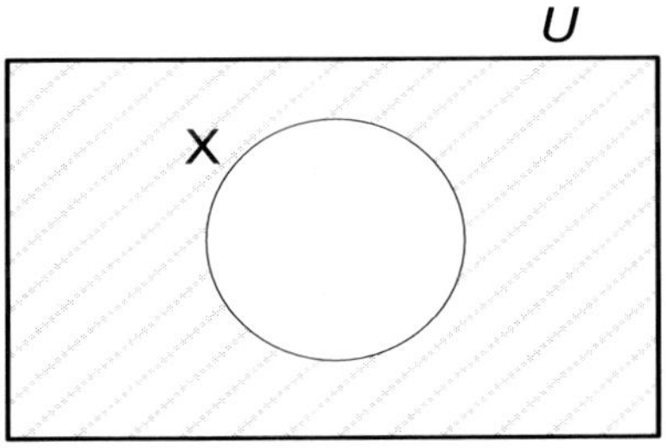

Fig.8.8 X^C or X'

Example 7

Given $U = \{1,2,3,4,5,6,7,8,9,10\}$ and $X = \{1,2,4,6,8\}$, find X'.

Solution

$X' = X^C = \{3,5,7,9,10\}$

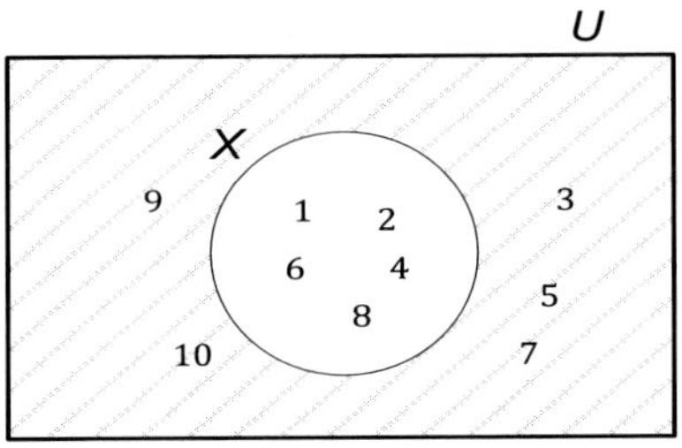

Fig.8.9 X^C

Example 8

Using a Venn diagram, shade the portion that satisfies the following:
(a) $A' \cap B$ (b) $(X \cup Y)'$

Solution

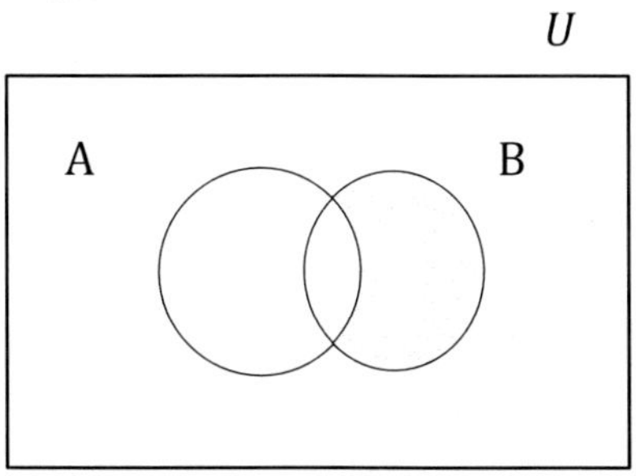

Fig. 8.10(a) $A' \cap B$

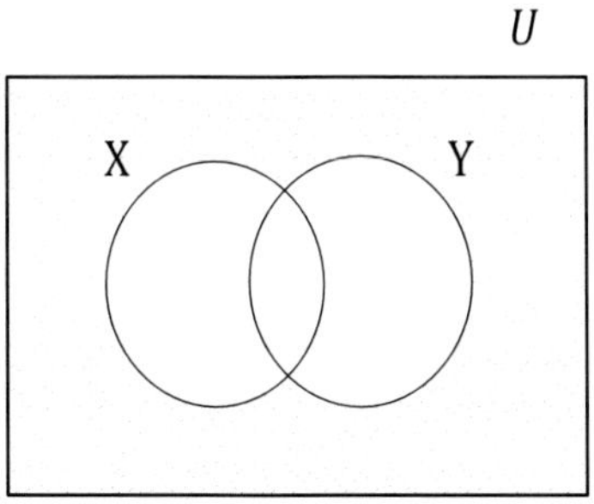

Fig. 8.10 (b) $(X \cup Y)'$

DIFFERENCE OF SETS: The difference of sets A and B (i.e., $A - B$), is the set of all the elements in set A which are not in another set B.

Example 9

Given the sets: $A = \{2, 5, 7, 8, 11, 12\}$ and $B = \{2, 5, 6, 7, 10\}$, find:
(a) $A - B$ (b) $B - A$

Solution

(a) $A - B = \{8, 11, 12\}$
(b) $B - A = \{6, 10\}$

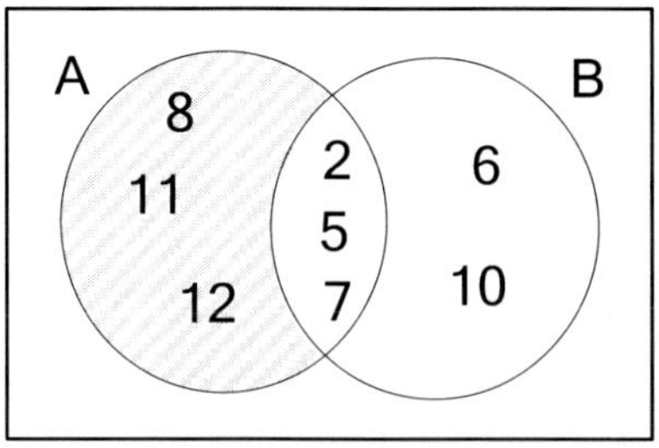

Fig. 8.11 (a) A - B

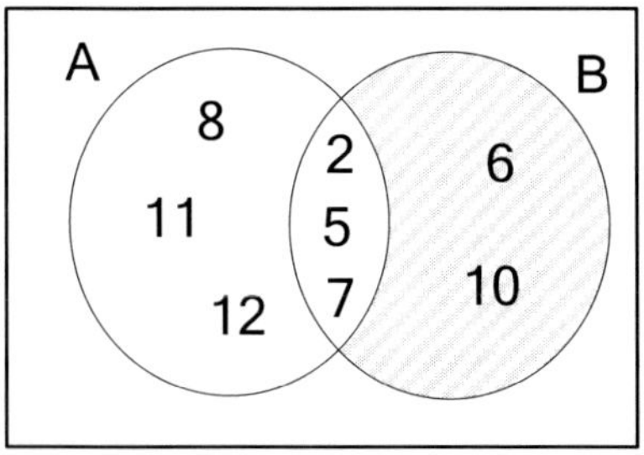

Fig. 8.11 (b) B - A

8.2 BASIC PROPERTIES/LAWS OF SET

DE MORGAN'S LAW:

(a) States that the complement of the union of two sets A and B, is equal to the complement of set A intersection the complement of set B.
Hence, $(A \cup B)' = A' \cap B'$

(b) The complement of the intersection of sets A and B, is equal to the complement of set A union the complement of set B.
Hence, $(A \cap B)' = A' \cup B'$

Example 10

Given the sets: $U = \{1, 2, 3, 4, 5, 6, 7, 8, 9, 10, 11, 12\}$, $A = \{2, 5, 7, 8, 11, 12\}$ and $B = \{2, 5, 6, 7, 10\}$. Show that: (i) $(A \cup B)' = A' \cap B'$ (ii) $(A \cap B)' = A' \cup B'$

Solution

(i) $(A \cup B)' = A' \cap B'$

$A \cup B = \{2, 5, 6, 7, 8, 10, 11, 12\}$

$(A \cup B)' = \{1, 3, 4, 9\}$

$A' = \{1, 3, 4, 6, 9, 10\}$ and $B' = \{1, 3, 4, 8, 9, 11, 12\}$

$A' \cap B' = \{1, 3, 4, 9\}$

Thus, it can be seen that $(A \cup B)' = A' \cap B' = \{1, 3, 4, 9\}$

(ii) $(A \cap B)' = A' \cup B'$

$A \cap B = \{2, 5, 7\}$

$(A \cap B)' = \{1, 3, 4, 6, 8, 9, 10, 11, 12\}$

$A' = \{1, 3, 4, 6, 9, 10\}$and $B' = \{1, 3, 4, 8, 9, 11, 12\}$

$A' \cup B' = \{1, 3, 4, 6, 8, 9, 10, 11, 12\}$

Thus, it can be seen that $(A \cap B)' = A' \cup B' = \{1, 3, 4, 6, 8, 9, 10, 11, 12\}$

COMMUTATIVE LAW: This is when two sets are combined and interposed such that the result is still the same after the inter-positioning.

Addition Rule:

From linear algebra we have: $a + b = b + a$
Similarly, in set theory $A \cup B = B \cup A$

The hint to the addition rule of set theory is that the plus sign (+) of the linear algebra is replaced with the union notation (∪) of set theory, e.g., $a + b \Rightarrow A \cup B$.

Multiplication Rule:

From linear algebra we have: $a \times b = b \times a$
Similarly, in set theory $A \cap B = B \cap A$

Similarly, the multiplication sign (×) of the linear algebra is replaced with the intersection notation (∩) of set theory, e.g., $a \times b \Rightarrow A \cap B$.

The set theory obeys the addition and multiplication rules thus it is said to be commutative.

Example 11

$A = \{a, b, c, d, e, f\}$ and $B = \{b, d, e, g\}$. Show that:
(a) $A \cup B = B \cup A$ (b) $A \cap B = B \cap A$

Solution

(a) $A \cup B = \{a, b, c, d, e, f, g\}$
$B \cup A = \{a, b, c, d, e, f, g\}$
$\therefore A \cup B = B \cup A = \{a, b, c, d, e, f, g\}$.
(b) $A \cap B = \{b, e\}$
$B \cap A = \{b, e\}$
$\therefore A \cap B = B \cap A = \{b, e\}$.

The set is commutative under addition and multiplication rules.

ASSOCIATIVE LAW: This is when three or more sets are combined to give the same result irrespective of the arrangement pattern.
Addition Rule:
From linear algebra, we have:
$a + b + c = (a + b) + c = (a + c) + b = a + (b + c)$
Similarly, in set theory $A \cup B \cup C = (A \cup B) \cup C = (A \cup C) \cup B = A \cup (B \cup C)$

Multiplication Rule:
From linear algebra, we have:
$a \times b \times c = (a \times b) \times c = (a \times c) \times b = a \times (b \times c)$
Similarly, in set theory, $A \cap B \cap C = (A \cap B) \cap C = (A \cap C) \cap B = A \cap (B \cap C)$

Therefore, set theory is associative under addition and multiplication rules.

DISTRIBUTIVE LAW: A set is only distributive for intersection ($\cap$) over union ($\cup$), and not otherwise.
Thus, from linear algebra: $a(b + c) = ab + ac = (a \times b) + (a \times c)$
Now, in set theory, $A \cap (B \cup C) = (A \cap B) \cup (A \cap C)$.

CLOSURE PROPERTY: If A and B are sets which are subsets of the universal set U, then it follows that $A \cup B \subset U$ and $A \cap B \subset U$. Then the union or intersection of A and B are subsets of the universal set U, hence, the operations are closed under such condition.

8.3 OTHER LAWS OF OPERATION OF SETS

(i) $A \cup \emptyset = A$ (This known as identity law)
(ii) $A \cup A = A$
(iii) $A \cup A' = A' \cup A = U$ (This is known as inverse law)
(iv) $A \cup U = U$
(v) $A \cap U = A$ (Identity law)
(vi) $A - A = \emptyset$
(vii) $A - \emptyset = A$
(viii) $A - B = A \cap B' = B' \cap A$
(ix) $(A')' = A$
(x) $\emptyset' = U$
(xi) $U' = \emptyset$
(xii) $A \cap A = A$
(xiii) $A \cap A' = \emptyset$ (Inverse law)
(xiv) $A - (B \cap C) = (A - B) \cup (A - C)$
(xv) $A - B = B - A = \emptyset$; if and only if A = B

Example 12

Given that A, B and C are subsets of the universal set U; each of which is defined as follows:

$U = \{x : 1 \leq x < 12, x \text{ is an integer}\}$
$A = \{x :\ 2 < x < 6\}$
$B = \{x :\ 3 < x \leq 8\}$
$C = \{x :\ x > 9\}$

(a) List the elements of sets U, A, B and C.
(b) (i) $(A \cup B) \cup C$
(ii) $(A \cup B) \cap C$
(iii) $A \cap B \cap C$
(iv) $A \cap (B \cup C)'$

Solution

(a) $U = \{1, 2, 3, 4, 5, 6, 7, 8, 9, 10, 11\}$; $A = \{3, 4, 5\}$; $B = \{4, 5, 6, 7, 8\}$; $C = \{10, 11\}$
(b) (i) $(A \cup B) \cup C = \{3, 4, 5, 6, 7, 8, 10, 11\}$
(ii) $(A \cup B) \cap C = \emptyset$
(iii) $A \cap B \cap C = \emptyset$
(iv) $A \cap (B \cup C)' = \{3\}$

Example 13

If $U = \{a, b, c, d, e, f, g, h, i, j, k\}$, $A = \{a, b, c, i, k\}$, $B = \{e, b, c, h, i, j, k\}$; $C = \{a, b, g, e, k\}$. Show that:

(a) $A \cap A' = \emptyset$
(b) $C \cap (A \cup B) = (C \cap A) \cup (C \cap B)$
(c) $A - (B \cap C) = (A - B) \cup (A - C)$

Solution

$U = \{a, b, c, d, e, f, g, h, i, j, k\}$; $A = \{a, b, c, i, k\}$; $B = \{e, b, c, h, i, j, k\}$; $C = \{a, b, g, e, k\}$.

(a) To show that: $A \cap A' = \emptyset$

$A = \{a, b, c, i, k\}$; $A' = \{d, e, f, g, h, j\}$

$\therefore\ A \cap A' = \{\ \} = \emptyset$

(b) To show that: $C \cap (A \cup B) = (C \cap A) \cup (C \cap B$

$A = \{a, b, c, i, k\}; B = \{e, b, c, h, i, j, k\}; C = \{a, b, g, e, k\}$.

$A \cup B = \{a, b, c, e, h, i, j, k\}$; $C = \{a, b, g, e, k\}$.

$\Rightarrow\ C \cap (A \cup B) = \{a, b, e, k\}$

Also, $C \cap A = \{a, b, k\}; C \cap B = \{e, b, k\}$

$\Rightarrow (C \cap A) \cup (C \cap B) = \{a, b, e, k\}$.

Thus, $C \cap (A \cup B) = (C \cap A) \cup (C \cap B) = \{a, b, e, k$

(c) $A - (B \cap C) = \{a, c, i\}$

$A - B = \{a\}$; $A - C = \{c, i\}$ $\quad \Rightarrow (A - B) \cup (A - C) = \{a, c, i\}$

$\therefore\ A - (B \cap C) = (A - B) \cup (A - C) = \{a, c, i\}$

8.4 SOLVING PROBLEMS USING VENN DIAGRAM

Example 14

In a survey of 200 students in a secondary school, 170 were listed as computer science students, and 150 as physics students; if 120 offer both subjects. Draw the Venn diagram, and find how many students that offer:

(a) physics only (b) computer science only.

Solution

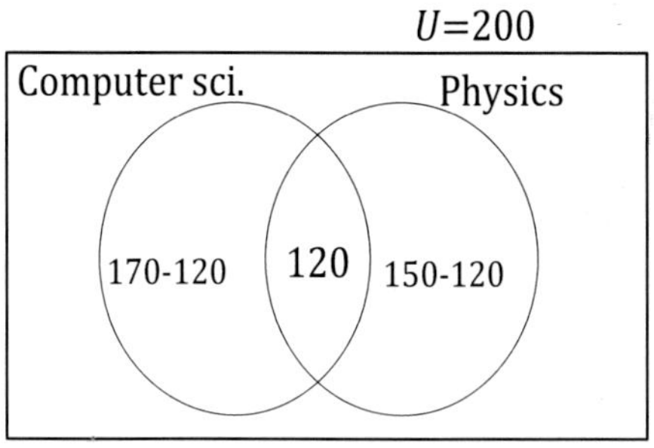

Fig. 8.12

Draw the Venn diagram. Now, subtract the total number of students listed in a particular subject from the number that offers both subjects to get the number of students that offer one subject only.

(a) Physics only $= 150 - 120 = 30$

$\therefore$ 30students offer physics only

(b) Computer Science only $= 170 - 120 = 50$

$\therefore$ 50 students offer Computer science only.

Example 15

Out of 50 candidates applying for a job, 30 have degree, 23 have diploma, and 5 neither have degree nor diploma. How many have both degree and diploma?

Solution

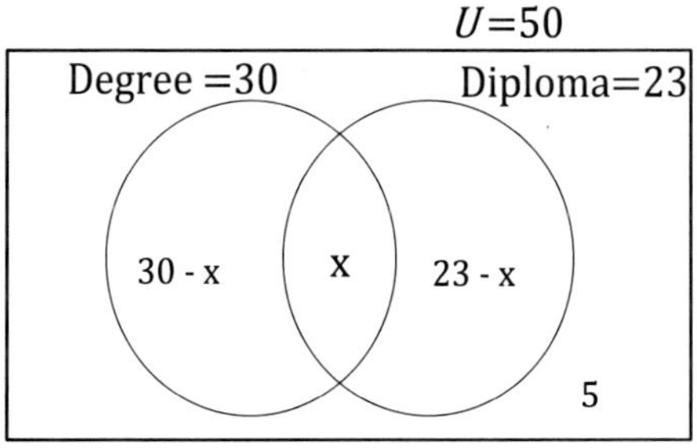

Fig. 8.13

Let x be the number of candidates applying for the job.
Then, $30 - x + x + 23 - x + 5 = 50$
$-x + 58 = 50$ Collecting like terms, then we have, $x = 58 - 50 = 8$
$\therefore$ the number of candidates that have both degree and diploma are 8.

Example 16

All the 82 students in SSS 3 class of a certain school take either mathematics(M), physics(P), or chemistry(C). If 60 take mathematics, 45 take physics, 38 take chemistry, and 30 take mathematics and physics, 25 take physics and chemistry, while 22 take mathematics and chemistry. How many students:

(a) Take all the three subjects;
(b) Take mathematics, but neither physics nor chemistry;
(c) Physics but neither mathematics nor chemistry;
(d) Chemistry but neither mathematics nor physics.

Solution

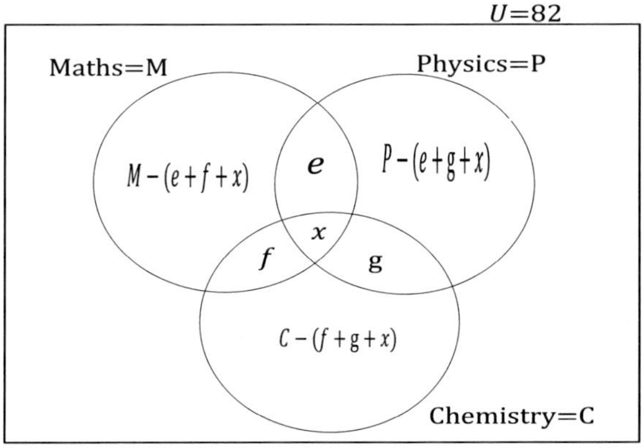

Fig. 8.14

Total number of SSS 3 students $= n(M \cup P \cup C) = 82$
Students offering math $= n(M) = 60$
Students offering physics $= n(P) = 45$
Students offering chemistry $= n(C) = 38$
Students offering both math and physics $= n(M \cap P) = e + x = 30$
Students offering both math and chemistry $= n(M \cap C) = f + x = 25$
Students offering both physic and chemistry $= n(P \cap C) = g + x = 22$
Students offering all the three subjects $= n(M \cap P \cap C) = x$

(a) $n(M) + n(P) + n(C) - n(M \cap P) - n(M \cap C) - n(P \cap C) + n(M \cap P \cap C)$
$= n(M \cup P \cup C)$
$\Rightarrow 60 + 45 + 38 - 30 - 25 - 22 + x = 82$
$66 + x = 82$
$x = 82 - 66 = 16$
$\therefore$ number of students that offer all the three subject $= x = 16$

(b) $n(M \cap P) = e + x = 30 \Rightarrow e = 30 - x = 30 - 16 = 14$
$n(M \cap C) = f + x = 25 \quad \Rightarrow f = 30 - x = 25 - 16 = 9$
Students that offer math only $= n(M) - (e + f + x)$
$= 60 - (14 + 9 + 16) = 60 - 39 = 21$ $\therefore$ students that offer math only $= 21$
(c) $n(P \cap C) = g + x = 22 \Rightarrow g = 22 - x = 22 - 16 = 6$
Students that offer physics only $= n(P) - (e + g + x)$
$= 45 - (14 + 6 + 16) = 45 - 36 = 9$ $\therefore$ students offering physic only $= 9$
(d) Students that offer chemistry only $= n(C) - (f + g + x)$
$= 38 - (9 + 6 + 16) = 38 - 31 = 7$ $\therefore$ students offering chemistry only $= 7$

Example 17

If the set $U = \{1, 2, 3, 4, \dots\}$, describe the following set by the listing method:
(a) $P = \{x : x^2 - 3x + 2 = 0\}$ (b) $P = \{x : x^2 - x + 2 = 0\}$

Solution

(a) $P = \{x : x^2 - 3x + 2 = 0\}$
Factorising $x^2 - 3x + 2 = 0 \quad \Rightarrow x = 1$ or 2
$\therefore P = \{1, 2\}$
P is a subset of the universal set U.

(b) $P = \{x : x^2 - x + 2 = 0\}$
Factorising $x^2 - x + 2 = 0$
$\Rightarrow x = \frac{1 \pm \sqrt{-7}}{2}$ this is a complex solution.
$\therefore Q = \emptyset$

OBJECTIVE TEST 8

1. Which subset is shaded in the figure?

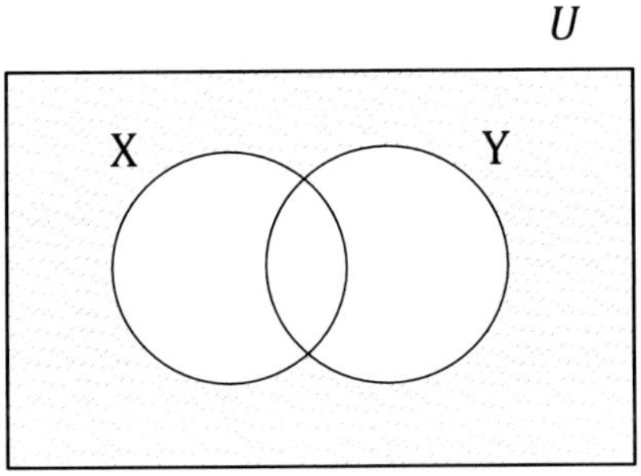

Fig 8.15

(A) $X' \cup Y'$ (B) $(X \cap Y)'$ (C) $(X \cup Y)'$ (D) $X \cup Y$

2. There are 50 students in a science class, 30 offer physics and 40 offer chemistry. How many students offer both subject?
   (A) 10 (B) 20 (C) 25 (D) 30
3. $A \cap (B \cup C) = (A \cap B) \cup (A \cap C)$
   (A) True (B) False
4. If a set is defined by $M = \{x: x \text{ is even numbers and } 0 \leq x \leq 10\}$, then the members of set M are:
   (A) $\{0, 2, 4, 6, 8, 10\}$ (B) $\{2, 4, 6, 8, 10\}$ (C) $\{2, 4, 6, 8\}$ (D) $\{0, 2, 4, 6, 8\}$
5. The shaded portion in figure 8.16 is represented by?

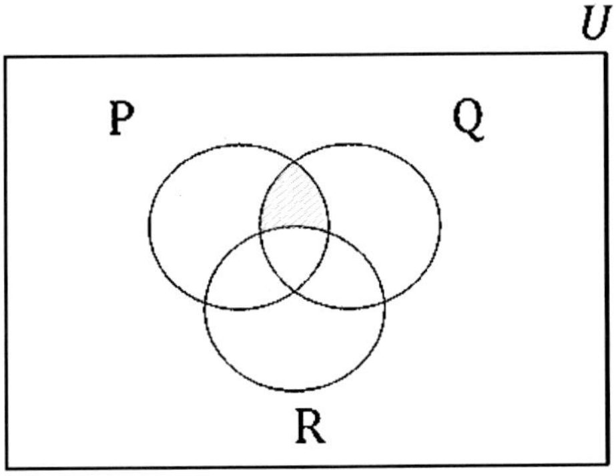

Fig. 8.16

(A) $P \cup Q$ (B) $P' \cap Q' \cap R$ (C) $P \cap Q \cap R'$ (D)) $P \cap Q' \cap R$

Let $X = \{a, e, i, o, u\}, Y = \{o, r, a, e, u\}$ and $Z = \{a, b, e, o\}$.
Use this information to answer question 6 and 7.

6. List $X \cup Y$
   (A) $\{a, e, b, i, o, u, r\}$ (B) $\{a, b, e, i, o, u\}$ (C) $\{a, e, i, o, u, r\}$ (D) $\{a, e, o, u\}$

7. List $X \cap Y \cap Z$
   (A) $\{a, e, o\}$ (B) $\{a, b, i, o, u\}$ (C) $\{a, e, i, o, u, r\}$ (D) $\{a, b, o, u\}$

8. A set $P = \{1, 3, 5, 6\}$. What is the power of the set?
   (A) 4 (B) 8 (C) 16 (D) 32

9. In a football team of 25 players; 12 are defenders and 18 attackers. If 3 neither play defense nor attack, how many play both defense and attack?
   (A) 5 (B) 7 (C) 8 (D) 9

10. $(A \cap B)' = A' \cup B'$. What law of set theory is illustrated here?
    (A) Commutative (B) Associative (C) De Morgan's law (D) Distributive

Exercise 8

1. List the following sets:
   (a) Set of days in a week
   (b) Set of five people that you know beginning with letter A.
2. Give two examples of the following:
   (a) Equivalent sets
   (b) Equal sets.
3. State whether each of the following statements is true or false:
   $A = \{1\};\ B = \{1, 3, 5\};\ C = \{2, 3, 5\};\ D = \{1, 3\};\ E = \{1, 2, 3, 5\}$
   $(i)\ A \subset B$ $(ii)\ A \subset D$ $(iii)\ D \subset B$ $(iv)\ D \not\subset B$ $(v) C \not\equiv E$
   $(vi) A \in E$ $(vii)\ C \in E$ $(viii)\ B \subset C$ $(ix)\ D \subset E$
4. State which of the following sets are comparable and give reasons for your answer:
   $A = \{1\};\ B = \{1, 3, 5\};\ C = \{2, 3, 5\};\ D = \{1, 3\};\ E = \{1, 2, 3, 5\}$
5. Let the universal set U be the set of all natural or counting numbers. Find the following sets by the listing method:
   (a) $A = \{x : x^2 - 5x + 6 = 0\}$
   (b) $B = \{x : x^2 + 8 = 0\}$
   (c) $C = \{x : x^2 \leq 3 + x\}$
   (d) $D = \{x : -2 \leq x \leq 10\}$

6. If $X = \{4, 6, 8\}$ and $Y = \{10, 12, 14, 16\}$. Find the following sets:
   (a) $X \cup Y$ (b) $X \cap Y$

7. Shade correctly the following sets:

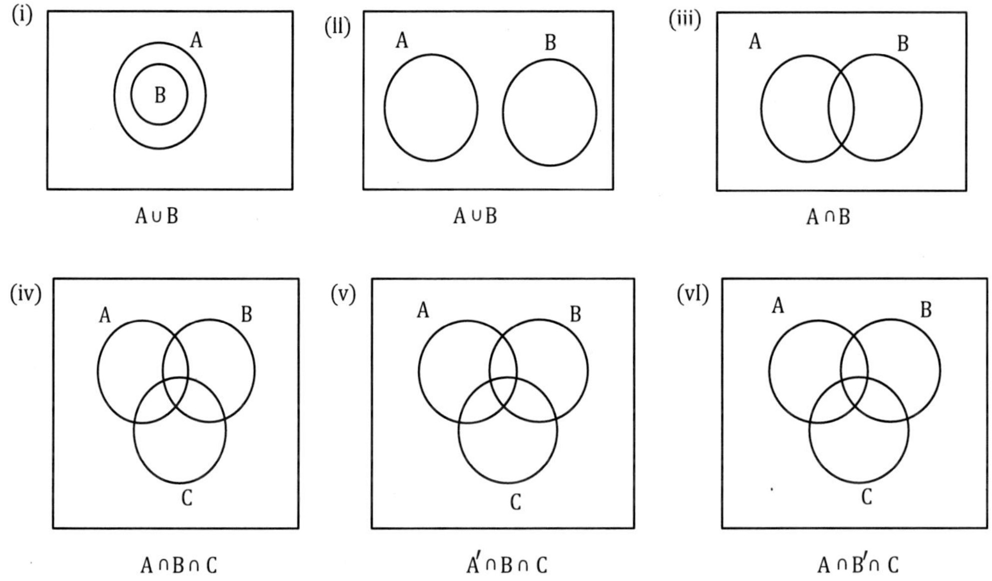

Fig. 8.17

8. Draw a Venn diagram for each of the following:
   (i) $X = \{a, e, i, o, u\}$ and $Y = \{a, b, c, d, e, f, u\}$
   (ii) $X = \{1, 2, 3, 4, 5, 6, 7\}$ and $Y = \{2, 3, 5, 6\}$

9. (a) Find the difference between the sets:
   $A = \{1, 2, 3, 4, 5, 6, 7, 8\}$ and $B = \{2, 1, 5, 3, 7\}$
   (b) If the sets in (a) above is to be represented in a Venn diagram, shade the area that corresponds to the solution.

10. How many subsets are there in the following sets:
    (i) $A = \{b\}$ (ii) $B = \{a, b\}$ (iii) $C = \{a, b, c\}$ (iv) $D = \{1, 2\}$ (v) $E = \{1, 2, 3, 4\}$

11. In a sport's club with 140 members, it was found that 60 played football; 56 played table tennis; 65 played badmintons. Then 35 played both football and table tennis; 28 played table tennis and badminton; and 30 played both football and badminton. It was also found that 35 members joined only for the social side of the club and did not play any of the three games:
    (a) How many played all the games?
    (b) How many played footballs only?
    (c) How many played badminton and table tennis but not football?

12. In a school interhouse sport, 80% of the students turned up at the athletic competition, 60% turned up at the football match. How many students attended both functions?

13. Show that the following laws of a set are true:
(a) $(A \cup B)' = A' \cap B'$ (b) $(A \cap B)' = A' \cup B'$

14. Find the solution set in each of the following:
(a) $\{x : 2x + 3 < 7,\ x \in R\}$ (b) $\{x : x^2 - 5x + 4 = 0,\ x \in R\}$
(c) $\{x : x^2 + 1 = 0, x \text{ is real}\}$

15. How many subsets are there in the following sets:
(a) $S = \{x : x^3 - 4x = 0\}$ (b) $S = \{a, \{b, c\}\}$ (c) $S = \{1, 2, \{3, 5, 7\}\}$

16. List the power set of the following:
(a) $X = \{1, 2, \{4, 7\}\}$ (b) $Y = \{a, b, c\}$

17. A given universal set U, is the set of real positive whole numbers less than 11. List the following sets:
(a) $S = \{x : x \in U\}$ (b) $S = \{x : x < 5\}$ (c) $S = \{x : x^2 - 4 = 0\}$
(d) $S = \{x : x - 3 \leq 4\}$ (e) $S = \{x : x \geq 8\}$

18. In a class of 45 students, 28 offer maths; 23 offer biology; 32 offer chemistry; and all the students take at least one of these subjects. If 6 offer biology and maths; 10 offer chemistry and maths; 15 offer biology and chemistry.
(a) Draw the Venn diagram
(b) Find the number of students who offer the three subjects.

19. Given U as the universal set consisting of all positive integers and X, Y and Z are subsets such that: $X = \{x : x \text{ are prime numbers}\}$;
$Y = \{x : x \text{ are even numbers}\}$;
$Z = \{x : 5 < x < 15\}$. List the elements of the following:
(a) $X \cap Y$ (b) $X' \cap Z$ (c) $X' \cap (Y' \cap Z)$

UNIT II ALGEBRA I & II

Specific Objectives:

At the end of this unit, the student should be able to:

(a) Simplify linear algebraic expressions;
(b) Solve problems in two variables (simultaneous equations);
(c) Make a given variable the subject formula of an expression;
(d) Solve quadratic equations using: Factorisation, formula method, completing the square and graphical method;
(e) Find the solution set that satisfies a given inequality;
(f) Add, subtract, multiply and divide polynomial functions;
(g) Use factor and remainder theorems to evaluate and factorise polynomial functions;
(h) Resolve functions into partial fractions.
(i) Solve problem on mapping and Functions
(j) Grasp basic statistical idea such as: Classifying and representing data in table and charts;
(k) Calculate the mean, mode and median of ungrouped and grouped data;
(l) Find the range, mean deviation, variance and standard deviation of ungrouped and grouped data;
(m) Know the basic laws of probability and Solve problems;
(n) Solve permutation and combination problems;
(o) Understand in depth of Complex number
(p) Understand Matrices and perform operations on matrices
(q) Solve simultaneous equation problems using Cramer's rule and Inverse matrix methods.

9 ALGEBRA PROCESS 1

Algebra is the branch of mathematics that uses letters to represent numbers or real-life objects. For instance, assume that a man is represented by x, while a woman by y; then the addition of 4 men and 6 women can be written as $4x + 6y$. This expression is called an algebraic expression.

9.1 SIMPLIFICATION OF ALGEBRAIC EXPRESSION

Simplifying an algebraic expression means to represent or reduce a complex expression into its simplest form.

9.1.1 LINEAR EXPRESSIONS

Example 1

Simplify:

(a) $10x + 6y - 3x + y$

(b) $4x^2 - 3xy + x^2 - 5xy - 5x^2$

(c) $3a^2 - 4ab + 5a^2 - 6ab + 5$

Solution

(a) $10x + 6y - 3x + y$

$= 10x - 3x + 6y + y = 7x + 7y$

$= 7(x + y)$

(b) $4x^2 - 3xy + x^2 - 5xy - 5x^2$

$= 4x^2 + x^2 - 5x^2 - 3xy - 5xy = 0 - 8xy$

$= -8xy$

(c) $3a^2 - 4ab + 5a^2 - 6ab + 5$

$= 3a^2 + 5a^2 - 4ab - 6ab + 5 = 8a^2 - 10ab + 5$

$= 8a^2 - 5(2ab - 1)$

9.1.2 ALGEBRAIC FRACTION

Example 2

Simplify:

(a) $\frac{3x}{5} + \frac{2x}{5}$ (b) $\frac{3}{2x} - \frac{5}{2x}$

(b) $\frac{3x-2}{2} + \frac{x+1}{5}$ (d) $\frac{x}{x+1} - \frac{x+2}{x-1}$

Solution

(a) $\frac{3x}{5}+\frac{2x}{5}=\frac{5\left(\frac{3x}{5}+\frac{2x}{5}\right)}{5}=\frac{3x+2x}{5}$

$=\frac{5x}{5}=x$

(b) $\frac{3}{2x}-\frac{5}{2x}=\frac{2x\left(\frac{3}{2x}-\frac{5}{2x}\right)}{2x}=\frac{3-5}{2x}=\frac{-2}{2x}$

$=-\frac{1}{x}$

(c) $\frac{3x-2}{2}+\frac{x+1}{5}=\frac{10\left(\frac{3x-2}{5}+\frac{x+1}{5}\right)}{10}=\frac{2(3x-2)+2(x+1)}{10}$

$=\frac{6x-4+2x+2}{10}=\frac{8x-2}{10}=\frac{2(4x-1)}{10}$

$=\frac{4x-1}{5}$

In each solution, find the LCM then multiply the LCM to the numerator and denominator of each fraction and divide again by the LCM. After which you can simplify to reduce to the simplest expression.

(d) $\frac{x}{x+1}-\frac{x+2}{x-1}=\frac{(x+1)(x-1)\left(\frac{x}{x+1}-\frac{x+2}{x-1}\right)}{(x+1)(x-1)}=\frac{x(x-1)-(x+1)(x+2)}{(x+1)(x-1)}$

$=\frac{x^2-x-(x^2+2x+x+2)}{(x+1)(x-1)}=\frac{x^2-x-x^2-2x-x-2}{(x+1)(x-1)}=\frac{-4x-2}{(x+1)(x-1)}$

$=\frac{-2(2x+1)}{(x+1)(x-1)}$

Example 3

Express each of the following as a single fraction:

(a) $\frac{2}{a}-\frac{5}{b}-\frac{1}{c}$ (b) $\frac{a-5}{2b}-\frac{b-2}{3b}$

Solution

(a) $\frac{2}{a}-\frac{5}{b}-\frac{1}{c}$

Take the LCM and multiply it through each fraction as follows:

$\frac{2}{a}-\frac{5}{b}-\frac{1}{c}=\frac{abc\left(\frac{2}{a}-\frac{5}{b}-\frac{1}{c}\right)}{abc}=\frac{2bc-5ac-ab}{abc}$

$=\frac{2bc-5ac-ab}{abc}$

(b) $\frac{a-5}{2b}-\frac{b-2}{3b}=\frac{6b\left(\frac{a-5}{2b}-\frac{b-2}{3b}\right)}{6b}=\frac{3(a-5)-2(b-2)}{6b}$

$=\frac{3a-15-2b+4}{6b}=\frac{3a-2b-11}{6b}$

9.1.3 SUBSTITUTION

Example 4

Find the value of the following expressions if $x = 1$ and $y = -5$

(a) $3x + 5y$ (b) $\frac{x}{3} + \frac{2y}{8}$ (c) $\frac{x+3}{2x-5}$

Solution

$x = 1$ and $y = -5$

(a) $3x + 5y = 3(1) + 5(-5) = 3 + (-25)$
$= 3 - 25 = -22$

(b) $\frac{x}{3} + \frac{2y}{8} = \frac{1}{3} + \frac{2(-5)}{8} = \frac{1}{3} - \frac{10}{8}$
$= \frac{24\left(\frac{1}{3} - \frac{10}{8}\right)}{24} = \frac{8(1) - 3(10)}{24} = \frac{8-30}{24} = \frac{-22}{24} = -\frac{11}{12}$

(c) $\frac{x+3}{2x-5} = \frac{1+3}{2(1)-5} = \frac{4}{2-5} = \frac{4}{-3} = -1\frac{1}{3}$

Example 5

Given that $x:y = 3:5$, evaluate $\frac{7x+y}{x-\frac{y}{2}}$

Solution

$x:y = 3:5 \quad \Rightarrow \frac{x}{y} = \frac{3}{5}$

$$\frac{7x+y}{x-\frac{y}{2}} = \frac{7x+y}{x-\frac{y}{2}} \times \frac{1/y}{1/y} = \frac{7\left(\frac{x}{y}\right) + 1}{\frac{x}{y} - \frac{1}{2}}$$

$$\frac{7\left(\frac{3}{5}\right) + 1}{\frac{3}{5} - \frac{1}{2}} = \frac{26/5}{1/10} = 52$$

Substitute $\frac{x}{y} = \frac{3}{5}$ into the expression $\frac{7x+y}{x-\frac{y}{2}}$, then simplify.

Example 6

If $x = \frac{a+1}{3a-1}$, express $\frac{2x+3}{x+1}$ in terms of a.

Solution

$$\frac{2x+3}{x+1} = \frac{2\left(\frac{a+1}{3a-1}\right)+3}{\frac{a+1}{3a-1}+1} = \frac{2(a+1)+3(3a-1)/3a-1}{a+1+3a-1/3a-1}$$

$$= \frac{2(a+1)+3(3a-1)}{3a-1} \times \frac{3a-1}{a+1+3a-1} = \frac{2a+2+9a-3}{3a-1} \times \frac{3a-1}{a+1+3a-1}$$

$$= \frac{11a-1}{3a-1} \times \frac{3a-1}{4a} = \frac{11a-1}{4a}$$

9.1.4 EXPANSION AND FACTORISATION

Example 7

Remove the brackets and simplify the following:

(a) $x(a+b)+b(x-a)$ (b) $5x+2(x+y-6z)$

Solution

(a) $x(a+b)+b(x-a) = ax+bx+bx-ab$
$= ax+2bx-ab$

(b) $5x+2(x+y-6z) = 5x+2x+2y-12z$
$= 7x+2y-12z$

In each case, multiply out the bracket, then collect like terms and reduce the equation to its simplest form.

Example 8

Expand the following:

(a) $(2x+3)(4x-5)$ (b) $(3x+y)^2$
(c) $(a+2)(a-2)$ (d) $(a-b)(3a+4b)$

Solution

(a) $(2x+3)(4x-5) = 2x(4x-5)+3(4x-5)$
$= 8x^2-10x+12x-15 = 8x^2+2x-15$

(b) $(3x+y)^2 = (3x+y)(3x+y) = 3x(3x+y)+y(3x+y)$
$= 9x^2+3xy+3xy+y^2 = 9x^2+6xy+y^2$

(c) $(a+2)(a-2) = a(a-2)+2(a-2) = a^2-2a+2a-4$
$= a^2-4$

(d) $(a-b)(3a+4b) = a(3a+4b)-b(3a+4b)$
$= 3a^2+4ab-3ab-4b^2$
$= 3a^2+ab-4b^2$

Example 9

Factorise the following:

(a) $6x-2y$ (b) $5x^2+20x$ (c) $2xy+4x^2y-6xy^2$
(d) $(b-x)(a+2x)-(b-x)^2$

Solution

In each case, find the HCF, then divide each term by the HCF so as to get the terms in the bracket.

(a) $6x-2y = 2(3x-y)$

(b) $5x^2+20x = 5x(x+4)$

(c) $2xy+4x^2y-6xy^2 = 2xy(1+2x-3y)$

(d) $(b-x)(a+2x)-(b-x)^2$
$= b-x\big((a+2x)-(b-x)\big) = b-x(a+2x-b+x)$
$= b-x(a-b+3x) = (b-x)(a-b+3x)$

1. Simplify the following:
   (a) $10x - 15 + 8x + 6y - 3$ (b) $x^3 + y^3 + 3x^2y + 3xy^2$ (c) $\frac{1}{2}x + \frac{2}{3}x$
   (d) $\frac{2}{3}x - \frac{1}{5}x + \frac{1}{7}x$ (e) $\frac{3x-1}{2} + \frac{x-2}{2}$ (f) $\frac{x}{x+1} - \frac{x-2}{x-1}$
   (g) $\frac{x^2-y^2}{y^2-x^2}$ (h) $\frac{2}{a} - \frac{2}{b} + \frac{1}{c}$ (i) $\frac{x-1}{2y} - \frac{y+3}{3y}$
2. Given that $x = 1,\ y = -2$. Evaluate the following:
   (a) $x + y$ (b) $3x - y^2 + \frac{x}{y}$ (c) $\frac{x+3}{2(x-5)}$ (d) $\frac{(y+5)^2}{(y-1)^3}$
3. Given that $x : y = 1 : 2$. Evaluate the following:
   (a) $\frac{x+y}{x-y}$ (b) $\frac{3x-y}{x+\frac{1}{2}y}$
4. Given that $p = 2,\ q = -3$ and $r = 1$. Evaluate $\frac{P^2-qr}{2pqr}$
5. Factorise the following:
   (a) $xy - y + zy - zx$ (b) $a^2m + am^2 - mn - an$ (c) $xy - x + y - 1$
   (d) $ax - y^2 - ay + x^2$ (e) $a(x - y) + b(y - x)$ (f) $12t^2 - t - 1$
   (g) $6 - x - x^2$ (h) $x^2 + x - \frac{3}{4}$ (i) $3k^2 + 2k - 1$ (j) $10x - 231 + x^2$
   (k) $a^2 + 10a - 24$ (l) $1 - 6a - 9a^2$ (m) $b^2 - 6b + 9$ (n) $r^2 + 2r + 1$
6. If $x = \frac{a+2}{3a+1}$, express $\frac{5x+2}{x-1}$ in terms of a.
7. Change the following expression into a single fraction:
   (a) $\frac{x}{y^2} + \frac{y}{x^2}$ (b) $\frac{x+1}{y+1} + \frac{y+1}{x+1}$
8. Expand the following:
   (a) $(x + 1)(x + 2)$ (b) $(x - 1)(x + 1)$ (c) $(2y + 10)(3 - y)$ (d) $(2x + y)^2$ (e) $(x + y)^3$
9. Factorise the following:
   (a) $2x - 10$ (b) $10x^2 + 30x$ (c) $3xy + 5x^2y - 6xy^2$ (d) $(a - x)(b + 2x) - (a + x)^2$
10. Find the coefficient of xy in the expansion of $(5x + 2y)(x - 6y)$
11. Find the coefficient of x in each of the following:
    (a) $(2x + 3)(x - 6) = 2(x - 3)$ (b) $(25x - 3)(x + 2) = 10(x + 4) - 55$
12. Simplify the following:
    (a) $\frac{a-b}{2} - \frac{5a+b}{2}$ (b) $\frac{7a+b}{3} + \frac{2a-b}{2}$ (c) $\frac{b}{a-2} + \frac{a}{2-b}$ (d) $\frac{b-2}{a+1} - \frac{a-3}{1+a}$
13. Simplify each of the following:
    (a) $\frac{x}{2} - \frac{x}{3}$ (b) $\frac{x}{5} + \frac{y}{2}$ (c) $\frac{y}{x-3} + \frac{y}{x+1}$ (d) $\frac{x-2}{x} - \frac{y-3}{y}$

14. Solve for x in each case: (a) $\frac{2x-1}{3} - \frac{3-x}{2} = \frac{x}{5}$ (b) $\frac{10-2x}{5-x} - \frac{1}{10} = 0$

(c) $2 - \frac{x+1}{5-x} = 0$ (d) $\frac{x+5}{x^2-4} \times \frac{x^2+2x}{x^2-25}$

15. Find the missing part of the fraction in each of the following:

(a) $\frac{2m+5}{10} - \frac{n+3}{5} = \frac{\square}{10}$ (b) $\frac{r-p}{3} - \frac{\square}{5} = \frac{p+5r}{15}$

16. Simplify the following:

(a) $\frac{2x}{5} + \frac{3x}{5} - \frac{10x}{3}$ (b) $\frac{2x+1}{x} + \frac{3y-2}{y} - 5$

17. Given $\frac{y}{x} = \frac{7}{2}$, evaluate $\frac{7x+y}{xy}$

18. If $p = 2, q = -3$, and $r = 1$. Evaluate $\frac{p^2-qr}{2pqr}$

19. Factorise the following:

(a) $xy - y^2 + 8y - xz$ (b) $a^2m + am^2 - mn - an$ (c) $xy - x + y - 1$

(d) $x^2y^2 + k^2x^2 + x^2a^2 + a^2y^2$ (e) $ax - ay + x^2$ (f) $a(x-y) + b(y-x)$

(g) $12t^2 - t - 1$ (*h*) $6 - x - x^2$ (i) $x^2 + x - \frac{3}{4}$ (j) $3k^2 + 2k - 1$

(k) $10x - 231 + x^2$ (l) $a^2 + 10a - 24$ (m) $1 - 6a + 9a^2$ (n) $b^2 - 6b + 9$

9.2 LINEAR EQUATION IN ONE VARIABLE

An equation is in one variable if the highest index power of a variable e.g., x^n or y^n is one (i.e., $n = 1$).

Example 10

Evaluate the following:

(a) $3y + 8 = 20$ (b) $\frac{3}{4}(x-1) + \frac{1}{3}(x+5) = 5$ (c) $3(2y-5) - 2(4y+3) = 0$

Solution

Collect like terms by putting the **unknown** variable on the lefthand side, while the other numerical values on the right.

(a) $3y + 8 = 20$

$3y = 20 - 8$

$3y = 12$

$\frac{3y}{3} = \frac{12}{3} = 4$

$\therefore y = 4$

(b) $\frac{3}{4}(x-1) + \frac{1}{3}(x+5) = 5$

$= \frac{12\left(\frac{3}{4}(x-1) + \frac{1}{3}(x+5)\right)}{12} = 5$

$\Rightarrow 9(x-1)+4(x+5)=5\times 12$

$9x-9+4x+20=60$

$9x+4x=60-20+9=49$

$13x=49$

$\frac{13x}{13}=\frac{49}{13}=3\frac{10}{13}$

$\therefore x=3\frac{10}{13}$

(c) $3(2y-5)-2(4y+3)=0$

$6y-15-8y-6=0$

$6y-8y=0+15+6=21$

$-2y=21$

$\frac{-2y}{-2}=\frac{21}{-2}=-10\frac{1}{2}$

$\therefore y=-10\frac{1}{2}$

Open the brackets and then collect like terms.

Example 11

A certain number is added to 3 and the result is 10. What is the number?

Solution

Let the number be x

Then, when the number is added to 3 implies $3+x$.

Thus, the result of adding $3+x$ equals 10 $\Rightarrow 3+x=10$

making the unknown variable the subject formula, we have: $x=10-3=7$

$\therefore x=7$

Example 12

A trader bought x mangoes at the rate of 5 mangoes for ₦15. If 3 of the mangoes were bad and he sold the remaining at the rate of 6 mangoes for ₦25 and made a gain of ₦12, find x.

Solution

Gain=₦12

x mangoes at the rate of 5 mangoes for ₦15.

$\therefore$ cost of 1 mango $=\frac{15}{5}\times 1$

and cost price of x mangoes $=\frac{15}{5}\times x$

Thus, if 3 mangoes were bad, the remaining mangoes will be $x-3$; and it was sold at the rate of 6 for ₦25.

$\therefore$ the selling price $=\frac{x-3}{6}\times 25$. Hence, gain = selling price − cost price

$=\frac{x-3}{6}\times 25-\frac{15}{5}\times x=12$

$$\frac{30\left(\frac{x-3}{6}\times 25-\frac{15}{5}\times x\right)}{30}=12$$

Find the LCM, and then simplify. Here, the LCM is 30.

$125x - 375 - 90x = 360$
$35x - 375 = 360$
$35x = 375 + 360 = 735$
$x = \frac{735}{35} = 21$
$\therefore$ The trader bought 21 mangoes.

9.3 LINEAR EQUATION IN TWO VARIABLES (SIMULTANEOUS EQUATION)

Let the linear equations be of the form:
$ax + by = c$ ....(i)
$dx + ey = f$ ....(ii)
Where $a, b, c, d, e,$ and f are constants. The unknown variables are x and y. Then, the equations (i) and (ii) exist simultaneously, if x and y for (i) is equal and equivalent to x and y in equation (ii).

9.3.1 METHODS OF SOLVING SIMULTANEOUS EQUATION

(1) Substitution method
(2) Elimination method
(3) Graphical method

SUBSTITUTION METHOD

Example 13
Solve the simultaneous equation using substitution method:
$3x + y = 1$ and $2x - 3y = 8$

Solution
$3x + y = 1$ ....(i)
$2x - 3y = 8$ ....(ii)
Make y the subject formula in equation (i)
Thus, $3x + y = 1 \quad \Rightarrow y = 1 - 3x$
Now, put $y = 1 - 3x$ into equation (ii)
$2x - 3y = 8$
$\Rightarrow 2x - 3(1 - 3x) = 8$
$2x - 3 + 9x = 8$
$2x + 9x = 8 + 3 = 11$
$11x = 11$
$x = \frac{11}{11} = 1 \quad \therefore x = 1$

You can make one of the variables the subject of the formula either from equation (i) or (ii), depending on which one is most convenient and easier to use.

Now, put the value of $x = 1$ in either equation (i) or (ii).
Let's use equation (i) in this case:
thus, $3x + y = 1 \quad \Rightarrow 3(1) + y = 1$
$y = 1 - 3 = -2$
$\Rightarrow (x, y) = (1, -2)$

Example 14

Solve for x and y:
$x + y = 4$
$2x - y = 5$

Solution

$x + y = 4$ ....(i)
$2x - y = 5$ ....(ii)

Make x the subject formula in equation (i):
$x + y = 4 \quad \Rightarrow x = 4 - y$
Now, put this value of x into equation (ii), which is $2x - y = 5$
So, $2(4 - y) - y = 5$
$8 - 2y - y = 5$
$8 - 3y = 5$
$-3y = 5 - 8 = -3$
$-3y = -3$
$y = \frac{-3}{-3} = 1 \quad \therefore y = 1$
Put this value of y into equation (i) i.e., $x + y = 4$
$x + 1 = 4 \quad \Rightarrow x = 4 - 1 = 3$
$\therefore x = 3 \quad \Rightarrow (x, y) = (3, 1)$

ELIMINATION METHOD

This method is often used when none of the coefficients of the unknown variables is unity. Thus, in order to find the solution of the simultaneous equation, one of the unknown variables is first eliminated so as to find the other variable.

Steps to follow when using Elimination method:

(1) Choose a specific variable to eliminate;
(2) Make the coefficients of this variable equal by multiplying the first and second equations by convenient factors;
(3) Then add or subtract the two equations. If addition will eliminate one of the variables, then add, else subtract to eliminate.
(4) Solve for the remaining variable and then substitute the answer to any of the equations to find the other unknown variable.

Example 15

Solve for x and y in the simultaneous equations:

$x - 5y = 1$

$x + 4y = 28$

Solution

$x - 5y = 1$ .... (i)

$-\ x + 4y = 28$ .... (ii)

$0 + 9y = 27$

> The coefficients of x in equation (i) and (ii) are equal. So, x can be easily eliminated by subtracting (ii) from (i).

$9y = 27$

$y = \frac{27}{9} = 3$

$\therefore y = 3$

Now, put $y = 3$ into equation (i) i.e., $x - 5y = 1$

$\Rightarrow x - 5(3) = 1$

$x - 15 = 1$

$x = 1 + 15 = 16$

$\therefore x = 16$

Example 16

$2x - 3y = -2$

$3x + 2y = -6$

> **Step 1**: Choose the variable to eliminate, e.g., x.
> **Step 2**: Multiply equation (i) by factor of 3.
> Then multiply equation (ii) by factor of 2.
> **Step 3**: Subtract equation (iv) from (iii).
> **Step 4**: Solve for y and x respectively.

Solution

$2x - 3y = -2$ .... (i)

$3x + 2y = -6$ .... (ii)

$3 \times (2x - 3y) = 3 \times (-2)$

$2 \times (3x + 2y) = 2 \times (-6)$

$6x - 9y = -6$ .... (iii)

$-\ 6x + 4y = -12$ .... (iv)

$0 + 13y = -6$

$\Rightarrow 13y = -6$

Then, $y = -\frac{6}{13}$

Put $y = -\frac{6}{13}$ into equation (i), i.e., $2x - 3y = -2$

$\Rightarrow 2x - 3\left(-\frac{6}{13}\right) = -2$

$2x + \frac{18}{13} = -2 \quad \Rightarrow 2x = -2 - \frac{18}{13} = -\frac{(26+18)}{13} = -\frac{44}{13}$

$x = -\frac{44}{2(13)} = -\frac{22}{13}$

$\therefore x = -1\frac{9}{13}$ Hence, $(x, y) = \left(-1\frac{9}{13}, -\frac{6}{13}\right)$

Example 17

Given $\frac{5y}{2} - 5x = 5$ and $\frac{y}{3} - \frac{x}{2} = \frac{17}{6}$

Find the values of x and y.

Solution

Step 1: Choose the variable to eliminate e.g., y.

Step 2: Multiply equation (i) by factor of 2. Then, multiply equation (ii) by factor of 15.

Step 3: Subtract equation (iv) from (iii).

Step 4: Solve for x and y.

$\frac{5y}{2} - 5x = 5$ ....(i)

$\frac{y}{3} - \frac{x}{2} = \frac{17}{6}$ ....(ii)

$2 \times \left(\frac{5y}{2} - 5x\right) = 2 \times 5$

$5y - 10x = 10$ ....(iii)

Similarly,

$6 \times \left(\frac{y}{3} - \frac{x}{2}\right) = 6 \times \frac{17}{6}$

$2y - 3x = 17$ ....(iv)

Choosing to eliminate y, multiply equation iii by 2 and equation (iv) by 5.

$$\begin{array}{l} 2 \times (5y - 10x) - 2 \times 10 \\ - \\ 5 \times (2y - 3x) = 5 \times 17 \end{array}$$

$10y - 20x = 20$ ....(v)

$-$

$10y - 15x = 85$ ....(vi)

$0 + 5y = 65$

$5y = 65 \quad \Rightarrow y = 13$

Now, put y into any of the equations; for instance (iii) i.e., $5y - 10x = 10$

$5(13) - 10x = 5 \quad \Rightarrow -10x = 5 - 65 = -60$

$x = \frac{-60}{-10} \quad \Rightarrow x = 6$

$\therefore (x, y) = (6, 13)$

GRAPHICAL METHOD

This requires that we draw the graph of two equations using the same axes and scale. Then, the point of intersection between the two lines depicts the required solution of the simultaneous equation.

Example 18

Solve graphically the simultaneous equations:

$x + y = 3$

$3x - y = 1$

Solution

Table 9.1(a) $x + y = 3$

x	-2	0	2	4
y	5	3	1	-1

Table 9.1(b) $3x - y = 1$

x	-2	0	2	4
y	-7	-1	5	11

Use any convenient interval for x and y of your choice (at least an interval of $-2 \leq x \leq 2$).

Now, plot the graph using convenient scale and the same axes for both table 9.1(a) and table 9.1(b). Thus, the solution lies at the point of intersection $(x, y) = (1, 2)$

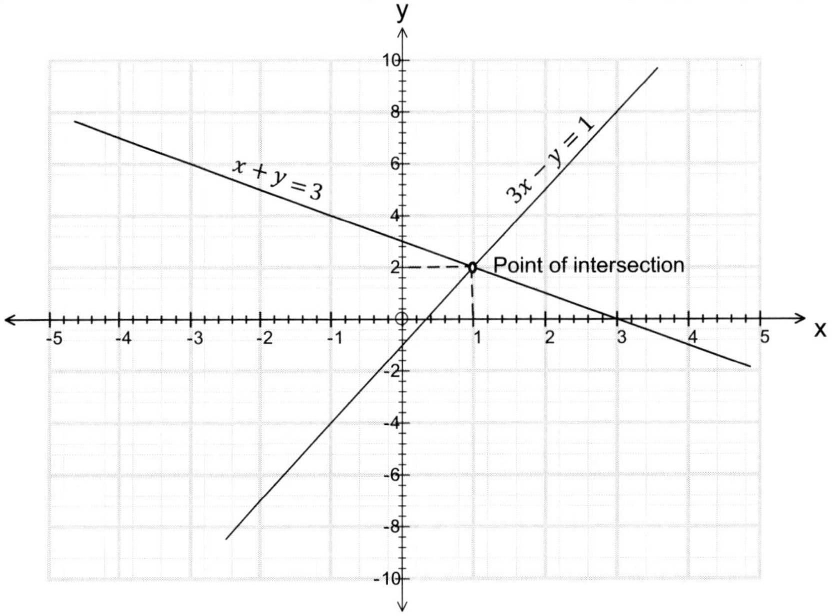

Fig. 9.1 Graphical solution of linear simultaneous equation

9.3.2 WORD PROBLEMS LEADING TO SIMULTANEOUS LINEAR EQUATION

This is the process of translating English statements into mathematical forms (i.e., linear equations). Then, the unknown variables are found using any method of solving simultaneous linear equation.

Example 19

A father is three times as old as his son, and twelve years ago he was six times as old as his son. How old are the son and the father?

Solution

Let the father's age be y, and let the son's age be x.

$\therefore y = 3x$ ....(i)

$y - 12 = 6(x - 12)$ ....(ii)

Using substitution method:

Put $y = 3x$ from equation (i) into (ii):

$\therefore y - 12 = 6(x - 12) \Rightarrow 3x - 12 = 6(x - 12)$

$3x - 12 = 6x - 72$

$6x - 3x = 72 - 12 = 60$

$3x = 60$

$x = \frac{60}{3} = 20$

$\therefore$ The son's age is 20 years.

Now, put $x = 20$ into equation (i), $y = 3x = 3 \times 20 = 60$.

$\therefore$ The father's age is 60 years.

Twelve years ago, the father's age will be twelve years less than his present age (i.e., $y - 12$). Likewise, that of the son ($x - 12$).

Example 20

The perimeter of a rectangle is 28cm, while the difference between the length and the width is 2cm. Find the length and width of the rectangle.

Solutions

Let the length be L, and the width W.

$\therefore$ Perimeter $= 2(L + W) = 28$

$L + W = 14$ ....(i)

$L - W = 2$ ....(ii)

Using elimination method, add equation (i) and (ii) to eliminate W:

$L + W = 14$

$L - W = 2$

$2L + 0 = 16$

$\therefore L = \frac{16}{2} = 8$

Put $L = 8$ into equation (i), i.e. $L + W = 14$

thus, $8 + W = 14$

$W = 14 - 8 = 6$

$\Rightarrow$ Length of rectangle $=$ 8cm and width of rectangle$=$ 6cm.

9.4 CHANGE OF SUBJECT OF FORMULA

A formula is simply a mathematical equation consisting of letters which represent real life quantities. In many cases, it will be necessary to change the subject of a formula; and this is achieved by collecting like terms, and putting the letter to be made the subject of the formula on one side of the equation while the other variables (letters) are kept on the opposite side of the equation.

Example 21

(a) $y = mx + c$; make x the subject of the formula.
(b) $v^2 = u^2 + 2as$; make a the subject formula.
(c) $\frac{1}{f} = \frac{1}{u} + \frac{1}{v}$; make v the subject formula.

Solution

(a) $y = mx + c$

Rearrange the equation:

$\therefore y - c = mx$

Divide through by m

$\frac{y-c}{m} = \frac{mx}{m} \quad \Rightarrow x = \frac{y-c}{m}$

(b) $v^2 = u^2 + 2as$

Rearrange the equation:

$\therefore v^2 - u^2 = 2as$

Divide through by $2s$

$\frac{v^2-u^2}{2s} = \frac{2as}{2s} \quad \Rightarrow a = \frac{v^2-u^2}{2s}$

(c) $\frac{1}{f} = \frac{1}{u} + \frac{1}{v}$

Rearranging the equation so that $\frac{1}{v}$ is left alone on one side of the equation.

$\therefore \frac{1}{f} - \frac{1}{u} = \frac{1}{v} \quad \Rightarrow \frac{u-f}{uf} = \frac{1}{v}$

Take the reciprocal of both sides of the equation,

$\therefore \frac{uf}{u-f} = v$

Thus, $v = \frac{uf}{u-f}$

Example 22

(a) $\frac{x^2}{a} + \frac{y^2}{b} = 1$ Make y the subject formula.

(b) $ax^2 + bx + c = 0$ Make x the subject formula.

Solution

(a) $\frac{x^2}{a} + \frac{y^2}{b} = 1$

$\frac{y^2}{b} = 1 - \frac{x^2}{a} = \frac{a-x^2}{a}$

Leave $\frac{y^2}{b}$ on the LHS, while other terms are kept on the RHS.

Thus, $\frac{y^2}{b} = \frac{a-x^2}{a} \quad \Rightarrow y^2 = \frac{b(a-x^2)}{a}$

Hence, $y = \sqrt{\frac{b(a-x^2)}{a}}$

(b) $ax^2 + bx + c = 0$

Divide through by the coefficient of x^2

$$\frac{ax^2}{a} + \frac{bx}{a} + \frac{c}{a} = 0$$

$$\therefore x^2 + \frac{b}{a}x = -\frac{c}{a}$$

$$x^2 + \frac{b}{a}x + \left(\frac{1}{2} \times \frac{b}{a}\right)^2 = -\frac{c}{a} + \left(\frac{1}{2} \times \frac{b}{a}\right)^2$$

$$\left(x + \frac{b}{2a}\right)^2 = -\frac{c}{a} + \frac{b^2}{4a^2} = \frac{b^2-4ac}{4a^2}$$

$$x + \frac{b}{2a} = \sqrt{\frac{b^2-4ac}{4a^2}} = \frac{\sqrt{b^2-4ac}}{2a}$$

$$x = -\frac{b}{2a} \pm \frac{\sqrt{b^2-4ac}}{2a} = \frac{-b \pm \sqrt{b^2-4ac}}{2a}$$

$$\therefore x = \frac{-b \pm \sqrt{b^2-4ac}}{2a}$$

Half the coefficient of x (i.e., $\frac{1}{2} \times \frac{b}{a}$); then square it, $\left(\frac{1}{2} \times \frac{b}{a}\right)^2$. Now, add this squared value to both sides of the equation, then simplify to make x the subject formula.

9.5 QUADRATIC EQUATION

This is any equation of the form, $ax^2 + bx + c = 0$; where a, b and c are constants such that $a \neq 0$.

Methods of Solving Quadratic Equation

(1) Factorisation method
(2) Completing the square method
(3) Quadratic formula method
(4) Graphical method

9.5.1 FACTORISATION METHOD

This involves finding two linear factors. And to do this, it is advisable to first test if the equation is factorisable using the discriminant, given by, $D = b^2 - 4ac$. If the discriminant is a perfect square, then you can continue to use factorisation method. But if it is not a perfect square, then use other methods like completing the square or formula method.

Steps to follow when using factorisation method:

Given $ax^2 + bx + c = 0$, then:

(1) Find two numbers (i.e., factors f_1 and f_2) by trial and error such that their sum is equal to "b". Thus, $f_1 + f_2 = b$

(2) If $a = 1$, multiply f_1 and f_2 to see if the product is equal "c".
thus, $f_1 f_2 = c$
and if $a > 1$, then $f_1 f_2 = ac$.

When conditions (1) and (2) are satisfied, then f_1 and f_2 are correct factors.

(3) Then replace "b" in the quadratic equation with $f_1 + f_2$ from step (1).
Thus, $ax^2 + (f_1 + f_2)x + c = 0$

(4) Solve for x by simplifying the equation into its factored form (i.e. find the common factors, then completely factorise).

Example 23

Solve for x, if (a) $x^2 + 5x + 6 = 0$ (b) $2x^2 - 3x - 5 = 0$

Solution

(a) $x^2 + 5x + 6 = 0$

Here we have: $a = 1, \; b = 5, \; c = 6$

Now, use the aforementioned steps:

step 1:

$b = 5$, possible trials (factors) are: $1 + 4 = 5$, or $2 + 3 = 5$

step 2 :

$c = 6$, possible trials are $1 \times 4 = 4 \neq 6$, or $2 \times 3 = 6$.

Thus, the correct factor is when the sum of the factors is equal to b, and their product equal c. Therefore, $f_1 = 2$ and $f_2 = 3$.

step 3 :

$x^2 + 5x + 6 = 0 \quad \Rightarrow x^2 + (f_1 + f_2)x + 6 = 0$

$x^2 + (2 + 3)x + 6 = 0$

$\Rightarrow x^2 + 2x + 3x + 6 = 0$

step 4 :

$x(x + 2) + 3(x + 2) = 0$

$(x + 2)(x + 3) = 0 \quad \Rightarrow (x + 2) = 0$ and $(x + 3) = 0$

$\therefore x = -2$ or -3

(b) $2x^2 - 3x - 5 = 0$

$a = 2, b = -3, c = -5$

$b = -3$, possible factors are: $-4 + 1 = -3$ or $-5 + 2 = -3$

Since, $a > 1$ (i.e., $a = 2$) $\quad \Rightarrow ac = 2 \times (-5) = -10$

$\therefore \; -4 \times 1 = -4 \neq -10$, and $-5 \times 2 = -10$ Correct factors are: -5 and 2

$\therefore 2x^2 - 3x - 5 = 0 \quad \Rightarrow 2x^2 + 2x - 5x - 5 = 0$

$2x(x+1) - 5(x+1) = 0$

$(x+1)(2x-5) = 0$

$\Rightarrow (x+1) = 0$ and $(2x-5) = 0$

$\therefore x = -1$ or $2\frac{1}{2}$

9.5.2 COMPLETING THE SQUARE METHOD

This method solves quadratic problems that cannot be solved by factorisation method.

Example 2

(a) $2x^2 + 3x - 5 = 0$, solve for x

(b) $y^2 - 5y + 6 = 0$, solve for y

Solution

Half the coefficient of x $\left(\text{i. e., } \frac{1}{2} \times \frac{3}{2} = \frac{3}{4}\right)$ then square it, $\left(\frac{3}{4}\right)^2$. Now, add this squared value to both sides of the equation, then simplify to solve for x.

(a) $2x^2 + 3x - 5 = 0$

$\frac{2}{2}x^2 + \frac{3}{2}x - \frac{5}{2} - 0 \quad \rightarrow x^2 + \frac{3}{2}x = \frac{5}{2}$

$x^2 + \frac{3}{2}x + \left(\frac{3}{4}\right)^2 = \frac{5}{2} + \left(\frac{3}{4}\right)^2$

$\left(x + \frac{3}{4}\right)^2 = \frac{5}{2} + \frac{9}{16} = \frac{40+9}{16} = \frac{49}{16}$

$x + \frac{3}{4} = \sqrt{\frac{49}{16}} = \pm\frac{7}{4}$

$x = \pm\frac{7}{4} - \frac{3}{4}$

$x = +\frac{7}{4} - \frac{3}{4}$ or $-\frac{7}{4} - \frac{3}{4}$

$x = \frac{+7-3}{4}$ or $\frac{-7-3}{4} \quad \Rightarrow x = \frac{+4}{4}$ or $\frac{-10}{4}$

$x = \frac{4}{4} = 1$ or $\frac{-10}{4} = -2\frac{1}{2}$

$\therefore x = 1$ or $-2\frac{1}{2}$

(b) $y^2 - 5y + 6 = 0$

$y^2 - 5y = -6$

$y^2 - 5y + \left(-\frac{5}{2}\right)^2 = -6 + \left(-\frac{5}{2}\right)^2$

$$\left(y-\frac{5}{2}\right)^2 = -6+\left(-\frac{5}{2}\right)^2 = -6+\frac{25}{4}$$

$$\left(y-\frac{5}{2}\right)^2 = \frac{-24+25}{4} = \frac{1}{4}$$

$$y-\frac{5}{2} = \sqrt{\frac{1}{4}} = \pm\frac{1}{2}$$

$$y = \pm\frac{1}{2}+\frac{5}{2} = \frac{+1+5}{2} \text{ or } \frac{-1+5}{2}$$

$$y = \frac{6}{2} = 3 \quad \text{or } y = \frac{4}{2} = 2$$

$\Rightarrow y = 3$ or 2

Example 25

What must be added to each of the following equations to make it a perfect square?

(a) x^2+3x (b) $2x^2-4x$

Solution

(a) x^2+3x

Use the discriminant, $D = b^2-4ac$

$\Rightarrow x^2+3x+c=0$

where c is the required value.

Thus, $D = b^2-4ac = 0$

$a = 1, \quad b = 3$

$\therefore 3^2-4\times 1\times c = 0 \quad \Rightarrow 9-4c = 0$

$\therefore c = \frac{9}{4}$

The discriminant, D, must be equal to zero so as to obtain the value needed to make the equation a perfect square.

(b) $2x^2-4x$

Thus, $D = b^2-4ac = 0$

$a = 2, \ b = -4$

$\therefore (-4)^2-4\times 2\times c = 0 \quad \Rightarrow 16-8c = 0$

$\therefore c = \frac{16}{8}$

9.5.3 QUADRATIC FORMULA METHOD

This formula is derived by applying the method of completing the square on a quadratic equation, $ax^2+bx+c=0$.

Proof:

$ax^2 + bx + c = 0$, then divide through by a to have:

$\frac{ax^2}{a} + \frac{bx}{a} + \frac{c}{a} = 0 \quad \therefore\ x^2 + \frac{b}{a}x = -\frac{c}{a}$

$$x^2 + \frac{b}{a}x + \left(\frac{b}{2a}\right)^2 = -\frac{c}{a} + \left(\frac{b}{2a}\right)^2$$

$$\left(x + \frac{b}{2a}\right)^2 = -\frac{c}{a} + \frac{b^2}{4a^2} = \frac{b^2-4ac}{4a^2}$$

$$x + \frac{b}{2a} = \sqrt{\frac{b^2-4ac}{4a^2}} = \frac{\sqrt{b^2-4ac}}{2a}$$

$$x = -\frac{b}{2a} \pm \frac{\sqrt{b^2-4ac}}{2a} = \frac{-b\pm\sqrt{b^2-4ac}}{2a}$$

$$\therefore\ x = \frac{-b\pm\sqrt{b^2-4ac}}{2a} = \frac{-b\pm\sqrt{D}}{2a}$$

$D = b^2 - 4ac$ this is called the discriminant.

Thus,

if $D > 0$, then D =positive $\therefore$ roots are real and distinct.

if $D < 0$, then D = negative $\therefore$ roots are imaginary and complex.

if $D = 0$, then D = zero $\therefore$ roots are equal, thus they are coincident roots.

Example 26

Use the quadratic formula to find the solutions for x in each of the given equations:

(a) $2x^2 - 3x + 1 = 0$ (b) $3x^2 + 8x + 2 = 0$

Solution

(a) $2x^2 - 3x + 1 = 0$

Comparing the coefficients with that of the form: $ax^2 + bx + c = 0$

$a = 2,\ b = -3,\ c = 1$; substitute these values into the quadratic formula.

$$\therefore\ x = \frac{-b\pm\sqrt{b^2-4ac}}{2a} \quad \Rightarrow x = \frac{-(-3)\pm\sqrt{(-3)^2-4\times2\times1}}{2\times2}$$

$$x = \frac{+3\pm\sqrt{9-8}}{2\times2} = \frac{+3\pm\sqrt{1}}{4} = \frac{+3\pm1}{4}$$

$$= \frac{+3+1}{4} \text{ or } \frac{-3+1}{4} \quad \Rightarrow \frac{4}{4} = 1, \text{ or } \frac{-2}{4} = -\frac{1}{2}$$

$$\therefore x = 1 \text{ or } -\frac{1}{2}$$

(b) $3x^2 + 8x + 2 = 0$

$a = 3, \quad b = 8, \quad c = 2$

$$\text{Using, } x = \frac{-b\pm\sqrt{b^2-4ac}}{2a}$$

$$x = \frac{-8 \pm \sqrt{8^2 - 4 \times 3 \times 2}}{2 \times 3} = \frac{-8 \pm \sqrt{64 - 24}}{6} = \frac{-8 \pm \sqrt{40}}{6}$$

$$x = \frac{-8 \pm 6.33}{6} = \frac{-8 + 6.33}{6} \text{ or } \frac{-8 - 6.33}{6}$$

$$= \frac{-1.67}{6} \text{ or } \frac{-14.33}{6} = -0.28 \text{ or } -2.39$$

$$x = -0.28 \text{ or } -2.39$$

Formulating Quadratic Equation From Given Roots

A quadratic equation can be formed if the roots are known. So, let $x = \alpha$ and $x = \beta$ represent the roots of a quadratic expression, $ax^2 + bx + c = 0$.
then, $(x - \alpha)(x - \beta) = 0$
$x^2 - \alpha x - \beta x + \alpha\beta = 0$
$\Rightarrow x^2 - (\alpha + \beta)x + \alpha\beta = 0$

Compare this equation with $ax^2 + bx + c = 0$

$a = 1, \quad b = -(\alpha + \beta), \quad c = \alpha\beta$

$\therefore \quad x^2 - (Sum\ of\ roots)x + Product\ of\ roots = 0$

However, the sum of roots and product of roots can be found using the quadratic formula:

$$x = \frac{-b \pm \sqrt{b^2 - 4ac}}{2a}$$

where the roots are $x_1 = \alpha = \frac{-b + \sqrt{b^2 - 4ac}}{2a}$ or $x_2 = \beta = \frac{-b - \sqrt{b^2 - 4ac}}{2a}$

Thus, sum of roots $= \alpha + \beta = \frac{-b + \sqrt{b^2 - 4ac}}{2a} + \frac{-b - \sqrt{b^2 - 4ac}}{2a}$

$$= \frac{-2b}{2a} = -\frac{b}{a}$$

Also, product of roots $= \alpha\beta = \left(\frac{-b + \sqrt{b^2 - 4ac}}{2a}\right)\left(\frac{-b - \sqrt{b^2 - 4ac}}{2a}\right)$

$$= \frac{b^2 + b\sqrt{b^2 - 4ac} - b\sqrt{b^2 - 4ac} - b^2 + 4ac}{4a^2} = \frac{4ac}{4a^2} = \frac{c}{a}$$

Thus, sum of roots $= \alpha + \beta = -\frac{b}{a}$ and product of roots $= \alpha\beta = \frac{c}{a}$

Example 27

Find the quadratic equation with the following roots:

(a) 5 and 3 (b) $-\frac{1}{3}$ and 2 (c) 2 twice (d) $\frac{1}{2}$ and $-\frac{3}{4}$

Solution

The quadratic equation whose roots are α and β is given by $x^2 - (\alpha + \beta)x + \alpha\beta = 0$

(a) 5 and 3 : let $\alpha = 5$ and $\beta = 3$

$x^2 - (5+3)x + 5 \times 3 = 0 \quad \Rightarrow x^2 - 8x + 15 = 0$

(b) $-\frac{1}{3}$ and 2: let $\alpha = -\frac{1}{3}$ and $\beta = 2$

$x^2 - (-\frac{1}{3} + 2)x + -\frac{1}{3} \times 2 = 0$

$x^2 - \frac{5}{3}x - \frac{2}{3} = 0 \quad \Rightarrow 3x^2 - 5x - 2 = 0$

(c) 2 twice: let $\alpha = 2$ and $\beta = 2$

$x^2 - (2+2)x + 2 \times 2 = 0$

$x^2 - 4x + 4 = 0$

(d) $\frac{1}{2}$ and $-\frac{3}{4}$: let $\alpha = \frac{1}{2}$ and $\beta = -\frac{3}{4}$

$x^2 - \left(\frac{1}{2} + \left(-\frac{3}{4}\right)\right)x + \frac{1}{2} \times (-\frac{3}{4}) = 0$

$x^2 - \left(-\frac{1}{4}\right)x + (-\frac{3}{8}) = 0 \quad \Rightarrow \quad x^2 + \frac{1}{4}x - \frac{3}{8} = 0$

$\therefore \quad 8x^2 + 2x - 3 = 0$

Example 28

Find the sum and product of the following roots:

(a) $x^2 - 4x - 3 = 0$ (b) $2x^2 + 11x - 10 = 0$

(c) $3x^2 - \frac{1}{2}x - 1 = 0$ (d) $\frac{1}{3}x - \frac{2}{x-1} - x = 0$

Solution

(a) $x^2 - 4x - 3 = 0$

Quadratic equation is of the form $ax^2 + bx + c$

Thus, $a = 1,\ b = -4,\ c = -3$

sum of roots, $\alpha + \beta = -\frac{b}{a} = -\left(\frac{-4}{1}\right) = 4$

product of roots, $\alpha\beta = \frac{c}{a} = \frac{-3}{1} = -3$

$\therefore \alpha + \beta = 4$, and $\alpha\beta = -3$

(b) $2x^2 + 11x - 10 = 0$

Thus, $a = 2,\ b = 11,\ c = -10$

sum of roots, $\alpha + \beta = -\frac{b}{a} = -\left(\frac{11}{2}\right) = -5\frac{1}{2}$

product of roots, $\alpha\beta = \frac{c}{a} = \frac{-10}{2} = -5$

$\therefore \alpha + \beta = -5\frac{1}{2}$ and $\alpha\beta = -5$

(c) $3x^2 - \frac{1}{2}x - 1 = 0$

Thus, $a = 3,\ b = -\frac{1}{2},\ c = -1$

Sum of roots, $\alpha + \beta = -\frac{b}{a} = -\frac{(-1/2)}{3} = \frac{1}{6}$

Product of roots, $\alpha\beta = \frac{c}{a} = \frac{-1}{3}$

$\therefore \alpha + \beta = \frac{1}{6}$ and $\alpha\beta = \frac{-1}{3}$

(d) $\frac{1}{3}x - \frac{2}{x-1} - x = 0$

> Simplify this expression so as to obtain it in its quadratic form.

$3(x-1)\frac{1}{3}x - 3(x-1)\frac{2}{x-1} - 3(x-1)x = 0$

$x(x-1) - 6 - 3x(x-1) = 0$

$x^2 - x - 6 - 3x^2 + 3 = 0$

$-2x^2 + 2x - 6 = 0$

Thus, $a = -2,\ b = 2,\ c = -6$

Sum of roots, $\alpha + \beta = -\frac{b}{a} = -\frac{2}{-2} = 1$

Product of roots, $\alpha\beta = \frac{c}{a} = \frac{-6}{-2} = 3$

$\therefore \alpha + \beta = 1$ and $\alpha\beta = 3$

Example 29

Given $5x^2 + 10x + 20$, evaluate the following:

(a) $\alpha^2 + \beta^2$ (b) $\frac{1}{\alpha} + \frac{1}{\beta}$ (c) $\alpha^3 + \beta^3$

Solution

$5x^2 + 10x + 20$

$a = 5, \quad b = 10, \quad c = 20$

(a) $\alpha^2 + \beta^2 = (\alpha + \beta)^2 - 2\alpha\beta = (-\frac{b}{a})^2 - 2(\frac{c}{a})$

$= \left(-\frac{10}{5}\right)^2 - 2\left(\frac{20}{5}\right) = (-2)^2 - 2(4) = 4 - 8 = -4$

(b) $\frac{1}{\alpha} + \frac{1}{\beta} = \frac{\alpha+\beta}{\alpha\beta} = \frac{-b/a}{c/a} = \frac{-10/5}{20/5}$

$\frac{-2}{4} = -\frac{1}{2}$

(c) $\alpha^3 + \beta^3 = (\alpha + \beta)^3 - 3\alpha\beta(\alpha + \beta) = \left(-b/a\right)^3 - 3(c/a)(-b/a)$

$= \left(-\frac{10}{5}\right)^3 - 3\left(\frac{20}{5}\right)\left(-\frac{10}{5}\right) = (-2)^3 - 3(4)(-2) = -8 + 24 = 16$

Word Problems Leading to Quadratic Equation

This is the translation of statements into mathematical (quadratic) equations, then the rule of quadratic equation is applied to solve the given problem.

Example 30

Find two numbers whose difference is 5 and product is 50.

Solution

Let the smaller number be x, and the larger number be y.
$y - x = 5$ ....(i)
$xy = 50$ ....(ii)
From equation (i), y can be made the subject of the formula,
$\Rightarrow y = x + 5$
Put this value of y into equation (ii):
$x(x+5) = 50 \quad \Rightarrow x^2 + 5x = 50$ Now, factorise the quadratic expression:
$x^2 + 5x - 50 = 0 \quad \therefore x^2 + 10x - 5x - 50 = 0$
$x(x+10) - 5(x+10) = 0$ Hence, we have $(x-5)(x+10) = 0$
$\therefore x + 10 = 0 \quad \Rightarrow x = -10$
When $x = 5,\ y = x + 5 = 5 + 5 = 10$
When $x = -10,\ y = x + 5 = -10 + 5 = -5$
Since x is the smaller number, while y is the larger number.
then, $x = -10$ or 5, $y = -5$ or 10

Example 31

The ages of two brothers are 11 and 8 years. In how many years' time will the product of their ages be 208?

Solution

Let the age of the younger brother now be 8 years,
and the age of the elder brother now be 11 years
Thus, in x time their ages will be $x + 8$, and $x + 11$
Then the product of their ages in x time $= (x+8)(x+11) = 208$
$(x+8)(x+11) = 208$
$x^2 + 11x + 8x + 88 = 208$
Re-arranging $x^2 + 11x + 8x + 88 - 208 = 0$
$x^2 + 19x - 120 = 0$, factorise the quadratic expression
$x^2 + 24x - 5x - 120 = 0 \quad \Rightarrow x(x+24) - 5(x+24) = 0$
$(x-5)(x+24) = 0$
$\therefore x - 5 = 0 \quad \Rightarrow x = 5$
$x + 24 = 0 \quad \Rightarrow x = -24$
Since time cannot be negative, therefore, the correct answer is $x = 5$.
Hence, in 5 years' time, the product of their ages will be 208.

Example 32

The length of a rectangle is twice its width. If the width is reduced by 1cm while the length is reduced by 2cm, then the area will be 15cm². Find the dimensions of the original rectangle.

Solution

Let width $= w$, length$= L$

Where $L = 2w$

When the length and width were reduced, then we have dimensions as,

$L - 2$ and $w - 1$. Hence, $(L - 2)(w - 1) = 15$

But, $L = 2w$

thus, $(2w - 2)(w - 1) = 15 \quad \Rightarrow 2w^2 - 2w - 2w + 2 = 15$

$2w^2 - 4w - 13 = 0$

Now, use the quadratic formula to solve the equation:

$$w = \frac{-b \pm \sqrt{b^2 - 4ac}}{2a}$$

$a = 2, \qquad b = -4, \qquad c = -13$

$$w = \frac{-(-4) \pm \sqrt{(-4)^2 - 4 \times 2 \times (-13)}}{2 \times 2} = \frac{4 \pm \sqrt{16 + 120}}{4}$$

$$w = \frac{4 \pm \sqrt{136}}{4} = \frac{4 \pm 10.95}{4} = \frac{4 + 10.95}{4} \text{ or } \frac{4 - 10.95}{4}$$

$$w = \frac{14.95}{4} = 3.74\text{cm} \text{ or } \frac{-6.94}{4} = -1.74\text{cm}$$

So, $w = 3.74$cm, which is the positive solution.

$\therefore \; L = 2w = 2 \times 3.74 = 7.48$ cm

$\therefore \; w = 3.74$cm, $L = 7.48$ cm

9.5.4 GRAPHICAL SOLUTION OF QUADRATIC EQUATION

The graphical equation, $ax^2 + bx + c = 0$ can be solved graphical by drawing the graph of the function $y = F(x) = ax^2 + bx + c$ with interval of $m \le x \le n$, where m and n are integers. Hence, the point(s) where the graph cuts the x axis are the roots/solution of the quadratic equation.

Also, the graph of a quadratic function is always a curve. And the shape of the curve depends on the value of the constant, a. Thus, if $a > 0$, then the curve is a ∪ −shape; but when $a < 0$, then the curve is an ∩ −shape curve. Thus, the lowest point on the ∪ −shape where the curve abruptly changes direction is called the "minimum point," and it occurs at a value of $x = -\frac{b}{2a}$; while that of the ∩ −shape is called the "maximum point," and it also occurs at a value of $x = -\frac{b}{2a}$.

The value of $x = -\frac{b}{2a}$ is referred to as the "line of Symmetry." And at this point the corresponding value of y is $\frac{4ac-b^2}{4a}$.

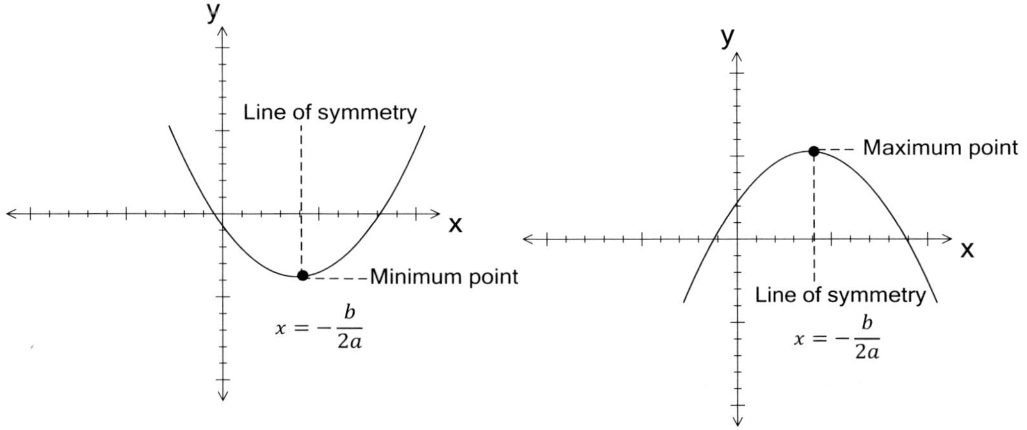

Fig. 9.2 Minimum and Maximum point

<u>Steps to Follow When Using Graphical Method</u>

(1) Construct a table of value for x and y using convenient interval, $m \leq x \leq n$;
(2) Choose convenient scales for both x and y axes, and indicate it on the top right-hand corner of your graph sheet;
(3) Plot the tabulated values on the graph;
(4) Join the points by using free hand sketch, or with a "French curve" if available;
(5) Title your graph;
(6) Deduce the roots (i.e. the solution) from the point(s) where the curve cuts the x axis.

Thus, the following curves show the variation of the roots of a quadratic equation:

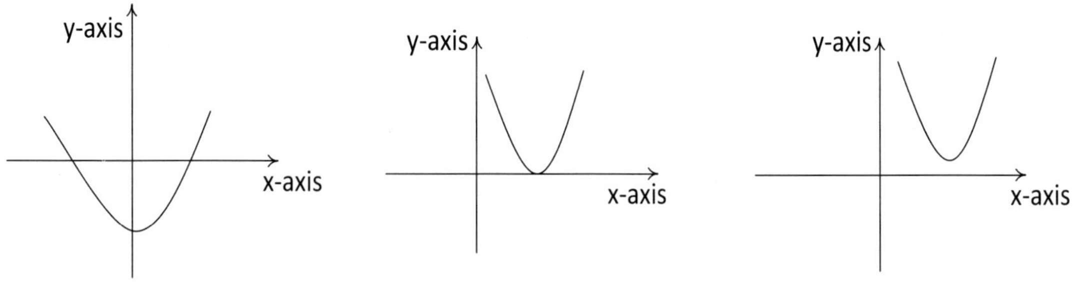

(i) Distinct Roots (ii) Repeated/Coincidental Roots (iii) Complex/Imaginary Roots

Fig. 9.3 Forms of Quadratic Roots

Example 33

Draw the graph of $2x^2 + 3x - 2 = 0$ using interval of $-3 \le x \le 3$. Take a scale of 1cm to represent 1 unit on x-axis, and 5 units on y-axis. Use the graph to find the following:

(a) the roots of the equation $2x^2 + 3x - 2 = 0$

(b) the values of x when $y = 5$

(c) the minimum value of the graph, $y = 2x^2 + 3x - 2$

(d) the line of symmetry (i.e., the value of x for which y is minimum).

Solution

Table 9.2

x	-3	-2	-1	0	1	2	3
$2x^2$	18	8	2	0	2	8	18
$+3x$	-9	-6	-3	0	3	6	9
-2	-2	-2	-2	-2	-2	-2	-2
y	7	0	-3	-2	3	12	25

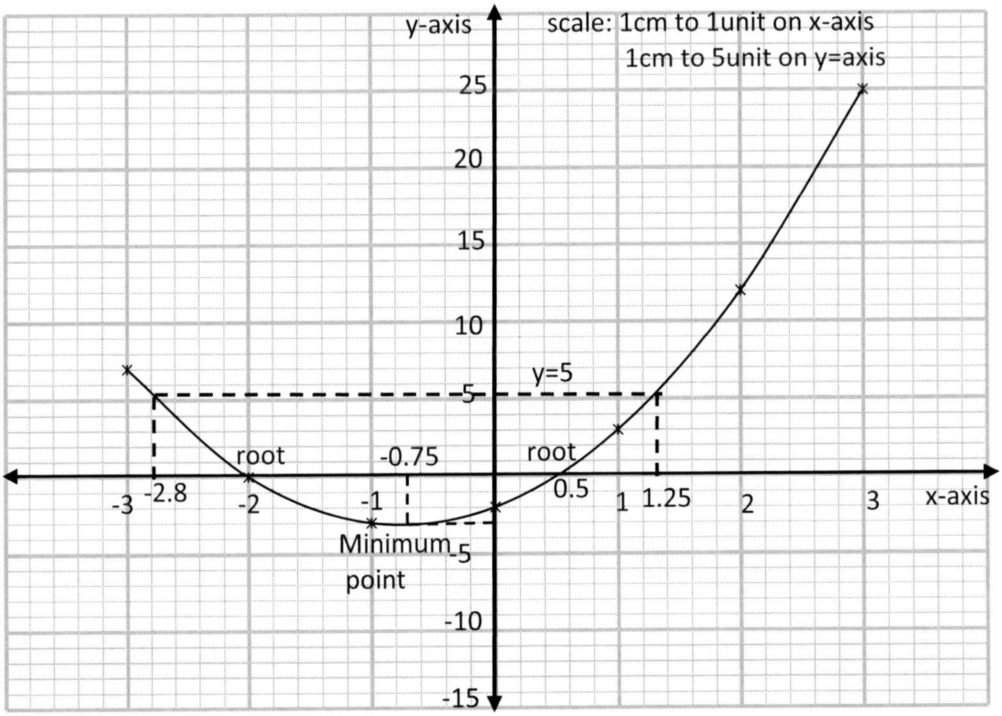

Fig. 9.4 Quadratic graph of $y = 2x^2 + 3x - 2$

(a) The roots of the equation $2x^2 + 3x - 2 = 0$ are the points where the curve cuts the $x -$ axis. Hence, from the graph above, the points are -2 and 0.5

$\therefore\ x = -2$ and 0.5

(b) When $y = 5$; then the values of x are gotten by drawing a horizontal line to touch the x axis. Then the corresponding values of x will give the solution when $y = 5$. Hence, $x = -2.8$ and 1.25

(c) The minimum value of the graph, $y = 2x^2 + 3x - 2$ is the point when the curve begins to have a turning point. Hence $y = -3$

(d) The line of symmetry is the corresponding value of x where the minimum point occurs. Hence, $x = -0.75$

Example 34

Plot the graph of the quadratic equation $5 + 2x - x^2 = 0$, If $-4 \leq x \leq 4$. Hence, use the graph to evaluate the following:

(i) the roots of the equation, $5 + 2x - x^2 = 0$

(ii) the maximum value of y

Solution

Table 9.3

x	-4	-3	-2	-1	0	1	2	3	4
5	5	5	5	5	5	5	5	5	5
$+2x$	-8	-6	-4	-2	0	2	4	6	8
$-x^2$	-16	-9	-4	-1	0	-1	-4	-9	-16
y	-19	-10	-3	2	5	6	5	2	-3

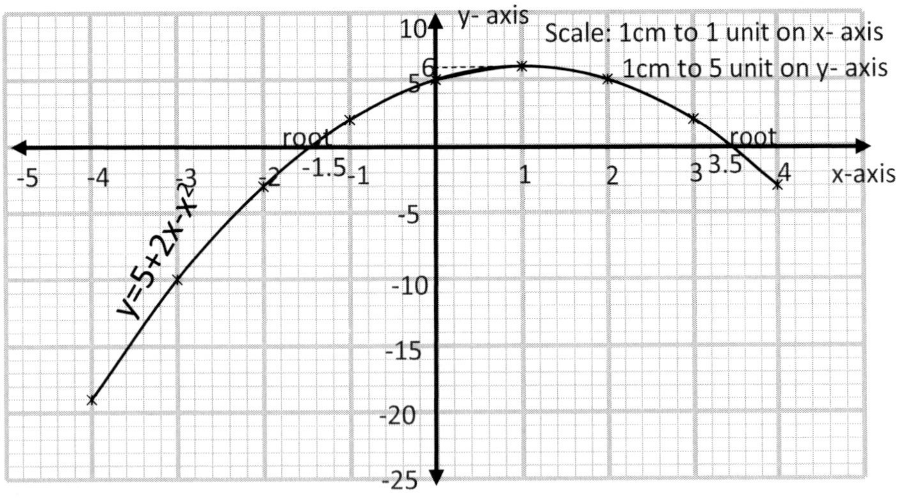

Fig. 9.5 Quadratic Graph

9.6 SIMULTANEOUS QUADRATIC AND LINEAR EQUATION

This is the combination of a quadratic and a linear equation both existing simultaneously for two variables x and y.

Example 35

Solve the following simultaneous equations:

(a) $x^2 + y^2 = 5$
$x + y = 1$

(b) $3x + y = 25$
$xy = 8$

Solution

(a) $x^2 + y^2 = 5$ ....(i)
$x + y = 1$ ....(ii)
From equation (ii), make y the subject of the formula
$x + y = 1 \quad \Rightarrow y = 1 - x$
Now, substitute $y = 1 - x$ into equation (i)
$x^2 + y^2 = 5 \quad \Rightarrow x^2 + (1 - x)^2 = 5$
$x^2 + 1 - 2x + x^2 = 5$
$x^2 + x^2 - 2x + 1 - 5 = 0$
$2x^2 - 2x - 4 = 0$ Factorise the quadratic expression:
$2(x + 1)(x - 2) = 0 \quad \therefore x = -1 \text{ or } 2$
Put the values of x into equation (ii):
Thus, when $x = -1$
$x + y = 1 \quad \Rightarrow -1 + y = 1$
$\therefore y = 1 + 1 = 2$
Thus, when $x = 2$
$x + y = 1 \quad \Rightarrow 2 + y = 1$
$\therefore y = 1 - 2 = -1$
Hence, $y = 2 \text{ or } -1$
$\Rightarrow (x, y) = (-1, 2) \text{ or } (2, -1)$

(b) $3x + y = 25$ ....(i)
$xy = 8$ ....(ii)
From equation (i), make y the subject of the formula.
$3x + y = 25 \quad \Rightarrow y = 25 - 3x$
Now, substitute $y = 25 - 3x$ into equation (ii):
$xy = 8 \quad \Rightarrow x(25 - 3x) = 8$
$25x - 3x^2 = 8$
$3x^2 - 25x + 8 = 0$ Factorise the quadratic expression:
$(3x - 1)(x - 8) = 0 \quad \therefore x = \frac{1}{3} \text{ or } 8$

Put the values of x into equation (i); therefore, when $x = \frac{1}{3}$

then, $3x + y = 25 \quad \Rightarrow 3\left(\frac{1}{3}\right) + y = 25$

$1 + y = 25 \quad \therefore y = 25 - 1 = 24$

when $x = 8$

$3x + y = 25 \quad \Rightarrow 3(8) + y = 25$

$24 + y = 25 \quad \therefore y = 25 - 24 = 1$

Hence, $y = 24$ or $1 \quad \Rightarrow (x, y) = \left(\frac{1}{3}, 24\right)$ or $(8, 1)$

Example 36

The area of a rectangle is 72cm^2 and its length is twice the width. Find the length and width of the rectangle.

Solution

Let the length of the rectangle and width be l and b.

$\therefore$ Area $= lb = 72\text{cm}^2$ ....(i)

$l = 2b$ ....(ii)

Put equation (ii) into (i)

$(2b)b = 72 \quad \Rightarrow 2b^2 = 72$

$b^2 = \frac{72}{2} = 36$

$b = \sqrt{36} = 6$

$\therefore b = 6\text{cm}$

Put the value of b into equation (ii)

thus $l = 2b \quad \Rightarrow l = 2 \times 6 = 12$

$\therefore l = 12\text{cm}$

Hence, the length is 12cm and the width 6cm.

Example 37

Solve for x and y: $\frac{1}{x^2} + \frac{1}{y^2} = 1$ and $\frac{1}{x} + \frac{1}{y} = 1$

Solution

$\frac{1}{x^2} + \frac{1}{y^2} = 2$ ....(i)

$\frac{1}{x} + \frac{1}{y} = 1$ ....(ii)

Using equation (ii), make $\frac{1}{y}$ the subject formula.

$\frac{1}{x} + \frac{1}{y} = 1 \quad \Rightarrow \frac{1}{y} = 1 - \frac{1}{x}$

Now, substitute $\frac{1}{y} = 1 - \frac{1}{x}$ into equation (i)

$\frac{1}{x^2} + \frac{1}{y^2} = 2$, this can be re-written as $\left(\frac{1}{x}\right)^2 + \left(\frac{1}{y}\right)^2 = 2$

$$\left(\frac{1}{x}\right)^2 + (1 - \frac{1}{x})^2 = 2$$

Open the bracket and simplify the expression.

$$\left(\frac{1}{x}\right)^2 + \left(1 - \frac{1}{x}\right)^2 = 2$$

$$\left(\frac{1}{x}\right)^2 + 1 - 2\left(\frac{1}{x}\right) + \left(\frac{1}{x}\right)^2 = 2 \quad \Rightarrow \left(\frac{1}{x}\right)^2 - 2\left(\frac{1}{x}\right) + \left(\frac{1}{x}\right)^2 + 1 - 2 = 0$$

$$2\left(\frac{1}{x}\right)^2 - 2\left(\frac{1}{x}\right) - 1 = 0$$

Now, assume $\frac{1}{x} = m \quad \therefore \; 2\left(\frac{1}{x}\right)^2 - \frac{2}{x} - 1 = 0 \quad \Rightarrow 2m^2 - 2m - 1 = 0$

$$2m^2 - 2m - 1 = 0$$

This expression cannot be factorised. So, the best approach is to use the quadratic formula method.

$$m = \frac{-b \pm \sqrt{b^2 - 4ac}}{2a}$$

Where $a = 2,\ b = -2,\ c = -1$

$$\frac{-(-2) \pm \sqrt{(-2)^2 - 4 \times 2 \times (-1)}}{2 \times 2} = \frac{2 \pm \sqrt{4 + 8}}{4} = \frac{2 \pm \sqrt{12}}{4}$$

$$m = \frac{2 \pm 3.46}{4} \quad \Rightarrow \frac{2 + 3.46}{4} \text{ or } \frac{2 - 3.46}{4}$$

$$m = \frac{5.46}{4} = 1.37 \text{ or } \frac{-1.46}{4} = -0.37$$

$$\therefore m = 1.37 \text{ or } -0.37$$

Remember that $\frac{1}{x} = m \quad \Rightarrow \frac{1}{x} = 1.37$ and $\frac{1}{x} = -0.37$

$$\therefore x = \frac{1}{1.37} = 0.73 \text{ and } x = \frac{1}{-0.37} = -2.7$$

Finally, use equation (ii) to find y :

$$\frac{1}{x} + \frac{1}{y} = 1 \quad \Rightarrow \frac{1}{y} = 1 - \frac{1}{x} = 1 - m$$

when $m = 1.37$

$$\frac{1}{y} = 1 - 1.37 = -0.37 \quad \Rightarrow y = -\frac{1}{0.37} = -2.7$$

Also, when $m = -0.37$

$$\frac{1}{y} = 1 - (-0.37) = 1 + 0.37 = 1.37 \quad \Rightarrow y = \frac{1}{1.37} = 0.73$$

Hence, $(x, y) = (0.73, -0.37)$ or $(-2.7,\ 0.73)$

Example 38

A number of two digits is such that the sum of its digits is 7. When the digits are interchanged the number is decreased by 45, find the number.

Solution

Let the digits be x and y, where x and y are the tens and unit digits respectively.
Therefore, the number is $10x + y$
Now, the sum of the digits is $x + y = 7$ ....(i)
and the number when the digits are interchanged is $10y + x$.
So, when the digits are interchanged, the number decreased by 45. This means that:
$(10x + y) - 45 = (10y + x)$, re-arranging this equation, we have:
$10x + y - 10y - x = 45$
$\therefore \quad 9x - 9y = 45$ ....(ii)
Make y the subject formula from (i), then put it into (ii)
$x + y = 7 \quad \Rightarrow y = 7 - x$
then putting this value into equation (ii), we have:
$9x - 9(7 - x) = 45 \quad \Rightarrow 9x - 63 + 9x = 45$
$\therefore 18x = 45 + 63 = 108$
$\therefore x = \dfrac{108}{18} = 6$
This follows that: $y = 7 - x = 7 - 6 = 1$
$\therefore x = 6, y = 1$ where the number is $10x + y = 10(6) + 1 = 61$
So, the number is 61

Example 39

Solve for x and y in each of the following simultaneous inequality:
(a) $\log_2(x - 4y) = 5$ and $\log_2\left(\frac{x}{y}\right) = 3$ (b) $\dfrac{9^x}{27^{x+2y}} = 81$ and $2x - y = 10$

Solution

(a) $\log_2 x - 4y = 5$ and $\log_2\left(\frac{x}{y}\right) = 3$
So, $\log_2(x - 4y) = 5 \quad \Rightarrow x - 4y = 2^5 = 32$
and $\log_2\left(\frac{x}{y}\right) = 3 \quad \Rightarrow \frac{x}{y} = 2^3 = 8 \quad$ (i.e. $x = 8y$)
$\therefore \; x - 4y = 32$ ....(i)
$x = 8y$ ....(ii)
Put $x = 8y$ from (ii) into (i)
$\Rightarrow x - 4y = 32$ becomes $(8y) - 4y = 32$
$\Rightarrow 4y = 32 \quad \therefore \; y = \dfrac{32}{4} = 8$
then, $x = 8y \quad \Rightarrow x = 8(8) = 64$
$\therefore \; (x, y) = (64, 8)$

(b) $\frac{9^x}{27^{x+2y}} = 81$ and $2x - y = 10$

$\frac{9^x}{27^{x+2y}} = 81 \quad \Rightarrow \frac{3^{2x}}{3^{3(x+2y)}} = 3^4$

This becomes, $3^{(2x)-3(x+2y)} = 3^4$

Remember that from the law of indices, we have $\frac{a^m}{a^n} = a^{m-n}$

Comparing the index powers of the RHS and LHS, we have that:

$(2x) - 3(x + 2y) = 4$

$\Rightarrow 2x - 3x - 6y = 4 \quad$ i.e., $-x - 6y = 4$

So, $-x - 6y = 4$ ....(i)

and $2x - y = 10$ ....(ii)

Make x the subject formula from (i), then put it into equation (ii)

Thus, $-x - 6y = 4 \quad \Rightarrow x = -6y - 4$

then, $2x - y = 10 \quad \Rightarrow 2(-6y - 4) - y = 10$

$\therefore -12y - 8 - y = 10$

$-13y = 10 + 8 = 18$

$\therefore y = -\frac{18}{13} = -1\frac{5}{13}$

$x = -6y - 4 = -6\left(-\frac{18}{13}\right) - 4$

$= \frac{108-52}{13} = \frac{56}{13} = 4\frac{5}{13}$

$\therefore (x, y) = \left(4\frac{5}{13}, -1\frac{5}{13}\right)$

GRAPHICAL SOLUTION OF SIMULTANEOUS QUADRATIC AND LINEAR EQUATION

To solve a problem involving a quadratic and linear equation, first the quadratic graph of $ax^2 + bx + c = 0$ is drawn to obtain either a ∪ or ∩ shape curve depending on the value of a ($i.e.$ $a > 0$, or $a < 0$). Then the linear graph of $y = mx + c$ is drawn on the same graph, using same axes. Thus, the points of intersection of the two graphs are the solution of the simultaneous quadratic and linear equation.

Example 40

Use graphical method to solve the following simultaneous equations:

(a) $y = x^2 - 2x - 3$ and $y = 5 - 2x$

(b) $y = 10 + 2x - x^2$ and $y = 2x - 3$

(a) Construct tables for $y = x^2 - 2x - 3$ and $y = 5 - 2x$ using a range of $-4 \leq x \leq 4$.

Table 9.4(i) $y = x^2 - 2x - 3$

x	-4	-3	-2	-1	0	1	2	3	4
x^2	16	9	4	1	0	1	4	9	16
$-2x$	8	6	4	2	0	-2	-4	-6	-8
-3	-3	-3	-3	-3	-3	-3	-3	-3	-3
y	21	12	5	0	-3	-4	-3	0	5

Table 9.4(ii) $y = 5 - 2x$

x	-4	-3	-2	-1	0	1	2	3	4
5	5	5	5	5	5	5	5	5	5
$-2x$	8	6	4	2	0	-2	-4	-6	-8
y	13	11	9	7	5	3	1	-1	-3

The solution of the graph are the points where the curve and the line intersect.

$\therefore Roots = -2.8$ and 2.8

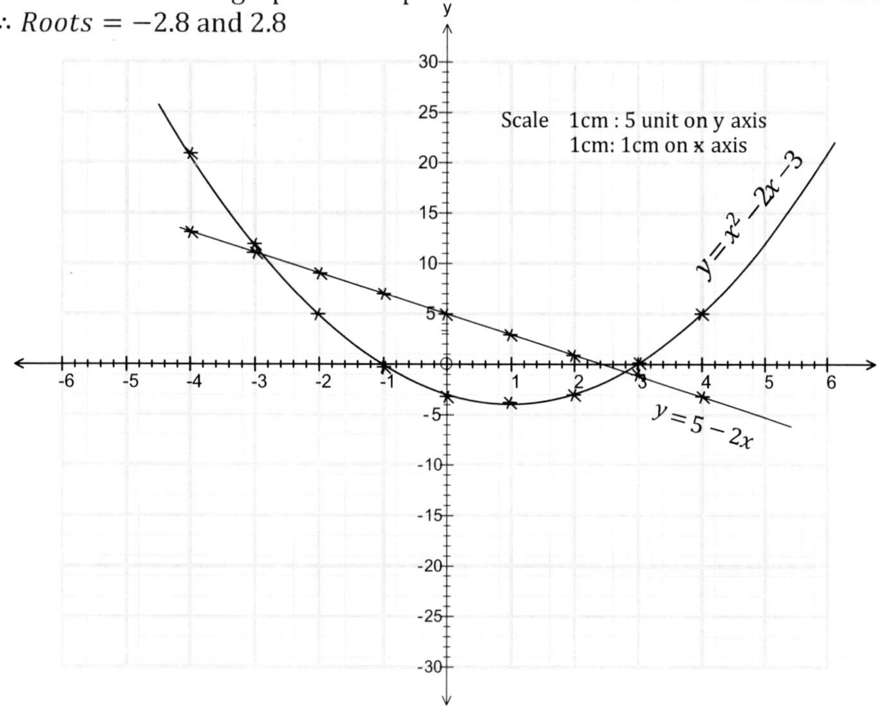

Fig. 9.6

(b) Construct tables of values for $y = 10 + 3x - 2x^2$ and $y = 2x - 3$ using a range of $-4 \leq x \leq 4$.

Table 9.5 (i) $y = 10 + 3x - 2x^2$

x	-4	-3	-2	-1	0	1	2	3	4
10	10	10	10	10	10	10	10	10	10
$+3x$	-12	-9	-6	-3	0	3	6	9	12
$-2x^2$	-32	-18	-8	-2	0	-2	-8	-18	-32
y	-34	-19	-4	5	10	11	8	1	-10

Table 9.5 (ii) $y = 2x - 3$

x	-4	-3	-2	-1	0	1	2	3	4
$2x$	-8	-6	-4	-2	0	2	4	6	8
-3	-3	-3	-3	-3	-3	-3	-3	-3	-3
y	-11	-9	-7	-5	-3	-1	1	3	5

The solution of the graph are the points where the curve and the line intersect.

$\therefore$ *Roots* are approximately -2.4 and 2.7

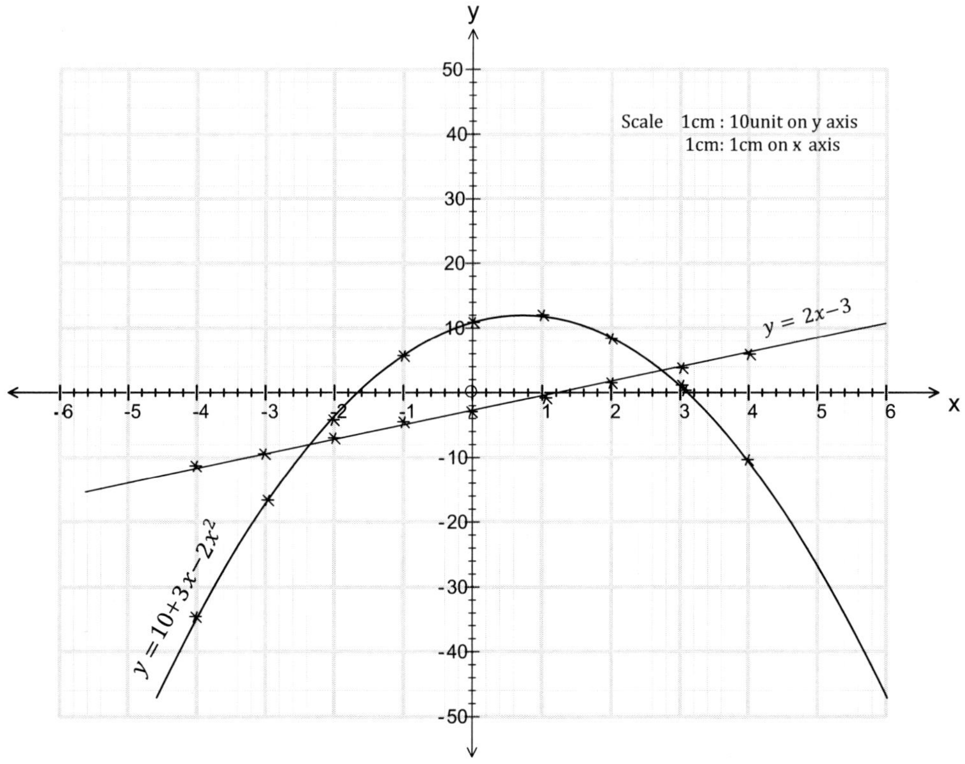

Fig. 9.7 Simultaneous Linear Quadratic Graph

OBJECTIVE TEST 9

1. Simplify $\frac{1}{a-b} + \frac{2}{a}$
   (A) $\frac{3a+2b}{a(a+b)}$ (B) $\frac{3a-2b}{a(a-b)}$ (C) $\frac{a+2}{a-b}$ (D) $\frac{2b-3a}{a(a-b)}$

2. For what value of x is the expression not defined $\frac{7}{x-2} + 1$?
   (A) $x = 1$ (B) $x = 2$ (C) $x = 7$ (D) $x = 5$

3. Given $y = \frac{x-2a}{x+a}$ make x the subject formula
   (A) $\frac{a(2-y)}{1-y}$ (B) $\frac{a(1-y)}{2+y}$ (C) $\frac{a(2+y)}{1-y}$ (D) $\frac{a(2-y)}{1+y}$

4. Factorise $25(x + y)^2 - 16(3x - y)^2$
   (A) $17x(2y - 7x)$ (B) $y(8x + 9y)$
   (C) $(17x + y)(7x - 9y)$ (D) $(17x + y)(9y - 7x)$

5. Find the equation whose roots are -2 and 1
   (A) $x^2 + x - 2$ (B) $x^2 - x + 2$ (C) $x^2 + 3x + 2$ (D) $2x^2 + x - 1$

6. A man is 5 years times as old as his child. In 5 years time the product of their ages will be 300, how old is the child now ?
   (A) 10 years (B) 4 years (C) 5 years (D) 11 years

7. Find the solution of the simultaneous linear quadratic equation:
   $x^2 - 4y^2 = 9$ and $x + 2y = 1$
   (A) $(5, 2)$ (B) $(2, 5)$ (C) $(5, -2)$ (D) $(-5, -2)$

8. Solve the simultaneous linear equation: $\frac{1}{x} - \frac{1}{2y} = 3$, and $\frac{1}{3x} + \frac{1}{y} = 2$
   (A) $\left(\frac{7}{24}, \frac{7}{6}\right)$ (B) $\left(\frac{1}{6}, 3\frac{3}{7}\right)$ (C) $\left(\frac{7}{6}, \frac{7}{24}\right)$ (D) $\left(3\frac{3}{7}, \frac{1}{6}\right)$

9. A number of two digits is increased by 18 when the digits are reversed, if the sum of the digits is 10, find the number.
   (A) 46 (B) 56 (C) 64 (D) 55

10. The sum of two numbers is 13 and their product is -30, find the numbers.
    (A) 10 and 3 (B) 10 and -3 (C) 15 and -2 (D) -15 and 2

Exercise 9b

1. Solve for x in each of the following equations:
   (a) $5x - 10 = 35$ (b) $2x^2 - 50 = 0$ (c) $\frac{1}{4}(x-1) + \frac{1}{2}(x+2) = 9$
   (d) $7(10x - 3) - 25(11x + 5) - 1 = 0$ (e) $\frac{2}{3}(5x+4) + \frac{1}{7}(x-2) = 0$
2. A certain number is added to 5 and the result is twice their difference. What is the number?
3. A man is three times as old as his son, if in ten years' time he will be twice as old as the son. Find the man's age and that of the son.
4. The ages of two sisters differ by 3. If the sum of their ages is 37 years, what are their ages now?
5. Think of a number and take away 10 from it, if the result is 40. What value is needed to change this result to 70?
6. Solve the following pairs of simultaneous equation using substitution method:
   (a) $x + y = 1$, $x - y = 1$ (b) $2x + y = 10$, $x + 3y = 50$ (c) $m + n = 2$, $2m - n = 5$ (d) $2a + 10b = 8$, $5a + 3b = 15$
7. Use elimination method to solve for x and y:
   (a) $x - y = 1$, $2x + y = 5$ (b) $2x - 3y = 5$, $3x + 2y = 10$ (c) $\frac{1}{2}x - \frac{1}{3}y = 4$, $\frac{1}{3}x - \frac{1}{2}y = 13$ (d) $x^2 + y = 32$, $y^2 - x = 42$
8. Solve graphically the following pairs of simultaneous equation:
   (a) $x + y = 1$, $3x - y = 2$ (b) $5x + 3y = 10$, $3x + 5y = 2$ (c) $a + b = 5$, $2a - b = 4$ (d) $x - y = 1$, $2x + y = 5$
9. Solve either with substitution or elimination method where necessary to find the solution of the following pairs of simultaneous equation:
   (a) $2^{x+y} = 1$, $3^{2x-y} = 27$ (b) $\log_2 x + 2y = 2$, $\log_3 2x - y = 2$ (c) $\log 5x + y = 2$, $\log 6x - y = 1$ (d) $3^{5x-y} = 9^{x+y+3}$, $16^{x+y} = 4^{x-y+1}$
10. The perimeter of a rectangle is 26cm and the difference between the length and the width is 3cm. Find the length and width of the rectangle.
11. The ages of a man and his son are now in the ratio 7: 3, if the son is 15 years old now. What will be the ratio of their ages in 5 years' time?
12. Solve the following quadratic simultaneous equations:
   (a) $\frac{1}{2x^2} + y = 20$, $\frac{1}{x^2} - y = 30$ (b) $x^2 + y^2 = 81$, $x^2 - y^2 = 17$ (c) $2x^2 - y^2 = 8$, $3x^2 + 2y^2 = 12$ (d) $\frac{1}{x^2} + \frac{1}{y^2} = 5$, $\frac{3}{x^2} + \frac{7}{y^2} = 24$
13. Find the values of x and y in each of the following simultaneous equations:
   (a) $\frac{1}{3x} + \frac{1}{y} = 15$, $\frac{1}{x} - \frac{1}{2y} = 10$ (b) $\frac{1}{x} + \frac{1}{y} = \frac{1}{2}$, $\frac{1}{2x} - \frac{1}{y} = 3$ (c) $\frac{1}{3x} - \frac{1}{y} = \frac{1}{3}$, $\frac{1}{2x} + \frac{1}{2y} = \frac{1}{2}$ (d) $x + \frac{1}{y} = \frac{1}{2}$, $\frac{1}{x} - \frac{y}{3} = 1$
14. Given that $x - y + 1 = 2x + 3y - 5 = 5x + 2y$, find x and y.

10 ALGEBRA PROCESS II

10.1 LINEAR INEQUALITY

When two or more quantities are to be compared, they are thus connected by inequality signs.

10.1.1 INEQUALITY NOTATIONS

Table 10.1 Inequality Notations and Meaning

NOTATIONS	MEANING
$<$	Less than, e.g. $2 < 3$
$>$	Greater than, e.g. $10 > 7$
$\leq$	Less than or equal to, e.g. $-2 \leq x$; where x can be $-2, -1, 0, \ldots$
$\geq$	Greater than or equal to, e.g. $-2 \geq x$; where x can be $-2, -3, -4, \ldots$
$\nless$	Not less than, e.g. $5 \nless y$, where y can be $5, 4, 3, \ldots$
$\ngtr$	Not greater than, e.g. $5 \ngtr y$; where y can be $5, 6,7, \ldots$
$\neq$	Not equal to, e.g. $5 \neq 6$
	Greater than or equal to
	Greater than
	Less than or equal to
	Less than

10.1.2 NUMBER LINE

A number line is used for representing an inequality statement. The number zero, is the reference point such that numbers to the right of zero are termed "positive number," while numbers to the left are termed "negative numbers."

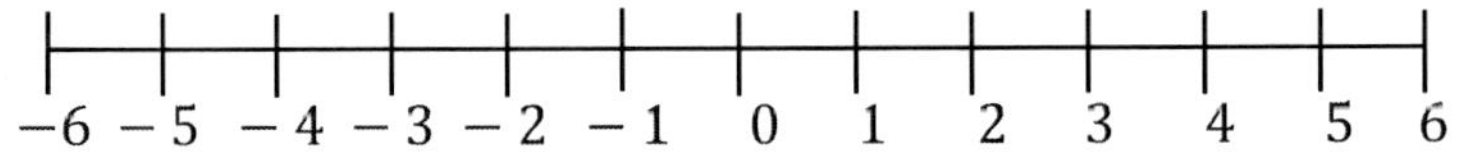

Fig. 10.1 Illustration of number Line

Example 1

Represent the following on a number line

(a) $x \geq -6$ (b) $x \leq -1$ (c) $x < 3$ (d) $x > 2$

Solution

(a)

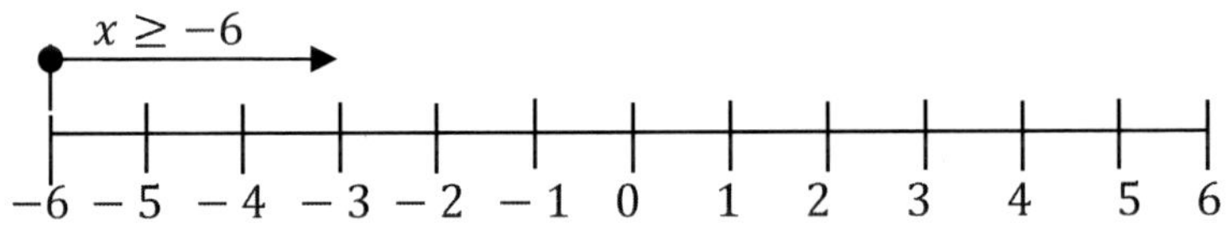

Fig. 10.2 (a)

(b)

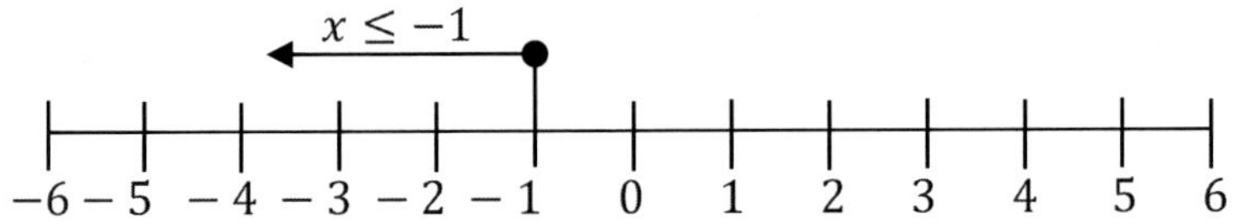

Fig. 10.2 (b)

(c)

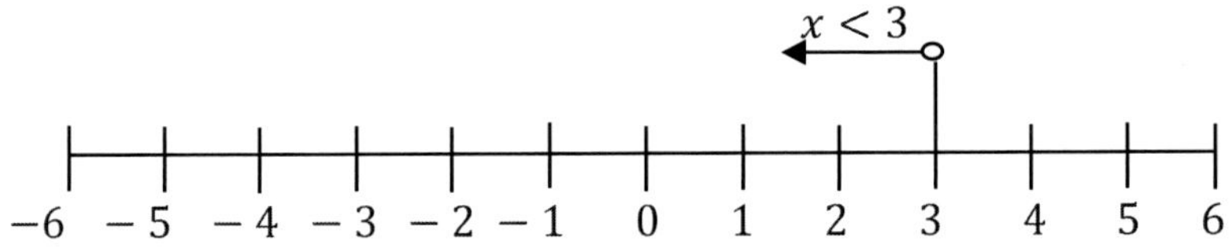

Fig. 10.2 (c)

(d)

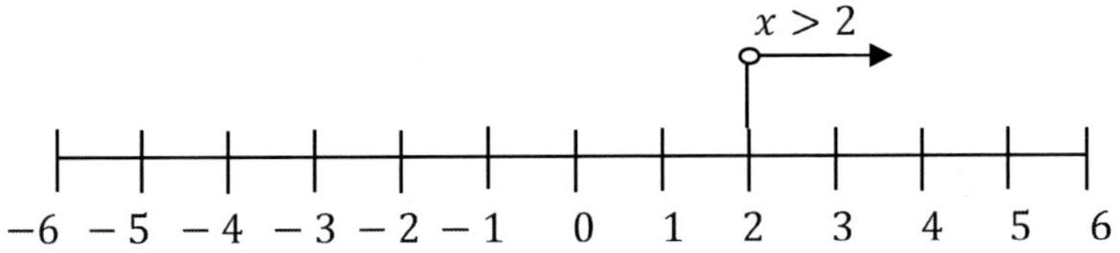

Fig. 10.2 (d)

10.1.3 RULES OF INEQUALITY

1. (a) If $x > y$, then $y < x$
   (b) If $x < y$, then $y > x$

2. (a) If $x > y$ and $y > z$, then $x > z$
   (b) If $x < y$ and $y < z$, then $x < z$
3. (a) If $x > y$, then $x + z > y + z$ and $x - z > y - z$
   (b) If $x < y$, then $x + z < y + z$ and $x - z < y - z$

4. (a) If $x > y$ and $z > 0$, then $xz > yz$ and $\frac{x}{z} > \frac{y}{z}$
   (b) $x < y$ and $z > 0$, then $xz < yz$ and $\frac{x}{z} < \frac{y}{z}$

5. (a) If $x > y$ and $z < 0$, then $xz < yz$ and $\frac{x}{z} < \frac{y}{z}$
   (b) If $x < y$ and $z < 0$, then $xz > yz$ and $\frac{x}{z} > \frac{y}{z}$

6. (a) If $x > y$ and $z > m$, then $x + z > y + m$
   (b) If $x < y$ and $z < m$, then $x + z < y + m$

7. (a) If $x > y$ and $z > m$, then $x - z > y - m$
   (b) If $x < y$ and $z < m$, then $x - z < y - m$

8. (a) If $x > y$, $y > 0$ and $z > m$, $m > 0$; then $xz > ym$
   (b) If $x < y$, $y < 0$ and $z < m$, $m < 0$; then $xz < ym$

9. (a) If $x > y$ and $a > 0$, then $x^a > y^a$
   (b) If $x < y$ and $a < 0$, then $x^a < y^a$

10. (a) If $-x > y$, then $x < -y$
    (b) If $-x < y$, then $x > -y$
    (c) If $-x < -y$, then $x > y$
    (d) If $-x > -y$, then $x < y$

10.1.4 LINEAR INEQUALITY IN ONE VARIABLE

A linear inequality in one variable is an inequality expressed in just one unknown term, e.g., $x + 1 > 0$, $x > 20$, $3 \geq x - 6$, or $3x - 5 \leq 7$. Thus, it follows that only the unknown, x, is to be found.

Example 2

Find the solution for the set of the following inequalities and represent them on a number line:
(a) $2x - 5 > 7$ (b) $5x - 8 \leq 2 + 7x$

Solution

(a) $2x - 5 > 7$

$2x > 7 + 5$

$x > \frac{12}{2}$

$\therefore\ x > 6$

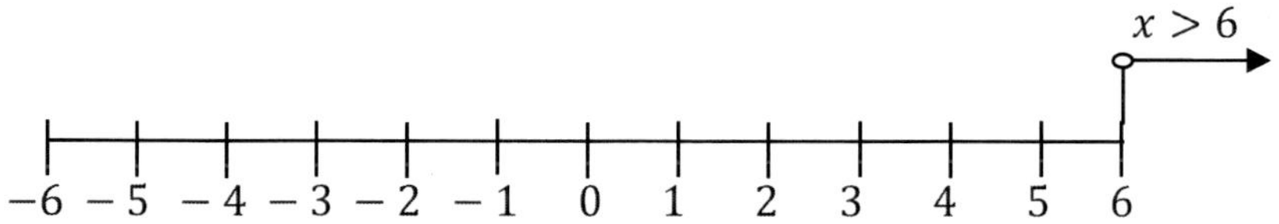

Fig. 10.3(a)

(b) $5x - 8 \leq 2 + 7x$

$5x - 7x \leq 2 + 8$

$-2x \leq 10$

$-x \leq \frac{10}{2} \quad \Rightarrow -x \leq 5$

$x \geq -5$

Notice here, that the inequality sign has changed; this is because the negative sign on x was removed by dividing it through by -1. And this will likewise change the inequality sign.

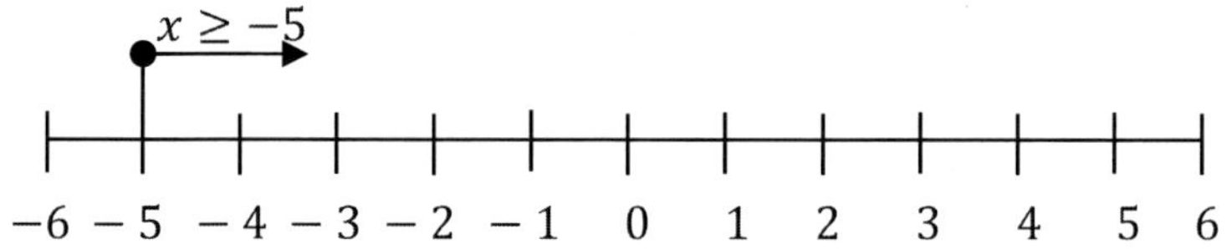

Fig. 10.4 (b)

Example 3

Find the solution for the set of the following inequalities and represent them on a number line.

(a) $\frac{x+1}{2} - \frac{x}{3} \leq \frac{1}{4}$

(b) $\frac{x}{2} - 1 > \frac{2x}{3} + \frac{1}{2}$

(a) $\frac{x+1}{2} - \frac{x}{3} \leq \frac{1}{4}$

$12\left(\frac{x+1}{2}\right) - 12\left(\frac{x}{3}\right) \leq 12 \times \frac{1}{4}$

$6(x + 1) - 4x \leq 3$

$6x + 6 - 4x \leq 3$

$6x - 4x \leq 3 - 6$

The inequality is multiplied through by its LCM (i.e., 12). Then, the value of x is found by simplifying the expression.

Hence, $2x \leq -3$

$x \leq \frac{-3}{2}$

$\therefore x \leq -1.5$

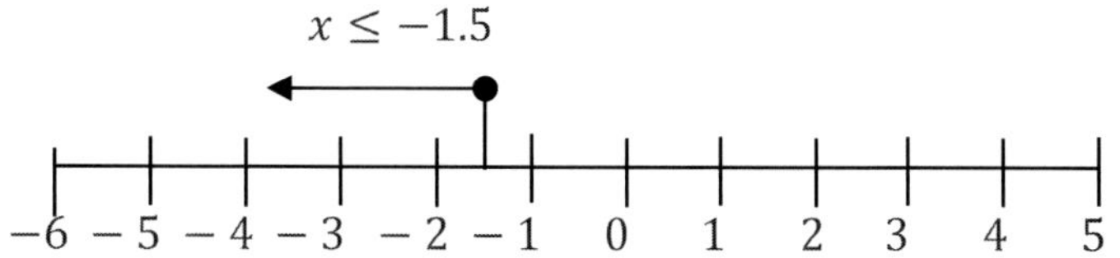

Fig. 10.5(a)

(b) $\frac{x}{2} - 1 > \frac{2x}{3} + \frac{1}{2}$

$6\left(\frac{x}{2}\right) - 6(1) > 6\left(\frac{2x}{3} + \frac{1}{2}\right) \quad \Rightarrow 3x - 6 > 4x + 3$

$3x - 4x > 6 + 3$

$-x > 9$

$\therefore x < -9$

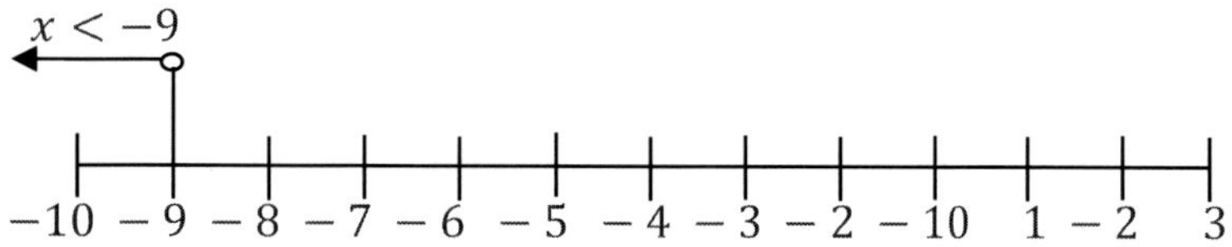

Fig.10.5(b)

Example 4

Find the solution set for x in the following expressions and represent the solution on a number line.

(a) $3 + x \leq 4$ and $7 + x > 3$

(b) $(5 - x)(2 + x) \geq 0$

(c) $\frac{1}{x-3} < -1$

(d) $\frac{(x-1)(x-3)}{(x+1)(x-2)} > 0$

Solution

(a) $3 + x \leq 4$ and $7 + x > 3$

For: $3 + x \leq 4$

$x \leq 4 - 3$

$\therefore x \leq 1$

For: $7 + x > 3$

$x > 3 - 7$

$\therefore x > -4$

The range that satisfies the inequality is $-4 < x \leq 1$

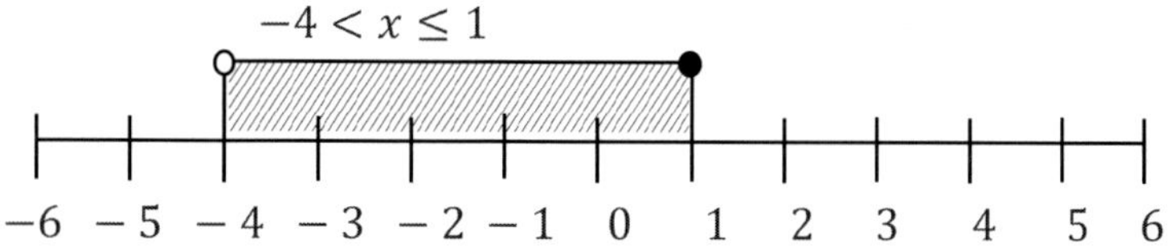

Fig. 10.6 (a)

(b) $(5 - x)(2 + x) \geq 0$

$\Rightarrow 5 - x \geq 0$ and $2 + x \geq 0$ Or $5 - x \leq 0$ and $2 + x \leq 0$

Then either : $x \le 5$ and $x \ge -2$ OR $x \ge 5$ and $x \le -2$
Now, solve the two pairs of inequality using a "sign table" to find the correct solution that satisfies: $(5-x)(2+x) \ge 0$

Table 10.1

	$x \le -2$	$-2 \le x \le 5$	$x \ge 5$
$5-x$	+	+	−
$2+x$	−	+	+
$(5-x)(2+x)$	−	+	−

The range that satisfies the inequality, $(5-x)(2+x) \ge 0$ is the product with positive sign because of the "≥" in the inequality.

$\therefore\ -2 \le x \le 5$
The number line is shown in fig. 10.6(b)

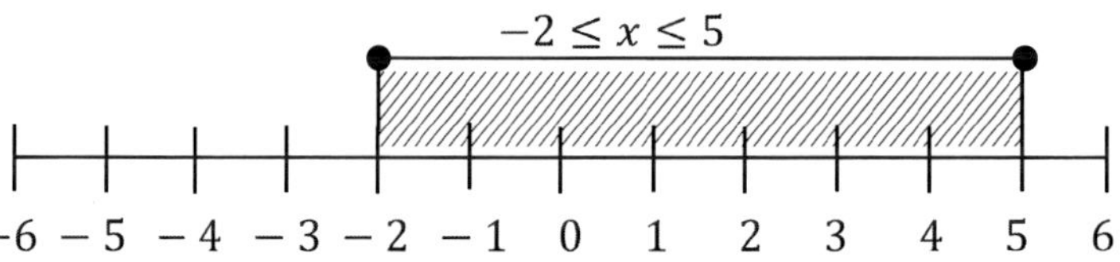

Fig. 10.6 (b)

Example 5

Find the solution set of each of the following inequalities:

(a) $\frac{1}{x-3} < -1$ (b) $\frac{(x-1)(x-3)}{(x+1)(x-2)} < 0$

Solution

(a) $\frac{1}{x-3} < -1$

$(x-3)^2\left(\frac{1}{x-3}\right) < -1(x-3)^2$

Multiplied through by $(x-3)^2$

$(x-3) < -1(x^2-6x+9)$
$(x-3) + 1(x^2-6x+9) < 0$
$x-3+x^2-6x+9 < 0$
$\Rightarrow x^2-5x+6 < 0$
$(x-3)(x-2) < 0$
$x < 2$ or $x > 3$ and $x > 2$ or $x < 3$

In this case, the inequality is satisfied when $x > 2$ or $x < 3$ which can be verified using the "sign table" in table 10.2.

Table 10.2

	$x < 2$	$2 < x < 3$	$x > 3$
$x - 3$	−	−	+
$x - 2$	−	+	+
$(x - 3)(x - 2)$	+	−	+

$\therefore$ the inequality is satisfied when $2 < x < 3$

The range that satisfies the inequality, $(x - 3)(x - 2) < 0$ is the product with "negative sign" because of the "<" in the inequality.

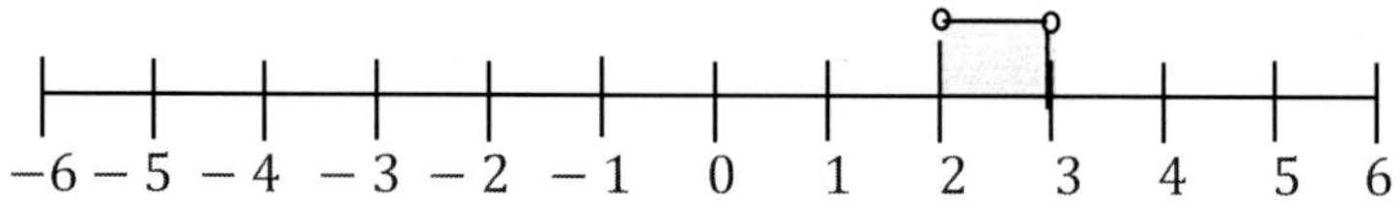

Fig. 10.6 (c)

(b) $\frac{(x-1)(x-3)}{(x+1)(x-2)} < 0$

Multiply through by $(x + 1)^2(x - 2)^2$

$$\frac{(x-1)(x-3)}{(x+1)(x-2)}(x + 1)^2(x - 2)^2 < 0(x + 1)^2(x - 2)^2$$

$\Rightarrow (x - 1)(x - 3)(x + 1)(x - 2) < 0$

$x < 1,\ x < 3,\ x < -1,\ x < 2$ and $x > 1,\ x > 3,\ x > -1,\ x > 2$

Table 10.3

	$x < -1$	$-1 < x < 1$	$1 < x < 2$	$2 < x < 3$	$x > 3$
$x - 1$	−	−	+	+	+
$x - 3$	−	−	−	−	+
$x + 1$	−	+	+	+	+
$x - 2$	−	−	−	+	+
$(x - 1)(x - 3)(x + 1)(x - 2)$	+	−	+	−	+

$\frac{(x-1)(x-3)}{(x+1)(x-2)} < 0$ then the solution that satisfies the inequality from the sign table is the product with "negative signs." $\therefore$ $-1 < x < 1$ and $2 < x < 3$

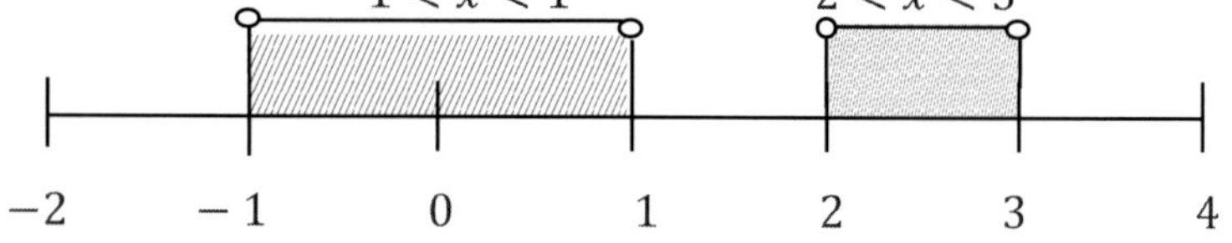

Fig. 10.6 (d)

10.1.5 LINEAR INEQUALITY IN TWO VARIABLE

This inequality has two unknowns (i.e., x and y); and the solution set of this inequality is found by graphical representation on the Cartesian-plane at points (x, y) of the two variables x and y.

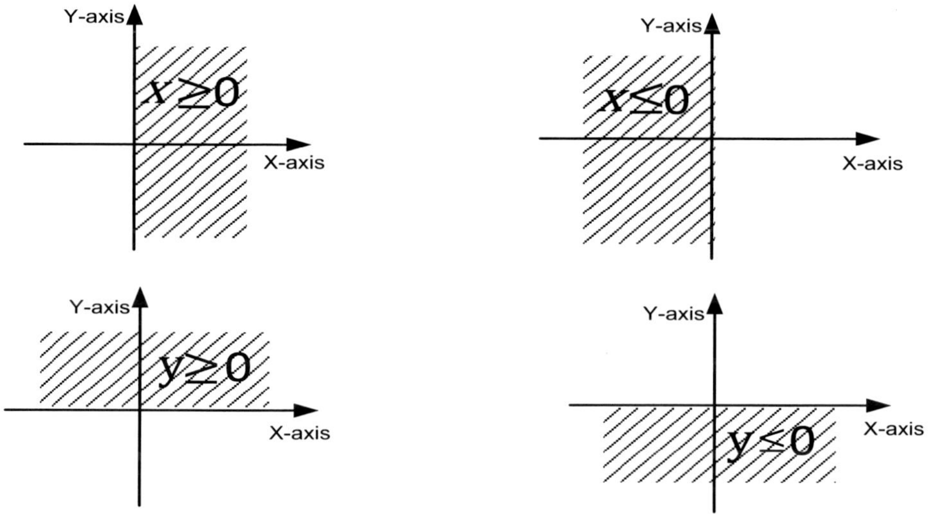

Fig. 10.7

Example 6

Show graphically the region that satisfies each of the following inequalities:

(a) $2x + y \geq -1$

(b) $x + y > 2$ and $x - 3y \geq 3$

(c) $y \geq 3$ and $x \geq 2$

Solution

(a) $\mathbf{2x + y \geq -1}$

Put $y = 0$ into the equation, so as to find x

$\Rightarrow 2x + (0) = -1$

$x = -\frac{1}{2}, \quad \therefore x = -0.5$

$(x_1, y_1) = (-0.5,\ 0)$

A line which equates the L.H.S with the R.H.S is drawn by putting $y = 0$, then solve for x. Also, putting a value of $x = 0$, then solve for y.

Similarly, put $x = 0$ into the equation so as to find y.

$2x + y = -1$

$\Rightarrow 2(0) + y = -1$

$\therefore y = -1 \quad (x_2, y_2) = (0, -1)$

Thus, the shaded region satisfies the inequality.

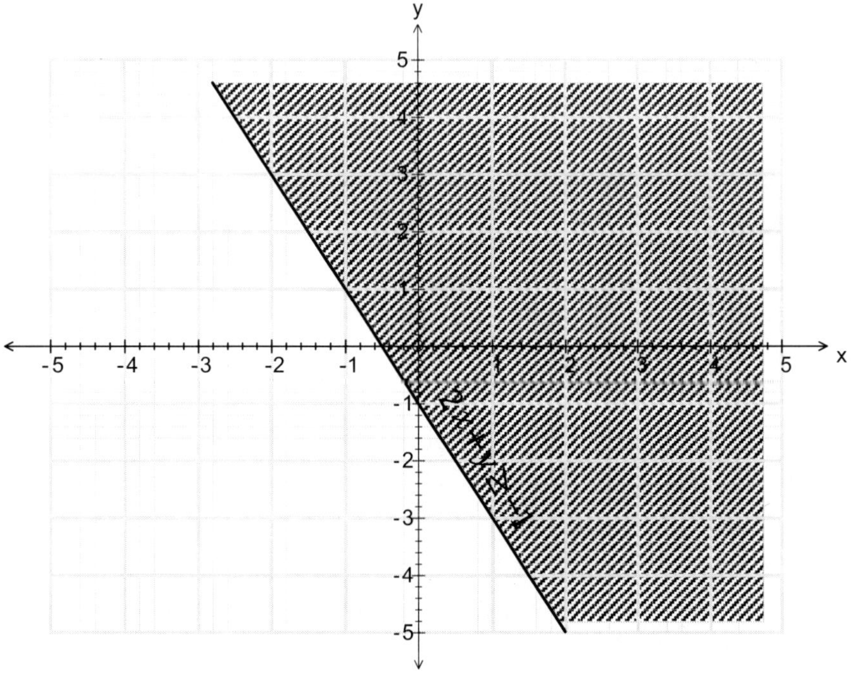

Fig. 10.8(a)

(b) $\boldsymbol{x + y > 2}$ and $\boldsymbol{x - 3y \geq 3}$

For $x + y = 2$

When $x = 0 \quad \Rightarrow (0) + y = 2$

$\therefore y = 2$

When $y = 0 \quad \Rightarrow x + (0) > 2$

$\therefore x > 2 \qquad (x_1, y_1) = (0, 2)$ and $(x_2, y_2) = (2, 0)$

Similarly,

$x - 3y \geq 3$

When $x = 0 \quad \Rightarrow (0) - 3y = 3$

$\therefore y = -1$ When $y = 0 \quad \Rightarrow x - 3(0) = 3$

$\therefore x = 3$

$(x_3, y_3) = (0, -1)$ and $(x_4, y_4) = (3, 0)$

The shaded region satisfies the inequality

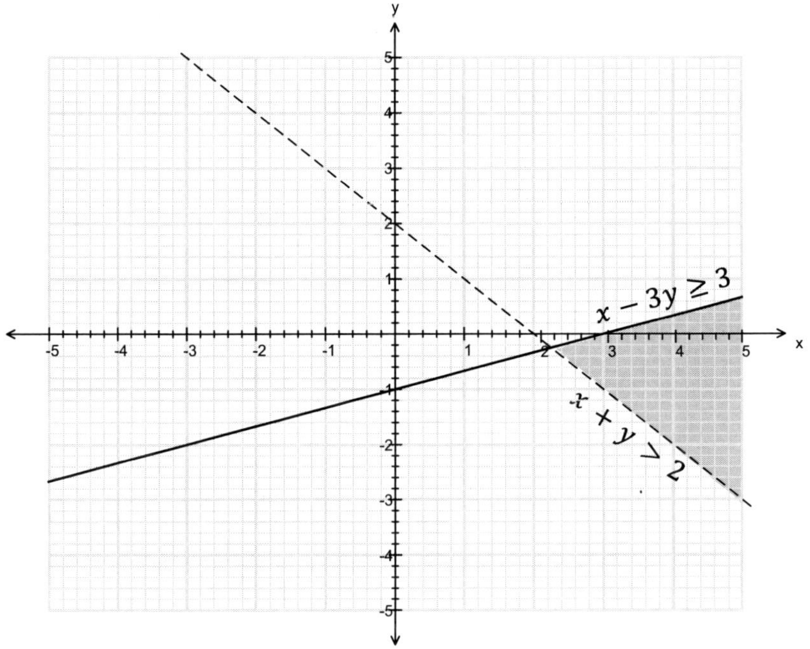

Fig. 10.8(b)

(c) $y > 3$ and $x > 2$

The shaded region satisfies the inequality

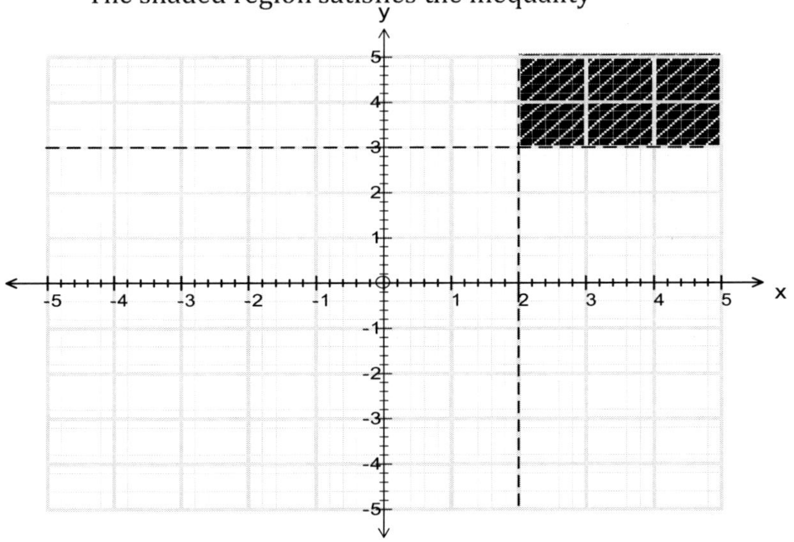

Fig. 10.8(c)

Example 7

Show the region that satisfies simultaneously the following inequalities:

(a) $2x + y \leq 3$
$x - 2y \geq -3$
$x \geq 0$
$y \geq 0$

(b) $x + y \geq 3$
$x - y \geq 1$
$x \geq 0$
$y \geq 0$

Solution

(a) $\mathbf{2x + y \leq 3}$

For $2x + y = 3$
When $x = 0 \quad \Rightarrow 2(0) + y = 3$
$\therefore\ y = 3$
When $y = 0 \quad \Rightarrow 2x + (0) = 3$
$\therefore\ x = \frac{3}{2}$
$(x_1, y_1) = (0, 3)$ and $(x_2, y_2) = (\frac{3}{2}, 0)$

$\mathbf{x - 2y \geq -3}$
For $x - 2y = -3$
When $x = 0 \quad \Rightarrow (0) - 2y = -3$
$\therefore y = \frac{3}{2}$
When $y = 0 \quad \Rightarrow x - 2(0) = -3$
$\therefore\ x = -3$
$(x_3, y_3) = \left(0, \frac{3}{2}\right)$ and $(x_4, y_4) = (-3, 0)$

If $\mathbf{x \geq 0} \quad \Rightarrow x = 0$
and $\mathbf{y \geq 0} \quad \Rightarrow y = 0$

See fig. 10.9(a) in the next page for the graphical solution.

The shaded region satisfies the inequality

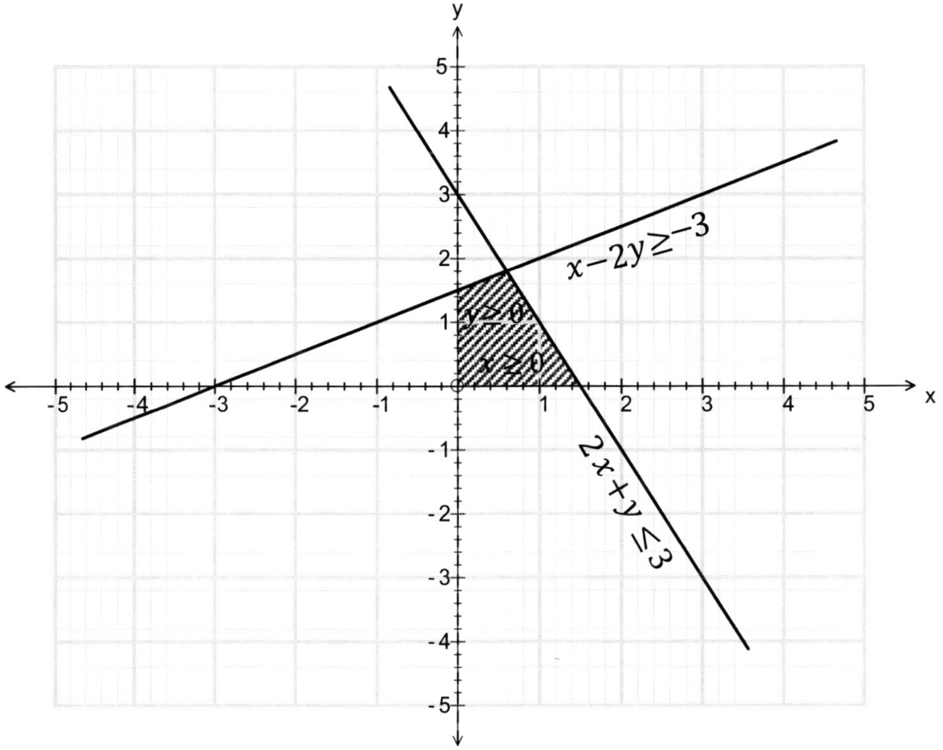

Fig. 10.9(a)

(b) $x + y \geq 3$
$x - y \geq 1$
$x \geq 0$
$y \geq 0$

If $\boldsymbol{x + y \geq 3}$

Then when $x = 0 \quad \Rightarrow (0) + y = 3$

$\therefore y = 3$

When $y = 0 \quad \Rightarrow x + (0) = 3$

$\therefore x = 3$

$(x_1, y_1) = (0, 3)$ and $(x_2, y_2) = (3,0)$

If $\boldsymbol{x - y \geq 1}$
When $x = 0 \quad \Rightarrow (0) - y = 1$
$\therefore y = -1$
When $y = 0 \quad \Rightarrow x - (0) = 1$
$\therefore x = 1$
$(x_3, y_3) = (0, -1)$ and $(x_4, y_4) = (1, 0)$

If $\boldsymbol{x \geq 0} \quad \Rightarrow x = 0$

If $\boldsymbol{y \geq 0} \quad \Rightarrow y = 0$

The shaded region satisfies the inequality

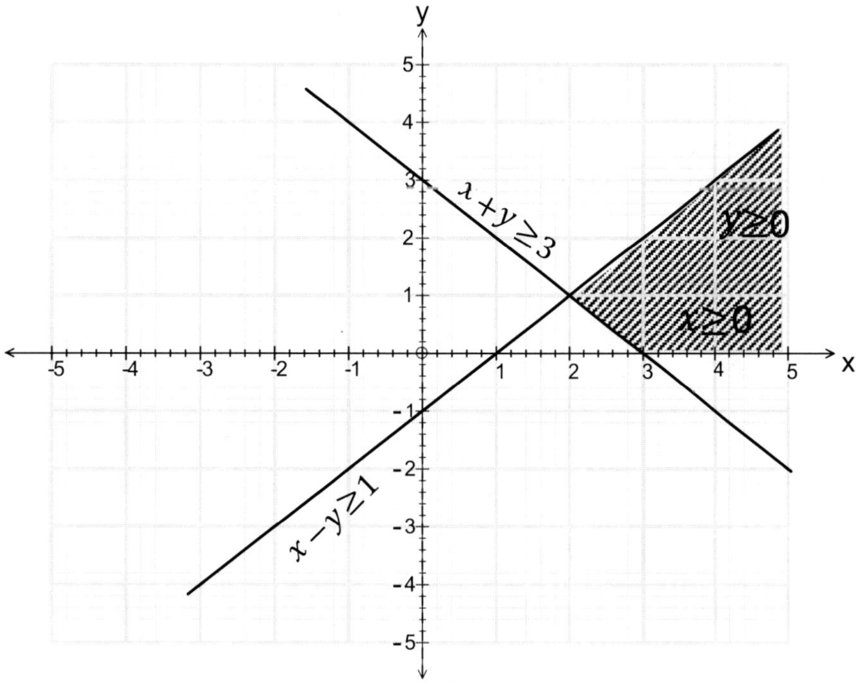

Fig. 10.9(b)

10.2 INEQUALITY WITH MODULUS

An inequality with modulus of the form:
(i) $|x - b| < a$ is written as $-a < x - b < a$;
(ii) $|x - b| > a$ is writen as $-a > x - b > a$ (i.e. $x - b < -a$ and $x - b > a$);
(iii) $|x - b| < |x - a|$ is written as $-|x - a| < x - b < |x - a|$;
(iv) $|x - b| > |x - a|$ is written as $-|x - a| > x - b > |x - a|$
OR $(x - b)^2 = (x - a)^2$

Example 8

Solve the following inequalities: (a) $|2x+3|<1$ (b) $|2x-3|<5+x$

Solution

(a) $|2x+3|<1 \quad \Rightarrow -1<2x+3<1$

Firstly consider: $2x+3<1$

$\therefore\ 2x<1-3 \quad \Rightarrow 2x<-2$

$\therefore\ x<-1$

Secondly consider: $-1<2x+3$

$\therefore\ -3-1<2x \quad \Rightarrow -4<2x \quad$ i.e. $2x>-4$

$\therefore\ x>-2$

Thus, the solution that satisfies the inequality are: $x<-1$ and $x>-2$

(b) $|2x-3|<5+x \quad \Rightarrow -5-x<2x-3<5+x$

Firstly consider: $2x-3<5+x$

$\therefore\ 2x-x<5+3 \quad \Rightarrow x<8$

$\therefore\ x<8$

Secondly consider: $-5-x<2x-3$

$\therefore -x-2x<5-3 \quad \Rightarrow -3x<2$

$\therefore x>\frac{-2}{3}$

Thus, the solution that satisfies the inequality are: $x>\frac{-2}{3}$ and $x<8$

Example 9

Solve for x if $\left|\frac{1}{x+2}\right|<1$

Solution

$\left|\frac{1}{x+2}\right|<1 \quad \Rightarrow -1<\frac{1}{x+2}<1$

Firstly consider: $\frac{1}{x+2}<1$ ....(i)

$\therefore\ \frac{1}{x+2}-1<0 \quad \Rightarrow \frac{-x-1}{x+2}<0$

Multiply through by $(x+2)^2$, so that the equation becomes $-(x+1)(x+2)<0$

$-(x+1)(x+2)<0$

$\Rightarrow (x+1)(x+2)>0$

This means: $(x+1)>0$ and $(x+2)>0$ **OR** $(x+1)<0$ and $(x+2)<0$

i.e. $x>-1$ and $x>-2$ **OR** $x<-1$ and $x<-2$

Thus, we have to test for the correct solution using the sign table (see table 10.4(i)).

Table 10.4(i)

	$x<-2$	$-2<x<-1$	$x>-1$
$x+1$	−	−	+
$x+2$	−	+	+
$(x+1)(x+2)$	+	−	+

Since the equation is $(x+1)(x+2) > 0$, such that the inequality notation joining the RHS and LHS is " > " ; then the solution that satisfies the inequality is the product with "plus" sign from table 10.1(i). Thus, $x < -2$ and $x > -1$.

Secondly consider: $-1 < \frac{1}{x+2}$ ....(ii)

$$0 < \frac{1}{x+2} + 1 \quad \Rightarrow 0 < \frac{1}{x+2} + 1$$

$\therefore\ 0 < \frac{x+3}{x+2}$ which can be rewritten as $\frac{x+3}{x+2} > 0$

Multiplying through by $(x+2)^2$, then we have $(x+3)(x+2) > 0$

This means either: $(x+3) > 0$ and $(x+2) > 0$

OR $(x+3) < 0$ and $(x+2) < 0$ i.e. $x > -3$ and $x > -2$ OR $x < -3$ and $x < -2$

Therefore, create a sign table as shown in table 10.4(ii):

Table 10.4(ii)

	$x < -3$	$-3 < x < -2$	$x > -2$
$x+3$	−	+	+
$x+2$	−	−	+
$(x+3)(x+2)$	+	−	+

Since the equation is $(x+3)(x+2) > 0$, such that the inequality notation joining the RHS and LHS in this case is " > ", then the solution that satisfies the inequality is the product with "positive" signs from table 10.4(ii).

$\therefore x > -2$ and $x < -3$

Finally, combine equations (i) and (ii) on a number line to find the solution of the inequality,

$\left|\frac{1}{x+2}\right| < 1$

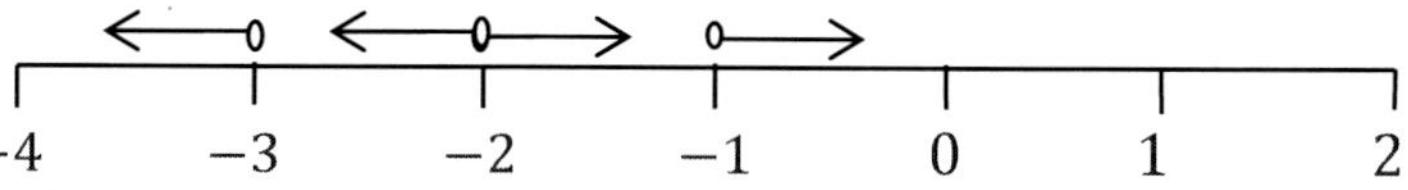

Therefore, the inequality solutions are $x < -3$ and $x > -1$

Example 10

$$\left|\frac{2x-3}{3x+1}\right| \leq 5$$

Solution

$$\left|\frac{2x-3}{3x+1}\right| \leq 5 \qquad \Rightarrow -5 \leq \frac{2x-3}{3x+1} \leq 5$$

Firstly consider: $\frac{2x-3}{3x+1} \leq 5$ ....(i)

$$\therefore \frac{2x-3}{3x+1} - 5 \leq 0 \quad \Rightarrow \quad \frac{2x-3-15x-5}{3x+1} \leq 0$$

i.e. $\frac{-13x-8}{3x+1} \leq 0$

$\therefore (-13x-8)(3x+1) \leq 0$

Multiply through by $(3x+1)^2$, so that the equation becomes $(-13x-8)(3x+1) < 0$

Thus, to remove the minus sign of the equation,$(-13x-8)(3x+1)\leq 0$, we rewrite this as $(13x+8)(3x+1)\geq 0$

This means either: $(13x+8)\leq 0$ and $(3x+1)\geq 0 \quad \Rightarrow x\leq -\frac{8}{13}$ and $x\geq -\frac{1}{3}$

OR $(13x+8)\geq 0$ and $(3x+1)\leq 0 \quad \Rightarrow \; x\geq -\frac{8}{13}$ and $x\leq -\frac{1}{3}$

Using the sign table to find the correct solution, we have:

Table 10.5(i)

	$x<-\frac{1}{3}$	$-\frac{1}{3}<x<-\frac{8}{13}$	$x>-\frac{8}{13}$
$13x+8$	−	+	+
$3x+1$	−	−	+
$(13x+8)(3x+1)$	+	−	+

Since, $(13x+8)(3x+1)\geq 0$ that means that the "positive sign" gives the correct solution, hence $x\leq -\frac{8}{13}$ and $x\geq -\frac{1}{3}$

Secondly consider: $-5\leq \frac{2x-3}{3x+1}$ ....(ii)

$\therefore\; 0\leq \frac{2x-3}{3x+1}+5 \quad \Rightarrow \quad \frac{2x-3}{3x+1}+5\geq 0$

i.e. $\frac{17x+2}{3x+1}\geq 0 \quad \therefore\; (17x+2)(3x+1)\geq 0$

Thus $(17x+2)\leq 0$ and $(3x+1)\geq 0$ **OR** $(17x+2)\geq 0$ and $(3x+1)\leq 0$

$\Rightarrow\; x\leq -\frac{2}{17}$ and $x\geq -\frac{1}{3}$ **OR** $x\geq -\frac{2}{17}$ and $x\leq -\frac{1}{3}$

Using the sign table to find the correct solution, we have:

Table 10.5(ii)

	$x<-\frac{1}{3}$	$-\frac{1}{3}<x<-\frac{2}{17}$	$x>-\frac{2}{17}$
$17x+2$	−	−	+
$3x+1$	−	+	+
$(17x+2)(3x+1)$	+	−	+

Since, $(17x+2)(3x+1)\geq 0$, that means the "positive sign" gives the correct solution, hence $x\leq -\frac{1}{3}$ and $x\geq -\frac{2}{17}$.

Therefore, combining equations (i) and (ii) on a number line, we have:

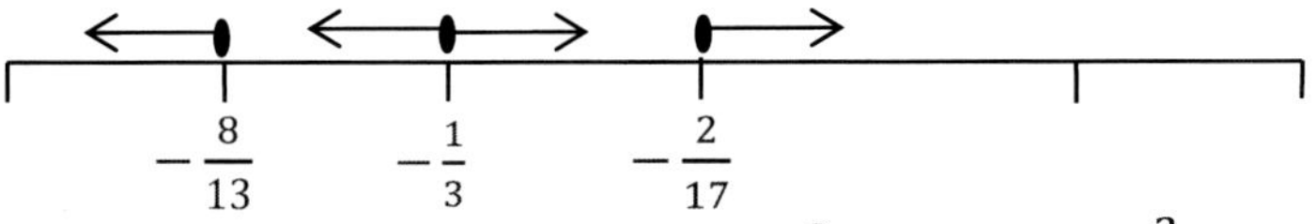

So, the solution of the inequality is: $x\leq -\frac{8}{13}$ and $x\geq -\frac{2}{17}$

Example 11

Find the solution set of the inequality $|2y| > |5 - 3y|$

Solution

$|2y| > |5 - 3y| \quad \Rightarrow -|5 - 3y| > 2y > |5 - 3y|$

Thus,

$2y > |5 - 3y|$ ....(i)

And $-|5 - 3y| > 2y$ which can be rewritten as $2y < -|5 - 3y|$

$2y < -|5 - 3y|$ ....(ii)

From equation (i) we have:

$2y > |5 - 3y|$ which means $|5 - 3y| < 2y$

$\Rightarrow -2y < 5 - 3y < 2y$

Using $-2y < 5 - 3y$ we have that: $3y - 2y < 5$

$\therefore y < 5$

Also, using $5 - 3y < 2y$ we have that: $-3y - 2y < -5$

$\therefore y > 1$

Hence, $2y > |5 - 3y|$ is $1 < y < 5$

While equation (ii), $2y < -|5 - 3y|$, has no solution.

Therefore, the solution of the given inequality, $|2y| > |5 - 3y|$, is

$1 < y < 5$

Alternatively,

$|2y| > |5 - 3y|$ can be solved as $(2y)^2 > (5 - 3y)^2$

then, $4y^2 > 25 - 30y + 9y^2$ re-arranging the inequality, we have:

$5y^2 - 30y + 25 < 0 \quad \Rightarrow y^2 - 6y + 5 < 0$

$(y - 5)(y - 1) < 0$

Therefore, the solution is $1 < y < 5$

Example 12

Find the solution of the given inequality: $|2x^2 - 7x - 5| \leq 10$

Solution

$|2x^2 - 7x - 5| \leq 10 \quad \Rightarrow -10 \leq 2x^2 - 7x - 5 \leq 10$

Hence, $-10 \leq 2x^2 - 7x - 5$ ....(i)

$2x^2 - 7x - 5 \leq 10$ ....(ii)

From equation(i) we have that: $-10 \leq 2x^2 - 7x - 5$

$\Rightarrow 0 \leq 2x^2 - 7x - 5 + 10$

and $0 \leq 2x^2 - 7x - 5 + 10$ means $2x^2 - 7x - 5 + 10 \leq 0$

$\therefore 2x^2 - 7x + 5 \leq 0$

$\therefore (x - 5)(x - 2) \leq 0$

$x \geq 2$ and $x \leq 5$ i.e. $2 \leq x \leq 5$

Solve the quadratic equation using factorisation method.

Similarly, from equation (ii) we have that: $2x^2 - 7x - 5 \leq 10$

$\Rightarrow\ 2x^2 - 7x - 5 - 10 \leq 0$

$\therefore\ 2x^2 - 7x - 5 - 15 \leq 0$

Thus, $(2x + 3)(x - 5) \leq 0 \qquad \Rightarrow x \geq -\frac{3}{2}$ and $x \leq 5$

i.e. $-\frac{3}{2} \leq x \leq 5$

Now combining equation (i) and (ii), we have that the solution of the inequality are:

$2 \leq x \leq 5$ OR $-\frac{3}{2} \leq x \leq 5$

Example 13

Solve and shade the region which satisfies the inequality $|x| - y \leq 5$

Solution

$|x| - y \leq 5 \quad \Rightarrow |x| \leq y + 5$

$\therefore\ -y - 5 \leq x \leq y + 5$

Thus, $x \leq y + 5$ ....(i)

$-y - 5 \leq x$ ....(ii)

Using equation (i) we have that: $x \leq y + 5 \quad \Rightarrow x - y - 5 \leq 0$

The line is at: $x - y - 5 = 0$

Put $x = 0 \quad \therefore (0) - y - 5 = 0$

$\therefore\ y = -5$

Also, put $y = 0 \quad \therefore x - (0) - 5 = 0$

$\therefore x = 5$

$(x_1, y_1) = (5, -5)$

Using equation (ii), we have that: $-y - 5 \leq x \quad \Rightarrow x + y + 5 \geq 0$

But the boundary line is at $x + y + 5 = 0$

put $x = 0 \quad \Rightarrow (0) + y + 5 = 0$

$\therefore y = -5$

Also, put $y = 0$

then, $x + (0) + 5 = 0$

$\therefore\ x = -5$

$(x_2, y_2) = (-5, -5)$

See fig. 10.10 in the next page for the region of the graph that satisfies the inequality:

The shaded region gives the solution of the inequality

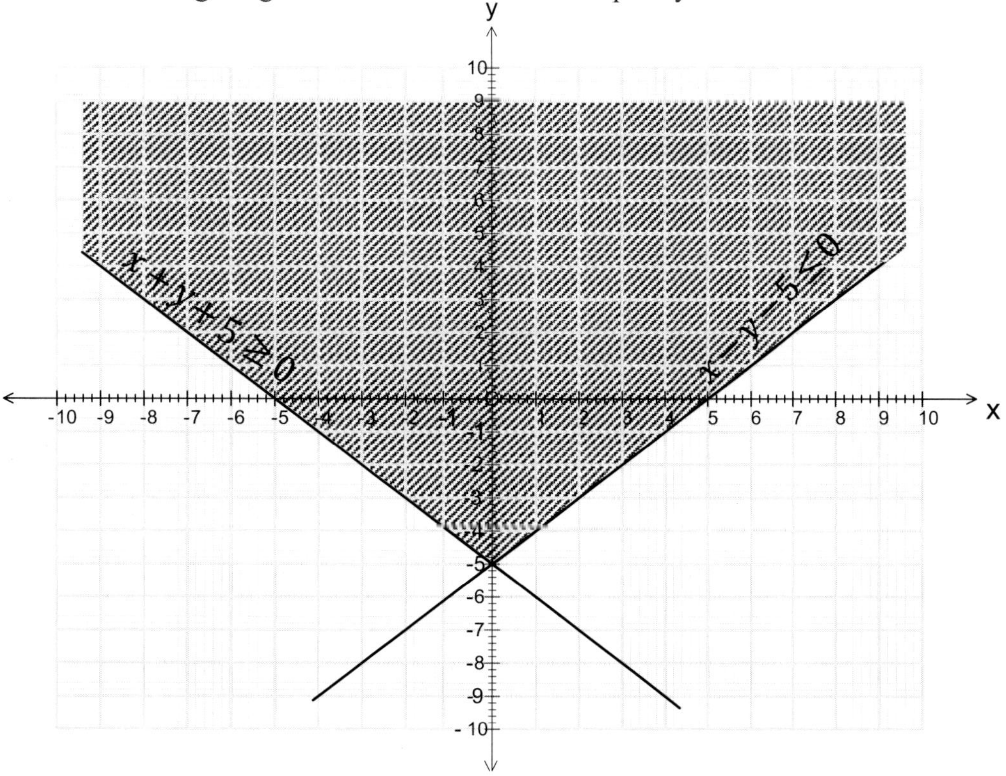

Fig. 10.10

Example 14

Solve graphically the inequality $|y| \leq x^2 - 9$

Solution

The inequality can be expressed as $9 - x^2 \leq y \leq x^2 - 9$

Thus, $y \leq x^2 - 9$ ....(i)

$9 - x^2 \leq y$ ....(ii)

Using equation (i) we have that: $y \leq x^2 - 9 \quad \Rightarrow y - x^2 + 9 \leq 0$

The boundary curve is at $y - x^2 + 9 = 0$

Put $x = 0 \quad \therefore\ y - (0)^2 + 9 = 0$

$\therefore y = -9$

Also, put $y = 0 \quad \therefore (0) - x^2 + 9 = 0$

$\therefore x = \sqrt{9} = \pm 3$

$(x_1, y_1) = (\pm 3, -9)$

Using equation (ii), we have that: $9 - x^2 \leq y \quad \Rightarrow 9 - y - x^2 \leq 0$
The boundary curve is at: $9 - y - x^2 = 0$
Put $x = 0 \quad \therefore\ 9 - y - (0)^2 = 0$
$\therefore\ y = 9$
Also, put $y = 0 \quad \therefore\ 9 + (0) - x^2 = 0$
$\therefore\ x = \sqrt{9} = \pm 3$
$(x_2,\ y_2) = (\pm 3,\ 9)$

The shaded regions give the solution set of the inequality

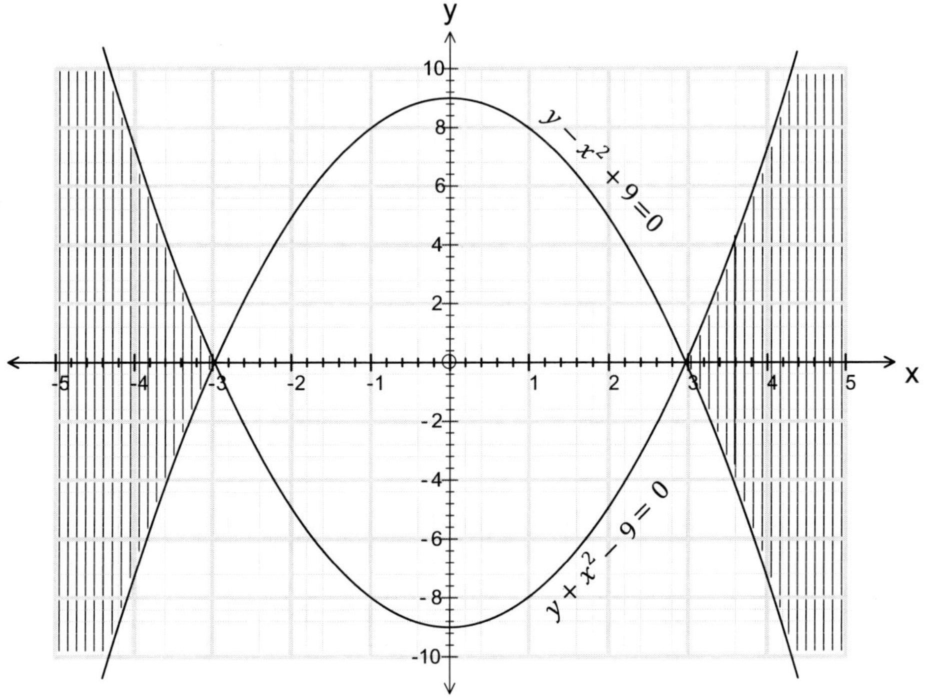

Fig. 10.11

OBJECTIVE TEST 10

1. Solve for x if: $\frac{x-2}{5} + 3 \leq 0$
   (A) $x \geq 13$ (B) $x \leq -13$ (C) $x < -13$ (D) $x \geq -13$

2. Find the true solution that satisfies the inequality: $(x+4)(x-1) \geq 0$
   (A) $-4 \leq x \leq 1$ (B) $x \geq -4$ or $x \geq 1$
   (C) $x \leq -4$ or $x \geq 1$ (D) $x \geq -4$ or $x \leq 1$

3. What is the range of value of x in $|x|<5$
   (A) $-5 \leq x < 5$ (B) $-5 \leq x \leq 5$ (C) $5 < x < -5$ (D) $-5 < x < 5$

4. Find the value of p for which the equation $2x^2 + px + 4 = 0$ will have real roots.
   (A) $-4\sqrt{2} \leq p \leq 4\sqrt{2}$ (B) $p \leq -4\sqrt{2}$ or $p \geq 4\sqrt{2}$
   (C) $p \geq 4\sqrt{2}$ or $p \leq 4\sqrt{2}$ (D) $p = 4\sqrt{2}$

5. Find the solution for y in the equation $|y-5| \leq 1$
   (A) $4 \leq y \leq 6$ (B) $-4 \leq y \leq 6$ (C) $y \leq 4$ or $y \geq 6$ (D) $4 < y < 6$

6. Find the true solution of $(x+3)(x-2) \leq 0$
   (A) $-2 \leq x \leq 3$ (B) $-3 \leq x \leq 2$ (C) $x \geq -3$ or $x \geq 2$ (D) $x \leq -3$ or $x \geq 2$

7. What is the equation that satisfies the shaded region of the function represented by the graph?

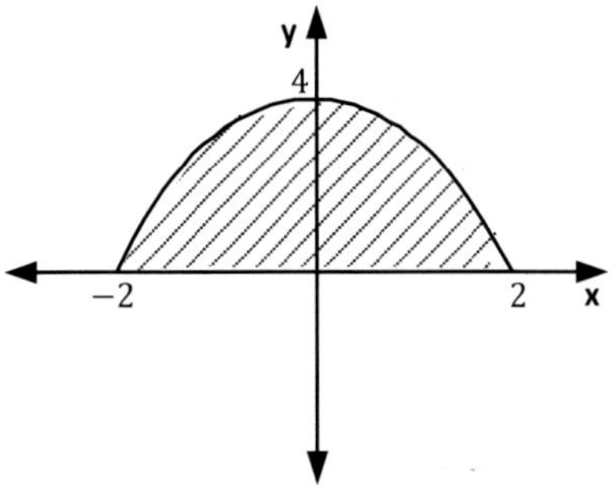

(A) $x^2 + y - 4 \leq 0$ (B) $x + y^2 - 4 \leq 0$ (C) $x^2 - y - 4 \leq 0$ (D) $x^2 + y - 4 \geq 0$

8. The inequality, $-2 \le x \le 3$, is represented on a number line as:

(A) 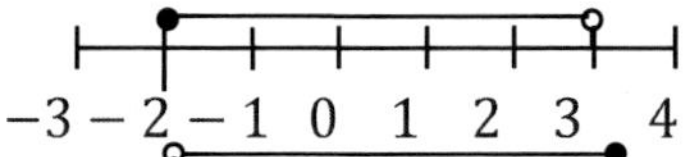

(B) 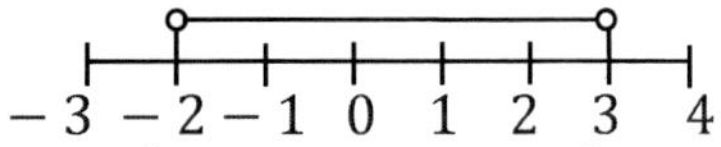

(C) 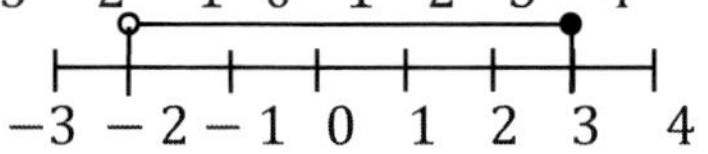

(D) 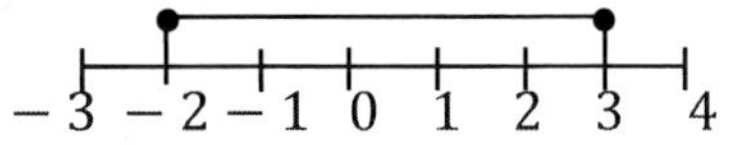

9. Find the solution set of the inequality $\frac{x-3}{x+1} \le 0$

(A) $-1 \le x \le 3$ (B) $1 \le x \le -3$ (C) $x \le -1$ or $x \ge 3$ (D) $x < -1$ or $x < 3$

10. Which of the shaded graphs represent the correct solution of the inequalities: $2x - 2 \le y$ and $2y - x \le 2$

(A) 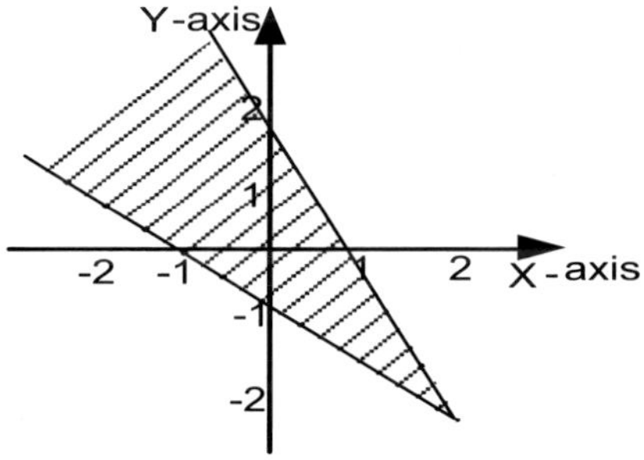

(B) 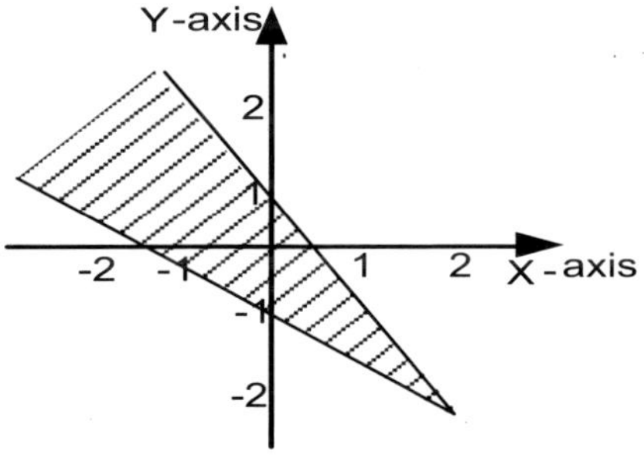

(C) 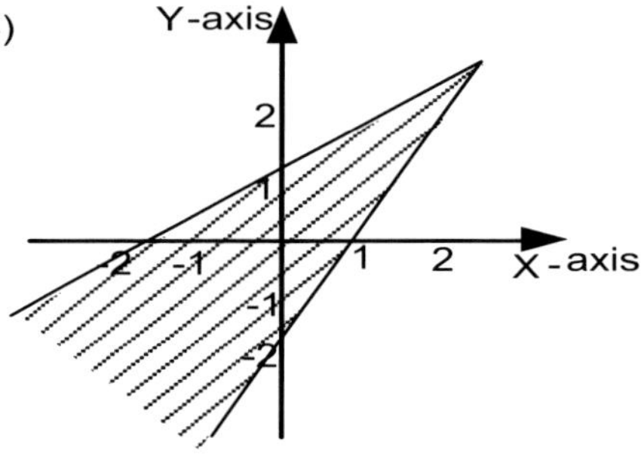

(D) 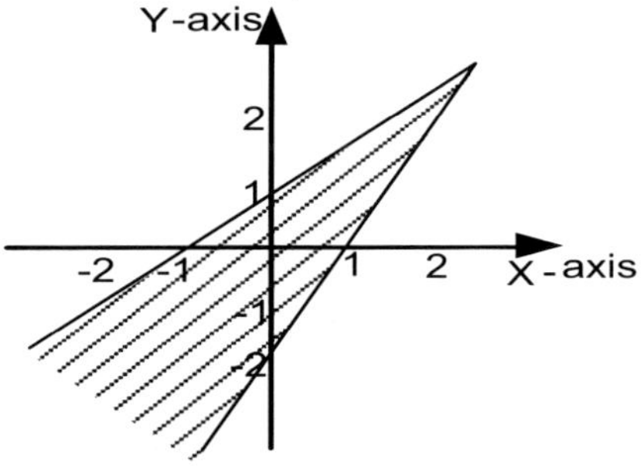

Exercise 10

1. What are the integral values of x that satisfy the following inequalities:
   (a) $1 \leq 5 - 2x \leq 3$ (b) $-3 \leq 2x + 7 < 3$

2. If $y = x^2 - x - 6$, find the range of values of x for which $y \geq 0$

3. Solve the following inequalities:
   (a) $2x + 3 < 18 - x$ (b) $2 - 5x > 7$ (c) $3x - 4 < 2 - x$
   (d) $9x - (2x + 3) \geq 0$ (e) $-3(x + 1) < 0$

4. Evaluate each of the following systems of inequality:
   (a) $2 - x \geq 7$ and $\frac{x-1}{2} - x < 5\frac{1}{3}$ (b) $4 \leq \frac{1}{3}(x - 5) \leq 2$
   (c) $3x + 1 \leq 10 - 3x \leq 6x + 1$

5. Represent the solution set of each of the following inequalities on a number line:
   (a) $2(x + 1) \leq x + 1$ (b) $x + 1 \leq 3$
   (c) $3x - 5 \leq 2x + 1$ and $x + 2 < 2x + 5$ (d) $2x - 3 < x + 2 < 2x + 1$

6. Solve the following inequalities:
   (a) $(3 - x)(x + 1) \geq 0$ (b) $(x - 1)(x + 2) \leq 0$
   (c) $\frac{3x-10}{x+1} < 0$ (d) $\frac{x-10}{x+2} > 3$

7. Shade the region that satisfies each of the solution set of the following pairs of simultaneous linear inequalities:
   (a) $x + y < 1$, $2x - y \leq 5$, $x \geq 0$ (b) $2x + 3y \leq 5$, $x - 5y \geq 1$, $y \geq 0$

8. Solve for x in the following inequalities with modulus:
   (a) $|2x + 1| > 7$ (b) $|x + 3| \geq 5$ (c) $|2x - 3| \leq 1$ (d) $|10x - 7| < 13$
   (e) $|5x + 2| > 3 - x$ (f) $|3x| > |7 - 4x|$ (g) $|6 - x| \leq |5x + 2|$

9. Solve each of the inequalities:
   (a) $\frac{2x}{3-x} > -1$ (b) $\frac{x+1}{x-2} < 5$ (c) $\frac{2x-3}{5x+2} \leq \frac{1}{3}$ (d) $\frac{7x-1}{10+x} > \frac{1}{5}$

10. Find the solution set of the following inequalities:
   (a) $3 \geq |2x^2 - 9x - 8|$ (b) $\left|\frac{2}{3x}\right| < 5$ (c) $|x - 5| \geq 2$ (d) $|3x + 1| > |x - 1|$

11. Solve graphically the following inequalities with modulus:
   (a) $|x| - y \geq 3$ (b) $x - |2y| \leq 5$ (c) $|x - y| \geq 2$ (d) $|y| - 1 \leq 2$ (e) $|3x| - y \leq 10$

12. Show the region that satisfies the following inequalities:
   (a) $|x| - 2y \geq 10$ (b) $|x - y| \leq 5$ (c) $x^2 + y \geq 25,\ y \geq 0,\ x \geq 0$ (d) $|x| - 1 \leq |y|$

11 POLYNOMIALS AND PARTIAL FRACTION

11.1 POLYNOMIALS

The general form of a polynomial is given as:

$$a_n x^n + a_{n-1}x^{n-1} + a_{n-2}x^{n-2} + \cdots + a_2 x^2 + a_1 x^1 + a_0\, x^0$$

Where $a_0 \neq 0$, and $a_n, a_{n-1}, a_{n-2}, a_2, a_1, a_0$ are numerical constants which are the coefficients of x^n, x^{n-1}, $x^{n-2}, \ldots x^2, x, x^0$ respectively. Also, n is the highest power of the polynomial, and it is called the "degree" of the polynomial.

11.1.1 ADDITION OF POLYNOMIALS

The addition of polynomials is achieved by combining the like terms of the polynomial with respect to their degree of power. To do this, the functions may have to be arranged in ascending or descending order of degree of power of x.

Example 1

Given that $P_{1(x)} = 7x^3 - 4x^2 + 5x + 8$, $\quad P_{2(x)} = 5x^3 + 6x + 10$
and $P_{3(x)} = 4x^3 + 2x^2 - x + 6$. Find:
(a) $P_{1(x)} + P_{2(x)}$ (b) $P_{1(x)} + P_{3(x)}$ (c) $P_{1(x)} + P_{2(x)} + P_{3(x)}$

Solution

(a) $P_{1(x)} + P_{2(x)} = (7x^3 - 4x^2 + 5x + 8) + (5x^3 + 6x + 10)$
$= 7x^3 + 5x^3 - 4x^2 + 5x + 6x + 8 + 10$
$= 12x^3 - 4x^2 + 11x + 18$

(b) $P_{1(x)} + P_{3(x)} = (7x^3 - 4x^2 + 5x + 8) + (4x^3 + 2x^2 - x + 6)$
$= 7x^3 + 4x^3 - 4x^2 + 2x^2 + 5x - x + 8 + 6$
$= 11x^3 - 2x^2 + 4x + 14$

(c) $P_{1(x)} + P_{2(x)} + P_{3(x)} = (7x^3 - 4x^2 + 5x + 8) + (5x^3 + 6x + 10) + (4x^3 + 2x^2 - x + 6)$
$= 7x^3 + 5x^3 + 4x^3 - 4x^2 + 2x^2 + 5x + 6x - x + 8 + 10 + 6$
$= 16x^3 - 2x^2 + 10x + 24$

Example 2

Given that $P_{1(x)} = 3x^3 + 4x^2 - x + 1$, $P_{2(x)} = x^4 + x^3 + 2x - 5$ and $P_{3(x)} = 15x^3 - 2x + 4$
Find: (a) $2P_{1(x)} + P_{2(x)}$ (b) $3P_{1(x)} + P_{3(x)}$ (c) $5P_{1(x)} + 3P_{2(x)} + 2P_{3(x)}$

Solution

(a) $2P_{1(x)} + P_{2(x)} = 2(3x^3 + 4x^2 - x + 1) + (x^4 + x^3 + 2x - 5)$
$= (6x^3 + 8x^2 - 2x + 2) + (x^4 + x^3 + 2x - 5)$
$= x^4 + 6x^3 + x^3 + 8x^2 - 2x + 2x + 2 - 5 = x^4 + 7x^3 + 8x^2 - 3$

(b) $3P_{1(x)} + P_{3(x)} - 3(3x^3 + 4x^2 - x + 1) + (15x^3 - 2x + 4)$
$= (9x^3 + 12x^2 - 3x + 3) + (15x^3 - 2x + 4)$
$= 15x^3 + 9x^3 + 12x^2 - 3x - 2x + 3 + 4$
$= 24x^3 + 12x^2 - 5x + 7$

(c) $5P_{1(x)} + 3P_{2(x)} + 2P_{3(x)}$
$= 5(3x^3 + 4x^2 - x + 1) + 3(x^4 + x^3 + 2x - 5) + 2(15x^3 - 2x + 4)$
$= (15x^3 + 20x^2 - 5x + 5) + (3x^4 + 3x^3 + 6x - 15) + (30x^3 - 4x + 8)$
$= 3x^4 + 30x^3 + 15x^3 + 3x^3 + 20x^2 + 6x - 5x - 4x + 5 + 8 - 15$
$= 3x^4 + 48x^3 + 20x^2 - 3x - 2$

11.1.2 SUBTRACTION OF POLYNOMIALS

The subtraction of polynomials uses the same method as addition of polynomials; however, the only difference is that the "+" sign is changed to "−".

Example 3

Given that $P_{1(x)} = x^6 - 5x^5 + 2x^3 + 10$, $P_{2(x)} = 5x^4 + 2x^3 - x^2 + 1$, and $P_{3(x)} = 1 - 2x^2 + 3$, find:
(a) $P_{1(x)} - P_{2(x)}$ (b) $3P_{1(x)} - 2P_{3(x)}$ (c) $5P_{1(x)} - P_{2(x)} - P_{3(x)}$

Solution

(a) $P_{1(x)} - P_{2(x)} = (x^6 - 5x^5 + 2x^3 + 10) - (5x^4 + 2x^3 - x^2 + 1)$
$= x^6 - 5x^5 + 2x^3 + 10 - 5x^4 - 2x^3 + x^2 - 1$
$= x^6 - 5x^5 - 5x^4 + 2x^3 - 2x^3 + x^2 + 10 - 1 = x^6 - 5x^5 - 5x^4 + x^2 + 9$
$\therefore P_{1(x)} - P_{2(x)} = x^6 - 5x^5 - 5x^4 + x^2 + 9$

(b) $3P_{1(x)} - 2P_{3(x)} = 3(x^6 - 5x^5 + 2x^3 + 10) - 2(1 - 2x^2 + 3)$
$= (3x^6 - 15x^5 + 6x^3 + 30) - (2 - 4x^2 + 6)$
$= 3x^6 - 15x^5 + 6x^3 + 30 - 2 + 4x^2 - 6$
$= 3x^6 - 15x^5 + 6x^3 + 4x^2 + 22$
$\therefore 3P_{1(x)} - 2P_{3(x)} = 3x^6 - 15x^5 + 6x^3 + 4x^2 + 22$

(c) $5P_{1(x)} - P_{2(x)} - P_{3(x)}$
$= 5(x^6 - 5x^5 + 2x^3 + 10) - (5x^4 + 2x^3 - x^2 + 1) - (1 - 2x^2 + 3)$
$= (5x^6 - 25x^5 + 10x^3 + 50) - (5x^4 + 2x^3 - x^2 + 1) - (1 - 2x^2 + 3)$
$= 5x^6 - 25x^5 + 10x^3 + 50 - 5x^4 - 2x^3 + x^2 - 1 - 1 + 2x^2 - 3$
$= 5x^6 - 25x^5 - 5x^4 + 10x^3 - 2x^3 + 2x^2 + x^2 - 1 - 1 - 3$
$= 5x^6 - 25x^5 - 5x^4 + 8x^3 + 3x^2 - 5$
$\therefore 5P_{1(x)} - P_{2(x)} - P_{3(x)} = 5x^6 - 25x^5 - 5x^4 + 8x^3 + 3x^2 - 5$

11.1.3 MULTIPLICATION OF POLYNOMIALS

The product of two polynomials of degrees, m and n, is another polynomial of degree $m + n$.

Example 4

Given that $P_{1(x)} = x^3 + 10x^2 + 3$ and $P_{2(x)} = 3x^4 + 2x^2 - 1$, find:

(a) $P_{1(x)}P_{2(x)}$ (b) $P_1{}^2{}_{(x)}P_{2(x)}$

Solution

(a) $P_{1(x)}\,P_{2(x)} = (x^3 + 10x^2 + 3)(3x^4 + 2x^2 - 1)$

$= 3x^4(x^3 + 10x^2 + 3) + 2x^2(x^3 + 10x^2 + 3) - 1(x^3 + 10x^2 + 3)$

$= 3x^7 + 30x^6 + 9x^4 + 2x^5 + 20x^4 + 6x^2 - x^3 - 10x^2 - 3$

$= 3x^7 + 30x^6 + 2x^5 + 9x^4 + 20x^4 - x^3 + 6x^2 - 10x^2 - 3$

$= 3x^7 + 30x^6 + 2x^5 + 29x^4 - x^3 - 4x^2 - 3$

$\therefore P_{1(x)}P_{2(x)} = 3x^7 + 30x^6 + 2x^5 + 29x^4 - x^3 - 4x^2 - 3$

(b) $P_1{}^2{}_{(x)}P_{2(x)} = (x^3 + 10x^2 + 3\,)^2(3x^4 + 2x^2 - 1)$

$= (x^6 + 20x^5 + 100x^4 + 6x^3 + 60x^2 + 9)(3x^4 + 2x^2 - 1)$

$= (x^6 + 20x^5 + 100x^4 + 6x^3 + 60x^2 + 9)(3x^4 + 2x^2 - 1)$

$= 3x^{10} + 2x^8 - x^6 + 60x^9 - 140x^7 - 20x^5 + 300x^8 + 200x^6$
$-100x^4 + 18x^7 + 12x^5 - 6x^3 + 180x^6 + 120x^4 - 60x^2 + 27x^4 + 18x^2 - 9$

$= 3x^{10} + 60x^9 + 302x^8 + 58x^7 + 379x^6 - 8x^5 + 47x^4 - 6x^3 - 42x^2 - 9$

$\therefore P_1{}^2{}_{(x)}P_{2(x)} = 3x^{10} + 60x^9 + 302x^8 + 58x^7 + 379x^6 - 8x^5 + 47x^4 - 6x^3 - 42x^2 - 9$

Example 5

Find the values of the unknown constants given that:

$(x^3 + 2x^2 + x - 1)(x^2 + x + 2) = ax^5 - bx^4cx^3 - dx^2 + ex - f$

Solution

$(x^3 + 2x^2 + x - 1)(x^2 + x + 2)$

$= x^{3+2} + x^{3+1} + 2x^3 + 2x^{2+2} + 2x^{2+1} + 4x^2 + x^{1+2} + x^{1+1} + 2x - x^2 - x - 2$

$= x^5 + x^4 + 2x^3 + 2x^4 + 2x^3 + 4x^2 + x^3 + x^2 + 2x - x^2 - x - 2$

$= x^5 + 3x^4 + 5x^3 + 4x^2 + x - 2 = ax^5 - bx^4 + cx^3 - dx^2 + ex - f$

$\therefore a = 1, \quad b = -3, \quad c = 5, \quad d = -4, \quad e = 1, \quad f = 2$

11.1.4 DIVISION OF POLYNOMIALS

A polynomial of degree n can be divided by another polynomial of degree m if and only if $n \geq m$.

Let: $P_{1(x)} = a_nx^n + a_{n-1}x^{n-1} + a_{n-2}x^{n-2} + \cdots + a_2x^2 + a_1x^1 + a_0\,x^0$

$P_{2(x)} = a_mx^m + a_{m-1}x^{m-1} + a_{m-2}x^{m-2} + \cdots + a_2x^2 + a_1x^1 + a_0\,x^0$

Then, $\frac{P_{1(x)}}{P_{2(x)}} = Q_{(x)} + R_{(x)}$

Where: $P_{1(x)}$ is the dividend, $P_{2(x)}$ is the divisor, $Q_{(x)}$ is the quotient, and $R_{(x)}$ is the remainder.

Thus, the result of the division of $P_{1(x)}$ and $P_{2(x)}$ results to the "quotient" (i. e. $Q_{(x)}$); and any left-over value after the division is called the "remainder" (i.e., $R_{(x)}$).

Example 6

Divide the following polynomials:

(a) $3x^2 - 2x - 4$ by $x + 2$ (b) $2x^3 + 3x + 1$ by $2x + 1$

Solution

(a) Let $P_{1(x)} = 3x^2 - 2x - 4$, and $P_{2(x)} = x + 2$

$$\begin{array}{r|l} & \quad 3x - 8 \\ \hline x + 2 & 3x^2 - 2x - 4 \\ & -\ 3x^2 + 6x + 0 \\ \hline & \quad -8x - 4 \\ & -\ -8x - 16 \\ \hline & \qquad 12 \end{array}$$

First, divide $3x^2$ by x and that will give $3x$; then write it out on the top bar. Now, multiply $(x + 2)$ by $3x$ to get $3x^2 + 6x + 0$; then subtract this result from $P_{1(x)}$ to have $-8x - 4$. Similarly, divide again $-8x$ by x to get -8, write it on the top bar as shown. Also, multiply $(x + 2)$ by -8 to get $-8x - 16$. Finally, subtract $-8x - 16$ from $-8x - 4$ to give a remainder, 12.

$$\frac{P_{1(x)}}{P_{2(x)}} = Q_{(x)} + R_{(x)}$$

$$= \frac{3x^2 - 2x - 4}{x + 2} = 3x - 8 + \frac{12}{x + 2}$$

Where the quotient, $Q_{(x)} = 3x - 8$; and the remainder, $R_{(x)} = \frac{12}{x+2}$.

(b) $2x^3 + 3x + 1$ by $2x + 1$

Let $P_{1(x)} = 2x^3 + 3x + 1$ and $P_{2(x)} = 2x + 1$

$$\begin{array}{r|l} & \qquad x^2 - {}^{x}/_{2} + {}^{7}/_{4} \\ \hline & -\quad 2x^3 + 3x + 1 \\ & \quad\ \ \underline{2x^3 + x^2 + 0} \\ & \qquad -\quad x^2 + 3x + 1 \\ 2x + 1 & \qquad\quad\ \ -x^2 - {}^{x}/_{2} \\ \hline & \qquad\quad -\ \ {}^{7}/_{4}x + 1 \\ & \qquad\qquad {}^{7}/_{2}x + {}^{7}/_{4} \\ \hline & \qquad\qquad\quad -{}^{3}/_{4} \end{array}$$

$$\frac{P_{1(x)}}{P_{2(x)}} = Q_{(x)} + R_{(x)} = \frac{2x^3+3x+1}{2x+1} = x^2 - {}^{x}/_{2} + {}^{7}/_{4} + \frac{-{}^{3}/_{4}}{2x+1}$$

11.1.5 ZEROES OF POLYNOMIALS

If $P_{(x)} = a_nx^n + a_{n-1}x^{n-1} + a_{n-2}x^{n-2} + \cdots + a_2x^2 + a_1x^1 + a_0\,x^0$ is a polynomial of degree n, then the zeroes of the polynomial are the solution for x when $P_{(x)} = 0$.

Example 7

Find the zeroes of the following polynomials:

(a) $x^2 - 5x + 6$ (b) $2x^3 + 5x^2 - 3x - 4$ (c) $\frac{x^3-2x^2-5x+6}{3x^2-4x+5}$

Solution

(a) The zeroes of the polynomial $x^2 - 5x + 6$, are the solution for x when the polynomial is equated to zero.

$\Rightarrow x^2 - 5x + 6 = 0$

$(x-3)(x-2) = 0 \quad \therefore x = 3$ or 2

Hence, the zeroes of the polynomial are 3 and 2.

(b) $2x^3 + 5x^2 - 3x - 4$

$\Rightarrow 2x^3 + 5x^2 - 3x - 4 = 0$

$2x^3 + 5x^2 - 3x - 4 = (x-1)(2x^2 + 7x + 4)$

$\Rightarrow (x-1)(2x^2 + 7x + 4) = 0$

Solving for x in $2x^2 + 7x + 4$ using the quadratic formula method:

then, $2x^2 + 7x + 4 = (x - 0.72)(x + 2.78) = 0$

$2x^3 + 5x^2 - 3x - 4 = (x-1)(x-0.72)(x+2.78)$

$\therefore$ zeroes of the polynomial are: $x = 1,\ 0.72$ and -2.78

(c) $\frac{x^3-2x^2-5x+6}{3x^2-4x+5} = 0 \quad \Rightarrow \quad x^3 - 2x^2 - 5x + 6 = 0 \times (3x^2 - 4x + 5)$

Thus, $x^3 - 2x^2 - 5x + 6 = 0$

Factorising, we have: $(x-1)(x+2)(x-3) = 0$

Therefore, the zeroes of the polynomial are $x = 1, x = -2, x = 3$

11.1.6 POLYNOMIAL SOLUTION BY SUBSTITUTION

Given a polynomial function, then a given solution of the polynomial can be determined when we substitute a defined value for x into the given function.

Example 8

Given that a function is defined as $P_{(x)} = 2x^3 + 5x^2 - 9x - 18$, find:

(a) $P(1)$ (b) $P(2)$ (c) $P(0)$ (d) $P(3)$

Solution

(a) $P(1) = 2(1)^3 + 5(1)^2 - 9(1) - 18 = 2 \times 1 + 5 \times 1 - 9 \times 1 - 18$

$= 2 + 5 - 9 - 18 = -20$

$\therefore P(1) = -20$

(b) $P(2) = 2(2)^3 + 5(2)^2 - 9(2) - 18 = 2 \times 8 + 5 \times 4 - 9 \times 2 - 18$
$= 16 + 20 - 18 - 18 = 0 \quad \therefore P(2) = 0$

(c) $P(0) = 2(0)^3 + 5(0)^2 - 9(0) - 18 = 2 \times 0 + 5 \times 0 - 9 \times 0 - 18$
$= 0 + 0 - 0 - 18 = -18 \therefore P(0) = -18$

(d) $P(3) = 2(3)^3 + 5(3)^2 - 9(3) - 18 = 2 \times 27 + 5 \times 9 - 9 \times 3 - 18$
$= 54 + 45 - 18 - 18 = 54 \quad \therefore P(3) = 54$

Example 9

If $f_{(x)} = 5x^3 - 2x^2 + x + 3$, find (a) $f(-1)$ (b) $f(0)$ (c) $f(-6)$ (d) $f(2)$

Solution

(a) $f(-1) = 5(-1)^3 - 2(-1)^2 + (-1) + 3 = 5 \times (-1) - 2 \times 1 - 1 + 3$
$= -5 - 2 - 1 + 3 = -5$
$\therefore P(-1) = -5$
(b) $f(0) = 5(0)^3 - 2(0)^2 + (0) + 3 = 5 \times 0 - 2 \times 0 + 0 + 3$
$= 0 - 0 - 0 + 3 = 3$
$\therefore P(0) = 3$

(c) $f(-6) = 5(-6)^3 - 2(-6)^2 + (-6) + 3 = 5 \times (-216) - 2 \times 36 - 6 + 3$
$= -1080 - 72 - 6 + 3 = -1155$
$\therefore P(-6) = -1155$

(d) $f(2) = 5(2)^3 - 2(2)^2 + (2) + 3 = 5 \times 8 - 2 \times 4 + 2 + 3$
$= 40 - 8 + 2 + 3 = 37$
$\therefore P(2) = 37$

11.1.7 REMAINDER THEOREM AND FACTOR THEOREM

Remainder's theorem: States that when a function $f(x)$ is divided by $x - a$, the remainder is $f(a)$. Similarly, when $f(x)$ is divided by $ax - b$, then the remainder is found when we substitute $f\left(\frac{b}{a}\right)$ into the function.

Example 10

Given that $f(x) = 5x^3 + 3x^2 - 2x + 1$
Find the remainder when $f(x)$ is divided by $x - 1$

Solution

Let R be the remainder; then from remainder's theorem $f(a) = R$
$\Rightarrow f(1) = R$
$f(1) = 5(1)^3 + 3(1)^2 - 2(1) + 1$
$= 5 + 3 - 2 + 1 = 7$
$\therefore \quad R = 7$

Since the divisor is $x - 1$, i.e., $x - a$
$\Rightarrow x - 1 = x - a$
$\therefore a = 1$

Example 11

Find the remainder when $f(x) = (x+2)(x-3)(x+4)$ is divided by $x+1$

Solution

Let R be the remainder when $f(x)$ is divided by $x+1$

$\Rightarrow f(-1) = R$

The divisor is $x+1$, i.e., $x-a$
$\Rightarrow x+1 = x-a \quad \therefore a=-1$

$$f(-1) = (-1+2)(-1-3)(-1+4)$$
$$= (1)(-4)(3) = -12$$
$$\therefore \quad R = -12$$

Example 12

Find the remainder when $f(x) = x^4 + 3x^3 - 6x + 10$ is divided by $2x+1$.

Solution

Let R be the remainder when $f(x)$ is divided by $2x+1$

$\Rightarrow f\left(-\frac{1}{2}\right) = R$

The divisor is $2x+1$, i.e. $ax-b$
$\Rightarrow 2x+1 = ax-b$
$\therefore a=2$ and $b=-1$
Then, $f\left(\frac{b}{a}\right) = f\left(-\frac{1}{2}\right)$

$$f\left(-\frac{1}{2}\right) = \left(-\frac{1}{2}\right)^4 + 3\left(-\frac{1}{2}\right)^3 - 6\left(-\frac{1}{2}\right) + 10$$

$$= \left(\frac{1}{16}\right) + 3\left(-\frac{1}{8}\right) - 6\left(-\frac{1}{2}\right) + 10 = \frac{1}{16} - \frac{3}{8} + 3 + 10 = \frac{1}{16} - \frac{3}{8} + 13$$

$$= \frac{1-6+208}{16} = \frac{203}{16} = 12\frac{11}{16} \quad \therefore R = 12\frac{11}{16}$$

Factor's theorem: States that when a function $f(x)$ is divided by $x-a$, and leaves no remainder; then $x-a$ is a factor of the polynomial.
i.e. $f(a) = R = 0$; then $x-a$ is a factor of the polynomial, $f(x)$.

Example 13

Show that the remainder of the function $f(x) = 3x^5 + 2x^3 - x + 4$ is zero when divided by $x+1$.

Solution

$f(x) = 3x^5 + 2x^3 - x + 4$ and the divisor is $x+1$

$\Rightarrow f(-1) = 3(-1)^5 + 2(-1)^3 - 1 + 4 = 3(-1) + 2(-1) + 1 + 4$

$= -3 - 2 + 5 = -5 + 5 = 0$

$\therefore R = 0$

Example 14

Show that $x-2$ is a factor of $f(x) = 2x^6 - 3x^4 - x^2 - 76$

Solution

To show that $x-2$ is a factor; then make $x-2=0$, such that $x=2$

$\Rightarrow f(2) = 2(2)^6 - 3(2)^4 - 2^2 - 76 = 2(64) - 3(16) - 4 - 76$

$= 128 - 48 - 4 - 76 = 128 - 128 = 0$

$\therefore R = 0 \quad$ Hence, $x-2$ is a factor because $R = 0$

Example 15

Factorise $x^3 - 8x^2 - x + 8$

Solution

Using trial method, test with $x = \pm 1$

When $x = 1$

$\Rightarrow f(1) = x^3 - 8x^2 - x + 8 = (1)^3 - 8(1)^2 - (1) + 8 = 2$

$2 \neq 0$, so it is not a factor.

Now, using $x = -1 \quad \Rightarrow f(-1) = x^3 - 8x^2 - x + 8 = (-1)^3 - 8(-1)^2 - (-1) + 8$

$= -1 - 8 + 1 + 8 = 0$

Since, $f(-1) = 0$ then $x = -1$ (i.e., $x + 1 = 0$) is a factor.

Now, use polynomial division to find the quotient when $x^3 - 8x^2 - x + 8$ is divided by $x + 1$.

Thus,

$$
\begin{array}{r|l}
 & x^2 - 9x + 8 \\
\hline
x + 1 & x^3 - 8x^2 - x + 8 \\
 & -\underline{(x^3 + x^2)} \\
 & -9x^2 - x + 8 \\
 & -\underline{(-9x^2 - 9x)} \\
 & 8x + 8 \\
 & \underline{8x + 8} \\
 & 0
\end{array}
$$

The quotient is $x^2 - 9x + 8 \quad \Rightarrow x^2 - 9x + 8 = (x - 8)(x - 1)$

$\therefore$ the factorisation of $x^3 - 8x^2 - x + 8$ is $(x + 1)(x - 8)(x - 1)$

11.1.8 POLYNOMIAL GRAPHS

Here, only the graphs of polynomial of degree 2 and degree 3 shall be considered:

Example 16

Sketch the following graphs (a) $y = x^2 + 4x - 21$ (b) $y = 6 - x - x^2$

Solution

(a)

This equation $y = x^2 + 4x - 21$ is of degree 2 and it will have a U-shape because $a > 0$. Therefore, the roots are $x = (-7, 0)$ and $(3, 0)$. While the y intercept is $y = (0, -21)$.

The minimum value occurs at the vertex, (h, k)) where $h = \frac{-b}{2a} = \frac{-4}{2(1)} = -2$

And $k = \frac{4ac - b^2}{4a} = \frac{4*1*(-21) - 4^2}{4*1} = -25$

$y = x^2 + 4x - 21 = (x + 2)^2 - 25$

thus, at points $(-2, -25)$

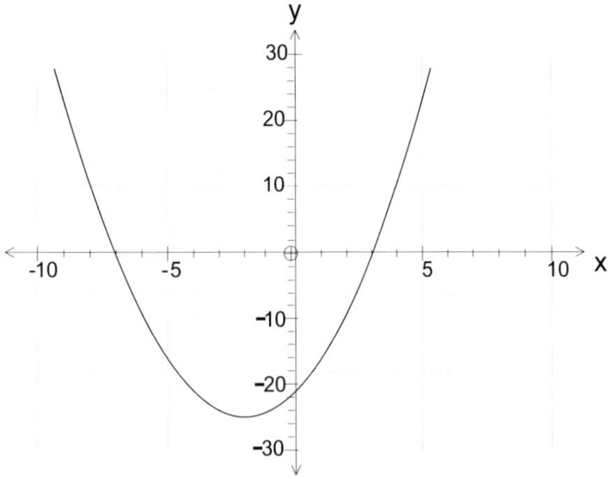

Fig. 11.1 (a) Graph of $y = x^2 + 4x - 21$

(b)
This equation $y = 6 - x - x^2$ will have an ∩-shape because $a < 0$. Therefore, the roots are $x = (-3, 0)$ and $(2, 0)$. While the y intercept is $y = (0, 6)$. And the maximum value occurs when $y = 6 - x - x^2 = \frac{25}{4} - \left(x + \frac{1}{2}\right)^2$; thus, at points $\left(-\frac{1}{2}, \frac{25}{4}\right)$.

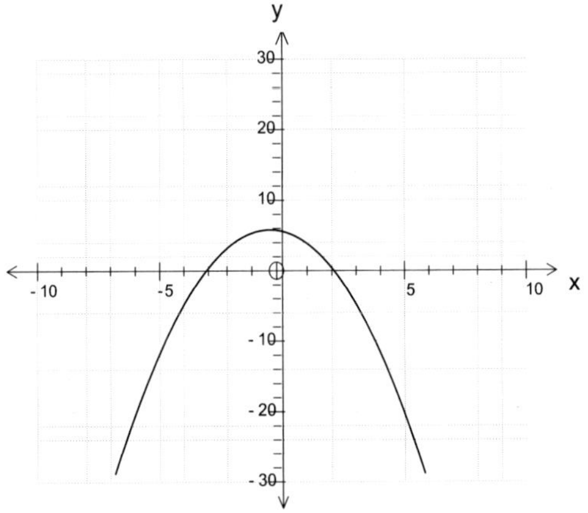

Fig. 11.1 (b) Graph of $y = 6 - x - x^2$

Example 17

Sketch the graph $x^3 - 2x^2 - 19x + 20$

Solution

This equation $x^3 - 2x^2 - 19x + 20$ is of degree 3, hence it has just three roots. They are $x = (1, 0)$; $(5, 0)$ and $(-4, 0)$. the y intercept is $y = (0, 20)$. And the maximum value occurs at approximate points $(-1.9, 42)$; while the minimum value occurs at approximate points $(3.3, -28)$.

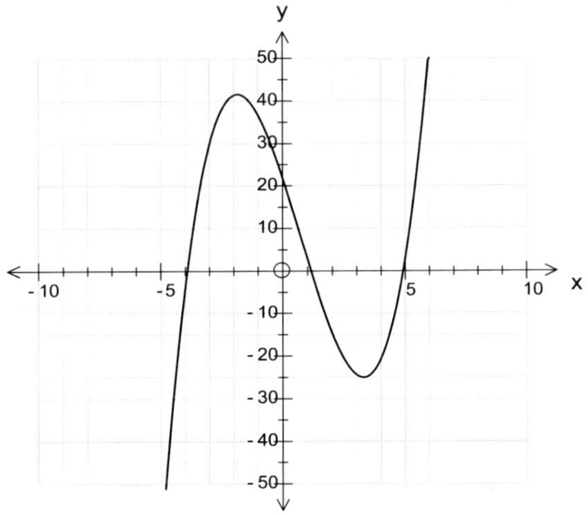

Fig. 11.2 Graph of $x^3 - 2x^2 - 19x + 20$

Exercise 11a

1. Given that $P_{(x)} = x^3 + 4x^2 - 3x + 1$, find:
   (a) $P(1)$ (b) $P(-2)$ (c) $P\left(\frac{1}{2}\right)$ (d) $P(0)$
2. Given $P_{1(x)} = 2x^3 - 3x^2 + 4x - 1$, $P_{2(x)} = 4x^3 + 7x^2 - 3x + 8$ and $P_{3(x)} = x^3 - 5x^2 + 2x + 6$, find:
   (a) $P_{1(x)} + P_{2(x)}$ (b) $P_{2(x)} + P_{3(x)}$ (c) $P_{1(x)} + P_{3(x)}$ (d) $P_{1(x)} + P_{2(x)} + P_{3(x)}$
3. If $P_{1(x)} = 2x^3 + 5x^2 - 3x + 3$, $P_{2(x)} = x^3 - 3x^2 + 2x + 1$ and $P_{3(x)} = 4x^3 + 2x^2 - 3x + 4$. Find (a) $2P_{1(x)}$ (b) $3P_{2(x)}$ (c) $4P_{3(x)}$
   (d) $2P_{1(x)} + 3P_{2(x)}$ (e) $3P_{1(x)} + 4P_{3(x)}$ (f) $2P_{1(x)} - 4P_{3(x)}$
4. Show that $(x - 3)$ is a factor of the polynomial $f_{(x)} = x^3 - x^2 + 4x + 24$

5. When $x^4 + Px^3 + Qx^2 + R$ is divided by $(x - 1), (x + 1)$ and $(x - 2)$, the remainder are 12, 6 and 54 respectively. Find the values of P, Q and R.
6. Draw the graph of the function $y = F_{(x)} = 2x^3 + 3x^2 - 4x - 6$, hence determine the approximate location of the real zeroes.
7. If $(2x - 1)$ is a factor of $8x^4 - ax^2 + 4x - 1$, find the value of a.
8. Given that $P_{1(x)} = 5x^3 + 3x^2 - 2x + 6$, $P_{2(x)} = x^3 + 4x^2 - 3x + 1$, and $P_{3(x)} = 2x^3 - 3x^2 + 3x + 2$. Find the value of the following:
   (a) P_1P_2 (b) P_2P_3 (c) P_1P_3 (d) $P_3(P_1 + P_2)$ (e) $P_2(P_1 + P_3)$ (f) $(P_2 - P_3)\,P_1$
9. Find the remainder of the following functions:
   (a) When $x^4 - 3x^3 - x^2 + 2x + 1$ is divided by $x - 1$
   (b) When $x^3 + 5x^2 - 3x + 1$ is divided by $x + 1$
   (c) When $x^5 - 3x^3 + x - 3$ is divided by $x - 2$
   (d) When $2x^3 - 4x^2 + x - 3$ is divided by $x + 1$
10. If when $g_{(x)} = f_{(x+1)}$ and $f_{(x)} = x^2 + 3x + 2$. Find the remainder when $g_{(x)}$ is divided by $x - 2$.
11. The polynomial $f_{(x)} = 2x^3 - 36kx + 4k^2$ where k is a constant, has $(x - 3)$ as a factor; find the possible values of k.
12. Factorise $x^3 - 2x^2 - 5x + 6$
13. Find the zeroes of the polynomial $P_{(x)} = x^3 + 4x^2 + x - 6$
14. If $f_{(x)} = 6x^3 + 13x^2 + 2x - 5$ (a) Show that $f(-1) = 0$ (b) Find the factors of $f_{(x)}$
15. The remainder when $f_{(x)} = ax^3 + 4b + 12x + 7$ is divided by $(x + 1)$ is 5, and when it is divided by $(x - 1)$ the remainder is -17. Determine the values of a and b.
16. When the expression $x^3 + kx^2 + 2$ is divided by $(x + 2)$ the remainder is 1 less than when it is divided by $(x + 1)$, find k.
17. What is the remainder when $x^4 - 6x^2 + 1$ is divided by $(x + 2)$?
18. If $x + 2y$ is a factor of $x^3 + 6x^2y + 11xy^2 + 6y^3$; find the other factors.
19. If $P_{1(x)} = 7x^3 - 4x^2 + 5x + 8$, $P_{2(x)} = 5x^3 + 6x + 10$ and $P_{3(x)} = 4x^3 + 2x^2 + 2$; find:
    (a) $P_{1(x)} + P_{2(x)}$ (b) $P_{1(x)} + P_{3(x)}$ (c) $P_{1(x)} - P_{2(x)}$
    (d) $P_{3(x)} - P_{1(x)}$ (e) $P_{1(x)} + P_{2(x)} + P_{3(x)}$
20. Given that:
    $P_{1(x)} = 3x^3 + 4x^2 - x + 1$, $P_{2(x)} = x^4 + x^3 + 2x - 5$, $P_{3(x)} = 4x^3 - 2x^2 + 4$, find:
    (a) $2P_{1(x)} + P_{2(x)}$ (b) $3P_{1(x)} + P_{3(x)}$ (c) $P_{3(x)} + 2P_{1(x)} - 3P_{2(x)}$ (d) $5P_{1(x)} - 3P_{2(x)}$
21. Find the quotient and the remainder when $x^3 + 10$ is divided by $x^2 + 2x + 4$
22. A function is defined as $f_{(x)} = 2x^7 + 5x^5 - x^2 + 5$, Find: (a) $f(0)$ (b) $f(2)$ (c) $f(-1)$

11.2 PARTIAL FRACTION

The partial fraction of a function is the part or sub – functions such that when summed together, it will give equivalent form of the single fraction/function.

e.g. $\frac{Ax+B}{(x+C)(x+D)} = \frac{M}{(x+C)} + \frac{N}{(x+D)}$

Where the expression on the RHS is the partial fraction of the "single fraction" on the LHS of the equation. Thus, by convention, partial fraction must be "proper," in the sense that the degree of x in the numerator must be less than the degree of x in the denominator. If otherwise, we use the technique of polynomial division to break it down into its quotient and remainder; then we convert the fraction with a degree of x in the numerator less than the degree of x in the denominator into its constituent partial fractions.

11.2.1 METHODS OF SOLVING PARTIAL FRACTION

There are two common methods used for solving partial fraction, they include:
(i) Cover-up rule;
(ii) Method of comparing coefficients

11.2.2 FORMS OF PARTIAL FRACTION

(a) Non- repeated *linear* factors at the denominator
(b) Non- repeated with *nonlinear* factors at the denominator
(c) Repeated factors at the denominator
(d) Improper rational function

NON-REPEATED *LINEAR* FACTORS AT THE DENOMINATOR

A non – repeated linear factor is of the form $\frac{f_{(x)}}{(a_1x+b_1)(a_2x+b_1)\dots(a_nx+b_n)}$

Where $f_{(x)}$ is a polynomial (function) of degree less than the degree of the product of the factors in the denominator.

$$\Rightarrow \frac{f_{(x)}}{(a_1x+b_1)(a_2x+b_1)\dots(a_nx+b_n)} = \frac{A}{(a_1x+b_1)} + \frac{B}{(a_1x+b_1)} + \dots + \frac{Z}{(a_nx+b_n)}$$

The best and fastest method of finding the partial fraction of this form is by using "cover-up rule."

Example 18

Resolve the following into partial fraction:

(a) $\frac{3x+2}{(x+2)(x-5)}$ (b) $\frac{2x+3}{(2x-1)(x+2)}$ (c) $\frac{x^2+2x-3}{(x+4)(x-7)(x+1)}$

Solution

(a) $\frac{3x+2}{(x+2)(x-5)} = \frac{A}{(x+2)} + \frac{B}{(x-5)}$

Cover-up $x+2$:

$\Rightarrow x = -2$

$\therefore A = \frac{3x+2}{(x-5)} = \frac{3(-2)+2}{(x-5)} = \frac{-6+2}{-2-5}$

$= \frac{-4}{-7} = \frac{4}{7} \quad \therefore A = \frac{4}{7}$

Using cover-up rule, one of the factors in the denominators of the RHS fraction is equated to 0; e.g., $x+2=0$ where $x=-2$, then this value is substituted in $\frac{3x+2}{(x-5)}$ to find the value of A. Similarly, $x-5=0$ is covered-up by putting $x=5$ in $\frac{3x+2}{(x+2)}$ to find B.

Similarly, cover-up $x-5$ (i.e., put $x=5$ into the other factor of the denominator)

$\Rightarrow B = \frac{3x+2}{(x+2)} = \frac{3(5)+2}{(5+2)} = \frac{15+2}{7} = \frac{17}{7} \qquad \therefore B = \frac{17}{7}$

Thus, $\frac{A}{(x+2)} + \frac{B}{(x-5)} = \frac{4/7}{(x+2)} + \frac{17/7}{(x-5)} = \frac{4}{7(x+2)} + \frac{17}{7(x-5)}$

(b) $\frac{2x+3}{(2x-1)(x+2)} = \frac{A}{(2x-1)} + \frac{B}{(x+2)}$

Using Cover-up rule, we have:

$A = \frac{2x+3}{(x+2)_{\text{when } x=\frac{1}{2}}} = \frac{2(\frac{1}{2})+3}{\frac{1}{2}+2} = \frac{1+3}{5/2} = \frac{4}{5/2}$

$\therefore A = \frac{8}{5}$

$B = \frac{2x+3}{(2x-1)_{\text{when } x=-2}} = \frac{2(-2)+3}{2(-2)-1} = \frac{-4+3}{-4-1} = \frac{-1}{-5} = \frac{1}{5}$

$\therefore B = \frac{1}{5}$

Thus, $\frac{A}{(2x-1)} + \frac{B}{(x+2)} = \frac{8/5}{(2x-1)} + \frac{1/5}{(x+2)} = \frac{8}{5(2x-1)} + \frac{1}{5(x+2)}$

(c) $\frac{x^2+2x-3}{(x+4)(x-7)(x+1)} = \frac{A}{(x+4)} + \frac{B}{(x-7)} + \frac{C}{(x+1)}$

Using Cover-up rule:

$A = \frac{x^2+2x-3}{(x-7)(x+1)_{x=-4}} = \frac{(-4)^2+2(-4)-3}{(-4-7)(-4+1)} = \frac{16-8-3}{-11(-3)} = \frac{5}{33}$

$\therefore A = \frac{5}{33}$

$B = \frac{x^2+2x-3}{(x+4)(x+1)_{x=7}} = \frac{(7)^2+2(7)-3}{(7+4)(7+1)} = \frac{49+14-3}{11(8)}$

$$= \frac{60}{88} = \frac{15}{22} \quad \therefore B = \frac{15}{22}$$

$$C = \frac{x^2 + 2x - 3}{(x+4)(x-7)_{x=-1}} = \frac{(-1)^2 + 2(-1) - 3}{(-1+4)(-1-7)} = \frac{1-2-3}{-3(-8)}$$

$$= \frac{-4}{-24} = \frac{1}{6} \quad \therefore C = \frac{1}{6}$$

Thus, $\frac{A}{(x+4)} + \frac{B}{(x-7)} + \frac{C}{(x+1)} = \frac{5/33}{(x+4)} + \frac{15/22}{(x-7)} + \frac{1/6}{(x+1)}$

$$= \frac{5}{22(x+4)} + \frac{15}{22(x-7)} + \frac{1}{6(x+1)}$$

(b) NON- REPEATED WITH *NON-LINEAR* FACTORS AT THE DENOMINATOR

If a non-factorisable polynomial of degree, n, occurs at the denominator of a compound fraction, then assign a polynomial of degree, $n-1$, to the denominator of the corresponding partial fraction.

Example 19

Resolve into partial fraction:

(a) $\frac{2x^2-3x+1}{(x+1)(x^2+5x+1)}$ (b) $\frac{5}{(x-2)(x^2+5)}$

> $x^2 + 5x + 1$ cannot be factorised into linear factor; so we write its partial fraction in the form $\frac{Bx+C}{x^2+5x+1}$, then use the "method of comparing coefficients" to find the values of B and C; then the value of A is determined.

Solution

(a) $\frac{2x^2-3x+1}{(x+1)(x^2+5x+1)} = \frac{A}{x+1} + \frac{Bx+C}{x^2+5x+1}$

Now,

$$\frac{A}{x+1} + \frac{Bx+C}{x^2+5x+1} = \frac{A(x^2+5x+1)+(Bx+C)(x+1)}{(x+1)(x^2+5x+1)}$$

$$= \frac{Ax^2+5Ax+A+Bx^2+Bx+Cx+C}{(x+1)(x^2+5x+1)} = \frac{x^2(A+B)+x(5A+B+C)+A+C}{(x+1)(x^2+5x+1)}$$

$$\therefore \frac{2x^2 - 3x + 1}{(x+1)(x^2+5x+1)} = \frac{x^2(A+B) + x(5A+B+C) + A + C}{(x+1)(x^2+5x+1)}$$

Now, compare the coefficients of the RHS and LHS:

$\Rightarrow A + B = 2$ ....(i)

$5A + B + C = -3$ ....(ii)

$A + C = 1$ ....(iii)

From Equation (iii), Put $A = 1 - C$ into equation (ii):
$\therefore 5(1 - C) + B + C = -3$
$\Rightarrow 5 - 5C + B + C = -3$
$\therefore B - 4C = -8$ ....(iv)
Similarly, put $A = 1 - C$ into equation (i):
$\therefore (1 - C) + B = 2$
$\Rightarrow B - C = 2 - 1 = 1$
$\therefore B - C = 1$ ....(v)
Now, make B subject formula (i.e., $B = 1 + C$) from equation (v) and put into equation(iv).
$\therefore B - 4C = -8 \quad \Rightarrow (1 + C) - 4C = -8$
$-3C = -8 - 1 = -9$
$C = \frac{-9}{-3} = 3$
$\therefore C = 3$
Thus, $B = 1 + C = 1 + 3$
$\therefore B = 4$

Note that you can solve the simultaneous equations, i.e., equations i, ii, and iii, using elimination method, matrix method or by using Crammer's rule. Visit chapter 18 of this book on matrix to learn how to solve simultaneous equations of three variables.

Similarly, from equation (iii), we have:
$A + C = 1 \quad \Rightarrow A + 3 = 1$
$\therefore A = 1 - 3 = -2$

Hence, $\frac{A}{x+2} + \frac{Bx+C}{x^2+5x+1} = \frac{-2}{x+2} + \frac{4x+3}{x^2+5x+1}$

This can be rewritten as, $\frac{4x+3}{x^2+5x+1} - \frac{2}{x+2}$

(b) $\frac{5}{(x-2)(x^2+5)} = \frac{A}{x-2} + \frac{Bx+C}{x^2+5}$

$\therefore \quad \frac{A}{x-2} + \frac{Bx+C}{x^2+5} = \frac{A(x^2+5)+(Bx+C)(x-2)}{(x-2)(x^2+5)}$

$= \frac{Ax^2+5A+Bx^2-2Bx+Cx-2C}{(x-2)(x^2+5)} = \frac{x^2(A+B)+x(C-2B)+5A-2C}{(x-2)(x^2+5)}$

Now, use the method of comparing coefficients to find the unknown constants (i.e., A, B and C):

$$\frac{5}{(x-2)(x^2+5)} = \frac{x^2(A+B)+x(C-2B)+5A-2C}{(x-2)(x^2+5)}$$

$A + B = 0$ ....(i)
$C - 2B = 0$ ....(ii)
$5A - 2C = 5$ ....(iii)

From equation (i), put $A = -B$ into equation (iii):

$5A - 2C = 5 \quad \Rightarrow 5(-B) - 2C = 5$

$\therefore 5B + 2C = -5$ ....(iv)

Now, using equation (ii), put $C = 2B$ into equation (iv):

$5B + 2C = -5 \quad \Rightarrow 5B + 2(2B) = -5$

$\therefore 9B = -5$

So, $B = \frac{-5}{9}$

Thus, $C = 2B = 2\left(\frac{-5}{9}\right)$

$\therefore C = -\frac{10}{9}$

Also, using equation (i) $A = -B$

$\Rightarrow A = -\left(\frac{-5}{9}\right) = \frac{5}{9}$

$$\frac{A}{x-2} + \frac{Bx+C}{x^2+5} = \frac{5/9}{x-2} + \frac{-5/9 x + (-10/9)}{x^2+5x} = \frac{5}{9(x-2)} - \frac{5x+10}{9(x^2+5)}$$

Example 20

Find the partial fraction of the expression,

$$\frac{2x^4-6x^3+4x^2-x}{(x-1)(x^2+1)}$$

Solution

(a) $\frac{2x^4-6x^3+4x^2-x}{(x-1)(x^2+1)} = \frac{2x^4-6x^3+4x^2-x}{x^3-x^2+x-1}$

Since the degree of power of the numerator(i.e. $n = 4$) is greater than the denominator (i.e. $m = 3$), the best approach is to use polynomial division to reduce it. Then, when $m > n$, we can now resolve the given term into partial fraction.

$$\therefore \quad x^3 - x^2 + x - 1 \;\overline{\big)\; 2x^4 - 6x^3 + 4x^2 - x}\quad \text{quotient: } 2x - 4$$

$$\begin{array}{r} 2x^4 - 6x^3 + 4x^2 - x \\ -\;(2x^4 - 2x^3 + 2x^2 - 2x) \\ \hline -4x^3 + 2x^2 + x \\ -\;(-4x^3 + 4x^2 - 4x + 4) \\ \hline -2x^2 + 5x - 4 \end{array}$$

This is the remainder of the polynomial division which can now be resolved into partial fraction.

$$\Rightarrow \frac{2x^4-6x^3+4x^2-x}{(x-1)(x^2+1)} = 2x - 4 + \frac{-2x^2+5x-4}{(x-1)(x^2+1)}$$

$$\frac{-2x^2+5x-4}{(x-1)(x^2+1)} = \frac{A}{x-1} + \frac{Bx+C}{x^2+1} = \frac{A(x^2+1)+(Bx+C)(x-1)}{(x-1)(x^2+1)}$$

$$\frac{-2x^2+5x-4}{(x-1)(x^2+1)} = \frac{x^2(A+B)+x(C-B)+A-C}{(x-1)(x^2+1)}$$

Using the method of comparing coefficients, we have:

$A + B = -2$ ....(i)

$C - B = 5$ ....(ii)

$A - C = -4$ ....(iii)

From (ii) put $C = 5 + B$ into (iii)

$\therefore\ A - C = -4 \quad \Rightarrow A - (5 + B) = -4$

$A - B = -4 + 5 = 1$

$A - B = 1$ ....(iv)

Subtract (iv) from (i):

$$\begin{array}{r} A + B = -2 \\ -\quad A - B = 1 \\ \hline 0 + 2B = -3 \end{array}$$

$\therefore B = -\frac{3}{2}$

Now, put $B = -\frac{3}{2}$ into (iv) to find A

$\Rightarrow A - \left(-\frac{3}{2}\right) = 1 \quad \Rightarrow A + \frac{3}{2} = 1$

$\therefore\ A = 1 - \frac{3}{2} = -\frac{1}{2}$

Similarly, from (ii) we have that $C = 5 + B$

$\Rightarrow C = 5 + \left(-\frac{3}{2}\right) = 5 - \frac{3}{2} = \frac{7}{2}$

$\therefore A = -\frac{1}{2}, B = -\frac{3}{2}$ and $C = \frac{7}{2}$

$$\frac{A}{x-1} + \frac{Bx+C}{x^2+1} = -\frac{1}{2(x-1)} + \frac{-3x+7}{2(x^2+1)} = -\frac{1}{2(x-1)} + \frac{7-3x}{2(x^2+1)}$$

Thus, $\frac{2x^4-6x^3+4x^2-x}{(x-1)(x^2+1)} = 2x - 4 - \frac{1}{2(x-1)} + \frac{7-3x}{2(x^2+1)}$

Example 21

Express $\frac{1}{x^3-1}$ as a partial fraction.

Solution

$$\frac{1}{x^3-1} = \frac{1}{(x-1)(x^2+x+1)} = \frac{A}{x-1} + \frac{Bx+C}{x^2+x+1}$$

Where $\frac{A}{x-1} + \frac{Bx+C}{x^2+x+1} = \frac{A(x^2+x+1)+(Bx+C)(x-1)}{(x-1)(x^2+x+1)}$

$$= \frac{x^2(A+B) + x(A-B+C) + A - C}{(x-1)(x^2+x+1)}$$

$$\Rightarrow \frac{1}{x^3-1} = \frac{x^2(A+B)+x(A-B+C)+A-C}{(x-1)(x^2+x+1)}$$

Now, applying the method of comparing coefficients, we have:

$A + B = 0$ ....(i)

$A - B + C = 0$ ....(ii)

$A - C = 1$ ....(iii)

From (i), $A + B = 0$, $\Rightarrow B = -A$, then put into (ii)

$A - B + C = 0 \Rightarrow A - (-A) + C = 0$

$\therefore A + A + C = 0$

$2A + C = 0$ ....(iv)

Then from (iv), make C subject formula (i.e., $C = -2A$) and put into (iii):

$\therefore A - C = 1 \quad \Rightarrow A - (-2A) = 1$

$A + 2A = 1 \quad \Rightarrow 3A = 1$

$\therefore A = \frac{1}{3}$

Then, $B = -A \quad \Rightarrow B = -\frac{1}{3}$

Also, $C = -2A$

$= -2\left(\frac{1}{3}\right) = -\frac{2}{3}$

$\therefore A = \frac{1}{3}$, $B = -\frac{1}{3}$ and $C = -\frac{2}{3}$

$$\frac{A}{x-1} + \frac{Bx+C}{x^2+x+1} = \frac{\frac{1}{3}}{x-1} + \frac{-\frac{1}{3}x+\left(-\frac{2}{3}\right)}{x^2+x+1} = \frac{1}{3(x-1)} - \frac{x+2}{3(x^2+x+1)}$$

$$\therefore \frac{1}{x^3-1} = \frac{1}{3(x-1)} - \frac{x+2}{3(x^2+x+1)}$$

(c) REPEATED FACTORS AT THE DENOMINATOR

If the denominator has repeated factor(s) of the form $(ax+b)^n$, where a and b are constants, and n is the index power which represents the "number of repetitions," then the partial fraction is depicted as:

$$\frac{f_{(x)}}{(ax+b)^n} = \frac{A}{ax+b} + \frac{B}{(ax+b)^2} + \frac{C}{(ax+b)^3} + \cdots + \frac{Z}{(ax+b)^n}$$

Though, at this level, we are going to restrict our calculation to $n \leq 3$

$$\Rightarrow \frac{f_{(x)}}{(ax+b)^3} = \frac{A}{ax+b} + \frac{B}{(ax+b)^2} + \frac{C}{(ax+b)^3}$$

Example 72

Resolve the following into partial fraction:

(a) $\frac{x+5}{(2x+1)^2}$ (b) $\frac{7x+2}{(2x-3)(x+1)^2}$

Solution

(a) $\frac{x+5}{(2x+1)^2} = \frac{A}{2x+1} + \frac{B}{(2x+1)^2} = \frac{2Ax+A+B}{(2x+1)^2}$

This expression is gotten when we find the LCM of $\frac{A}{2x+1} + \frac{B}{(2x+1)^2}$ and then express as a single fraction.

Using the method of comparing coefficients, we have:

$2A = 1$ ....(i)

$A + B = 5$ ....(ii)

From equation (i), $A = \frac{1}{2}$

Now, putting this value in equation (ii) to find B, we have:

$A + B = 5 \quad \Rightarrow \frac{1}{2} + B = 5$

$\therefore B = 5 - \frac{1}{2} = \frac{9}{2}$

Thus, $\frac{A}{2x+1} + \frac{B}{(2x+1)^2} = \frac{1}{2(2x+1)} + \frac{9}{2(2x+1)^2}$

So, $\frac{x+5}{(2x+1)^2} = \frac{1}{2(2x+1)} + \frac{9}{2(2x+1)^2}$

(b) $\frac{7x+2}{(2x-3)(x+1)^2} = \frac{A}{2x-3} + \frac{B}{x+1} + \frac{C}{(x+1)^2}$

$= \frac{A(x+1)^2+B(2x-3)(x+1)+C(2x-3)}{(2x-3)(x+1)^2} = \frac{(A+B)x^2+(2A-B+2C)x+A-3B-3C}{(2x-3)(x+1)^2}$

Using the method of comparing coefficients:

$\frac{7x+2}{(2x-3)(x+1)^2} = \frac{(A+2B)x^2+(2A-B+2C)x+A-3B-3C}{(2x-3)(x+1)^2}$

$\therefore A + 2B = 0$ ....(i)

$2A - B + 2C = 7$ ....(ii)

$A - 3B - 3C = 2$ ....(iii)

From equation (i) $A = -2B$, put into equation (ii) i.e. $2A - B + 2C = 7$

$2(-2B) - B + 2C = 7 \quad \Rightarrow -4B - B + 2C = 7$

$\therefore -5B + 2C = 7$ ....(iv)

Also, put $A = -2B$ into equation (iii)

$A - 3B - 3C = 2 \quad \Rightarrow -2B - 3B - 3C = 2$

$\therefore 5B + 3C = -2$ ....(v)

Now, use equation (iv) and (v) to find the values of B and C by solving simultaneously:

$$\begin{array}{r} -5B + 2C = 7 \\ + \quad 5B + 3C = -2 \\ \hline 0 + 5C = 5 \end{array}$$

$\therefore C = 1$

Hence, $5B + 3C = -2$

$\Rightarrow 5B + 3(1) = -2$

$5B = -2 - 3 = -5$

$\therefore B = -1$

But, $A = -2B \quad \Rightarrow A = -2(-1) = 2$

$\therefore A = 2\,, B = -1\,, C = 1$

$$\frac{A}{2x-3} + \frac{B}{x+1} + \frac{C}{(x+1)^2} = \frac{2}{2x-3} + \frac{-1}{x+1} + \frac{1}{(x+1)^2} = \frac{2}{2x-3} - \frac{1}{x+1} + \frac{1}{(x+1)^2}$$

$$\therefore \quad \frac{7x+2}{(2x-3)(x+1)^2} = \frac{2}{2x-3} - \frac{1}{x+1} + \frac{1}{(x+1)^2}$$

Example 23

Resolve into partial fraction

$$\frac{2x^3+2x^2+2}{(x+1)^2(x^2+1)}$$

Solution

$$\frac{2x^3+2x^2+2}{(x+1)^2(x^2+1)} = \frac{A}{x+1} + \frac{B}{(x+1)^2} + \frac{Cx+D}{x^2+1}$$

$$= \frac{A(x+1)(x^2+1)+B(x^2+1)+(Cx+D)(x^2+1)^2}{(x+1)^2(x^2+1)}$$

$$= \frac{Ax^3+Ax^2+A+Bx^2+B+Cx^3+2Cx^2+Cx+Dx^2+2Dx+1}{(x+1)^2(x^2+1)}$$

$$= \frac{(A+C)x^3+(A+B+2C+D)x^2+(A+C+2D)x+A+B+D}{(x+1)^2(x^2+1)}$$

Using the method of comparing coefficients:

$A + C = 2$ ....(i)

$A + B + 2C + D = 2$ ....(ii)

$A + C + 2D = 0$ ....(iii)

$A + B + D = 2$ ....(iv)

From equation (i), $C = 2 - A$, put this value into equation (ii)

$A + B + 2C + D = 2 \Rightarrow A + B + 2(2 - A) + D = 2$

$\therefore B + D - A = 2$ ....(v)

Also, put $C = 2 - A$, into equation (iii)

$A + C + 2D = 0 \quad \Rightarrow A + 2 - A + 2D = 0$

$\therefore 2D = -2$

$\therefore D = \frac{-2}{2} = -1$

Now, from equation (iv) we have:

$A + B + D = 2 \quad \Rightarrow A + B + (-1) = 2$

$A + B = 2 + 1 = 3$

$\therefore A + B = 3$..(vi)

Put $A + B = 3$, and $D = -1$ into equation (ii)

$A + B + 2C + D = 2 \quad \Rightarrow (3) + 2C + (-1) = 2$

$\therefore 2C = 2 - 3 + 1 = 0$

$\therefore C = \frac{0}{2} = 0$

Using equation (i), $A + C = 2$

$A + 0 = 2 \quad \therefore A = 2$

Also, from equation (vi), $A + B = 3$

$\therefore 2 + B = 3 \quad \Rightarrow B = 3 - 2 = 1$

Thus, $A = 2,\ B = 1,\ C = 0, D = -1$

$$\frac{A}{x+1} + \frac{B}{(x+1)^2} + \frac{Cx+D}{x^2+1} = \frac{2}{x+1} + \frac{1}{(x+1)^2} + \frac{0x+(-1)}{x^2+1}$$

$$= \frac{2}{x+1} + \frac{1}{(x+1)^2} + \frac{-1}{x^2+1} = \frac{2}{x+1} + \frac{1}{(x+1)^2} - \frac{1}{x^2+1}$$

IMPROPER RATIONAL FUNCTION

This is called "improper rational function" because the highest number of degrees of the numerator is always equal or greater than that of the denominator. Thus, when the highest degree of the numerator is equal or greater than that of the denominator, then use polynomial division method to divide the numerator by the denominator before proceeding to resolving into partial fraction.

Example 24

Find the partial fraction of the following expressions

(a) $\frac{2x^2+3x+10}{(x+1)(x+2)}$ (b) $\frac{x^4+8}{x(x^3+3)}$

Solution

$$\frac{2x^2+3x+10}{(x+1)(x+2)} = \frac{2x^2+3x+10}{x^2+3x+2}$$

Observe here, that the degree of the numerator and that of the denominator are equal. So, we have to use polynomial division to reduce it, so that the highest degree of the numerator will be less than that of the denominator before proceeding to resolving it into partial fraction.

$$\therefore \quad x^2+3x+2 \enclose{longdiv}{2x^2+3x+10} \quad \text{quotient } 2$$

$$\begin{array}{r} 2 \\ 2x^2+3x+10 \\ -\underline{\left(2x^2+6x+4\right)} \\ -3x+6 \end{array}$$

$$\frac{2x^2+3x+10}{(x+1)(x+2)} = 2 + \frac{-3x+6}{x^2+3x+2}$$

$$\therefore \frac{-3x+6}{x^2+3x+2} = \frac{-3x+6}{(x+1)(x+2)} = \frac{A}{(x+1)} + \frac{B}{(x+2)}$$

Using cover-up rule, we have that:

$$A = \frac{-3x+6}{(x+2)_{x=-1}} = \frac{-3(-1)+6}{-1+2} = \frac{3+6}{1}$$

$$\therefore A = 9$$

$$B = \frac{-3x+6}{(x+1)_{x=-2}} = \frac{-3(-2)+6}{-2+1} = \frac{6+6}{-1}$$

$$B = -12$$

$$\frac{A}{(x+1)} + \frac{B}{(x+2)} = \frac{9}{(x+1)} + \frac{-12}{(x+2)} = \frac{9}{(x+1)} - \frac{12}{(x+2)}$$

$$\frac{-3x+6}{x^2+3x+2} = \frac{9}{(x+1)} - \frac{12}{(x+2)}$$

$$\therefore \frac{2x^2+3x+10}{(x+1)(x+2)} = 2 + \frac{9}{(x+1)} - \frac{12}{(x+2)}$$

(b) $\frac{x^4+8}{x(x^3+3)} = \frac{x^4+8}{x^4+3x}$

$$\therefore \quad x^4+3x \enclose{longdiv}{x^4+8} \quad \text{quotient } 1$$

$$\begin{array}{r} 1 \\ x^4+8 \\ -\underline{\left(x^4+3x\right)} \\ -3x+8 \end{array}$$

$$\frac{x^4+8}{x(x^3+3)} = 1 + \frac{-3x+8}{x(x^3+3)}$$

Expressing $\frac{-3x+8}{x^4+3x}$ into partial fraction, we have:

$$\frac{-3x+8}{x(x^3+3)} = \frac{A}{x} + \frac{Bx^2+Cx+D}{x^3+3}$$

$$\frac{A}{x} + \frac{Bx^2+Cx+D}{x^3+3} = \frac{A(x^3+3)+x(Bx^2+Cx+D)}{x^3+3} = \frac{x^3(A+B)+x^2C+Dx+3A}{x(x^3+3)}$$

Using the method of comparing coefficients, we have:

$$\frac{-3x+8}{x(x^3+3)} = \frac{x^3(A+B)+x^2C+Dx+3A}{x(x^3+3)}$$

$\therefore A+B=0$ ....(i)

$C=0$ ....(ii)

$D=-3$ ....(iii)

$3A=8$ ....(iv)

From equation (iv), $A=\frac{8}{3}$

Put this value of A into equation (i) to find B :

$A+B=0 \quad \Rightarrow B=-A=-\frac{8}{3}$

$\therefore\ A=\frac{8}{3},\ B=-\frac{8}{3},\ C=0,\ D=-3$

$$\frac{x^4+8}{x(x^3+3)} = 1+\frac{A}{x}+\frac{Bx^2+Cx+D}{x^3+3} = 1+\frac{8}{3x}+\frac{-\frac{8}{3}x^2+0x-3}{x^3+3}$$

$$= 1+\frac{8}{3x}-\frac{8x^2+9}{3(x^3+3)}$$

Example 25

Resolve into partial fraction $\frac{4x^3-11x^2+8x+1}{x^3-1}$

Solution

(a) $\frac{4x^3-11x^2+8x+1}{x^3-1}$

Use polynomial division to reduce this single fraction.

$$\begin{array}{r|l} & \quad 4 \\ x^3-1 & 4x^3-11x^2+8x+1 \\ & -\underline{4x^3+0x^2+0x-4} \\ & \quad -11x^2+8x+5 \end{array}$$

$$\therefore \frac{(4x+1)(x-2)(x-3)}{x^3-1} = 4+\frac{-11x^2+8x+5}{x^3-1}$$

$$\Rightarrow \frac{-11x^2+8x+5}{x^3-1} = \frac{-11x^2+8x+5}{(x-1)(x^2+x+1)} = \frac{A}{x-1}+\frac{Bx+C}{x^2+x+1}$$

$$\frac{A}{x-1}+\frac{B}{x^2+x+1} = \frac{A(x^2+x+1)+(Bx+C)(x-1)}{(x-1)(x^2+x+1)} = \frac{(A+B)x^2+(A-B+C)x+A-C}{(x-1)(x^2+x+1)}$$

Thus, using the method of comparing coefficients, we have:

$A+B=-11$ ....(i)

$A-B+C=8$ ....(ii)

$A-C=5$ ....(iii)

From equation (iii), make A subject formula (i.e., $A = 5 + C$), then put into equation (ii):

$(5 + C) - B + C = -8 \quad \Rightarrow 2C - B = -8 - 5 = -13$

$2C - B = -13$ ....(iv)

Similarly, put $A = 5 + C$ into equation (i):

$(5 + C) + B = -11 \quad \Rightarrow C + B = -11 - 5 = -16$

$\therefore\ C + B = -16$ ....(v)

Now, combining equation (iv) and (v)

$$\begin{array}{r} 2C - B = -13 \\ + \quad C + B = -16 \\ \hline 3C = -29 \end{array}$$

$\therefore C = -\dfrac{29}{3}$

Add equations (iv) and (v) to eliminate B, then find the solution to C. Finally, put the value of C into equation (iv) or (v) to find B.

Put C into equation (v)

$\Rightarrow -\dfrac{29}{3} + B = -16$

$\therefore B = \dfrac{29}{3} - 16 = -\dfrac{19}{3}$

Now, from equation (iii), $A - C = 5$

$\Rightarrow A = 3 + \left(-\dfrac{29}{3}\right) = -\dfrac{20}{3}$

$\therefore A = -\dfrac{20}{3}, B = -\dfrac{19}{3}, C = -\dfrac{29}{3}$

$$\frac{A}{x-1} + \frac{Bx+C}{x^2+x+1} = \frac{-20/3}{(x-1)} + \frac{-19/3x + (-29/3)}{x^2+x+1}$$

$$= -\frac{20}{3(x-1)} - \frac{19x+29}{3(x^2+x+1)}$$

$$\therefore \frac{4x^3-11x^2+8x+1}{x^3-1} = 4 - \frac{20}{3(x-1)} - \frac{19x+29}{3(x^2+x+1)}$$

Example 26

Find the partial fraction of the given expression:

$$\frac{x^4+1}{(x+2)^2}$$

Solution

$$\frac{x^4+1}{(x+2)^2} = \frac{x^4+1}{x^2+4x+4}$$

The degree of the numerator is greater than that of the denominator. So, we must first reduce it using "polynomial division" before resolving it to its partial fraction.

$$\begin{array}{r|l} & x^2 - 4x + 12 \\ \hline x^2 + 4x + 4 & x^4 + 0x^3 + 0x^2 + 0x + 1 \\ & -\;x^4 + 4x^3 + 4x^2 + 0x + 0 \\ \hline & -4x^3 - 4x^2 + 0x + 1 \\ & -\;-4x^3 - 16x^2 - 16x \\ \hline & 12x^2 + 16x + 1 \\ & -\;12x^2 + 48x + 48 \\ \hline & -32x - 47 \end{array}$$

For convenience, $x^4 + 1$ is replaced with $x^4 + 0x^3 + 0x^2 + 0x + 1$ this is because they are still same and equal.

$$\frac{x^4+1}{(x+2)^2} = x^2 + 4x + 12 + \frac{-32x-47}{(x+2)^2}$$

$$\frac{-32x-47}{(x+2)^2} = \frac{A}{x+2} + \frac{B}{(x+2)^2} = \frac{Ax+2A+B}{(x+2)^2}$$

Now, using the method of comparing coefficient, we have:

$A = -32$ ....(i)

$2A + B = -47$..............................(ii)

Now, put the value of $A = -32$, into (ii):

$2(-32) + B = -47$

$B = -47 + 64 = 17$

$\therefore A = -32, \; B = 17$

$$\frac{A}{x+2} + \frac{B}{(x+2)^2} = \frac{-32}{x+2} + \frac{17}{(x+2)^2}$$

$$\therefore \frac{x^4+1}{(x+2)^2} = x^2 + 4x + 12 - \frac{32}{x+2} + \frac{17}{(x+2)^2}$$

Note that in the method of comparing coefficients, we consider the RHS and LHS equations, i.e., $\frac{-32x-47}{(x+2)^2} = \frac{Ax+2A+B}{(x+2)^2}$, then compare the corresponding terms in the numerator.

Example 27

Resolve the following into partial fraction: (a) $\frac{2x^2-x+1}{(x^2+2)^2}$ (b) $\frac{10+2x^2}{(x+1)^3}$

Solution

(a) $$\frac{2x^2-x+1}{(x^2+2)^2} = \frac{Ax+B}{x^2+2} + \frac{Cx+D}{(x^2+2)^2} = \frac{(Ax+B)(x^2+2)+(Cx+D)}{(x^2+2)^2}$$

$$\frac{2x^2-x+1}{(x^2+2)^2} = \frac{Ax^3+2Ax+Bx^2+2B+Cx+D}{(x^2+2)^2} = \frac{Ax^3+Bx^2+(2A+C)x+2B+D}{(x^2+2)^2}$$

Using the method of comparing coefficients, we have:

$\therefore A = 0, \; B = 2, \; 2A + C = -1, \; 2B + D = 1 \quad \Rightarrow C = -1$ and $D = -3$;

$$Thus, \; \frac{Ax+B}{x^2+2} + \frac{Cx+D}{(x^2+2)^2} = \frac{0x+2}{x^2+2} + \frac{-x-3}{(x^2+2)^2} = \frac{0x+2}{x^2+2} - \frac{x+3}{(x^2+2)^2} = \frac{2}{x^2+2} - \frac{x+3}{(x^2+2)^2}$$

(b) $\frac{10+2x^2}{(x+1)^3} = \frac{A}{x+1} + \frac{B}{(x+1)^2} + \frac{C}{(x+1)^3} = \frac{A(x+1)^2+B(x+1)+C}{(x+1)^3}$

$= \frac{Ax^2+(2A+B)x+A+B+C}{(x+1)^3}$

Comparing coefficients, we have that:

$\frac{10+2x^2}{(x+1)^3} = \frac{Ax^2+(2A+B)x+A+B+C}{(x+1)^3}$

$\Rightarrow \; A = 2$ ....(i)

$2A + B = 0$ ....(ii)

$A + B + C = 10$ ....(iii)

Put $A = 2$ into (ii)

$2A + B = 0 \quad \Rightarrow 2(2) + B = 0$

$\therefore B = -4$

Now, put $A = 2$, and $B = -4$ into (iii)

$A + B + C = 10 \quad \Rightarrow 2 + (-4) + C = 10$

$\therefore C = 10 - 2 + 4 = 12$

$\therefore A = 2, B = -4$ and $C = 12$

$\Rightarrow \frac{A}{x+1} + \frac{B}{(x+1)^2} + \frac{C}{(x+1)^3} = \frac{2}{x+1} + \frac{-4}{(x+1)^2} + \frac{12}{(x+1)^3}$

$\therefore \frac{10+2x^2}{(x+1)^3} = \frac{2}{x+1} - \frac{4}{(x+1)^2} + \frac{12}{(x+1)^3}$

OBJECTIVE TEST 11

1. Express in partial fraction $\frac{11x+2}{6x^2-x-1}$

   (A) $\frac{1}{3x-1}+\frac{3}{2x+1}$ (B) $\frac{3}{3x-1}+\frac{1}{2x+1}$ (C) $\frac{1}{3x+1}+\frac{3}{2x-1}$ (D) $\frac{1}{2x+3}+\frac{3}{2x-1}$

2. Express $\frac{1}{x+1}-\frac{1}{x-2}$ as a single algebraic fraction

   (A) $\frac{1}{(x+1)(x-2)}$ (B) $\frac{3}{(x+1)(2-x)}$ (C) $\frac{-3}{(x+1)(x+2)}$ (D) $\frac{1}{(x+1)(x-2)}$

3. If $f(x-4)=x^2+2x+3$, find $f(2)$.

   (A) 6 (B) 11 (C) 27 (D) 51

4. Divide x^3-2x^2-5x+6 by $x-1$

   (A) x^2-x-6 (B) x^2-3x+6 (C) x^2-7x+6 (D) x^2-5x-6

5. If $(y-1),(y+1)$ and $(y-2)$ are factors of the polynomial ay^3+by^2+cy-1. Find a, b and c respectively.

   (A) $\frac{1}{2}$, 1 and $-\frac{1}{2}$ (B) $\frac{1}{2}$, -1 and $\frac{1}{2}$ (C) $\frac{1}{2}$, 1 and $\frac{1}{2}$ (D) $-\frac{1}{2}$, 1 and $\frac{1}{2}$

6. Resolve into partial fraction $\frac{5x-12}{(x-2)(x-3)}$

   (A) $\frac{2}{x+2}-\frac{3}{x-3}$ (B) $\frac{2}{x-2}+\frac{3}{x-3}$ (C) $\frac{2}{x-3}-\frac{3}{x-2}$ (D) $\frac{5}{x-3}+\frac{4}{x-2}$

7. Find the value of P and Q if $\frac{5+2x}{(x+1)(x-2)}$ expressed in partial fraction is $\frac{P}{x-2}+\frac{Q}{x+1}$

   (A) $P=3, Q=-1$ (B) $P=1, Q=3$ (C) $P=-1, Q=3$ (D) $P=-3, Q=1$

8. If $f(x)=2x^2-5x+3$, find $f(x+1)$.

   (A) $2x^2-x$ (B) $2x^2-x+10$ (C) $4x^2+3x+2$ (D) $x^2+3x+12$

9. When the expression Px^2+Qx+1 is divided by $(x+1)$, the remainder is 2. And when divided by $(x-1)$, the remainder becomes 4. Find the values of P and Q respectively. (A) 2, 1 (B) -1, 2 (C) 3, -2 (D) -2, 3

10. Express $\frac{1}{1-x^3}$ in partial fraction

    (A) $\frac{1}{3(x-1)}-\frac{x+2}{3(x^2+x+1)}$ (B) $\frac{1}{3(1-x)}+\frac{x+2}{3(x^2+x+1)}$

    (C) $\frac{1}{3(1-x)}-\frac{x+2}{3(x^2+x+1)}$ (D) $\frac{1}{3(x-1)}+\frac{x+2}{3(x^2+x+1)}$

Exercise 11b

1. State the form of partial fraction of the following, then resolve them into partial fraction:

(a) $\frac{1}{x^2-1}$ (b) $\frac{1}{(x+1)(x-3)}$ (c) $\frac{3x+10}{2x(x^3-1)}$

(d) $\frac{x+1}{(x-1)(x^2+x+1)}$ (e) $\frac{x^3+1}{(x+1)(x-2)}$

2. Resolve the following into their respective partial fraction:

(a) $\frac{2x-1}{(x+1)(x+2)}$ (b) $\frac{5x+6}{x^2-4}$ (c) $\frac{3}{x(x^2+1)}$ (d) $\frac{x^3-1}{x^2+1}$

3. The following expressions have non repeated linear factor at the denominator. Resolve them to their respective partial fractions:

(a) $\frac{1}{(x-2)(x+3)}$ (b) $\frac{3x^2-2x+1}{(x+2)(x-1)(x+3)(x-6)}$

(c) $\frac{7x-2}{(x+5)(x-2)}$ (d) $\frac{10x^2-6x+1}{(x-5)(x+1)(x-3)}$

4. The following are non-repeated with nonlinear factor in the denominator. Express each of them in partial fraction.

(a) $\frac{1}{x(x^2+1)}$ (b) $\frac{3x+1}{(x-1)(2x^2-5x+1)}$ (c) $\frac{x^2-x+1}{x^3-x^2-2x+2}$ (d) $\frac{x^2+3x-2}{x(x^3+1)}$

5. Find the single equivalent fraction of each of the following partial fractions:

(a) $\frac{1}{x-1}+\frac{1}{x-3}-\frac{1}{2x-1}$ (b) $\frac{3}{x-1}+\frac{2}{x^2+2x+5}$ (c) $\frac{2}{2x-3}-\frac{1}{x+1}-\frac{1}{(x+1)^2}$

6. Find the partial fraction of the each of the following:

(a) $\frac{2x^3+2x^2+2}{(x+1)^2(x^2+1)}$ (b) $\frac{x^2+5}{(x+2)^2(x^2+1)}$ (c) $\frac{8x+15}{(x+4)^2(x-3)}$ (d) $\frac{1-2x^2}{(x+1)^3}$

7. Resolve into partial fraction:

(a) $\frac{1}{(x+3)(x-1)(x-5)}$ (b) $\frac{2x+3}{x^2(x-1)}$ (c) $\frac{2x^4+5x^3-3x^2+8}{x^3+x^2-5x-2}$

12 MAPPING AND FUNCTIONS

12.1 DEFINITION

Mapping is a mathematical relation which associates each element of a given set (called the domain of the function) to another element of another set (called the co-domain of the function).

A mapping is denoted by letters such as $f, g, h, \theta, \beta, \alpha,$ and so on. Thus, if f is a mapping from a set P to set Q, it is written as $f: P \rightarrow Q$; where the arrow, $\rightarrow$, indicates that the mapping is from the domain, P, to the co-domain, Q.

12.2 PROPERTIES AND NOTATIONS OF MAPPING

The properties of a mapping are the attributes shared by all the members of a set that describe the mapping. They include:

12.2.1 DOMAIN

The "domain" of a mapping is the set, say, P which is mapped to another set, say, Q (see fig. 12.1);

where P is the domain of the function.

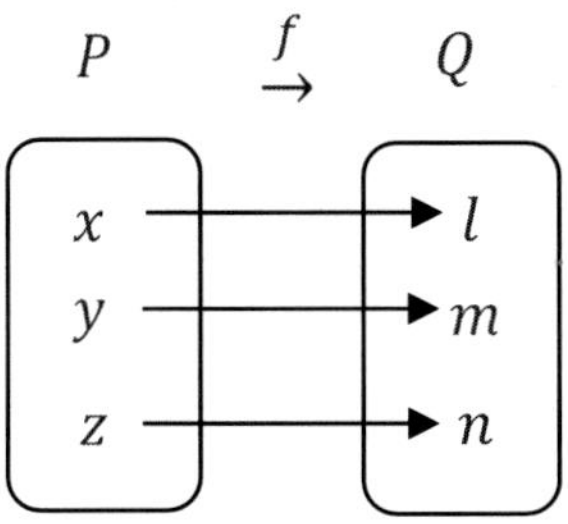

Fig. 12.1

12.2.2 CO – DOMAIN

The "co-domain" of a mapping is the set, say, Q which contains the images of the domain of the mapping (see fig. 12.1); where Q is the co-domain

12.2.3 IMAGE

This is the set of values of the dependent variable (i.e., elements of the co-domain) for which a function is defined. So, for a function to be called a mapping, every element of the domain must have an image in the co-domain. Thus, from fig. 12.1, $\{l, m, n\}$ are the images of the co-domain.

Therefore, the image of an element, $x \in P$, under a mapping $f: P \rightarrow Q$, is denoted as $f(x)$; where $f(x) \in Q$.

12.2.4 RANGE

This is the set of the images of a mapping. Hence, it is a subset of the co-domain. Thus, the range of a mapping, $f: P \rightarrow Q$, is denoted as $f(P)$; where $f(P) \subset Q$.

12.3 REPRESENTATION OF MAPPING

Mapping can be represented in four basic forms, namely:

(i) Arrow Diagram
(ii) Parallel Axes
(iii) Cartesian Graph
(iv) Set of Ordered Pair

12.3.1 ARROW DIAGRAM

The arrow diagram uses an arrow, →, to indicate a mapping/function. The following steps are rules to follow when using arrow diagram:

(i) The arrow starts from members of the domain, then ends with members of the co-domain;
(ii) Every element of the domain has an arrow starting from it;
(iii) Just one arrow starts from each element of the domain, and the end of the arrow terminates at the co-domain;
(iv) It is possible that more than one arrow ends with some members of the co-domain, for instance, in fig. 12.2, the elements b and c have the same image, q.

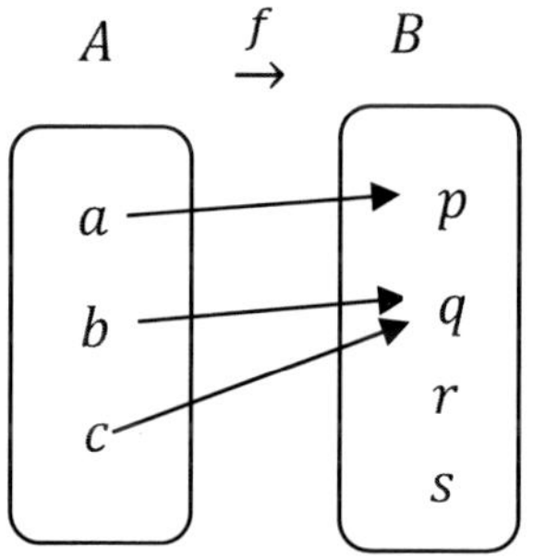

Fig. 12.2 Arrow Diagram

Thus, the mapping in fig. 12.2 is defined by $f: A \to B$, where $A = \{a, b, c\}$ is the domain and $B = \{p, q, r, s\}$ is the co-domain. The images are $f(a) = p$; $f(b) = q$; $f(c) = q$. Hence, the range $f(A) = \{p, q\}$ which is a proper subset of B.

12.3.2 PARALLEL AXES

This method uses real number lines that are parallel to each other. One of the lines represents the domain of the mapping, and the other line represents the range of the mapping.

Example 1

Using parallel axes representation, find the mapping of the function, $f: A \to 2x + 1$, if the domain, $A = \{5, 3, 1, -1, -3, -5\}$.

Solution

The images of the mapping:

$f(5) = 2(x) + 1 = 2(5) + 1 = 11$; $f(3) = 2(x) + 1 = 2(3) + 1 = 7$;
$f(1) = 2(x) + 1 = 2(1) + 1 = 3$; $f(-1) = 2(x) + 1 = 2(-1) + 1 = -1$;

$f(-3) = 2(x) + 1 = 2(-3) + 1 = -5;\ \ f(-5) = 2(x) + 1 = 2(-5) + 1 = -9.$
$\therefore$ *the range*, $f(A) = \{11, 7, 3, -1, -5, -9\}$

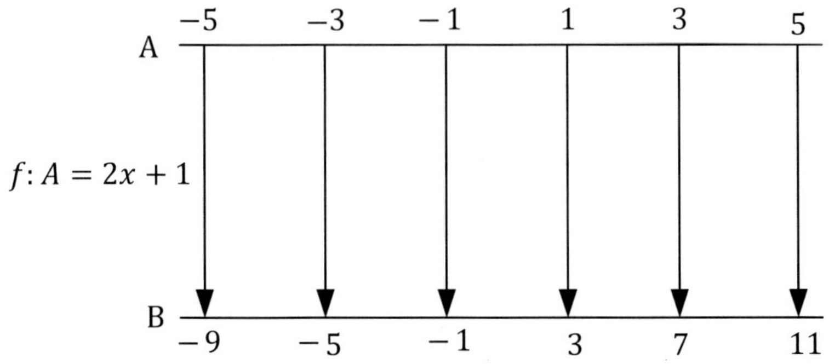

Fig. 12.3 Parallel Axes

12.3.3 CARTESIAN GRAPH

This uses the Cartesian plane (X-Y), with the axes perpendicular to each other. The vertical axis is Y, which represents the co-domain; the horizontal axis is X, which represents the domain of the mapping.

Example 2

Draw the graph of the mapping $f: x \rightarrow x^2 - 5$ using an interval of $-3 \leq x \leq 3$.

Solution

The interval $-3 \leq x \leq 3$.

Table 12.1 Cartesian Table

Domain (x)	−3	−2	−1	0	1	2	3
x^2	9	4	1	0	1	4	9
−5	−5	−5	−5	−5	−5	−5	−5
Co-domain (y)	4	−1	−4	−5	−4	−1	4

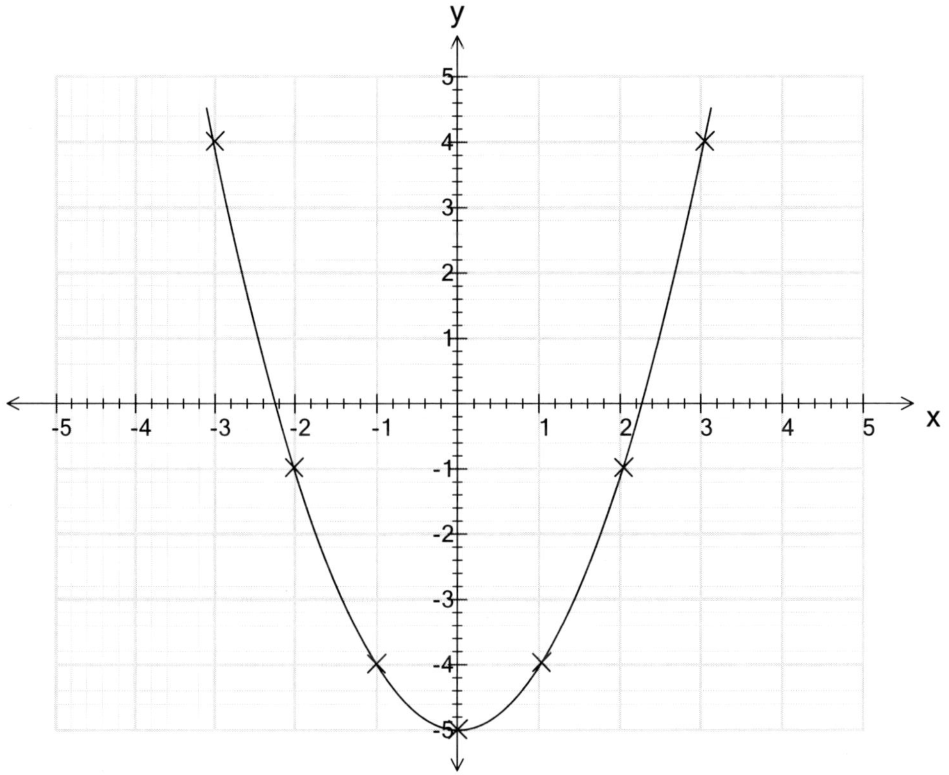

Fig. 12.4 Cartesian Graph of $f: x \to x^2 - 5$

12.3.4 SET OF ORDERED PAIR

This is the Cartesian Product Set of two non-empty sets X and Y.
$X * Y = \{(x, y): x \varepsilon X,\ y \varepsilon Y\}$

Example 3

Write each mapping as a set of ordered pair:
(a) $f: \{-2, -1, 0, 1, 2\};\ f(x) \to x^2$
(b) $g: \{-\frac{1}{3}, 0, 1\};\ g(x) \to 3x^2 - x + 1$

Solution

(a) $f: \{-2, -1, 0, 1, 2\};\ f(x) \to x^2$
The domain is $\{-2, -1, 0, 1, 2\}$
$\Rightarrow$ the images will be found using $f(x) \to x^2$
$\Rightarrow f(-2) = (-2)^2 = 4;\quad f(-1) = (-1)^2 = 1;\quad f(0) = (0)^2 = 0$
$f(1) = (1)^2 = 1;\quad f(2) = (2)^2 = 4$
$\therefore$ Set of ordered pair $X * Y = \{(-2, 4), (-1, 1), (0, 0), (1, 1), (2, 4)\}$

(b) $g:\{-\frac{1}{3},0,1\};\ g(x) \to 3x^2 - x + 1$

The domain is $\{-\frac{1}{3},0,1\}$

$\Rightarrow$ the images will be found using $f(x) \to 3x^2 - x + 1$

$\Rightarrow f\left(-\frac{1}{3}\right) = 3\left(-\frac{1}{3}\right)^2 - \left(-\frac{1}{3}\right) + 1 = \frac{5}{3}$; $f(0) = 3(0)^2 - (0) + 1 = 1$;

$f(1) = 3(1)^2 - (1) + 1 = 3$

$\therefore$ Set of ordered pair $X * Y = \left\{\left(-\frac{1}{3}, \frac{5}{3}\right), (0,1), (1,3)\right\}$

Example 4

If $x = \{p,q,r,s\}$ and $y = \{5,3,1,-1\}$, write down the product set of the following:

(i) $X * X$ (ii) $X *Y$ (iii) $Y * Y$ (iv) $Y * X$

Solution

(i) $X * X = \{(p,p),(q,q),(r,r),(s,s)\}$

(ii) $X * Y = \{(p,5),(q,3),(r,1),(s,1)\}$

(iii) $Y * Y = \{(5,5),(3,3),(1,1),(-1,-1)\}$

(iv) $Y * X = \{(5,p),(3,q),(1,r),(-1,s)\}$

Example 5

Find the range of the function $f: R \to R$ defined by the rule $f: x \to 3x^2$.

Solution

Taking points with coordinates (x,y) which satisfies the equation, $y = 3x^2$

Then, $f(-2) = 3(-2)^2 = 12$; $f(-1) = 3(-1)^2 = 3$; $f(0) = 3(0)^2 = 0$;

$f(1) = 3(1)^2 = 3$; $f(2) = 3(2)^2 = 12$

$f: X * Y = \{(-2,12),(-1,3),(0,0),(1,3),(2,12)\}$

$\therefore$ Range $= \{y \in R: y \geq 0\}$

12.4 TYPES OF MAPPING

There are three types of mapping, namely:

(i) One - to - One Mapping (Injective Mapping)
(ii) Onto Mapping (Surjective Mapping)
(iii) Bijective Mapping

12.4.1 ONE – TO – ONE MAPPING (INJECTIVE MAPPING)

It is a mapping in which each element in the domain has one and only one image in the co-domain. That is to say, each element in the domain is mapped to just a unique single element (image) in the co-domain.

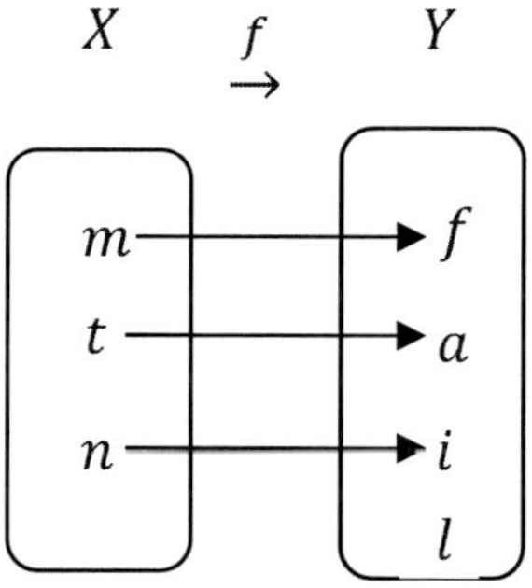

Fig. 12.5 One-to-One Mapping

Example 6

Represent the mapping $f(x) \to x^3$ using arrow graph, and write out:
(a) the elements of the domain (b) the images (c) elements of the co-domain
(d) range of the mapping.

Solution

(a) Let the domain be A, then the elements of $A = \{-2, -1, 0, 1, 2\}$

(b) Images: $f(x) \to x^3$

$\Rightarrow f(-2) = (-2)^3 = -8$

$\Rightarrow f(-1) = (-1)^3 = -1$

$\Rightarrow f(0) = (0)^3 = 0$

$\Rightarrow f(1) = (1)^3 = 1$

$\Rightarrow f(2) = (2)^3 = 8$

(c) Let the co-domain be B, then the elements/images of the co-domain $B = \{-8, -1, 0, 1, 8\}$

(d) The range, $f(A) = \{-8, -1, 0, 1, 8\}$

Thus, the mapping is one-to-one mapping because each element in the domain is mapped to just one element in the co-domain.

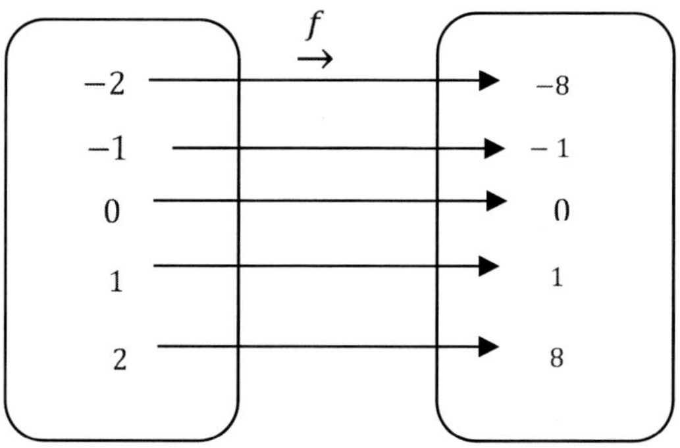

Fig. 12.6 $f(x) \to x^3$

12.4.2 ONTO – MAPPING (SURJECTIVE MAPPING)

It is a mapping in which every element in the co-domain is the image of at least one element in the domain, and the range is equal to its co-domain.
Thus, for a mapping to be called an onto-mapping, one of these conditions must hold:

(i) The range must be equal to the co-domain; this means that every element in the co-domain must be an image.

(ii) At least one element of the domain will be mapped to an image in the co-domain; this means that two or more element in the domain can have the same element in the co-domain.

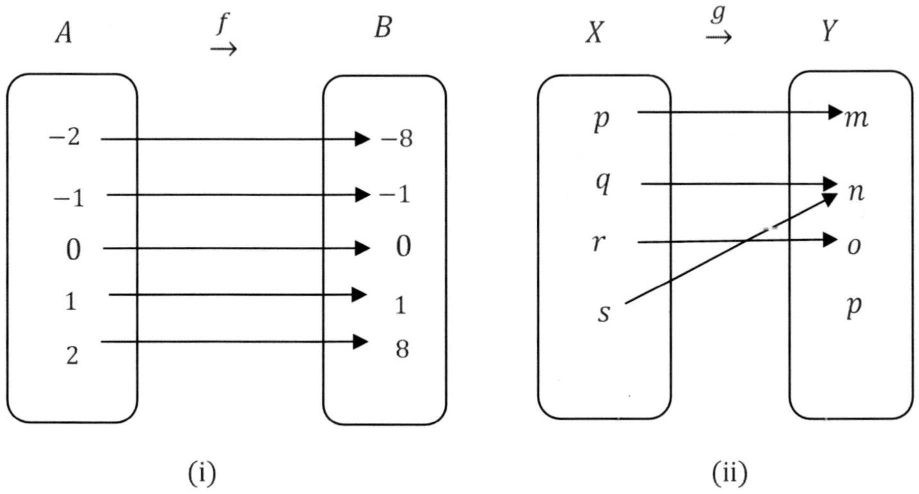

Fig. 12.7 Onto-Mapping

In fig. 12.7(i), the function is onto-mapping because no element in the co-domain is left out; and the range is equal to the co-domain.
While in fig. 12.7(ii), the function is onto-mapping because q and s have the same image, n, in the co-domain.

12.4.3 BIJECTIVE MAPPING

This is when a mapping (function) is both one-to-one and onto. Therefore, from fig. 12.7(i) the mapping $A \xrightarrow{f} B$ is bijective.

12.5 IDENTITY MAPPING

This is a mapping which takes an element onto itself; the domain, co-domain and range are equal. E.g., if $f_I: A \to B$ is a mapping defined by $f_I: \{x, y, z\} = \{x, y, z\}$, then elements of the domain and co-domain are equal, then the mapping is an identity mapping.

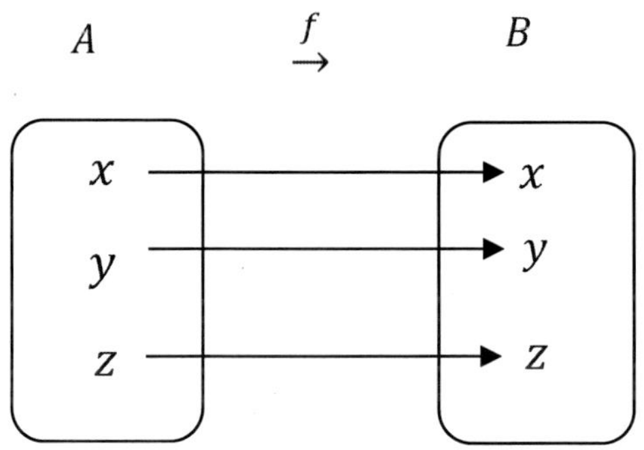

Fig. 12.8 Identity Mapping

Identity mapping is both one-to-one-mapping and onto-mapping.

12.6 CONSTANT MAPPING

It is a mapping that assigns every element in the domain with the same image. For instance, if $f_I: A \to B$ is a mapping defined by $f_I : \{x, y, z\} = \{k\}$, then all the elements of the domain is mapped to the image, k, of the co-domain, then the mapping is constant mapping.

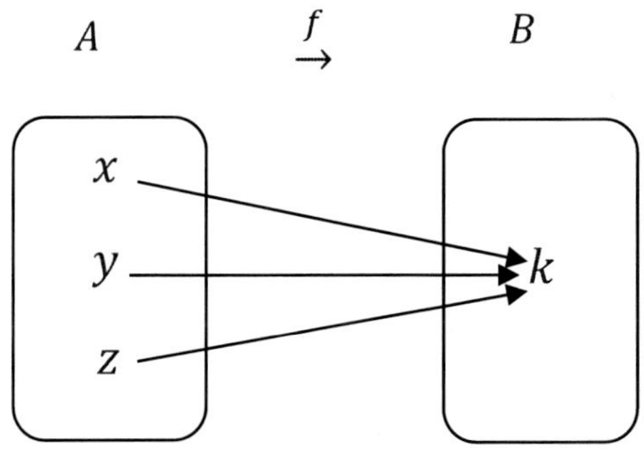

Fig. 12.9 Constant Mapping

Constant mapping is neither one-to-one-mapping nor onto-mapping.

Example 7

Identify the function that is a mapping, hence state the type of mapping of the function.

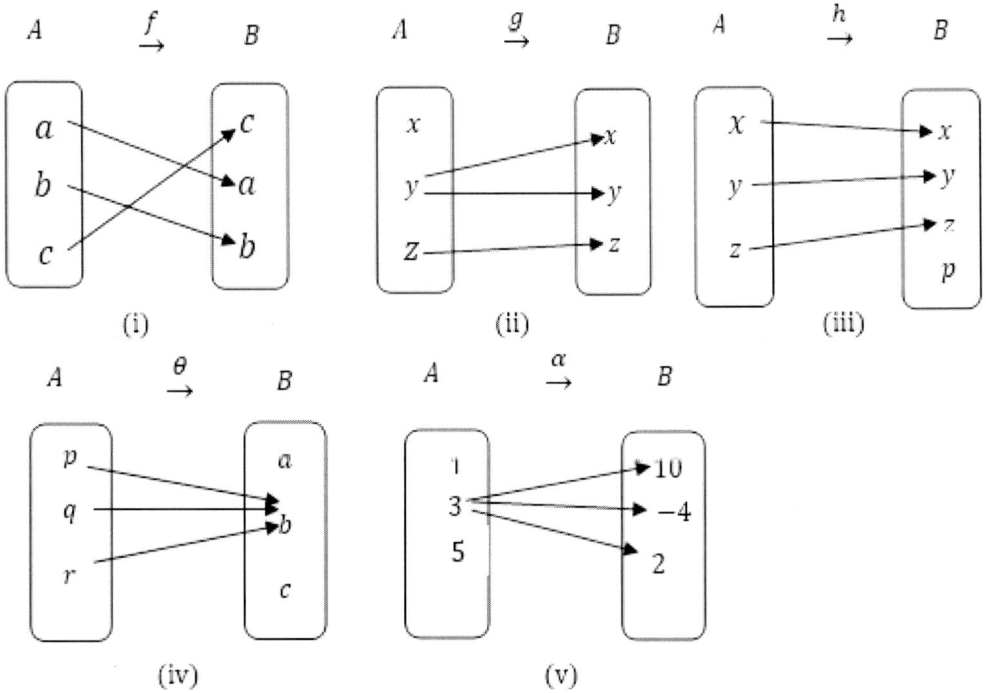

Fig. 12.10

Solution

From fig. 12.10 (i-v) the following analysis can be made:

(i) The function $f: A \to B$ is an identity mapping because the domain, co-domain and range of the function are equal.

(ii) The function $g: A \to B$ is not a mapping because an element in the domain has two images; thus, a function is a mapping when the element(s) in the domain has one image, but not otherwise.

(iii) The function $h: A \to B$ is a mapping. And it is a one-mapping because each element in the domain has just a single unique element in in the co-domain.

(iv) The function $\theta: A \to B$ is a constant mapping because the elements p, q, r in the domain have one image, b, in the co-domain.

(v) The function $\alpha: A \to B$ is not a mapping because an element, 3, in the domain has three images, $10, -4,\ 2$ in the co-domain.

12.7 COMPOSITION OF MAPPING

This is a mapping of two functions in which the co-domain of the first function becomes the domain of the second function, i.e., the output of the first function is used as the input of the second function. Thus, let P, Q and R be any non-empty sets such that $f: P \rightarrow Q$ and $g: Q \rightarrow R$ are functions of a mapping; then the co-domain of f is equal to the domain of g, which implies that the " composition of f and g" denoted by gof will be the mapping of g which takes an element in P via Q to give an image in R(see fig. 12.11(a) and (b)).

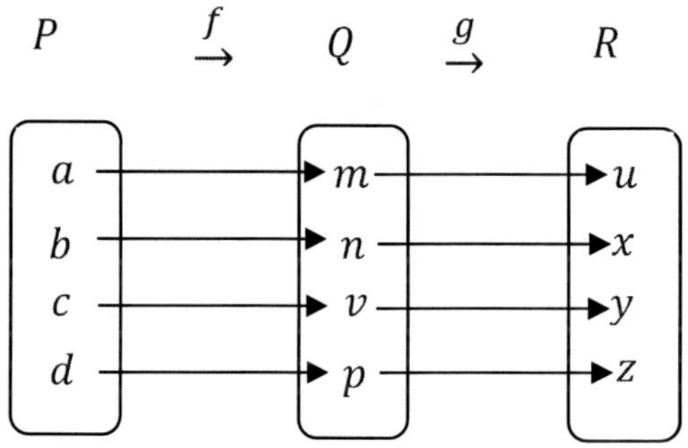

(a)

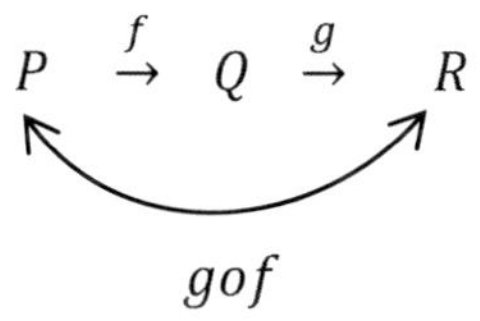

(b)

Fig. 12.11 Composite Mapping

Hence, the function of the composite mapping is written as $gof: P \rightarrow Q$. For instance, in fig. 12.11(a), we have that $gof(a) = g(f(a)) = g(m) = u$;

$gof(b) = g(f(b)) = g(n) = x$; $gof(c) = g(f(c)) = g(v) = y$;

$gof(d) = g(f(d)) = g(p) = z$.

Example 8

Given that $f: x \rightarrow 2x - 1$ and $g: x \rightarrow x^2 + 2x + 3$, find:

(a) $f^2(x)$ (b) $g^2(x)$ (c) fog (d) gof

Solution

(a) $f^2(x) = fof(x) = f(f(x)) = f(2x - 1)$

$= 2(2x - 1) - 1 = 4x - 2 - 1 = 4x - 3$

$\therefore f^2(x) = 4x - 3$

(b) $g^2(x) = gog(x) = g(g(x)) = g(x^2 + 2x + 3)$

$= (x^2 + 2x + 3)^2 + 2(x^2 + 2x + 3) + 3 = x^4 + 4x^3 + 11x^2 + 16x + 18$

$\therefore g^2(x) = x^4 + 4x^3 + 12x^2 + 16x + 18$

(c) $fog = fog(x) = f(g(x)) = f(x^2 + 2x + 3)$

$= 2(x^2 + 2x + 3) - 1 = 2x^2 + 4x + 6 - 1 = 2x^2 + 4x + 5$

$\therefore fog = 2x^2 + 4x + 5$

(d) $gof = gof(x) = g(f(x)) = g(2x-1)$
$= (2x-1)^2 + 2(2x-1) + 3 = 4x^2 + 2$
$\therefore gof = 4x^2 + 2$

Example

If $f(x) = x - 1,\ g(x) = x^2 + 4$ and $h(x) = 5x + 3$. Show that:
(a) $gof \neq fog$ (b) $gofoh \neq hogof$

Solution

(a) To show that $gof \neq fog$
$gof = gof(x) = g(f(x)) = g(x-1)$
$= (x-1)^2 + 4 = x^2 - 2x + 1 + 4 = x^2 - 2x+5$
$\Rightarrow gof = x^2 - 2x+5$
And, $fog = fog(x) = f(g(x)) = f(x^2+4)$
$= (x^2+4) - 1 = x^2 + 3 \quad \Rightarrow fog = x^2 + 3$
Since $x^2 - 2x+5 \neq x^2 + 3$
$\therefore gof \neq fog$

(b) $go(foh) = go(foh)(x) = go(foh(x)) = go(f(5x+3))$
$= g((5x+3) - 1) = g(5x+2)$
Now, put this value into $x^2 + 4$
$= (5x+2)^2 + 4 = 25x^2 + 20x + 8$
$\Rightarrow fogoh = 25x^2 + 20x + 8$
And,
$hogof = h(x^2 - 2x + 5)$
Now, put this value into $= 5x + 3$
$= 5(x^2 - 2x + 5) + 3 = 5x^2 - 10x + 28$
$\Rightarrow hogof = 5x^2 - 10x + 28$
Since $25x^2 + 20x + 8 \neq 5x^2 - 10x + 28$
$\therefore\ gofoh \neq hogof$

Example 11

Let $f(x) = \frac{x^2-2}{x+1}$ and $g(x) = 2x - 1$. Find (a) gof (b) fog

Solution

(a) $gof = g(f(x)) = g\left(\frac{x^2-2}{x+1}\right)$

$\Rightarrow gof = 2\left(\frac{x^2-2}{x+1}\right) - 1 = \frac{2x^2-4}{x+1} - 1 = \frac{2x^2-x-5}{x+1}$

$\therefore gof = \frac{2x^2-x-5}{x+1}$

Example 12

A function is given as, $R(x) = (x^2 + 5x - 1)^3$, write it as a composition of:
(i) two functions (ii) three functions

Solution

Given $R(x) = (x^2 + 5x - 1)^3$
(i) Thus, let the composition of two functions for $R(x)$ be $f(x)$ and $g(x)$
$\Rightarrow f(x) = x^2 + 5x - 1$ and $g(x) = x^3$
That means $R(x) = (fog)(x)$
(ii)Let the composition of three functions for $R(x)$ be $f(x)$, $g(x)$ and $h(x)$:
Thus, $f(x) = x^2 + 5x$; $g(x) = x - 1$; $h(x) = x^3$
That means, $R(x) = (fogoh)(x)$

12.8 FUNCTION

A function is a rule that maps a given number to a unique number. The mapped input number is referred to as the independent variable, while the unique output is the dependent variable. Thus, a function must never have two different outputs for the same input argument. If a graph is drawn and one is to determine if it's a valid function, then a vertical line test which is to draw a vertical straight line with a ruler to see if it touches the graph at most one point then it is valid function otherwise it is not a valid function.

Example 13

A function is defined by $f: x \rightarrow 2x^2 + 3x - 2$, find the images at the following points:
(a) $f(0)$ (b) $f(-1)$ (c) $f(3)$ (d) $f(-5)$; hence represent the information using arrow graph.

Solution

Given that the function is defined by:
$f: x \rightarrow 2x^2 + 3x - 2$,
the images $y = f(x)$ is thus:
(a) $f(0) = 2(0)^2 + 3(0) - 2 = 0 + 0 - 2 = -2$
(b) $f(-1) = 2(-1)^2 + 3(-1) - 2 = 2 - 3 - 2 = -3$
(c) $f(3) = 2(3)^2 + 3(3) - 2 = 18 + 9 - 2 = 25$
(d) $f(-5) = 2(-5)^2 + 3(-5) - 2 = 50 - 15 - 2 = 33$

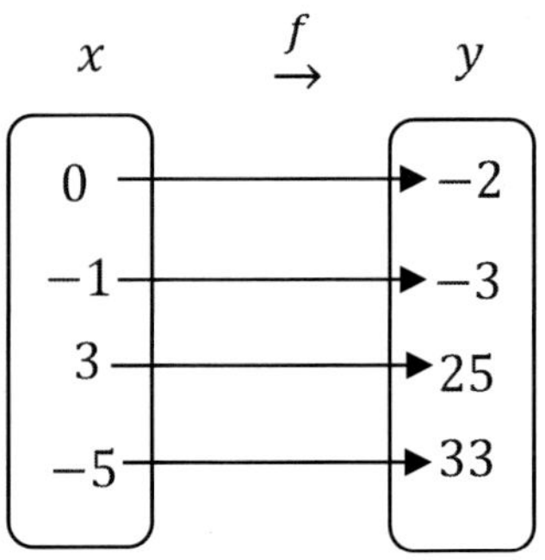

Fig. 12.12

Example 14

The parallel axes in fig. 12.13 show the domain and co-domain of a function, $f: x \to ax^2 + bx + c$. Find the values of a, b and c.

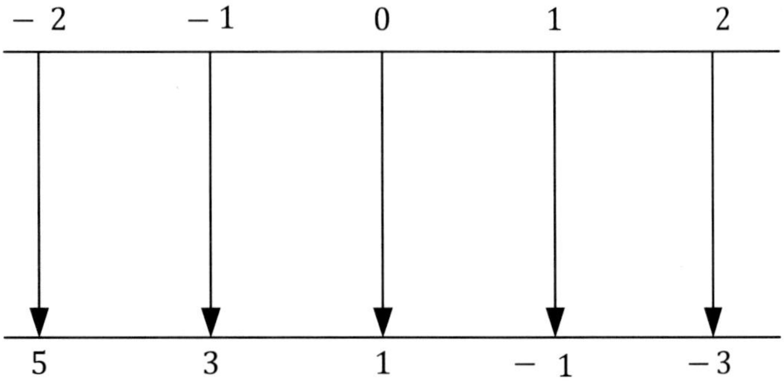

$f: x \to ax^2 + bx + c$

Fig. 12.13

Solution

The function is given as, $f: x \to ax^2 + bx + c$

When $x = -2,\ y = 5 \quad \Rightarrow f: x = y = a(-2)^2 + b(-2) + c = 5$

$4a - 2b + c = 5$ ....(i)

When $x = -1,\ y = 3 \Rightarrow f: x = y = a(-1)^2 + b(-1) + c = 3$

$a - b + c = 3$ ....(ii)

When $x =,\ y = 1 \quad \Rightarrow f: x = y = a(0)^2 + b(0) + c = 1$

$c = 1$ ....(iii)

Substitute the value of $c = 1$ from equation (iii) into equation (i) and (ii).

Thus, from (i) $4a - 2b + 1 = 5$

$\Rightarrow 4a - 2b = 4$ ....(iv)

Similarly, from (ii) $a - b + 1 = 3$

$\Rightarrow a - b = 2$ ....(v)

Solving simultaneously equations (iv) and (v), we have that:

$4(b + 2) - 2b = 4$

$4b + 8 - 2b = 4$

$2b = -4 \quad \Rightarrow b = \frac{-4}{2} = -2$

$\therefore b = -2$

Then, $a = b + 2 = -2 + 2 = 0$

So, $a = 0, \quad b = -2, \quad c = 1$

the function $f: x \to ax^2 + bx + c \quad \Rightarrow f: x \to 1 - 2x$

The variable a was made the subject formula in equation (v), i.e., $a = b + 2$, then put this value into equation (iv) in order to solve for b.

12.9 INVERSE OF A FUNCTION

This is a function derived when the domain and the co-domain of a function are interchanged.
The inverse of a function, $f(x)$, is denoted as $f^{-1}(x)$.
Hence, it is defined as $(fof^{-1})(x) = x$.
Thus, a function can have an inverse when the following conditions are satisfied:

(i) When the function is one-to-one and onto-mapping;
(ii) The domain of the function $f(x)$ is the range of the inverse function $f^{-1}(x)$;
(iii) The domain of $f^{-1}(x)$ is the range of $f(x)$.

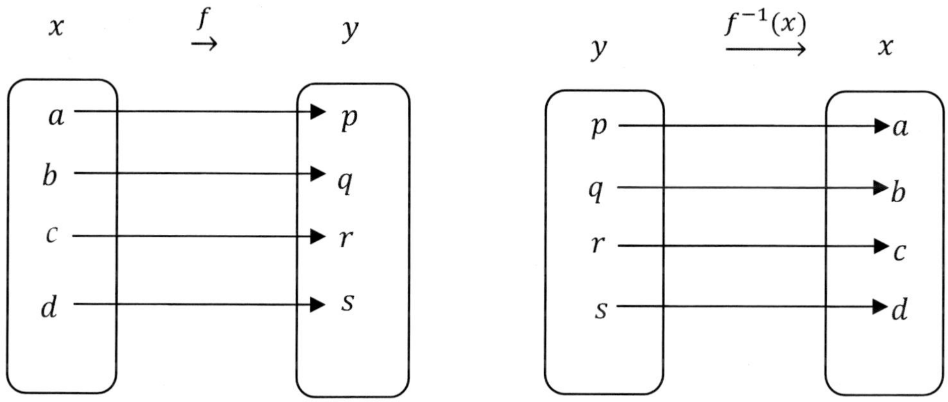

(i) Function (ii) Inverse function
Fig.12.14 Function and Inverse Function

Example 15

Find the inverse of the function $f(x) = 5x$

Solution

The inverse relation is given as $(fof^{-1})(x) = x$
where f^{-1} is the inverse function of f.
then, $(fof^{-1})(x) = x \quad \Rightarrow 5f^{-1}(x) = x$
$\therefore\ f^{-1}(x) = \frac{x}{5}$

Example 16

Find the inverse of the function $g(x) = 5x^3$

Solution

The inverse relation is given as $(gog^{-1})(x) = x$
where g^{-1} is the inverse function of g
then $(gog^{-1})(x) = x \quad \Rightarrow g(g^{-1}(x)) = x$
$5\,[\,g^{-1}(x)\,]^3 = x \quad \Rightarrow [\,g^{-1}(x)\,]^3 = \frac{x}{5}$
$\therefore\ g^{-1}(x) = \left(\frac{x}{5}\right)^{\frac{1}{3}}$

Example 17

If a function is defined by $f(x) = x^3 - 8$, find the inverse of the function.

Solution

The inverse relation is given as $(fof^{-1})(x) = x$
where f^{-1} is the inverse function of f
then $(fof^{-1})(x) = x \Rightarrow f(f^{-1}(x)) = x$
$([f^{-1}(x)]^3 - 8) = x \Rightarrow [f^{-1}(x)]^3 = x + 8$
$\therefore f^{-1}(x) = (x+8)^{\frac{1}{3}}$

Alternative method:

Let the function $f(x) = y = x^3 - 8$, then the inverse is found by making x the subject formula from the given equation.

Thus, $y = x^3 - 8 \quad \Rightarrow x = (y+8)^{\frac{1}{3}}$

So, the inverse function, $f^{-1}(x) = (x+8)^{\frac{1}{3}}$

The y in the equation is later replaced with x.

Example 18

Given that $g: x \rightarrow 3 + \frac{2}{x-1}$, find $g^{-1}(x)$.

Solution

Let $g: x \rightarrow y$, then $y = 3 + \frac{2}{x-1}$

$\Rightarrow x = 1 + \frac{2}{y-3} = \frac{y-1}{y-3}$

$\therefore \; g^{-1}(x) = \frac{x-1}{x-3}$

Make x the subject formula.

Example 19

Given the functions $f: x \rightarrow 7x$ and $g: x \rightarrow 2x - 1$, find: (a) $f^{-1}o\, g^{-1}$ (b) $g^{-1}o\, f^{-1}$

Solution

The inverse of the function $f: x \rightarrow 7x$ is $f^{-1}(x) = \frac{x}{7}$

And the inverse of the function $g: x \rightarrow 2x - 1$ is $g^{-1}(x) = \frac{x+1}{2}$

(a) $f^{-1}o\, g^{-1} = f^{-1}(g^{-1}(x)) = f^{-1}\left(\frac{x+1}{2}\right)$

$\Rightarrow f^{-1}\left(\frac{x+1}{2}\right) = \frac{\left(\frac{x+1}{2}\right)}{7} = \frac{x+1}{14}$

(b) $g^{-1}o\, f^{-1} = g^{-1}(f^{-1}(x)) = g^{-1}\left(\frac{x}{7}\right)$

$\Rightarrow g^{-1}\left(\frac{x}{7}\right) = \frac{\left(\frac{x}{7}+1\right)}{2} = \frac{x+7}{14}$

Example 20

Given the following functions: $f: x \rightarrow \frac{x+1}{2x+3}$ and $g: x \rightarrow 1 + \frac{3}{2-x}$, find:

(a) f^{-1} (b) g^{-1} (c) $f^{-1} o\, g^{-1}$

Solution

(a) $f: x \rightarrow \frac{x+1}{2x+3}$

Let $f: x \rightarrow y$, then $y = \frac{x+1}{2x+3}$.

Make x the subject formula.

$\Rightarrow 2xy + 3y = x + 1$

Then, $2xy - x = 1 - 3y \Rightarrow x(2y-1) = 1 - 3y$

$x = \frac{1-3y}{2y-1}$

Replace the letter y with x in order to finally get the inverse function, $f^{-1}(x)$.

(b) $g: x \rightarrow 1 + \frac{3}{2-x}$

Let $g: x \rightarrow y$, then $y = 1 + \frac{3}{2-x}$

Using the same step as in solution (a) above, it can be deduced that:

$y = 1 + \frac{3}{2-x}$ then making x subject formula, we have $x = 2 - \frac{3}{y-1} = \frac{2y-5}{y-1}$

$\therefore\ g^{-1}(x) = \frac{2x-5}{x-1}$

(c) $f^{-1} o\, g^{-1} = f^{-1}(g^{-1}(x)) = f^{-1}\left(\frac{2x-5}{x-1}\right)$

Put this value into the function, $f^{-1}(x) = \frac{1-3x}{2x-1}$.

$= \frac{1-3\left(\frac{2x-5}{x-1}\right)}{2\left(\frac{2x-5}{x-1}\right)-1} = \frac{x-1-6x+15}{4x-10-x+1} = \frac{14-5x}{3x-9}$

$\therefore f^{-1} o\, g^{-1} = \frac{14-5x}{3x-9}$

OBJECTIVE TEST 12

1. The mapping of all the elements in the domain to a single image in the co-domain is called ....
   (A) Identity mapping (B) One-to-One mapping
   (C) Onto-mapping (D) Constant mapping

2. An example of a mapping which assigns every student to the teacher in a class is .....
   (A) Onto mapping (B) One-to-One mapping
   (C) Constant mapping (D)Identity mapping.

3. State if the function, f, in fig. 12.15 is a mapping.
   (A) f is a mapping
   (B) f is not a mapping
   (C) f is a constant mapping
   (D) f is a one-to-one mapping

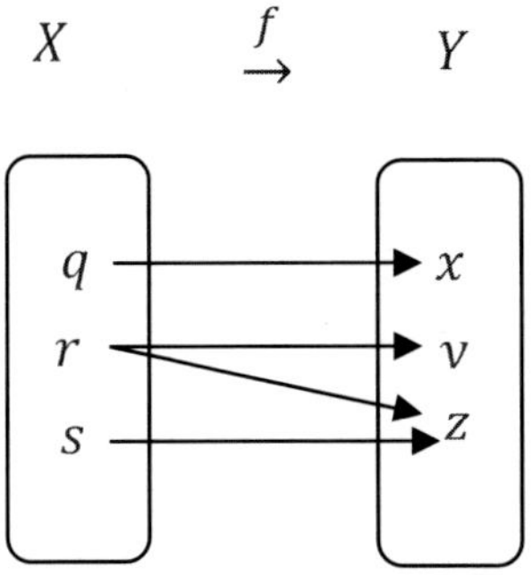

Fig. 12.15

4. A mapping that is both one-to-one mapping and onto - mapping is called ....
   (A) Surjective mapping (B) Injective mapping
   (C) Bijective mapping (D) None of these

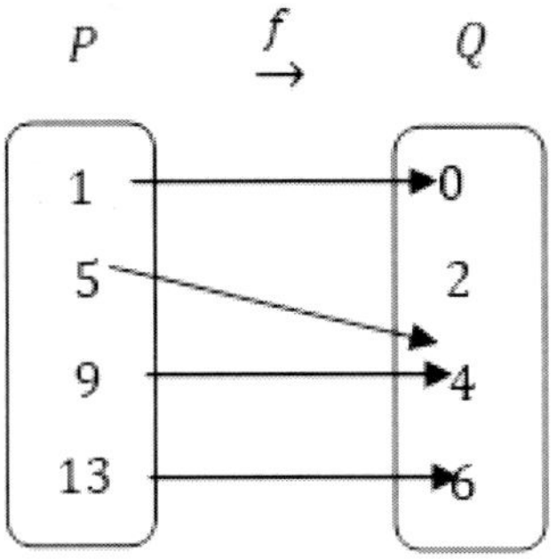

Fig. 12.16

5. Find the range of the function in fig. 12.16.
   (A) $\{1, 5, 9, 13\}$ (B) $\{0, 2, 4\}$
   (C) $\{0, 2, 4, 6\}$ (D) $\{0, 4, 6\}$

6. If a mapping is defined by $f: x \rightarrow 2x^2 + 3$, find the image of the point -5 under the function, f.
   (A) 53 (B) -53 (C) 13 (D) 23

7. A mapping is defined by the rule $f:(x,y) \rightarrow (-y, 10x+y)$ where $x, y \in R$. Find the image of the co-ordinate points $(-2,-1)$.
   (A) $(1,21)$ (B) $(21,1)$ (C) $(-1,-21)$ (D) $(1,-21)$

8. Find the inverse of the function $f(x) = \frac{x-2}{3-x}$
   (A) $\frac{3-x}{x-2}$ (B) $\frac{x-3}{x-2}$ (C) $\frac{3x+2}{x+1}$ (D) $\frac{x+1}{3x+2}$

9. A composite function is defined as: $f:x \rightarrow x+1$ and $g:x \rightarrow 3x^2-2$, find gof
   (A) $3x^2-6x+2$ (B) $3x^2+6x+1$ (C) $3x^2-1$ (D) $6x^2+x+1$

10. Given $A = \{2,3,8,11,14\}$ and $B = \{-1,4,59,x,191\}$, such that the mapping is one-to-one, find the value of x.
   (A) 104 (B) 132 (C) 116 (D) 121

Exercise 12

1. What do you understand by the terms: (i) mapping (ii) function
2. State if there is/are any difference between mapping and function
3. State three types of mapping; hence explain each of them.
4. If $f:R \rightarrow R$ is a function defined by $f:x \rightarrow 2x^2-1$, find:
   (a) the image of the point (i) 2 (ii) 1 (iii) -3
   (b) the value of f under the point -5.
5. Find the image under the function:
   (i) $f:x \rightarrow 3x^2-2$ at point -1 (ii) $g:y \rightarrow \frac{y^2}{2}+1$ at point -3
   (iii) $h:z \rightarrow \frac{z+1}{z^2-1}$ at point 2 (iv) $G:\theta \rightarrow sin\theta$ at point $\frac{\pi}{6}$
   (v) $H:\theta \rightarrow \frac{sin2\theta}{cos\theta}$ at point $\frac{\pi}{3}$
6. Find the range of the function:
   (i) $f:x \rightarrow (x-1)(x+2)$ if the domain is $\{-\frac{1}{2}, 0, 1, \frac{3}{2}\}$
   (ii) $g:y \rightarrow y^4-6$ if the domain is $\{-2,-1,0,1,2\}$
   (iii) $h:z \rightarrow z^2-2z+1$ if the domain is $\{\frac{1}{2}, \frac{1}{3}, 1, 3, 5\}$
   (iv) $G:\theta \rightarrow sin^2\theta + 3cos\theta$ if the domain is $\{0, \frac{\pi}{6}, \frac{\pi}{3}, \frac{\pi}{2}, \pi\}$
7. Draw an arrow graph for each solution in question (6).
8. A mother, m, has four sons p, q, r and s. Assuming the sons as a set X and the mother as a singleton set Y, represent this statement as a set of ordered pair; and draw the arrow graph given that the statement is a mapping.
9. Let $f:R \rightarrow R$ defined by $f(x) = x+1$ and $g:R \rightarrow R$ defined by $g(x) = x^2$, find:
   (i) $(gof)(x)$ (ii) $(fog)(x)$

10. Using fig. 12.17, find (i) $f^{-1}(0)$ (ii) $f^{-1}(2)$ (iii) $f^{-1}(4)$

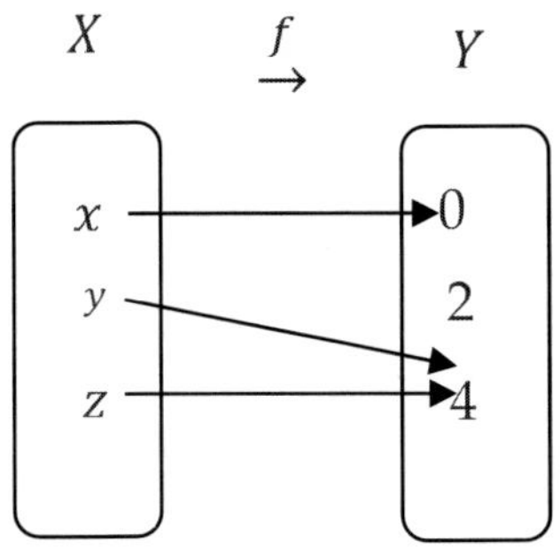

Fig 12.17

11. Identify if the following is a mapping; state what type of mapping.

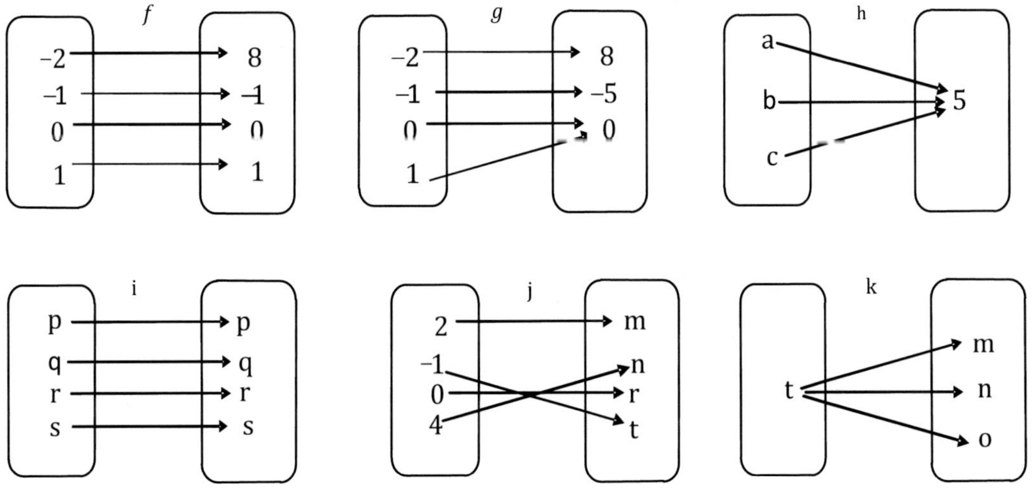

Fig. 12.18

12. Given $f: x \to x + 1$ and $g: x \to 5x^2 - 1$, find the solution of $fog(x) = gof(x)$

13. If $f(x) = 3x + 2$, $g(x) = x^2 - 1$ and $h(x) = x + 7$, show that:
  (a) $gof \neq fog$ (b) $gofoh \neq hogof$

14. Let $f(x) = \frac{x^2-1}{x+3}$ and $g(x) = x - 5$. Calculate (a) gof (b) fog

15. If a function is defined by $f(x) = x^3 - 125$, find the inverse of the function,

16. Given that $f: x \to 1 - 4x$ and $g: x \to x^2 + 2x - 1$, find:
  (a) $f^2(x)$ (b) $g^2(x)$ (c) fog (d) gof

17. Given the following functions: $f: x \to \frac{x+1}{2x+3}$ and $g: x \to 1 + \frac{3}{2-x}$, find:
  (a) f^{-1} (b) g^{-1} (c) $f^{-1}o\, g^{-1}$

13 BINARY OPERATION

Binary operation is a rule of combination for any two elements of a non-empty set. Also, a binary operation on a set, S, is a binary relation that maps elements of the Cartesian product $S \times S$ to S. Thus,

$$f: S \times S \rightarrow S$$

13.1 OPERATORS/NOTATIONS

Operators: The following operators are commonly used in binary operation:

$*, \ \oplus, \otimes, \ominus, \circ, \ \triangle, \ \nabla.$

Notations: The following notations are used in binary operation:

Notation	Meaning
R	The set of real numbers; any rational or irrational number
N	The set of natural numbers (i.e., 1, 2, 3, 4, 5, 6, ...)
Z	The set of all integers (i.e. ..., $-3, -2, -1, 0, 1, 2, 3, 4, 5$, ...)
Z^+	The set of positive integers; natural numbers (i.e., 1, 2, 3, 4, 5, ...)
Z^-	The set of negative integers (i.e. $-1, -2, -3, -4$, ...)
Q	The set of rational numbers (i.e., $\frac{p}{q}$ where $p, q \in Z, \ q \neq 0$)

13.2 PROPERTIES OF BINARY OPERATION

The following are the properties that describe a binary operation:

13.2.1 CLOSURE PROPERTY

A non-empty set, S, is closed under a binary operation $*$, if for all $x, y \in S$, then the operation $x * y \in S$. Therefore, the set of all integers, Z, is closed under addition $(+)$, subtraction $(-)$, and multiplication $(\times)$. However, Z is not closed under division $(\div)$.

Example 1

If a set of all integers is denoted as Z. Check if addition, subtraction, multiplication and division is closed under the operation.

Solution

Use any integers, say, 7, 4. $\therefore$ 7, 4 $\in Z$
For **Addition:** $7 + 4 = 11$ and $4 + 7 = 11$; thus addition is closed under Z.
For **Subtraction:** $7 - 4 = 3$ and $4 - 7 = -3$; thus subtraction is closed under Z.
(because -3 is an integer, i.e. negative integer)
For **Multiplication:** $7 \times 4 = 28$ and $4 \times 7 = 28$; thus multiplication is closed under Z.
For **Division:** $\frac{7}{4} = 1.75$ and $\frac{4}{7} = 0.5714$; thus division is not closed under Z (this is because 1.75 and 0.5714 have decimal point, i.e. it is a fraction).

Example 2

An operation $a * b = \frac{a+b}{3}$ is defined on the set of real numbers R, such that $a, b, \in R$.
Find: (i) $2 * 3$ (ii) $-7 * 4$ (iii) $3 * 2$ (iv) $\frac{1}{5} * \frac{1}{3}$
hence state if the solution is closed under R.

Solution

(i) $a * b = 2 * 3 \quad \Rightarrow 2 * 3 = \frac{2+3}{3} = \frac{5}{3}$

(ii) $a * b = -7 * 4 \quad \Rightarrow -7 * 4 = \frac{-7+4}{3} = \frac{-3}{3} = -1$

(iii) $a * b = 3 * 2 \quad \Rightarrow 3 * 2 = \frac{3+2}{3} = \frac{5}{3}$

(iv) $a * b = \frac{1}{5} * \frac{1}{3} \quad \Rightarrow \frac{1}{5} * \frac{1}{3} = \frac{\frac{1}{5}+\frac{1}{3}}{3} = \frac{\frac{8}{15}}{3} = \frac{8}{5}$

Since $a, b \in R$ and all the solution of $\frac{a+b}{3} \in R$, the set R is closed under the operation $*$

Example 3

An operation $\oplus$ is defined on the set, $A = \{-3, -1, -1, 0, 1, 2\}$ by $x \oplus y = x + y - xy$. Find:
(a) $-2 \oplus 3$ (b) $0 \oplus -3$ (c) $-3 \oplus -2$ (d) $3 \oplus 1$

Solution

(a) $x \oplus y = x + y - xy \quad \Rightarrow -2 \oplus 3 = -2 + 3 - (-2)(3) = 1 + 6 = 7$
where $7 \notin A$
(b) $x \oplus y = x + y - xy \quad \Rightarrow 0 \oplus -3 = 0 + (-3) - (0)(-3) = -3 + 0 = -3$
where $-3 \in A$
(c) $x \oplus y = x + y - xy \quad \Rightarrow -3 \oplus -2 = (-3) + (-2) - (-3)(-2)$
$= -5 - 6 = -11$; where $-11 \notin A$
(d) $x \oplus y = x + y - xy \quad \Rightarrow 3 \oplus 1 = 3 + 1 - (3)(1) = 4 - 3 = 1$
where $1 \in A$

Thus, from solution (a) and (c) it shows that the set is not closed under the operation $\oplus$ because $7, -11 \notin A$ while in solution (b) and (d) the set is closed under the operation $\oplus$ because $-3, 1 \in A$.

Example 4

(a) Construct an operation table for addition in modulo 5
(b) From the table in (a), determine whether the closure property is satisfied.

Solution

(a) Using the set, $S_5 = \{0, 1, 2, 3, 4\}$

Table 13.1

+ (*mod* 5)	0	1	2	3	4
0	0	1	2	3	4
1	1	2	3	4	0
2	2	3	4	0	1
3	3	4	0	1	2
4	4	0	1	2	3

(b) From the table, it can be observed that each element in the column and row is contained in the set $S_5 = \{0, 1, 2, 3, 4\}$; therefore, the closure property is satisfied.

Example 5

(a) Construct an operation table for multiplication in modulo 5.
(b) From the table in (a) above, determine whether the closure property is satisfied.

Solution

Using the set, $S_5 = \{0, 1, 2, 3, 4\}$
(a)

Table 13.2

× (*mod* 5)	0	1	2	3	4
0	0	0	0	0	0
1	0	1	2	3	4
2	0	2	4	1	3
3	0	3	1	4	2
4	0	4	3	2	1

(b) From the table, it can be observed that each element in the column and row is contained in the set $S_5 = \{0, 1, 2, 3, 4\}$; therefore, the closure property is satisfied.

Example 6

Obtain the operation table of the binary operation*, where $a * b$ is the remainder when ab is divided by 8; hence state whether the set is closed under the operation$^{\otimes}$

(i) $\{1, 4\}$ (ii) $\{1, 3, 5\}$ (iii) $\{1, 2, 4, 6\}$ (iv) $\{1, 5, 9\}$

Solution

If $a * b$ is the remainder when ab is divided by 8, then the statement implies that we should create a table for multiplication (i.e., ab) under modulo 8.

Table 13.3 (i-iv)

$\otimes$	1	4
1	1	4
4	4	0

(i)

$\otimes$	1	3	5
1	1	3	5
3	3	1	7
5	5	7	1

(ii)

$\otimes$	1	2	4	6
1	1	2	4	6
2	2	4	0	4
4	4	0	0	0
6	6	4	0	4

(ii)

$\otimes$	1	3	9
1	1	3	9
3	3	1	3
9	1	3	1

(iv)

Thus from the given operation tables, we have that:

(i) $\{1, 4\}$ is not closed under the operation$^{\otimes}$, because 0 is not an element in the set.

(ii) $\{1, 3, 5\}$ is not closed under the operation$^{\otimes}$, because 7 is not an element in the set

(iii) $\{1, 2, 4, 6\}$ is not close under the operation$^{\otimes}$, because 0 is not an element in the set.

(iv) $\{1, 5, 9\}$ is close under the operation$^{\otimes}$, because each element in the rows and columns is contained in the set.

Example 7

Test whether or not the closure property is satisfied in each of the following operations:

(i) $(Z, \otimes)$ where $x \otimes y = x^2y$ (ii) $(Q, *)$ where $x * y = xy + x - y$

(iii) $(R, \triangle)$ where $x \triangle y = x + y\sqrt{2}$ (iv) $(N, *)$ where $x * y = x + y - 1$

Solution

(i) $(Z, \otimes)$ where $x \otimes y = x^2y$; then choose any integer, say, 5 and 3.

$\Rightarrow x \otimes y = 5 \otimes 3 = 5^2(3)$

$= 25(3) = 75$ $\quad \therefore 75 \in Z$, hence it is closed under $\otimes$.

(ii) $(Q, *)$ where $x * y = xy + x - y$; then choose any rational numbers, say, $\frac{2}{3}$ and $\frac{1}{4}$.

$\Rightarrow x * y = \frac{2}{3} \times \frac{1}{4} + \frac{2}{3} - \frac{1}{4} = \frac{7}{12}$

$\therefore \frac{7}{12} \in Q$, hence it is closed under $*$.

(iii) $(R, \triangle)$ where $x \triangle y = x + \sqrt{y}$; then choose any real number, say, $\frac{2}{5}$ and 9.

$\Rightarrow x \triangle y = \frac{2}{5} \times \sqrt{9} = \frac{2}{5} \times 3 = \frac{6}{5}$ $\quad \therefore \frac{6}{5} \in R$, hence it is closed under $\triangle$.

(iv) $(N, *)$ where $x * y = x + y - 1$; then choose any natural number, say, 10 and 7
$\Rightarrow x * y = 10 * 7 = 10 + 7 - 1$
$= 17 - 1 = 16$
$\therefore 16 \in N$, hence it is closed under $*$.

13.2.2 COMMUTATIVE PROPERTY

A binary operation is commutative if the operation of any two elements of a set upon which it is defined gives the same result even if the order of operation is reversed. Thus, if $x, y \in R$ where R is the set of real numbers, then
$x + y = y + x$; and $xy = yx$. Therefore, addition and multiplication of real numbers are commutative. On the other hand, $x - y \neq y - x$; and $\frac{x}{y} \neq \frac{y}{x}$ except when $x = y$. Therefore, subtraction and division are not commutative.

Example 8

If a binary operation $*$ is defined on set of real numbers, show that addition and multiplication are commutative, while subtraction and division are not commutative.

Solution

Let $x, y \in R$ where R is the set of real numbers.
Now using arbitrary values, $x = 14$, $y = 2$

(i) For **Addition**: $x + y = 14 + 2 = 16$, and $y + x = 2 + 14 = 16$
since $14 = 14$, so addition is commutative for the set of real numbers.

(ii) For **Subtraction**: $x - y = 14 - 2 = 12$, and $y - x = 2 - 14 = -12$
since $12 \neq -12$, so subtraction is not is commutative for the set of real numbers.

(iii) For **Multiplication**: $x \times y = 14 \times 2 = 28$ and $y \times x = 2 \times 14 = 28$

Since $xy = yx$, so multiplication is commutative for the set or real numbers.

(iv) For **Division**: $\frac{x}{y} = \frac{14}{2} = 7$ and $\frac{y}{x} = \frac{2}{7}$

Since $\frac{x}{y} \neq \frac{y}{x}$ (i.e., $7 \neq \frac{2}{7}$) that means that division is not commutative for the set of real numbers.

Example 9

Examine whether the property of commutativity is satisfied under the following conditions: (i) $a * b = 5a - b$ (ii) $a \oplus b = a + b - 2ab$

Solution

(i) Choosing the integers $a = 5$ and $b = 3$.
$\Rightarrow a * b = 5 * 3 = 5(5) - 3 = 22$; $b * a = 3 * 5 = 3(3) - 5 = 4$
since $22 \neq 4$ under the operation $*$, then it is not commutative.
(ii) Choosing the integers $a = 5$ and $b = 3$.
$a \oplus b = a + b - 2ab = 5 + 3 - 2(5)(3) = -22$
And, $b \oplus a = b + a - 2ba = 3 + 5 - 2(3)(5) = -22$
$\Rightarrow a \oplus b = b \oplus a$ under the operation $\oplus$, thus it is commutative.

13.2.3 ASSOCIATIVE PROPERTY

If x, y and z are elements of a set, then the set are associative when:

$$(x * y) * z = x * (y * z)$$

This implies that the position of the bracket does not affect the result obtained from the operation. Thus, addition and multiplication are associative under the set of real numbers, while subtraction and division are not associative under the set of real numbers.

Example 10

State whether each of the following operations is associative or not:
(i) $x * y = x - 2y$ (ii) $a \oplus b = 2a - b + ab$

Solution

(i) The operation is defined as, $x * y = x - 2y$
then, $(x * y) * z = (x - 2y) * z \Rightarrow (x - 2y) * z = (x - 2y) - 2z = x - 2y - 2z$
And, $x * (y * z) = x * (y - 2z) \Rightarrow x * (y - 2z) = x - 2(y - 2z) = x - 2y + 4z$
Therefore, $x - 2y - 2z \neq x - 2y + 4z$,
thus, the operation is not associative under the given operation.
(ii) The operation is defined as, $a \oplus b = 2a - b + ab$
then, $(a \oplus b) \oplus c = (2a - b + ab) \oplus c$
$\Rightarrow (2a - b + ab) \oplus c = 2(2a - b + ab) - c + (2a - b + ab)c$
$= 4a - 2b + 2ab - c + 2ac - bc + abc$
And, $a \oplus (b \oplus c) = a \oplus (2b - c + bc)$
$\Rightarrow a \oplus (2b - c + bc) = 2a - (2b - c + bc) + a(2b - c + bc)$
$= 2a - 2b + c - bc + 2ab - ac + abc$
Since, $4a - 2b + 2ab - c + 2ac - bc + abc \neq 2a - 2b + c - bc + 2ab - ac + abc$,
then it is not associative.

Example 11

Show that the binary operation defined by $p \nabla q = p + q - pq$ is commutative.

Solution

$p\nabla q = p + q - pq$
then $(p\nabla q) \nabla r = (p + q - pq) \nabla r = (p + q - pq) + r - (p + q - pq)r$
$=(p + q - pq)(1 - r) + r$
then, choosing arbitrary values: $p = 10,\ q = 4,\ r = 6$
$\therefore\ (p + q - pq)(1 - r) + r = (10 + 4 - 10 \times 4)(1 - 6) + 6$
$= (14 - 40)(-5) + 6 = (-26)(-5)+6$
$= 130 + 6 = \mathbf{136}$
And, $p\nabla(q\nabla r) = p\nabla(q + r - qr) = p + (q + r - qr) - p(q + r - qr)$
$(q + r - qr)(1 - p) + p$
Also, using the same arbitrary values: $p = 10,\ q = 4,\ r = 6$
$(q + r - qr)(1 - p) + p = (4 + 6 - 4 \times 6)(1 - 10) + 10$

$= (10 - 24)(-9) + 10 = (-14)(-9) + 10$
$= 126 + 10 = \mathbf{136}$
Hence it can be seen that, $(p + q - pq)(1 - r) + r = (q + r - qr)(1 - p) + p$
this means that it is associative.

13.2.4 DISTRIBUTIVE PROPERTY

If a set $(S,^{*,}\ ^{\oplus})$ is defined on two binary operators, then the operation will be distributive when:
(i) $x * (y \oplus z) = (x * y) \oplus (x * z)$
(ii) $(x \oplus y) * z = (x * z) \oplus (y * z)$
where $x, y, z \in S$.

Thus from (i), the binary operation * is distributive from the left over the operation $^{\oplus}$ (i.e., it is left distributive); whereas in (ii), the binary operation * is distributive from the right over operation $^{\oplus}$ (i.e., it is right distributive).

Example 12

Show that multiplication is distributive over addition for all $x, y, z \in R$

Solution

To show that $(R,^{+,}\ ^{\times})$
$\Rightarrow x \times (y + z) = (x \times y) + (x \times z) = xy + xz$
From the LHS, $x \times (y + z) = xy + xz$
and from the RHS $(x \times y) + (x \times z) = xy + xz$
Because the LHS = RHS, thus multiplication is distributive over addition.

Example 13

Given a set $(Q,^{-,}\ ^{\div})$ show that division is right distributive and not left distributive over subtraction.

Solution

To check if division is right distributive over subtraction, we have:
$(x - y) \div z = (x \div z) - (y \div z)$
Choose arbitrary values for x, y and z; let say, $x = \frac{1}{2}$, $y = \frac{1}{3}$ and $z = \frac{1}{5}$
On the LHS, $(x - y) \div z = \left(\frac{1}{2} - \frac{1}{3}\right) \div \frac{1}{5} = \frac{1}{6} \div \frac{1}{5} = \mathbf{\frac{5}{6}}$
Now on the RHS, $(x \div z) - (y \div z) = \left(\frac{1}{2} \div \frac{1}{5}\right) - \left(\frac{1}{3} \div \frac{1}{5}\right)$
$= \frac{5}{2} - \frac{5}{3} = \mathbf{\frac{5}{6}}$
Thus, LHS and RHS are equal, then division is right distributive over subtraction in $(Q,^{-,}\ ^{\div})$
However, to show that division is not left distributive, we have:
$x \div (y - z) \neq (x \div y) - (x \div z)$

On the LHS, $x \div (y - z) = \frac{1}{2} \div \left(\frac{1}{3} - \frac{1}{5}\right) = \frac{1}{2} \div \frac{2}{15} = \mathbf{\frac{15}{4}}$

Now on the RHS, $(x \div y) - (x \div z) = \left(\frac{1}{2} \div \frac{1}{3}\right) - \left(\frac{1}{2} \div \frac{1}{5}\right)$

$= \frac{3}{2} - \frac{5}{2} = \frac{2}{2} = 1$

Thus, LHS$\neq$ RHS (i.e., $\mathbf{\frac{15}{4}} \neq 1$) which means that division is not left distributive over subtraction.

13.3 IDENTITY ELEMENT

An identity element is an element that does not change the outcome of a binary operation. Thus, when a binary operation* is defined on a set, S, such that $a \in S$, then the identity element, e, which exists in S is given as:

$$a * e = e * a = a$$

Where: $a * e = a$ is called the right identity element;
$e * a = a$ is called the left identity element.

Remarks on Identity Element

(i) The identity element of an operation is called two - sided identity if the right and left identities exist and are equal (i.e., $a * e = e * a = a$).

(ii) The identity element on addition operation for all integers is two - sided identity, which is zero. If $a \in Z$, thus $a + 0 = 0 + a = a$ {where $0 \in Z$}.

(iii) The identity element on multiplication operation is unity (i.e., 1). If $a \in R$, thus $a \times 1 = 1 \times a = a$ {where $1 \in R$ }.

(iv) Subtraction operation has only right identity, which is zero. Thus, $a - 0 = a$ which is right identity but no left identity, since $0 - a \neq a$.

(v) Division operation has only right identity element which is unity (i.e., 1). Thus, if $a, b \in Q$ and $a, b \neq 0$, then $\frac{a}{b} \div 1 = \frac{a}{b}$ {where $1 \in Q$}. However, there is no left identity element for division because $1 \div \frac{a}{b} \neq \frac{a}{b}$.

(vi) The identity element of union of sets is $\emptyset$. Thus, if $a \in \mathcal{U}$ then $a \cup \emptyset = \emptyset \cup a = a$ {where $\emptyset \in \mathcal{U}$}

Example 14

Find the identity element of each binary operation:
(i) $a \oplus b = a + b$ (ii) $a \oplus b = 2a - b$

Solution

Let the identity element be given as, e.
Thus, by definition, $a \oplus e = e \oplus a = a$
(i) $a \oplus b = a + b$
$\Rightarrow a \oplus e = a + e = a$
$e = a - a = 0 \quad \therefore e = 0$

(ii) $a \oplus e = 2a - b$

$\Rightarrow \; a \oplus e = 2a - e = a$

$e = 2a - a = a$

$\therefore\; e = a$ this means that the identity element depends on "a"

Example 15

Find the identity element defined by the binary operation $x * y = xy$

Solution

Let the identity element be given as, e.

Thus, by definition, $x * e = e * x = x$

$x * y = xy \quad \Rightarrow x * e = xe = x$

$\therefore\; e = \frac{x}{x} = 1$

Example 16

Find the identity element if it exists in $x * y = x + y - 3xy$, defined under the set of rational numbers.

Solution

Let the identity element be given as, e.

This implies that by definition, $x * e = e * x = x$

$x * y = x + y - 3xy$

$\Rightarrow x * e = x + e - 3xe = x$

So, $x + e - 3xe = x$

$e(1 - 3x) = x - x$ Solve for e.

$e = \frac{0}{1-3x} = 0$

$\therefore\; e = 0 \qquad$ (provided $x \neq \frac{1}{3}$)

Example 17

Obtain the identity element from the table below:

Table 13.4 Addition table on modulo 5

$\oplus$ (*mod* 5)	0	1	2	3	4
0	0	1	2	3	4
1	1	2	3	4	0
2	2	3	4	0	1
3	3	4	0	1	2
4	4	0	1	2	3

Solution

In table 13.4, it can be seen that the set of integers on modulo 5 is,

$S = \{0, 1, 2, 3, 4\}$.

This is the identity element because $0 \oplus 0 = 0$, this satisfies the law of identity element, i.e., $m \oplus m = m$.

Thus, the row and column whose operations present the correct order as given in the set, $S = \{0, 1, 2, 3, 4\}$, are the entries in the first row and first column; this implies that the common element (i.e., the intersection point between that row and column) is the element, 0. Hence, the identity element is 0.

Example 18

Create an operation$^{\otimes}$ table for multiplication on modulo 14 using the set of integers, $S = \{\,2,\ 4,\ 6, 8, 10\}$. Hence, from the table determine:
(i) if the operation$^{\otimes}$ is closed (ii) the identity element if it exists.

Solution

Table 13.5

$\otimes$ (*mod* 14)	2	4	6	8	10
2	4	8	12	2	8
4	8	2	10	4	12
6	12	10	8	6	4
8	2	4	6	8	10
10	6	12	4	10	2

(i) The operation is not closed because the operation table contains an entry of 12, which is not a member of the set, $S = \{\,2, 4, 6, 8, 10\}$.

(ii) The identity element is found by checking the row and column that contains the correct order of elements as found in the set. Thus, by inspection on table 13.5, it is observed that the fourth row and fourth column satisfy this condition; and the element of their point of intersection gives the identity element.

$\therefore$ the identity element is 8 (i.e., $8 \otimes 8 = 8$)

13.4 INVERSE OF AN ELEMENT

The inverse of an element operates on the element to give the identity element, e. This implies that if $x \in S$ defined under a binary operation *such that S is a set containing also the identity element, e, then the inverse will be defined by the relation:

$$x * x^{-1} = x^{-1} * x = e$$

Thus, the above equation can be reduced in two forms:

$x * x^{-1} = e$ ....(i) where x^{-1} is the right inverse of x.
$x^{-1} * x = e$ .... (ii) where x^{-1} is the left inverse of x.

Remarks on Inverse

(a) If the inverse exists in any operation, then it is always unique.
(b) Either equation (i) or (ii) above is used for finding the inverse of x.
(c) The identity element of an operation must be known before it can be possible to calculate the inverse of the operation.

Example 19

If a binary operation$*$ is defined by $x * y = x + y + 1$, find the inverse of x .

Solution

Let the identity element be represented as, e.
Thus, by definition: $x * e = e * x = x$
But, $x * y = x + y + 1$
$\Rightarrow x * e = x + e + 1 = x$
So, $x + e + 1 = x$
$e + 1 = x - x = 0$
$\therefore e = -1$
Now, let the inverse be x^{-1}
$\Rightarrow x * x^{-1} = x^{-1} * x = e$ {inverse property defined}
And, $x * y = x + y + 1$ $\Rightarrow x * x^{-1} = x + x^{-1} + 1 = e$ { binary operation defined}
But, $e = -1$ $\Rightarrow x + x^{-1} + 1 = -1$
$x^{-1} = -1 - 1 - x = -2 - x$
$\therefore x^{-1} = -2 - x$

Example 20

A binary operation* is defined by $a * b = a^2 + b^2 - 3a$ where $a, b \in R$, find:
(i) the identity element (ii) the inverse of an element $a \in R$ under the operation.

Solution

(i) $a * e = e * a = a$ {property of an identity element defined}
$a * e = a^2 + e^2 - 3a = a$ {binary operation defined}
$\Rightarrow e^2 = a + 3a - a^2 = 4a - a^2$
$\therefore e = \sqrt{a(4 - a)}$

(ii) Now, let the inverse be given as a^{-1}
$\Rightarrow a * a^{-1} = a^{-1} * a = e$ {inverse property defined}
and, $a * a^{-1} = a^2 + (a^{-1})^2 - 3a = e$ {binary operation defined}
But, $e = \sqrt{a(4 - a)}$ $\Rightarrow a^2 + (a^{-1})^2 - 3a = \sqrt{a(4 - a)}$
$(a^{-1})^2 = \sqrt{a(4 - a)} + 3a - a^2$
$\therefore a^{-1} = \sqrt{\sqrt{a(4 - a)} + 3a - a^2}$
Therefore, the inverse a^{-1} depends on the given value of a.

Example 21

Find the inverse of an element in a set $(S, ^{\oplus})$ defined under $x \oplus y = x + y - xy$

Solution

$x \oplus e = e \oplus x = x$ {identity element defined}
$\Rightarrow x \oplus e = x + e - xe = x$ {binary operation defined}
$e(1 - x) = x - x = 0$
$\therefore\ e = 0$
Now, let the inverse be given as x^{-1}
$\Rightarrow x \oplus x^{-1} = x^{-1} \oplus x = e$ {inverse property defined}
and, $x \oplus x^{-1} = x + x^{-1} - xx^{-1} = e$ { binary operation defined}
But, $e = 0 \quad \Rightarrow x + x^{-1} - xx^{-1} = 0$
$x^{-1}(1 - x) = 0 - x$
$x^{-1} = \frac{-x}{1-x} = \frac{x}{x-1}\{x \neq 1\}$
Thus, the inverse of the set $(S, ^{\oplus})$ is $x^{-1} = \frac{x}{x-1}$

Example 22

A binary operation is defined by $a * b = a^2 b + a - b$ for any real number a and b. If the identity element is zero, find the inverse of 3.

Solution

The binary operation is defined by $a * b = a^2 b + a - b$
Let the x represent the inverse of 3
$\Rightarrow 3 * x = 3^2(x) + 3 - x = e$ {But the identity element, $e = 0$)
$= 9x + 3 - x = 0 \qquad \therefore\ 9x - x = 0 - 3$
$8x = -3$
$\therefore\ x = -\frac{3}{8}$
Therefore, the inverse of 3 is $-\frac{3}{8}$

Example 23.

Two binary operations $\oplus$ and $*$ are defined by:
$x * y = x^2 + y^2 - 2xy$ and $x \oplus y = x^y - 10$, find the value of $5 \oplus (2 * 3)$.

Solution

The binary operations are defined by:
$x * y = x^2 + y^2 - 2xy$ and $x \oplus y = x^y - 10$
then $x * y = 2 * 3 = 2^2 + 3^2 - 2(2)(3) = 4 + 9 - 12$
$\Rightarrow 2 * 3 = 13 - 12 = 1$
Therefore, $5 \oplus (2 * 3) = 5 \oplus 1$
and $x \oplus y = 5 \oplus 1 = 5^1 - 10 \quad \Rightarrow 5 \oplus 1 = 5 - 10 = -5$
$\therefore\ 5 \oplus (2 * 3) = -5$

Example 24

The binary operation $*$, is defined by $x * y - xy - x - y$ for $x, y \in R$, if $x * 2 = 3 * x$, find x.

Solution

The operation is defined by, $x * y = xy - x - y$ and $x * 2 = 3 * x$

$\Rightarrow x * 2 = x(2) - x - 2 = x - 2$

And, $3 * x = 3x - x - 3 = 2x - 3$

Then, $x * 2 = 3 * x \quad \Rightarrow \quad x - 2 = 2x - 3$

$2x - x = 3 - 2 \quad \therefore \ x = 1$

Example 25

Create an operation table for the set $A\{-2, 0, 2, 4, 6\}$ on addition under modulo 10. Hence, from the table find: (a) the identity element (b) the inverse of (i) -2 (ii) 2 (iii) 4 (iv) 6

Solution

Table 13.6

$\oplus$ (*mod* 10)	-2	0	2	4	6
-2	-4	-2	0	2	4
0	-2	0	2	4	6
2	0	2	4	6	8
4	2	4	6	8	0
6	4	6	8	0	2

(a) The identity element from the table is 0.

(b) The inverse property is defined by $x \oplus x^{-1} = e$

Where x is the element, x^{-1} is the inverse of the element, and e is the identity element.

Thus:

(i) $(-2)^{-1} = 2$ (ii) $(2)^{-1} = -2$ (iii) $(4)^{-1} = 6$ (iv) $(6)^{-1} = 4$

Example 26

Table 13.7

$*$	x	y	z
x	y	x	z
y	x	y	z
z	x	z	y

(a)

$\oplus$	a	b	c
a	b	a	c
b	a	b	c
c	b	c	a

(b)

Use table 13.7 (a) and (b) to answer the following:

(i) Given $x * q = x$, find q (ii) $b \oplus m = a$, find m

(iii) evaluate $(c \oplus c) \oplus a$ (iv) evaluate $(a \oplus b) * x$

Solution

(i) From table (a), $x * y = x$ can be compared with the operation $x * q = x$.
Thus, $q = y$

(ii) In table (b), $a \oplus b = a$ can be compared with the operation $m \oplus b = a$.
Thus, $m = a$

(iii) Given $(c \oplus c) \oplus a = p$; then from table (b), we have that:
$c \oplus c = a \quad \Rightarrow (c \oplus c) \oplus a = a \oplus a = b$
$\therefore \quad (c \oplus c) \oplus a = b$

(iv) $(a \oplus b) * x = a * x$ Thus, the operation $a * x$ does not exist in the given table.

OBJECTIVE TEST 13

1. Two binary operations $*$ and $\wedge$ are defined as $p * q = pq + q + 1$ and $p \wedge q = pq + 2q - 1$ for all real numbers p and q. What is the value of $2 \wedge (3 * 4)$?
(A) 68 (B) −1 (C) 67 (D) 1
2. Determine the inverse for each element of a set $(S,*)$ where $x * y = x + y - xy$.
(A) $\frac{1}{x-1}$ (B) $\frac{x}{1-x}$ (C) $\frac{x}{x-1}$ (D) $\frac{1-x}{x}$
3. $x * y = x + y - 2xy$, find x when $(x * 3) - (x * 2) = 29$
(A) 14 (B) −14 (C) 28 (D) −7
4. Find the identity element in table 13.8.

Table 13.8

*	x	y	z
x	y	x	z
y	x	y	z
z	x	z	y

(A) z (B) x (C) y (D) No identity element

5. A binary operation* is defined by $p * q = pq - p - q$ for all real numbers p and q, find the identity element under the operation when $p = 2$
(A) 5 (B) 4 (C) −4 (D) −5

6. The binary operation* on the set of integers is defined by $a * b = (ab)^{\frac{1}{3}}$, find the value of $2 * (4 * 16)$.
   (A) 16 (B) 8 (C) −2 (D) 2

7. Which of these is not a property of a binary operation?
   (A) closure (B) commutativity (C) associativity (D) inverse

8. Compute $\frac{2}{3} \nabla \frac{3}{7}$ if a binary operation is defined as $\frac{a}{b} \nabla \frac{c}{d} = \frac{a+c}{bd}$
   (A) $\frac{21}{5}$ (B) $\frac{5}{21}$ (C) $-\frac{5}{21}$ (D) $-\frac{21}{5}$

9. A binary operation* is defined by $x * y = xy + x + y$ where $x, y \in R$. If $x * 2 = 5 * x$, find the value of x.
   (A) 0 (B) 1 (C) −1 (D) 2

10. Table 13.9 shows a multiplication table on modulo 10. Find the inverse of 4.

Table 13.9

*	2	4	6
2	4	8	2
4	8	6	4
6	2	4	6

   (A) 2(B) 4 (C) 6 (D) 8

Exercise 13

1. Given $(Z, *)$ where $*$ is defined for all integers x and y as:
   $x * y = x + y + xy$, evaluate:
   (a) $2 * 5$ (b) $3 * 7$ (c) $(-5) * 12$ (d) $(-3) * (-1)$

2. In a set $(Q^+, *)$ where $*$ is defined as $\frac{x}{y} * \frac{m}{n} = \frac{x^2 n}{m+y}$, compute:
   (i) $\frac{2}{3} * \frac{1}{2}$ (ii) $5\frac{1}{3} * 2\frac{1}{2}$ (iii) $\frac{1}{4} * \frac{2}{5}$ (iv) $\frac{1}{7} * \frac{1}{3}$

3. Evaluate each of the following:
   (i) $2 \triangle 3$ if $a \triangle b = 5a - 2b$ in $(N, \triangle)$ (ii) $3 * 2$ if $a * b = a - 2b$ in $(Z, *)$
   (iii) $-3 \oplus -1$ if $a \oplus b = (a+b)^2$ in $(Z, \oplus)$ (iv) $\frac{1}{2} \otimes \frac{3}{4}$ if $a \otimes b = a^3 + b^3$ in $(Z, \otimes)$

4. A binary operation ^{o}, is defined as a composition of function, compute:
   (a) fog (b) gof
   If (i) $f(x) = 2x - 1,\ g(x) = x + 3$ (ii) $f(x) = 1 - x,\ g(x) = 5x + 2$
   (iii) $f(x) = x^2 - 1,\ g(x) = 5 - x + x^2$ (iv) $f(x) = 10 + 3x,\ g(x) = 2(x + 1)$

5. Create an operation table on modulo 3 for each of the following:
   (a) $(\{0, 1\}, +)$ (b) $(\{0, 1, 2\}, \times)$ (c) $(\{-1, 0, 1\}, \times)$

6. Obtain the operation table of the binary operation*, where $a * b$ is the remainder when ab is divided by 6. Hence, state whether the set is closed under the operation*.
   (a)$\{1, 4\}$ (b) $\{1, 2, 3\}$ (c) $\{3, 5, 7\}$ (d) $\{1, 2, 3, 4\}$ (e) $\{1, 2, 4\}$ (f) $\{1, 3, 5\}$

7. Determine whether or not if the closure property is satisfied:
   (i) $(Z, \otimes\}$ where $a \otimes b = a + b$
   (ii) $(Q, *\}$ where $a \otimes b = ab + a - b$
   (iii)$(N, \circ)$ where $a \circ b = (ab)^2 - 4$
   (iv) $(R^+, *)$ where $a * b = 2\log_a b$

8. Determine whether or not $(S, *)$ is commutative in each of the following:
   (i) $(Z, -)$ (ii) $(N, +)$ (iii) $(R, \times)$ (iv) $(Q^+, \div)$

9. Check if the property of commutativity is satisfied:
   (i) $(N, -), a * b = a^2 b$ (ii) $(R, \triangle), a \triangle b = b^a$
   (iii) $(Z, \oplus), a \oplus b = a^2 - b^2$ (iv) $(Q, \otimes), a \otimes b = ab$
   (v) $(Q, *), \frac{a+b}{ab}$ (vi) $(N, *), a * b = a + b - 5$

10. Examine whether or not the property of associativity is satisfied in each of the following:
   (a) $(Z, *), x * y = a + b - ab$ (b) $(Q, \otimes), x^2(y^2 - 1)$

11. State the identity element in each of the following:
   (i) $(N, -)$ (ii) $(Z, +)$ (iii) $(R, \times)$, (iv) $(Q^*, \div)$ (v) $(Z^+, \times)$ (vi) $(E, \cup)$

12. Find the identity element if it exists in each of the following:
   (i) $(\{1, 6\}, +)$ under modulo 8 (ii) $(\{2, 4\}, \times)$ under modulo 6
   (iii) $(\{1, 3, 7, 10, 12, 6\}, \times)$ under modulo 20.

13. Show that $\div$ is right distributive over $+$ but not left distributive in $(R^*+, \div)$

14. Show whether $+$ is distributive over $\times$ in each of the following:
   (i) $(R, +, \times)$ (ii) $(Z, +, \times)$ (iii) $(Q, +, \times)$

15. Create an operation$^{\oplus}$ table on addition under modulo 6. Hence, from the table find:
   (a) the identity element
   (b) the inverse of (i) 0 (ii) 1 (iii) 5 (iv) 2 (iii) 3

16. Test whether $*$ is distributive over $\oplus$ in $(R, \oplus, *)$ if:
   (a) $x \oplus y = \frac{1}{2}(x - y), \quad x * y = \frac{1}{2}xy$
   (b)$x \oplus y = x + y + 2, \quad x * y = xy + x + y$

17. Given the set $(Z, \oplus)$ where $\oplus$ is defined for all integers x and y as $x \oplus y = xy - x + y$, evaluate:
   (i) $4 \oplus 6$ (ii) $-7 \oplus 8$ (iii) $x \oplus 5 = -11$, find x
   (iv) $x \oplus - 3 = 1$, find x.

18. A binary operation is defined by $x * y = x^2 y - 3xy$, evaluate:
   (i) $-2 * 5$ (ii) $x \oplus 5 = 20$ (iii) $-3 * -1 = x$ (iv) $x * 2 = 8 * x$

14 STATISTICS

Statistics is a branch of mathematics that deals with the collection, organisation, interpretation and storage of large amount of data for a specific purpose.

14.1 DATA COLLECTION, CLASSIFICATION AND PRESENTATION

Data means raw facts that can be measured, quantified, and processed to meaningful information.

14.1.1 DATA COLLECTION

Data can be collected by sampling, measurement, or selection. Thus, a sampled data is a data selected at random from a population, and used to test hypotheses about the population.

14.1.2 DATA CLASSIFICATION

Data are classified as discrete and continuous data.

Discrete Data: These are data obtained by "counting." Examples are the number of students that offer mathematics; the number of pages of a book; the number of bottled drinks in a refrigerator, etc. These data are called discrete because they involve discrete numbers (i.e., absolute whole numbers) such as 1, 2, 3, 4, 5, ...

Continuous Data: These are data obtained by measurement, observation, and assumption. Examples are volume of water in a tank; area of football field; mass of a car in kilogram; height of a student in a certain class.

14.1.3 DATA PRESENTATION

This is the format or way data are arranged for easy interpretation. There are two common forms of data presentation: tabular form and charts/graph.

Tabular Form: This uses frequency table which is known as frequency distribution table for the presentation of data so that it can be easily read and analysed.

The frequency table has a row and column for data entry. The row contains each possible data entry, while the column defines the number of occurrence (i.e., frequency) of each of these data. Thus, the frequency table has the advantage of presenting data in an orderly manner; also, information can be easily read just by mere glance at the table.

Charts/Graph: This is a visual display of data on a graph. The common charts used for data presentation are: pie chart, bar chart, histogram, line graph, and frequency polygon.

14.2 GROUPED FREQUENCY DISTRIBUTION

Grouped frequency distribution is associated with data that have several data entry repetition/recurrence. The following terms are used for describing grouped frequency distribution:

Class Interval: In order to classify data in groups, the number of groups (called classes) must be decided at the beginning of data entry. Each of these groups is given an interval called the "class interval." Thus, *class interval* can be defined as the number of classes in a frequency table. Suppose we wish to classify a data into 10 equal intervals, then the class interval is 10 and the difference between the least entry and the highest entry is divided by 10, to give the width.

Also, class intervals have limits. The beginning unit of a class interval is called the "Lower Class Limit," while the maximum end unit is called the "Upper Class Limit." Take for instance, if we have class interval $5 - 9$; then 5 is the lower limit and 9 is the upper limit of the class, while the class width is the difference between the upper limit and lower limit, i.e., $9 - 5 = 4$.

Class Boundary: It is used to account for the possibilities of rounded figures within a class interval in a grouped frequency distribution.

Class Mark: Another term associated with grouped frequency distribution is "class mark." The *class mark* is the mid-point of any class interval. For instance, a class interval of $70 - 75$ will have a class mark of,

$$x = \frac{70+75}{2} = \frac{145}{2} = 72.5$$

Cumulative frequency: This is a frequency distribution that has its frequencies added up successively.

Example 1

1	5	3	5	7	2	12	4	2	10
2	8	5	3	2	3	14	7	6	2
10	3	1	14	11	9	7	6	3	8

From the given data, construct a frequency table with a class interval of $1 - 5, 6 - 10, \ldots$ and determine the class mark of each class.

Solution

Table 14.1

Class Interval	Class Boundary	Class Mark (x)	Frequency
1 – 5	0.5 – 5.5	3	16
6 – 10	5.5 – 10.5	8	10
11 – 15	10.5 – 15.5	13	4
			$\sum F = 30$

Example 2

14	20	30	23	5	19	1	4	42	10
12	28	32	25	2	15	8	43	6	2
16	15	34	27	3	4	48	36	3	8
7	49	28	17	18	2	35	23	41	17
46	11	38	24	19	5	47	25	34	21

The above distribution is the marks of 50 students in a mathematics examination. Using ten equal class intervals, construct:
(a) a grouped frequency table;
(b) cumulative frequency table.

Solution

The lowest mark is 1, and the highest mark is 49.
$\Rightarrow$ the range $= 49 - 1 = 48$

So, class width $= \dfrac{\text{range}}{\text{class interval}}$

$= \dfrac{48}{10} = 4.8$

$\Rightarrow 4.8 \triangleq 5$

This number is approximated to 5 for conveniency. Thus, we are going to use a class interval of $1–5, 6 - 10,\ 11 - 15, \dots, 46 - 50$; see table 14.2 and 14.3 on next page.

(a)

Table 14.2 Grouped Frequency Table

Class Interval	Class Boundary	Class Mark (x)	Frequency
1 – 5	0.5 – 5.5	3	10
6 – 10	5.5 – 10.5	8	5
11 – 15	10.5 – 15.5	13	5
16 – 20	15.5 – 20.5	18	7
21 – 25	20.5 – 25.5	23	6
26 – 30	25.5 – 30.5	28	4
31 – 35	30.5 – 35.5	33	4
36 – 40	35.5 – 40.5	38	2
41 – 45	40.5 – 45.5	43	3
46 – 50	45.5 – 50.5	48	4
			$\sum F = 50$

(b)

Table 14. 3 Cumulative Frequency Table

Class Interval	Class Boundary	Class Mark (x)	Cumulative Frequency
1 – 5	0.5 – 5.5	3	10
6 – 10	5.5 – 10.5	8	15
11 – 15	10.5 – 15.5	13	20
16 – 20	15.5 – 20.5	18	27
21 – 25	20.5 – 25.5	23	33
26 – 30	25.5 – 30.5	28	37
31 – 35	30.5 – 35.5	33	41
36 – 40	35.5 – 40.5	38	43
41 – 45	40.5 – 45.5	43	46
46 – 50	45.5 – 50.5	48	50

14.3 CHARTS

This is the pictorial /graphical representation of data.

14.3.1 PICTOGRAM

Pictogram uses simple and clear symbols to represent data, and a key is then used to define the quantity of each symbol.

Example 3

The total bags of rice imported to Nigeria from 2020 to 2024 are given in the table below:

Table 14.4 Imported Goods

Year	Rice (bags)
2020	10000
2021	8000
2022	6000
2023	5000
2024	3000

Draw a pictogram to show the total import for each given year.

Solution

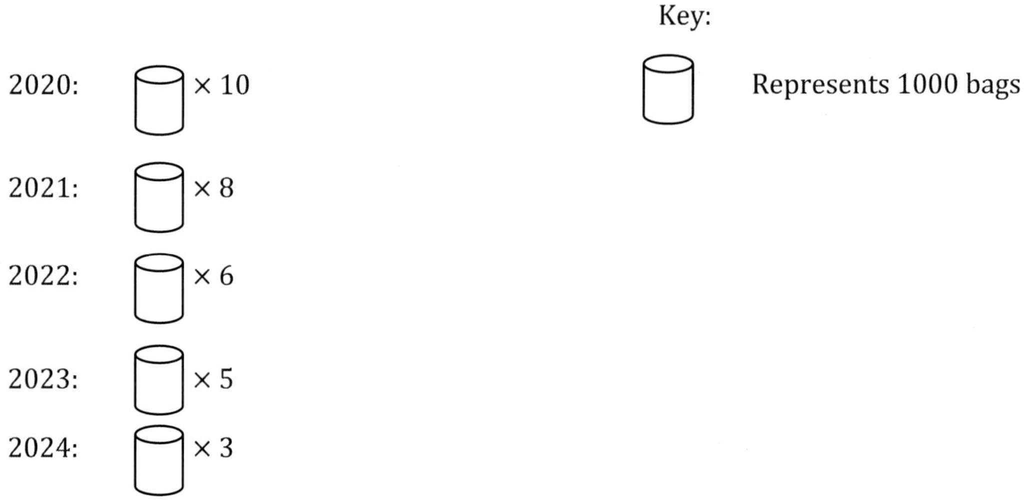

Example 4

In the Unified Tertiary Matriculation Examination (UTME) conducted in 2024, if it was recorded that 100,000 students wrote mathematics; 60,000 wrote physics; 50,000 wrote chemistry; and 1,200,000 wrote English language. Represent the given information using a pictogram.

Solution

Table 14.5

Subject	Number of Students
English language	☺ × 120
Mathematics	☺ × 10
Physics	☺ × 6
Chemistry	☺ × 5

Key: ☺ represents 10,000 students.

14.3.2 BAR CHART

Bar chart uses rectangular bars of equal widths, such that the heights of the rectangles are proportional to the sizes of the number of items they represent. Bar chart is classified into simple and component/composite bar chart. Hence for simple bar chart, the bar represents just a single item. Whereas, in component/composite bar chart, each bar represents more than one item at a time.

Example 5

The number of graduates in 2024 by a country's statistical record is given below:

Table 14.6

Profession	Number of Graduates
Engineering	10,000
Medical Science	7,000
Accountancy	8,000
Law	12,000
Economics	5,000
Psychology	3,000
Philosophy	1,500

Use the information to construct a bar chart which will represent the above data.

Solution

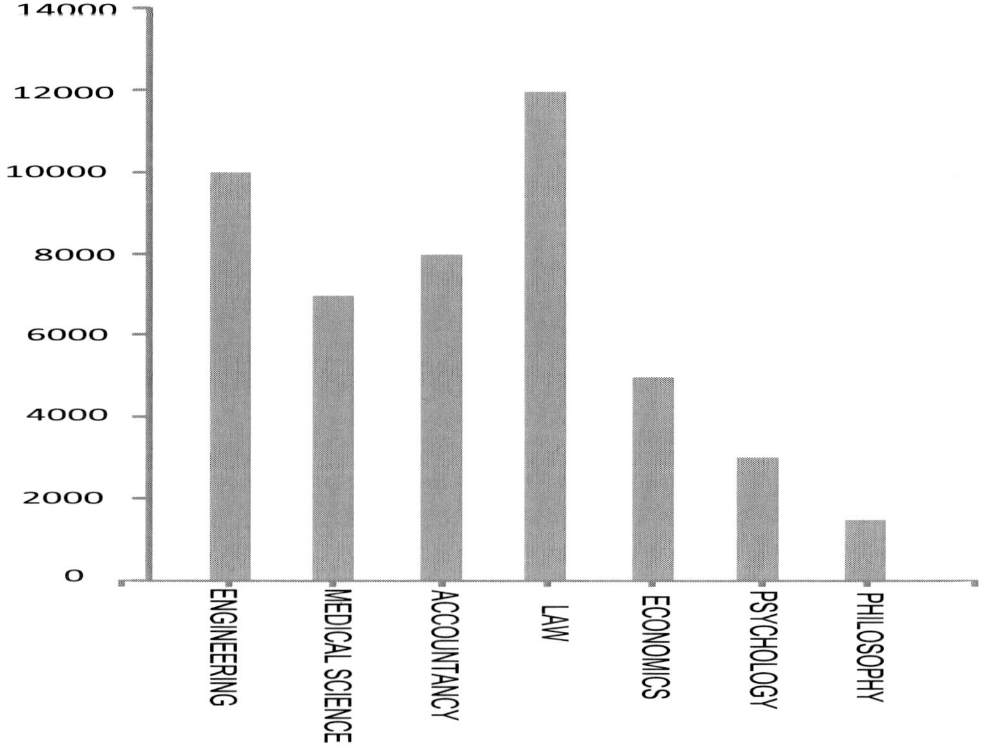

Fig. 14.1 Simple Bar Chart

Example 6

The total number of male and female students in a certain school that enrolled for the Senior Secondary School Certificate Examination (SSCE) from 2005 to 2010 are represented below:

Table 14.7

YEAR	MALE STUDENTS	FEMALE STUDENTS
2005	80	50
2006	70	45
2007	90	60
2008	50	70
2009	100	90
2010	120	120

Draw a composite bar chart to represent the given information.

Solution

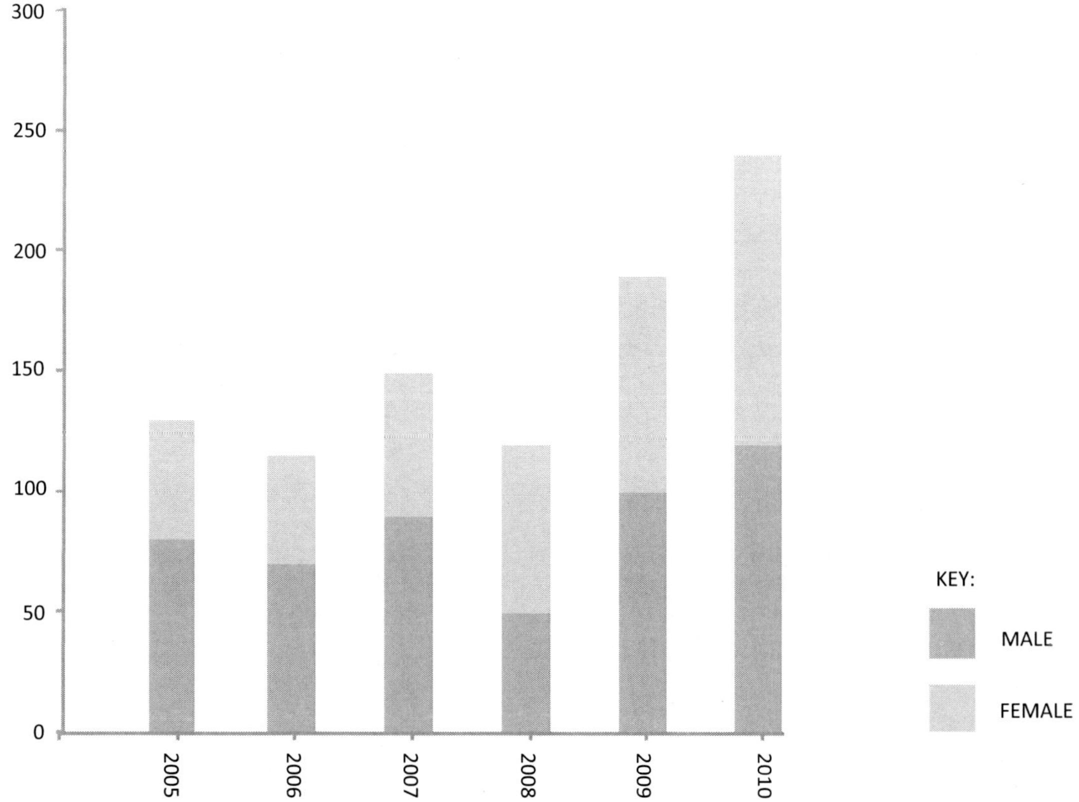

Fig. 14.2 Composite Bar Chart

14.3.3 HISTOGRAM

A histogram is a graphical representation of data (frequency distribution) using equal rectangular bars joined together (i.e., no space between them); it is constructed in similar way as bar chart, except that each bar is not separated/spaced.

Example 7

Construct a histogram using similar data in table 14.6 of example 5.

Solution

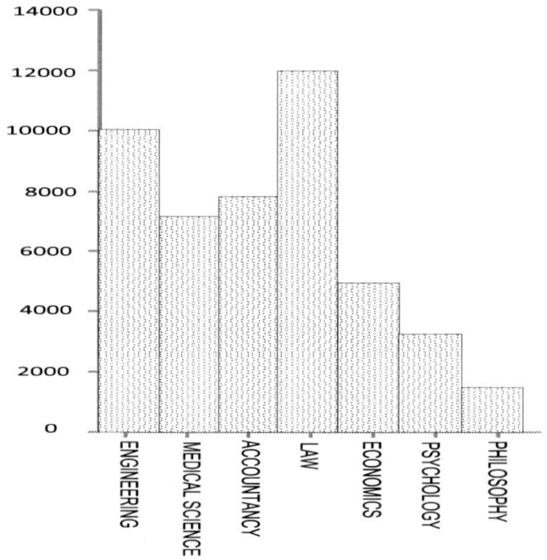

Fig. 14.3 Histogram

14.3.4 LINE GRAPH

This uses lines instead of rectangles to represent frequency distribution. The advantage is that it is simplest to construct than bar chart and histogram.

Example 8

Construct a line graph using similar data in table 14.6 of example 5.

Solution

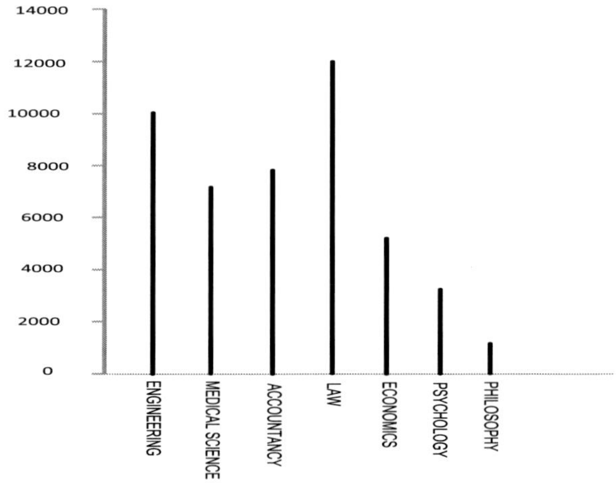

Fig. 14.4 Line Graph

14.3.5 PIE CHART

A pie chart is a circular chart divided into sectors, and the area of each sector represents the magnitude of a given data which is also a percentage of the whole data.

Example 9

The prices of items bought from a shop is given below:

Table 14.8

Item	Price(₦)
Shoe	2,000
T- shirt	1,000
Wrist- watch	500
Eye-glass	1,500
Ear-rings	1000

Solution

Total price = 2000 + 1000 + 500 + 1500 + 1000 = ₦6,000

The angle for each sector is given by:

$$\angle\, for\ an\ item, x = \frac{price\ of\ x}{total\ price} \times 360°$$

$$\Rightarrow \angle\, for\ Shoe = \frac{2000}{6000} \times 360° = 120°$$

$$\Rightarrow \angle\, for\ Tshirt = \frac{1000}{6000} \times 360° = 60°$$

$$\Rightarrow \angle\, for\ Wrist\ watch = \frac{500}{6000} \times 360° = 30°$$

$$\Rightarrow \angle\, for\ Eye\ glass = \frac{1500}{6000} \times 360° = 90°$$

$$\Rightarrow \angle\, for\ Ear\ rings = \frac{1000}{6000} \times 360° = 60°$$

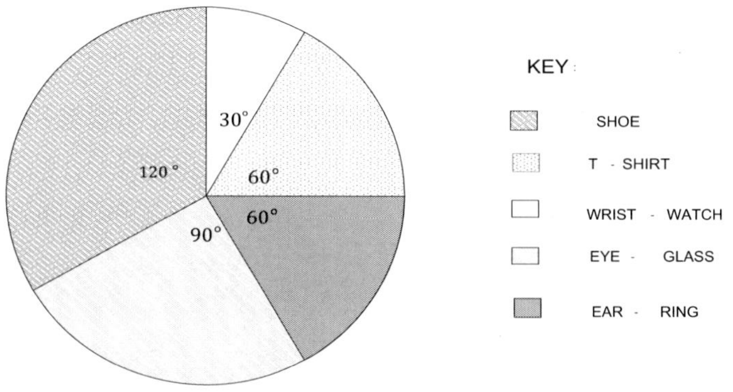

Fig. 14.5 Pie–Chart

14.3.6 FREQUENCY POLYGON

A frequency polygon is a line graph generated by joining each of the class- mid points of the top of the rectangular bars of a histogram. Thus, each of the mid-points must corresponds to their respective height (frequency); then each point is joined together by drawing a straight line to obtain the frequency polygon. However, when the points are joined by drawing a smooth curve, it is called a "frequency curve."

Example 10

Construct a frequency polygon of table 14.9.

Table 14.9

age (yrs)	*Class boundary*	*Class mark*	*Frequency*
11 – 15	10.5 – 15.5	13	2
16 – 20	15.5 – 20.5	18	5
21 – 25	20.5 – 25.5	23	8
26 – 30	25.5 – 30.5	28	4
31 – 35	30.5 – 35.5	33	7
35 – 40	35.5 – 40.5	38	1

Solution

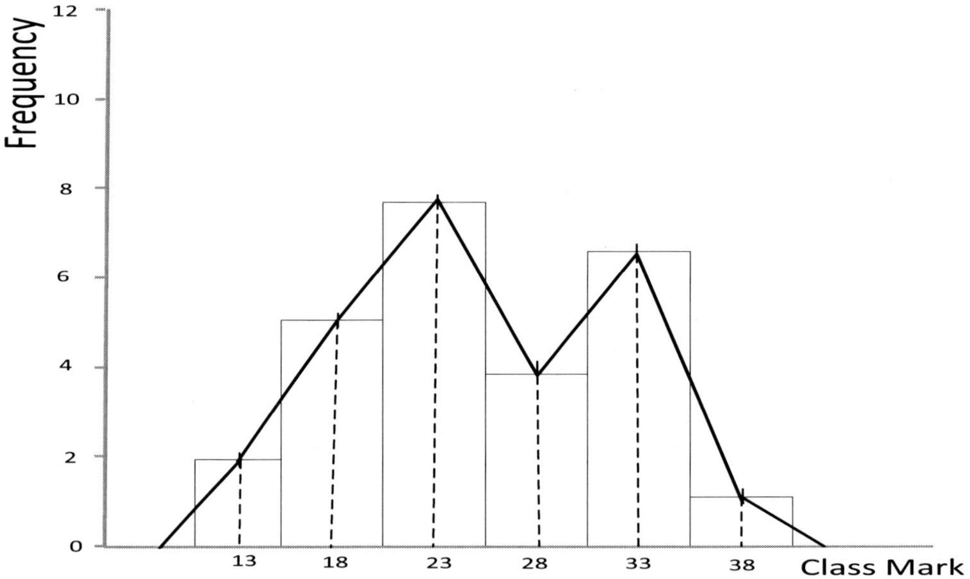

Fig. 14.6 Frequency Polygon

14.3.7 CUMULATIVE FREQUENCY CURVE (OGIVE)

This is the graph of cumulative frequency against the upper limit of the class boundaries. Thus, the ogive helps us to estimate median, percentile, and interquartile range of a given distribution.

Example 11

Use the data in table 14.10 to draw the cumulative frequency curve (ogive).

Table 14.10

Mass(kg)	Class boundary	*Frequency*	*Cumulative Frequency*
10 – 14	9.5–14.5	4	4
15 – 19	14.5 – 19.5	2	6
20 – 24	19.5 – 24.5	3	9
25 – 29	24.5 – 29.5	5	14
30 – 34	29.5 – 34.5	8	22
35 – 39	34.5 – 39.5	10	32
40 – 44	39.5 – 44.5	9	41
45 – 49	44.5 – 49.5	11	52
50 – 54	49.5 – 54.5	5	57
55 – 59	54.5 – 59.5	3	60

From the ogive, determine:

(a) the 25th percentile
(b) 75th percentile
(c) 50th percentile
(d) inter-quartile range

Solution

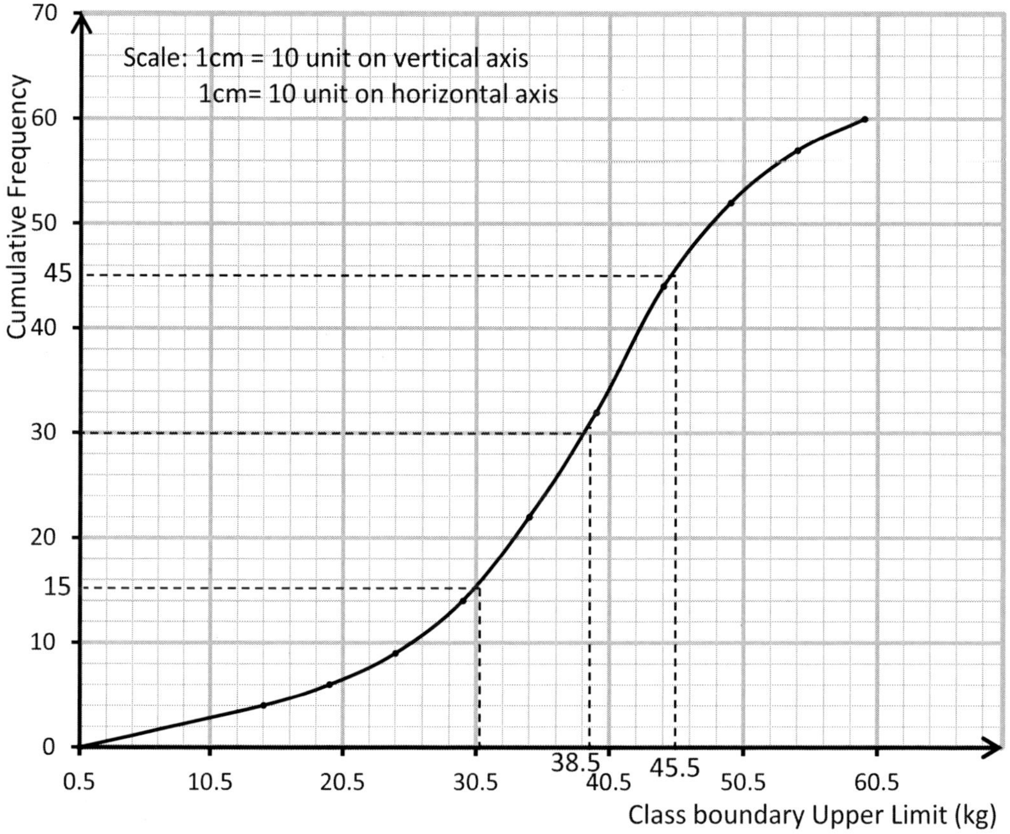

Fig. 14.7 Ogive

(a) From the ogive drawn in fig. 14.7, the 25th percentile is given by:

$\frac{25}{100} \times \sum F$ where $\sum F$ is the sum of all the frequency.

$= \frac{25}{100} \times 60 = 15$ C.F term

$= 30.5$kg.

From the ogive, trace the **15 C.F** term on the cumulative frequency axis to touch the ogive. Then trace down to the class boundary axis to find the corresponding value.

(b) The 75th percentile is given by:

$\frac{75}{100} \times \sum F \Rightarrow \frac{75}{100} \times 60 = 45$ C.F term

$= 45.5$kg

From the ogive, the **45 C.F** term is **45.5kg**

(c) The 50th percentile is given by:

$Q_2 = \frac{1}{2} \times \sum F = \frac{1}{2} \times 60 = 30$ C.F term

$= 38.5$kg. From the ogive, the 30 C.F term is 38.5kg

(d) The interquartile range is given by:

$Q_3 - Q_1 = 45.5 - 30.5 = 15$

$= 15$kg.

14.4 MEASURES OF CENTRAL TENDENCY

These are statistical measures that describe the middle/central element(s) of a set of data. Thus, the measures of central tendency include mean, mode and median.

14.4.1 MEAN

This is sometimes called "arithmetic mean," it is the average of a data. Thus, the mean of ungrouped or grouped data is the ratio of the total sum of items to the total number of the given items. It is the best and most accurate method of obtaining the correct representation of the center of a set of data.

Mathematically,

$\bar{x} = \frac{\sum x_i}{N}$ .... {for ungrouped data}

Where: $\bar{x}$= Arithmetic mean;

x_i= Individual item;

$\sum x_i$= Sum of the individual item from $i = 1$ to $i = n$

$\bar{x} = \frac{\sum Fx_i}{\sum F}$ .... {for grouped data}

Where : $\bar{x}$ = Arithmetic mean;

x_i =Individual item;

$\sum Fx_i$ = Sum of the product of the corresponding frequency and the individual item;

$\sum F$ = Total frequency of data

However, the arithmetic mean of a given data can also be determined using "Assumed mean method" also called guessed mean.

$\bar{x} = A + \frac{\sum d_i}{N}$ .... {for ungrouped data}

$\bar{x} = A + \frac{\sum Fd_i}{\sum F}$ .... {for grouped data}

Where $\bar{x}$= Arithmetic mean; A = Assumed mean d_i = Deviation of x_i from A;

$\sum Fd_i$ = Sum of the product of the corresponding frequency and the deviation;

$\sum F$ = Total frequency of data

Example 12

Find the mean of the given data 1, 2, 3, 5, 7, 8, 9, 12, 18, 15.

Solution

This is an ungrouped data. So, the best formula to use is $\bar{x} = \frac{\sum x_i}{N}$

$\Rightarrow \bar{x} = \frac{1+2+3+5+7+8+9+12+18+15}{10} = \frac{80}{10} = 8$

$\therefore \bar{x} = 8$

Example 13

Consider the score of 40 students in a physics test.

3	8	10	3	9	5	5	9
10	8	9	3	5	9	9	5
9	5	9	5	9	5	8	3
8	5	5	9	5	10	8	10
5	10	9	5	8	5	9	8

Obtain the mean score of the data.

Solution

First, obtain the frequency distribution table; then apply the formula for finding mean of grouped data.

Table 14.11

Score(x)	*Frequency(F)*	*Fx*
3	4	12
8	7	56
10	5	50
9	11	99
5	13	65
...	$\sum F = 40$	$\sum Fx = 282$

$\Rightarrow \bar{x} = \frac{\sum Fx}{\sum F} = \frac{282}{40} = 7.05$ $\quad\therefore$ the mean $\bar{x} = 7.05$

Example 14

The table below shows the distribution of weekly wages earned by 50 employees in SICON construction company in Abuja. Using an assumed mean of ₦3245, calculate the mean of the distribution.

Table 14.12

Wages(₦)	Number of Employees
2000 – 2490	20
2500 – 2990	15
3000 – 3490	4
3500 – 3990	8
4000 – 4490	2
4500 – 4990	1

Solution

Table 14.13 *Assumed Mean* = ₦3245

Wages (₦)	*Class Mark*	*Frequency*(F)	$d = x - A$	Fd
2000 – 2490	2245	20	−1000	−20000
2500 – 2990	2745	15	−500	−7500
3000 – 3490	3245	4	0	0
3500 – 3990	3745	8	500	4000
4000 – 4490	4245	2	1000	2000
4500 – 4990	4745	1	1500	1500
…	…	$\sum F = 50$	…	$\sum Fd = -20000$

$$\Rightarrow \bar{x} = A + \frac{\sum Fd}{\sum F} = 3245 + \left(\frac{-20000}{50}\right)$$
$$= 3245 - 400 = 2845$$

$\therefore$ the mean $\bar{x} = 2845$

14.4.2 MEDIAN

The median of a set of data is the middle (i.e., center) value when the set of data is orderly arranged in ascending or descending order.

Mathematically,

$M_e = \left(\frac{N+1}{2}\right)^{th}$ term .... {for ungrouped data or frequency distribution when N is odd}

$M_e = \frac{\left(\frac{N}{2}\right)^{th} + \left(\frac{N}{2}+1\right)^{th}}{2}$ term ...... {for grouped data or frequency distribution when N is even}

$M_e = L_1 + \left(\frac{\frac{N}{2} - f_1}{f_m}\right) C$ .... {for grouped data}

Where $M_e = Median$; $L_1 = Lower\ Class\ Boundary$;
$f_1 = Cumulative\ Frequency\ of\ the\ Class\ just\ before\ the\ Median\ Class$;
$f_m = Frequency\ of\ the\ Median\ Class$; $C = Class\ Interval(Width)$;
$N = Total\ Frequency\ (i.e. Cumulative\ Frequency)$

Example 15

Find the median of each of the following data:
(a) 5, 4, 8, 9, 7, 6, 3, 2, 12
(b) 2, 3, 4, 6, 9, 8, 7, 2, 9, 13

Solution

(a) Given 5, 4, 8, 9, 7, 6, 3, 2, 12

Rearrange the numbers in ascending order of magnitude:

since $N = 9$ then $= \left(\frac{N+1}{2}\right)^{th} = \left(\frac{9+1}{2}\right)^{th} = \left(\frac{10}{2}\right)^{th} = 5^{th}$ term

2, 3, 4, 5, 6, 7, 8, 9, 12

$\Rightarrow M_e = 6$

This number is the 5th term after counting-off four numbers to the RHS or LHS.

(b) Given 2, 3, 4, 6, 9, 8, 7, 2, 9, 13
Rearrange the numbers in ascending order of magnitude:
2, 2, 3, 4, 6, 7, 8, 9, 9, 13

since $N = 10$ then $M_e = \frac{\left(\frac{N}{2}\right)^{th} + \left(\frac{N}{2}+1\right)^{th}}{2} = \frac{\left(\frac{10}{2}\right)^{th} + \left(\frac{10}{2}+1\right)^{th}}{2} = \frac{5^{th}+6^{th}}{2}$ term

$\Rightarrow \frac{5^{th}+6^{th}}{2}$ term lies between 6 and 7

$\Rightarrow M_e = \frac{6+7}{2} = \frac{13}{2} = 6.5 \qquad \therefore M_e = 6.5$

Example 16

Estimate the median of the frequency distribution in table 14.14.

Table 14.14

Mass(kg)	5	15	25	35	45
Frequency	3	6	10	5	1

Solution

The data is a frequency distribution without class interval. So, a cumulative frequency table is prepared for the distribution.

Table 14.15

Mass(kg)	5	15	25	35	45
Frequency	3	6	10	5	1
Cumulative Freq.	3	9	19	24	25

Cumulative Freq. $= N = 25$

$\Rightarrow M_e = \left(\frac{N+1}{2}\right)^{th} = \left(\frac{25+1}{2}\right)^{th} = \left(\frac{26}{2}\right)^{th} = 13^{th}$ term

Thus, the 13th term is 25kg

Example 17

Table. 14.16 shows the age distribution of young people in a polygamous family.

Table 14.16

Age (yrs)	1 – 5	6 – 10	11 – 15	16 – 20	21 – 25
Number	10	15	30	5	25

Hence, from the table estimate the median of the age distribution.

Solution

In order to estimate the median, create a cumulative frequency table and class boundary as shown in table 14.17.

Table 14.17

Age (yrs)	*Class Boundary*	*Number(Frequency)*	*Cumulative Frequency*
1 – 5	0.5 – 5.5	10	10
6 – 10	5.5 – 10.5	15	25
11 – 15	10.5 – 15.5	30	55
16 – 20	15.5 – 20.5	5	60
21 – 25	20.5 – 25.5	25	85

The frequency of the median class will be located at, $M_e = \left(\frac{N+1}{2}\right)^{th}$ term

$\Rightarrow M_e = \frac{85+1}{2} = \frac{86}{2} = 43^{rd}$ term; hence frequency of the median class, $F_m = 30$

Then, sum of frequency above F_m is given as $f_1 = 25$; Class width $C=5$;

Lower class boundary of the median class, $L_1 = 10.5$; $N = 85$

$$\Rightarrow M_e = L_1 + \left(\frac{\frac{N}{2}-f_1}{f_m}\right)C = 10.5 + \left(\frac{\frac{85}{2}-25}{30}\right) \times 5 = 10.5 + \left(\frac{42.5-25}{30}\right) \times 5$$

$$= 10.5 + \left(\frac{17.5}{30}\right) \times 5 = 10.5 + \frac{87.5}{30} = 10.5 + 2.9 = 13.5$$

$\therefore M_e = 13.5$

14.4.3 MODE

The mode of distribution of a data is the value with the most frequent occurrence.

Mathematically,

$$M_o = L_1 + \left(\frac{f_m - f_1}{(f_m - f_1) + (f_m - f_2)}\right)C \qquad \text{\{for grouped data\}}$$

Where: M_o = mode of the Distribution

L_1 = Lower Class boundary of the modal class

f_m = Frequency of the modal class

f_1 = Frequency before the modal frequency

f_2 = Frequency after the modal frequency

C = Class width

However, for an ungrouped data, the mode is found by inspection. That is, looking directly at the number that occurred most in the list of distribution.

Example 18

Estimate the mode of each of the following distribution:

(a) 2, 1, 1, 3, 2, 4, 5, 8, 2, 3, 3, 2.

(b)

Table 14.18

Graduates	1000	2800	500	3000
Number of Schools	5	20	15	17

Solution

(a) 2, 1, 1, 3, 2, 4, 5, 8, 2, 3, 3, 2:

This is an ungrouped data; by inspection the mode is 2 because it occurs most than any other number.

$\therefore Mode = 2$

(b) From table 14.18, by inspection, it can be seen that the highest number of schools is 20. Thus, the corresponding number of graduates is 2800.

$\therefore Mode = 2800$

Example 19

Estimate the mode of the distribution in table 14.19.

Table 14.19

Age (yrs)	*Class Boundary*	*Number(Frequency)*
1 – 5	0.5 – 5.5	12
6 – 10	5.5 – 10.5	9
11 – 15	10.5 – 15.5	20
16 – 20	15.5 – 20.5	7
21 – 25	20.5 – 25.5	15

Solution

From table 14.19, we have that:

L_1 = Lower Class boundary of the modal class =10.5

f_m = Frequency of the modal class = 20

f_1 = Frequency before the modal frequency =9

f_2 = Frequency after the modal class =7

C = Class Width = 15.5−10.5 = 5

Now, using $M_o = L_1 + \left(\frac{f_m - f_1}{(f_m - f_1)+(f_m - f_2)}\right) C$

$$\Rightarrow M_o = 10.5 + \left(\frac{20-9}{(20-9)+(20-7)}\right) \times 5$$

$$M_o = 10.5 + \left(\frac{11}{11+13}\right) \times 5 = 10.5 + \left(\frac{11}{24}\right) \times 5 = 10.5 + \frac{55}{24}$$

$$= 10.5 + 2.3 = 12.8$$

$\therefore Mode = 12.8$ yrs

MODE FROM HISTOGRAM

The mode of a distribution can be calculated from the histogram of the given distribution. Thus, in order to obtain the mode, use the following steps:

(i) Locate the modal class by choosing the tallest rectangle.

(ii) Draw a line from the top left corner of this rectangle to touch the top left corner of the rectangle just to its right.

(iii) Similarly, draw a line from the top right corner of the tallest rectangle to touch the right top corner of the rectangle just to its left.

(iv) Now, from the intersection point of the lines, trace a perpendicular line to the horizontal (i.e., class boundary) axis.

(v) Finally, read-off the modal class at the point where the perpendicular trace touches the horizontal axis.

Example 20

Draw a histogram using table 14.20, hence estimate the mode of the distribution.

Table 14.20

Age (yrs)	*Class Boundary*	*Number(Frequency)*
1 – 5	0.5 – 5.5	12
6 – 10	5.5 – 10.5	9
11 – 15	10.5 – 15.5	20
16 – 20	15.5 – 20.5	7
21 – 25	20.5 – 25.5	15

Solution

The estimated mode from the histogram is approximately 12.8 years. However, when this answer is compared with the formula method as in example 19, it is observed that they are equal.

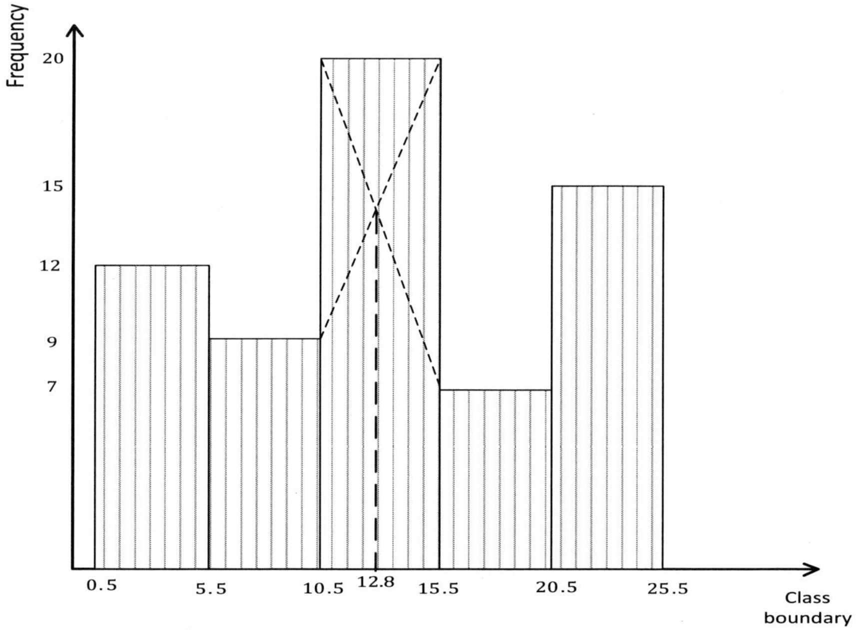

Fig. 14.8 Mode from Histogram

14.5 MEASURE OF DISPERSION

Dispersion refers to the spread/scattering of figures or data over a range. Therefore, the measure of dispersion are techniques employed for measuring the dispersion of set of data. These measures include: range, mean deviation, variance and standard deviation.

14.5.1 RANGE

This is the difference between the largest and smallest observation (data points) in a set. Thus, the range takes account of the two extreme values in a distribution; because of this, it does not really give an accurate representation of the distribution. But on the contrary, it is very simple to calculate, for this reason it is widely used.

Example 21

2, 3, 7, 9, 15, 20, 4, 3, 5, 8. Find the range of the distribution.

Solution

Range = Highest datum – Lowest datum $= 20 - 2 = 18$

$\therefore$ Range $= 18$

14.5.2 MEAN DEVIATION

This is the average deviation from the mean of a distribution.

$M_d = \frac{\sum|x-\bar{x}|}{N}$ .... { for ungrouped data}

$M_d = \frac{\sum F|x-\bar{x}|}{\sum F}$ or $M_d = \frac{\sum F|d|}{\sum F}$ .... { for grouped data}

Where $\bar{x} =$ Mean of the distribution;

$x =$ Datapoint (i.e., a member of a set of data);

$|d| = |x - \bar{x}| =$ Absolute values of the difference between $\bar{x}$ and x;

$N =$ Total number of data points, i.e., for ungrouped data;

$\sum F =$ Summation of the frequency of the distribution.

Example 22

Given the distribution: 2, 3, 6, 6, 7, 9, 9, 10, 11, 17.

Find (a) the mean (b) the mean deviation

Solution

(a) $\bar{x} = \frac{\sum x_i}{N} = \frac{2+3+6+6+7+9+9+10+11+17}{10}$

$= \frac{80}{10} = 8$

$\therefore$ mean, $\bar{x} = 8$

(b) The mean deviation is given by, $M_d = \frac{\sum|x-\bar{x}|}{N}$

$$\Rightarrow M_d = \frac{|2-8|+|3-8|+|6-8|+|6-8|+|7-8|+|9-8|+|9-8|+|10-8|+|11-8|+|17-8|}{10}$$

$$= \frac{|-6|+|-5|+|-2|+|-2|+|-1|+|1|+|1|+|2|+|3|+|9|}{10} = \frac{6+5+2+2+1+1+1+2+3+9}{10}$$

$$= \frac{32}{10} = 3.2$$

$\therefore$ mean deviation, $M_d = 3.2$

Example 23

Find the mean deviation of the distribution of marks obtained by 30 students in a chemistry test.

Table 14.21 Chemistry test score

Marks	5	6	7	8	9
Students	13	7	5	3	2

Solution

To find the mean deviation, create a table for the deviation of each datum.

Table 14.22

$Marks(x)$	$Frequency(F)$	Fx	$d = x - \bar{x}$	$\|d\| = \|x - \bar{x}\|$	$F\|d\|$
5	13	65	−1.13	1.13	14.69
6	7	42	−0.13	0.13	0.91
7	5	35	0.87	0.87	4.35
8	3	24	1.87	1.87	5.61
9	2	18	2.87	2.87	5.74
...	$\sum F = 30$	$\sum Fx = 184$	...	...	$\sum F\|d\| = 31.3$

$$\bar{x} = \frac{\sum Fx}{\sum F} = \frac{184}{30} = 6.13$$

$$\Rightarrow M_d = \frac{\sum F|x-\bar{x}|}{\sum F} = \frac{\sum F|d|}{\sum F}$$

$$= \frac{31.3}{30} = 1.04$$

$\therefore M_d = 1.04$

14.5.3 VARIANCE AND STANDARD DEVIATION

Variance: This is the value of the square of the deviations of random variables from its mean value. Thus, the variance can be calculated from any of the following formulas:

$$\delta^2 = \frac{\sum|x-\bar{x}|^2}{N} \quad \text{or} \quad \delta^2 = \frac{\sum x^2 - \frac{(\sum x)^2}{N}}{N}$$ ....{ for ungrouped data}

$$\delta^2 = \frac{\sum F|x-\bar{x}|^2}{\sum F} \quad \text{or} \quad \delta^2 = \frac{\sum Fx^2 - \frac{(\sum Fx)^2}{\sum F}}{\sum F}$$ ....{ for grouped data}

Where δ^2 = Variance

$\bar{x}$ = Mean of the distribution;

x = Datapoint (i.e., a member of a set of data);

$|x-\bar{x}|$ = Absolute values of the difference between $\bar{x}$ and x;

N = Total number of data points (i.e., for ungrouped data);

$\sum F$ = Summation of the frequency of the distribution.

Standard Deviation: This is the square root of the mean square deviation of a random variable from its mean value. Hence, it is simply the square root of the variance.

i.e. $standard\ deviation = \sqrt{Variance}$

Thus,

$$\delta = \sqrt{\frac{\sum|x-\bar{x}|^2}{N}} \quad \text{or} \quad \delta = \sqrt{\frac{\sum x^2 - \frac{(\sum x)^2}{N}}{N}}$$ ....{ for ungrouped data}

$$\delta = \sqrt{\frac{\sum F|x-\bar{x}|^2}{\sum F}} \quad \text{or} \quad \delta = \sqrt{\frac{\sum Fx^2 - \frac{(\sum Fx)^2}{\sum F}}{\sum F}}$$ ....{ for grouped data}

Where δ = Standard Deviation;

$\bar{x}$ = Mean of the distribution;

x = Datapoint (i.e., a member of a set of data);

$|x-\bar{x}|$ = Absolute values of the difference between $\bar{x}$ and x;

N = Total number of data points (i.e., for ungrouped data);

$\sum F$ = Summation of the frequency of the distribution.

Example 24

Calculate the variance and standard deviation of the distribution 5, 6, 7, 8, 9, 10

Solution

5, 6, 7, 8, 9, 10

$$\bar{x} = \frac{5+6+7+8+9+10}{6} = \frac{45}{6} = 7.5$$

$\Rightarrow$ Mean, $\bar{x} = 7.5$

Since this is an ungrouped data, then using the formula for variance, $\delta^2 = \frac{\sum|x-\bar{x}|^2}{N}$

$$\text{Thus, } \delta^2 = \frac{\sum|x-\bar{x}|^2}{N} = \frac{|5-7.5|^2+|6-7.5|^2+|7-7.5|^2+|8-7.5|^2+|9-7.5|^2+|10-7.5|^2}{6}$$

$$= \frac{2.5^2+1.5^2+0.5^2+0.5^2+1.5^2+2.5^2}{6} = \frac{6.25+2.25+0.25+0.25+2.25+6.25}{6}$$

$$= \frac{17.5}{6} = 2.92$$

$\therefore$ Variance, $\delta^2 = 2.92$

The standard deviation is calculated as, $\delta = \sqrt{2.92} \triangleq 1.71$

$\therefore$ Standard deviation, $\delta = 1.71$

Example 25

Using the formula, $\delta = \sqrt{\frac{\sum x^2 - \frac{(\sum x)^2}{N}}{N}}$

Calculate the standard deviation of the distribution 2, 5, 8, 7, 3, 2, 3.

Solution

2, 5, 8, 7, 3, 2, 3 is an ungrouped data.

$\sum x = 2+5+8+7+3+2+3 = \mathbf{30}$

$(\sum x)^2 = 30^2 = \mathbf{900}$

$\sum x^2 = 2^2+5^2+8^2+7^2+3^2+2^2+3^2$

$= 4+25+64+49+9+4+9 = \mathbf{164}$

$N = \mathbf{7}$

Thus, using $\delta = \sqrt{\frac{\sum x^2 - \frac{(\sum x)^2}{N}}{N}}$ we have that:

$$\sqrt{\frac{164 - \frac{900}{7}}{7}} = \sqrt{\frac{164-128.6}{7}} = \sqrt{\frac{35.4}{7}} = \sqrt{5.06} = 2.25$$

$\therefore$ the standard deviation, $\delta = 2.25$

Example 26

Use the information in table 14.23 to find the (i) variance (ii) standard deviation

Table 14.23

x	2	4	6	8	10
F	5	3	10	7	5

Solution

Table 14.24

x	F	Fx	$\lvert x-\bar{x}\rvert$	$\lvert x-\bar{x}\rvert^2$	$F\lvert x-\bar{x}\rvert^2$
2	5	10	4.27	18.23	91.15
4	3	12	2.27	5.15	15.45
6	10	60	0.27	0.073	0.73
8	7	56	1.73	2.99	20.93
10	5	50	3.73	13.91	69.55
...	$\sum F = 30$	$\sum Fx = 188$	...	...	$\sum F\lvert x-\bar{x}\rvert^2 = 197.81$

$\bar{x} = \frac{\sum Fx}{\sum F} = \frac{188}{30} = 6.27$ Therefore, the mean is 6.27

(i) The variance, $\delta^2 = \frac{\sum F|x-\bar{x}|^2}{\sum F} = \frac{197.81}{30} \triangleq 6.6$

$\therefore\ \delta^2 \triangleq 6.6$

(ii) The standard deviation, $\delta = \sqrt{\frac{\sum F|x-\bar{x}|^2}{\sum F}} = \sqrt{6.6} = 2.57$

$\therefore\ \delta \triangleq 2.57$

Alternative method:

The variance and standard deviation can be computed using another approach as shown in table 14.25.

Table 14.25

x	F	Fx	x^2	Fx^2
2	5	10	4	20
4	3	12	16	48
6	10	60	36	360
8	7	56	64	448
10	5	50	100	500
...	$\sum F = 30$	$\sum Fx = 188$	...	$\sum Fx^2 = 1376$

(i) $\delta^2 = \dfrac{\Sigma Fx^2 - \frac{(\Sigma Fx)^2}{\Sigma F}}{\Sigma F} = \dfrac{1376 - \frac{(188)^2}{30}}{30} = \dfrac{1376 - \frac{35344}{30}}{30}$

$= \dfrac{1376-1178}{30} = \dfrac{198}{30} = 6.6$

$\therefore$ The variance, $\delta^2 = 6.6$

(ii) Standard deviation, $\delta = \sqrt{6.6} = 2.57$

$\therefore$ $\delta = 2.57$

Example 27

The scores of 50 students in a mathematics examination are listed below:

3	12	11	18	20	31	33	35	58	72
36	38	40	43	70	95	84	86	61	63
42	45	21	23	36	22	37	50	39	43
21	23	25	28	31	25	29	27	28	59
29	27	39	38	48	69	78	65	53	54

Using a class interval of 1 – 10, 11–20, 21–30, . . . calculate:
(a) the mean (b) variance (c) standard deviation

Solution

Table 14.26

Class Interval	F	Class mark (x)	Fx	$\lvert x - \bar{x}\rvert$	$\lvert x - \bar{x}\rvert^2$	$F\lvert x - \bar{x}\rvert^2$
1 – 10	1	5.5	5.5	32.14	1032.98	1032.98
11 – 20	4	15.5	62.0	22.14	490.18	1960.72
21 – 30	13	25.5	331.5	12.14	147.38	1915.94
31 – 40	11	35.5	390.5	2.14	4.58	50.38
41 – 50	6	45.5	50	7.86	61.78	370.68
51 – 60	4	55.5	222	17.86	318.98	1275.92
61 – 70	5	65.5	327.5	27.86	776.18	3880.90
71 – 80	3	75.5	226.5	37.86	1433.38	4300.14
81 – 90	2	85.5	171	47.86	2290.58	4581.16
91 – 100	1	95.5	95.5	57.86	3347.78	3347.78
…	$\sum F = 50$	…	$\sum Fx = 1882$	…	…	$\sum F\lvert x - \bar{x}\rvert^2 = 22716.6$

(a) $\bar{x} = \dfrac{\sum Fx}{\sum F} = \dfrac{1882}{50} = 37.64$

$\therefore$ the mean $\bar{x} = 37.64$

(b) the variance, $\delta^2 = \dfrac{\sum F|x-\bar{x}|^2}{\sum F} = \dfrac{22716.6}{50} \triangleq 454.33$

$\therefore \delta^2 = 454.33$

(c) Standard deviation, $\delta = \sqrt{variance} = \sqrt{454.33}$

$\therefore \delta = 21.32$

OBJECTIVE TEST 14

Table 14.27

Score (x)	0	1	2	3	4	5	6	7	8	9	10
Number of students	5	10	3	6	4	2	1	8	3	7	1

The table above shows the scores obtained by students in a computer science test. Use this to answer the question 1−3.

1. How many students took the test?
   (A) 25 (B) 40 (C) 50 (D) 60
2. What is the mean score?
   (A) 2.38 (B) 4.38 (C) 3.65 (D) 2
3. Calculate the mean deviation.
   (A) 2.89 (B) 2.98 (C) 4.38 (D) 3.89
4. The mean score of 5 students is 80. If another student's score is added so that the new mean score becomes 82, what is the value of the added score?
   (A) 10 (B) 70 (C) 80 (D) 92

5.

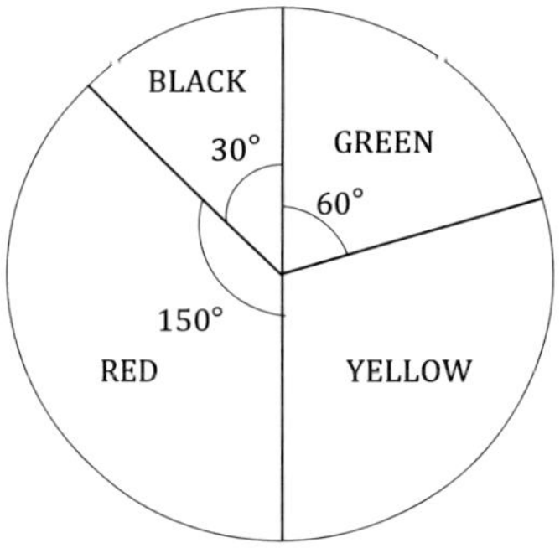

Fig. 14.9

A 60kg sample is a composite of four colors as shown in fig. 14.9. What is the weight of yellow in the sample?

(A) 10kg (B) 15kg (C) 20kg (D) 30kg

6. Find the difference between the range and the variance of the set of numbers 3, 5, 6, 8, 10, 4, 3, 8, 5, 8
   (A) 7 (B) 5.2 (C) 12.2 (D) 1.8

7. Which of the following is not a measure of dispersion?
   (A) Range (B) Standard deviation (C) Mode (D) Semi- interquartile range

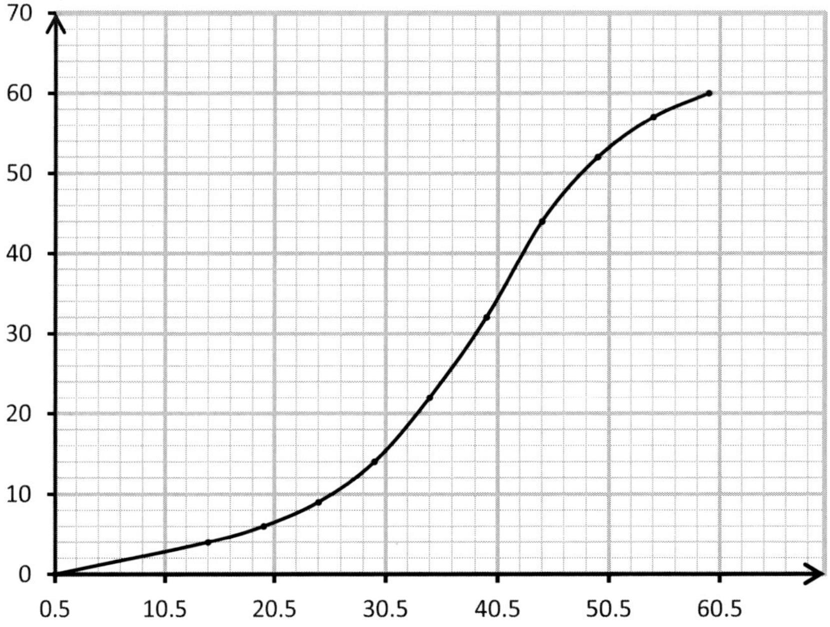

Fig. 14.10

8. Find the interquartile range in fig. 14.10 above.
   (A) 6.5 (B) 22.5 (C)28.5 (D) 15.0

9. What is the 90th percentile?
   (A) 51.5 (B) 30 (C) 25 (D)50.0

10. The formula for determining the mode of a grouped frequency distribution is represented by:
   (A) $M_o = L_1 + \left(\frac{f_m - f_1}{(2f_m - f_1) + (f_m - f_2)}\right) C$ (B) $M_o = L_1 + \left(\frac{f_1 - f_m}{(f_m - f_1) + (f_m - f_2)}\right) C$
   (C) $M_o = L_1 + \left(\frac{f_m - f_1}{(f_m - f_1) + (f_m - f_2)}\right) C$ (D) $M_o = L_1 + \left(\frac{f_m - f_1}{(f_m - f_1) + (f_2 - f_m)}\right) C$

Exercise 14

1. Define the following terms:
   (i) mean (ii) mode (iii) median (iv) range
   (v) variance (vi) standard deviation

2. List five methods of data representation.

3. Table 14.28 shows the marks obtained by students in a mathematics test.

Table 14.28

Score (x)	0	1	3	5	7	8
Number of Students	2	1	3	3	6	1

Find:
(i) the modal number of students in the test
(ii) how many students scored between 5 and 8?
(iii) the modal score
(iv) the mean score

4. A die is thrown and the outcome of each throw is recorded as: 2, 4, 2, 3, 3, 3, 1, 6, 4, 3.
   Find the:
   (a) mode (b) median (c) mean

5. The weights in kilograms recorded by 60 students in a science class is shown in table 14.29:

Table 14.29

Weight (x)	50 – 54	55 – 59	60 – 64	65 – 69	70 – 74
Number of students	10	18	15	5	12

(a) Draw a histogram to represent the given information.

(b) Calculate the: (i) mean (ii) mode (iii) median

6. The data below is the score of 90 students in a test:

Table 14.30

Marks	Frequency
1 – 10	15
11 – 20	18
21 – 30	12
31 – 40	10
41 – 50	6
51 – 60	5
61 – 70	8
71 – 80	4
81 – 90	5
91 – 100	7

Using the data in table 14.30:
(a) draw the ogive (b) estimate the median (c) find the upper and lower quartile.

7. The marks of 45 applicants for an aptitude test are listed below:

65	60	54	40	53	62	91	67	58
58	46	58	23	45	43	98	32	69
51	44	59	26	61	39	83	52	78
50	65	69	39	40	80	75	48	82
42	70	51	67	38	70	40	43	50

(a) Using a class interval of 21 – 30, 31 – 40, 41 – 50, . . . prepare the frequency distribution table.
(b) Calculate the mean mark of the distribution

(c) Estimate the (i)mode (ii) median.
(d) What percentage of students scored more than 60?

8. The mean score of 20 students in a physics test is 70%. If one student's score, x, was later added and computed, the new mean score was recalculated as 71%, find x.

9. Find y if the mean of the data 14, 38, 25, 2 y, 10, 4 is 20.

10. A research was carried out in a small town to find out the number of people within certain age group that are unemployed. If the frequency distribution is represented in table 14.31:

Table 14.31

Age	20 –29	30 – 39	40 – 49	50 – 59	60 – 70
Frequency	800	500	400	100	200

Represent the given information using a pie chart.

11. The table below shows the distribution of the heights of players in a football team.

Table 14.32

Height (cm)	150 – 169	170 – 189	190 – 209	210 – 229	230 – 249
Players	1	5	8	10	6

From the table, find:
(a) the modal class and its corresponding mode
(b) the mean height

12. Find the mean and median of the distribution: 3, 5, 3, 5, 7, 7, 3, 5, 7

13. The average weight of ten boys is 50.3kg; and that of fifteen girls is 38kg. What is the average weight both boys and girls?

14. Calculate the (i) mean and (ii) standard deviation of the distribution

Table 14.33

x	5	6	7	8	9
F	4	9	12	7	3

15. Table 14.34 shows the mark distribution of 70 students in a biology examination.

Table 14.34

Marks	Frequency
1 – 10	1
11 – 20	3
21 – 30	5
31 – 40	6
41 – 50	15
51 – 60	20
61 – 70	8
71 – 80	4
81 – 90	5
91 – 100	3

Calculate:

(i) the mean (ii) mean deviation (iii) variance (iv) standard deviation

15 PROBABILITY

Probability is a mathematical concept which shows the degree of possibility of an event occurring under certain conditions.

15.1 DEFINITION AND TERMS USED IN PROBABILITY

(a) **Experiment**: This is the act of performing a controlled test. Examples are tossing a coin, throwing a pair of ludo dice for a number of times, picking a ball at random from a basket containing many balls.

(b) **Trial**: This is a single act performed in an experiment.

(c) **Outcome**: It is the possible result that can occur in a trial of an experiment.

(d) **Sample Space**: This is the set of all possible outcomes or results obtained in an experiment. For instance, when a dice is thrown, the possible outcomes are, S = {1, 2, 3, 4, 5 and 6}; this is a sample space of the outcome. Thus, a sample space is termed the "universal set of an experiment."

(e) **Sample Point**: A sample point is each outcome in a sample space. For instance, in a sample space of a dice {1, 2, 3, 4, 5, and 6}, each element in the set is called a "sample point."

(f) **Event**: An event is the collection of outcomes (i.e., sample points) that are related or common with one another. Hence, an event is a subset of a sample space.

(g) **Mutually Exclusive Event**: This is when two events have no outcome in common. Thus, we say that an event is mutually exclusive when no two events in a sample space can occur simultaneously, e.g., events $A\,\{2, 4\}$ and $B\,\{1, 3, 5, 6, 10, 12\}$ are mutually exclusive because no sample point in A is repeated in B. While, the events $A\,\{2, 4\}$ and $C\,\{1, 2, 4, 12\}$ are not mutually exclusive because the sample points,
$\{2, 4\}$, occur in both A and C.

(h) **Independent Events**: This is the occurrence of two or more events in an experiment without any influence on one another. Example is getting a six from a thrown dice been absolutely independent from getting a head from a tossed coin in an experiment.

(i) **At least**: This means the probability of an event or more of it occurring.

(J) **At most**: This means the probability of an event or less of it occurring.

(k) **Or**: This means "additive rule" of the probabilities of events. Thus, it is the union of two events that occur if at least one of the events occurs. For instance, assumming we have the probability of two events occurring as $P(A)$ and $P(B)$; then the probability of A or B occurring is given as:
$P(A \cup B) = P(A) + P(B)$ { for mutually exclusive event}

$$P(A \cup B) = P(A) + P(B) - P(AB) \quad \{\text{for not mutually exclusive events}\}$$

This additive rule can be extended to three or more events, thus:

$P(A \cup B \cup C \cup ...) = P(A) + P(B) + P(C) + \cdots$ { for mutually exclusive event)

$P(A \cup B \cup C) = P(A) + P(B) + P(C) - P(AB) - P(BC) - P(AC) + P(ABC)$ { for not mutually exclusive event}

(l) **Not**: This is the probability of an event not occurring; it is the complement of an event. Thus, $Q(A) = 1 - P(A)$; where $Q(A)$ is the probability of an event not occurring.

(m) **And**: This means "multiplication or intersection rule" of the probabilities of at least two events. Assuming we have the probability of two events occurring as $P(A)$ and $P(B)$, then the probability of A and B occurring is given as:

$P(A \cap B) = P(AB) = P(A) \times P(B)$ {for independent events}

$P(AB) = 0$ {for mutually exclusive event}

(n) **Conditional Probability**: This is the probability of one event occurring if another event occurred. Thus, the conditional probability of A given B is written as, $P(A/B)$.

$$\Rightarrow \quad P(A/B) = \frac{P(A.B)}{P(B)}$$

15.2 BASIC LAWS OF PROBABILITY

The following rules/laws are universally applied when solving problems on probability:

1. $0 \leq P(E) \leq 1$ {the probability of an event, $P(E)$, occurring lies between 0 and 1}
2. (a) If $P(E) = 1$ { then it is an absolute certainty of the event occurring}
   (b) If $P(E) = 0$ { then it is an absolute certainty of the event not occurring}
3. $P(E) = \dfrac{number\ of\ Sample\ point\ of\ an\ event}{Total\ number\ of\ sample\ points\ in\ a\ sample\ space} = \dfrac{n(E)}{n(S)}$
4. $Q(E) = 1 - P(E)$ { the complement of the probability of an event}

Example 1

In an experiment, a fair coin is tossed (i) once (ii) twice :

(a) represent the sample space in each case.

(b) find in each case the probability of obtaining a head.

Solution

(a) (i) When the coin is tossed once, the sample space is represented as:

$S = \{H, T\}$

(ii) When the coin is tossed twice, then the sample space is,

$S = \{HH,\ HT,\ TH,\ TT\}$

(b) (i) The probability of getting a head when the coin was tossed once is given as:

$$P_{(H)} = \frac{number\ of\ head\ in\ sample\ space}{total\ outcome\ of\ sample\ space} = \frac{n(E)}{n(S)}$$

$$\therefore\ P_{(H)} = \frac{1}{2}$$

(ii) The probability of getting a head when the coin was tossed twice is given as:

$$P_{(H)} = P_{(Head\ then\ tail)} + P_{(tail\ then\ head)} = \frac{n(HT)}{n(S)} + \frac{n(TH)}{n(S)}$$

$$P_{(H)} = \frac{1}{4} + \frac{1}{4} = \frac{1}{2} \qquad \therefore \quad P_{(H)} = \frac{1}{2}$$

Example 2

A dice is thrown twice, represent the sample space. Hence find the probability of getting: (i) same number (ii) sum of numbers less than 10.

Solution

When the die is thrown twice, the sample space is given as:

Table 15.1 Sample space

Second throw \ First throw	1	2	3	4	5	6
1	1, 1	2, 1	3, 1	4, 1	5, 1	6, 1
2	1, 2	2, 2	3, 2	4, 2	5, 2	6, 2
3	1, 3	2, 3	3, 3	4, 3	5, 3	6, 3
4	1, 4	2, 4	3, 4	4, 4	5, 4	6, 4
5	1, 5	2, 5	3, 5	4, 5	5, 5	6, 5
6	1, 6	2, 6	3, 6	4, 6	5, 6	6, 6

(i) Thus, the probability of getting same number,

$$P_{(same\ number)} = \frac{total\ same\ number}{total\ outcome\ of\ Sample\ space} = \frac{6}{36}$$

$$\therefore \quad P_{(same\ number)} = \frac{1}{6}$$

(ii) Thus, the probability of getting the sum of numbers less than 10

$$P_{(sum\ less\ than\ 10)} = \frac{total\ sum\ less\ than\ 10}{total\ outcome\ of\ Sample\ space} = \frac{30}{36}$$

$$\therefore \quad P_{(sum\ less\ than\ 10)} = \frac{5}{6}$$

Example 3

A dice is thrown once, find the probability that the face up shows:
(i) 3 (ii) an even number (iii) an odd number (iv) a prime number

Solution

The sample space is given as, $S = \{1, 2, 3, 4, 5, 6\}$

(i) $P_{(3)} = \dfrac{number\ of\ three\ in\ sample\ space}{total\ outcome\ of\ sample\ space} = \dfrac{n(3)}{n(S)}$

$\therefore\ P_{(3)} = \dfrac{1}{6}$

(ii) $P_{(even)} = \dfrac{even\ numbers\ in\ sample\ space}{total\ outcome\ of\ Sample\ space} = \dfrac{n(2,\ 4,\ 6)}{n(S)}$

$P_{(even)} = \dfrac{3}{6} = \dfrac{1}{2} \quad \therefore P_{(even)} = \dfrac{1}{2}$

(iii) $P_{(odd)} = \dfrac{odd\ numbers\ in\ sample\ space}{total\ outcome\ of\ Sample\ space} = \dfrac{n(1,\ 3,\ 5)}{n(S)}$

$P_{(odd)} = \dfrac{3}{6} = \dfrac{1}{2} \quad \therefore P_{(odd)} = \dfrac{1}{2}$

(iv) $P_{(prime)} = \dfrac{prime\ numbers\ in\ sample\ space}{total\ outcome\ of\ Sample\ space} = \dfrac{n(2,\ 3,\ 5)}{n(S)}$

$P_{(prime)} = \dfrac{3}{6} = \dfrac{1}{2} \quad \therefore P_{(prime)} = \dfrac{1}{2}$

Example 4

A coin is thrown twice. If the sample space is given by $S = \{HH, HT, TH, TT\}$, such that $X = \{HH, HT\}$ and $Y = \{TH, TT\}$:
(a) Draw a Venn diagram to represent the above information.
(b) Find the probability of (i) X (ii) Y (iii) $X \cap Y$ (iv) $X \cup Y$.

Solution

(a) The Venn diagram is given in fig. 15.1.

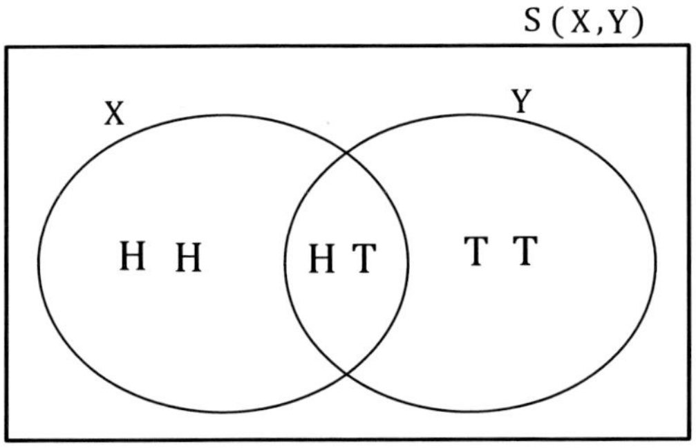

Fig. 15.1

(b) (i) $P_{(X)} = \frac{n(X)}{n(S)} = \frac{n(HH,\ HT)}{n(HH,\ HT,\ TH,\ TT)}$

$= \frac{2}{4} = \frac{1}{2} \quad \therefore\ P_{(X)} = \frac{1}{2}$

(ii) $P_{(Y)} = \frac{n(Y)}{n(S)} = \frac{n(TH,\ TT)}{n(HH,\ HT,\ TH,\ TT)}$

$= \frac{2}{4} = \frac{1}{2} \quad \therefore P_{(Y)} = \frac{1}{2}$

(iii) $P_{(X \cap Y)} = \frac{n(X \cap Y)}{n(S)} = \frac{n(HT)}{n(HH,\ HT,\ TH,\ TT)} = \frac{1}{4} \qquad \therefore P_{(X \cap Y)} = \frac{1}{4}$

(iv) $P_{(X \cup Y)} = \frac{n(X \cup Y)}{n(S)} = \frac{n(HH)+n(HT)+n(TT)}{n(HH,\ HT,\ TH,\ TT)} = \frac{3}{4} \qquad \therefore P_{(X \cup Y)} = \frac{3}{4}$

Alternatively:

$P_{(X \cup Y)} = P_{(X)} + P_{(Y)} - P_{(X \cap Y)} = \frac{1}{2} + \frac{1}{2} - \frac{1}{4} = \frac{3}{4}$

$\therefore P_{(X \cup Y)} = \frac{3}{4}$

Example 5

A ball is drawn at random from a box containing 5 red balls (R), 4 white balls (W), and 11 blue balls (B). Find the probability that it is:
(i) Red (ii) White (iii) Blue (iv) not Blue (v) Red or Blue (vi) Red or White

Solution

$n(R) = 5, n(W) = 4,\ n(B) = 11$

$\Rightarrow n(S) = \{n(R) + n(W) + n(B)\} = \{5 + 4 + 11\} = 20$ balls

(i) $P_{(R)} = \frac{n(R)}{n(S)} = \frac{5}{20} = \frac{1}{4} \qquad \therefore P_{(R)} = \frac{1}{4}$

(ii) $P_{(W)} = \frac{n(W)}{n(S)} = \frac{4}{20} = \frac{1}{5} \qquad \therefore P_{(W)} = \frac{1}{5}$

(iii $P_{(B)} = \frac{n(B)}{n(S)} = \frac{11}{20} \qquad \therefore P_{(B)} = \frac{11}{20}$

(iv) Let $Q_{(B)} = 1 - P_{(B)}$; where $Q_{(B)}$ probability of not Blue

$\Rightarrow Q_{(B)} = 1 - \frac{11}{20} = \frac{9}{20} \qquad \therefore Q_{(B)} = \frac{9}{20}$

(v) Probability of Red or Blue $= P_{(R)} + P_{(B)} = \frac{1}{4} + \frac{11}{20} = \frac{4}{5}$

$\therefore\ P_{(R)\ or\ (B)} = \frac{4}{5}$

(v) Probability of Red or White $= P_{(R)} + P_{(W)} = \frac{1}{4} + \frac{1}{5} = \frac{9}{20}$

$\therefore P_{(R)\ \text{or}\ (W)} = \frac{9}{20}$

Example 6

A basket contains 5 oranges and 3 mangoes. If two fruits are drawn successfully at random from the bag (a) with replacement (b) without replacement.
What is the probability of obtaining an orange and then mango in that order for case (a) and (b) respectively?

Solution

(a) **With replacement**:

Orange $= n(O) = 5$, Mango $= n(M) = 3$, Sample space $= n(S) = \{n(O) + n(M)\}$

Then, the probability of the fruit drawn first is Orange,$P_{(O)} = \frac{n(O)}{n(S)}$

$\Rightarrow P_{(O)} = \frac{5}{8}$

And, the probability of the second fruit drawn is Mango, $P_{(M)} = \frac{n(M)}{n(S)}$

$\Rightarrow P_{(M)} = \frac{3}{8} \quad \therefore P_{(O \text{ and then } M)} = \frac{5}{8} \times \frac{3}{8} = \frac{15}{64}$

(b) **Without replacement**:

Orange $= n(O) = 5$, Mango $= n(M) = 3$, Sample space $= n(S) = \{n(O) + n(M)\} = 8$

Then, the probability of the fruit drawn first is Orange, $P_{(O)} = \frac{n(O)}{n(S)}$

$\Rightarrow P_{(O)} = \frac{5}{8}$

And, the probability of the second fruit drawn is Mango, $P_{(M/O)} = \frac{n(M)}{n(S-1)}$

$\Rightarrow P_{(M/O)} = \frac{3}{7}$

$\therefore P_{(O \text{ and then } M)} = \frac{5}{8} \times \frac{3}{7} = \frac{15}{56}$

Example 7

A bag contains 3 apples and 2 mangoes. If two fruits are drawn one after the other without replacement. What is the probability of:

(i) taking an apple first?
(ii) taking a mango second?
(iii) taking an apple and a mango in that order?
(iv) first and second selection are mangoes?
(v) first and second selection are apples?

Solution

Apple $= n(A) = 3$, Mango $= n(M) = 2$, Sample space $= n(S) = \{n(A) + n(M)\} = 5$

(i) The probability that the first fruit drawn is Apple, $P_{(A)} = \frac{n(A)}{n(S)}$

$\Rightarrow P_{(A)} = \frac{3}{5}$

(ii) The probability that the second fruit drawn is Mango, $P_{(M/A)} = \frac{n(M)}{n(S-1)}$

$\Rightarrow P_{(M)} = \frac{2}{5-1} = \frac{2}{4} \quad \therefore \quad P_{(M/A)} = \frac{2}{4} = \frac{1}{2}$

(iii) The probability of drawing an apple and mango in that order,

$P_{(AM)} = P_{(A)} \times P_{(M/A)}$

$\Rightarrow P_{(AM)} = P_{(A)} \times P_{(M/A)} = \frac{3}{5} \times \frac{1}{2} = \frac{3}{10} \quad \therefore \ P_{(AM)} = \frac{3}{10}$

(iv) The probability that the first and second fruits drawn are mangoes,

$P_{(MM)} = P_{(M)} \times P_{(M/M)} = \frac{n(M)}{n(S)} \times \frac{n(M-1)}{n(S-1)} \quad \Rightarrow P_{(MM)} = \frac{2}{5} \times \frac{2-1}{5-1} = \frac{2}{5} \times \frac{1}{4}$

$\therefore P_{(MM)} = \frac{2}{20} = \frac{1}{10}$

(v) The probability that the first and second fruits drawn are apples,

$P_{(AA)} = P_{(A)} \times P_{(A/A)}$

$P_{(AA)} = P_{(A)} \times P_{(A/A)} = \frac{n(A)}{n(S)} \times \frac{n(A-1)}{n(S-1)} \quad \Rightarrow P_{(AA)} = \frac{3}{5} \times \frac{3-1}{5-1} = \frac{3}{5} \times \frac{2}{4}$

$\therefore P_{(AA)} = \frac{3}{5} \times \frac{2}{4} = \frac{6}{20} = \frac{3}{10}$

Example 8

Three balls are drawn successfully from a box containing 5 red balls, 3 blue balls, and 2 green balls. Find the probability that they are drawn in the order of green, red, and blue on the condition that each ball is (a) replaced (b) not replaced.

Solution

Let: Red $= n(R) = 5$, Blue $= n(B) = 3$, Green $= n(G) = 2$,

Sample space $= n(S) = \{ n(R) + n(B) + n(G)\} = \{5 + 3 + 2\} = 10$

Then, the probability of taking Green, Red, and Blue on the condition that:

(a) Ball replaced: $P_{(GRB)} = \frac{n(G)}{n(S)} \times \frac{n(R)}{n(S)} \times \frac{n(B)}{n(S)} = \frac{2}{10} \times \frac{5}{10} \times \frac{3}{10}$

$\Rightarrow P_{(GRB)} = \frac{30}{1000} = \frac{3}{100} \quad \therefore P_{(GRB)} = \frac{3}{100}$

(b) Ball not replaced: $P_{(GRB)} = \frac{n(G)}{n(S)} \times \frac{n(R)}{n(S-1)} \times \frac{n(B)}{n(S-2)} = \frac{2}{10} \times \frac{5}{9} \times \frac{3}{8}$

$\Rightarrow P_{(GRB)} = \frac{30}{720} = \frac{1}{24} \quad \therefore P_{(GRB)} = \frac{1}{24}$

Example 9

A dice is thrown twice:
(a) list all the possible outcomes (b) find the probability of obtaining (i) sum of 8; (ii) the same number; (iii) only even numbers.

Solution

(a)

Table 15.2

Second throw \ First throw	1	2	3	4	5	6
1	1,1	2,1	3,1	4,1	5,1	6,1
2	1,2	2,2	3,2	4,2	5,2	6,2
3	1,3	2,3	3,3	4,3	5,3	6,3
4	1,4	2,4	3,4	4,4	5,4	6,4
5	1,5	2,5	3,5	4,5	5,5	6,5
6	1,6	2,6	3,6	4,6	5,6	6,6

(b) (i) Sample space $= n(S) = 36, n(\text{sum of } 8) = 5$

$$\Rightarrow P = \frac{n(sum\ of\ 8)}{n(S)} = \frac{5}{36} \quad \therefore P_{(sum\ of\ 8)} = \frac{5}{36}$$

(ii) $n(\text{same number}) = 6$

$$\Rightarrow P = \frac{n(same\ number)}{n(S)} = \frac{6}{36} = \frac{1}{6} \quad \therefore P_{(same\ number)} = \frac{1}{6}$$

(iii) $n(\text{only even number}) = 9$

$$\Rightarrow P = \frac{n(only\ even\ numbers)}{n(S)} = \frac{9}{36} = \frac{1}{3} \quad \therefore P_{(only\ even\ numbers)} = \frac{1}{4}$$

Example 10

A committee is made up of 4 males and 6 females. Two members are selected successfully at random without replacement. What is the probability that the selected members are:
(a) both males (b) both females (c) first is a female and the second is a male.

Solution

Male $= n(M) = 4$, Female $= n(F) = 6, n(S) = \{\text{Sample space} = n(M) + n(F)\} = 10$

(a) Probability that both are male: $P_{(MM)} = \frac{n(M)}{n(S)} \times \frac{n(M-1)}{n(S-1)} = \frac{4}{10} \times \frac{3}{9}$

$\Rightarrow P_{(MM)} = \frac{4}{10} \times \frac{3}{9} = \frac{12}{90} = \frac{2}{15} \quad \therefore P_{(both\ male)} = \frac{2}{15}$

(b) Probability that both are female: $P_{(FF)} = \frac{n(F)}{n(S)} \times \frac{n(F-1)}{n(S-1)} = \frac{6}{10} \times \frac{5}{9}$

$\Rightarrow P_{(FF)} = \frac{6}{10} \times \frac{5}{9} = \frac{30}{90} = \frac{1}{3} \quad \therefore P_{(both\ female)} = \frac{1}{3}$

(c)The probability that the first is female, and the second male:

$P_{(FM)} = \frac{n(F)}{n(S)} \times \frac{n(M)}{n(S-1)} = \frac{6}{10} \times \frac{4}{9}$

$\Rightarrow P_{(FF)} = \frac{6}{10} \times \frac{5}{9} = \frac{24}{90} = \frac{4}{15} \quad \therefore P_{(both\ female)} = \frac{4}{15}$

Example 11

In a Chess competition, the probability that Musa wins $= \frac{1}{3}$, Ade wins $= \frac{1}{5}$, and Emeka wins $= \frac{3}{7}$. What is the probability that:

(a) all three wins? (b) only Ade and Emeka wins? (c) Musa loses? (d) only two wins? (e) all loses the game?

Solution

Let: Musa = M$= \frac{1}{3}$, Ade = A$= \frac{1}{5}$, Emeka = E $= \frac{3}{7}$

(a) All three wins: $P_{(M\cap A\cap E)} = \frac{1}{3} \times \frac{1}{5} \times \frac{3}{7} = \frac{1}{35}$

(b) Only Ade and Emeka win: this implies that Ade and Emeka win while Musa loses

$\Rightarrow \frac{1}{5} \times \frac{3}{7} \times \left(1 - \frac{1}{3}\right) = \frac{1}{5} \times \frac{3}{7} \times \frac{2}{3} = \frac{2}{35}$

(c) Musa loses: $(1 - Musa\ wins)$

$\Rightarrow 1 - \frac{1}{3} = \frac{2}{3}$

(d) Probability that only two win the game: (Musa and Ade win while Emeka loses) or (Musa and Emeka win while Ade loses) or (Ade and Emeka win while M=Musa loses)

$\Rightarrow \frac{1}{3} \times \frac{1}{5} \times \left(1 - \frac{3}{7}\right) + \frac{1}{3} \times \frac{3}{7} \times \left(1 - \frac{2}{5}\right) + \frac{1}{5} \times \frac{3}{7} \times \left(1 - \frac{1}{3}\right)$

$= \left(\frac{1}{3} \times \frac{1}{5} \times \frac{4}{7}\right) + \left(\frac{1}{3} \times \frac{3}{7} \times \frac{3}{5}\right) + \left(\frac{1}{5} \times \frac{3}{7} \times \frac{2}{3}\right) = \frac{4}{105} + \frac{9}{105} + \frac{6}{105}$

$= \frac{19}{105}$

(e) Probability that all loses the game:

$\Rightarrow$ Musa loses and Ade loses and Emeka loses$= \left(1 - \frac{1}{3}\right)\left(1 - \frac{1}{5}\right)\left(1 - \frac{3}{7}\right)$

$= \frac{2}{3} \times \frac{2}{5} \times \frac{3}{7} = \frac{12}{105} = \frac{4}{15} \quad \therefore$ Probability that all loses the game $= \frac{4}{15}$

OBJECTIVE TEST 15

1. What is the probability that a number chosen at random from the integers between 11 and 20 inclusive is either a prime or a multiple of 3?

   (A) $\frac{1}{2}$ (B) $\frac{2}{5}$ (C) $\frac{3}{10}$ (D) $\frac{7}{10}$

2. 10 identical balls are contained in a bag; 4 are red, while 6 are black. If two balls are taken one after the other without replacement, what is the probability of getting a red and a black ball in that order?

   (A) $\frac{6}{25}$ (B) $\frac{2}{15}$ (C) $\frac{4}{15}$ (D) $\frac{1}{5}$

U_1 and U_2 are mutually exclusive events with $P_{(U_1)} = \frac{1}{3}$ and $P_{(U_2)} = \frac{2}{5}$, use the information to answer question 3 and 4.

3. What is the probability of U_1 or U_2?

   (A) $\frac{1}{2}$ (B) $\frac{2}{15}$ (C) $\frac{11}{15}$ (D) $\frac{5}{6}$

4. By how much is the probability of U_2 less than the probability of not U_1?

   (A) $\frac{1}{2}$ (B) $\frac{2}{3}$ (C) $\frac{1}{15}$ (D) $\frac{4}{15}$

5. Two fair dice are tossed simultaneously, find the probability of obtaining the same scores in both dice.

   (A) $\frac{1}{6}$ (B) $\frac{1}{3}$ (C) $\frac{5}{6}$ (D) $\frac{1}{2}$

6. A number is selected at random from 10 to 30. What is the probability that it is an odd number?

   (A) $\frac{2}{7}$ (B) $\frac{10}{21}$ (C) $\frac{11}{21}$ (D) $\frac{11}{15}$

7. A squad of 5 team players were selected for a penalty kick. Three of the players have $\frac{1}{7}$ chance each of scoring and the fourth player has $\frac{1}{5}$ chance of scoring. What is the probability that the last player will score?

   (A) $\frac{3}{7}$ (B) $\frac{22}{35}$ (C) $\frac{13}{35}$ (D) 1

8. A bag contains 15 red balls, 10 white balls, and x blue balls. If the probability of picking a red ball is $\frac{1}{3}$, find x.
(A) 5 (B) 12 (C) 20 (D) 25

9. A bag contains 8 white balls, 5 red balls, and 2 blue balls. If two balls are drawn from the bag without replacement, find the probability of obtaining one red and one blue ball in that order.
(A) $\frac{2}{45}$ (B) $\frac{1}{3}$ (C) $\frac{1}{21}$ (D) $\frac{20}{21}$

10. Table 15.3

Age	15 – 20	21 – 25	26 – 30	31 – 55
Number	12	5	8	10

Table 15.3 shows the age distribution of students in a department selected for an I.Q test. Find the probability that the number of students chosen in the department is at least 21 years.
(A) $\frac{23}{35}$ (B) $\frac{6}{7}$ (C) $\frac{1}{7}$ (D) $\frac{12}{35}$

Exercise 15

1. Define the following terms:
(a) probability (b) experiment (c) event (d) outcome
(e) mutually exclusive event (f) independent event (g) conditional probability
2. Give the condition(s) when the following statements are used in probability:
(a) Not (b) Or (c) And (d) Either (e) Neither
3. A coin is thrown twice:
(a) list all the possible outcome
(b) what is the probability of obtaining:
(i) two head (ii) a head (iii) a tail (iv) two tail
4. A fair die is thrown once:
(a) create a sample space of the experiment
(b) find the probability of obtaining (i) a 1 (ii) a 3 (iii) 2 or 3 (iv) at least 4
5. Two ludo die are thrown: (a) create a sample space for the possible outcome
(b) find the probability of obtaining (i) double – six (ii) sum of 8
(iii) two even numbers (iv) two prime numbers.

6. A bag contains 5 black balls and 3 red balls. If one ball is chosen at random, what is the probability that: (a) it is black (b) it is red

7. State the sample space that are mutually exclusive:
A={ 2, 3}; B= { 2, 3, 5, 6}; C= { H, T}; D={ HT, TH, HH,TT }

8. A box contains coloured beads: 5 red, 10 blue, 3 black, 12 yellow and 6 green. If three beads are selected at random (a) without replacement (b) with replacement. Find in each case, the probability of obtaining:
(i) three red (ii) red, black, and yellow in that order
(iii) two green and one black (iv) first and second are black and the third is not black.

9.

Table 15.4

Age	16	17	18
Number	12	5	8

The table above shows the age and number of students in a class, what is the probability that a student chosen at random is:
(i) 16 years? (ii)18 years? (iii) at most 17 years? (iv) at least 17 years?

10. $S = \{ HT, TH, HH, TT \}$ find the probability of obtaining:
(a) two tail (b) two head (c) a head (d) a tail

11. Given that $S = \{ x\text{: } 0 \leq x < 30,$ where x is an integer$\}$
(a) List the elements of the set S (b) determine the probability of obtaining even numbers (c) odd numbers (d) odd numbers

12. A box contains 10 red balls, 8 blue balls, and x white balls. If the probability of obtaining red balls is $\frac{2}{5}$, find x.

13. A basket contains 18 mangoes and 10 oranges. If some quantities of apples were introduced in order to make the probability of picking apples from the basket $\frac{1}{3}$, find the number of apples introduced.

14. A bag contains 4 black balls, 3 yellow balls and 5 white balls. If a ball is chosen at random from the bag, what is the probability that it is (i) black (ii) white (iii) yellow (iv) not yellow

15. A basket contains 2 red balls, 3 black balls and 7 green balls. It two balls are picked at random one after the other (a) with replacement (b) without replacement.
Find the probability that:
(i) both are red (ii) first is black and the second is red (iii) one is green and the other is black (iv) they are all the same colour.

16. Two soldiers aim at a target. The probability that the first will hit the target is $\frac{1}{3}$, while the probability that the second hits the target is $\frac{2}{5}$. Find the probability that:
(a) both will hit the target (b) just one hits the target (c) both will miss
(d) only the second soldier hits the target.

17. The probability of three independent events A, B and C occurring are $\frac{1}{4}$, $\frac{1}{3}$ and $\frac{2}{7}$ respectively. Find the probability that:
(i) All will occur (ii) A and B only will occur (iii) C only occurs
(iv) two of them only occur (v) one occurs (vi) none occurs.

18. A fair dice is rolled once while a coin is tossed once. Find the probability of getting:
(a) a 6 and a tail (b) an even number and a head

16 PERMUTATION AND COMBINATION

16.1 PERMUTATION

Permutation is an ordered arrangement of objects; it shows all the possible ways of arranging objects or doing things in an orderly manner.

16.1.1 GENERAL FORMULA OF PERMUTATION

Mathematically, the general equation of permutation is given as:

$$^{n}P_r = \frac{n!}{(n-r)!}$$

The above equation can be expanded as:

$$\frac{n!}{(n-r)!} = \frac{n(n-2)(n-3)\ldots(n-r+1)(n-r)(n-r-1)\ldots 3\times 2\times 1\times 0!}{(n-r)(n-r-1)\ldots 3\times 2\times 1\times 0!}$$

$$= n(n-2)(n-3)\ldots(n-r+1)$$

Where: n = number of objects;
r = objects taken at a time;
$!$ = factorial function which means to multiply a series of descending natural number.

(N/B: $0! = 1$)

Example 1

Find the factorial functions of each of the following:

(a) $5!$ (b) $12!$ (c) $4!\,6!$ (d) $\frac{7!}{4!}$

Solution

(a) $5! = 5\times 4\times 3\times 2\times 1\times 0! = 120$

(b) $12! = 12\times 11\times 10\times 9\times 8\times 7\times 6\times 5\times 4\times 3\times 2\times 1\times 0! = 47901600$

(c) $6!\,4! = (6\times 5\times 4\times 3\times 2\times 1\times 0!)(4\times 3\times 2\times 1\times 0!)$
$= 720\times 24 = 17280$

(d) $\frac{7!}{4!} = \frac{7\times6\times5\times4\times3\times2\times1\times0!}{4\times3\times2\times1\times0!} = \frac{5040}{24} = 210$

Example 2

In how many ways can a student arrange the answers of two out of five essay questions in an examination?

Solution

$n = 5, r = 2$

Then using the formula ${}^{n}P_r = \frac{n!}{(n-r)!}$

$\Rightarrow {}^{5}P_2 = \frac{5!}{(5-2)!} = \frac{5!}{3!} = \frac{5\times4\times3\times2\times1\times0!}{3\times2\times1\times0!}$

$\therefore {}^{5}P_2 = \frac{120}{6} = 20$

Thus, two out of five essay questions can be arranged in 20 ways.

Example 3

In how many ways can one gold, one silver, and one bronze be awarded to 12 contestants in a competition?

Solution

$n = 12, r = one\ gold + one\ silver + one\ bronze = 3$ prizes to be awarded.

Then using the formula ${}^{n}P_r = \frac{n!}{(n-r)!}$

$\rightarrow {}^{12}P_3 = \frac{12!}{(12-3)!} = \frac{12!}{9!} = \frac{12\times11\times10\times9!}{9!}$

$\therefore {}^{12}P_3 = 12 \times 11 \times 10 = 1320$

The first, second, and third prizes can be awarded to 12 contestants in 1320 ways.

16.1.2 SPECIAL CASES/FORMULAS OF PERMUTATION

The following formulas can also be used for solving problems on permutation depending on the nature of the problem:

CASE 1 ${}^{n}P_r = n^r$ { Used when r objects can be chosen from n objects with "repetition allowed"}

Example 4

In how many ways can a red ball, black ball, and white ball be chosen from 10 different coloured balls, if repetition is allowed.

Solution

$n = 10, \ r = 3$

Then, using the formula ${}^{n}P_r = n^r$

$\Rightarrow {}^{10}P_3 = 10^3 = 10 \times 10 \times 10 = 1000$ ways

$\therefore {}^{10}P_3$ with replacement allowed=1000 ways

Since repetition is allowed, then in the first choice there are n possibilities; in the second choice there are still n possibilities; and finally in the third choice there are n possibilities.

Example 5

In how many ways can a student fail 5 objective questions, each with options A, B, C and D.

Solution

The student can fail the first objective question in 3 ways. This is because, if the correct answer is A, then choosing B, C or D will be wrong.
$\Rightarrow$ for five objective questions, the permutation will be repeated as:
$3 \times 3 \times 3 \times 3 \times 3 = 243$ ways

Example 6

The 11- digit phone number in a country has 081 as the first three digits. How many different possible phone numbers can be generated?

Solution

Let the first number be 081-00000000
then each digit in 00000000 has ten (10) possible numbers (i.e, 0 to 9), thus for the eight digits:
the total arrangement becomes, n^r
$\Rightarrow 10^8 = 10 \times 10 \times 10 \times 10 \times 10 \times 10 \times 10 \times 10$
$= 100{,}000{,}000$ possible phone numbers

Since 081 is fixed for all phone numbers, then we can only permute the remaining eight digits, thus $r = 8$ objects with each n possible numbers.

CASE 2 $^nP_n = n!$ { This is used when n objects are to be arranged "with all taken at a time with no repetition of any object"}

Example 7

Find the number of ways each of the following can be arranged:
(a) BOY (b) GIRL (c) MONEY (d) ABSCUND

Solution

(a) BOY:
The word has $n = 3$ letters, and all are taken at a time with no repetition of any letter.
$^3P_3 = 3! = 3 \times 2 \times 1 \times 0! = 6$ possible ways of arrangement.

(b) GIRL:
The word has $n = 4$ letters, and all are taken at a time.
$^4P_4 = 4! = 4 \times 3 \times 2 \times 1 \times 0! = 24$ possible ways of arrangement.

(c) MONEY:
The word has $n = 5$ letters, and all are taken at a time with no repetition of any letter.
$^5P_5 = 5! = 5 \times 4 \times 3 \times 2 \times 1 \times 0! = 120$ possible ways of arrangement.

(d) ABSCUND:
The word has $n = 7$ letters, and all are taken at a time with no repetition of any letter.
$^7P_7 = 7! = 7 \times 6 \times 5 \times 4 \times 3 \times 2 \times 1 \times 0! = 5040$ possible ways of arrangement.

CASE 3 $N_{ways} = \frac{n!}{r_1!\,r_2!\,r_3! \ldots}$ {This is used when some objects are the same}

Example 8

In how many ways can each of the following words be arranged:
(a) RICHARD (b) TOMORROW (c) ABDULLAZEEZ (d) MISSISSIPPI

Solution

(a) RICHARD:
The word has $n = 7$ letters; with repetition of letter R twice (i.e., $r = 2$)
$n = 7, r = 2$
$N = \frac{n!}{r!} = \frac{7!}{2!} = \frac{7\times6\times5\times4\times3\times\cancel{2!}}{\cancel{2!}}$
$\Rightarrow N = 7 \times 6 \times 5 \times 4 \times 3 = 2520$ possible ways.

(b) TOMORROW:
The word has $n = 8$ letters; with repetition of letters O and R.
$\Rightarrow r_1 =$ letter O $= 3$, $r_2 =$ letter R -2
$N = \frac{n!}{r_1!r_2!} = \frac{8!}{3!2!} = \frac{8\times7\times6\times5\times4\times3\times\cancel{2!}}{3\times2\times1\times\cancel{2!}}$
$\Rightarrow N = \frac{8\times7\times6\times5\times4\times3}{3\times2\times1} = \frac{20160}{6} = 3360$ possible ways

(c) ABDULLAZEEZ:
The word has $n = 11$ letters; with repetition of letter A, L, E and Z.
$\Rightarrow r_1 =$ letter A $= 2$, $r_2 =$ letter L $=2$, $r_3 =$ letter E $=2$, $r_4 =$ letter Z $=2$
Thus, $N = \frac{n!}{r_1!r_2!r_3!r_4!} = \frac{11!}{2!2!2!2!} = \frac{11\times10\times9\times8\times7\times6\times5\times4\times3\times2\times1\times0!}{2!2!2!2!}$
$\Rightarrow N = \frac{39916800}{16} =2494800$ possible ways

(d) MISSISSIPPI:
The word has $n = 11$ letters; with repitition of letter I, S and P
$\Rightarrow r_1 =$ letter I $=4$, $r_2 =$ letter S $= 4$, $r_3 =$ letter P $=2$
$N = \frac{n!}{r_1!r_2!r_3!} = \frac{11!}{4!4!2!} = \frac{11\times10\times9\times8\times7\times6\times5\times4\times3\times2\times1\times0!}{(4\times3\times2\times1\times0!)(4\times3\times2\times1\times0!)(2\times1\times0!)}$
$\Rightarrow N = \frac{39916800}{1152} = 34650$ possible ways

Example 9

In how many ways can 4 men, 3 women, 2 boys and 5 girls be orderly arranged in a row.

Solution

$n = 4$ men $+ 3$ women $+ 2$ boys $+ 5$ girls $= 14$ people
$r_1 =$ men $= 4$, $r_2 =$ women $=3$, $r_3 =$ boys $=2$, $r_4 =$ girls $= 5$

Thus, $N = \frac{n!}{r_1!r_2!r_3!r_4!} = \frac{14!}{4!3!2!5!} = \frac{14\times13\times12\times11\times10\times9\times8\times7\times6\times5\times4\times3\times2\times1\times0!}{(4\times3\times2\times1\times0!)(3\times2\times1\times0!)(2\times1\times0!)(5\times4\times3\times2\times1\times0!)}$

$\Rightarrow N = \frac{87178291200}{34560}$ =2522520 possible ways

CASE 4 (a) $N_{ways} = (n-1)!$ .... { This is called cyclic permutation; it is used for arrangement of objects in a circular platform, e.g. round table }

(b) $N_{ways} = \frac{(n-1)!}{2}$ .... { This is called cyclic permutation; it is used for arrangement of objects on a ring }

Example 10

In how many ways can six people labelled, A, B, C, D, E, F be orderly arranged around a circular table?

Solution

n =6 people

$N_{ways} = (n-1)! = (6-1)! = 5!$

$\Rightarrow 5! = 5 \times 4 \times 3 \times 2 \times 1 \times 0! = 120$ possible ways

Example 11

Using example 10 above, what is the total possible arrangement assuming two people: (a) A and D must sit together (b) A and D must not sit together?

Solution

(a) If A and D must sit together, then we have two ways of arranging A and D (i.e, AD or DA). Also, AD and DA are viewed as one to be arranged together with B, C, E, F;

thus, $N_{ways} = (5-1) = 4! = 24$ways.

Therefore, total possible arrangement is $2 \times 24 = 48$ ways

(b) If A and D must not sit together, then A can be fixed to sit in between BC, FB or EF; thus we have 3 ways to fix A far away from D.

Therefore, total possible arrangement $= 3 \times 24 = 72$ ways

Example 12

In how many ways can 5 brothers and 4 sisters be arranged to eat on a dining table?

Solution

$n = 5\ brothers + 4\ sisters = 9$ people

$N_{ways} = (n-1)! = (9-1)! = 8!$

$\Rightarrow 8! = 8 \times 7 \times 6 \times 5 \times 4 \times 3 \times 2 \times 1 \times 0! = 40{,}320$ possible ways

Example 13

In how many ways can 7 different keys be arranged on a key ring?

Solution

$N_{ways} = \frac{(n-1)!}{2} = \frac{(7-1)!}{2} = \frac{6!}{2} = \frac{6\times5\times4\times3\times2\times1}{2} = 360$ possible ways

16.2 COMBINATION

Combination is the selection of objects (r) from a set of data (n) without considering the order of arrangement. Thus, combination is denoted as, ${}^{n}C_r$ or $\binom{n}{r}$.

Mathematically,

$$ {}^{n}C_r = \frac{nP_r}{r!} = \frac{\frac{n!}{(n-r)!}}{r!} $$

$$ \therefore \quad {}^{n}C_r = \frac{n!}{(n-r)!r!} $$

Example 14

In how many ways can 3 people be chosen from a group of 10.

Solution

$n = 10,\ r = 3$

$$ {}^{n}C_r = \frac{n!}{(n-r)!r!} \Rightarrow \frac{10!}{(10-3)!3!} = \frac{10!}{7!3!} $$

$$ \therefore \frac{10!}{7!3!} = \frac{10\times9\times8\times7!}{7!\times3\times2\times1\times0!} = \frac{720}{6} = 120 \text{ ways} $$

Example 15

Evaluate each of the following:

(i) ${}^{8}C_2$ (ii) ${}^{5}C_5$ (iii) ${}^{5}C_0$ (iv) ${}^{5}C_2 \times {}^{10}C_4$ (v) ${}^{7}C_5 \times {}^{12}C_{10}$

Solution

(i) $${}^{8}C_2 = \frac{8!}{(8-2)!2!} = \frac{8\times7\times6!}{6!2!} \Rightarrow \frac{8\times7}{2\times1\times0!} = \frac{56}{2} = 28$$

(ii) $${}^{5}C_5 = \frac{5!}{(5-5)!5!} = \frac{5!}{0!5!} \Rightarrow \frac{1}{0!} = \frac{1}{1} = 1$$

(iii) $${}^{5}C_0 = \frac{5!}{(5-0)!0!} = \frac{5!}{5!0!} \Rightarrow \frac{1}{0!} = \frac{1}{1} = 1$$

(iv) $${}^{5}C_2 \times {}^{10}C_4 = \frac{5!}{(5-2)!2!} \times \frac{10!}{(10-4)!4!} = \frac{5!}{3!2!} \times \frac{10!}{6!4!}$$

$$ = \frac{5\times4\times3!}{3!2!} \times \frac{10\times9\times8\times7\times6!}{6!4!} = \frac{5\times4}{2} \times \frac{10\times9\times8\times7}{4\times3\times2\times1\times0!} = \frac{100800}{48} $$

$= 2100$

(v) $${}^{7}C_5 \times {}^{12}C_{10} = \frac{7!}{(7-5)!5!} \times \frac{12!}{(12-10)!10!} = \frac{7!}{2!5!} \times \frac{12!}{2!10!}$$

$$ = \frac{7\times6\times5!}{2!5!} \times \frac{12\times11\times10!}{2!10!} = \frac{42}{2} \times \frac{12\times132}{2} = \frac{66582}{4} $$

$= 16632$

Example 16

In how many ways can a committee of 10, comprising of 7 males and 3 females be formed from a group that consists of 12 males and 8 females.

Solution

males = 12 from which 7 will be selected; females = 8 from which 3 will be selected.

$\Rightarrow$ number of possible combinations, $^{12}C_7 \times {}^8C_3 = \frac{12!}{(12-7)!7!} \times \frac{8!}{(8-3)!3!}$

$= \frac{12!}{5!7!} \times \frac{8!}{5!3!} = \frac{12\times11\times10\times9\times8\times\cancel{7}!}{5!\cancel{7}!} \times \frac{8\times7\times6\times\cancel{5}!}{\cancel{5}!3!} = \frac{12\times11\times10\times9\times8}{5!} \times \frac{8\times7\times6}{3!}$

$= \frac{12\times11\times10\times9\times8}{5\times4\times3\times2\times1\times0!} \times \frac{8\times7\times6}{3\times2\times1\times0!} = \frac{95040}{120} \times \frac{336}{6}$

$= 792 \times 56 = 44352$

Example 17

In the "2023 Africa nation cup finals," 5 players from a team of 10 players in the Super Eagles of Nigeria were to be chosen for penalty kick. If 3 players of the team have been chosen, what is the total possible ways of selecting the remaining 2 players from the team?

Solution

Number of players in the team = 10, number of players to be selected = 5,
Number of already selected players =3,
Number of remaining players to be selected =2
$\Rightarrow$ number of team players remaining =10 −3 = 7 from which 2 remaining will be selected
$\therefore$ Total possible ways of selecting the remaining 2 players = $^7C_2$

$^7C_2 = \frac{7!}{(7-2)2!} = \frac{7!}{5!2!} = \frac{7\times6\times\cancel{5}!}{\cancel{5}!2!}$

$= \frac{7\times6}{2!} = \frac{42}{2} = 21$ ways

Example 18

Given that $^5C_{r+1} = {}^5C_r$, find r.

Solution

$^5C_{r+1} = {}^5C_r \quad \Rightarrow \frac{5!}{(5-(r+1))!(r+1)!} = \frac{5!}{(5-r)!r!}$

$\cancel{5!}(5-r)!\,r! = \cancel{5!}\,(5-(r+1))!\,(r+1)!$

$\Rightarrow (5-r)!\,r! = (4-r)!\,(r+1)!$

$(5-r)\cancel{(4-r)!}\,\cancel{r!} = \cancel{(4-r)!}\,(r+1)\cancel{r!}$

$\Rightarrow 5 - r = r + 1 \quad$ {Collect like terms}

$r + r = 5 - 1 = 4$

$2r = 4$

$\therefore r = \frac{4}{2} = 2$

OBJECTIVE TEST 16

1. Evaluate ${}^{10}P_2$
   (A) 90 (B) 45 (c) 900 (d) 72

2. Evaluate ${}^{6}P_2 \times {}^{10}P_3$
   (A) 720 (B) 30 (c) 10800 (d) 21600

3. In how many ways can the word LINGUISTIC be arranged?
   (A) 10! (B) 9! (c) $\frac{10!}{2!}$ (d) $\frac{10!}{3!}$

4. If there are 50 questions, each with options A, B, C and D in the UTME 2010 mathematics examination, in how many ways can a student fail the questions?
   (A) 50^3 (B) 3^4 (C) 3^{50} (D) $3 \times 50!$

5. A male student has 5 shoes, 3 trousers, 4 shirts and 2 belts. How many combinations of these can he wear?
   (A) 60 (B) 605404800 (C) 120 (D) 240

6. A committee of four is to be formed consisting of 3 men and 1 woman from a group of 5 men and 4 women, in how many ways can the members be selected?
   (A) 40 (B) 20 (C) 126 (D) 63

7. There are five pens on the table, in how many ways can a student arrange them if two of the pens are of the same colour?
   (A) 120 (B) 60 (C) 24 (D) 48

8. Solve for x in the equation $4({}^{3}P_x) - 6({}^{4}C_x) = 0$
   (A) 2 (B) 3 (C) 4 (D) 5

9. Evaluate ${}^{10}C_3 \times {}^{8}C_4 \times {}^{12}C_{10}$
   (A) 250 (B) 554400 (C) 252000 (D) 1108800

10. A die is thrown twice and a coin is tossed eight times, what is the total possible combination between the die and coin?
    (A) 256 (B) 36 (C) 9216 (D) 292

Exercise 16

1. What do you understand by the terms (i) Permutation (ii) Combination
2. State the difference between permutation and combination?
3. State whether each of the following terms is a permutation or combination:
   (a) Selecting 5 players from a team for penalty kick.
   (b) Awarding three medal to 12 participants in a competition.
   (c) Appointing the president, vice president and secretary from a group of 50.
   (d) Choosing 2 universities out of 52 institutions for admission.
   (e) Choosing F.U.T.O, U.N.N and I.M.S.U from 60 universities in Nigeria.
   (f) Choosing three favorite dishes from a list of menu.
4. In how many ways can each of the following words be arranged:
   (a) ABCDEFG (b) MOTOROLA (c) HOUSE (d) IMMEDIATELY (e) LANGUAGE
   (f) YESTERDAY (g) MATHEMATICS (h) INNOVATION (i) EMPIRE
5. In how many ways can 5 physics, 4 chemistry and 3 mathematics textbooks be arranged by a science student in a book shelve?
6. Evaluate the following factorials:
   (i) 0! (ii) 1! (iii) 5! (iv) 10! (v) 6! (vi) $\frac{8!}{3!}$ (vii) $\frac{12!}{3!5!10!}$ (viii) 8! 3! (ix) 2! 6! (x) $\frac{10!3!}{4!8!}$
7. In how many ways can 5 men and 3 women be arranged round a circular table?

8. In how many ways can the letters A, M, P, Q, T and U be arranged on a circular table.
9. If $\frac{^8C_r}{^8P_r} = \frac{1}{24}$, find r.
10. Find the number of ways 9 subjects can be selected from 12 subjects for IGCSE.
11. If $^8P_r = 8$, find: (a) $^8P_{r+1}$ (b) $^8P_{r-1}$
12. A coin is tossed 10 times, how many combinations do we have?
13. What is the number of combination of getting at least 8 answers correct out of 10 questions in a test?
14. In how many ways can a committee of 6 be selected from 15 members of a group?
15. A bag contains 4 blue balls, 3 red balls and 5 white balls. If two balls are selected at random, find the number of possible ways of doing it.
16. Evaluate: (a) $^3C_1 \times {^8C_1}$ (b) $^{10}C_3 \times {^8C_5}$ (c) $^{10}P_2 \times {^5C_3}$ (d) $\frac{^6C_3}{^6P_3}$ (e) $\frac{^5P_r}{^4C_r} = \frac{5}{2}$
17. A man was asked to select one wife from 4 beautiful ladies. How many possible choices does he have?
18. Three medals are to be given to three winners in a competition consisting of 20 participants. In how many ways can the prizes be awarded?
19. Using question 18 above, if the medals are gold, silver, and bronze; in how many ways can these be given to 20 people in that order?
20. The first two digits of 9-digit telephone numbers in a state is 08, how many different telephone numbers are possible?

17 COMPLEX NUMBER

17.1 DEFINITION AND ALGEBRAIC OPERATION

17.1.1 DEFINITION AND PROPERTIES

These are numbers that are expressed in the form of $a + ib$ where, a, b are real numbers and 'i' is an imaginary number called "iota"; then the value of $i = \sqrt{-1}$. The idea of complex number is applied on periodic functions as wave forms which has its large applications in engineering: Waves, Electricity, Magnetism, and many more.

Given, $i = \sqrt{-1}$ then

- $i^2 = \sqrt{-1}^2 = -1$
- $i^3 = \left(\sqrt{-1}^2\right).\sqrt{-1} - -1.\sqrt{-1} - -i$
- $i^4 = \left(\sqrt{-1}^2\right).\left(\sqrt{-1}^2\right) = -1.-1 = 1$

The powers of i has a repeating form, hence:

- $i = \sqrt{-1} = i^{4n+1} = i$
- $i^2 = i^{4n+2} = -1$
- $i^3 = i^{4n+3} = -i$
- $i^4 = i^{4n+4} = 1$

Complex numbers can be represented as a two-dimensional complex plane. The horizontal axis is real number and the vertical axis is the imaginary number.

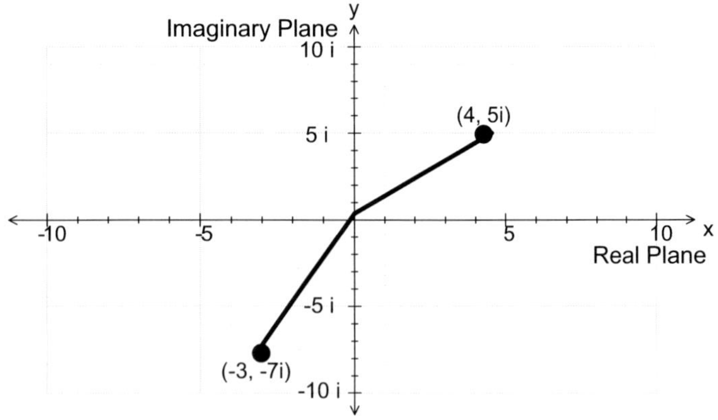

Figure 17.1: Complex Number Representation

17.1.2 ALGEBRAIC COMPLEX OPERATION

Given that $x = a + ib$ and $y = c + id$ then we can say that the two complex numbers are equal when $x = y \quad \Rightarrow a + ib = c + id$ then comparing both sides of the equation, $a = c \; and \; b = d$.

Meanwhile, let's take a look at operations that can be performed on complex numbers examples,

Example 1

If $x = a + ib$ and $y = c + id$. Find (a) $x + y$ (b) $x - y$ (c) $x \times y$ (d) $\frac{x}{y}$

Solution

(a) Addition: $x + y = (a + ib) + (c + +id) = (a + b) + i(b + d)$

(b) Subtraction: $x - y = (a + ib - (c + +id) = (a - b) + i(b - d)$

(c) Multiplication $x \times y = (a + ib) \times (c + +id) = a.c + ia.d + ib.c + i^2 b.d$

$$= a.c + ia.d + ib.c - b.d = (a.c - b.d) + i(a.d - b.c)$$

Remember $i^2 = (\sqrt{-1})^2 = -1$, then the equation becomes, $a.c + ia.d + ib.c - b.d$

(d) Division $\frac{x}{y} = \frac{a+ib}{c+id} = \frac{a+ib}{c+id} \times \frac{c-id}{c-id}$

$$= \frac{a.c-ia.d+ib.c-i^2 b.d}{c^2-ic.d+ic.d-i^2 d^2}$$

$$= \frac{a.c+i(b.c-a.d)+b.d}{c^2-i(c.d-c.d)+d^2}$$

$$= \frac{(a.c+b.d)+i(b.c-a.d)}{c^2-i0+d^2}$$

$$= \frac{(a.c+b.d)+i(b.c-a.d)}{c^2+d^2}$$

In division, you multiply both the numerator and denominator by the conjugate of the complex number. The conjugate of the complex number, $c + id$ is $c - id$. It is just a change in the operation sign.

Example 2

If $x = 3 + i4$ and $y = 6 + i8$, Find (a) $x + y$ (b) $x - y$ (c) $x \times y$ (d) $\frac{x}{y}$

Solution

(a) $x + y = (3 + i4) + (6 + i8) = (3 + 6) + i(4 + 8) = 9 + 12i$

(b) $x - y = (3 + i4) - (6 + i8) = (3 - 6) + i(4 - 8) = -3 - 4i$

(c) $x \times y = (3 + i4) \times (6 + i8) = 3 * 6 + 3 * i8 + i4 * 6 + 4 * 8i^2$

$= 18 + 24i + 24i - 32 = (18 - 32) + i(24 + 24) = -14 + 28i$

(d) $\frac{x}{y} = \frac{3+4i}{6+8i} = \frac{3+4i}{6+8i} \times \frac{6-8i}{6-8i}$

$= \frac{(3*6+4*8)+i(4*6-3*8)}{6^2+8^2}$

$= \frac{(18+32)+i(24-24)}{6^2+8^2} = \frac{50+0i}{100} = \frac{50}{100} = \frac{1}{2}$

In division, you multiply both the numerator and denominator by the conjugate of the complex number as shown previously in example 1d.

Hence, $\frac{a+ib}{c+id} = \frac{(a.c+b.d)+i(b.c-a.d)}{c^2+d^2}$

Example 3

Given $z = 3 - 2i$, Find (a) $5z$ (b) z^2 (c) z^4

Solution

(a) $5z = 5(3 - 2i) = 5 * 3 - 5 * 2i = 15 - 10i$

(b) $z^2 = (3 - 2i)^2 = (3 - 2i) * (3 - 2i) = 3 * 3 - 3 * 2i - 2 * 3i + 2i * 2i$

$= 9 - 6i - 6i + 4i^2 = 9 - 12 + 4(-1) = 9 - 12 - 4$

Remember,
$i^2 = (\sqrt{-1})^2 = -1$

$= -7$

(c) $z^4 = (z^2)^2 = (-7)^2 = 49$

We have already solved in example 3b for $z^2 = -7$

17.2 MODULUS/MAGNITUDE AND CONJUGATE

17.2.1 MODULUS/MAGNITUDE: The modulus of a number is referred to as its magnitude. Thus, a complex number, $z = x + iy$; then, modulus of z, will be the resolution of the equation using Pythagoras theorem to evaluate the line from the origin to the point in space of the complex number.

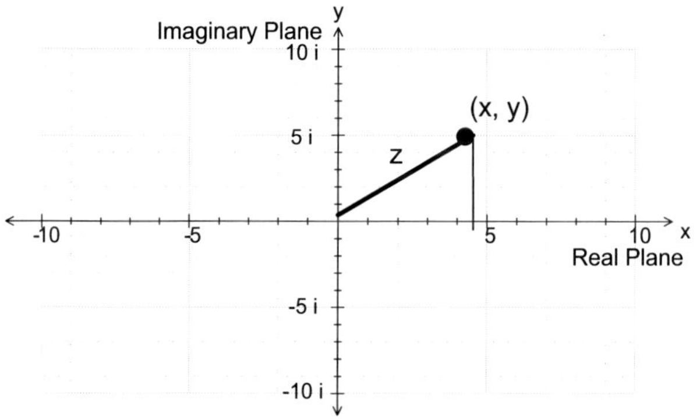

Figure 17.2: Modulus of a Complex Number

The modulus, of $z = x + iy$ is written as $|z| = |x + iy|$

$$|z| = \sqrt{x^2 + y^2}$$

Example 4

Given $z = 2i$, Find (a) $|z|$ (b) $|z^2|$

Solution

(a) If $z = 2i = 0 + 2i$; hence x=0 and y=2

$|z| = \sqrt{x^2 + y^2} = \sqrt{0^2 + 2^2} = \sqrt{4} = 2$

$\therefore |z| = 2$

(b) $|z^2| = |2i.2i| = |4i^2| = |-4|$

Now, $z^2 = -4 = -4 + 0i$ then x=−4 and y=0

$|z^2| = \sqrt{(-4)^2 + 0^2} = \sqrt{16 + 0} = \sqrt{16} = 4$

$\therefore |z^2| = 4$

Example 5

Given $x = 3 + i4$ and $y = 6 + i8$, Find (a) $|x|$ (b) $|y|$

Solution

(a) If $z = 3 + 4i$; then a=3 and b=4

$|x| = \sqrt{a^2 + b^2} = \sqrt{3^2 + 4^2} = \sqrt{9 + 16}$

$= \sqrt{25} = 5$

$\therefore |x| = 5$

(b) If $z = 6 + 8i$; then a=6 and b=8

$|z| = \sqrt{a^2 + b^2} = \sqrt{6^2 + 8^2} = \sqrt{36 + 64}$

$= \sqrt{100} = 10$

$\therefore |z| = 10$

Alternatively, $y = 6 + i8 = 2(3 + 4i) = 2|3 + 4i| = 2 * 5 = 10$

Example 6

Using the values of example 5, find: (a) $|x + y|$ (b) $|x - y|$

Solution

(a) If $x + y = (3 + 4i) + (6 + 8i) = 9 + 12i$

then a=9 and b=12

$|x + y| = \sqrt{9^2 + 12^2} = \sqrt{81 + 144}$

$= \sqrt{225} = 15$

$\therefore |x + y| = 15$

(b) If $x - y = (3 + 4i) - (6 + 8i) = -3 - 4i$

then a= -3 and b=-4

$|x - y| = \sqrt{-3^2 + -4^2} = \sqrt{9 + 16}$

$= \sqrt{25} = 5$

$\therefore |x - y| = 5$

17.2.2 CONJUGATE OF A COMPLEX NUMBER:

Given a complex number, $z = x + iy$, then the conjugate is denoted as $\bar{z}$.
Where, $\bar{z} = x - iy$

Example 7

Given $x = 3 + 4i$, $y = -1 + 7i$, , $z = 1 - 2i$ Find (a) $\bar{x}$ (b) $\bar{y}$ (c) $\bar{z}$

Solution

(a) $\bar{x} = 3 - 4i$
(b) $\bar{y} = -1 - 7i$
(c) $\bar{z} = 1 + 2i$

Example 8

Given $z = 2 - 3i$ find (a) $\bar{z}$ (b) $z + \bar{z}$ (c) $z\bar{z}$

Solution

(a) $\bar{z} = 2 + 3i$
(b) $z + \bar{z} = 2 - 3i + 2 + 3i = 4$
(c) $z\bar{z} = (2 - 3i)(2 + 3i) = 4 - 6i + 6i - 9i^2 = 4 + 9 = 13$

Example 9

If $z = \frac{5-2i}{1-i} = a + bi$, find the value of a and b

Solution

Given, $z = \frac{5-2i}{1-i}$ then we have $\frac{5-2i}{1-i} * \frac{1+i}{1+i} = \frac{(5-2i)(1+i)}{(1-i)(1+i)}$

$\Rightarrow \frac{5+5i-2i-2i^2}{1+i-i-i^2} = \frac{5+3i+2}{1+0+1} = \frac{7+3i}{2} = \frac{7}{2} + \frac{3}{2}i$

Hence, $\frac{7}{2} + \frac{3}{2}i = a + bi$ Comparing both sides of the equation

$a = \frac{7}{2}$ and $b = \frac{3}{2}$

17.3 MORE EXAMPLES ON COMPLEX NUMBER

In this subsection, more examples will be solved on complex number

Example 10

Evaluate the following (a) $(1+i)^{10}$ (b) $i^{2020} + i^{2023} + i^{2025}$

Solution

(a) Given, $(1+i)^{10}$ then we have

$$\Rightarrow (1+i)^{10} = [(1+i)^2]^5 = (2i)^5$$
$$(2)^5 * i^5 = 32 * i^4.i = 32i$$

Solving for $(1+i)^2$:

$$(1+i)^2 = (1+i)(1+i)$$
$$= 1 + 2i - 1 = 2i$$

(b) $i^{2020} + i^{2023} + i^{2025}$

$$i^{2020} = i^{4(505)} = i^{4(n)} = 1$$
$$i^{2023} = i^{4(505)+3} = i^{4(n)+3} = -i$$
$$i^{2025} = i^{4(506)+1} = i^{4(n)+1} = i$$

Thus, $i^{2020} + i^{2023} + i^{2025} = 1 + (-i) + i$

$$i^{2020} + i^{2023} + i^{2025} = 1$$

Remember this:

- $i = \sqrt{-1} = i^{4n+1} = i$
- $i^2 = i^{4n+2} = -1$
- $i^3 = i^{4n+3} = -i$
- $i^4 = i^{4n} = 1$

Example 11

Evaluate the following $5 - \sqrt{-9} + 3\sqrt{-4}$

Solution

$$5 - \sqrt{-9} + 3\sqrt{-4} = 5 - \sqrt{9} * \sqrt{-1} + 3(\sqrt{4} * \sqrt{-1})$$

$$= 5 - 3 * \sqrt{-1} + 3(2 * \sqrt{-1}) = 5 - 3i + 3(2i) = 5 - 3i + 6i$$

$$= 5 + 3i$$

Example 12

If $x = 1 - i$ then evaluate $3x^2 + 2x - 4$

Solution

Substitute the value of $x = 1 - i$, into the equation $3x^2 + 2x - 4$

$3(1 - i)^2 + 2(1 - i) - 4 = 3(1 - 2i - 1) + (2 - 2i) - 4$

Simplifying, we have:

$3(-2i) + (2 - 2i) - 4 = -6i - 2i - 2$

$= -8i - 2$

Solving for $(1 - i)^2$:

$(1 - i)^2 = (1 - i)(1 - i)$

$= 1 - i - i + i^2 = 1 - 2i - 1 = -2i$

Example 13

A polynomial function is defined by $f(x) = 7x^3 - 4x^2 + 2x + 3$ find $f(i)$

Solution

Substitute the value of $x = i$, into the equation $f(x) = 7x^3 - 4x^2 + 2x + 3$

$7(i)^3 - 4(i)^2 + 2(i) + 3 = 7(i^2 * i) - 4(-1) + 2i + 3$

$-7i + 4 + 2i + 3 = -7i + 2i + 4 + 3 = -5i + 7$

Example 14

What is the value of $\left|\frac{2-i}{3-i}\right|$

Solution

Let $\left|\frac{2-i}{3-i}\right| = |z|$, then we first find the conjugate resolution of z in the form $x + iy$

Hence,

$\frac{2-i}{3-i}$ becomes $\frac{2-i}{3-i} * \frac{3+i}{3+i} = \frac{(2-i)(3+i)}{(3-i)(3+i)} \Rightarrow \frac{6+2i-3i-i^2}{9-i^2} = \frac{6-i+1}{9+1} = \frac{7-i}{10}$

$= \frac{7}{10} - \frac{1}{10}i = \frac{1}{10}(7 - i)$

Then we find the magnitude, $|z| = \sqrt{x^2 + y^2} = \frac{1}{10} * \sqrt{7^2 + (-1)^2} = \frac{1}{10} * \sqrt{49 + 1}$

$\frac{1}{10} * \sqrt{50} = \frac{1}{10} * \sqrt{25} * \sqrt{2} = \frac{1}{10} * 5\sqrt{2} = \frac{1}{2}\sqrt{2}$

17.4 POLAR AND EXPONENIAL REPRESENTATION

17.4.1POLAR FORM: So far, we have been dealing with complex number representation in the Cartesian form. However, we can represent it in Polar form.

Given that $z = x + iy$ in the Cartesian form then the polar representation will become:

$$z = r(cos\theta + isin\theta)$$

Where, $r = \sqrt{x^2 + y^2} = |z|$; $x = rcos\theta$; $y = rsin\theta$;
$\theta = \arg(z)$ *called argument of z*; and $tan\theta = \frac{y}{x}$

Example 15

Convert the following complex number, (a) $z = 1 - i$ (b) $z = -7$

Solution

(a) $z = x + iy = 1 - i$ then x=1 and y=−1

Hence to change to polar form, $z = r(cos\theta + sin\theta)$

Where, $r = \sqrt{x^2 + y^2} = |z| = \sqrt{1^2 + (-1)^2} = \sqrt{2}$

$x = rcos\theta \Rightarrow 1 = \sqrt{2}cos\theta \therefore cos\theta = \frac{1}{\sqrt{2}} \quad \theta = \frac{\pi}{4}$ or $\frac{7\pi}{4}$

Similarly, $y = rsin\theta$; $\Rightarrow -1 = \sqrt{2}sin\theta$
$\therefore sin\theta = \frac{-1}{\sqrt{2}} \quad \theta = \frac{3\pi}{4}$ or $\frac{7\pi}{4}$

Sine is negative and must be in third or fourth quadrant; while cosine is positive and must be in first or fourth quadrant. Thus, both are equal in the fourth quadrant so we choose the fourth quadrant value, $2\pi - \frac{\pi}{4} = \frac{7\pi}{4}$

Then the polar form is $z = \sqrt{2}(cos\frac{7\pi}{4} + isin\frac{7\pi}{4})$
(b) $z = x + iy = -7 = -7 + 0i$ then x= -7 and y=0

Hence to change to polar form, $z = r(cos\theta + isin\theta)$

Where, $r = \sqrt{x^2 + y^2} = |z| = \sqrt{-7^2 + 0^2} = 7$

$x = rcos\theta \Rightarrow -7 = 7cos\theta \therefore cos\theta = -1$

Similarly, $y = rsin\theta$; $\Rightarrow 0 = 7cos\theta$
$\therefore sin\theta = 0$ thus, $\theta = \pi$ Then the polar form is, $z = 7(cos\pi + isin\pi)$

Example 16

Convert from polar to Cartesian form

(a) $z = 1(cos\pi + isin\pi)$ (b) $z = 3\left(cos\frac{\pi}{4} + isin\frac{\pi}{4}\right)$

Solution

(a) $z = 1(cos\pi + isin\pi) = r(cos\theta + isin\theta)$

Hence to change to polar form, $z = r(cos\theta + isin\theta) = 1.(cos\pi + isin\pi)$

Where, $r = \sqrt{x^2 + y^2} = |z| = \sqrt{x^2 + y^2} = 1$ and $\theta = \pi$

$x = 1.cos\pi \Rightarrow x = 1.(-1) = -1$
$y = 1.sin\pi \Rightarrow y = 1.(0) = 0$
$z = x + iy = -1 + i0 = -1$

(b) $z = 3\left(cos\frac{\pi}{4} + isin\frac{\pi}{4}\right) = r(cos\theta + isin\theta)$

Hence to change to Cartesian form, $z = 3\left(cos\frac{\pi}{4} + isin\frac{\pi}{4}\right) = r(cos\theta + isin\theta)$

Where, $r = \sqrt{x^2 + y^2} = |z| = \sqrt{x^2 + y^2} = 3$ and $\theta = \frac{\pi}{4}$

$x = 3.cos\frac{\pi}{4} \Rightarrow x = 3.\left(\frac{1}{\sqrt{2}}\right) = \frac{3}{\sqrt{2}}$
$y = 3.sin\pi \Rightarrow y = 3.\left(\frac{1}{\sqrt{2}}\right) = \frac{3}{\sqrt{2}}$
$z = x + iy = \frac{3}{\sqrt{2}} + i\frac{3}{\sqrt{2}}$

17.4.2 EXPONENTIAL FORM:

Given that $z = x + iy$ in the Cartesian form then the exponential representation will become:

$$z = re^{i\theta}$$

Where, $r = \sqrt{x^2 + y^2} = |z|$; and $tan\theta = \frac{y}{x}$

Example 12

Convert the following complex numbers, (a) $z = 1 - i$ (b) $z = -7$ to exponential form

Solution

(a) Hence to change to exponential form, $z = re^{i\theta}$

Where, $r = \sqrt{x^2 + y^2} = |z| = \sqrt{1^2 + 1^2} = \sqrt{2}$

$z = re^{i\theta} = \sqrt{2}e^{i\frac{7\pi}{4}}$ or $\sqrt{2}e^{i\frac{-\pi}{4}}$

(b) $z = x + iy = -7 = -7 + 0i$ then x= -7 and y=0

Hence to change to exponential form, $z = re^{i\theta}$

Where, $r = \sqrt{x^2 + y^2} = |z| = \sqrt{-7^2 + 0^2} = 7$

$tan\theta = \frac{y}{x} = \frac{0}{7} = 0$ therefore, $\theta = \pi$

$= 7e^{i\pi}$

OBJECTIVE TEST 17

1. If $x = 3 + i4$ and $y = 6 - i8$ find $2x - 3y$
   (A) $32i + 12$ (B) $16 - 32i$ (C) $-12 + 32i$ (D) $12 - 32i$

2. If $x = 1 + i$ and $y = i + 2$ find xy
   (A) $i - 3$ (B) $1 + 3i$ (C) $3 + i$ (D) $3i$

3. If $x = 2 - i$ and $y = 2 + 3i$ find $\frac{y}{x}$
   (A) $\frac{1+8i}{5}$ (B) $\frac{1-2i}{5}$ (C) $5(1 + 2i)$ (D) $\frac{1+5i}{2}$

4. If $z = 3 - 2i$ find z^2
   (A) $13 - 12i$ (B) $5 - 12i$ (C) $12 + 5i$ (D) $5 + 12i$

5. What is the modulus of the complex number $5 - 12i$
   (A) 60 (B) 12 (C) 7 (D) 13

6. Simplify, $i^{2019} + i^{2020} + i^{2022}$
   (A) i (B) 2 (C) $-i$ (D) $i - 2$

7. If $\frac{2+i}{3-2i} = a + bi$, find the values of a and b
   (A) $a = \frac{4}{13}$, $b = \frac{7}{13}$ (B) $a = \frac{-4}{13}$, $b = -\frac{7}{13}$
   (C) $a = \frac{7}{13}$, $b = \frac{4}{13}$ (D) $a = \frac{8}{13}$, $b = \frac{7}{13}$

8. Given that $x = 2i$ is a solution of $4x^3 + 8kx = 0$. Find k
   (A) -2 (B) -16 (C) 16 (D) 2

9. Change the complex number to polar coordinate, $z = 3 + 4i$
   (A) $5(cos37^0, isin37^0)$ (B) $5(cos53^0, isin53^0)$
   (C) $5(sin37^0, icos37^0)$ (D) $5(cos53^0, sin53^0)$

10. What is the exponential form of $z = 3$
    (A) $3e^{i\pi}$ (B) $-3e^{i\pi}$ (C) $3e^{i\frac{\pi}{2}}$ (D) $-3e^{i\frac{\pi}{2}}$

Exercise 17

1. What is the value of each of the following expressions?
   (a) $\sqrt{i} * \sqrt{i}$ (b) i^6 (c) i^9 (d) i^{11} (e) i^{3001}

2. What is the imaginary part for each of the following complex numbers:
   (a) $4 - 3i$ (b) $-3 + 5i$ (c) -9 (d) 0

3. Simplify the following:
   (a) $(2 + 5i) - (-6 + 5i)$ (b) $(1 + i) + (3 - i)$ (c) $i^4 - i^7$ (d) $3i^9 + 5i^{12} + 8i^{13}$

4. Solve for each of the following:
   (a) $i^{22} + i^{23} + i^{24} + i^{25}$ (b) $(i^{2020} - i^{2023}) * (i^{2025} + i^{2032})$

5. For $i = \sqrt{-1}$ Simplify the following:
   (a) $(-7 + 5\sqrt{-1})(2\sqrt{-1})$ (b) $(5 - \sqrt{-4}) + (7 - 4\sqrt{-9})$ (c) $(2 - 3\sqrt{-1}) - (3 + 2\sqrt{-1})$

6. State the conjugate of each of the following:
   (a) $1 + i$ (b) $33i$ (c) $-1 - i$ (d) $3i - 5$ (e) $2 - i^3$

7. Calculate the modulus of the following:
   (a) $5i$ (b) $2 - i$ (c) $4 - \sqrt{-9}$ (d) $\frac{i+3}{i-3}$ (e) $\frac{i-1}{i+1}$

8. Given that $a - bi = \frac{i+2}{i-1}$ Find a and b

9. Given that $a + bi = \frac{3-4i}{2i-1}$ Find $a + b$

10. A polynomial function is defined by $f(x) = 3x^3 + 16x^2 - 2x + 16$, find:
    (a) $f(i)$ (b) $f(-i)$ (c) $f(2i)$

11. If $x = 1 - i$, find $2x^2 - 5x + 4$

12. Given that $x = -5i$ is a solution of $2x^3 + kx = 0$. Find k

13. Simplify the following complex numbers:
    (a) $\frac{i-1}{i+1}$ (b) $\frac{3-2i^3}{4+3i}$

14. From figure 17.3, (a) trace and determine the complex number (b) find the magnitude

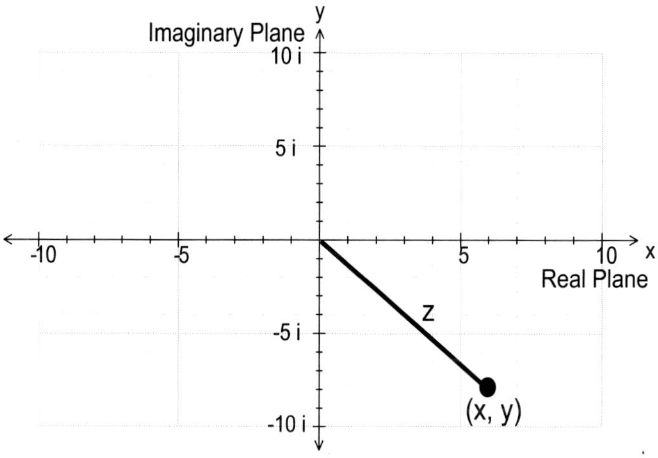

Figure 17.3

15. Change the following complex number to polar coordinate form:
(a) $z = 3 + 4i$ (b) $z = 3 - 4i$ (c) $z = -i$

16. Convert the following Complex number in polar form to Cartesian form:
(a) $z = 1(cos\pi + isin\pi)$ (b) $z = 5\left(cos\frac{\pi}{4} + isin\frac{\pi}{4}\right)$

18 MATRICES

A matrix is a rectangular array of numbers enclosed by pair of brackets and governed by a rule of operation. So, if a matrix is denoted by A with $(M \times N)$ as the size or order of the matrix, then:

$$A = \begin{pmatrix} a_{11} & a_{12} & \cdots & a_{1n} \\ a_{21} & a_{22} & \cdots & a_{2n} \\ \vdots & \vdots & \ddots & \vdots \\ a_{m1} & a_{m2} & \cdots & a_{mn} \end{pmatrix}$$

Where: a_{ij} are the components of the matrix showing its ith row and jth column; and m and n represent the space size (i.e. $i = 1,2,3, \dots m$ and $j = 1, 2, 3, \dots n$)

18.1 CLASSIFICATIONS OF MATRICES

Matrix can be classified as square, triangular, diagonal, identity, zero, equal, similar, symmetric, skewed symmetric, idempotent and nilpotent matrices.

18.1.1 SQUARE MATRIX

A square matrix is a matrix with the same number of rows and columns (i.e., $m = n$)

Example $A = \begin{pmatrix} a_{11} & a_{12} & a_{13} \\ a_{21} & a_{22} & a_{23} \\ a_{31} & a_{32} & a_{33} \end{pmatrix}$ is a 3×3 matrix, thus it is a square matrix.

18.1.2 TRIANGULAR MATRIX

This is made up of upper triangular and lower triangular matrices.

For instance, $A = \begin{pmatrix} a_{11} & a_{12} & a_{13} \\ 0 & a_{22} & a_{23} \\ 0 & 0 & a_{33} \end{pmatrix}$ this is upper triangular matrix.

$B = \begin{pmatrix} a_{1} & 0 & 0 \\ a_{21} & a_{22} & 0 \\ a_{31} & a_{32} & a_{33} \end{pmatrix}$ this is lower triangular matrix.

18.1.3 DIAGONAL MATRIX

A matrix is called a "diagonal matrix" if it has all their elements equal zeroes, except possibly, on its "main diagonal".

Thus, $A = \begin{pmatrix} a_{11} & 0 & 0 \\ 0 & a_{22} & 0 \\ 0 & 0 & a_{33} \end{pmatrix}$ is a diagonal matrix because the main or leading diagonal have possibly non-zero elements.

18.1.4 IDENTITY MATRIX

This is also called a "unit matrix" because it has each entry in the main diagonal as equal to one.

$I = \begin{pmatrix} 1 & 0 \\ 0 & 1 \end{pmatrix}$ for a 2×2 matrix; and $I = \begin{pmatrix} 1 & 0 & 0 \\ 0 & 1 & 0 \\ 0 & 0 & 1 \end{pmatrix}$ for a 3×3 matrix.

18.1.5 ZERO MATRIX

It is a matrix that has all its elements as zeroes.

$$X = \begin{pmatrix} 0 & 0 & 0 \\ 0 & 0 & 0 \\ 0 & 0 & 0 \end{pmatrix}$$

18.1.6 EQUAL MATRICES

If two matrices A and B with the same order of entry and each element of one is equal to the corresponding element of the other, then they are said to be equal matrices.

$$A = \begin{pmatrix} 1 & 3 & 8 \\ 3 & 2 & 7 \\ 5 & 4 & -1 \end{pmatrix} \text{ and } B = \begin{pmatrix} 1 & 3 & 8 \\ 3 & 2 & 7 \\ 5 & 4 & -1 \end{pmatrix}$$

Hence, $A = B$

18.1.7 SIMILAR MATRICES

Two matrices are similar if they have equal traces. Whereby, the trace of the matrix is the sum of the elements in its main diagonals.

Given, $A = \begin{pmatrix} 1 & 9 & 8 \\ 4 & 2 & 7 \\ 5 & 4 & 5 \end{pmatrix}$ and $B = \begin{pmatrix} 3 & 7 & 0 \\ 2 & 7 & 6 \\ -1 & 4 & -2 \end{pmatrix}$

Thus, trace of $A = 1 + 2 + 5 = 8$

And , trace of $B = 3 + 7 - 2 = 8$

Hence, trace A is equal to trace B; then the matrices are similar.

18.1.8 SYMMETRIC MATRICES

A matrix X is symetrical if it is equal to its transpose X.
Hence, $X = X^T$

18.1.9 SKEWED SYMMETRIC MATRIX

This is when a matrix X is equal to the negative of its transpose X^T.
Hence, $X = -X^T$

18.1.10 IDEMPOTENT MATRIX

If a matrix has unchanged value when it multiplies itself, then the matrix is said to be idempotent.
Hence, $A \times A = A$ where A is the idempotent matrix.

18.1.11 NILPOTENT MATRIX

A matrix, A, is nilpotent when all its result is equal to zero when raised to a certain index power.
For instance, $A^n = 0$, where n is a positive integer.

18.2 BASIC OPERATIONS OF MATRICES

18.2.1 ADDITION OF MATRICES

If $A = a_{ij}$ and $B = b_{ij}$ are two matrices, then their sum is given by:
$S_{ij} = A + B = a_{ij} + b_{ij}$
Therefore,

$$A = \begin{pmatrix} a_{11} & a_{12} & \cdots & a_{1n} \\ a_{21} & a_{22} & \cdots & a_{2n} \\ \vdots & \vdots & \ddots & \vdots \\ a_{m1} & a_{m2} & \cdots & a_{mn} \end{pmatrix} \text{ and } B = \begin{pmatrix} b_{11} & b_{12} & \cdots & b_{1n} \\ b_{21} & b_{22} & \cdots & b_{2n} \\ \vdots & \vdots & \ddots & \vdots \\ b_{m1} & b_{m2} & \cdots & b_{mn} \end{pmatrix}$$

$$\text{Then, } S = A + B = \begin{pmatrix} a_{11} + b_{11} & a_{12} + b_{12} & \cdots & a_{1n} + b_{1n} \\ a_{21} + b_{21} & a_{22} + b_{22} & \cdots & a_{2n} + b_{23} \\ \vdots & \vdots & \ddots & \vdots \\ a_{m1} + b_{m1} & a_{m2} + b_{m2} & \cdots & a_{mn} + b_{mn} \end{pmatrix}$$

Example 1

Given $X = \begin{pmatrix} 4 & 7 \\ 10 & 1 \end{pmatrix}$, $Y = \begin{pmatrix} 3 & 5 \\ 6 & 2 \end{pmatrix}$ and $Z = \begin{pmatrix} -1 & 4 \\ 3 & 12 \end{pmatrix}$
Find:
(i) $X + Y$ (ii) $Y + Z$ (iii) $X + Z$

Solution

(i) $X + Y = \begin{pmatrix} 4 & 7 \\ 10 & 1 \end{pmatrix} + \begin{pmatrix} 3 & 5 \\ 6 & 2 \end{pmatrix} = \begin{pmatrix} 4+3 & 7+5 \\ 10+6 & 1+2 \end{pmatrix} = \begin{pmatrix} 7 & 12 \\ 16 & 3 \end{pmatrix}$

$\therefore\ X + Y = \begin{pmatrix} 7 & 12 \\ 16 & 3 \end{pmatrix}$

(ii) $Y + Z = \begin{pmatrix} 3 & 5 \\ 6 & 2 \end{pmatrix} + \begin{pmatrix} -1 & 4 \\ 3 & 12 \end{pmatrix} = \begin{pmatrix} 3+(-1) & 5+4 \\ 6+3 & 2+12 \end{pmatrix} = \begin{pmatrix} 2 & 9 \\ 9 & 14 \end{pmatrix}$

$\therefore\ Y + Z = \begin{pmatrix} 2 & 9 \\ 9 & 14 \end{pmatrix}$

(iii) $X + Z = \begin{pmatrix} 4 & 7 \\ 10 & 1 \end{pmatrix} + \begin{pmatrix} -1 & 4 \\ 3 & 12 \end{pmatrix} = \begin{pmatrix} 4+(-1) & 7+4 \\ 10+3 & 1+12 \end{pmatrix} = \begin{pmatrix} 3 & 11 \\ 13 & 13 \end{pmatrix}$

$\therefore\ X + Z = \begin{pmatrix} 3 & 11 \\ 13 & 13 \end{pmatrix}$

Example 2

Given $A = \begin{pmatrix} 2 & 3 & 4 \\ 3 & 0 & -1 \\ -2 & 5 & -5 \end{pmatrix}$, $B = \begin{pmatrix} 1 & 0 & 4 \\ 0 & 2 & 4 \end{pmatrix}$ and $C = \begin{pmatrix} -1 & 3 & 2 \\ -2 & 1 & 5 \\ 0 & 5 & 8 \end{pmatrix}$

Find:

(i) $A + B$ (ii) $A + C$ (iii) $B + C$

Solution

(i) $A + B = \begin{pmatrix} 2 & 3 & 4 \\ 3 & 0 & -1 \\ -2 & 5 & -5 \end{pmatrix} + \begin{pmatrix} 1 & 0 & 4 \\ 0 & 2 & 4 \end{pmatrix}$ this is indeterminable.

> This is because a 3×3 cannot be added to 2×3 matrix. Thus, for matrices to add, they must be of equal order.

(ii) $A + C = \begin{pmatrix} 2 & 3 & 4 \\ 3 & 0 & -1 \\ -2 & 5 & -5 \end{pmatrix} + \begin{pmatrix} -1 & 3 & 2 \\ -2 & 1 & 5 \\ 0 & 5 & 8 \end{pmatrix}$

$= \begin{pmatrix} 2+(-1) & 3+3 & 4+2 \\ 3+(-2) & 0+1 & -1+5 \\ -2+0 & 5+5 & -5+8 \end{pmatrix} = \begin{pmatrix} 1 & 6 & 6 \\ 1 & 1 & 4 \\ -2 & 10 & 3 \end{pmatrix}$

$\therefore\ A + C = \begin{pmatrix} 1 & 6 & 6 \\ 1 & 1 & 4 \\ -2 & 10 & 3 \end{pmatrix}$

(iii) $B + C = \begin{pmatrix} 1 & 0 & 4 \\ 0 & 2 & 4 \end{pmatrix} + \begin{pmatrix} -1 & 3 & 2 \\ -2 & 1 & 5 \\ 0 & 5 & 8 \end{pmatrix}$ this is indeterminable

18.2.2 SUBTRACTION OF MATRICES

The subtraction of matrices employs the same principle as addition of matrices. If $A = a_{ij}$ and $B = b_{ij}$ are two matrices, then their difference/subtraction is given by:

$S_{ij} = A - B = a_{ij} - b_{ij}$

Therefore,

$$A = \begin{pmatrix} a_{11} & a_{12} & \cdots & a_{1n} \\ a_{21} & a_{22} & \cdots & a_{2n} \\ \vdots & \vdots & \ddots & \vdots \\ a_{m1} & a_{m2} & \cdots & a_{mn} \end{pmatrix} \text{and } B = \begin{pmatrix} b_{11} & b_{12} & \cdots & b_{1n} \\ b_{21} & b_{22} & \cdots & b_{2n} \\ \vdots & \vdots & \ddots & \vdots \\ b_{m1} & b_{m2} & \cdots & b_{mn} \end{pmatrix}$$

$$\text{Then, } S = A - B = \begin{pmatrix} a_{11} - b_{11} & a_{12} - b_{12} & \cdots & a_{1n} - b_{1n} \\ a_{21} - b_{21} & a_{22} - b_{22} & \cdots & a_{2n} - b_{23} \\ \vdots & \vdots & \ddots & \vdots \\ a_{m1} - b_{m1} & a_{m2} - b_{m2} & \cdots & a_{mn} - b_{mn} \end{pmatrix}$$

Example 3

Given $X = \begin{pmatrix} 3 & 8 \\ 2 & 10 \end{pmatrix}$, $Y = \begin{pmatrix} 5 & 4 \\ 1 & 3 \end{pmatrix}$ and $Z = \begin{pmatrix} 9 & -2 \\ 3 & 1 \end{pmatrix}$

Find:

(i) $X - Y$ (ii) $Y - Z$ (iii) $X - Z$ (iv) $Z - X$

Solution

(i) $X - Y = \begin{pmatrix} 3 & 8 \\ 2 & 10 \end{pmatrix} - \begin{pmatrix} 5 & 4 \\ 1 & 3 \end{pmatrix} = \begin{pmatrix} 3-5 & 8-4 \\ 2-1 & 10-3 \end{pmatrix} = \begin{pmatrix} -2 & 4 \\ 1 & 7 \end{pmatrix}$

$\therefore X - Y = \begin{pmatrix} -2 & 4 \\ 1 & 7 \end{pmatrix}$

(ii) $Y - Z = \begin{pmatrix} 5 & 4 \\ 1 & 3 \end{pmatrix} - \begin{pmatrix} 9 & -2 \\ 3 & 1 \end{pmatrix} = \begin{pmatrix} 5-9 & 4-(-2) \\ 1-3 & 3-1 \end{pmatrix} = \begin{pmatrix} -4 & 6 \\ -2 & 2 \end{pmatrix}$

$\therefore Y - Z = \begin{pmatrix} -4 & 6 \\ -2 & 2 \end{pmatrix}$

(iii) $X - Z = \begin{pmatrix} 3 & 8 \\ 2 & 10 \end{pmatrix} - \begin{pmatrix} 9 & -2 \\ 3 & 1 \end{pmatrix} = \begin{pmatrix} 3-9 & 8-(-2) \\ 2-3 & 10-1 \end{pmatrix} = \begin{pmatrix} -6 & 10 \\ -1 & 9 \end{pmatrix}$

$\therefore X - Z = \begin{pmatrix} -6 & 10 \\ -1 & 9 \end{pmatrix}$

(iv) $Z - X = \begin{pmatrix} 9 & -2 \\ 3 & 1 \end{pmatrix} - \begin{pmatrix} 3 & 8 \\ 2 & 10 \end{pmatrix} = \begin{pmatrix} 9-3 & -2-8 \\ 3-2 & 1-10 \end{pmatrix} = \begin{pmatrix} 6 & -10 \\ 1 & -9 \end{pmatrix}$

$\therefore Z - X = \begin{pmatrix} 6 & -10 \\ 1 & -9 \end{pmatrix}$

This follows that from solution (iii) and (iv), it can be seen that $X - Z \neq Z - X$.

Example 4

Given $A = \begin{pmatrix} 25 & 7 & 4 \\ 4 & 1 & 5 \\ 8 & 6 & 9 \end{pmatrix}$, $B = \begin{pmatrix} 3 & 12 & 5 \\ 15 & 2 & 4 \end{pmatrix}$ and $C = \begin{pmatrix} 10 & 13 & 1 \\ 15 & 32 & 71 \\ 4 & 16 & 8 \end{pmatrix}$

Find:

(i) $A - B$ (ii) $A - C$

Solution

(i) $A - B = \begin{pmatrix} 25 & 7 & 4 \\ 4 & 1 & 5 \\ 8 & 6 & 9 \end{pmatrix} - \begin{pmatrix} 3 & 12 & 5 \\ 15 & 2 & 4 \end{pmatrix}$ this is indeterminable.

(ii) $A - C = \begin{pmatrix} 25 & 7 & 4 \\ 4 & 1 & 5 \\ 8 & 6 & 9 \end{pmatrix} - \begin{pmatrix} 10 & 13 & 1 \\ 15 & 32 & 71 \\ 4 & 16 & 8 \end{pmatrix}$

$$= \begin{pmatrix} 25-10 & 7-13 & 4-1 \\ 4-15 & 1-32 & 5-71 \\ 8-4 & 6-16 & 9-8 \end{pmatrix} = \begin{pmatrix} 15 & -6 & 3 \\ -11 & -31 & -66 \\ 4 & -10 & 1 \end{pmatrix}$$

$$\therefore A - C = \begin{pmatrix} 15 & -6 & 3 \\ -11 & -31 & -66 \\ 4 & -10 & 1 \end{pmatrix}$$

18.2.3 MULTIPLICATION OF MATRICES

The multiplication of matrices is of two forms, namely:
(i) scalar-matrix multiplication and (ii) matrix-matrix multiplication.

(i) **Scalar-Matrix multiplication**

This is the multiplication of a given matrix, A, by a scalar, K.

If, $A = \begin{pmatrix} a_{11} & a_{12} & \cdots & a_{1n} \\ a_{21} & a_{22} & \cdots & a_{2n} \\ \vdots & \vdots & \ddots & \vdots \\ a_{m1} & a_{m2} & \cdots & a_{mn} \end{pmatrix}$ then $K \times A = KA$

then,

$$KA = K\begin{pmatrix} a_{11} & a_{12} & \cdots & a_{1n} \\ a_{21} & a_{22} & \cdots & a_{2n} \\ \vdots & \vdots & \ddots & \vdots \\ a_{m1} & a_{m2} & \cdots & a_{mn} \end{pmatrix} = \begin{pmatrix} Ka_{11} & Ka_{12} & \cdots & Ka_{1n} \\ Ka_{21} & Ka_{22} & \cdots & Ka_{2n} \\ \vdots & \vdots & \ddots & \vdots \\ Ka_{m1} & Ka_{m2} & \cdots & Ka_{mn} \end{pmatrix}$$

Example 5

Given, $A = \begin{pmatrix} 3 & 4 \\ -1 & 2 \end{pmatrix}$

Find: (i) $2A$ (ii) $-5A$ (iii) $\frac{2}{3}A$

Solution

(i) $2A = 2\begin{pmatrix} 3 & 4 \\ -1 & 2 \end{pmatrix} = \begin{pmatrix} 2(3) & 2(4) \\ 2(-1) & 2(2) \end{pmatrix} = \begin{pmatrix} 6 & 8 \\ -2 & 4 \end{pmatrix}$

(ii) $-5A = -5\begin{pmatrix} 3 & 4 \\ -1 & 2 \end{pmatrix} = \begin{pmatrix} -5(3) & -5(4) \\ -5(-1) & -5(2) \end{pmatrix} = \begin{pmatrix} -15 & -20 \\ 5 & -10 \end{pmatrix}$

(iii) $\frac{2}{3}A = \frac{2}{3}\begin{pmatrix} 3 & 4 \\ -1 & 2 \end{pmatrix} = \begin{pmatrix} \frac{2}{3}(3) & \frac{2}{3}(4) \\ \frac{2}{3}(-1) & \frac{2}{3}(2) \end{pmatrix} = \begin{pmatrix} 2 & \frac{8}{3} \\ \frac{-2}{3} & \frac{4}{3} \end{pmatrix}$

Example 6

Given, $B = \begin{pmatrix} 2 & 8 & -3 \\ 10 & 12 & 5 \\ 8 & 7 & 4 \end{pmatrix}$

Find: (i) $3B$ (ii) $-B$ (iii) $\frac{1}{5}B$

Solution

(i) $3B = 3\begin{pmatrix} 2 & 8 & -3 \\ 10 & 12 & 5 \\ 8 & 7 & 4 \end{pmatrix} = \begin{pmatrix} 3(2) & 3(8) & 3(-3) \\ 3(10) & 3(12) & 3(5) \\ 3(8) & 3(7) & 3(4) \end{pmatrix}$

$= \begin{pmatrix} 6 & 24 & -9 \\ 30 & 36 & 15 \\ 24 & 21 & 12 \end{pmatrix}$

(ii) $-B = -\begin{pmatrix} 2 & 8 & -3 \\ 10 & 12 & 5 \\ 8 & 7 & 4 \end{pmatrix} = \begin{pmatrix} -(2) & -(8) & -(-3) \\ -(10) & -(12) & -(5) \\ -(8) & -(7) & -(4) \end{pmatrix} = \begin{pmatrix} -2 & -8 & 3 \\ -10 & -12 & -5 \\ -8 & -7 & -4 \end{pmatrix}$

(iii) $\frac{1}{5}B = \frac{1}{5}\begin{pmatrix} 2 & 8 & -3 \\ 10 & 12 & 5 \\ 8 & 7 & 4 \end{pmatrix} = \begin{pmatrix} \frac{1}{5}(2) & \frac{1}{5}(8) & \frac{1}{5}(-3) \\ \frac{1}{5}(10) & \frac{1}{5}(12) & \frac{1}{5}(5) \\ \frac{1}{5}(8) & \frac{1}{5}(7) & \frac{1}{5}(4) \end{pmatrix} = \begin{pmatrix} \frac{2}{5} & \frac{8}{5} & -\frac{3}{5} \\ 2 & \frac{12}{5} & 1 \\ \frac{8}{5} & \frac{7}{5} & \frac{4}{5} \end{pmatrix}$

(i) **Matrix-Matrix Multiplication**

This is the multiplication of two matrices. Thus, when two matrices, A and B, are multiplied, then the resulting element of their operation is the sum of the product of row $i = 1, 2, 3, \ldots, m$ and the corresponding column $j = 1, 2, 3, \ldots, n$.

For instance:

$A = a_{ij}$ with size $m \times n$

$B = b_{ij}$ with size $n \times r$

then, the elements $C = AB = c_{ij}$ with size $m \times r$

where $c_{ij} = \sum_{k=1}^{n} a_{ik}b_{kj} = a_{i1}b_{1j} + a_{i2}b_{2j} + a_{i3}b_{3j} + \cdots + a_{in}b_{nj}$

For instance, $A = \begin{pmatrix} a_{11} & a_{12} \\ a_{21} & a_{22} \end{pmatrix}$ and $B = \begin{pmatrix} b_{11} & b_{12} \\ b_{21} & b_{22} \end{pmatrix}$

Therefore, $C = \begin{pmatrix} a_{11}\,b_{11} + a_{12}\,b_{21} & a_{11}\,b_{12} + a_{12}\,b_{22} \\ a_{21}\,b_{11} + a_{22}\,b_{21} & a_{21}\,b_{12} + a_{22}\,b_{22} \end{pmatrix}$

Example 6

$A = (7 \quad 3)$, $B = \begin{pmatrix} 2 \\ 5 \end{pmatrix}$ and $C = \begin{pmatrix} 0 & 2 \\ 5 & 3 \end{pmatrix}$

Find: (i) AB (ii) BA (iii) AC (iv) BC

Solution

(i) $AB = (7 \quad 3)\begin{pmatrix} 2 \\ 5 \end{pmatrix} = (7(2) + 3(5)) = (14 + 15) = (29)$

$\therefore\ AB = (29)$

(ii) $BA = \begin{pmatrix} 2 \\ 5 \end{pmatrix}(7 \quad 3) = \begin{pmatrix} 2(7) & 2(3) \\ 5(7) & 5(3) \end{pmatrix} = \begin{pmatrix} 14 & 6 \\ 35 & 15 \end{pmatrix}$

(iii) $AC = (7 \quad 3)\begin{pmatrix} 0 & 2 \\ 5 & 3 \end{pmatrix} = (7(0) + 3(5) \quad 7(2) + 3(3))$

$= (0 + 15 \quad 14 + 9) = (15 \ \ 23)$

$\therefore\ AC = (15 \ \ 23)$

(iv) $BC = \begin{pmatrix} 2 \\ 5 \end{pmatrix}\begin{pmatrix} 0 & 2 \\ 5 & 3 \end{pmatrix}$ this cannot be possible because B is 2×1 matrix with "one" column, while C is 2×2 matrix with "two" rows (i.e., $i \neq j$).

Example 7

Given $P = (3 \quad 2 \quad -5)$, $Q = \begin{pmatrix} 2 \\ 6 \\ 1 \end{pmatrix}$, $R = \begin{pmatrix} 1 & 3 & 1 \\ 2 & 0 & 3 \end{pmatrix}$ and $S = \begin{pmatrix} 1 & 3 & 1 \\ 2 & 2 & -3 \\ 3 & 0 & -4 \end{pmatrix}$

Find: (i) PQ (ii) QP (iii) PR (iv) RP (v) PS (vi) RS

Solution

(i) $PQ = (3 \quad 2 \quad -5)\begin{pmatrix} 2 \\ 6 \\ 1 \end{pmatrix} = \big(3(2) + 2(6) - 5(1)\big)$

$= (6 + 12 - 5) = (13)$

(ii) $QP = \begin{pmatrix} 2 \\ 6 \\ 1 \end{pmatrix}(3 \quad 2 \quad -5) = \begin{pmatrix} 2(3) & 2(2) & 2(-5) \\ 6(3) & 6(2) & 6(-5) \\ 1(3) & 1(2) & 1(-5) \end{pmatrix}$

$= \begin{pmatrix} 6 & 4 & -10 \\ 18 & 12 & -30 \\ 3 & 2 & -5 \end{pmatrix}$

(iii) $PR - (3 \quad 2 \quad -5)\begin{pmatrix} 1 & 3 & 1 \\ 2 & 0 & 3 \end{pmatrix}$

Note: for multiplication to be possible, the number of columns in P must be equal to the number of rows in R.

This is not possible because P **is order** 1×3 matrix, while R **is order** 2×3. Thus, P has "three" columns, and R has "two" rows; then this makes the multiplication impossible.

(iv) $RP = \begin{pmatrix} 1 & 3 & 1 \\ 2 & 0 & 3 \end{pmatrix}(3 \quad 2 \quad -5)$

This is not possible because R **is order** 2×3 matrix while P **is order** 1×3. Thus, R has "three" columns and Q has "one" row; then this makes the multiplication impossible.

(v) $PS = (3 \quad 2 \quad -5)\begin{pmatrix} 1 & 3 & 1 \\ 2 & 2 & -3 \\ 3 & 0 & -4 \end{pmatrix}$

$= (3(1) + 2(2) - 5(3) \quad 3(3) + 2(2) + 3(0) \quad 3(1) + 2(-3) - 5(-4))$

$= (-8 \quad 13 \quad 17)$

(v) $RS = \begin{pmatrix} 1 & 3 & 1 \\ 2 & 0 & 3 \end{pmatrix}\begin{pmatrix} 1 & 3 & 1 \\ 2 & 2 & -3 \\ 3 & 0 & -4 \end{pmatrix}$

$= \begin{pmatrix} 1(1) + 3(2) + 1(3) & 1(3) + 3(2) + 1(0) & 1(1) + 3(-3) + 1(-4) \\ 2(1) + 0(2) + 3(3) & 2(3) + 0(2) + 3(0) & 2(1) + 0(-3) + 3(-4) \end{pmatrix}$

$= \begin{pmatrix} 10 & 9 & -12 \\ 11 & 6 & -10 \end{pmatrix}$

Example 8

Given $A = \begin{pmatrix} 5 & 2 & 1 \\ 3 & 2 & 4 \\ 6 & 3 & 0 \end{pmatrix}$ and $B = \begin{pmatrix} 6 & 4 & 3 \\ 9 & 7 & 12 \\ 1 & 3 & 8 \end{pmatrix}$

Find: (i) AB (ii) BA

Solution

(i) $AB = \begin{pmatrix} 5 & 2 & 1 \\ 3 & 2 & 4 \\ 6 & 3 & 0 \end{pmatrix}\begin{pmatrix} 6 & 4 & 3 \\ 9 & 7 & 12 \\ 1 & 3 & 8 \end{pmatrix}$

$$= \begin{pmatrix} 5(6)+2(9)+1(1) & 5(4)+2(7)+1(3) & 5(3)+2(12)+1(8) \\ 3(6)+2(9)+4(1) & 3(4)+2(7)+4(3) & 3(3)+2(12)+4(8) \\ 6(6)+3(9)+0(1) & 6(4)+3(7)+0(3) & 6(3)+3(12)+0(8) \end{pmatrix}$$

$$= \begin{pmatrix} 30+18+1 & 20+14+3 & 15+24+8 \\ 18+18+4 & 12+14+12 & 9+24+32 \\ 36+27+0 & 24+21+0 & 18+36+0 \end{pmatrix} = \begin{pmatrix} 49 & 37 & 47 \\ 40 & 38 & 63 \\ 63 & 45 & 54 \end{pmatrix}$$

$$\therefore\ AB = \begin{pmatrix} 49 & 37 & 47 \\ 40 & 38 & 63 \\ 63 & 45 & 54 \end{pmatrix}$$

(ii) $BA = \begin{pmatrix} 6 & 4 & 3 \\ 9 & 7 & 12 \\ 1 & 3 & 8 \end{pmatrix}\begin{pmatrix} 5 & 2 & 1 \\ 3 & 2 & 4 \\ 6 & 3 & 0 \end{pmatrix}$

$$= \begin{pmatrix} 6(5)+4(3)+3(6) & 6(2)+4(2)+3(3) & 6(1)+4(4)+3(0) \\ 9(5)+7(3)+12(6) & 9(2)+7(2)+12(3) & 9(1)+7(4)+12(0) \\ 1(5)+3(3)+8(6) & 1(2)+3(2)+8(3) & 1(1)+3(4)+8(0) \end{pmatrix}$$

$$= \begin{pmatrix} 30+12+18 & 12+8+9 & 6+16+0 \\ 45+21+72 & 18+14+36 & 9+28+0 \\ 5+9+48 & 2+6+24 & 1+12+0 \end{pmatrix} = \begin{pmatrix} 60 & 29 & 22 \\ 138 & 68 & 37 \\ 62 & 32 & 13 \end{pmatrix}$$

$$\therefore\ BA = \begin{pmatrix} 60 & 29 & 22 \\ 138 & 55 & 37 \\ 62 & 32 & 13 \end{pmatrix}$$

Hence, from the answers in (i) and (ii), we can see that $AB \neq BA$.

18.3 PROPERTIES OF MATRIX OPERATIONS

Matrices can exhibit the following properties: commutativity, associativity, and distributivity.

18.3.1 COMMUTATIVITY OF MATRICES

A matrix, A, is said to be commutative with another matrix B, if:

$$A + B = B + A$$

Example

Given $A = \begin{pmatrix} 2 & -1 \\ 5 & 0 \end{pmatrix}$ and $B = \begin{pmatrix} 3 & 2 \\ 1 & 4 \end{pmatrix}$

Check if the matrix is commutative under:

(i) addition (ii) subtraction (iii) multiplication

Solution

(I) $A + B = \begin{pmatrix} 2 & -1 \\ 5 & 0 \end{pmatrix} + \begin{pmatrix} 3 & 2 \\ 1 & 4 \end{pmatrix} = \begin{pmatrix} 5 & 1 \\ 6 & 4 \end{pmatrix}$

and, $B + A = \begin{pmatrix} 3 & 2 \\ 1 & 4 \end{pmatrix} + \begin{pmatrix} 2 & -1 \\ 5 & 0 \end{pmatrix} = \begin{pmatrix} 5 & 1 \\ 6 & 4 \end{pmatrix}$

Hence $A + B = B + A$, then it is commutative under addition.

(ii) $A - B = \begin{pmatrix} 2 & -1 \\ 5 & 0 \end{pmatrix} - \begin{pmatrix} 3 & 2 \\ 1 & 4 \end{pmatrix} = \begin{pmatrix} -1 & -3 \\ 4 & -4 \end{pmatrix}$

and $B - A = \begin{pmatrix} 3 & 2 \\ 1 & 4 \end{pmatrix} - \begin{pmatrix} 2 & -1 \\ 5 & 0 \end{pmatrix} = \begin{pmatrix} 1 & 3 \\ -4 & 4 \end{pmatrix}$

Hence $A - B \neq B - A$, then it is not commutative under subtraction.

(iii) $AB = \begin{pmatrix} 2 & -1 \\ 5 & 0 \end{pmatrix}\begin{pmatrix} 3 & 2 \\ 1 & 4 \end{pmatrix} = \begin{pmatrix} 5 & 0 \\ 15 & 10 \end{pmatrix}$

And $BA = \begin{pmatrix} 3 & 2 \\ 1 & 4 \end{pmatrix}\begin{pmatrix} 2 & -1 \\ 5 & 0 \end{pmatrix} = \begin{pmatrix} 16 & -3 \\ 22 & -1 \end{pmatrix}$

Hence $AB \neq BA$, then it is commutative under multiplication.

18.3.2 ASSOCIATIVITY OF MATRICES

A matrix is associative if any of these conditions holds:

$$A + (B + C) = (A + B) + C$$

and/or $$A(BC) = (AB)C$$

Example 10

Given $A = \begin{pmatrix} 1 & 0 & 2 \\ 0 & 3 & 1 \\ 2 & 5 & 8 \end{pmatrix}$, $B = \begin{pmatrix} 1 & 2 & 2 \\ 2 & 1 & 1 \\ 0 & -3 & 4 \end{pmatrix}$ and $C = \begin{pmatrix} 3 & 1 & -2 \\ 2 & 1 & 7 \\ 1 & 1 & 6 \end{pmatrix}$

Test whether the matrix operation is associative under:

(i) addition (ii) subtraction (iii) multiplication

Solution

(i) $A + B = \begin{pmatrix} 1 & 0 & 2 \\ 0 & 3 & 1 \\ 2 & 5 & 8 \end{pmatrix} + \begin{pmatrix} 1 & 2 & 2 \\ 2 & 1 & 1 \\ 0 & -3 & 4 \end{pmatrix} = \begin{pmatrix} 2 & 2 & 4 \\ 2 & 4 & 2 \\ 2 & 2 & 12 \end{pmatrix}$

then $(A + B) + C = \begin{pmatrix} 2 & 2 & 4 \\ 2 & 4 & 2 \\ 2 & 2 & 12 \end{pmatrix} + \begin{pmatrix} 3 & 1 & -2 \\ 2 & 1 & 7 \\ 1 & 1 & 6 \end{pmatrix} = \begin{pmatrix} 5 & 3 & 2 \\ 4 & 5 & 9 \\ 3 & 3 & 18 \end{pmatrix}$

Now using $A + (B + C)$, we have that:

$B + C = \begin{pmatrix} 1 & 2 & 2 \\ 2 & 1 & 1 \\ 0 & -3 & 4 \end{pmatrix} + \begin{pmatrix} 3 & 1 & -2 \\ 2 & 1 & 7 \\ 1 & 1 & 6 \end{pmatrix} = \begin{pmatrix} 4 & 3 & 0 \\ 4 & 2 & 8 \\ 1 & -2 & 10 \end{pmatrix}$

then, $A + (B + C) = \begin{pmatrix} 1 & 0 & 2 \\ 0 & 3 & 1 \\ 2 & 5 & 8 \end{pmatrix} + \begin{pmatrix} 4 & 3 & 0 \\ 4 & 2 & 8 \\ 1 & -2 & 10 \end{pmatrix} = \begin{pmatrix} 5 & 3 & 2 \\ 4 & 5 & 9 \\ 3 & 3 & 18 \end{pmatrix}$

Hence, $(A + B) + C = A + (B + C)$; that means addition is associative.

(ii) Testing for subtraction, we have that: $A - B = \begin{pmatrix} 1 & 0 & 2 \\ 0 & 3 & 1 \\ 2 & 5 & 8 \end{pmatrix} - \begin{pmatrix} 1 & 2 & 2 \\ 2 & 1 & 1 \\ 0 & -3 & 4 \end{pmatrix}$

$= \begin{pmatrix} 0 & -2 & 0 \\ -2 & 2 & 0 \\ 2 & 8 & 4 \end{pmatrix}$

then $(A - B) - C = \begin{pmatrix} 0 & -2 & 0 \\ -2 & 2 & 0 \\ 2 & 8 & 4 \end{pmatrix} - \begin{pmatrix} 3 & 1 & -2 \\ 2 & 1 & 7 \\ 1 & 1 & 6 \end{pmatrix} = \begin{pmatrix} -3 & -3 & 2 \\ -4 & 1 & -7 \\ 1 & 7 & 2 \end{pmatrix}$

Now using $A - (B - C)$, we have that:

$B - C = \begin{pmatrix} 1 & 2 & 2 \\ 2 & 1 & 1 \\ 0 & -3 & 4 \end{pmatrix} - \begin{pmatrix} 3 & 1 & -2 \\ 2 & 1 & 7 \\ 1 & 1 & 6 \end{pmatrix} = \begin{pmatrix} -2 & 1 & 4 \\ 0 & 0 & -6 \\ -1 & -4 & -2 \end{pmatrix}$

then, $A - (B - C) = \begin{pmatrix} 1 & 0 & 2 \\ 0 & 3 & 1 \\ 2 & 5 & 8 \end{pmatrix} - \begin{pmatrix} -2 & 1 & 4 \\ 0 & 0 & -6 \\ -1 & -4 & -2 \end{pmatrix} = \begin{pmatrix} 3 & -1 & -2 \\ 0 & 3 & 7 \\ 3 & 9 & 10 \end{pmatrix}$

Hence, $(A - B) - C \neq A - (B - C)$; that means subtraction is not associative.

(ii) Testing for multiplication, we have that:

$$AB = \begin{pmatrix} 1 & 0 & 2 \\ 0 & 3 & 1 \\ 2 & 5 & 8 \end{pmatrix}\begin{pmatrix} 1 & 2 & 2 \\ 2 & 1 & 1 \\ 0 & -3 & 4 \end{pmatrix} = \begin{pmatrix} 1 & -4 & 10 \\ 6 & 0 & 7 \\ 12 & -15 & 41 \end{pmatrix}$$

$$\text{then } (AB)C = \begin{pmatrix} 1 & -4 & 10 \\ 6 & 0 & 7 \\ 12 & -15 & 41 \end{pmatrix}\begin{pmatrix} 3 & 1 & -2 \\ 2 & 1 & 7 \\ 1 & 1 & 6 \end{pmatrix} = \begin{pmatrix} 5 & 7 & 30 \\ 25 & 13 & 30 \\ 25 & 68 & 327 \end{pmatrix}$$

Now using $A(BC)$, we have that:

$$BC = \begin{pmatrix} 1 & 2 & 2 \\ 2 & 1 & 1 \\ 0 & -3 & 4 \end{pmatrix}\begin{pmatrix} 3 & 1 & -2 \\ 2 & 1 & 7 \\ 1 & 1 & 6 \end{pmatrix} = \begin{pmatrix} 9 & 5 & 24 \\ 9 & 4 & 9 \\ -2 & 1 & 3 \end{pmatrix}$$

$$\text{Then, } A(BC) = \begin{pmatrix} 1 & 0 & 2 \\ 0 & 3 & 1 \\ 2 & 5 & 8 \end{pmatrix}\begin{pmatrix} 9 & 5 & 24 \\ 9 & 4 & 9 \\ -2 & 1 & 3 \end{pmatrix} = \begin{pmatrix} 5 & 7 & 30 \\ 25 & 13 & 30 \\ 25 & 68 & 327 \end{pmatrix}$$

Hence, $(AB)C = A(BC)$; that means multiplication is associative.

18.3.3 DISTRIBUTIVITY OF MATRICES

A matrix is distributive if this condition is satisfied:

$$A(B + C) = AB + AC$$

Example 11

Given $A = \begin{pmatrix} 2 & -1 & 2 \\ 0 & 5 & -1 \\ 1 & 1 & 4 \end{pmatrix}$, $B = \begin{pmatrix} 1 & 0 & -3 \\ 1 & 1 & 1 \\ 0 & 2 & 4 \end{pmatrix}$ and $C = \begin{pmatrix} 1 & 0 & 1 \\ 2 & 1 & 0 \\ 5 & 4 & 3 \end{pmatrix}$ show that the matrix operation is distributive.

Solution

$$B + C = \begin{pmatrix} 1 & 0 & -3 \\ 1 & 1 & 1 \\ 0 & 2 & 4 \end{pmatrix} + \begin{pmatrix} 1 & 0 & 1 \\ 2 & 1 & 0 \\ 5 & 4 & 3 \end{pmatrix} = \begin{pmatrix} 2 & 0 & -2 \\ 3 & 2 & 1 \\ 5 & 6 & 7 \end{pmatrix}$$

$$\text{then } A(B + C) = \begin{pmatrix} 2 & -1 & 2 \\ 0 & 5 & -1 \\ 1 & 1 & 4 \end{pmatrix}\begin{pmatrix} 2 & 0 & -2 \\ 3 & 2 & 1 \\ 5 & 6 & 7 \end{pmatrix} = \begin{pmatrix} 11 & 10 & 9 \\ 10 & 4 & -2 \\ 25 & 26 & 27 \end{pmatrix}$$

$$\text{If } AB = \begin{pmatrix} 2 & -1 & 2 \\ 0 & 5 & -1 \\ 1 & 1 & 4 \end{pmatrix}\begin{pmatrix} 1 & 0 & -3 \\ 1 & 1 & 1 \\ 0 & 2 & 4 \end{pmatrix} = \begin{pmatrix} 1 & 3 & 1 \\ 5 & 3 & 1 \\ 2 & 9 & 14 \end{pmatrix}$$

$$\text{and } AC = \begin{pmatrix} 2 & -1 & 2 \\ 0 & 5 & -1 \\ 1 & 1 & 4 \end{pmatrix}\begin{pmatrix} 1 & 0 & 1 \\ 2 & 1 & 0 \\ 5 & 4 & 3 \end{pmatrix} = \begin{pmatrix} 10 & 7 & 8 \\ 5 & 1 & -3 \\ 23 & 17 & 13 \end{pmatrix}$$

$$\text{then } AB + AC = \begin{pmatrix} 1 & 3 & 1 \\ 5 & 3 & 1 \\ 2 & 9 & 14 \end{pmatrix} + \begin{pmatrix} 10 & 7 & 8 \\ 5 & 1 & -3 \\ 23 & 17 & 13 \end{pmatrix} = \begin{pmatrix} 11 & 10 & 9 \\ 10 & 4 & -2 \\ 25 & 26 & 27 \end{pmatrix}$$

Hence, $A(B + C) = AB + AC$ as shown; then the matrix is distributive.

18.4 TRANSPOSE OF A MATRIX

The transpose of a given matrix, A, is denoted as A^T; it is obtained by interchanging the rows and columns of the given matrix A.

If, $A = \begin{pmatrix} a_{11} & a_{12} & \cdots & a_{1n} \\ a_{21} & a_{22} & \cdots & a_{2n} \\ \vdots & \vdots & \ddots & \vdots \\ a_{m1} & a_{m2} & \cdots & a_{mn} \end{pmatrix}$

then, $A^T = \begin{pmatrix} a_{11} & a_{21} & \cdots & a_{m1} \\ a_{12} & a_{22} & \cdots & a_{m2} \\ \vdots & \vdots & \ddots & \vdots \\ a_{1n} & a_{2n} & \cdots & a_{mn} \end{pmatrix}$

Rules of Transpose of a matrix

(i) $(A+B)^t = A^t + B^t$

(ii) $(A-B)^t = A^t - B^t$

(iii) $(A^t)^t = A$

(iv) $KA^t = KA^t$

(iv) $(AB)^t = B^t A^t$

Example 12

Find the transpose of each of the following matrices:

(i) $A = \begin{pmatrix} 1 & 1 \\ -1 & 1 \end{pmatrix}$ (ii) $B = \begin{pmatrix} 2 & -1 \\ 3 & 5 \end{pmatrix}$ (iii) $C = \begin{pmatrix} 3 & 2 & -1 \\ 1 & 0 & 4 \\ 7 & 2 & 3 \end{pmatrix}$

(ii) $D = \begin{pmatrix} 2 & 3 & 5 \\ 1 & -1 & 4 \end{pmatrix}$ (v) $\begin{pmatrix} 1 & 2 \\ 3 & 4 \\ 5 & -3 \end{pmatrix}$

Solution

(i) If $A = \begin{pmatrix} 1 & 1 \\ -1 & 1 \end{pmatrix} \Rightarrow A^T = \begin{pmatrix} 1 & -1 \\ 1 & 1 \end{pmatrix}$

(ii) If $B = \begin{pmatrix} 2 & -1 \\ 3 & 5 \end{pmatrix} \Rightarrow B^T = \begin{pmatrix} 2 & 3 \\ -1 & 5 \end{pmatrix}$

(iii) If $C = \begin{pmatrix} 3 & 2 & -1 \\ 1 & 0 & 4 \\ 7 & 2 & 3 \end{pmatrix} \Rightarrow C^T = \begin{pmatrix} 3 & 1 & 7 \\ 2 & 0 & 2 \\ -1 & 4 & 3 \end{pmatrix}$

(iv) If $D = \begin{pmatrix} 2 & 3 & 5 \\ 1 & -1 & 4 \end{pmatrix} \Rightarrow D^T = \begin{pmatrix} 2 & 1 \\ 3 & -1 \\ 5 & 4 \end{pmatrix}$

(v) If $E = \begin{pmatrix} 1 & 2 \\ 3 & 4 \\ 5 & -3 \end{pmatrix} \Rightarrow E^T = \begin{pmatrix} 1 & 3 & 5 \\ 2 & 4 & -3 \end{pmatrix}$

Example 13

Given $A = \begin{pmatrix} 7 & -1 \\ 3 & 2 \end{pmatrix}$ and $B = \begin{pmatrix} 0 & 2 \\ 1 & -5 \end{pmatrix}$

Find: (i) $(A+B)^t$ (ii) $A^t + B^t$ (iii) $(AB)^t$ (iv) $(BA)^t$ (v) A^tB^t

Solution

(i) $A + B = \begin{pmatrix} 7 & -1 \\ 3 & 2 \end{pmatrix} + \begin{pmatrix} 0 & 2 \\ 1 & -5 \end{pmatrix} = \begin{pmatrix} 7 & 1 \\ 4 & -3 \end{pmatrix}$

then, $(A+B)^t = \begin{pmatrix} 7 & 4 \\ 1 & -3 \end{pmatrix}$

(ii) $A^t = \begin{pmatrix} 7 & 3 \\ -1 & 2 \end{pmatrix}$ and $B^t = \begin{pmatrix} 0 & 1 \\ 2 & -5 \end{pmatrix}$

then, $A^t + B^t = \begin{pmatrix} 7 & 3 \\ -1 & 2 \end{pmatrix} + \begin{pmatrix} 0 & 1 \\ 2 & -5 \end{pmatrix} = \begin{pmatrix} 7 & 4 \\ 1 & -3 \end{pmatrix}$

(iii) $AB = \begin{pmatrix} 7 & -1 \\ 3 & 2 \end{pmatrix}\begin{pmatrix} 0 & 2 \\ 1 & -5 \end{pmatrix} = \begin{pmatrix} -1 & 19 \\ 2 & -4 \end{pmatrix}$

then, $(AB)^t = \begin{pmatrix} -1 & 2 \\ 19 & -4 \end{pmatrix}$

(iv) $BA = \begin{pmatrix} 0 & 2 \\ 1 & -5 \end{pmatrix}\begin{pmatrix} 7 & -1 \\ 3 & 2 \end{pmatrix} = \begin{pmatrix} 6 & 4 \\ -8 & -11 \end{pmatrix}$

then, $(BA)^t = \begin{pmatrix} 6 & -8 \\ 4 & -11 \end{pmatrix}$

Thus, it can be seen from (iii) and (iv) that $(BA)^t \neq (AB)^t$

(v) $A^t = \begin{pmatrix} 7 & 3 \\ -1 & 2 \end{pmatrix}$ and $B^t = \begin{pmatrix} 0 & 1 \\ 2 & -5 \end{pmatrix}$

then, $A^tB^t = \begin{pmatrix} 7 & 3 \\ -1 & 2 \end{pmatrix}\begin{pmatrix} 0 & 1 \\ 2 & -5 \end{pmatrix} = \begin{pmatrix} 6 & -8 \\ 4 & -11 \end{pmatrix}$

Thus, it can be seen from (iv) and (v) that $(BA)^t = A^tB^t$

18.5 MINORS AND COFACTORS OF A MATRIX

18.5.1 MINORS OF A MATRIX

The minor of an element of a matrix is the value of the determinant obtained by covering up the row and column containing the given element.

Thus, minor of a 3×3 matrix $A = \begin{pmatrix} a_{11} & a_{12} & a_{13} \\ a_{21} & a_{22} & a_{23} \\ a_{31} & a_{32} & a_{33} \end{pmatrix}$ is given by:

$$M = \begin{pmatrix} M_{11} & M_{12} & M_{13} \\ M_{21} & M_{22} & M_{23} \\ M_{31} & M_{32} & M_{33} \end{pmatrix} \quad \text{Where: } M_{11} = \begin{vmatrix} a_{22} & a_{23} \\ a_{32} & a_{33} \end{vmatrix}, M_{12} = \begin{vmatrix} a_{21} & a_{23} \\ a_{31} & a_{33} \end{vmatrix}, M_{13} = \begin{vmatrix} a_{21} & a_{22} \\ a_{31} & a_{32} \end{vmatrix},$$

$$M_{21} = \begin{vmatrix} a_{12} & a_{13} \\ a_{32} & a_{33} \end{vmatrix}, M_{22} = \begin{vmatrix} a_{11} & a_{13} \\ a_{31} & a_{33} \end{vmatrix}, M_{23} = \begin{vmatrix} a_{11} & a_{12} \\ a_{31} & a_{32} \end{vmatrix},$$

$$M_{31} = \begin{vmatrix} a_{12} & a_{13} \\ a_{22} & a_{23} \end{vmatrix}, M_{32} = \begin{vmatrix} a_{11} & a_{13} \\ a_{21} & a_{23} \end{vmatrix}, M_{33} = \begin{vmatrix} a_{11} & a_{12} \\ a_{21} & a_{22} \end{vmatrix}$$

Example 14

Given the matrices $A = \begin{pmatrix} 7 & -1 \\ 3 & 2 \end{pmatrix}$ and $B = \begin{pmatrix} 3 & 2 & -1 \\ 1 & 0 & 4 \\ 7 & 6 & 5 \end{pmatrix}$

Find the (i) minors of A (ii) minors of B

Solution

(i) $A = \begin{pmatrix} 7 & -1 \\ 3 & 2 \end{pmatrix}$ Then the minor of element 7 is obtained by covering the row $(7 \ -1)$ and column $\begin{pmatrix} 7 \\ 3 \end{pmatrix}$; thus leaving the value (**2**) as the minor of 7.

Similarly, the minor of -1 is achieved by covering the row $(7 \ -1)$ and column $\begin{pmatrix} -1 \\ 2 \end{pmatrix}$; thus leaving the value (**3**) as the minor of -1.

Hence, repeating for the elements 3 and 2, their minors are (**−1**) and (**7**) respectively.

Therefore, the minor matrix of A is $= \begin{pmatrix} 2 & 3 \\ -1 & 7 \end{pmatrix}$

(ii) $B = \begin{pmatrix} 3 & 2 & -1 \\ 1 & 0 & 4 \\ 7 & 6 & 5 \end{pmatrix}$ the minor of an element of a 3×3 matrix is the value the determinant of the 2×2 matrix obtained by covering the row and column of the given element.

Thus, the minor of 3 is obtained when the row $(3 \ 2 \ -1)$ and column $\begin{pmatrix} 3 \\ 1 \\ 7 \end{pmatrix}$ are covered, leaving

the 2×2 determinant $\begin{vmatrix} 0 & 4 \\ 6 & 5 \end{vmatrix} = 5(0) - 6(4) = -24$

Therefore, the minor of element 3 is $M_{11} = \mathbf{-24}$

similarly, the minor of the following are:

element 2 is $M_{12} = \begin{vmatrix} 1 & 4 \\ 7 & 5 \end{vmatrix} = 5(1) - 7(4) = 5 - 28 = \mathbf{-23}$

element -1 is $M_{13} = \begin{vmatrix} 1 & 0 \\ 7 & 6 \end{vmatrix} = 6(1) - 7(0) = 6 - 0 = \mathbf{6}$

element 1 is $M_{21} = \begin{vmatrix} 2 & -1 \\ 6 & 5 \end{vmatrix} = 5(2) - 6(-1) = 10 + 6 = \mathbf{18}$

element 0 is $M_{22} = \begin{vmatrix} 3 & -1 \\ 7 & 5 \end{vmatrix} = 5(3) - 7(-1) = 15 + 7 = \mathbf{22}$

element 4 is $M_{23} = \begin{vmatrix} 3 & 2 \\ 7 & 6 \end{vmatrix} = 6(3) - 7(2) = 18 - 14 = \mathbf{4}$

element 7 is$M_{31} = \begin{vmatrix} 2 & -1 \\ 0 & 4 \end{vmatrix} = 4(2) - 0(-1) = 8 + 0 = \mathbf{8}$

element 6 is $M_{32} = \begin{vmatrix} 3 & -1 \\ 1 & 4 \end{vmatrix} = 4(3) - 1(-1) = 12 + 1 = \mathbf{13}$

element 5 is $M_{33} = \begin{vmatrix} 3 & 2 \\ 1 & 0 \end{vmatrix} = 3(0) - 1(2) = 0 - 2 = \mathbf{-2}$

Hence, the matrix of the minors is $\begin{pmatrix} M_{11} & M_{12} & M_{13} \\ M_{21} & M_{22} & M_{23} \\ M_{31} & M_{32} & M_{33} \end{pmatrix} = \begin{pmatrix} -24 & -23 & 6 \\ 18 & 22 & 4 \\ 8 & 13 & -2 \end{pmatrix}$

18.5.2 COFACTORS OF A MATRIX

The cofactor of a matrix is the signed-minor of the matrix. Thus, the signed-minor matrix is given as:

$$C = \begin{pmatrix} +M_{11} & -M_{12} \\ -M_{21} & +M_{22} \end{pmatrix} \text{ for a } 2 \times 2 \text{ matrix}$$

$$C = \begin{pmatrix} +M_{11} & -M_{12} & +M_{13} \\ -M_{21} & +M_{22} & -M_{23} \\ +M_{31} & -M_{32} & +M_{33} \end{pmatrix} \text{ for a } 3 \times 3 \text{ matrix}$$

Example 15

Given the matrices $A = \begin{pmatrix} 7 & -1 \\ 3 & 2 \end{pmatrix}$ and $B = \begin{pmatrix} 3 & 2 & -1 \\ 1 & 0 & 4 \\ 7 & 6 & 5 \end{pmatrix}$

Find:
(i) cofactor of A (ii) cofactor of B

Solution

(i) The matrix of the minors of A is $\begin{pmatrix} 2 & 3 \\ -1 & 7 \end{pmatrix}$

Therefore, the cofactors become:

$$\begin{pmatrix} +2 & -3 \\ -(-1) & +7 \end{pmatrix} = \begin{pmatrix} 2 & -3 \\ 1 & 7 \end{pmatrix}$$

Revisit example 14 on how to obtain the minor of a matrix.

(ii) The matrix of the minors of B is $\begin{pmatrix} -24 & -23 & 6 \\ 18 & 22 & 4 \\ 8 & 13 & -2 \end{pmatrix}$

Therefore, the cofactors become:

Revisit example 14 on how to obtain the minor of a 3×3 matrix.

$$\begin{pmatrix} +(-24) & -(-23) & +6 \\ -18 & +22 & -4 \\ +8 & -13 & +(-2) \end{pmatrix}$$
$$= \begin{pmatrix} -24 & 23 & 6 \\ -18 & 22 & -4 \\ 8 & -13 & -2 \end{pmatrix}$$

18.6 ADJOINT OF A MATRIX

The adjoint of a matrix is the transpose of the cofactors of the matrix.

Thus, when given the cofactor of a matrix A, as $C = \begin{pmatrix} +M_{11} & -M_{12} & +M_{13} \\ -M_{21} & +M_{22} & -M_{23} \\ +M_{31} & -M_{32} & +M_{33} \end{pmatrix}$

then the adjoint of A is achieved by changing the first row to become first column, second row to second column, and third row to third column.

$$C^T = \begin{pmatrix} +M_{11} & -M_{21} & +M_{31} \\ -M_{12} & +M_{22} & -M_{32} \\ +M_{13} & -M_{23} & +M_{33} \end{pmatrix}$$

Example 16

Find the adjoint of the following matrices:

(i)$A = \begin{pmatrix} 7 & -1 \\ 3 & 2 \end{pmatrix}$ (ii) $B = \begin{pmatrix} 3 & 2 & -1 \\ 1 & 0 & 4 \\ 7 & 6 & 5 \end{pmatrix}$

See example 14 and 15 on how to obtain the minor and cofactor of a matrix.

Solution

(i) Given $A = \begin{pmatrix} 7 & -1 \\ 3 & 2 \end{pmatrix}$, then the cofactor is $C = \begin{pmatrix} 2 & -3 \\ 1 & 7 \end{pmatrix}$

Hence, adjoint $A = C^T = \begin{pmatrix} 2 & 1 \\ -3 & 7 \end{pmatrix}$

(ii) Given $B = \begin{pmatrix} 3 & 2 & -1 \\ 1 & 0 & 4 \\ 7 & 6 & 5 \end{pmatrix}$

See example 14 and 15 on how to obtain the minor and cofactor of a matrix.

then the cofactor is $C = \begin{pmatrix} -24 & 23 & 6 \\ -18 & 22 & -4 \\ 8 & -13 & -2 \end{pmatrix}$

Hence, adjoint $A = C^T = \begin{pmatrix} -24 & -18 & 8 \\ 23 & 22 & -13 \\ 6 & -4 & -2 \end{pmatrix}$

18.7 DETERMINANT OF A MATRIX

If a matrix is $A = \begin{pmatrix} a_{11} & a_{12} & \cdots & a_{1n} \\ a_{21} & a_{22} & \cdots & a_{2n} \\ \vdots & \vdots & \ddots & \vdots \\ a_{m1} & a_{m2} & \cdots & a_{mn} \end{pmatrix}$

then the determinant is thus written as:

$$A = \begin{vmatrix} a_{11} & a_{12} & \cdots & a_{1n} \\ a_{21} & a_{22} & \cdots & a_{2n} \\ \vdots & \vdots & \ddots & \vdots \\ a_{m1} & a_{m2} & \cdots & a_{mn} \end{vmatrix}$$

18.7.1 DETERMINANT OF TWO BY TWO (2 × 2) MATRICES

A two by two (2 × 2) matrix, $A = \begin{pmatrix} a_{11} & a_{12} \\ a_{21} & a_{22} \end{pmatrix}$ has its determinant as:

$$|A| = \begin{vmatrix} a_{11} & a_{12} \\ a_{21} & a_{22} \end{vmatrix} = a_{11}a_{22} - a_{12}a_{21}$$

Example 17

Find the determinant of each of the following matrices:

(a) $\begin{vmatrix} 1 & 0 \\ 0 & 1 \end{vmatrix}$ (b) $\begin{vmatrix} 2 & 3 \\ -1 & 5 \end{vmatrix}$ (c) $\begin{vmatrix} 5 & 4 \\ 7 & 2 \end{vmatrix}$ (d) $\begin{vmatrix} -3 & 10 \\ 12 & -5 \end{vmatrix}$

Solution

(a) $\begin{vmatrix} 1 & 0 \\ 0 & 1 \end{vmatrix} = 1(1) - 0(0) = 1 - 0 = 1$

(b) $\begin{vmatrix} 2 & 3 \\ -1 & 5 \end{vmatrix} = 2(5) - 3(-1) = 10 + 3 = 13$

(c) $\begin{vmatrix} 5 & 4 \\ 7 & 2 \end{vmatrix} = 5(2) - 4(7) = 10 - 28 = -18$

(d) $\begin{vmatrix} -3 & 10 \\ 12 & -5 \end{vmatrix} = -3(-5) - 10(12) = 15 - 120 = -105$

18.7.2 DETERMINANT OF THREE BY THREE (3 × 3) MATRICES

The determinant of a 3×3 matrix is the sum of the product of the elements and their cofactors of any row or any column of the given matrix.

Then using the first row, the determinant of the 3×3 matrix of A is, $|A| = \begin{vmatrix} a_{11} & a_{12} & a_{13} \\ a_{21} & a_{22} & a_{23} \\ a_{31} & a_{32} & a_{33} \end{vmatrix}$

$$= a_{11}\begin{vmatrix} a_{22} & a_{23} \\ a_{32} & a_{33} \end{vmatrix} - a_{12}\begin{vmatrix} a_{21} & a_{23} \\ a_{31} & a_{33} \end{vmatrix} + a_{13}\begin{vmatrix} a_{21} & a_{22} \\ a_{31} & a_{32} \end{vmatrix}$$

$$= a_{11}(a_{22}a_{33} - a_{23}a_{32}) - a_{12}(a_{21}a_{33} - a_{23}a_{31}) + a_{13}(a_{21}a_{32} - a_{22}a_{31})$$

Note that there are six possible ways (i.e., using row 1, row 2, row 3, column 1, column 2, and column 3) to find the determinant, and each of them will give the same result.

Example 18

Evaluate $\begin{vmatrix} 4 & 3 & 2 \\ 2 & 7 & 5 \\ 1 & 8 & 9 \end{vmatrix}$

Solution

Using the first row, we have that:

$$= a_{11}\begin{vmatrix} a_{22} & a_{23} \\ a_{32} & a_{33} \end{vmatrix} - a_{12}\begin{vmatrix} a_{21} & a_{23} \\ a_{31} & a_{33} \end{vmatrix} + a_{13}\begin{vmatrix} a_{21} & a_{22} \\ a_{31} & a_{32} \end{vmatrix}$$

$$= 4\begin{vmatrix} 7 & 5 \\ 8 & 9 \end{vmatrix} - 3\begin{vmatrix} 2 & 5 \\ 1 & 9 \end{vmatrix} + 2\begin{vmatrix} 2 & 7 \\ 1 & 8 \end{vmatrix}$$

$= 4(7 \times 9 - 5 \times 8) - 3(2 \times 9 - 5 \times 1) + 2(2 \times 8 - 7 \times 1)$

$= 4(63 - 40) - 3(18 - 5) + 2(16 - 7) = 4(23) - 3(13) + 2(9)$

$= 92 - 39 + 18 = 71$

Alternatively, using the third column, we have that:

$$= a_{13}\begin{vmatrix} a_{21} & a_{23} \\ a_{31} & a_{33} \end{vmatrix} - a_{23}\begin{vmatrix} a_{11} & a_{12} \\ a_{31} & a_{32} \end{vmatrix} + a_{33}\begin{vmatrix} a_{11} & a_{12} \\ a_{21} & a_{22} \end{vmatrix}$$

$$= 2\begin{vmatrix} 2 & 7 \\ 1 & 8 \end{vmatrix} - 3\begin{vmatrix} 4 & 3 \\ 1 & 8 \end{vmatrix} + 9\begin{vmatrix} 4 & 3 \\ 2 & 7 \end{vmatrix}$$

$= 2(2 \times 8 - 7 \times 1) - 5(4 \times 8 - 3 \times 1) + 9(4 \times 7 - 3 \times 2)$

$= 2(16 - 7) - 5(32 - 3) + 9(28 - 6) = 2(9) - 5(29) + 9(22)$

$= 18 - 145 + 198 = 71$

Thus, it can be seen that the determinant is the same regardless of the row or column used.

Example 19

Evaluate $\begin{vmatrix} -1 & 5 & 2 \\ 0 & 4 & -3 \\ 2 & 3 & 6 \end{vmatrix}$

Solution

Let's use the first row, thus:

$$= a_{11}\begin{vmatrix} a_{22} & a_{23} \\ a_{32} & a_{33} \end{vmatrix} - a_{12}\begin{vmatrix} a_{21} & a_{23} \\ a_{31} & a_{33} \end{vmatrix} + a_{13}\begin{vmatrix} a_{21} & a_{22} \\ a_{31} & a_{32} \end{vmatrix}$$

$$= -1\begin{vmatrix} 4 & -3 \\ 3 & 6 \end{vmatrix} - 5\begin{vmatrix} 0 & -3 \\ 2 & 6 \end{vmatrix} + 2\begin{vmatrix} 0 & 4 \\ 2 & 3 \end{vmatrix}$$

$$= -(4 \times 6 - (-3) \times 3) - 5(0 \times 6 - (-3) \times 2) + 2(0 \times 3 - 4 \times 2)$$

$$= -(24 + 9) - 5(0 + 6) + 2(0 - 8) = -33 - 30 - 16$$

$$= -79$$

PROPERTIES OF DETERMINANTS

(i) **If a determinant has a row of zeroes, or columns of zeroes; then the determinant is equal to zero.**

Example 20

Find the determinant of each of the following matrices:

(a) $\begin{pmatrix} 0 & 2 \\ 0 & 1 \end{pmatrix}$ (b) $\begin{pmatrix} 3 & 4 & 5 \\ 3 & 2 & 8 \\ 0 & 0 & 0 \end{pmatrix}$

Solution

(a) The determinant of the matrix $\begin{pmatrix} 0 & 2 \\ 0 & 1 \end{pmatrix}$ is written as $\begin{vmatrix} 0 & 2 \\ 0 & 1 \end{vmatrix}$

then $\begin{vmatrix} 0 & 2 \\ 0 & 1 \end{vmatrix} = 0$ From property (i) all the elements of the first column is zero, so the determinant is zero.

(b) The determinant of the matrix $\begin{pmatrix} 3 & 4 & 5 \\ 3 & 2 & 8 \\ 0 & 0 & 0 \end{pmatrix}$ is written as $\begin{vmatrix} 3 & 4 & 5 \\ 3 & 2 & 8 \\ 0 & 0 & 0 \end{vmatrix}$

then, $\begin{vmatrix} 3 & 4 & 5 \\ 3 & 2 & 8 \\ 0 & 0 & 0 \end{vmatrix} = 0$ from property (i) all the elements of the third row is zero, so the determinant is zero.

(ii) **If a matrix has identical rows and column, then the determinant is zero.**

Example 21

Show that the determinant of each of the matrices is zero:

(a) $\begin{pmatrix} 2 & 2 \\ 1 & 1 \end{pmatrix}$ (b) $\begin{pmatrix} 4 & 6 & 4 \\ 3 & -1 & 3 \\ 2 & 5 & 2 \end{pmatrix}$ (c) $\begin{pmatrix} 1 & 0 & 3 \\ 4 & 3 & -2 \\ 1 & 0 & 3 \end{pmatrix}$

Solution

(a) $\begin{vmatrix} 4 & 4 \\ 3 & 3 \end{vmatrix} = 4(3) - 4(3) = 0$

From property (ii) column 1 and 2 have the same value, hence the determinant is zero.

(b) $\begin{vmatrix} 4 & 6 & 4 \\ 3 & -1 & 3 \\ 2 & 5 & 2 \end{vmatrix}$

$= 4(-1(2) - 3(5)) - 6(3(2) - 3(2)) + 4(3(5) - (-1)(2))$

$= 4(-2 - 15) - 6(6 - 6) + 4(15 + 2) = 4(-17) - 6(0) + 4(15 + 2)$

$= -68 + 0 + 68 = 0$

From property (ii) , column 1 and 3 have the same value, hence the determinant is zero.

(c) $\begin{vmatrix} 1 & 0 & 3 \\ 4 & 3 & -2 \\ 1 & 0 & 3 \end{vmatrix} = 0$

From property (ii) , row 1 and 3 have the same value, hence the determinant is zero.

(iii) **The determinant of product of two matrices A and B is equal to the product of the determinant of the individual matrix. That is, $|AB| = |A||B|$.**

Example 22

If $A = \begin{pmatrix} 5 & 2 \\ 1 & 7 \end{pmatrix}$ and $B = \begin{pmatrix} 3 & 4 \\ 8 & 7 \end{pmatrix}$ show that $|AB| = |A||B|$

Solution

$A = \begin{vmatrix} 5 & 2 \\ 1 & 7 \end{vmatrix} = 7(5) - 2(1) = 35 - 2 = 33$

$B = \begin{vmatrix} 3 & 4 \\ 8 & 7 \end{vmatrix} = 3(7) - 4(8) = 21 - 32 = -11$

then, $|A||B| = (33) \times (-11) = -363$

And, $AB = \begin{pmatrix} 5 & 2 \\ 1 & 7 \end{pmatrix}\begin{pmatrix} 3 & 4 \\ 8 & 7 \end{pmatrix} = \begin{pmatrix} 31 & 34 \\ 59 & 53 \end{pmatrix}$

then $|AB| = \begin{vmatrix} 31 & 34 \\ 59 & 53 \end{vmatrix} = 31(53) - 34(59)$

$= 1643 - 2006 = -363$

Thus, $|AB| = |A||B|$

(iv) **The transpose of a matrix and the matrix itself have equal determinants. That is, $|A| = |A^T|$.**

Example 23

Given $X = \begin{pmatrix} 1 & 0 & 3 \\ 4 & 2 & 1 \\ -1 & 5 & 7 \end{pmatrix}$ show that $|X| = |X^T|$

Solution

$$|X| = \begin{vmatrix} 1 & 0 & 3 \\ 4 & 2 & 1 \\ -1 & 5 & 7 \end{vmatrix} = 1\begin{vmatrix} 2 & 1 \\ 5 & 7 \end{vmatrix} - 0\begin{vmatrix} 4 & 1 \\ -1 & 7 \end{vmatrix} + 3\begin{vmatrix} 4 & 2 \\ -1 & 5 \end{vmatrix}$$

$= 1(2 \times 7 - 1 \times 5) - 0(4 \times 7 - (-1) \times 1) + 3(4 \times 5 - 2 \times (-1))$

$= 1(14 - 5) - 0(28 + 1) + 3(20 + 2)$

$= 1(9) - 0(29) + 3(22) = 9 - 0 + 66 = 75$

$\Rightarrow |X| = 75$

Given $X = \begin{pmatrix} 1 & 0 & 3 \\ 4 & 2 & 1 \\ -1 & 5 & 7 \end{pmatrix}$ then $X^T = \begin{pmatrix} 1 & 4 & -1 \\ 0 & 2 & 5 \\ 3 & 1 & 7 \end{pmatrix}$

So, $|X^T| = \begin{vmatrix} 1 & 4 & -1 \\ 0 & 2 & 5 \\ 3 & 1 & 7 \end{vmatrix} = 1\begin{vmatrix} 2 & 5 \\ 1 & 7 \end{vmatrix} - 4\begin{vmatrix} 0 & 5 \\ 3 & 7 \end{vmatrix} - 1\begin{vmatrix} 0 & 2 \\ 3 & 1 \end{vmatrix}$

$= 1(2 \times 7 - 5 \times 1) - 4(0 \times 7 - 5 \times 3) - 1(0 \times 1 - 2 \times 3)$

$= 1(9) - 4(-15) - 1(-6) = 9 + 60 + 6 = 75$

$\Rightarrow |X^T| = 75$

Therefore, $|X| = |X^T| = 75$

18.8 INVERSE OF A MATRIX

The inverse of a matrix A is written as A^{-1}, and it is given as:

$$A^{-1} = \frac{Adjoint\ A}{|A|} \qquad \text{(where } |A| \neq 0\text{)}$$

Steps to follow when finding the inverse of a matrix

(ii) First, find the determinant of the given matrix A in order to get $|A|$;

(iii) Then find the minors of A, and call the matrix of the minors B.

(iv) Find the cofactors of the matrix B. That is, take the signed minors of the matrix B.

(v) Take the transpose of the cofactor, i.e. B^T. This is now called the adjoint of the matrix A. Thus, $Adj\ A = B^T$.

(vi) Finally, divide the adjoint of the matrix A by the determinant $|A|$; this is the inverse of the matrix.

18.8.1 INVERSE OF TWO BY TWO (2 × 2) MATRICES

The inverse of a matrix A is A^{-1} such that:

$$AA^{-1} = I \qquad \text{(where } I \text{ is a unit matrix)}$$

Example 24

Find the inverse of the matrix $A = \begin{pmatrix} 5 & 2 \\ 1 & 7 \end{pmatrix}$.

Solution

There are two approaches that can be employed here. First, we can use $AA^{-1} = I$; then solving for the elements in A^{-1}. The second method is to use inverse method,

$A^{-1} = \frac{Adjoint\ A}{|A|}$.

Thus, using $AA^{-1} = I$ we have that $\begin{pmatrix} 5 & 2 \\ 1 & 7 \end{pmatrix}\begin{pmatrix} a & b \\ c & d \end{pmatrix} = \begin{pmatrix} 1 & 0 \\ 0 & 1 \end{pmatrix}$

Where $A^{-1} = \begin{pmatrix} a & b \\ c & d \end{pmatrix}$

Then, $\begin{pmatrix} 5 & 2 \\ 1 & 7 \end{pmatrix}\begin{pmatrix} a & b \\ c & d \end{pmatrix} = \begin{pmatrix} 1 & 0 \\ 0 & 1 \end{pmatrix}$

$\Rightarrow \begin{pmatrix} 5a + 2c & 5b + 2d \\ a + 7c & b + 7d \end{pmatrix} = \begin{pmatrix} 1 & 0 \\ 0 & 1 \end{pmatrix}$

Equating the corresponding elements of both sides of the equation, we have:

$5a + 2c = 1$ ....(i)

$5b + 2d = 0$ ....(ii)

$a + 7c = 0$ ....(iii)

$b + 7d = 1$ ....(iv)

Solving for a, b, c and d, we have that:

From equation (iii) $a = -7c$, put this value into equation(i)

$\Rightarrow 5(-7c) + 2c = 1$ then $-35c + 2c = 1$

$c = -\frac{1}{33}$

And, $a = -7c = -7\left(-\frac{1}{33}\right) = \frac{7}{33}$

Similarly, from equation (ii) $b = -\frac{2}{5}d$, put this value into equation (iv)

Thus, $-\frac{2}{5}d + 7d = 1 \quad \Rightarrow d = \frac{5}{33}$

And, $b = -\frac{2}{5}d \quad \Rightarrow b = -\frac{2}{5}\left(\frac{5}{33}\right) = -\frac{2}{33}$

Therefore, the inverse $A^{-1} = \begin{pmatrix} a & b \\ c & d \end{pmatrix} = \begin{pmatrix} \frac{7}{33} & -\frac{2}{33} \\ -\frac{1}{33} & \frac{5}{33} \end{pmatrix}$

However, an alternative method of finding the inverse is $A^{-1} = \frac{Adjoint\ A}{|A|}$.

If $A = \begin{pmatrix} 5 & 2 \\ 1 & 7 \end{pmatrix}$, then the determinant $|A| = \begin{vmatrix} 5 & 2 \\ 1 & 7 \end{vmatrix}$

$= 5(7) - 2(1) = 35 - 2 = 33$

Next, the minor of $A = \begin{pmatrix} 5 & 2 \\ 1 & 7 \end{pmatrix}$ is the matrix $B = \begin{pmatrix} 7 & 1 \\ 2 & 5 \end{pmatrix}$

then, the cofactor of B is $\begin{pmatrix} 7 & -1 \\ -2 & 5 \end{pmatrix}$

Now take the transpose of the cofactor of B, $\Rightarrow B^T = \begin{pmatrix} 7 & -2 \\ -1 & 5 \end{pmatrix}$

So, $Adjoint\ A = \begin{pmatrix} 7 & -2 \\ -1 & 5 \end{pmatrix}$

Thus, $A^{-1} = \frac{Adjoint\ A}{|A|} = \frac{1}{33}\begin{pmatrix} 7 & -2 \\ -1 & 5 \end{pmatrix} = \begin{pmatrix} \frac{7}{33} & -\frac{2}{33} \\ -\frac{1}{33} & \frac{5}{33} \end{pmatrix}$

Therefore, it can be concluded that the quickest way of obtaining the inverse of a 2×2 matrix $A = \begin{pmatrix} a & b \\ c & d \end{pmatrix}$ is by:

(i) Interchanging the positions of a and d in the main diagonal;
(ii) Changing the operation signs of b and c in the minor diagonal – that is, multiplying b and c by -1;
(iii) Then, dividing the outcome of (i) and(ii) by the determinant.

Thus, the inverse of $A = \begin{pmatrix} a & b \\ c & d \end{pmatrix}$ is:

$$A^{-1} = \frac{1}{ad - bc}\begin{pmatrix} d & -b \\ -c & a \end{pmatrix}$$

Example 25

Find the inverse of the matrix (i) $A = \begin{pmatrix} 2 & 5 \\ -9 & 4 \end{pmatrix}$ (ii) $B = \begin{pmatrix} 25 & 20 \\ 3 & 6 \end{pmatrix}$

Solution

(i) Let $A = \begin{pmatrix} 2 & 5 \\ -9 & 4 \end{pmatrix} = \begin{pmatrix} a & b \\ c & d \end{pmatrix}$ then using $A^{-1} = \frac{1}{ad-bc}\begin{pmatrix} d & -b \\ -c & a \end{pmatrix}$

Thus, $\frac{1}{2(4)-5(-9)}\begin{pmatrix} 4 & -5 \\ -(-9) & 2 \end{pmatrix} = \frac{1}{8+45}\begin{pmatrix} 4 & -5 \\ 9 & 2 \end{pmatrix}$

$\frac{1}{53}\begin{pmatrix} 4 & -5 \\ 9 & 2 \end{pmatrix} = \begin{pmatrix} \frac{4}{53} & -\frac{5}{53} \\ \frac{9}{53} & \frac{2}{53} \end{pmatrix} \therefore A^{-1} = \begin{pmatrix} \frac{4}{53} & -\frac{5}{53} \\ \frac{9}{53} & \frac{2}{53} \end{pmatrix}$

(ii) Let $B = \begin{pmatrix} 25 & 20 \\ 3 & 6 \end{pmatrix} = \begin{pmatrix} a & b \\ c & d \end{pmatrix}$

then using $B^{-1} = \frac{1}{ad-bc}\begin{pmatrix} d & -b \\ -c & a \end{pmatrix}$

Thus, $\frac{1}{25(6)-20(3)}\begin{pmatrix} 6 & -20 \\ -3 & 25 \end{pmatrix} = \frac{1}{150-60}\begin{pmatrix} 6 & -20 \\ -3 & 25 \end{pmatrix}$

$$\frac{1}{90}\begin{pmatrix} 6 & -20 \\ -3 & 25 \end{pmatrix} = \begin{pmatrix} \frac{6}{90} & -\frac{20}{90} \\ -\frac{3}{90} & \frac{25}{90} \end{pmatrix} = \begin{pmatrix} \frac{1}{15} & -\frac{2}{9} \\ -\frac{1}{30} & \frac{5}{18} \end{pmatrix}$$

$$\therefore B^{-1} = \begin{pmatrix} \frac{1}{15} & -\frac{2}{9} \\ -\frac{1}{30} & \frac{5}{18} \end{pmatrix}$$

18.8.2 INVERSE OF THREE BY THREE (3 × 3) MATRICES

The best approach to finding the inverse of a 3 × 3 matrix is using:

$$A^{-1} = \frac{Adjoint\ A}{|A|}$$

Example 26

Find the inverse of the matrix $P = \begin{pmatrix} 2 & 3 & 7 \\ 6 & 4 & 1 \\ 5 & 8 & 9 \end{pmatrix}$

Solution

Given $P = \begin{pmatrix} 2 & 3 & 7 \\ 6 & 4 & 1 \\ 5 & 8 & 9 \end{pmatrix}$

then to find P^{-1} use the following steps:

Step 1: Find the determinant $|P| = \begin{vmatrix} 2 & 3 & 7 \\ 6 & 4 & 1 \\ 5 & 8 & 9 \end{vmatrix}$

Using the first row, $|P| = 2\begin{vmatrix} 4 & 1 \\ 8 & 9 \end{vmatrix} - 3\begin{vmatrix} 6 & 1 \\ 5 & 9 \end{vmatrix} + 7\begin{vmatrix} 6 & 4 \\ 5 & 8 \end{vmatrix}$

$= 2(36-8) - 3(54-5) + 7(48-20) = 2(28) - 3(49) + 7(28)$

$= 56 - 147 + 196 = 105$

$\therefore |P| = 105$

Step 2: Find the minors of the matrix $P = \begin{pmatrix} 2 & 3 & 7 \\ 6 & 4 & 1 \\ 5 & 8 & 9 \end{pmatrix}$

Let the minors of $P = \begin{pmatrix} M_{11} & M_{12} & M_{13} \\ M_{21} & M_{22} & M_{23} \\ M_{31} & M_{32} & M_{33} \end{pmatrix}$

Where: $M_{11} = \begin{vmatrix} 4 & 1 \\ 8 & 9 \end{vmatrix} = 4(9) - 1(8) = 36 - 8 = 28$

$$M_{12} = \begin{vmatrix} 6 & 1 \\ 5 & 9 \end{vmatrix} = 6(9) - 1(5) = 54 - 5 = 49$$

$$M_{13} = \begin{vmatrix} 6 & 4 \\ 5 & 8 \end{vmatrix} = 6(8) - 5(4) = 48 - 20 = 28$$

$$M_{21} = \begin{vmatrix} 3 & 7 \\ 8 & 9 \end{vmatrix} = 3(9) - 7(8) = 27 - 56 = -29$$

$$M_{22} = \begin{vmatrix} 2 & 7 \\ 5 & 9 \end{vmatrix} = 2(9) - 5(7) = 18 - 35 = -17$$

$$M_{23} = \begin{vmatrix} 2 & 3 \\ 5 & 8 \end{vmatrix} = 2(8) - 3(5) = 16 - 15 = 1$$

$$M_{31} = \begin{vmatrix} 3 & 7 \\ 4 & 1 \end{vmatrix} = 3(1) - 7(4) = 3 - 28 = -25$$

$$M_{32} = \begin{vmatrix} 2 & 7 \\ 6 & 1 \end{vmatrix} = 2(1) - 7(6) = 2 - 42 = -40$$

$$M_{33} = \begin{vmatrix} 2 & 3 \\ 6 & 4 \end{vmatrix} = 2(4) - 3(6) = 8 - 18 = -10$$

$\therefore$ the minors of the matrix $P = \begin{pmatrix} M_{11} & M_{12} & M_{13} \\ M_{21} & M_{22} & M_{23} \\ M_{31} & M_{32} & M_{33} \end{pmatrix} = \begin{pmatrix} 28 & 49 & 28 \\ -29 & -17 & 1 \\ -25 & -40 & -10 \end{pmatrix}$

Step 3: Find the cofactor of the matrix P,

$$\text{Cofactor of } P = \begin{pmatrix} +M_{11} & -M_{12} & +M_{13} \\ -M_{21} & +M_{22} & -M_{23} \\ +M_{31} & -M_{32} & +M_{33} \end{pmatrix} = \begin{pmatrix} +28 & -49 & +28 \\ -(-29) & +(-17) & -1 \\ +(-25) & -(-40) & -10 \end{pmatrix}$$

$$= \begin{pmatrix} 28 & -49 & 28 \\ 29 & -17 & -1 \\ -25 & 40 & -10 \end{pmatrix}$$

Step 4: Now, find the transpose (P^T) of the cofactor; this is now called the adjoint of the matrix.

$$\text{Adjoint } P = P^T = \begin{pmatrix} 28 & 29 & -25 \\ -49 & -17 & 40 \\ 28 & -1 & -10 \end{pmatrix}$$

Step 5: Finally, take the ratio of the "adjoint of the matrix to the determinant" in order to get the inverse of the matrix.

$$P^{-1} = \frac{Adjoint\ P}{|P|} = \frac{1}{105}\begin{pmatrix} 28 & 29 & -25 \\ -49 & -17 & 40 \\ 28 & -1 & -10 \end{pmatrix} = \begin{pmatrix} \frac{28}{105} & \frac{29}{105} & -\frac{25}{105} \\ -\frac{49}{105} & -\frac{17}{105} & \frac{40}{105} \\ \frac{28}{105} & -\frac{1}{105} & -\frac{10}{105} \end{pmatrix}$$

$$\therefore\ P^{-1} = \begin{pmatrix} \frac{28}{105} & \frac{29}{105} & -\frac{25}{105} \\ -\frac{49}{105} & -\frac{17}{105} & \frac{40}{105} \\ \frac{28}{105} & -\frac{1}{105} & -\frac{10}{105} \end{pmatrix} = \begin{pmatrix} \frac{4}{15} & \frac{29}{105} & -\frac{5}{21} \\ -\frac{7}{15} & -\frac{17}{105} & -\frac{8}{21} \\ \frac{4}{15} & -\frac{1}{105} & -\frac{2}{21} \end{pmatrix}$$

Example 27

Find the inverse of the matrix, $Q = \begin{pmatrix} 0 & 3 & 1 \\ 2 & -5 & 6 \\ -7 & 15 & -12 \end{pmatrix}$

Solution

The determinant is, $|Q| = \begin{vmatrix} 0 & 3 & 1 \\ 2 & -5 & 6 \\ -7 & 15 & -12 \end{vmatrix} = 0\begin{vmatrix} -5 & 6 \\ 15 & -12 \end{vmatrix} - 3\begin{vmatrix} 2 & 6 \\ -7 & -12 \end{vmatrix} + 1\begin{vmatrix} 2 & -5 \\ -7 & 15 \end{vmatrix} = -59$

The matrix of the minors is given by:

$$\begin{pmatrix} \begin{vmatrix} -5 & 6 \\ 15 & -12 \end{vmatrix} & \begin{vmatrix} 2 & 6 \\ -7 & -12 \end{vmatrix} & \begin{vmatrix} 2 & -5 \\ -7 & 15 \end{vmatrix} \\ \begin{vmatrix} 3 & 1 \\ 15 & -12 \end{vmatrix} & \begin{vmatrix} 0 & 1 \\ -7 & -12 \end{vmatrix} & \begin{vmatrix} 0 & 3 \\ -7 & 15 \end{vmatrix} \\ \begin{vmatrix} 3 & 1 \\ -5 & 6 \end{vmatrix} & \begin{vmatrix} 0 & 1 \\ 2 & 6 \end{vmatrix} & \begin{vmatrix} 0 & 3 \\ 2 & -5 \end{vmatrix} \end{pmatrix} = \begin{pmatrix} -30 & 18 & -5 \\ -51 & 7 & 21 \\ 23 & -2 & -6 \end{pmatrix}$$

Then the cofactor matrix becomes $\begin{pmatrix} -30 & -18 & -5 \\ 51 & 7 & -21 \\ 23 & 2 & -6 \end{pmatrix}$

And the adjoint matrix of $Q = Q^T = \begin{pmatrix} -30 & 51 & 23 \\ -18 & 7 & 2 \\ -5 & -21 & -6 \end{pmatrix}$

Hence, $Q^{-1} = \frac{Adjoint\ Q}{|Q|} = \frac{Q^T}{|Q|} = -\frac{1}{59}\begin{pmatrix} -30 & 51 & 23 \\ -18 & 7 & 2 \\ -5 & -21 & -6 \end{pmatrix}$

$$= \begin{pmatrix} \frac{30}{59} & -\frac{51}{59} & -\frac{23}{59} \\ \frac{18}{59} & -\frac{7}{59} & -\frac{2}{59} \\ \frac{5}{59} & \frac{21}{59} & \frac{6}{59} \end{pmatrix}$$

18.9 SOLUTION OF SIMULTANEOUS EQUATION BY DETERMINANT AND INVERSE MATRIX METHODS

18.9.1 DETERMINANT METHOD (CRAMER'S RULE)

The determinant method sometimes called "Cramer's rule" can be used for solving simultaneous equation in two or three variables:
Thus, $AX = C$ is the form of the simultaneous equation clearly written:

For two-by-two matrices as,

$$a_{11}x + a_{12}y = c_1$$

$$a_{21}x + a_{22}y = c_2 \qquad \Rightarrow \begin{pmatrix} a_{11} & a_{12} \\ a_{21} & a_{22} \end{pmatrix} \begin{pmatrix} x \\ y \end{pmatrix} = \begin{pmatrix} c_1 \\ c_2 \end{pmatrix}$$

where A is the coefficients of the variables, $= \begin{pmatrix} a_{11} & a_{12} \\ a_{21} & a_{22} \end{pmatrix}$,

X is the variables, $= \begin{pmatrix} x \\ y \end{pmatrix}$ and C is the RHS constants $= \begin{pmatrix} c_1 \\ c_2 \end{pmatrix}$

Then, from Cramer's rule:

$$x = \frac{\begin{vmatrix} c_1 & a_{12} \\ c_2 & a_{22} \end{vmatrix}}{\begin{vmatrix} a_{11} & a_{12} \\ a_{21} & a_{22} \end{vmatrix}} \text{ and } y = \frac{\begin{vmatrix} a_{11} & c_1 \\ a_{21} & c_2 \end{vmatrix}}{\begin{vmatrix} a_{11} & a_{12} \\ a_{21} & a_{22} \end{vmatrix}}$$

For three-by-three matrices as:

$$a_{11}x + a_{12}y + a_{13}z = c_1$$

$$a_{21}x + a_{22}y + a_{23}z = c_2$$

$$a_{31}x + a_{32}y + a_{33}z = c_3$$

$$\Rightarrow \begin{pmatrix} a_{11} & a_{12} & a_{13} \\ a_{21} & a_{22} & a_{23} \\ a_{31} & a_{32} & a_{33} \end{pmatrix} \begin{pmatrix} x \\ y \\ z \end{pmatrix} = \begin{pmatrix} c_1 \\ c_2 \\ c_3 \end{pmatrix} \qquad \text{Where } A = \begin{pmatrix} a_{11} & a_{12} & a_{13} \\ a_{21} & a_{22} & a_{23} \\ a_{31} & a_{32} & a_{33} \end{pmatrix},\ X = \begin{pmatrix} x \\ y \\ z \end{pmatrix} \text{ and } C = \begin{pmatrix} c_1 \\ c_2 \\ c_3 \end{pmatrix}$$

Then, from Cramer's rule:

$$x = \frac{\begin{vmatrix} c_1 & a_{12} & a_{13} \\ c_2 & a_{22} & a_{23} \\ c_3 & a_{32} & a_{33} \end{vmatrix}}{\begin{vmatrix} a_{11} & a_{12} & a_{13} \\ a_{21} & a_{22} & a_{23} \\ a_{31} & a_{32} & a_{33} \end{vmatrix}},\quad y = \frac{\begin{vmatrix} a_{11} & c_1 & a_{13} \\ a_{21} & c_2 & a_{23} \\ a_{31} & c_3 & a_{33} \end{vmatrix}}{\begin{vmatrix} a_{11} & a_{12} & a_{13} \\ a_{21} & a_{22} & a_{23} \\ a_{31} & a_{32} & a_{33} \end{vmatrix}} \text{ and } z = \frac{\begin{vmatrix} a_{11} & a_{12} & c_1 \\ a_{21} & a_{22} & c_2 \\ a_{31} & a_{32} & c_3 \end{vmatrix}}{\begin{vmatrix} a_{11} & a_{12} & a_{13} \\ a_{21} & a_{22} & a_{23} \\ a_{31} & a_{32} & a_{33} \end{vmatrix}}$$

It is clearly observed that the determinant method (i.e., Cramer's rule) replaces the first column in A with C when looking for x, while second column of A is replaced with C when looking for y, and third column of A is replaced with C when looking for z. After which their respective determinants are found, then divided by the determinant of matrix A to get the solution of $x, y,$ and z.

Example 28

Solve for x and y in the system of linear equation using Cramer's rule:

(a) $x + y = 2$ (b) $\frac{1}{2}x + \frac{2}{3}y = 1$

$x - y = 3$ $\quad x + \frac{1}{5}y = 2$

Solution

(a) $x + y = 2$

$x - y = 3$

The first approach is to write this equation out in matrix notation.

Then, $\begin{pmatrix} 1 & 1 \\ 1 & -1 \end{pmatrix}\begin{pmatrix} x \\ y \end{pmatrix} = \begin{pmatrix} 2 \\ 3 \end{pmatrix}$

From Cramer's rule, we have that:

$$x = \frac{\begin{vmatrix} c_1 & a_{12} \\ c_2 & a_{22} \end{vmatrix}}{\begin{vmatrix} a_{11} & a_{12} \\ a_{21} & a_{22} \end{vmatrix}} = \frac{\begin{vmatrix} 2 & 1 \\ 3 & -1 \end{vmatrix}}{\begin{vmatrix} 1 & 1 \\ 1 & -1 \end{vmatrix}} = \frac{2(-1)-3(1)}{1(-1)-1(1)} = \frac{-5}{-2} \qquad \therefore x = \frac{5}{2}$$

$$\text{and } y = \frac{\begin{vmatrix} a_{11} & c_1 \\ a_{21} & c_2 \end{vmatrix}}{\begin{vmatrix} a_{11} & a_{12} \\ a_{21} & a_{22} \end{vmatrix}} = \frac{\begin{vmatrix} 1 & 2 \\ 1 & 3 \end{vmatrix}}{\begin{vmatrix} 1 & 1 \\ 1 & -1 \end{vmatrix}} = \frac{3(1)-2(1)}{1(-1)-1(1)} = \frac{1}{-2} \qquad \therefore y = -\frac{1}{2}$$

Hence, $(x, y) = \left(\frac{5}{2}, -\frac{1}{2}\right)$ (b) $\frac{1}{2}x + \frac{2}{3}y = 1$

$x + \frac{1}{5}y = 2$

Then, $\begin{pmatrix} \frac{1}{2} & \frac{2}{3} \\ 1 & \frac{1}{5} \end{pmatrix}\begin{pmatrix} x \\ y \end{pmatrix} = \begin{pmatrix} 1 \\ 2 \end{pmatrix}$

From Cramer's rule, we have that:

$$x = \frac{\begin{vmatrix} c_1 & a_{12} \\ c_2 & a_{22} \end{vmatrix}}{\begin{vmatrix} a_{11} & a_{12} \\ a_{21} & a_{22} \end{vmatrix}} = \frac{\begin{vmatrix} 1 & \frac{2}{3} \\ 2 & \frac{1}{5} \end{vmatrix}}{\begin{vmatrix} \frac{1}{2} & \frac{2}{3} \\ 1 & \frac{1}{5} \end{vmatrix}} = \frac{\frac{1}{5}(1)-2(\frac{2}{3})}{\frac{1}{2}(\frac{1}{5})-\frac{2}{3}(1)} = \frac{-17/15}{-17/30}$$

$$= \frac{-17}{15} \times \frac{-30}{17} = 2$$

$$\therefore x = 2 \text{ and } y = \frac{\begin{vmatrix} a_{11} & c_1 \\ a_{21} & c_2 \end{vmatrix}}{\begin{vmatrix} a_{11} & a_{12} \\ a_{21} & a_{22} \end{vmatrix}} = \frac{\begin{vmatrix} \frac{1}{2} & 1 \\ 1 & 2 \end{vmatrix}}{\begin{vmatrix} \frac{1}{2} & \frac{2}{3} \\ 1 & \frac{1}{5} \end{vmatrix}} = \frac{\frac{1}{2}(2)-1(1)}{\frac{1}{2}(\frac{1}{5})-\frac{2}{3}(1)} = \frac{0}{-17/30}$$

$\therefore y = 0$

Hence, $(x, y) = (2, 0)$

Example 29

Find the values of x, y and z using Cramer's rule:

$$x + y + z = 6$$
$$2x + y - z = 5$$
$$3x - 2y + 5z = 4$$

Solution

Write out the system of equation in matrix form:

$$\begin{pmatrix} a_{11} & a_{12} & a_{13} \\ a_{21} & a_{22} & a_{23} \\ a_{31} & a_{32} & a_{33} \end{pmatrix}\begin{pmatrix} x \\ y \\ z \end{pmatrix} = \begin{pmatrix} c_1 \\ c_2 \\ c_3 \end{pmatrix}$$

$$\Rightarrow \begin{pmatrix} 1 & 1 & 1 \\ 2 & 1 & -1 \\ 3 & -2 & 5 \end{pmatrix}\begin{pmatrix} x \\ y \\ z \end{pmatrix} = \begin{pmatrix} 6 \\ 5 \\ 4 \end{pmatrix}$$

Then, from Cramer's rule:

$$x = \frac{\begin{vmatrix} c_1 & a_{12} & a_{13} \\ c_2 & a_{22} & a_{23} \\ c_3 & a_{32} & a_{33} \end{vmatrix}}{\begin{vmatrix} a_{11} & a_{12} & a_{13} \\ a_{21} & a_{22} & a_{23} \\ a_{31} & a_{32} & a_{33} \end{vmatrix}} = \frac{\begin{vmatrix} 6 & 1 & 1 \\ 5 & 1 & -1 \\ 4 & -2 & 5 \end{vmatrix}}{\begin{vmatrix} 1 & 1 & 1 \\ 2 & 1 & -1 \\ 3 & -2 & 5 \end{vmatrix}} = \frac{6\begin{vmatrix} 1 & -1 \\ -2 & 5 \end{vmatrix} - 1\begin{vmatrix} 5 & -1 \\ 4 & 5 \end{vmatrix} + 1\begin{vmatrix} 5 & 1 \\ 4 & -2 \end{vmatrix}}{1\begin{vmatrix} 1 & -1 \\ -2 & 5 \end{vmatrix} - 1\begin{vmatrix} 2 & -1 \\ 3 & 5 \end{vmatrix} + 1\begin{vmatrix} 2 & 1 \\ 3 & -2 \end{vmatrix}}$$

$$= \frac{6(5-2)-1(25+4)+1(-10-4)}{1(5-2)-1(10+3)+1(-4-3)} = \frac{18-29-14}{3-13-7} = \frac{-25}{-17}$$

$$\therefore x = \frac{25}{17} = 1\frac{8}{17}$$

$$\text{Also, } y = \frac{\begin{vmatrix} a_{11} & c_1 & a_{13} \\ a_{21} & c_2 & a_{23} \\ a_{31} & c_3 & a_{33} \end{vmatrix}}{\begin{vmatrix} a_{11} & a_{12} & a_{13} \\ a_{21} & a_{22} & a_{23} \\ a_{31} & a_{32} & a_{33} \end{vmatrix}} = \frac{\begin{vmatrix} 1 & 6 & 1 \\ 2 & 5 & -1 \\ 3 & 4 & 5 \end{vmatrix}}{\begin{vmatrix} 1 & 1 & 1 \\ 2 & 1 & -1 \\ 3 & -2 & 5 \end{vmatrix}} = \frac{1\begin{vmatrix} 5 & -1 \\ 4 & 5 \end{vmatrix} - 6\begin{vmatrix} 2 & -1 \\ 3 & 5 \end{vmatrix} + 1\begin{vmatrix} 2 & 5 \\ 3 & 4 \end{vmatrix}}{1\begin{vmatrix} 1 & -1 \\ -2 & 5 \end{vmatrix} - 1\begin{vmatrix} 2 & -1 \\ 3 & 5 \end{vmatrix} + 1\begin{vmatrix} 2 & 1 \\ 3 & -2 \end{vmatrix}}$$

$$= \frac{1(25+4)-6(10+3)+1(8-15)}{1(5-2)-1(10+3)+1(-4-3)} = \frac{29-78-7}{3-13-7} = \frac{-56}{-17}$$

$$\therefore y = \frac{56}{17} = 3\frac{5}{17}$$

$$\text{And, } z = \frac{\begin{vmatrix} a_{11} & a_{12} & c_1 \\ a_{21} & a_{22} & c_2 \\ a_{31} & a_{32} & c_3 \end{vmatrix}}{\begin{vmatrix} a_{11} & a_{12} & a_{13} \\ a_{21} & a_{22} & a_{23} \\ a_{31} & a_{32} & a_{33} \end{vmatrix}} = \frac{\begin{vmatrix} 1 & 1 & 6 \\ 2 & 1 & 5 \\ 3 & -2 & 4 \end{vmatrix}}{\begin{vmatrix} 1 & 1 & 1 \\ 2 & 1 & -1 \\ 3 & -2 & 5 \end{vmatrix}} = \frac{1\begin{vmatrix} 1 & 5 \\ -2 & 4 \end{vmatrix} - 1\begin{vmatrix} 2 & 5 \\ 3 & 4 \end{vmatrix} + 6\begin{vmatrix} 2 & 1 \\ 3 & -2 \end{vmatrix}}{1\begin{vmatrix} 1 & -1 \\ -2 & 5 \end{vmatrix} - 1\begin{vmatrix} 2 & -1 \\ 3 & 5 \end{vmatrix} + 1\begin{vmatrix} 2 & 1 \\ 3 & -2 \end{vmatrix}}$$

$$= \frac{1(4+10)-1(8-15)+6(-4-3)}{1(5-2)-1(10+3)+1(-4-3)} = \frac{14+7-42}{3-13-7} = \frac{-21}{-17}$$

$$\therefore z = \frac{21}{17} = 1\frac{4}{17} \quad \text{Hence, } (x, y, z) = \left(1\frac{8}{17}, 3\frac{5}{17}, 1\frac{4}{17}\right)$$

18.9.2 INVERSE MATRIX METHOD

The inverse matrix method of solving simultaneous equation uses the concept that:

$$IX = A^{-1}C$$

where A^{-1} is the inverse of the coefficient matrix A of variables X, while I is the unit matrix

Example 30

Find the solution of the simultaneous equation using inverse matrix method.

$3x + y = 1$
$x - 2y = 5$

Solution

Writing it out in matrix form, we have:

$$\begin{pmatrix} 3 & 1 \\ 1 & -2 \end{pmatrix}\begin{pmatrix} x \\ y \end{pmatrix} = \begin{pmatrix} 1 \\ 5 \end{pmatrix}$$

This is now in the form $IX = A^{-1}C$, where $A^{-1} = \begin{pmatrix} 3 & 1 \\ 1 & -2 \end{pmatrix}^{-1}$, $C = \begin{pmatrix} 1 \\ 5 \end{pmatrix}$, $I = \begin{pmatrix} 1 & 0 \\ 0 & 1 \end{pmatrix}$ and $X = \begin{pmatrix} x \\ y \end{pmatrix}$

Then, $\begin{pmatrix} 1 & 0 \\ 0 & 1 \end{pmatrix}\begin{pmatrix} x \\ y \end{pmatrix} = \begin{pmatrix} 3 & 1 \\ 1 & -2 \end{pmatrix}^{-1}\begin{pmatrix} 1 \\ 5 \end{pmatrix}$

$$A^{-1} = \frac{Adjoint\ A}{|A|}$$

Thus, $A^{-1} = \frac{1}{3(-2)-1(1)}\begin{pmatrix} -2 & -1 \\ -1 & 3 \end{pmatrix}$

$$= \frac{1}{-7}\begin{pmatrix} -2 & -1 \\ -1 & 3 \end{pmatrix} = \begin{pmatrix} \frac{2}{7} & \frac{1}{7} \\ \frac{1}{7} & -\frac{3}{7} \end{pmatrix}$$

So, $\begin{pmatrix} 1 & 0 \\ 0 & 1 \end{pmatrix}\begin{pmatrix} x \\ y \end{pmatrix} = \begin{pmatrix} \frac{2}{7} & \frac{1}{7} \\ \frac{1}{7} & -\frac{3}{7} \end{pmatrix}\begin{pmatrix} 1 \\ 5 \end{pmatrix} = \begin{pmatrix} \frac{2}{7} + \frac{5}{7} \\ \frac{1}{7} + \left(-\frac{15}{7}\right) \end{pmatrix} = \begin{pmatrix} 1 \\ -2 \end{pmatrix}$

$$\begin{pmatrix} x \\ y \end{pmatrix} = \begin{pmatrix} 1 \\ -2 \end{pmatrix}$$

$\therefore (x,\ y) = (1, -2)$

Example 31

Find x, y and z in the equation:

$$x + 3y - z = -3$$
$$2x - y + 3z = 4$$
$$5x + y + 2z = -1$$

Solution

First convert the equation into the form, $AX = C$

$$\begin{pmatrix} 1 & 3 & -1 \\ 2 & -1 & 3 \\ 5 & 1 & 2 \end{pmatrix}\begin{pmatrix} x \\ y \\ z \end{pmatrix} = \begin{pmatrix} -3 \\ 4 \\ -1 \end{pmatrix}$$

But, $IX = A^{-1}C$

So, $\begin{pmatrix} 1 & 0 & 0 \\ 0 & 1 & 0 \\ 0 & 0 & 1 \end{pmatrix}\begin{pmatrix} x \\ y \\ z \end{pmatrix} = \begin{pmatrix} 1 & 3 & -1 \\ 2 & -1 & 3 \\ 5 & 1 & 2 \end{pmatrix}^{-1}\begin{pmatrix} -3 \\ 4 \\ -1 \end{pmatrix}$

Next, find the inverse matrix $A^{-1} = \frac{Adjoint\ A}{|A|}$.

But, cofactor of $A = \begin{pmatrix} +\begin{vmatrix} -1 & 3 \\ 1 & 2 \end{vmatrix} & -\begin{vmatrix} 2 & 3 \\ 5 & 2 \end{vmatrix} & +\begin{vmatrix} 2 & -1 \\ 5 & 1 \end{vmatrix} \\ -\begin{vmatrix} 3 & -1 \\ 1 & 2 \end{vmatrix} & +\begin{vmatrix} 1 & -1 \\ 5 & 2 \end{vmatrix} & -\begin{vmatrix} 1 & 3 \\ 5 & 1 \end{vmatrix} \\ +\begin{vmatrix} 3 & -1 \\ -1 & 3 \end{vmatrix} & -\begin{vmatrix} 1 & -1 \\ 2 & 3 \end{vmatrix} & +\begin{vmatrix} 1 & 3 \\ 2 & -1 \end{vmatrix} \end{pmatrix}$

$$- \begin{pmatrix} -5 & 11 & 7 \\ -7 & 7 & 14 \\ 8 & -5 & -7 \end{pmatrix}$$

then the transpose of the cofactor matrix becomes the adjoint matrix of A,

Adjoint $A = \begin{pmatrix} -5 & -7 & 8 \\ 11 & 7 & -5 \\ 7 & 14 & -7 \end{pmatrix}$

Next, $A^{-1} = \frac{Adjoint\ A}{|A|} = \frac{\begin{pmatrix} -5 & -7 & 8 \\ 11 & 7 & -5 \\ 7 & 14 & -7 \end{pmatrix}}{\begin{vmatrix} 1 & 3 & -1 \\ 2 & -1 & 3 \\ 5 & 1 & 2 \end{vmatrix}} = \frac{\begin{pmatrix} -5 & -7 & 8 \\ 11 & 7 & -5 \\ 7 & 14 & -7 \end{pmatrix}}{1(-2-3)-3(4-15)+(-1)(2+5)}$

$$= \frac{\begin{pmatrix} -5 & -7 & 8 \\ 11 & 7 & -5 \\ 7 & 14 & -7 \end{pmatrix}}{-5+33-7} = \frac{1}{21}\begin{pmatrix} -5 & -7 & 8 \\ 11 & 7 & -5 \\ 7 & 14 & -7 \end{pmatrix}$$

Thus, $\begin{pmatrix} 1 & 0 & 0 \\ 0 & 1 & 0 \\ 0 & 0 & 1 \end{pmatrix}\begin{pmatrix} x \\ y \\ z \end{pmatrix} = \frac{1}{21}\begin{pmatrix} -5 & -7 & 8 \\ 11 & 7 & -5 \\ 7 & 14 & -7 \end{pmatrix}\begin{pmatrix} -3 \\ 4 \\ -1 \end{pmatrix} = \frac{1}{23}\begin{pmatrix} 15 - 28 - 8 \\ -33 + 28 + 5 \\ -21 + 56 + 7 \end{pmatrix}$

$$\begin{pmatrix} 1 & 0 & 0 \\ 0 & 1 & 0 \\ 0 & 0 & 1 \end{pmatrix}\begin{pmatrix} x \\ y \\ z \end{pmatrix} = \frac{1}{21}\begin{pmatrix} -21 \\ 0 \\ 42 \end{pmatrix} = \begin{pmatrix} -1 \\ 0 \\ 2 \end{pmatrix}$$

$\therefore \begin{pmatrix} x \\ y \\ z \end{pmatrix} = \begin{pmatrix} -1 \\ 0 \\ 2 \end{pmatrix}$ Hence, the unknown variables are: $x = -1$, $y = 0$, and $z = 2$

OBJECTIVE TEST 18

1. Given that $X = \begin{pmatrix} 8 & -3 \\ -7 & 5 \end{pmatrix}$ and $Y = \begin{pmatrix} 4 & 10 \\ 2 & 3 \end{pmatrix}$, find $3X + 2Y$

   (A) $\begin{pmatrix} 32 & 11 \\ -17 & 21 \end{pmatrix}$ (B) $\begin{pmatrix} 25 & 11 \\ 24 & 21 \end{pmatrix}$ (C) $\begin{pmatrix} 24 & 11 \\ 25 & 21 \end{pmatrix}$ (D) $\begin{pmatrix} 24 & 21 \\ 25 & 11 \end{pmatrix}$

2. If $A = \begin{pmatrix} 3 & -5 \\ 2 & 8 \end{pmatrix}$ and $I = \begin{pmatrix} 1 & 0 \\ 0 & 1 \end{pmatrix}$, find $A^2 + A - 5I$

   (A) $\begin{pmatrix} 2 & -60 \\ 24 & 62 \end{pmatrix}$ (B) $\begin{pmatrix} -3 & -60 \\ 24 & 59 \end{pmatrix}$ (C) $\begin{pmatrix} 3 & -60 \\ 24 & 57 \end{pmatrix}$ (D) $\begin{pmatrix} -60 & 3 \\ 24 & 57 \end{pmatrix}$

3. If $\begin{vmatrix} x & 2 \\ 5 & 8 \end{vmatrix} = 14$, find the value of x.

   (A) 2 (B) 3 (C) 4 (D) 5

4. Evaluate $\begin{vmatrix} 3 & 2 & 0 \\ 5 & 4 & -2 \\ 7 & 6 & 9 \end{vmatrix}$

   (A) 26 (B) −26 (C) 10 (D) −10

5. Given that $M = \begin{vmatrix} 3 & 4 \\ -1 & 2 \end{vmatrix}$, find M^{-1}

   (A) $\begin{pmatrix} \frac{3}{10} & \frac{2}{5} \\ -\frac{1}{10} & -\frac{1}{5} \end{pmatrix}$ (B) $\begin{pmatrix} \frac{3}{10} & -\frac{2}{5} \\ \frac{1}{10} & -\frac{1}{5} \end{pmatrix}$ (C) $\begin{pmatrix} \frac{3}{10} & -\frac{2}{5} \\ \frac{1}{10} & \frac{2}{5} \end{pmatrix}$ (D) $\begin{pmatrix} \frac{1}{5} & -\frac{2}{5} \\ \frac{1}{10} & \frac{3}{10} \end{pmatrix}$

6. Given that $\begin{vmatrix} x & 0 & -2 \\ 3 & x+2 & 5 \\ 5 & 3 & -1 \end{vmatrix} = 8$, find the possible values of x.

   (A) −1, −6 (B) −1, 6 (C) −7, 1 (D) 8, −1

7. The inverse of a matrix is given as $P^{-1} = \begin{pmatrix} \frac{1}{5} & \frac{1}{5} \\ \frac{2}{5} & -\frac{3}{5} \end{pmatrix}$, find the matrix P.

(A) $\begin{pmatrix} 3 & 1 \\ 2 & -1 \end{pmatrix}$ (B) $\begin{pmatrix} -1 & 3 \\ 2 & -1 \end{pmatrix}$ (C) $\begin{pmatrix} -3 & 1 \\ 2 & 1 \end{pmatrix}$ (D) $\begin{pmatrix} 3 & -1 \\ -2 & -1 \end{pmatrix}$

8. Solve for x and y, given that: $\begin{pmatrix} 3 & 5 \\ 2 & 1 \end{pmatrix}\begin{pmatrix} x \\ y \end{pmatrix} = \begin{pmatrix} 2 \\ -1 \end{pmatrix}$

(A) $x = -1, y = -1$ (B) $x = 1, y = 1$ (C) $x = 1, y = -1$ (D) $x = -1, y = 1$

9. Find the Adjoint of the matrix $X = \begin{pmatrix} 3 & 5 & 0 \\ 2 & 4 & 8 \\ -1 & 2 & 1 \end{pmatrix}$

(A) $\begin{pmatrix} -12 & -5 & 40 \\ -10 & 3 & -24 \\ 8 & -11 & 2 \end{pmatrix}$ (B) $\begin{pmatrix} -12 & -10 & 8 \\ -5 & 3 & -11 \\ 40 & -24 & -2 \end{pmatrix}$ (C) $\begin{pmatrix} 12 & 10 & 8 \\ 5 & 3 & 11 \\ 40 & 24 & 2 \end{pmatrix}$ (D) $\begin{pmatrix} 12 & 5 & 4 \\ 10 & -3 & 1 \\ 8 & 11 & 2 \end{pmatrix}$

10. Given that $P = \begin{pmatrix} 1 & 3 & 0 \\ 2 & -1 & 5 \\ 1 & 0 & 8 \end{pmatrix}$ and $Q = \begin{pmatrix} 5 & 3 & 4 \\ 1 & 6 & 2 \\ 0 & 7 & 1 \end{pmatrix}$, find $(PQ)^T$

(A) $\begin{pmatrix} 8 & 21 & 10 \\ 9 & 35 & 11 \\ 5 & 24 & 7 \end{pmatrix}$ (B) $\begin{pmatrix} 8 & 21 & -10 \\ 9 & -35 & 11 \\ 5 & 24 & 7 \end{pmatrix}$ (C) $\begin{pmatrix} 8 & 9 & 5 \\ 21 & 35 & 59 \\ 10 & 11 & 12 \end{pmatrix}$ (D) $\begin{pmatrix} 8 & 21 & 10 \\ 9 & 35 & 11 \\ 5 & 59 & 12 \end{pmatrix}$

Exercise 18

Given $P = \begin{pmatrix} 3 & 0 \\ 1 & 4 \end{pmatrix}$, $Q = \begin{pmatrix} 5 & 7 \\ 8 & -4 \end{pmatrix}$ and $R = \begin{pmatrix} 20 & 9 \\ 12 & 15 \end{pmatrix}$

Use this information to answer question 1 to 5:

1. Find (i) $P + Q + R$ (ii) $P - Q + 2R$ (iii) $2P - 3Q$ (iv) $Q - P$
2. Evaluate (i) $P^2 - Q^2$ (ii) PQ (iii) QP (iv) PR (v) RP (vi) QR
3. Determine (i) $|P|$ (ii) $|Q|$ (iii) $|R|$ (iv) $|PQ|$ (v) $|QR|$ (vi) $|PR|$
4. Find (i) P^T (ii) Q^T (iii) R^T (iv) P^TQ^T (v) $(PQ)^T$ (vi) Q^TR^T
5. Evaluate (i) P^{-1} (ii) Q^{-1} (iii) R^{-1} (iv) $(PQ)^{-1}$ (v) $(PR)^{-1}$ (vi) $(QR)^{-1}$

6. Calculate the trace of the following matrices:

(i) $\begin{pmatrix} 1 & 0 \\ 2 & 3 \end{pmatrix}$ (ii) $\begin{pmatrix} 5 & 20 \\ 3 & 4 \end{pmatrix}$ (iii) $\begin{pmatrix} -3 & 1 \\ 2 & 0 \end{pmatrix}$ (iv) $\begin{pmatrix} 1 & 0 \\ 0 & 1 \end{pmatrix}$ (v) $\begin{pmatrix} 1 & 4 & 7 \\ 2 & 5 & 8 \\ 3 & 6 & 9 \end{pmatrix}$

(vi) $\begin{pmatrix} 1 & 7 & 0 \\ 1 & 0 & 8 \\ -3 & 6 & -4 \end{pmatrix}$ (vii) $\begin{pmatrix} 5 & -8 & 2 \\ 6 & 0 & 3 \\ -2 & -9 & -3 \end{pmatrix}$ (viii) $\begin{pmatrix} 25 & 3 & -5 \\ 4 & 3 & 8 \\ 1 & 7 & -2 \end{pmatrix}$

7. Given the following matrices:

$A = \begin{pmatrix} 1 & 3 \\ 2 & 5 \end{pmatrix}, B = \begin{pmatrix} 0 & 4 \\ -5 & 3 \end{pmatrix}$ $C = \begin{pmatrix} 5 & 3 & 1 \\ 4 & 2 & 8 \end{pmatrix}$ and $D = \begin{pmatrix} -2 & 4 & 3 \\ -3 & 7 & 6 \end{pmatrix}$

Compute: (i) $A + B$ (ii) $C + D$ (iii) $A + C$ (iv) $A - B$ (v) $B - A$ (vi) $C - D$

8. Show that $AB \neq BA$ if $A = \begin{pmatrix} 6 & 1 & 5 \\ 0 & -4 & 1 \\ 2 & 2 & 1 \end{pmatrix}$ and $B = \begin{pmatrix} -5 & 4 & 3 \\ 1 & 0 & 0 \\ 3 & 5 & 4 \end{pmatrix}$

9. Three matrices are given as:

$X = \begin{pmatrix} 1 & 1 & 0 \\ 12 & 5 & 4 \\ 7 & 2 & 3 \end{pmatrix}, Y = \begin{pmatrix} 2 & 3 & 7 \\ 1 & 4 & -5 \\ -8 & 2 & 1 \end{pmatrix}$ and $Z = \begin{pmatrix} -4 & 1 & 9 \\ 8 & 5 & 2 \\ 1 & 0 & 0 \end{pmatrix}$

Find: (i) $X + Y + Z$ (ii) $X^2 + Y^2 + Z^2$ (iii) $(X + Y)(X - Z)$
(iv) XY (v) ZY (vi) XZ (vii) $(XY)Z$

10. If $P = (1 \quad 3)$, $Q = \begin{pmatrix} 0 & 4 \\ 5 & 9 \end{pmatrix}$ and $R = \begin{pmatrix} 2 \\ 7 \end{pmatrix}$

Find: (i) PQ (ii) PR (iii) $(PQ)R$ (iv) $(PR)Q$

11. Given $M = \begin{pmatrix} 2 & 5 & 1 \\ -1 & 4 & 3 \\ 0 & 3 & 6 \end{pmatrix}$ and the unit matrix $I = \begin{pmatrix} 1 & 0 & 0 \\ 0 & 1 & 0 \\ 0 & 0 & 1 \end{pmatrix}$

Evaluate: (i) IM (ii) $M^2 + 2M + I$

12. Given $A = \begin{pmatrix} 1 & -2 \\ 2 & 5 \end{pmatrix}$ and $B = \begin{pmatrix} 7 & 3 \\ 8 & 2 \end{pmatrix}$

Find: (i) A^T (ii) B^T (iii) A^TB^T (iv) $(AB)^T$ (v) $(BA)^T$

13. Given the matrices:

$A = \begin{pmatrix} 1 & -3 & 2 \\ -2 & 0 & 0 \\ 3 & -5 & 1 \end{pmatrix}, B = \begin{pmatrix} 1 & 0 & 1 \\ 2 & 3 & 2 \\ -1 & 1 & 5 \end{pmatrix}$ and $C = \begin{pmatrix} 3 & 4 & 6 \\ 2 & -1 & 7 \\ 1 & 2 & 9 \end{pmatrix}$

Show that:

(i) $A(B + C) = AB + AC$ (ii) $A(BC) = (AB)C$ (iii) $AB \neq BA$

14. Find the possible values of x given that:

$$\text{(i)}\ \begin{vmatrix} x & 0 & 0 \\ 0 & x+1 & 0 \\ 0 & 0 & x-4 \end{vmatrix} = 0 \quad \text{(ii)}\ \begin{vmatrix} x & 0 & -3 \\ 2 & x+1 & 0 \\ 1 & 5 & 1 \end{vmatrix} = 20$$

15. Evaluate the determinant of each of the matrices given below:

$$\text{(i)} \begin{pmatrix} -3 & 5 \\ 1 & 4 \end{pmatrix} \quad \text{(ii)} \begin{pmatrix} 2 & 7 \\ 3 & -6 \end{pmatrix} \quad \text{(iii)} \begin{pmatrix} 23 & 12 \\ 16 & 5 \end{pmatrix} \quad \text{(iv)} \begin{pmatrix} 2 & 8 \\ -5 & 4 \end{pmatrix} \quad \text{(v)} \begin{pmatrix} 0 & 26 \\ 0 & 12 \end{pmatrix}$$

$$\text{(vi)} \begin{pmatrix} 1 & 1 \\ 3 & 3 \end{pmatrix} \quad \text{(vii)} \begin{pmatrix} 2 & 3 \\ 2 & 3 \end{pmatrix} \quad \text{(viii)} \begin{pmatrix} e^x & 5 \\ 2e^{2x} & 3e^x \end{pmatrix} \quad \text{(ix)} \begin{pmatrix} 2x & x^3 \\ 1 & 3x^2 \end{pmatrix}$$

16. Find the determinant of each 3×3 matrices:

$$\text{(i)} \begin{pmatrix} 2 & 5 & -1 \\ 9 & 4 & 0 \\ 1 & -2 & 1 \end{pmatrix} \quad \text{(ii)} \begin{pmatrix} 5 & 0 & 0 \\ 4 & 1 & 12 \\ -3 & 5 & 2 \end{pmatrix} \quad \text{(iii)} \begin{pmatrix} 7 & 2 & 7 \\ 4 & 3 & 4 \\ -5 & 9 & -5 \end{pmatrix}$$

$$\text{(iv)} \begin{pmatrix} 1 & 3 & -5 \\ -8 & 2 & 7 \\ 3 & 4 & 2 \end{pmatrix} \quad \text{(v)} \begin{pmatrix} 3 & 7 & 4 \\ 11 & 1 & 2 \\ 6 & 25 & 5 \end{pmatrix} \quad \text{(vi)} \begin{pmatrix} 13 & 9 & 5 \\ 15 & 6 & 4 \\ 7 & 3 & 8 \end{pmatrix}$$

17. Given the matrices:

$$P = \begin{pmatrix} 4 & -3 \\ 2 & 1 \end{pmatrix} \text{ and } Q = \begin{pmatrix} 5 & 12 \\ 2 & -2 \end{pmatrix}$$

Find: (a) the minors of (i) P (ii) Q (iii) PQ (iv) QP
(b) the cofactor of (i) P (ii) Q (iii) PQ (iv) QP
(c) the adjoint of (i) P (ii) Q (iii) PQ (iv) QP
(d) the inverse of (i) P (ii) Q (iii) PQ (iv) QP

18. Find the (i) minors (ii) cofactors of the matrix $\begin{pmatrix} 5 & -2 & 4 \\ 7 & 1 & 3 \\ 10 & 26 & 8 \end{pmatrix}$

19. Find the inverse of each 3×3 matrices:

$$\text{(i)} \begin{pmatrix} 1 & 3 & 0 \\ 2 & 4 & 2 \\ 1 & 4 & 1 \end{pmatrix} \quad \text{(ii)} \begin{pmatrix} 5 & 0 & 0 \\ 0 & 1 & 12 \\ 0 & 5 & 2 \end{pmatrix} \quad \text{(iii)} \begin{pmatrix} 7 & 2 & 1 \\ 4 & 3 & 1 \\ 12 & 75 & 30 \end{pmatrix}$$

$$\text{(iv)} \begin{pmatrix} 20 & 3 & 5 \\ -8 & -1 & 4 \\ 3 & 6 & 2 \end{pmatrix} \quad \text{(v)} \begin{pmatrix} 2 & 0 & 3 \\ 10 & 1 & 2 \\ 5 & -2 & 7 \end{pmatrix} \quad \text{(vi)} \begin{pmatrix} \frac{3}{2} & \frac{1}{2} & 1 \\ 1 & 6 & 0 \\ -\frac{5}{2} & 2 & -1 \end{pmatrix}$$

$$\text{(vii)} \begin{pmatrix} -\frac{7}{2} & 1 & 0 \\ 3 & \frac{1}{2} & \frac{13}{5} \\ -\frac{5}{2} & 2 & -1 \end{pmatrix} \quad \text{(viii)} \begin{pmatrix} 0 & 0 & 0 \\ 3 & -4 & 1 \\ -1 & 5 & 2 \end{pmatrix} \quad \text{(ix)} \begin{pmatrix} 3 & 0 & 3 \\ 1 & 1 & 1 \\ -5 & 2 & -5 \end{pmatrix}$$

20. Find the solution to the simultaneous equations using Cramer's rule:

(i) $x + y = 2$, $2x - y = 5$ (ii) $5x + 3y = 1$, $2x + y = 10$ (iii) $\frac{1}{x} + \frac{1}{y} = 3$, $\frac{3}{x} + \frac{2}{y} = -2$

21. Solve the system of linear equations using Cramer's rule.

$$3x + 2y + z = -1$$
$$2x - y + 5z = 6$$
$$x + 3y + 2z = 5$$

22. Find the solution for x and y using inverse matrix method.

(i) $\begin{pmatrix} 2 & 1 \\ 3 & -2 \end{pmatrix}\begin{pmatrix} x \\ y \end{pmatrix} = \begin{pmatrix} 1 \\ 0 \end{pmatrix}$ (ii) $\begin{pmatrix} 12 & 5 \\ 3 & 4 \end{pmatrix}\begin{pmatrix} x \\ y \end{pmatrix} = \begin{pmatrix} 10 \\ -7 \end{pmatrix}$

23. Solve the following simultaneous equations using inverse matrix method:

(i) $3x + 4y = 0$, $2x + 5y = -7$ (ii) $-2x + 5y - 4 = 0$, $5x - y + 12 = 0$

24. Given the equations:

$$x + 2y + z + 5 = 0$$
$$5x + 3y + z - 1 = 0$$
$$x + y + 4z - 7 = 0$$

(a) Write the simultaneous equation in the form $AX = C$

(b) Hence, solve for $x, y,$ and z using either determinant method or inverse matrix method.

25. The inverse of a matrix is given as:

(i) $P^{-1} = \begin{pmatrix} 3 & 5 \\ 2 & -4 \end{pmatrix}$, find the matrix P (ii) $Q^{-1} = \begin{pmatrix} 1 & 3 & 0 \\ 2 & 4 & 2 \\ 1 & 4 & 1 \end{pmatrix}$, find the matrix Q.

26. Find the transpose of the following matrices:

(i) $A = \begin{pmatrix} 3 & 7 \\ 18 & -14 \end{pmatrix}$ (ii) $B = \begin{pmatrix} 40 & -5 \\ 30 & 17 \end{pmatrix}$ (iii) $\begin{pmatrix} 13 & 15 & -3 \\ 56 & 4 & -5 \\ 11 & 16 & 10 \end{pmatrix}$

(iv) $\begin{pmatrix} -85 & 105 & 122 \\ 43 & -38 & 90 \\ 132 & -160 & 58 \end{pmatrix}$ (v) $\begin{pmatrix} \frac{2}{3} & -\frac{3}{7} & 1 \\ -\frac{4}{5} & 0 & \frac{3}{8} \\ 1 & -1 & 0 \end{pmatrix}$

APPENDIX A

MATHEMATICAL TABLE

TABLE 1 : LOGARITHMS OF NUMBERS $x \longrightarrow \text{Log } x$

										Differences								
x	0	1	2	3	4	5	6	7	8	9	1	2	3	4	5	6	7	8	9
10	0000	0043	0086	0128	0170	0212	0253	0294	0334	0374	4	8	12	17	21	25	29	33	37
11	0414	0453	0492	0531	0569	0607	0645	0682	0719	0755	4	8	11	15	19	23	26	30	34
12	0792	0828	0864	0899	0934	0969	1004	1038	1072	1106	3	7	10	14	17	21	24	28	31
13	1139	1173	1206	1239	1271	1303	1335	1367	1399	1430	3	6	10	13	16	19	23	26	29
14	1461	1492	1523	1553	1584	1614	1644	1673	1703	1732	3	6	9	12	15	18	21	24	27
15	1761	1790	1818	1847	1875	1903	1931	1959	1987	2014	3	6	8	11	14	17	20	22	25
16	2041	2068	2095	2122	2148	2175	2201	2227	2253	2279	3	5	8	11	13	16	18	21	24
17	2304	2330	2355	2380	2405	2430	2455	2480	2504	2529	2	5	7	10	12	15	17	20	22
18	2553	2577	2601	2625	2648	2672	2695	2718	2742	2765	2	5	7	9	12	14	16	19	21
19	2788	2810	2833	2856	2878	2900	2923	2945	2967	2989	2	4	7	9	11	13	16	18	20
20	3010	3032	3054	3075	3096	3118	3139	3160	3181	3201	2	4	6	8	11	13	15	17	19
21	3222	3243	3263	3284	3304	3324	3345	3365	3385	3404	2	4	6	8	10	12	14	16	18
22	3424	3444	3464	3483	3502	3522	3541	3560	3579	3598	2	4	6	8	10	12	14	15	17
23	3617	3636	3655	3674	3692	3711	3729	3747	3766	3788	2	4	6	7	9	11	13	15	17
24	3802	3820	3838	3856	3874	3892	3909	3927	3945	3962	2	4	5	7	9	11	12	14	16
25	3979	3997	4014	4031	4048	4065	4082	4099	4116	4133	2	3	5	7	9	10	12	14	15
26	4150	4166	4183	4200	4216	4232	4249	4265	4281	4298	2	3	5	7	8	10	11	13	15
27	4314	4330	4346	4362	4378	4393	4409	4425	4440	4456	2	3	5	6	8	9	11	13	14
28	4472	4487	4502	4518	4533	4548	4564	4579	4594	4609	2	3	5	6	8	9	11	12	14
29	4624	4639	4654	4669	4683	4698	4713	4728	4742	4757	1	3	4	6	7	9	10	12	13
30	4771	4786	4800	4814	4829	4843	4857	4871	4886	4900	1	3	4	6	7	9	10	11	13
31	4914	4928	4942	4955	4969	4983	4997	5011	5024	5038	1	3	4	6	7	8	10	11	12
32	5051	5065	5079	5092	5105	5119	5132	5145	5159	5172	1	3	4	5	7	8	9	11	12
33	5185	5198	5211	5224	5237	5250	5263	5276	5289	5302	1	3	4	5	6	8	9	10	12
34	5315	5328	5340	5353	5366	5378	5391	5403	5416	5428	1	3	4	5	6	8	9	10	11
35	5441	5453	5465	5478	5490	5502	5514	5527	5539	5551	1	2	4	5	6	7	9	10	11
36	5563	5575	5587	5599	5611	5623	5635	5647	5658	5670	1	2	4	5	6	7	8	10	11
37	5682	5694	5705	5717	5729	5740	5752	5763	5775	5786	1	2	3	5	6	7	8	9	10
38	5798	5809	5821	5832	5843	5855	5866	5877	5888	5899	1	2	3	5	6	7	8	9	10
39	5911	5922	5933	5944	5955	5966	5977	5988	5999	6010	1	2	3	4	5	7	8	9	10
40	6021	6031	6042	6053	6064	6075	6085	6096	6107	6117	1	2	3	4	5	6	8	9	10
41	6128	6138	6149	6160	6170	6180	6191	6201	6212	6222	1	2	3	4	5	6	7	8	9
42	6232	6243	6253	6263	6274	6284	6294	6304	6314	6325	1	2	3	4	5	6	7	8	9
43	6335	6345	6355	6365	6375	6385	6395	6405	6415	6425	1	2	3	4	5	6	7	8	9
44	6435	6444	6454	6464	6474	6484	6493	6503	6513	6522	1	2	3	4	5	6	7	8	9
45	6532	6542	6551	6561	6571	6580	6590	6599	6609	6618	1	2	3	4	5	6	7	8	9
46	6628	6637	6646	6656	6665	6675	6684	6693	6702	6712	1	2	3	4	5	6	7	7	9
47	6721	6730	6739	6749	6758	6767	6776	6785	6794	6803	1	2	3	4	5	5	6	7	8
48	6812	6821	6830	6839	6848	6857	6866	6875	6884	6893	1	2	3	4	4	5	6	7	8
49	6902	6911	6920	6928	6937	6946	6955	6964	6972	6981	1	2	3	4	5	5	6	7	8
50	6990	6998	7007	7016	7024	7033	7042	7050	7059	7067	1	2	3	3	4	5	6	7	8
51	7076	7084	7093	7101	7110	7118	7126	7135	7143	7152	1	2	3	3	4	5	6	7	8
52	7160	7168	7177	7184	7193	7202	7210	7218	7226	7235	1	2	2	3	4	5	6	7	7
53	7243	7251	7259	7267	7275	7284	7292	7300	7308	7316	1	2	2	3	4	5	6	6	7
54	7324	7332	7340	7348	7356	7364	7372	7380	7388	7396	1	2	2	3	4	5	6	6	7
x	0	1	2	3	4	5	6	7	8	9	1	2	3	4	5	6	7	8	9

TABLE 1 (continued): LOGARITHMS OF NUMBERS $x \to \log x$

x	0	1	2	3	4	5	6	7	8	9	Differences 1	2	3	4	5	6	7	8	9
55	7404	7412	7419	7427	7435	7443	7451	7459	7466	7474	1	2	2	3	4	5	5	6	7
56	7482	7482	7490	7505	7513	7520	7528	7536	7543	7551	1	2	2	3	4	5	5	6	7
57	7559	7566	7574	7582	7589	7597	7604	7612	7619	7627	1	2	2	3	4	5	5	6	7
58	7634	7642	7649	7657	7664	7672	7679	7686	7694	7701	1	1	2	3	4	4	5	6	7
59	7709	7716	7723	7731	7738	7745	7752	7760	7767	7774	1	1	2	3	4	4	5	6	7
60	7782	7789	7796	7803	7810	7818	7825	7823	7839	7846	1	1	2	3	4	4	5	6	6
61	7853	7860	7868	7875	7882	7889	7876	7903	7910	7917	1	1	2	3	4	4	5	6	6
62	7924	7931	7938	7945	7952	7959	7966	7973	7980	7987	1	1	2	3	3	4	5	6	6
63	7993	8000	8007	8014	8021	8028	8035	8041	8048	8055	1	1	2	3	3	4	5	5	6
64	8062	8069	8075	8082	8089	8096	8102	8109	8116	8122	1	1	2	3	3	4	5	5	6
65	8129	8136	8142	8149	8156	8162	8169	8176	8182	8189	1	1	2	3	3	4	5	5	6
66	8195	8202	8209	8215	8222	8228	8235	8241	8248	8254	1	1	2	3	3	4	5	5	6
67	8261	8267	8274	8280	8287	8293	8299	8306	8312	8319	1	1	2	3	3	4	5	5	6
68	8325	8331	8338	8344	8351	8357	8363	8370	8376	8382	1	1	2	2	3	4	4	5	6
69	8388	8395	8401	8407	8414	8420	8426	8432	8439	8445	1	1	2	2	3	4	4	5	6
70	8451	8457	8463	8470	8476	8482	8488	8494	8500	8506	1	1	2	2	3	4	4	5	6
71	8513	8519	8525	8531	8537	8543	8549	8555	8561	8567	1	1	2	2	3	4	4	5	5
72	8573	8579	8585	8591	8597	8603	8609	8615	8621	8627	1	1	2	2	3	4	4	5	5
73	8633	8639	8645	8651	8657	8663	8669	8675	8681	8686	1	1	2	2	3	4	4	5	5
74	8692	8698	8704	8710	8716	8722	8727	8733	8739	8745	1	1	2	2	3	4	4	5	6
75	8751	8756	8762	8768	8774	8779	8785	8791	8797	8802	1	1	2	2	3	3	4	5	5
76	8808	8814	8820	8825	8831	8837	8842	8848	8854	8859	1	1	2	2	3	3	4	5	5
77	8865	8871	8876	8882	8887	8893	8899	8904	8910	8915	1	1	2	2	3	3	4	4	5
78	8921	8927	8932	8938	8943	8949	8954	8960	8965	8971	1	1	2	2	3	3	4	4	5
79	8976	8982	8987	8993	8998	9004	9009	9015	9020	9025	1	1	2	2	3	3	4	4	5
80	9031	9036	9042	9047	9053	9058	9063	9069	9074	9079	1	1	2	2	3	3	4	4	5
81	9085	9090	9096	9101	9106	9112	9117	9122	9128	9133	1	1	2	2	3	3	4	4	5
82	9138	9143	9149	9154	9159	9165	9170	9175	9180	9186	1	1	2	2	3	3	4	4	5
83	9191	9196	9201	9206	9212	9217	9222	9227	9232	9238	1	1	2	2	3	3	4	4	5
84	9243	9248	9253	9258	9263	9269	9274	9279	9284	9289	1	1	2	2	3	3	4	4	5
85	9295	9299	9304	9309	9315	9320	9325	9330	9335	9340	1	1	2	2	3	3	4	4	5
86	9345	9350	9355	9360	9365	9370	9375	9380	9385	9390	1	1	2	2	3	3	4	4	5
87	9395	9400	9405	9410	9415	9420	9424	9430	9435	9440	0	1	1	2	2	3	3	4	4
88	9445	9450	9455	9460	9465	9469	9474	9479	9484	9489	0	1	1	2	2	3	3	4	4
89	9494	9499	9504	9509	9513	9518	9523	9528	9533	9538	0	1	1	2	2	3	3	4	4
90	9542	9547	9552	9557	9562	9566	9571	9576	9581	9586	0	1	1	2	2	3	3	4	4
91	9590	9595	9600	9605	9609	9614	9619	9624	9628	9633	0	1	1	2	2	3	3	4	4
92	9638	9643	9647	9652	9657	9661	9666	9671	9675	9680	0	1	1	2	2	3	3	4	4
93	9685	9689	9694	9699	9703	9708	9713	9717	9722	9727	0	1	1	2	2	3	3	4	4
94	9731	9736	9741	9745	9750	9754	9759	9763	9768	9773	0	1	1	2	2	3	3	4	4
95	9777	9782	9786	9791	9795	9800	9805	9809	9814	9818	0	1	1	2	2	3	3	4	4
96	9823	9827	9832	9836	9841	9845	9850	9854	9859	9863	0	1	1	2	2	3	3	4	4
97	9868	9872	9877	9881	9886	9890	9894	9899	9903	9908	0	1	1	2	2	3	3	4	4
98	9912	9917	9921	9926	9930	9934	9939	9943	9948	9952	0	1	1	2	2	3	3	4	4
99	9956	9961	9965	9983	9974	9978	9983	9987	9991	9996	0	1	1	2	2	3	3	3	4
x	0	1	2	3	4	5	6	7	8	9	1	2	3	4	5	6	7	8	9

TABLE 2 : ANTILOGARITHMS OF NUMBERS $x \longrightarrow 10^x$

											Differences								
x	0	1	2	3	4	5	6	7	8	9	1	2	3	4	5	6	7	8	9
.00	1000	1002	1005	1007	1009	1012	1014	1016	1019	1021	0	0	1	1	1	1	2	2	2
.01	1023	1026	1028	1030	1033	1035	1038	1040	1042	1045	0	0	1	1	1	1	2	2	2
.02	1047	1050	1052	1054	1057	1059	1062	1064	1067	1069	0	0	1	1	1	1	2	2	2
.03	1072	1074	1076	1079	1081	1084	1086	1089	1091	1094	0	0	1	1	1	1	2	2	2
.04	1096	1099	1102	1104	1107	1109	1112	1114	1117	1119	0	1	1	1	1	2	2	2	2
.05	1122	1125	1127	1130	1132	1135	1138	1140	1143	1146	0	1	1	1	1	2	2	2	2
.06	1148	1151	1153	1156	1159	1161	1164	1167	1169	1172	0	1	1	1	1	2	2	2	2
.07	1175	1178	1180	1183	1186	1189	1191	1194	1197	1199	0	1	1	1	1	2	2	2	2
.08	1202	1205	1208	1211	1213	1216	1219	1222	1225	1227	0	1	1	1	1	2	2	2	3
.09	1230	1233	1236	1239	1242	1245	1247	1250	1253	1256	0	1	1	1	1	2	2	2	3
.10	1259	1262	1265	1268	1271	1274	1276	1279	1282	1285	0	1	1	1	1	2	2	2	3
.11	1288	1291	1294	1297	1300	1303	1306	1309	1312	1315	0	1	1	1	1	2	2	2	3
.12	1318	1321	1324	1327	1330	1334	1337	1340	1343	1346	0	1	1	1	1	2	2	2	3
.13	1349	1352	1355	1358	1361	1365	1368	1371	1374	1377	0	1	1	1	2	2	2	3	3
.14	1380	1384	1387	1390	1393	1396	1400	1403	1406	1409	0	1	1	1	2	2	2	3	3
.15	1413	1416	1419	1422	1426	1429	1432	1435	1439	1442	0	1	1	1	2	2	2	3	3
.16	1445	1449	1452	1455	1459	1462	1466	1469	1472	1476	0	1	1	1	2	2	2	3	3
.17	1479	1483	1486	1489	1493	1496	1500	1503	1507	1510	0	1	1	1	2	2	2	3	3
.18	1514	1517	1521	1524	1528	1531	1535	1538	1542	1545	0	1	1	1	2	2	2	3	3
.19	1549	1552	1556	1560	1563	1567	1570	1574	1578	1581	0	1	1	1	2	2	3	3	3
.20	1585	1589	1592	1596	1600	1603	1607	1611	1614	1618	0	1	1	1	2	2	3	3	3
.21	1622	1626	1629	1633	1637	1641	1644	1648	1652	1656	0	1	1	2	2	2	3	3	3
.22	1660	1663	1667	1671	1675	1679	1683	1687	1690	1694	0	1	1	2	2	2	3	3	3
23	1698	1702	1706	1710	1714	1718	1722	1726	1730	1734	0	1	1	2	2	2	3	3	4
.24	1738	1742	1746	1750	1754	1758	1762	1766	1770	1774	1	1	2	2	2	2	3	3	4
.25	1778	1782	1786	1791	1795	1799	1803	1807	1811	1816	1	1	2	2	2	2	3	3	4
.26	1820	1824	1828	1832	1837	1841	1845	1849	1854	1858	1	1	2	2	2	3	3	3	4
.27	1862	1866	1871	1875	1879	1884	1888	1892	1897	1901	1	1	2	2	2	3	3	3	4
.28	1905	1910	1914	1919	1923	1928	1932	1936	1941	1945	1	1	2	2	2	3	3	4	4
.29	1950	1954	1959	1963	1968	1972	1977	1982	1986	1991	1	1	2	2	2	3	3	4	4
.30	1995	2000	2004	2009	2014	2018	2023	2028	2032	2037	1	1	2	2	2	3	3	4	4
.31	2042	2046	2051	2056	2061	2065	2070	2075	2080	2084	1	1	2	2	2	3	3	4	4
.32	2089	2094	2099	2104	2109	2113	2118	2123	2128	2133	1	1	2	2	2	3	3	4	4
.33	2138	2143	2148	2153	2158	2163	2168	2173	2178	2183	1	1	2	2	2	3	3	4	4
.34	2188	2193	2198	2203	2208	2213	2218	2223	2228	2234	1	1	2	2	3	3	4	4	5
.35	2239	2244	2249	2254	2259	2265	2270	2275	2280	2286	1	1	2	2	3	3	4	4	5
.36	2291	2296	2301	2307	2312	2317	2323	2328	2333	2339	1	1	2	2	3	3	4	4	5
.37	2344	2350	2355	2360	2366	2371	2377	2382	2388	2393	1	1	2	2	3	3	4	4	5
.38	2399	2404	2410	2415	2421	2427	2432	2438	2443	2449	1	1	2	2	3	3	4	4	5
.39	2455	2460	2466	2472	2477	2483	2489	2495	2500	2506	1	1	2	2	3	3	4	5	5
.40	2512	2518	2523	2529	2535	2541	2547	2553	2559	2564	1	1	2	2	3	4	4	5	5
.41	2570	2576	2582	2588	2594	2600	2606	2612	2618	2624	1	1	2	2	3	4	4	5	5
.42	2630	2636	2642	2649	2655	2661	2667	2673	2679	2685	1	1	2	2	3	4	4	5	6
.43	2692	2698	2704	2710	2716	2723	2729	2735	2742	2748	1	1	2	3	3	4	4	5	6
.44	2754	2761	2767	2773	2780	2786	2793	2799	2805	2812	1	1	2	3	3	4	4	5	6
.45	2818	2825	2831	2838	2844	2851	2858	2864	2871	2877	1	1	2	3	3	4	5	5	6
.46	2884	2891	2897	2904	2911	2917	2924	2931	2938	2944	1	1	2	3	3	4	5	5	6
.47	2951	2958	2965	2972	2979	2985	2992	2999	3006	3013	1	1	2	3	3	4	5	5	6
.48	3020	3027	3034	3041	3048	3055	3062	3069	3076	3083	1	1	2	3	4	4	5	6	6
.49	3090	3097	3105	3112	3119	3126	3133	3141	3148	3155	1	1	2	3	4	4	5	6	6
x	0	1	2	3	4	5	6	7	8	9	1	2	3	4	5	6	7	8	9

TABLE 2 (continued): ANTILOGARITHMS OF NUMBERS $x \longrightarrow 10^x$

											Differences								
x	0	1	2	3	4	5	6	7	8	9	1	2	3	4	5	6	7	8	9
.50	3162	3170	3177	3184	3192	3199	3206	3214	3221	3228	1	1	2	3	4	4	5	6	7
.51	3236	3243	3251	3258	3266	3273	3281	3289	3296	3304	1	2	2	3	4	5	5	6	7
.52	3311	3319	3327	3334	3342	3350	3357	3365	3373	3381	1	2	2	3	4	5	5	6	7
.53	3388	3396	3404	3412	3420	3428	3436	3443	3451	3459	1	2	2	3	4	5	5	6	7
.54	3467	3475	3483	3491	3499	3508	3516	3524	3532	3540	1	2	2	3	4	5	6	6	7
.55	3548	3556	3565	3573	3581	3589	3597	3606	3614	3622	1	2	2	3	4	5	6	6	7
.56	3631	3639	3648	3656	3664	3673	3681	3690	3698	3707	1	2	3	3	4	5	6	7	8
.57	3715	3734	3733	3741	3750	3758	3767	3776	3784	3793	1	2	3	3	4	5	6	7	8
.58	3802	3811	3819	3828	3837	3846	3855	3864	3873	3882	1	2	3	4	4	5	6	7	8
.59	3890	3899	3908	3917	3926	3936	3945	3954	3963	3972	1	2	3	4	5	5	6	7	8
.60	3981	3990	3999	4009	4018	4027	4036	4046	4055	4064	1	2	3	4	5	6	6	7	8
.61	4074	4083	4093	4102	4111	4121	4130	4140	4150	4159	1	2	3	4	5	6	6	8	9
.62	4169	4178	4188	4198	4207	4217	4227	4236	4246	4256	1	2	3	4	5	5	7	8	9
.63	4266	4276	4285	4295	4305	4315	4325	4335	4345	4355	1	2	3	4	5	6	7	8	9
.64	4365	4375	4385	4396	4406	4416	4426	4436	4446	4457	1	2	3	4	5	6	7	8	9
.65	4467	4477	4487	4458	4508	4519	4529	4539	4550	4560	1	2	3	4	5	6	7	8	9
.66	4571	4581	4592	4603	4613	4624	4634	4645	4656	4667	1	2	3	4	5	6	7	8	10
.67	4677	4688	4699	4710	4721	4732	4742	4753	4764	4775	1	2	3	4	5	7	8	9	10
.68	4786	4797	4808	4819	4831	4842	4853	4864	4875	4887	1	2	3	4	6	7	8	9	10
.69	4898	4909	4920	4932	4943	4955	4966	4977	4989	5000	1	2	3	5	6	7	8	9	10
.70	5012	5023	5035	5047	5058	5070	5082	5093	5105	5117	1	2	4	5	6	7	8	9	11
.71	5129	5140	5152	5164	5176	5188	5200	5212	5224	5236	1	2	4	5	6	7	8	10	11
.72	5248	5260	5272	5284	5297	5309	5321	5333	5346	5358	1	2	4	5	6	7	9	10	11
.73	5370	5383	5395	5408	5420	5433	5445	5458	5470	5483	1	3	4	5	6	8	9	10	11
.74	5495	5508	5521	5534	5546	5559	5572	5585	5598	5610	1	3	4	5	6	8	9	10	12
.75	5623	5636	5649	5662	5675	5689	5702	5715	5728	5741	1	3	4	5	6	8	9	10	12
.76	5754	5768	5781	5794	5808	5821	5834	5848	5861	5875	1	3	4	5	7	8	9	11	12
.77	5888	5902	5916	5929	5943	5957	5970	5984	5998	6012	1	3	4	5	7	8	10	11	12
.78	6026	6039	6053	6067	6081	6095	6109	6124	6138	6152	1	3	4	6	7	8	10	11	13
.79	6166	6180	6194	6209	6223	6237	6252	6266	6281	6295	1	3	4	6	7	9	10	11	13
.80	6310	6324	6339	6353	6368	6383	6397	6412	6427	6442	1	3	4	6	7	9	10	12	13
.81	6457	6471	6486	6501	6516	6531	6546	6561	6577	6592	2	3	5	6	8	9	11	12	14
.82	6607	6622	6637	6653	6668	6683	6699	6714	6730	6745	2	3	5	6	8	9	11	12	14
.83	6761	6776	6792	6808	6823	6839	6855	6871	6887	6902	2	3	5	6	8	9	11	13	14
.84	6918	6934	6950	6966	6982	6998	7015	7031	7047	7063	2	3	5	6	8	10	11	13	15
.85	7079	7096	7112	7129	7145	7161	7178	7194	7211	7228	2	3	5	7	8	10	12	13	15
.86	7244	7261	7278	7295	7311	7328	7345	7362	7379	7396	2	3	5	7	8	10	12	13	15
.87	7413	7430	7447	7464	7482	7499	7516	7534	7551	7568	2	4	5	7	9	10	12	14	16
.88	7586	7603	7621	7638	7656	7674	7691	7709	7727	7745	2	4	5	7	9	11	12	14	16
.89	7762	7780	7798	7816	7834	7852	7870	7889	7907	7925	2	4	5	7	9	11	13	14	16
.90	7943	7962	7980	7998	8017	8034	8054	8072	8091	8110	2	4	6	7	9	11	13	15	17
.91	8128	8147	8166	8185	8204	8222	8241	8260	8279	8299	2	4	6	8	9	11	13	15	17
.92	8318	8337	8356	8375	8395	8414	8433	8453	8472	8492	2	4	6	8	10	12	14	15	17
.93	8511	8531	8551	8570	8590	8610	8630	8650	8670	8690	2	4	6	8	10	12	14	16	18
.94	8710	8730	8750	8770	8790	8810	8831	8851	8872	8892	2	4	6	8	10	12	14	16	18
.95	8913	8933	8954	8974	8995	9016	9036	9057	9078	9099	2	4	6	8	10	12	14	17	19
.96	9120	9141	9162	9183	9204	9226	9247	9268	9290	9311	2	4	6	8	11	13	15	17	19
.97	9333	9354	9376	9397	9419	9441	9462	9484	9506	9528	2	4	7	9	11	13	15	17	20
.98	9550	9572	9594	9616	9638	9661	9683	9705	9727	9750	2	4	7	9	11	13	16	18	20
.99	9772	9795	9817	9840	9863	9886	9908	9931	9954	9977	2	5	7	9	11	14	16	18	20
x	0	1	2	3	4	5	6	7	8	9	1	2	3	4	5	6	7	8	9

TABLE 3 : SINES OF ANGLES (x IN DEGREES) $x \longrightarrow \sin x$

x	·0	·1	.2	.3	·4	.5	.6	·7	.8	.9	1	2	3	4	5	6	7	8	9
											ADD Differences								
0	0.0000	.0017	.0035	.0052	.0070	.0087	.0105	.0122	.0140	.0157	2	3	5	7	9	10	12	14	16
1	0.0175	.0192	.0209	.0227	.0244	.0262	.0279	.0297	.0314	.0332	2	3	5	7	9	10	12	14	16
2	0.0349	.0366	.0384	.0401	.0419	.0436	.0454	.0471	.0488	.0506	2	3	5	7	9	10	12	14	16
3	0.0523	.0541	.0558	.0576	.0593	.0610	.0628	.0645	.0663	.0680	2	3	5	7	9	10	12	14	16
4	0.0698	.0715	.0732	.0750	.0767	.0785	.0802	.0819	.0837	.0854	2	3	5	7	9	10	12	14	16
5	0.0872	.0889	.0906	.0924	.0941	.0958	.0976	.0993	.1011	.1028	2	3	5	7	9	10	12	14	16
6	0.1045	.1063	.1080	.1097	.1115	.1132	.1149	.1167	.1184	.1201	2	3	5	7	9	10	12	14	16
7	0.1219	.1236	.1253	.1271	.1288	.1305	.1323	.1340	.1357	.1374	2	3	5	7	9	10	12	14	16
8	0.1392	.1409	.1426	.1444	.1461	.1478	.1495	.1513	.1530	.1547	2	3	5	7	9	10	12	14	16
9	0.1564	.1582	.1599	.1616	.1633	.1650	.1668	.1685	.1702	.1719	2	3	5	7	9	10	12	14	15
10	0.1736	.1754	.1771	.1788	.1805	.1822	.1840	.1857	.1874	.1891	2	3	5	7	9	10	12	14	15
11	0.1908	.1925	.1942	.1959	.1977	.1994	.2011	.2028	.2045	.2062	2	3	5	7	9	10	12	14	15
12	0.2079	.2096	.2113	.2130	.2147	.2164	.2181	.2198	.2215	.2233	2	3	5	7	9	10	12	14	15
13	0.2250	.2267	.2284	.2300	.2317	.2334	.2351	.2368	.2385	.2402	2	3	5	7	8	10	12	14	15
14	0.2419	.2436	.2453	.2470	.2487	.2504	.2521	.2538	.2554	.2571	2	3	5	7	8	10	12	14	15
15	0.2588	.2605	.2622	.2639	.2656	.2672	.2689	.2706	.2723	.2740	2	3	5	7	8	10	12	13	15
16	0.2756	.2773	.2790	.2807	.2823	.2840	.2857	.2874	.2890	.2907	2	3	5	7	8	10	12	13	15
17	0.2924	.2940	.2957	.2974	.2990	.3007	.3024	.3040	.3057	.3074	2	3	5	7	8	10	12	13	15
18	0.3090	.3107	.3123	.3140	.3156	.3173	.3190	.3206	.3223	.3239	2	3	5	7	8	10	12	13	15
19	0.3256	.3272	.3289	.3305	.3322	.3338	.3355	.3371	.3387	.3404	2	3	5	7	8	10	12	13	15
20	0.3420	.3437	.3453	.3469	.3486	.3502	.3518	.3535	.3551	.3567	2	3	5	7	8	10	11	13	15
21	0.3584	.3600	.3616	.3633	.3649	.3665	.3681	.3697	.3714	.3730	2	3	5	6	8	10	11	13	15
22	0.3746	.3762	.3778	.3795	.3811	.3827	.3843	.3859	.3875	.3891	2	3	5	6	8	10	11	13	15
23	0.3907	.3923	.3939	.3955	.3971	.3987	.4003	.4019	.4035	.4051	2	3	5	6	8	10	11	13	14
24	0.4067	.4083	.4099	.4115	.4131	.4147	.4163	.4179	.4195	.4210	2	3	5	6	8	10	11	13	14
25	0.4226	.4242	.4258	.4274	.4289	.4305	.4321	.4337	.4352	.4368	2	3	5	6	8	9	11	13	14
26	0.4384	.4399	.4415	.4431	.4446	.4462	.4478	.4493	.4509	.4524	2	3	5	5	8	9	11	12	14
27	0.4540	.4555	.4571	.4586	.4602	.4617	.4633	.4648	.4664	.4679	2	3	5	6	8	9	11	12	14
28	0.4695	.4710	.4726	.4741	.4756	.4772	.4787	.4802	.4818	.4833	2	3	5	6	8	9	11	12	14
29	0.4848	.4863	.4879	.4894	.4909	.4924	.4939	.4955	.4970	.4985	2	3	5	6	8	9	11	12	14
30	0.5000	.5015	.5030	.5045	.5060	.5075	.5090	.5105	.5120	.5135	2	3	5	6	8	9	11	12	14
31	0.5150	.5165	.5180	.5195	.5210	.5225	.5240	.5255	.5270	.5284	1	3	4	6	7	9	10	12	13
32	0.5299	.5314	.5329	.5344	.5358	.5373	.5388	.5402	.5417	.5432	1	3	4	6	7	9	10	12	13
33	0.5446	.5461	.5476	.5490	.5505	.5519	.5534	.5548	.5563	.5577	1	3	4	6	7	9	10	12	13
34	0.5592	.5606	.5621	.5635	.5640	.5664	.5678	.5693	.5707	.5721	1	3	4	6	7	9	10	12	13
35	0.5736	.5750	.5764	.5779	.5793	.5807	.5821	.5835	.5850	.5864	1	3	4	6	7	9	10	11	13
36	0.5878	.5892	.5906	.5920	.5934	.5948	.5962	.5976	.5990	.6004	1	3	4	6	7	8	10	11	13
37	0.6018	.6032	.6046	.6060	.6074	.6088	.6101	.6115	.6129	.6143	1	3	4	6	7	8	10	11	12
38	0.6157	.6170	.6184	.6198	.6211	.6225	.6239	.6252	.6266	.6280	1	3	4	5	7	8	10	11	12
39	0.6293	.6307	.6320	.6334	.6347	.6361	.6374	.6388	.6401	.6414	1	3	4	5	7	8	9	11	12
40	0.6428	.6441	.6455	.6468	.6481	.6494	.6508	.6521	.6534	.6547	1	3	4	5	7	8	9	11	12
41	0.6561	.6574	.6587	.6600	.6613	.6626	.6639	.6652	.6665	.6678	1	3	4	5	7	8	9	10	12
42	0.6691	.6704	.6717	.6730	.6743	.6756	.6769	.6782	.6794	.6807	1	3	4	5	6	8	9	10	12
43	0.6820	.6833	.6845	.6858	.6871	.6884	.6896	.6909	.6921	.6934	1	3	4	5	6	8	9	10	11
44	0.6947	.6959	.6972	.6984	.6997	.7009	.7022	.7034	.7046	.7059	1	2	4	5	6	7	9	10	11
x	.0	.1	.2	.3	.4	.5	.6	.7	.8	.9	1	2	3	4	5	6	7	8	9

TABLE 3 (continued): SINES OF ANGLES (x IN DEGREES) $x \longrightarrow sinx$

x	·0	·1	.2	.3	·4	.5	.6	·7	.8	.9	1	2	3	4	5	6	7	8	9
											ADD Differences								
45	0.7071	.7083	.7096	.7108	.7120	.7133	.7145	.7157	.7169	.7181	1	2	4	5	6	7	9	10	11
46	0.7193	.7206	.7218	.7230	.7242	.7254	.7266	.7278	.7290	.7302	1	2	4	5	6	7	8	10	11
47	0.7314	.7325	.7337	.7349	.7361	.7373	.7385	.7396	.7408	.7420	1	2	4	5	6	7	8	9	11
48	0.7431	.7443	.7455	.7466	.7478	.7490	.7501	.7513	.7524	.7536	1	2	3	5	6	7	8	9	10
49	0.7547	.7559	.7570	.7581	.7593	.7604	.7615	.7627	.7638	.7649	1	2	3	5	6	7	8	9	10
50	0.7660	.7672	.7683	.7694	.7705	.7716	.7727	.7738	.7749	.7760	1	2	3	4	6	7	8	9	10
51	0.7771	.7782	.7793	.7804	.7815	.7826	.7837	.7848	.7859	.7869	1	2	3	4	5	7	8	9	10
52	0.7880	.7891	.7902	.7912	.7923	.7934	.7944	.7955	.7965	.7976	1	2	3	4	5	6	7	8	10
53	0.7986	.7997	.8007	.8018	.8028	.8039	.8049	.8059	.8070	.8080	1	2	3	4	5	6	7	8	9
54	0.8090	.8100	.8111	.8121	.8131	.8141	.8151	.8161	.8171	.8181	1	2	3	4	5	6	7	8	9
55	0.8192	.8202	.8211	.8221	.8231	.8241	.8251	.8261	.8271	.8281	1	2	3	4	5	6	7	8	9
56	0.8290	.8300	.8310	.8320	.8329	.8339	.8348	.8358	.8368	.8377	1	2	3	4	5	6	7	8	9
57	0.8387	.8396	.8406	.8415	.8425	.8434	.8443	.8453	.8462	.8471	1	2	3	4	5	6	7	8	8
58	0.8480	.8490	.8499	.8508	.8517	.8526	.8536	.8545	.8554	.8563	1	2	3	4	5	5	6	7	8
59	0.8572	.8581	.8590	.8599	.8607	.8616	.8625	.8634	.8643	.8652	1	2	3	4	4	5	6	7	8
60	0.8660	.8669	.8678	.8686	.8695	.8704	.8712	.8721	.8739	.8738	1	2	3	3	4	5	6	7	8
61	0.8746	.8755	.8763	.8771	.8780	.8788	.8796	.8805	.8813	.8821	1	2	2	3	4	5	6	7	7
62	0.8829	.8838	.8846	.8854	.8862	.8870	.8878	.8886	.8894	.8902	1	2	2	3	4	5	6	6	7
63	0.8910	.8918	.8926	.8934	.8942	.8949	.8957	.8965	.8973	.8980	1	2	2	3	4	5	5	6	7
64	0.8988	.8996	.9003	.9011	.9018	.9026	.9033	.9041	.9048	.9056	1	2	2	3	4	5	5	6	7
65	0.9063	.9070	.9078	.9085	.9092	.9100	.9107	.9114	.9121	.9128	1	1	2	3	4	4	5	6	7
66	0.9135	.9143	.9150	.9157	.9164	.9171	.9178	.9184	.9191	.9198	1	1	2	3	3	4	5	6	6
67	0.9205	.9212	.9219	.9225	.9232	.9239	.9245	.9252	.9259	.9265	1	1	2	3	3	4	5	5	6
68	0.9272	.9278	.9285	.9291	.9298	.9304	.9311	.9317	.9323	.9330	1	1	2	3	3	4	4	5	6
69	0.9336	.9342	.9348	.9354	.9361	.9367	.9373	.9379	.9385	.9391	1	1	2	2	3	4	4	5	6
70	0.9397	.9403	.9409	.9415	.9421	.9426	.9432	.9438	.9444	.9449	1	1	2	2	3	3	4	5	5
71	0.9455	.9461	.9466	.9472	.9478	.9483	.9489	.9494	.9500	.9505	1	1	2	2	3	3	4	4	5
72	0.9511	.9516	.9521	.9527	.9532	.9537	.9542	.9548	.9553	.9558	1	1	2	2	3	3	4	4	5
73	0.9563	.9568	.9573	.9578	.9583	.9588	.9593	.9598	.9603	.9608	0	1	1	2	2	3	3	4	4
74	0.9613	.7617	.9622	.9627	.9632	.9636	.9641	.9646	.9650	.9655	0	1	1	2	2	3	3	4	4
75	0.9659	.9664	.9668	.9673	.9677	.9681	.9686	.9690	.9694	.9699	0	1	1	2	2	3	3	3	4
76	0.9703	.9707	.9711	.9715	.9720	.9724	.9728	.9732	.9736	.9740	0	1	1	2	2	2	3	3	4
77	0.9744	.9748	.9751	.9755	.9759	.9763	.9767	.9770	.9774	.9778	0	1	1	2	2	2	3	3	3
78	0.9781	.9785	.9789	.9792	.9796	.9799	.9803	.9806	.9810	.9813	0	1	1	1	2	2	2	3	3
79	0.9816	.9820	.9823	.9826	.9829	.9833	.9836	.9839	.9842	.9845	0	1	1	1	2	2	2	3	3
80	0.9848	.9851	.9854	.9857	.9860	.9863	.9866	.9869	.9871	.9874	0	1	1	1	1	2	2	2	3
81	0.9877	.9880	.9882	.9885	.9888	.9890	.9893	.9895	.9898	.9900	0	1	1	1	1	2	2	2	2
82	0.9903	.9905	.9907	.9910	.9912	.9914	.9917	.9919	.9921	.9923	0	0	1	1	1	1	2	2	2
83	0.9925	.9928	.9930	.9932	.9934	.9936	.9938	.9940	.9942	.9943	0	0	1	1	1	1	1	2	2
84	0.9945	.9947	.9949	.9951	.9952	.9954	.9956	.9957	.9959	.9960	0	0	1	1	1	1	1	1	2
85	0.9962	.9963	.9965	.9966	.9968	.9969	.9971	.9972	.9973	.9974	0	0	0	1	1	1	1	1	1
86	0.9976	.9977	.9978	.9979	.9980	.9981	.9982	.9983	.9984	.9985	0	0	0	0	1	1	1	1	1
87	0.9986	.9987	.9988	.9989	.9990	.9990	.9991	.9992	.9993	.9993	0	0	0	0	0	0	1	1	1
88	0.9994	.9995	.9995	.9996	.9996	.9997	.9997	.9997	.9998	.9998									
89	0.9998	.9999	.9999	.9999	.9999	1.000	1.000	1.000	1.000	1.000									
x	·0	·1	.2	.3	.4	.5	.6	·7	.8	.9	1	2	3	4	5	6	7	8	9

TABLE 4 : COSINES OF ANGLES (x IN DEGREES) $x \longrightarrow \cos x$

x	·0	·1	.2	.3	·4	.5	.6	·7	.8	.9	SUBTRACT Differences								
											1	2	3	4	5	6	7	8	9
0	1.0000	1.000	1.000	1.000	1.000	1.000	.9999	.9999	.9999	.9999									
1	0.9998	.9998	.9998	.9997	.9997	.9997	.9996	.9996	.9995	.9995									
2	0.9994	.9993	.9993	.9992	.9991	.9990	.9990	.9989	.9988	.9987	0	0	0	0	0	0	1	1	1
3	0.9986	.9985	.9984	.9983	.9982	.9981	.9980	.9979	.9978	.9977	0	0	0	0	1	1	1	1	1
4	0.9976	.9974	.9973	.9972	.9971	.9969	.9968	.9966	.9965	.9963	0	0	0	1	1	1	1	1	1
5	0.9962	.9960	.9959	.9957	.9956	.9954	.9952	.9951	.9949	.9947	0	0	1	1	1	1	1	1	2
6	0.9945	.9943	.9942	.9940	.9938	.9936	.9934	.9932	.9930	.9928	0	0	1	1	1	1	1	2	2
7	0.9925	.9923	.9921	.9919	.9917	.9914	.9912	.9910	.9907	.9905	0	0	1	1	1	1	2	2	2
8	0.9903	.9900	.9898	.9895	.9893	.9890	.9888	.9885	.9882	.9880	0	1	1	1	1	2	2	2	2
9	0.9877	.9874	.9871	.9869	.9866	.9863	.9860	.9857	.8954	.9851	0	1	1	1	1	2	2	2	3
10	0.9848	.9845	.9842	.9839	.9836	.9833	.9829	.9826	.9823	.9820	0	1	1	1	2	2	2	3	3
11	0.9816	.9813	.9810	.9806	.9803	.9799	.9796	.9792	.9789	.9785	0	1	1	1	2	2	2	3	3
12	0.9781	.9778	.9774	.9770	.9767	.9763	.9759	.9755	.9751	.9748	0	1	1	2	2	2	3	3	3
13	0.9744	.9740	.9736	.9732	.9728	.9724	.9720	.9715	.9711	.9707	0	1	1	2	2	2	3	3	4
14	0.9703	.9699	.9694	.9690	.9686	.9681	.9677	.9673	.9668	.9664	0	1	1	2	2	3	3	3	4
15	0.9659	.9655	.9650	.9646	.9641	.9636	.9632	.9627	.9622	.9617	0	1	1	2	2	3	3	4	4
16	0.9613	.9608	.9603	.9598	.9593	.9588	.9583	.9578	.9573	.9568	0	1	1	2	2	3	3	4	4
17	0.9563	.9558	.9553	.9548	.9542	.9537	.9532	.9527	.9521	.9516	1	1	2	2	3	3	4	4	5
18	0.9511	.9505	.9500	.9494	.9489	.9483	.9478	.9472	.9466	.9461	1	1	2	2	3	3	4	4	5
19	0.9455	.9449	.9444	.9438	.9432	.9426	.9421	.9415	.9409	.9403	1	1	2	2	3	3	4	5	5
20	0.9397	.9391	.9385	.9379	.9373	.9367	.9361	.9354	.9348	.9342	1	1	2	2	3	4	4	5	6
21	0.9336	.9330	.9323	.9317	.9311	.9304	.9298	.9291	.9285	.9278	1	1	2	3	3	4	4	5	6
22	0.9272	.9265	.9259	.9252	.9245	.9239	.9232	.9225	.9219	.9212	1	1	2	3	3	4	5	5	6
23	0.9205	.9198	.9191	.9184	.9178	.9171	.9164	.9157	.9150	.9143	1	1	2	3	3	4	5	6	6
24	0.9135	.9128	.9121	.9114	.9107	.9100	.9092	.9085	.9078	.9070	1	1	2	3	4	4	5	6	7
25	0.9063	.9056	.9048	.9041	.9033	.9026	.9018	.9011	.9003	.8996	1	2	2	3	4	5	5	6	7
26	0.8988	.8980	.8973	.8965	.8957	.8949	.8942	.8934	.8926	.8918	1	2	2	3	4	5	5	6	7
27	0.8910	.8902	.8894	.8886	.8878	.8870	.8862	.8854	.8846	.8838	1	2	2	3	4	5	6	6	7
28	0.8829	.8821	.8813	.8805	.8796	.8788	.8780	.8771	.8763	.8755	1	2	2	3	4	5	6	7	7
29	0.8746	.8738	.8729	.8721	.8712	.8704	.8695	.8686	.8678	.8669	1	2	3	3	4	5	6	7	8
30	0.8660	.8652	.8643	.8634	.8625	.8616	.8607	.8599	.8590	.8581	1	2	3	4	4	5	6	7	8
31	0.8572	.8563	.8554	.8545	.8536	.8526	.8517	.8508	.8499	.8490	1	2	3	4	5	5	6	7	8
32	0.8480	.8471	.8462	.8453	.8443	.8434	.8425	.8415	.8406	.8396	1	2	3	4	5	6	7	8	8
33	0.8387	.8377	.8368	.8358	.8348	.8339	.8329	.8320	.8310	.8300	1	2	3	4	5	6	7	8	9
34	0.8290	.8281	.8271	.8261	.8251	.8241	.8231	.8221	.8211	.8202	1	2	3	4	5	6	7	8	9
35	0.8192	.8181	.8171	.8161	.8151	.8141	.8131	.8212	.8111	.8100	1	2	3	4	5	6	7	8	9
36	0.8090	.8080	.8070	.8059	.8049	.8039	.8028	.8018	.8007	.7997	1	2	3	4	5	6	7	8	9
37	0.7986	.7976	.7965	.7955	.7944	.7934	.7923	.7912	.7902	.7891	1	2	3	4	5	6	7	8	10
38	0.7880	.7869	.7859	.7848	.7837	.7826	.7815	.7804	.7793	.7782	1	2	3	4	5	6	7	9	10
39	0.7771	.7760	.7749	.7738	.7727	.7716	.7705	.7694	.7683	.7672	1	2	3	4	6	7	8	9	10
40	0.7660	.7649	.7638	.7627	.7615	.7604	.7593	.7581	.7570	.7559	1	2	3	5	6	7	8	9	10
41	0.7547	.7536	.7524	.7513	.7501	.7490	.7478	.7466	.7455	.7443	1	2	3	5	6	7	8	9	10
42	0.7431	.7420	.7408	.7396	.7385	.7373	.7361	.7349	.7337	.7325	1	2	4	5	6	7	8	9	11
43	0.7314	.7302	.7290	.7278	.7266	.7254	.7242	.7230	.7218	.7206	1	2	4	5	6	7	8	10	11
44	0.7193	.7181	.7169	.7157	.7145	.7133	.7120	.7108	.7096	.7083	1	2	4	5	6	7	9	10	11
x	.0	·1	.2	.3	.4	.5	.6	·7	.8	.9	1	2	3	4	5	6	7	8	9

TABLE 4 (continued): COSINES OF ANGLES (x IN DEGREES) $x \longrightarrow cos x$

x	·0	·1	.2	.3	·4	.5	.6	·7	.8	.9	SUBTRACT Differences 1	2	3	4	5	6	7	8	9
45	0.7071	.7059	.7046	.7034	.7022	.7009	.6997	.6984	.6972	.6959	1	2	4	5	6	7	9	10	11
46	0.6947	.6934	.6794	.6789	.6896	.6884	.6871	.6858	.6845	.6833	1	3	4	5	6	8	9	10	11
47	0.6820	.6807	.6794	.6782	.6769	.6756	.6743	.6730	.6717	.6704	1	3	4	5	6	8	9	10	12
48	0.6891	.6678	.6665	.6652	.6639	.6626	.6613	.6600	.6587	.6574	1	3	4	5	7	8	9	10	12
49	0.6561	.6547	.6534	.6521	.6508	.6494	.6481	.6468	.6455	.6441	1	3	4	5	7	8	9	11	12
50	0.6428	.6414	.6401	.6388	.6374	.6361	.6347	.6334	.6320	.6307	1	3	4	5	7	8	9	11	12
51	0.6293	.6280	.6266	.6252	.6239	.6225	.6211	.6198	.6184	.6170	1	3	4	5	7	8	10	11	12
52	0.6157	.6143	.6129	.6115	.6101	.6088	.6074	.6060	.6046	.6032	1	3	4	5	7	8	10	11	12
53	0.6018	.6004	.5990	.5976	.5962	.5948	.5934	.5920	.5906	.5892	1	3	4	6	7	8	10	11	13
54	0.5878	.5864	.5850	.5830	.5821	.5807	.5793	.5779	.5764	.5750	1	3	4	6	7	9	10	11	13
55	0.5736	.5721	.5707	.5693	.5678	.5664	.5650	.5635	.5621	.5606	1	3	4	6	7	9	10	12	13
56	0.5592	.5577	.5563	.5548	.5534	.5519	.5505	.5490	.5476	.5461	1	3	4	6	7	9	10	12	13
57	0.5446	.5432	.5417	.5402	.5388	.5373	.5358	.5344	.5329	.5314	1	3	4	6	7	9	10	12	13
58	0.5299	.5284	.5270	.5255	.5240	.5225	.5210	.5195	.5180	.5165	1	3	4	6	7	9	10	12	13
59	0.5150	.5135	.5120	.5150	.5090	.5075	.5060	.5045	.5030	.5015	2	3	5	6	8	9	11	12	14
60	0.5000	.4985	.4970	.4955	.4939	.4924	.4909	.4894	.4879	.4863	2	3	5	6	8	9	11	12	14
61	0.4848	.4833	.4818	.4802	.4787	.4772	.4756	.4741	.4726	.4710	2	3	5	6	8	9	11	12	14
62	0.4695	.4679	.4664	.4648	.4633	.4617	.4602	.4586	.4571	.4555	2	3	5	6	8	9	11	12	14
63	0.4540	.4524	.4509	.4493	.4478	.4462	.4446	.4431	.4415	.4399	2	3	5	6	8	9	11	12	14
64	0.4384	.4368	.4352	.4337	.4321	.4305	.4289	.4274	.4258	.4242	2	3	5	6	8	9	11	13	14
65	0.4226	.4210	.4195	.4179	.4163	.4147	.4131	.4115	.4099	.4083	2	3	5	6	8	10	11	13	14
66	0.4067	.4051	.4035	.4019	.4003	.3987	.3971	.3955	.3939	.3923	2	3	5	6	8	10	11	13	14
67	0.3907	.3891	.3875	.3859	.3843	.3827	.3811	.3795	.3778	.3762	2	3	5	6	8	10	11	13	15
68	0.3746	.3730	.3714	.3697	.3681	.3665	.3649	.3633	.3616	.3600	2	3	5	6	8	10	11	13	15
69	0.3584	.3567	.3551	.3535	.3518	.3502	.3486	.3469	.3453	.3437	2	3	5	6	8	10	11	13	15
70	0.3420	.3404	.3387	.3371	.3355	.3338	.3322	.3305	.3289	.3272	2	3	5	7	8	10	12	13	15
71	0.3256	.3239	.3223	.3206	.3190	.3173	.3156	.3140	.3123	.3107	2	3	5	7	8	10	12	13	15
72	0.3090	.3074	.3057	.3040	.3024	.3007	.2990	.2974	.2957	.2940	2	3	5	7	8	10	12	13	15
73	0.2924	.2907	.2890	.2874	.2857	.2840	.2823	.2807	.2790	.2773	2	3	5	7	8	10	12	13	15
74	0.2756	.2740	.2723	.2706	.2689	.2672	.2656	.2639	.2622	.2605	2	3	5	7	8	10	12	13	15
75	0.2588	.2571	.2554	.2538	.2521	.2504	.2487	.2470	.2453	.2436	2	3	5	7	8	10	12	13	15
76	0.2419	.2402	.2385	.2368	.2351	.2334	.2317	.2300	.2284	.2267	2	3	5	7	8	10	12	13	15
77	0.2250	.2233	.2215	.2198	.2181	.2164	.2147	.2130	.2113	.2096	2	3	5	7	9	10	12	13	15
78	0.2079	.2062	.2045	.2028	.2011	.1994	.1977	.1959	.1942	.1925	2	3	5	7	9	10	12	13	15
79	0.1908	.1891	.1874	.1857	.1840	.1822	.1805	.1788	.1771	.1754	2	3	5	7	9	10	12	13	15
80	0.1736	.1719	.1702	.1685	.1668	.1650	.1633	.1616	.1599	.1582	2	3	5	7	9	10	12	13	15
81	0.1564	.1547	.1530	.1513	.1495	.1478	.1461	.1444	.1426	.1409	2	3	5	7	9	10	12	13	16
82	0.1392	.1374	.1357	.1340	.1323	.1305	.1288	.1271	.1253	.1236	2	3	5	7	9	10	12	13	16
83	0.1219	.1201	.1184	.1167	.1149	.1132	.1115	.1097	.1080	.1063	2	3	5	7	9	10	12	13	16
84	0.1045	.1028	.1011	.0471	.0976	.0958	.0941	.0924	.0906	.0889	2	3	5	7	9	10	12	13	16
85	0.0872	.0854	.0837	.0819	.0802	.0785	.0767	.0750	.0732	.0715	2	3	5	7	9	10	12	13	16
86	0.0698	.0680	.0663	.0645	.0628	.0610	.0593	.0576	.0558	.0541	2	3	5	7	9	10	12	13	16
87	0.0523	.0506	.0488	.0471	.0454	.0436	.0419	.0401	.0384	.0366	2	3	5	7	9	10	12	13	16
88	0.0349	.0332	.0314	.0297	.0279	.0262	.0244	.0227	.0209	.0192	2	3	5	7	9	10	12	13	16
89	0.0175	.0157	.0140	.0122	.0105	.0087	.0070	.0052	.0035	.0017	2	3	5	7	9	10	12	13	16
x	·0	·1	.2	.3	.4	.5	.6	·7	.8	.9	1	2	3	4	5	6	7	8	9

TABLE 5 : TANGENTS OF ANGLES (x IN DEGREES) $x \longrightarrow tanx$

											ADD Differences								
x	·0	·1	.2	.3	·4	.5	.6	·7	.8	.9	1	2	3	4	5	6	7	8	9
0	0.0000	.0017	.0035	.0052	.0070	.0087	.0105	.0122	.0140	.0157	2	3	5	7	9	10	12	14	16
1	0.0175	.0192	.0209	.0227	.0244	.0262	.0279	.0297	.0314	.0332	2	3	5	7	9	10	12	14	16
2	0.0349	.0367	.0384	.0402	.0419	.0436	.0454	.0472	.0489	.0507	2	3	5	7	9	10	12	14	16
3	0.0524	.0542	.0559	.0577	.0594	.0612	.0629	.0647	.0664	.0682	2	4	5	7	9	11	12	14	16
4	0.0699	.0717	.0734	.0752	.0769	.0787	.0805	.0822	.0840	.0857	2	4	5	7	9	11	12	14	16
5	0.0875	.0892	.0910	.0928	.0945	.0963	.0981	.0998	.1016	.1033	2	4	5	7	9	11	12	14	16
6	0.1051	.1069	.1086	.1104	.1122	.1139	.1157	.1175	.1192	.1210	2	4	5	7	9	11	12	14	16
7	0.1228	.1246	.1263	.1281	.1299	.1317	.1334	.1352	.1370	.1388	2	4	5	7	9	11	12	14	16
8	0.1405	.1423	.1441	.1459	.1477	.1495	.1512	.1530	.1548	.1566	2	4	5	7	9	11	12	14	16
9	0.1584	.1602	.1620	.1638	.1655	.1673	.1691	.1709	.1727	.1745	2	4	5	7	9	11	13	14	16
10	0.1763	.1781	.1799	.1817	.1835	.1853	.1871	.1890	.1908	.1926	2	4	5	7	9	11	13	14	16
11	0.1944	.1962	.1980	.1998	.2016	.2035	.2053	.2071	.2089	.2107	2	4	5	7	9	11	13	15	16
12	0.2126	.2144	.2162	.2180	.2199	.2217	.2235	.2254	.2272	.2290	2	4	5	7	9	11	13	15	16
13	0.2309	.2327	.2345	.2364	.2382	.2401	.2419	.2438	.2456	.2475	2	4	6	7	9	11	13	15	17
14	0.2493	.2512	.2530	.2549	.2568	.2586	.2605	.2623	.2642	.2661	2	4	6	7	9	11	13	15	17
15	0.2679	.2698	.2717	.2736	.2754	.2773	.2792	.2811	.2830	.2849	2	4	6	8	9	11	13	15	17
16	0.2867	.2886	.2905	.2924	.2943	.2962	.2981	.3000	.3019	.3620	2	4	6	8	9	11	13	15	17
17	0.3057	.3076	.3096	.3115	.3134	.3153	.3172	.3191	.3211	.3230	2	4	6	8	10	12	13	15	17
18	0.3249	.3269	.3288	.3307	.3327	.3346	.3365	.3385	.3404	.3424	2	4	6	8	10	12	14	16	17
19	0.3443	.3463	.3482	.3502	.3522	.3541	.3561	.3581	.3600	.3620	2	4	6	8	10	12	14	16	18
20	0.3640	.3659	.3679	.3699	.3719	.3739	.3759	.3779	.3799	.3819	2	4	6	8	10	12	14	16	18
21	0.3839	.3859	.3879	.3899	.3919	.3939	.3959	.3697	.4000	.4020	2	4	6	8	10	12	14	16	18
22	0.4040	.4061	.4081	.4101	.4122	.4142	.4163	.4183	.4204	.4224	2	4	6	8	10	12	14	16	18
23	0.4245	.4265	.4286	.4307	.4327	.4348	.4369	.4390	.4411	.4431	2	4	6	8	10	12	15	17	19
24	0.4452	.4473	.4494	.4515	.4536	.4557	.4578	.4599	.4621	.4642	2	4	6	8	11	13	15	17	19
25	0.4665	.4684	.4706	.4727	.4748	.4770	.4791	.4183	.4834	.4856	2	4	6	9	11	13	15	17	19
26	0.4877	.4899	.4921	.4942	.4964	.4986	.5008	.5029	.5051	.5073	2	4	7	9	11	13	15	17	20
27	0.5095	.5117	.5139	.5161	.5184	.5206	.5228	.5250	.5272	.5295	2	4	7	9	11	13	16	18	20
28	0.5317	.5340	.5362	.5384	.5407	.5430	.5452	.5475	.5498	.5520	2	5	7	9	11	14	16	18	20
29	0.5543	.5566	.5589	.5612	.5635	.5658	.5681	.5704	.5727	.5750	2	5	7	9	12	14	16	18	21
30	0.5774	.5797	.5820	.5844	.5867	.5890	.5914	.5938	.5961	.5135	2	5	7	9	12	14	16	19	21
31	0.6009	.6032	.6056	.6080	.6104	.6128	.6152	.6176	.6200	.6224	2	5	7	10	12	14	17	19	22
32	0.6249	.6273	.6297	.6322	.6346	.6371	.6395	.6420	.6445	.6469	2	5	7	10	12	15	17	20	22
33	0.6494	.6519	.6544	.6569	.6594	.6619	.6644	.6669	.6694	.6720	3	5	8	10	13	15	18	20	23
34	0.6745	.6771	.6796	.6822	.6847	.6873	.6899	.6924	.6950	.6976	3	5	8	10	13	15	18	21	23
35	0.7002	.7028	.7054	.7080	.7107	.7133	.7159	.7186	.7212	.7239	3	5	8	11	13	16	18	21	24
36	0.7265	.7292	.7319	.7346	.7373	.7400	.7427	.7454	.7481	.7508	3	5	8	11	14	16	19	22	24
37	0.7536	.7563	.7590	.7618	.7646	.7673	.7701	.7729	.7757	.7785	3	6	8	11	14	17	19	22	25
38	0.7813	.7841	.7869	.7898	.7926	.7954	.7983	.8012	.8040	.8069	3	6	9	11	14	17	20	23	26
39	0.8098	.8127	.8156	.8185	.8214	.8243	.8273	.8302	.8332	.8361	3	6	9	12	15	18	21	23	26
40	0.8391	.8421	.8451	.8481	.8511	.8541	.8571	.8601	.8632	.8662	3	6	9	12	15	18	21	24	27
41	0.8693	.8724	.8754	.8785	.8816	.8847	.8878	.8910	.8941	.8972	3	6	9	12	16	19	22	25	28
42	0.9004	.9036	.9067	.9099	.9131	.9163	.9195	.9228	.9260	.8662	3	6	10	13	16	19	22	26	29
43	0.9325	.9358	.9391	.9424	.9457	.9490	.9523	.9556	.9590	.9623	3	7	10	13	17	20	23	27	30
44	0.9657	.9691	.9725	.9759	.9793	.9827	.9861	.9896	.9930	.9965	3	7	10	14	17	21	24	27	31
x	.0	.1	.2	.3	.4	.5	.6	.7	.8	.9	1	2	3	4	5	6	7	8	9

TABLE 5 (continued): TANGENTS OF ANGLES (x IN DEGREES) $x \longrightarrow tan x$

											ADD Differences								
x	·0	·1	.2	.3	·4	.5	.6	·7	.8	.9	1	2	3	4	5	6	7	8	9
45	1.000	1.003	1.007	1.011	1.014	1.018	1.021	1.025	1.028	1.032	0	1	1	1	2	2	2	3	3
46	1.036	1.039	1.043	1.046	1.050	1.054	1.057	1.061	1.065	1.069	0	1	1	1	2	2	3	3	3
47	1.072	1.076	1.080	1.084	1.087	1.091	1.095	1.099	1.103	1.107	0	1	1	2	2	2	3	3	3
48	1.111	1.115	1.159	1.122	1.126	1.130	1.134	1.138	1.142	1.146	0	1	1	2	2	2	3	3	3
49	1.150	1.154	1.159	1.163	1.167	1.171	1.175	1.179	1.183	1.188	0	1	1	2	2	2	3	3	3
50	1.192	1.196	1.200	1.205	1.209	1.213	1.217	1.222	1.226	1.230	0	1	1	2	2	3	3	4	4
51	1.235	1.239	1.244	1.248	1.253	1.257	1.262	1.266	1.271	1.275	0	1	1	2	2	3	3	4	4
52	1.280	1.285	1.289	1.295	1.299	1.303	1.308	1.313	1.317	1.322	0	1	1	2	2	3	3	4	4
53	1.327	1.332	1.337	1.342	1.347	1.351	1.356	1.361	1.366	1.371	0	1	1	2	2	3	3	4	4
54	1.376	1.381	1.387	1.392	1.397	1.402	1.407	1.412	1.418	1.423	1	1	2	2	3	3	4	4	5
55	1.428	1.433	1.439	1.444	1.450	1.455	1.460	1.466	1.471	1.477	1	1	2	2	3	3	4	4	5
56	1.483	1.488	1.494	1.499	1.505	1.511	1.517	1.522	1.528	1.534	1	1	2	2	3	3	4	4	5
57	1.540	1.546	1.552	1.558	1.564	1.570	1.576	1.582	1.588	1.594	1	1	2	2	3	4	4	4	5
58	1.600	1.607	1.613	1.619	1.625	1.632	1.638	1.645	1.651	1.658	1	1	2	3	3	4	4	4	6
59	1.664	1.671	1.678	1.684	1.691	1.698	1.704	1.711	1.718	1.725	1	1	2	3	3	4	4	4	6
60	1.732	1.739	1.746	1.753	1.760	1.767	1.775	1.782	1.789	1.797	1	1	2	3	4	4	5	6	6
61	1.804	1.811	1.819	1.827	1.834	1.842	1.849	1.857	1.865	1.873	1	2	2	3	4	5	5	6	7
62	1.881	1.889	1.897	1.905	1.913	1.921	1.929	1.937	1.946	1.954	1	2	2	3	4	5	6	7	7
63	1.963	1.971	1.980	1.988	1.997	2.006	2.014	2.023	2.032	2.041	1	2	3	4	4	5	6	7	8
64	2.050	2.059	2.069	2.078	2.087	2.097	2.106	2.116	2.125	2.135	1	2	3	4	5	6	7	8	8
65	2.145	2.154	2.164	2.174	2.184	2.194	2.204	2.215	2.225	2.236	1	2	3	4	5	6	7	8	9
66	2.246	2.257	2.267	2.278	2.289	2.300	2.311	2.322	2.333	2.344	1	2	3	4	5	7	8	9	10
67	2.356	2.367	2.379	2.391	2.402	2.414	2.426	2.438	2.450	2.463	1	3	4	5	6	7	8	10	11
68	2.475	2.488	2.500	2.513	2.526	2.539	2.552	2.565	2.578	2.592	1	3	4	5	7	8	9	10	12
69	2.605	2.619	2.633	2.278	2.660	2.675	2.689	2.703	2.718	2.733	1	3	4	6	7	9	10	11	13
70	2.747	2.762	2.778	2.793	2.808	2.824	2.840	2.856	2.872	2.888	2	3	5	6	8	9	11	13	14
71	2.904	2.921	2.937	2.954	2.971	2.989	3.006	3.024	3.042	3.060	2	3	5	7	9	10	12	14	16
72	3.078	3.096	3.115	3.133	3.152	3.172	3.191	3.211	3.230	3.251	2	4	6	8	10	12	14	15	17
73	3.271	3.291	3.312	3.333	3.354	3.376	3.398	3.420	3.442	3.465	2	4	6	9	11	13	16	17	19
74	3.487	3.511	3.534	3.558	3.582	3.606	3.630	3.655	3.681	3.706	2	5	7	10	12	15	17	20	22
75	3.271	3.758	3.785	3.812	3.839	3.867	3.895	3.923	3.952	3.981	3	6	8	11	14	17	20	22	25
76	4.011	4.041	4.071	4.102	4.134	4.165	4.198	4.230	4.264	4.297	3	6	10	13	16	19	22	26	29
77	4.331	4.366	4.402	4.437	4.474	4.511	4.548	4.586	4.625	4.665									
78	4.705	4.745	4.787	4.829	4.872	4.915	4.959	5.005	5.050	5.097									
79	5.145	5.193	5.242	5.292	5.343	5.396	5.449	5.503	5.558	5.614									
80	5.671	5.730	5.789	5.850	5.912	5.976	6.041	6.107	6.174	6.243									
81	6.314	6.386	6.460	6.535	6.612	6.691	6.772	6.855	6.940	7.026									
82	7.115	7.207	7.300	7.396	7.495	7.596	7.700	7.806	7.916	8.028									
83	8.144	8.264	8.386	8.513	8.643	8.777	8.915	9.058	9.205	9.357									
84	9.514	9.677	9.845	10.02	10.20	10.39	10.58	10.78	10.99	11.20									
85	11.43	11.66	11.91	12.16	12.43	12.71	13.00	13.30	13.62	13.95									
86	14.30	14.67	15.06	15.46	15.89	16.35	16.83	17.34	17.89	18.46									
87	19.08	19.74	20.45	21.20	22.02	22.90	23.86	24.90	26.03	27.27									
88	28.64	30.14	31.82	33.69	35.80	38.19	40.92	44.07	47.74	52.08									
89	57.29	63.66	71.62	81.85	95.46	114.6	143.2	191.0	286.5	573.0									
x	·0	·1	.2	.3	·4	.5	.6	·7	.8	.9	1	2	3	4	5	6	7	8	9

TABLE 6: LOGARITHMS OF SINES OF ANGLES (x IN DEGREES) $x \longrightarrow \text{Log} \sin x$

x	·0	·1	.2	.3	·4	.5	.6	·7	.8	.9	1	2	3	4	5	6	7	8	9
											ADD Differences								
0	− ∞	$\bar{3}$.242	.543	.719	.844	$\bar{3}$.941	$\bar{2}$.020	.087	.145	.196									
1	$\bar{2}$.242	.283	.321	.356	.388	.418	.446	.472	.497	.521									
2	$\bar{2}$.543	.564	.584	.603	.622	.640					2	4	6	8	10	11	13	15	17
						.640	.657	.673	.689	.704	2	3	5	6	8	10	11	13	14
3	$\bar{2}$.719	.733	.747	.760	.773	.786	.798	.810	.821	.833	1	2	4	5	6	7	8	10	11
4	$\bar{2}$.844	.854	.865	.875	.885	.895					1	2	3	4	5	5	7	8	9
						.895	.904	.913	.923	.932	1	2	3	4	5	5	6	7	8
5	$\bar{2}$.940	.949	.957	.966	.974	.982	.989	$\bar{2}$.997	$\bar{1}$.005	.012	1	2	2	3	4	5	6	6	7
6	$\bar{1}$.019	.026	.033	.040	.047	.054	.060	.067	.073	.080	1	1	2	3	4	4	5	6	6
7	$\bar{1}$.0859	0920	0981								6	12	18	24	31	37	43	49	45
			0981	1040	1099	1157	1214	1271			6	12	17	23	29	35	41	46	52
								1271	1326	1381	6	11	17	22	28	33	39	44	50
8	$\bar{1}$.1436	1489	1542	1594	1646	1697					5	10	16	21	26	31	36	42	47
						1697	1747	1797	1847	1895	5	10	15	20	25	29	34	39	44
9	$\bar{1}$.1943	1991	2038	2085	2131						5	9	14	19	24	28	33	38	42
					2131	2176	2221	2266	2310	2353	4	9	13	18	22	26	31	35	40
10	$\bar{1}$.2397	2439	2482	2524	2565	2606					4	8	13	17	21	25	29	34	38
						2006	2647	2687	2727	2767	4	8	12	16	20	24	28	32	36
11	$\bar{1}$.2806	2845	2883	2921	2959	2997		3070	3107	3143	4	8	11	15	19	23	27	30	34
						2997	3034				4	7	11	14	18	22	25	29	32
12	$\bar{1}$.3179	3214	3250	3284	3319	3353	3387	3421	3455	3488	4	7	11	14	18	21	25	28	32
					3319	3353	3387				3	7	10	14	17	20	24	27	31
13	$\bar{1}$.3527	3554	3586	3618	3650	3682	3713	3745	3775	3806	3	6	10	13	16	19	22	26	29
						3682	3713				3	6	9	12	16	19	22	25	28
14	$\bar{1}$.3837	3867	3897	3927	3957	3986	4015	4044	4073	4102	3	6	9	12	15	17	20	23	26
15	$\bar{1}$.4130	4158	4186	4214	4242	4269	4296	4323	4350	4377	3	5	8	11	14	16	19	22	24
16	$\bar{1}$.4403	4430	4456	4482	4508	4533	4559	4584	4609	4634	3	5	8	10	13	16	18	21	23
17	$\bar{1}$.4659	4684	4709	4733	4757	4781	4805	4829	4853	4876	2	5	7	10	12	14	17	19	22
18	$\bar{1}$.4900	4923	4946	4969	4992	5015	5037	5060	5082	5104	2	4	7	9	12	14	16	18	21
19	$\bar{1}$.5126	5148	5170	5192	5213	5235	5256	5278	5299	5320	2	4	7	9	11	13	15	21	20
20	$\bar{1}$.5341	5361	5382	5402	5423	5443	5463	5484	5504	5523	2	4	6	8	10	12	14	16	18
21	$\bar{1}$.5543	5563	5583	5602	5621	5641	5660	5679	5698	5717	2	4	6	8	10	11	13	15	17
22	$\bar{1}$.5736	5754	5773	5792	5810	5828	5847	5865	5883	5901	2	4	5	7	9	11	13	14	16
23	$\bar{1}$.5919	5937	5954	5972	5990	6007	6024	6042	6059	6076	2	3	5	7	9	10	12	14	15
24	$\bar{1}$.6093	6110	6127	6144	6161	6177	6194	6210	6227	6243	2	3	5	7	9	10	12	14	15
25	$\bar{1}$.6259	6276	6292	6308	6324	6340	6356	6371	6387	6403	2	3	5	6	8	10	11	12	14
26	$\bar{1}$.6418	6434	6449	6465	6480	6495	6510	6526	6541	6556	2	3	5	6	8	9	11	12	14
27	$\bar{1}$.6570	6585	6600	6615	6629	6644	6659	6673	6687	6702	2	3	5	6	8	9	11	12	14
28	$\bar{1}$.6716	6730	6744	6759	6773	6787	6801	6814	6828	6842	1	3	4	6	7	8	10	11	13
29	$\bar{1}$.6856	6869	6883	6896	6910	6923	6937	6950	6963	6977	1	3	4	5	7	8	9	10	12
30	$\bar{1}$.6990	7003	7016	7029	7042	7055	7068	7080	7093	7106	1	3	4	5	7	8	9	10	12
31	$\bar{1}$.7118	7131	7144	7156	7168	7181	7193	7205	7218	7230	1	2	4	5	6	7	8	10	11
32	$\bar{1}$.7242	7254	7266	7278	7290	7302	7314	7326	7338	7349	1	2	4	5	6	7	8	10	11
33	$\bar{1}$.7361	7373	7384	7396	7407	7419	7430	7442	7453	7464	1	2	4	5	6	7	8	10	11
34	$\bar{1}$.7476	7487	7498	7509	7520	7531	7542	7553	7564	7575	1	2	3	4	6	7	8	9	10
35	$\bar{1}$.7586	7597	7609	7618	7629	7640	7650	7661	7671	7682	1	2	3	4	6	7	8	9	10
36	$\bar{1}$.7692	7703	7713	7723	7734	7744	7754	7764	7774	7785	1	2	3	4	5	6	7	8	9
37	$\bar{1}$.7795	7805	7815	7825	7835	7844	7854	7864	7874	7884	1	2	3	4	5	6	7	8	9
38	$\bar{1}$.7893	7903	7913	7922	7932	7941	7951	7960	7970	7979	1	2	3	4	5	6	7	8	9
39	$\bar{1}$.7989	7998	8007	8017	8026	8035	8044	8053	8063	8072	1	2	3	4	5	6	6	7	8
40	$\bar{1}$.8081	8090	8099	8108	8117	8125	8134	8143	8152	8161	1	2	3	4	5	5	6	7	8
x	.0	.1	.2	.3	.4	.5	.6	.7	.8	.9	1	2	3	4	5	6	7	8	9

$\text{Log cosec}x = -\text{Log} \sin x$

TABLE 6 (continued): LOGARITHMS OF SINES OF ANGLES (x IN DEGREES) $x \longrightarrow \text{Log} \sin x$

x	·0	·1	.2	.3	·4	.5	.6	·7	.8	.9	1	2	3	4	5	6	7	8	9
											ADD Differences								
41	$\bar{1}$.8169	8178	8187	8195	8204	8213	8221	8230	8238	8247	1	2	2	4	5	5	6	7	8
42	$\bar{1}$.8255	8264	8272	8280	8289	8297	8305	8313	8322	8330	1	2	2	3	4	5	6	6	7
43	$\bar{1}$.8338	8346	8354	8362	8370	8378	8386	8394	8302	8410	1	1	2	3	4	5	6	6	7
44	$\bar{1}$.8418	8426	8433	8441	8449	8457	8464	8472	8480	8487	1	2	2	3	4	5	6	6	7
45	$\bar{1}$.8495	8502	8510	8517	8525	8532	8540	8547	8555	8562	1	1	2	3	4	4	5	6	6
46	$\bar{1}$.8569	8577	8584	8591	8598	8606	8613	8620	8627	8634	1	1	2	3	4	4	5	6	6
47	$\bar{1}$.8641	8648	8655	8662	8669	8676	8683	8690	8697	8704	1	1	2	3	4	4	5	6	6
48	$\bar{1}$.8711	8718	8724	8731	8738	8745	8751	8758	8765	8771	1	1	2	3	4	4	5	6	6
49	$\bar{1}$.8778	8784	8791	8797	8804	8810	8817	8823	8830	8836	1	1	2	2	4	4	4	5	5
50	$\bar{1}$.8843	8849	8855	8862	8868	8874	8880	8887	8893	8899	1	1	2	2	3	4	4	5	5
51	$\bar{1}$.8905	8911	8917	8923	8929	8935	8941	8947	8953	8959	1	1	2	2	3	4	4	5	5
52	$\bar{1}$.8965	8971	8977	8983	8989	8995	9000	9006	9012	9018	1	1	2	2	3	4	4	5	5
53	$\bar{1}$.9023	9029	9035	9041	9046	9052	9057	9063	9069	9074	1	1	2	2	3	4	4	5	5
54	$\bar{1}$.9080	9085	9091	9096	9101	9107	9112	9118	9123	9128	1	1	2	2	3	3	4	4	5
55	$\bar{1}$.9134	9139	9144	9149	9155	9160	9165	9170	9175	9181	1	1	2	2	3	3	4	4	5
56	$\bar{1}$.9186	9191	9196	9201	9206	9211	9216	9221	9226	9231	1	1	2	2	3	3	4	4	5
57	$\bar{1}$.9236	9241	9246	9251	9255	9260	9265	9270	9275	9279	1	1	2	2	3	3	4	4	5
58	$\bar{1}$.9284	9289	9294	9298	9303	9308	0312	9317	9322	9326	1	1	2	2	3	3	4	4	5
59	$\bar{1}$.9331	9335	9340	9344	9349	9353	9358	9362	9367	9371	0	1	1	2	2	2	3	3	4
60	$\bar{1}$9375	9380	9384	9388	9393	9397	9401	9406	9410	9414	0	1	1	2	2	2	3	3	4
61	$\bar{1}$.9418	9422	9427	9431	9435	9439	9443	9447	9451	9455	0	1	1	2	2	2	3	3	4
62	$\bar{1}$.9459	9463	9467	9471	9475	9479	9483	9487	9491	9495	0	1	1	2	2	2	3	3	4
63	$\bar{1}$.9499	9503	9506	9510	9514	9518	9522	9525	9529	9533	0	1	1	2	2	2	3	3	4
64	$\bar{1}$.9537	9540	9544	9548	9551	9555	9558	9562	9566	9569	0	1	1	2	2	2	3	3	4
65	$\bar{1}$.9573	9576	9580	9583	9587	9590	9594	9597	9601	9604	0	1	1	1	2	2	2	2	3
66	$\bar{1}$.9607	9611	9614	9617	9621	9624	9627	9631	9634	9637	0	1	1	1	2	2	2	2	3
67	$\bar{1}$.9640	9643	9647	9650	9653	9656	9659	9662	9666	9669	0	1	1	1	2	2	2	2	3
68	$\bar{1}$.9672	9675	9678	9681	9684	9687	9690	9693	9696	9699	0	1	1	1	2	2	2	2	3
69	$\bar{1}$.9702	9704	9707	9710	9713	9716	9719	9722	9724	9727	0	1	1	1	2	2	2	2	3
70	$\bar{1}$.9730	9733	9735	9738	9741	9743	9746	9749	9751	9754	0	1	1	1	2	2	2	2	3
71	$\bar{1}$.9757	9759	9762	9764	9767	9770	9772	9775	9777	9780	0	0	1	1	2	1	1	2	2
72	$\bar{1}$.9782	9785	9787	9789	9792	9794	9797	9799	9801	9804	0	0	1	1	1	1	1	2	2
73	$\bar{1}$.9806	9808	9811	9813	9815	9817	9820	9822	9824	9826	0	0	1	1	1	1	1	2	2
74	$\bar{1}$.9828	9831	9833	9835	9837	9839	9841	9843	9845	9847	0	0	1	1	1	1	1	2	2
75	$\bar{1}$.9849	9851	9853	9855	9857	9859	9861	9863	9865	9867	0	0	1	1	1	1	1	2	2
76	$\bar{1}$.9869	9871	9873	9875	9876	9878	9880	9882	9884	9885	0	0	1	1	1	1	1	2	2
77	$\bar{1}$.9887	9889	9891	9892	9894	9896	9897	9899	9901	9902	0	0	1	1	1	1	1	2	2
78	$\bar{1}$.9904	9906	9907	9909	9910	9912	9913	9915	9916	9918	0	0	1	1	1	1	1	2	2
79	$\bar{1}$.9919	9921	9922	9924	9925	9927	9928	9929	9931	9932	0	0	1	1	1	1	1	2	2
80	$\bar{1}$.9934	9935	9936	9937	9939	9940	9941	9943	9944	9983									
81	$\bar{1}$.9946	9947	9949	9950	9951	9952	9953	9954	9955	9956									
82	$\bar{1}$.9958	9959	9960	9961	9962	9963	9964	9965	9966	9967									
83	$\bar{1}$.9968	9968	9969	9970	9971	9972	9973	9974	9975	9975									
84	$\bar{1}$.9976	9977	9978	9978	9979	9980	9981	9981	9982	9983									
85	$\bar{1}$.9983	9984	9985	9985	9986	9987	9987	9988	9988	9989									
86	$\bar{1}$.9989	9990	9990	9991	9991	9992	9992	9993	9993	9994									
87	$\bar{1}$.9994	9994	9995	9995	9996	9996	9996	9996	9997	9997									
88	$\bar{1}$.9997	9998	9998	9998	9998	9999	9999	9999	9999	9999									
89	$\bar{1}$.9999	9999	0000	0000	0000	0000	0000	0000	0000	0000									
x	·0	·1	.2	.3	·4	.5	.6	·7	.8	.9	1	2	3	4	5	6	7	8	9

$\text{Log cosec}x = -\text{Log} \sin x$ Way

TABLE 7 : LOGARITHMS OF COSINES OF ANGLES (x IN DEGREES) $x \longrightarrow \text{Log} \cos x$

x	·0	·1	.2	.3	·4	.5	.6	·7	.8	.9	SUBTRACT Differences 1	2	3	4	5	6	7	8	9
0	0.0000	0000	0000	0000	0000	0000	0000	0000	0000	9999									
1	$\bar{1}$.9999	9999	9999	9999	9999	9999	9998	9998	9998	9998									
2	$\bar{1}$.9997	9997	9997	9996	9996	9996	9996	9995	9995	9994									
3	$\bar{1}$.9994	9994	9993	9993	9992	9992	9991	9991	9990	9990									
4	$\bar{1}$.9989	9989	9988	9988	9987	9987	9986	9985	9985	9984									
5	$\bar{1}$.9983	9983	9982	9981	9981	9980	9979	9978	9978	9977									
6	$\bar{1}$.9976	9975	9975	9974	9973	9972	9971	9970	9969	9968									
7	$\bar{1}$.9968	9967	9966	9965	9964	9963	9962	9961	9960	9959									
8	$\bar{1}$.9958	9956	9955	9954	9953	9952	9951	9950	9949	9947									
9	$\bar{1}$.9946	9945	9944	9943	9941	9940	9939	9937	9936	9935									
10	$\bar{1}$.9934	9932	9931	9929	9928	9927	9925	9924	9922	9921	0	0	1	1	1	1	1	2	2
11	$\bar{1}$.9919	9918	9916	9915	9913	9912	9910	9909	9907	9906	0	0	1	1	1	1	1	2	2
12	$\bar{1}$.9904	9902	9901	9899	9897	9896	9894	9892	9891	9889	0	0	1	1	1	1	1	2	2
13	$\bar{1}$.9887	9885	9884	9882	9880	9878	9876	9875	9873	9871	0	0	1	1	1	1	1	2	2
14	$\bar{1}$.9869	9867	9865	9863	9861	9859	9857	9855	9853	9851	0	0	1	1	1	1	1	2	2
15	$\bar{1}$.9849	9847	9845	9843	9841	9839	9837	9835	9833	9831	0	0	1	1	1	1	1	2	2
16	$\bar{1}$.9828	9826	9824	9822	9820	9817	9815	9813	9811	9808	0	0	1	1	1	1	1	2	2
17	$\bar{1}$.9806	9804	9801	9799	9797	9794	9792	9789	9787	9785	0	0	1	1	1	1	1	2	2
18	$\bar{1}$.9782	9780	9777	9775	9772	9770	9767	9764	9762	9759	0	0	1	1	1	1	1	2	2
19	$\bar{1}$.9757	9754	9751	9749	9746	9743	9741	9738	9735	9733	0	0	1	1	2	2	2	2	3
20	$\bar{1}$.9730	9727	9724	9722	9719	9716	9713	9710	9707	9704	0	1	1	1	2	2	2	2	3
21	$\bar{1}$.9702	9699	9696	9693	9690	9687	9684	9681	9678	9675	0	1	1	1	2	2	2	2	3
22	$\bar{1}$.9672	9669	9666	9662	9659	9656	9653	9650	9647	9643	0	1	1	1	2	2	2	2	3
23	$\bar{1}$.9640	9637	9634	9631	9627	9624	9621	9617	9614	9611	0	1	1	1	2	2	2	2	3
24	$\bar{1}$.9607	9604	9601	9597	9594	9590	9587	9583	9580	9576	0	1	1	1	2	2	2	2	3
25	$\bar{1}$.9573	9569	9566	9562	9558	9555	9551	9548	9544	9540	0	1	1	2	2	2	3	3	4
26	$\bar{1}$.9537	9533	9529	9525	9522	9518	9514	9510	9506	9503	0	1	1	2	2	2	3	4	4
27	$\bar{1}$.9499	9495	9491	9487	9483	9479	9475	9471	9467	9463	0	1	1	2	2	2	3	3	4
28	$\bar{1}$.9459	9455	9451	9447	9443	9439	9435	9431	9427	9422	0	1	1	2	2	2	3	3	4
29	$\bar{1}$.9418	9414	9410	9406	9401	9397	9393	9388	9384	9380	0	1	1	2	2	2	3	3	4
30	$\bar{1}$.9375	9371	9367	9362	9358	9353	9349	9344	9340	9335	0	1	1	2	2	2	3	3	4
31	$\bar{1}$.9331	9326	9322	9317	9312	9308	0303	9298	9294	9289	1	1	2	2	3	3	4	4	5
32	$\bar{1}$.9284	9279	9275	9270	9265	9260	9255	9251	9246	9241	1	1	2	2	3	3	4	4	5
33	$\bar{1}$.9236	9231	9226	9221	9216	9211	9206	9201	9196	9191	1	1	2	2	3	3	4	4	5
34	$\bar{1}$.9186	9181	9175	9170	9165	9160	9155	9149	9144	9139	1	1	2	2	3	3	4	4	5
35	$\bar{1}$.9134	9128	9123	9118	9112	9107	9101	9096	9091	9085	1	1	2	2	3	3	4	4	5
36	$\bar{1}$.9080	9074	9069	9063	9057	9052	9046	9041	9035	9029	1	1	2	2	3	4	4	5	5
37	$\bar{1}$.9023	9018	9012	9006	9000	8995	8989	8983	8977	8971	1	1	2	2	3	4	4	5	5
38	$\bar{1}$.8965	8959	8953	8947	8941	8935	8929	8923	8917	8911	1	1	2	2	3	4	4	5	5
39	$\bar{1}$.8905	8899	8893	8887	8880	8874	8868	8862	8855	8849	1	1	2	2	3	4	4	5	5
40	$\bar{1}$.8843	8836	8830	8823	8817	8810	8804	8797	8791	8784	1	1	2	2	3	4	4	5	5
41	$\bar{1}$.8778	8771	8765	8758	8751	8745	8738	8731	8724	8718	1	1	2	3	4	4	5	6	6
42	$\bar{1}$.8711	8704	8697	8690	8683	8676	8669	8662	8655	8648	1	1	2	3	4	4	5	6	6
43	$\bar{1}$.8641	8634	8627	8620	8613	8606	8598	8591	8584	8577	1	1	2	3	4	4	5	6	6
44	$\bar{1}$.8569	8562	8555	8547	8540	8532	8525	8517	8510	8502	1	1	2	3	4	4	5	6	6
x	.0	.1	.2	.3	.4	.5	.6	.7	.8	.9	1	2	3	4	5	6	7	8	9

$\text{Log} \sec x = -\text{Log} \cos x$

TABLE 7 (continued): LOGARITHMS OF COSINES OF ANGLES (x IN DEGREES) $x \longrightarrow \text{Log} \cos x$

x	·0	·1	.2	.3	·4	.5	.6	·7	.8	.9	SUBTRACT Differences 1	2	3	4	5	6	7	8	9
45	$\bar{1}$.8495	8487	8480	8472	8464	8457	8449	8441	8433	8426	1	2	2	3	4	5	6	6	7
46	$\bar{1}$.8418	8410	8402	8394	8386	8378	8370	8362	8354	8346	1	2	2	3	4	5	6	6	7
47	$\bar{1}$.8338	8330	8322	8313	8305	8297	8289	8280	8272	8264	1	2	2	3	4	5	6	6	7
48	$\bar{1}$.8255	8247	8238	8230	8221	8213	8204	8195	8187	8178	1	2	3	4	5	5	6	7	8
49	$\bar{1}$.8169	8161	8152	8143	8134	8125	8117	8108	8099	8090	1	2	3	4	5	5	6	7	8
50	$\bar{1}$.8081	8072	8063	8053	8044	8035	8026	8017	8007	7998	1	2	3	4	5	5	6	7	8
51	$\bar{1}$.7989	7979	7970	7960	7951	7941	7932	7922	7913	7903	1	2	3	4	5	6	7	8	9
52	$\bar{1}$.7893	7884	7874	7864	7854	7844	7835	7825	7815	7805	1	2	3	4	5	6	7	8	9
53	$\bar{1}$.7795	7785	7774	7764	7754	7744	7734	7723	7713	7703	1	2	3	4	5	6	7	8	9
54	$\bar{1}$.7692	7682	7671	7661	7650	7640	7629	7618	7607	7597	1	2	3	4	6	7	8	9	10
55	$\bar{1}$.7586	7575	7564	7553	7542	7531	7520	7509	7498	7487	1	2	3	4	6	7	8	9	10
56	$\bar{1}$.7476	7464	7453	7442	7430	7419	7407	7396	7384	7373	1	2	4	5	6	7	8	10	11
57	$\bar{1}$.7361	7349	7338	7326	7314	7302	7290	7278	7266	7254	1	2	4	5	6	7	8	10	11
58	$\bar{1}$.7242	7230	7218	7205	7193	7181	7168	7156	7144	7131	1	2	4	5	6	7	8	10	11
59	$\bar{1}$.7118	7106	7093	7080	7068	7055	7042	7029	7016	7003	1	3	4	5	6	8	9	10	12
60	$\bar{1}$.6990	6977	6963	6950	6937	6923	6910	6896	6883	6869	1	3	4	5	7	8	9	10	12
61	$\bar{1}$.6856	6842	6828	6814	6801	6787	6773	6759	6744	6730	1	3	4	6	7	8	10	11	13
62	$\bar{1}$.6716	6702	6687	6673	6659	6644	6629	6615	6600	6585	2	3	5	6	8	9	11	12	14
63	$\bar{1}$.6570	6556	6541	6526	6510	6495	6480	6465	6449	6434	2	3	5	6	8	9	11	12	14
64	$\bar{1}$.6418	6403	6387	6371	6356	6340	6324	6308	6292	6276	2	3	5	6	8	10	11	13	14
65	$\bar{1}$.6259	6243	6227	6210	6194	6177	6161	6144	6127	6110	2	3	5	7	9	10	12	14	15
66	$\bar{1}$.6093	6076	6059	6042	6024	6007	5990	5972	5954	5937	2	3	5	7	9	10	12	14	15
67	$\bar{1}$.5919	5901	5883	5865	5847	5828	5810	5792	5773	5754	2	4	5	7	9	11	13	14	16
68	$\bar{1}$.5736	5717	5698	5679	5660	5641	5621	5602	5583	5563	2	4	6	8	10	11	13	15	17
69	$\bar{1}$.5543	5523	5504	5484	5463	5443	5423	5402	5382	5361	2	4	6	8	10	12	14	16	18
70	$\bar{1}$.5341	5320	5299	5278	5256	5235	5213	5192	5170	5148	2	4	7	9	11	13	15	18	20
71	$\bar{1}$.5126	5104	5082	5060	5037	5015	4992	4969	4946	4923	2	5	7	9	12	14	16	18	21
72	$\bar{1}$.4900	4876	4853	4829	4805	4781	4757	4733	4709	4684	2	5	7	10	12	14	17	19	22
73	$\bar{1}$.4659	4634	4609	4584	4559	4533	4508	4482	4456	4430	3	5	8	10	13	16	18	21	23
74	$\bar{1}$.4403	4377	4350	4323	4296	4269	4242	4214	4186	4158	3	5	8	11	14	16	19	22	24
75	$\bar{1}$.4130	4102	4073	4044	4015	3986	3957	3927	3897	3867	3	6	9	12	15	17	20	23	26
76	$\bar{1}$.3837	3806	3775	3745	3713	3682	3650	3618	3586	3554									
77	$\bar{1}$.5126	3488	3455	3421	3387	3353	3319	3284	3250	3214									
78	$\bar{1}$.3179	3143	3107	3070	3034	2997	2959	2921	2883	2845									
79	$\bar{1}$.2806	2767	2727	2687	2647	2606	2565	2524	2482	2439									
80	$\bar{1}$.2397	2353	2310	2266	2221	2176	2131	2085	2038	1991									
81	$\bar{1}$.1943	1895	1847	1797	1747	1697	1646	1594	1542	1489									
82	$\bar{2}$.1436	1381	1326	1271	1214	1157	1099	1040	0981	0920									
83	$\bar{2}$.0859	0797	0734	0640	0605	0539	0472	0403	0334	0264									
84	$\bar{1}$.0192	0210	0046	9970	9894	9816	9736	9655	9573	9489									
85	$\bar{1}$.9403	9315	9226	9135	9042	8946	8849	8749	8647	8543									
86	$\bar{1}$.8436	8326	8213	8098	7979	7857	7731	7602	7468	7330									
87	$\bar{2}$.719	.704	.689	.673	.657	.640	.622	.603	.584	.564									
88	$\bar{2}$.543	.521	.497	.472	.446	.418	.388	.356	.321	.283									
89	$\bar{2}$.242	.196	.145	.087	$\bar{2}$.020	$\bar{3}$.941	.844	.719	.543	.242									
x	·0	·1	.2	.3	·4	.5	.6	·7	.8	.9	1	2	3	4	5	6	7	8	9

$\text{Log} \sec x = -\text{Log} \cos x$

TABLE 8 : LOGARITHMS OF TANGENTS OF ANGLES (x IN DEGREES) $x \longrightarrow \text{Log} \tan x$

											ADD Differences								
x	·0	·1	.2	.3	·4	.5	.6	·7	.8	.9	1	2	3	4	5	6	7	8	9
0	$-\infty$	3.242	.543	.719	.844	3.941	2.020	.087	.145	.196									
1	$\bar{2}.242$	.283	.321	.356	.388	.418	.446	.472	.497	.521									
2	$\bar{2}.543$	.564	.585	.604	.622	.640					2	4	6	8	10	11	13	15	17
						.640	.657	.674	.689	.705	2	3	5	6	8	10	11	13	14
3	$\bar{2}.719$	.734	.747	.761	.774						1	3	4	6	7	7	10	11	13
					.774	.786	.799	.811	.822	.834	1	2	4	5	6	7	8	10	11
4	$\bar{2}.845$	.855	.866	.876	.886	.896	.906	.915	.924	.933	1	2	3	4	5	6	7	8	9
5	$\bar{2}.942$	.951	.959	.967	.976	.984	.991	2.999	1.007	.014	1	2	2	3	4	5	6	6	7
6	$\bar{1}.022$	.029	.036	.043	.050	.057	.063	.070	.073	.080	1	1	2	3	4	4	5	6	6
7	$\bar{1}.0891$	0954	1015	1076							6	12	19	25	31	37	43	50	56
				1076	1135	1194	1252	1310			6	12	18	24	30	35	41	47	53
								1310	1367	1423	6	11	17	22	28	34	39	45	50
8	$\bar{1}.1478$	1533	1587	1640	1693						5	11	16	22	27	32	38	43	49
					1693	1745	1797	1848	1898	1948	5	10	16	20	25	30	35	42	47
9	$\bar{1}.1997$	2046	2094	2142	2189	2236					5	10	14	19	24	29	34	40	43
						2176	2282	2382	2374	2419	5	9	14	18	23	27	32	36	41
10	$\bar{1}.2463$	2507	2551	2594	2637	2680					4	9	13	17	22	26	30	34	39
						2680	2722	2764	2805	2846	4	8	12	16	21	25	29	33	37
11	$\bar{1}.2887$	2927	2967	3006	3046	3085					4	8	12	16	20	24	28	32	36
						3085	3123	3162	3200	3237	4	8	11	15	19	23	27	30	34
12	$\bar{1}.3275$	3312	3349	3385	3422	3458	3493	3529	3564	3599	4	7	11	14	18	22	25	29	32
13	$\bar{1}.3634$	3668	3702	3736	3770	3804	3837	3870	3903	3935	3	7	10	13	17	20	23	26	30
14	$\bar{1}.3968$	4000	4032	4064	4095	4127	4158	4189	4220	4250	3	6	9	12	16	19	22	25	28
15	$\bar{1}.4281$	4311	4341	4371	4400	4430	4459	4488	4517	4546	3	6	9	12	15	17	20	23	26
16	$\bar{1}.4575$	4603	4632	4660	4688	4716	4744	4771	4799	4826	3	6	8	11	14	17	20	22	25
17	$\bar{1}.4853$	4880	4907	4934	4961	4987	5014	5040	5066	5092	3	5	8	10	13	16	18	21	23
18	$\bar{1}.5118$	5143	5169	5195	5220	5245	5270	5295	5320	5345	3	5	8	10	13	15	18	20	23
19	$\bar{1}.5370$	5394	5419	5443	5467	5491	5516	5539	5563	5587	2	5	7	10	12	14	17	19	22
20	$\bar{1}.5611$	5634	5658	5681	5704	5727	5750	5773	5796	5819	2	5	7	9	12	14	16	18	21
21	$\bar{1}.5842$	5864	5887	5909	5932	5954	5976	5998	6020	6042	2	4	7	9	11	13	15	18	20
22	$\bar{1}.6064$	6086	6108	6129	6151	6172	6194	6215	6236	6257	2	4	7	9	11	13	15	18	20
23	$\bar{1}.6279$	6300	6321	6341	6362	6383	6404	6424	6445	6465	2	4	6	8	11	13	15	17	19
24	$\bar{1}.6486$	6506	6726	6746	6567	6587	6607	6627	6647	6667	2	4	6	8	10	12	14	16	18
25	$\bar{1}.6687$	6706	6726	6746	6765	6785	6804	6824	6843	6863	2	4	6	8	10	12	14	1 6	18
26	$\bar{1}.6882$	6901	6920	6939	6958	6977	6996	7015	7034	7053	2	4	6	8	10	12	13	15	17
27	$\bar{1}.7072$	7090	7109	7128	7146	7165	7183	7202	7220	7238	2	4	5	7	9	11	13	14	16
28	$\bar{1}.7257$	7275	7293	7311	7330	7348	7890	7384	7402	7053	2	4	5	7	9	11	13	14	16
29	$\bar{1}.7438$	7455	7473	7491	7509	7526	7544	7562	7579	7597	2	4	5	7	9	11	13	14	16
30	$\bar{1}.7614$	7632	7649	7667	7684	7701	7719	7736	7753	7771	2	3	5	7	9	10	12	14	15
31	$\bar{1}.7788$	7805	7822	7839	7856	7873	7890	7907	7924	7941	2	3	5	7	9	10	12	14	15
32	$\bar{1}.7958$	7975	7992	8008	8025	8052	8059	8075	8092	8109	2	3	5	7	9	10	12	14	15
33	$\bar{1}.8125$	8142	8158	8175	8191	8202	8224	8241	8257	8274	2	3	5	6	8	10	11	13	14
34	$\bar{1}.8290$	8387	8803	8339	8355	8371	8388	8404	8420	8436	2	3	5	6	8	10	11	13	14
35	$\bar{1}.8452$	8468	8484	8660	8517	8533	8549	8565	8581	8597	2	3	5	6	8	10	11	13	14
36	$\bar{1}.8613$	8629	8644	8660	8676	8692	8708	8724	8740	8755	2	3	5	6	8	10	11	13	14
37	$\bar{1}.8771$	8787	8803	8818	8834	8850	8865	8881	8897	8812	2	3	5	6	8	10	11	13	14
38	$\bar{1}.8928$	8944	8959	8975	8990	9006	9022	9037	9053	9068	2	3	5	6	8	10	11	13	14
39	$\bar{1}.9084$	9099	9115	9130	9146	9161	9176	9192	9207	9223	2	3	5	6	8	9	11	12	14
40	$\bar{1}.9238$	9254	9269	9284	9300	9315	9330	9346	9361	9376	2	3	5	6	8	10	11	12	14
x	·0	·1	.2	.3	·4	.5	.6	·7	.8	.9	1	2	3	4	5	6	7	8	9

$\text{Log} \cot x = -\text{Log} \tan x$

TABLE 8
(continued):

LOGARITHMS OF TANGENTS OF ANGLES (x IN DEGREES) $x \longrightarrow \text{Log} \tan x$

x	·0	·1	.2	.3	·4	.5	.6	·7	.8	.9	ADD Differences 1	2	3	4	5	6	7	8	9
41	$\bar{1}.9392$	9407	9422	9438	9453	9422	9438	9499	9514	9529	2	3	5	6	8	9	11	12	14
42	$\bar{1}.9544$	9560	9575	9590	9560	9575	9590	9651	9666	9681	2	3	5	6	8	9	11	12	14
43	$\bar{1}.9697$	9712	9727	9742	9712	9727	9742	9803	9818	9833	2	3	5	6	8	9	11	12	14
44	$\bar{1}.9848$	9864	9879	9894	9909	9924	9939	9955	9970	9985	2	3	5	6	8	9	11	12	14
45	0.0000	0015	0030	0045	0061	0076	0091	0106	0121	0136	2	3	5	6	8	9	11	12	14
46	0.0152	0167	0182	0197	0212	0228	0243	0258	0273	0288	2	3	5	6	8	9	11	12	14
47	0.0303	0319	0334	0349	0364	0379	0395	0410	0425	0440	2	3	5	6	8	9	11	12	14
48	0.0456	0471	0486	0501	0517	0532	0547	0562	0578	0593	2	3	5	6	8	9	11	12	14
49	0.0608	0624	0639	0654	0670	0685	0700	0716	0731	0746	2	3	5	6	8	9	11	12	14
50	0.0762	0777	0793	0808	0824	0839	0854	0870	0885	0901	2	3	5	6	8	10	11	12	14
51	0.0916	0932	0947	0963	0978	0994	1010	1025	1041	1056	2	3	5	6	8	10	11	13	14
52	0.1072	1088	1103	1119	1135	1150	1166	1182	1197	1213	2	3	5	6	8	10	11	13	14
53	0.1229	1245	1260	1276	1292	1308	1324	1340	1356	1371	2	3	5	6	8	10	11	13	14
54	0.1387	1403	1419	1435	1451	1467	1483	1499	1516	1532	2	3	5	6	8	10	11	13	14
55	0.1548	1564	1580	1596	1612	1629	1645	1661	1677	1694	2	3	5	6	8	10	11	13	14
56	0.1710	1726	1743	1759	1776	1792	1809	1825	1842	1858	2	3	5	6	8	10	11	13	14
57	0.1875	1891	1908	1925	1941	1958	1975	1992	2008	2025	2	3	5	7	9	10	12	14	15
58	0.2042	2059	2076	2093	2110	2127	2144	2161	2178	2195	2	3	5	7	9	10	12	14	15
59	0.2212	2229	2247	2264	2281	2299	2316	2333	2351	2368	2	3	5	7	9	10	12	14	15
60	0.2386	2403	2421	2438	2456	2474	2491	2509	2527	2725	2	4	5	7	9	11	13	14	16
61	0.2562	2580	2598	2616	2634	2652	2670	2689	2707	2725	2	4	5	7	9	11	13	14	16
62	0.2743	2762	2780	2798	2817	2835	2854	2872	2891	2910	2	4	5	7	9	11	13	14	16
63	0.2928	2947	2966	2985	3004	3023	3042	3061	3080	3099	2	4	6	8	10	11	13	15	17
64	0.3118	3137	3157	3176	3196	3215	3235	3254	3274	3294	2	4	6	8	10	12	14	16	18
65	0.3313	3333	3353	3373	3393	3413	3433	3453	3473	3494	2	4	6	8	10	12	14	16	18
66	0.3514	3535	3555	3576	3596	3617	3638	3659	3679	3700	2	4	6	8	11	13	15	17	19
67	0.3721	3743	3764	3785	3806	3828	3849	3871	3892	3914	2	4	7	9	11	13	15	18	20
68	0.3936	3958	3980	4002	4024	4046	4068	4091	4113	4136	2	4	7	9	11	13	15	18	20
69	0.4158	4181	4204	4227	4250	4273	4296	4319	4342	4366	2	5	7	9	12	14	16	18	21
70	0.4389	4413	4437	4461	4484	4509	4533	4557	4581	4606	2	5	7	10	12	14	17	19	22
71	0.4630	4655	4680	4705	4730	4755	4780	4805	4831	4857	3	5	8	10	13	15	18	20	23
72	0.4882	4908	4934	4960	4986	5013	5039	5066	5093	5120	3	5	8	10	13	16	18	21	23
73	0.5147	5174	5201	5229	5256	5284	5312	5340	5368	5397	3	6	8	11	14	17	20	22	25
74	0.5425	5454	5483	5512	5541	5570	5600	5629	5659	5689	3	6	9	12	15	17	20	23	26
75	0.5719	5750	5780	5811	5842	5873	5905	5936	5968	6000	3	6	9	12	16	19	22	25	28
76	0.6032	6065	6097	6130	6163	6196	6230	6264	6298	6332	3	7	10	13	17	20	23	26	30
77	0.6366	6401	6436	6471	6507	6542	6578	6615	6651	6688	4	7	11	14	18	22	25	29	32
78	0.6725	6763	6800	6838	6877	6915	6954	6994	7033	7073									
79	0.7113	7154	7195	7236	7278	7320	7363	7406	7449	7493									
80	0.7537	7581	7626	7672	7718	7764	7811	7858	7906	7954									
81	0.8003	8052	8102	8152	8203	8255	8307	8360	8413	8467									
82	0.8522	8577	8633	8690	8748	8806	8865	8924	8985	9046									
83	0.9109	9172	9236	9301	9367	9433	9501	9570	9640	9711									
84	0.9784	9857	9932	0008	0085	0164	0244	0326	0409	0494									
85	1.0580	0669	0759	0850	0944	1040	1138	1238	1341	1446									
86	1.1554	1664	1777	1893	2012	2135	2261	2391	2525	2623									
87	1.281	.295	.311	.326	.343	.360	.378	.396	.415	.436									
88	1.457	.479	.503	.528	.554	.582	.612	.644	.679	.717									
89	1.758	.804	.855	.913	1.980	2.059	.158	.281	.457	.758									
x	·0	·1	.2	.3	·4	.5	.6	·7	.8	.9	1	2	3	4	5	6	7	8	9

$\text{Log} \cot x = -\text{Log} \tan x$

APPENDIX B

ANSWERS TO OBJECTIVE TESTS AND EXERCISES

OBJECTIVE TEST 1

1B 2B 3B 4C 5A 6D 7B 8B 9C 10A

Exercise 1

1. (a) C.F = 1×3 and H.C.F = 3 (b) C.F = 1×2 and H.C.F= 2

   (c) C.F = $1 \times 2 \times 3 \times 6$ and H.C.F= 6 (d) C.F = 1×3 and H.C.F= 3

   (e) No C.F and H.C.F (f) C.F = 3×5 and H.C.F= 15

   (g) C.F = $2^2 \times 3$ and H.C.F = 12 (h) C.F = $3^2 \times 7^2$ and H.C.F = 441

   (i) C.F = 4×5 and H.C.F = 20

2. (a) Prime factors of 126 = 2, 3 and 7;
   Prime factors of 288 = 2 and 3;
   Prime factors of 380 = 2 and 3
   (b) C.F = 2×3^2 and H.C.F= 18

3. (a) 6 (b) 315 (c) 36 (d) x^2 (e) $(x-1)(x+1)^2$
   (f) $(x-1)^3$ (g) $2^2 \times 3^2$ (h) $5^3 \times 7^3$

4. (a) 4.893 (b) 0.077 (c) 15.256 (d) 456.883

5. (a) 116.29 (b) 2.03 (c) 12.41 (d) 100.02

6. (a) 210 (b) 0.21 (c) 4100 (d) 3100

7. (a) 1 (b) 287 (c) 604 (d) 0

8. (a) 3.18×10^{-5} (b) 1.315×10^4 (c) 8.63×10^{-5} (d) 4.56756×10^5

9. (a) 2.9×10^2 (b) 8.3×10^{-8} (c) 7.1×10^3 (d) 4.2×10^3

10. (a) 5×10^3 (b) 1.35×10^{-2} (c) 7.78×10^2 (d) 2×10^{-1}

11. (a) 3.8×10^{-1} (b) 1.8×10^1 (c) 6.5×10^2 (d) 2.5×10^0 (e) 5.4×10^1

12. (a) 1.35×10^3 (b) 1.5×10^{-2} (c) 2.19×10^4
13. (a) 0.24 (b) 89000 (c) 0.44 (d) 0.33

OBJECTIVE TEST 2

1B 2D 3D 4B 5C 6B 7C 8C 9B 10D

Exercise 2a

1. $\frac{71}{5} = 14\frac{1}{5}$
2. $4\frac{11}{168}$
3. ₦36,540
4. (a) $\frac{3}{8}$ (b) $\frac{1}{4}$ (c) $\frac{1}{500}$
   (d) $\frac{3}{20}$ (e) $\frac{17}{20}$ (f) $\frac{3}{8}$
5. (a) 0.5 (b) 0.05 (c) 0.375
   (d) $0.6\dot{6}$ (e) 0.6
6. $\frac{4}{7} < \frac{3}{4} < \frac{7}{8} < \frac{8}{9}$
7. (a) $\frac{13}{4}$ (b) $\frac{127}{70}$
8. (a) $\frac{187}{140}$ (b) $-\frac{29}{8}$
9. $\frac{107}{15}$
10. $\frac{3}{5}$; 40
11. 240
12. (a) $\frac{8}{9}$ (b) $\frac{2}{3}$ (c) $61\frac{1}{4}$
    (d) $\frac{11}{20}$ (e) $17\frac{31}{2500}$
13. (a) $x = 3$ (b) $x = -1$ or 4 (c) $x = \pm 2$
    (d) $x = \pm 3$ (e) $x = -7, 1$ or $2\frac{1}{2}$
14. (a) $x = -1$ (b) $x = \frac{1}{2}$
    (c) $x = -1.76$ or 4.26 (d) $x = 3$ or 23

Exercise 2b

1. ₦29,040
2. ₦110.16
3. 9 yrs
4. ₦625
5. 1.5%
6. 50%
7. 6%
8. 4.6%
9. 4.8%
10. 8%
11. man's age $=$ 39 years
    wife's age$=$ 34 years

OBJECTIVE TEST 3

1 A 2 C 3 B 4 B 5 B 6 C 7 A 8 A 9 B 10 A

1. (a) 432_9 (b) 334_5 (c) 11010_2 (d) 11100000_2 (e) 6160_8 (f) 233_5

2. (a) 01_2 (b) 001_2 (c) 01000_2 (d) 030_4 (e) 130_8

   (f) 6622_7 (g) 111_8 (h) 490_{16} (i) $B7DD_{16}$ (j) $3FD86_{16}$

3. (a) 1111_2 (b) 1111001_2 (c) 100111111_2 (d) 100011.11_2 (e) 111.010_2

   (f) 101.1111_2 (g) 11010.111_2 (h) 103.322_4 (i) 24210.50_6 (j) 1264.734_8

4. (a) 101_2 (b) 011_2 (c) 1011_2 (d) 141_7 remainder 21_7
   (e) 53_8 remainder 7_8 (f) 24_5 remainder 01_5

5. (a)3 (b) 13 (c) 59 (d) 14 (e) 1837 (f) $3\frac{3}{8}$ (g) $2\frac{3}{4}$ (h) $4\frac{7}{8}$ (i) $13\frac{3}{4}$

6. (a) 44 (b) 467 (c) 24518 (d) $29\frac{23}{64}$ (e) $67\frac{11}{32}$ (f) $2502\frac{21}{64}$

7. (a) 43860 (b) 353 (c) 2419 (d) 35778 (e) 326

8. (a) 1111_2 (b) $0.\overline{1000010100011110100 1}_2$ (c) 1001000_2 (d) 1101001_2

   (e) 1011011001_2 (f) 1000000000_2 (g) 111001_2 (h) 1101010_2

   (i) $0.\overline{0110}_2$ (j) $10000.10\overline{10}_2$

9. (a) 328 (b)2746 (c)30223 (d) 4099

10. (a) 102_3 (b) 10011000_2 (c) 5317_8 (d) $1B_{16}$ (e) $132.\overline{15304}_6$

    (f) 521_9 (g) 211201_4 (h) 18995_{10} (i) $5t9_{12}$

11. (a) $X = 12$ (b) $X = 11000$ (c) $X = 11$ (d) $X = DD$

OBJECTIVE TEST 4

1C 2A 3B 4A 5C 6A 7D 8C 9B 10B

Exercise 4a

1. (a) 2 (b) $\frac{1}{169}$ (c) $\frac{1}{4}$ (d) $\frac{64}{125}$ (e) $\frac{3}{2}$

2. (a) $\frac{2}{125a^2}$ (b) $2a^2$

3. (a) $x = 2$ (b) $x = -4$ (c) $x = 2$ (d) $x = -\frac{10}{3}$
   (e) $x = -3$ (f) $x = \frac{7}{12}$ (g) $x = 2$

4. (a) $\frac{1}{25}$ (b) 100

5. (a) $x = -\frac{2}{3}$ (b) $x = -3$ (c) $x = 4$ (d) $x = \pm\frac{1}{2}$ (e) $x = -\frac{1}{2}$ (f) $x = -4$

6. (a) $x = 0$ or 1 (b) $x = 0, 2$ and 3

Exercise 4b

1. (a) $\frac{9}{25}$ (b) $112\frac{1}{2}$

2. $q = 100$

3. (a) $\frac{3}{10}$ (b) $\frac{9}{16}$

4. $x = 100$

5. (a) $3\frac{1}{2}$ (b) 7 (c) $\frac{2}{3}$ (d) -1

6. (a) 2 (b) -3 (c) 7 (d) 2 (e) 4 (f) -1
   (g) 2 (h) 5 (i) -4 (j) 6 (k) $\frac{5}{2}$ (l) 4 (m) $\sqrt{2}$

7. (a) 2 (b) $\frac{5}{3}$ (c) $\frac{1}{3}$ (d) $\frac{1}{2}$ (e) 4

8. (a) 5 (b) -2 (c) -4

9. (a) 4.289 (b) 0.006960 (c) 0.2733
   (d) 349.1 (e) 1.071 (f) 0.06682

10. (a) 945800 (b) 2.286
    (c) 406.6 (d) 2113000000
    (e) 3.509 (f) 44540000 (g) 2.6
    (h) 6.955 (i) 1.406

11. (a) $x = 5$ (b) $x = 2$

12. $n = 10$

13. (a) $x = 10$ (b) $x = 125$

OBJECTIVE TEST 5

1B 2A 3C 4A 5C 6A 7A 8A 9C 10C

Exercise 5

1. (a) $\sqrt{20}$ (b)$\sqrt{18}$ (c) $\sqrt{50}$ (d) $\sqrt{147}$ (e)$\sqrt{200}$ (f)$\sqrt{144}$ (g)$\sqrt{243}$

2. (a) $2\sqrt{2}$ (b) $3\sqrt{2}$ (c) $5\sqrt{2}$ (d) $5\sqrt{3}$

   (e) $2\sqrt{10}$ (f) $16\sqrt{2}$ (g) $10\sqrt{6}$ (h) $4\sqrt{3}$

3. (a) $3+2\sqrt{2}$ (b) $4+2\sqrt{3}$ (c) $13-4\sqrt{3}$ (d) $38+12\sqrt{10}$ (e) $5-2\sqrt{6}$

4. (a) $11\sqrt{3}$ (b) $4\sqrt{6}$ (c) $10\sqrt{7}$ (d) $-5\sqrt{2}$

5. (a) 8.660 (b) 2.828 (c) 12.31 (d) 2.464

6. (a) 3.54 (b) 0.28 (c) 5.484 (d)-8.14

7. (a) $2\sqrt{5}$ (b) $\frac{\sqrt{2}+\sqrt{6}}{2}$ (c) $\frac{3+\sqrt{6}}{5}$ (d) $\frac{\sqrt{15}}{3}$

8. (a) $\frac{\sqrt{7}-1}{3}$ (b) $2\sqrt{2}-3$ (c) $\frac{1}{2}(\sqrt{5}-3)$ (d) $\frac{8+\sqrt{5}}{59}$ (e) $\frac{1}{19}(21+8\sqrt{6})$

9. $a=-\frac{5}{13}, b=-\frac{2}{13}, n=3$

10. $-5-2\sqrt{6}$

11. $\frac{\sqrt{2}}{2}+\frac{\sqrt{7}}{7}$

12. $7\sqrt{2}$

13. (a) $\frac{1}{2}\left(\sqrt{14}+\sqrt{10}\right)$ twice (b) $1+2\sqrt{5}$ twice
    (c) $2\sqrt{3}-\sqrt{2}$ and $\sqrt{2}-2\sqrt{3}$ (d)$\left(\sqrt{5}+\sqrt{2}\right)$ twice

14. $2(x+\sqrt{x(x+1)}+1$

15. $24\sqrt{2}$

16. $\frac{5+4\sqrt{10}}{9}$

OBJECTIVE TEST 6

1D 2A 3A 4C 5B 6A 7D 8B 9B 10C

Exercise 6a

1. $y = 5x$, and $x = 6$
2. (a) $P = 2Q^2$ (b) $P = 50, Q = 4$
3. (a) $R = \frac{2}{3}\sqrt{M}$ (b) $R = 4, M = 324$
4. (a) $V = 4.25R^3$ (b) $V = 114.75\text{cm}^3$ (c) $R = 5\text{cm}$
5. $V = 4d$
6. $P = 100$ and $Q = 4\sqrt{2}$
7. 1,000,000
8. (a) $A = 0.8N$ (b) $A = 12$, and $N = 25$
9. ₦2,000
10. (a) $y = \frac{27}{8}x^{\frac{3}{2}}$ (b) $y = 421.875$

Exercise 6b

1. $y = \frac{36}{x}$
2. $y = 5,\ x = 7.5$
3. (a) $P = \frac{225}{Q^2}$ (b) 56.25 (c) 2
4. (a) $t = \frac{150}{v}$ (b) 7.5m/s
5. (a) $C = \frac{75000}{N}$ (b) 15,000
6. (a) $R = \frac{75}{\sqrt{M}}$ (b) 9.375

Exercise 6c

1. (a) $y = 0.6xz$ (b) $x = 10$
2. (a) $y = \frac{4x}{z}$ (b) $y = 10$
3. (a) $P = \frac{8Q^2}{3R}$ (b) $P = \frac{50}{3}$
4. (a) $V = \frac{100T}{P}$ (b) $V = 250\text{cm}^3$ (c) 400mmHg
5. (a) $C = \frac{24A}{L}$ (b) $C = 4\text{F}$
6. (a) $P = \frac{75H}{\sqrt{R}}$ (b) $H = 8\text{m}$
7. (a) $E = \frac{1}{2}mv^2$ (b) $4.5 \times 10^5\text{J}$
8. $x = 10$

Exercise 6d

1. (a) $y = 20 + 5x$ (b) $y = 70$
2. $k = 280$ and $c = -300$
3. (a) $M = 100 + 5.6N^2$ (b) $M = 150.9$ and $N = 1.9$
4. (a) $V = \frac{3}{11000}gR - \frac{170}{11}$ (b) $V = 7.46$ km/s

OBJECTIVE TEST 7

1C 2B 3D 4B 5A 6A 7B 8C 9D 10A

Exercise 7a

1. (a) 2 (b) 8 (c) 25 (d) 1 (e) $\frac{1}{2}$
2. $U_{10} = 28$
3. (a) $U_{10} = 41$ (b) $n = 25$
4. $\frac{7}{9}, \frac{9}{11}, \frac{11}{13}$
5. $d = 5$
6. (a) $U_n = 16 + 4n$ (b) $U_n = 8 - 3n$ (c) $U_n = 3n - 13$

7. (a) $U_7 = 35$ (c) $U_{12} = 60$, (c) $S_{10} = 550$

8. 26, 29, 32.

9. (a) $U_1 = 4$ (b) $U_3 = 32$

10. 19

11. $S_n = 442$

12. (a) $S_{15} = 360$ (b) $S_5 = 25$

13. Go to section 7.11 for the prove

14. $n = 8$

Exercise 7b

1. (a) 2 (b) 2 (c) -6 (d) $\frac{1}{8}$
2. (a) -2 (b) -3 (c) 63
3. (a) 2 (b) 3
4. $x = 6, y = 12$
5. ...
6. (a) 4 (b) $\frac{1}{6}$
7. (a) $\frac{5}{3}$ (b) $\frac{75}{2}$
8. $\frac{1}{24}, \frac{1}{48}, \frac{1}{96}, \frac{1}{192}$
9. $n = 6$
10. $n = 7$
11. geometric mean: 15, 25
12. ...
13. $S_\infty = \frac{8}{7}$
14. (a) $S_\infty = \frac{1}{15}$ (b) $S_\infty = \frac{43}{99}$
15. (a) $S_\infty = \frac{3}{2}$ (b) $2\frac{2}{7}$
16. (a) $S_\infty = \frac{6}{5}$ (b) $S'_\infty = \frac{5}{3}$

OBJECTIVE TEST 8

1C 2B 3A 4A 5C 6C 7A 8C 9C 10C

Exercise 8a

1. (a) {April, June, September, November}
   (b) {Sunday, Monday, Tuesday, Wednesday, Thursday, Friday, Saturday}
   (c) { Ade, Alice, Amodu, Anne, Abdullahi}

2. (a) {a,b,c} ≡ {1,2,3} and {a,e,i,o,u} ≡ {Chuma, Vitalis, Bede, Ayo, Adamu}
(b) {1,2,3,4} ={4, 2, 1, 3} and {1, 2, 3, 4}= {1, 2, 3, 2, 4, 4}

3. (a) True (b) True (c)True (d) False (e) False
(f) False (g) False (h) False (i)True

4. A is comparable to sets $B, D,$ and E

B is comparable to sets D and E

C is comparable to sets D and E

D is comparable to sets A, B and E

E is comparable to sets A, B, C and D

5. (a) $A = \{2, 3\}$ (b) $B = \emptyset$ (c) $C = \emptyset$ (d) $\{1, 2, 3, 4, 5, 6, 7, 8, 9, 10\}$

6. (a) $\{ 4, 6, 8, 10, 12, 14, 16 \}$ (b) $\emptyset$

7.

(i)

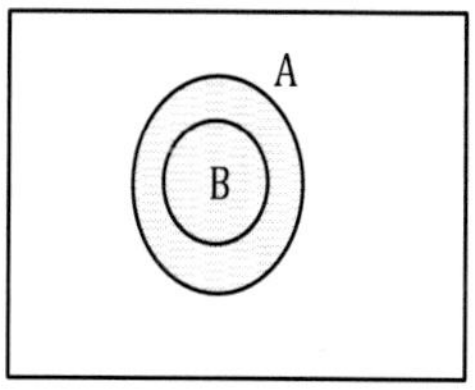

$A \cup B$

(ii)

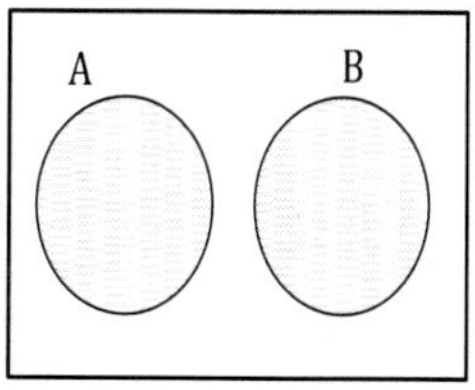

$A \cup B$

(iii)

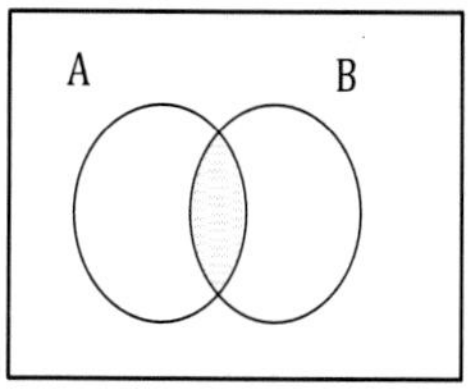

$A \cap B$

(iv)

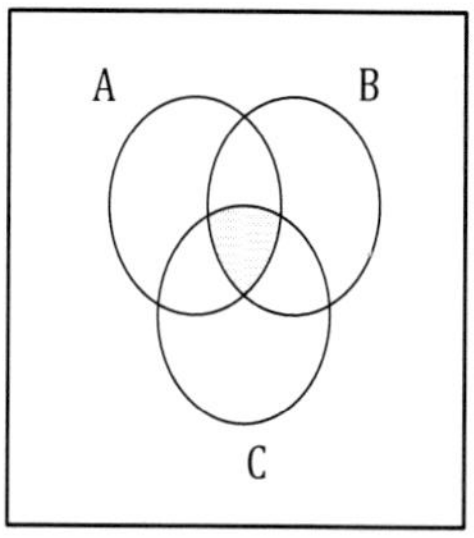

$A \cap B \cap C$

(v)

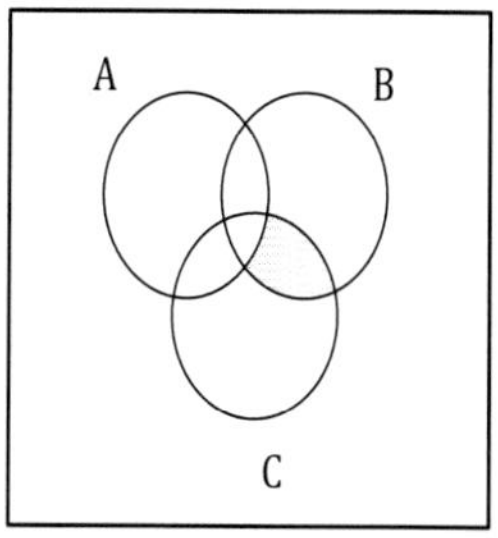

$A' \cap B \cap C$

(vI)

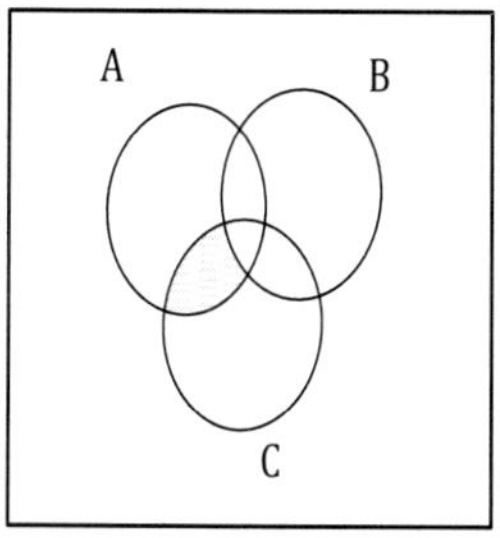

$A \cap B' \cap C$

8.

(i)

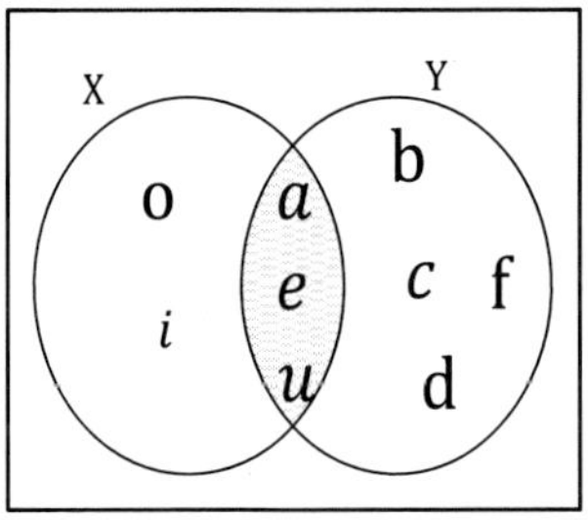

(ii)

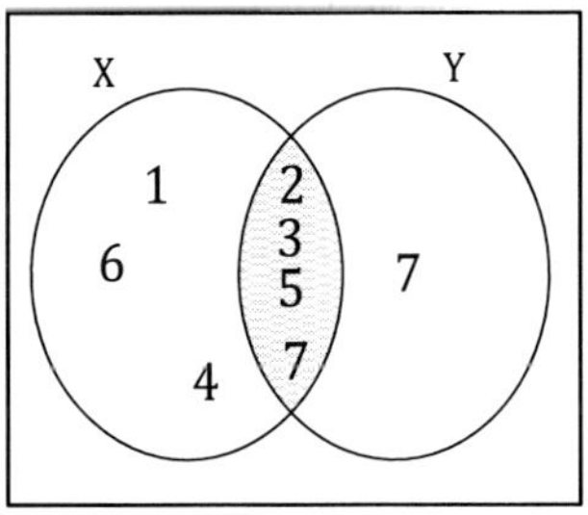

9. (a) $A - B = \{4, 6, 8\}$ and $B - A = \{5, 7\}$

(b)

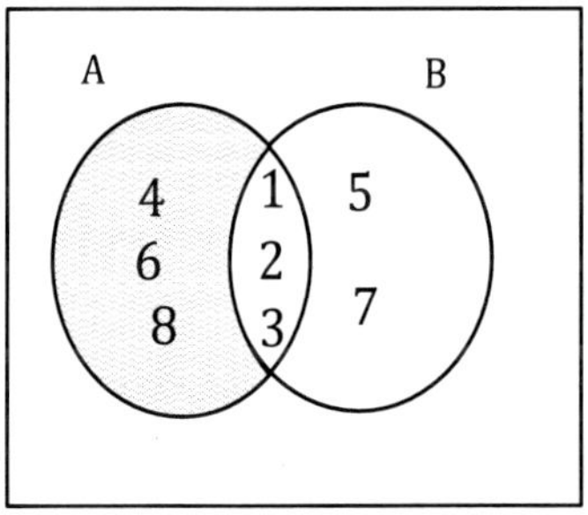

(i) $A - B$ Shaded

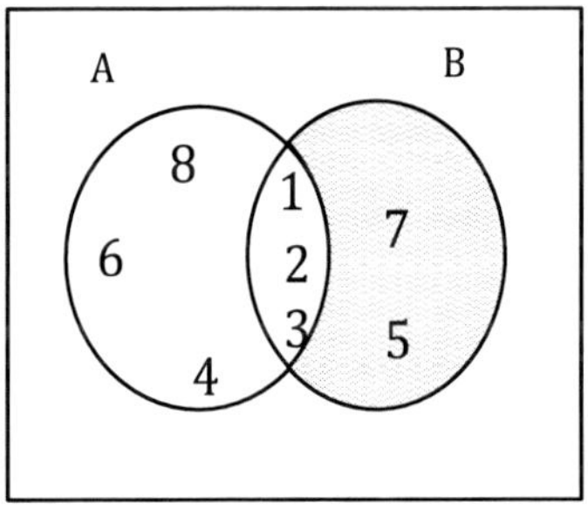

(ii) $B - A$ Shaded

10. (i) 2 (ii) 4 (iii) 8 (iv) 4 (v) 16

11. (a) 17 (b) 12 (c) 11

12. 40%

13. ...

14. (a) $\{x: x < 2, x = 1\}$ (b) $\{x: x = 1, x = 4\}$ (c) $\emptyset$

15. (a) 8 (b) 4 (c) 8

16. (a) $X = \{1, 2, \{4, 7\}$; $\{1\}$; $\{2\}$; $\{4, 7\}$; $\{1, 2\}$; $\{1, \{4, 7\}\}$; $\{2, \{4, 7\}\}$; $\{\ \}$

(b) $\{a, b, c\}$; $\{a, b\}$; $\{a, c\}$; $\{b, c\}$; $\{a\}$; $\{b\}$; $\{c\}$; $\{\ \}$

17. (a) $S = \{1, 2, 3, 4, \ldots, 10\}$ (b) $S = \{1, 2, 3, 4\}$ (c) $S = \{4\}$
(d) $S = \{1, 2, 3, 4, 5, 6, 7\}$ (e) $S = \{8, 9, 10\}$

18. (a)

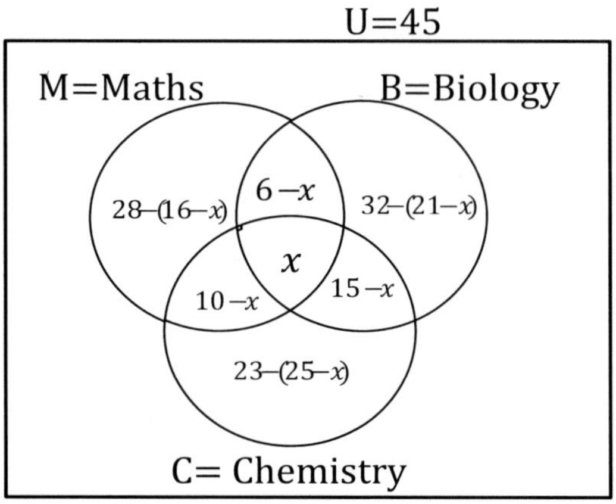

(b) $x = 3$

19. (a) $X \cap Y = \{2\}$ (b) $X' \cap Z = \{6, 8, 9, 10, 12, 14\}$ (c) $X' \cap (Y' \cap Z) = \{9\}$

OBJECTIVE TEST 9

1B 2B 3C 4D 5A 6C 7C 8A 9A 10C

Exercise 9a

1. (a) $3(6x - 2y - 1)$ (b) $(x + 3)^3$ (c) $\frac{7}{6}x$ (d) $\frac{64}{105}x$ (e) $x - 1$

   (f) $\frac{2}{x^2-1}$ (g) -1 (h) $\frac{2c(b-a)+ab}{abc}$ (i) $\frac{3x-2y-9}{6x}$

2. (a) -1 (b) $-\frac{3}{2}$ (c) $-\frac{1}{2}$ (d) $-\frac{1}{3}$

3. (a) -1 (b) $\frac{1}{2}$

4. $\frac{7}{12}$

5. (a) $(x - y)(y - z)$ (b) $(am - n)(a + m)$ (c) $(x + y)(y - 1)$

   (d) $(x - y)(a + x + y)$ (e) $(a - b)(x - y)$ (f) $(3t - 1)(4t + 1)$

   (g) $(3 + x)(2 - x)$ (h) $\left(x - \frac{1}{2}\right)\left(x + \frac{3}{2}\right)$ (i) $(k + 1)(3k - 1)$

   (j) $(x + 21)(x - 11)$ (k) $(a + 12)(a - 2)$ (l) $(1 - 3a)^2$
   (m) $(b - 3)^2$ (n) $(1 + r)^2$

6. $\frac{8-a}{1-a}$

7. (a) $\frac{x^3-y^3}{x^2y^2}$ (b) $\frac{x^2+y^2+2(x+y+1)}{(x+1)(y+1)}$

8. (a) x^2+3x+3 (b) x^2-1 (c) $2(15-2y-y^2)$

(d) $4x^2+4xy+y^2$ (e) $x^3+y^3+3x^2y+3xy^2$

9. (a) $2(x-5)$ (b) $10x(x+3)$ (c) $3xy(1+5x-2y)$ (d) $(a-x)(b-a+x)$

10. -28

11. (a) 1 (b) 37

12. (a) $-(2a+b)$ (b) $\frac{20a-b}{6}$ (c) $\frac{2(a+b)(a-b)^2}{(a-2)(b-2)}$ (d) $\frac{b-a+1}{a+1}$

13. (a) $\frac{x}{6}$ (b) $\frac{2x+5y}{10}$ (c) $2y(x-1)$ (d) $\frac{3x-2y}{xy}$

14. (a) $\frac{55}{29}$ (b) 5 (c) 3 (d) $\frac{x}{(x-2)(x-5)}$

15. (a) $2(m-n)-1$ (b) $-2p$

16. (a) $-\frac{7}{3}x$ (b) $\frac{y-2x}{xy}$

17. 21

18. $-\frac{7}{12}$

19. (a) $(x-y)(y-z)$ (b) $(am-n)(a+m)$ (c) $(x+1)(y-1)$
(d) $(k^2+a^2)(x^2+y^2)$ (e) $(x-y)(a+x+y)$ (f) $(a-b)(x-y)$
(g) $(3t-1)(4t+1)$ (h) $(3+x)(2-x)$ (i) $(x-\frac{1}{2})(x+\frac{3}{2})$
(j) $(k+1)(3k-1)$ (k) $(x+21)(x-11)$ (l) $(a+12)(a-2)$
(m) $(1-3a)^2$ (n) $(b-3)^2$

Exercise 9b

1. (a) $x = 9$ (b) $x = \pm 5$ (c) $x = 11$ (d) $x = 74$
2. $x = 15$
3. The man's age = 30, and the son's age = 10
4. The ages of the sisters are 17, and 20
5. The value is 20
6. (a) $(x, y) = (1, 0)$ (b) $(x, y) = (-4, 18)$
   (c) $(m, n) = (\frac{7}{3}, -\frac{1}{3})$ (d) $(a, b) = (\frac{63}{22}, \frac{5}{22})$
7. (a) $(x, y) = (2, 1)$ (b) $(x, y) = \left(\frac{65}{26}, \frac{5}{13}\right)$
   (c) $(x, y) = (5\frac{3}{13}, -4\frac{2}{13})$ (d) $(x, y) = (5, 7)$
8. (a) $(x, y) = \left(\frac{3}{4}, \frac{1}{4}\right)$ (b) $(x, y) = \left(6\frac{1}{9}, 3\frac{1}{3}\right)$
   (c) $(a, b) = (3, 2)$ (d) $(x, y) = (2, 1)$
9. (a) $(x, y) = (1, -1)$ (b) $(x, y) = (4, -1)$
   (c) $(x, y) = (10, 50)$ (d) $(x, y) = \left(-\frac{8}{5}, \frac{2}{5}\right)$
10. Length = 8cm, Width = 5cm
11. 2 : 1
12. (a) $(x, y) = \left(\frac{\sqrt{3}}{10}, \frac{10}{3}\right)$ (b) $(x, y) = \left(\sqrt{\frac{115}{3}}, \frac{8}{\sqrt{3}}\right)$
    (c) $(x, y) = (\pm 2, 0)$ (d) $(x, y) = \left(\pm \frac{2}{\sqrt{11}}, \pm \frac{2}{3}\right)$
13. (a) $(x, y) = \left(\frac{1}{15}, -\frac{1}{10}\right)$ (b) $(x, y) = \left(\frac{23}{22}, -\frac{6}{11}\right)$
    (c) $(x, y) = (1, 0)$ (d)$(x_1, y_1) = \left(1\frac{1}{2}, -1\right)$ and $(x_2, y_2) = \left(\frac{1}{3}, 6\right)$
14. $(x, y) = \left(-1\frac{6}{13}, \frac{8}{13}\right)$

OBJECTIVE TEST 10

1B 2C 3D 4B 5A 6B 7A 8D 9A 10C

Exercise 10

1. (a) $1 \leq x < 2$ (b) $-5 \leq x < -1$

2. $x \leq -2$, and $x \geq 3$

3. (a) $x < 5$ (b) $x < -1$ (c) $x < \frac{3}{2}$ (d) $x \geq \frac{3}{7}$ (e) $x \geq -1$

4. (a) $x \leq -5$ and $x \geq -11\frac{2}{3}$ (b) $x \geq 17$ and $x < 11$ (c) $x \leq \frac{3}{2}$ and $x \geq 1$

5. (a) $x \leq -1$ (b) $x \leq 2$

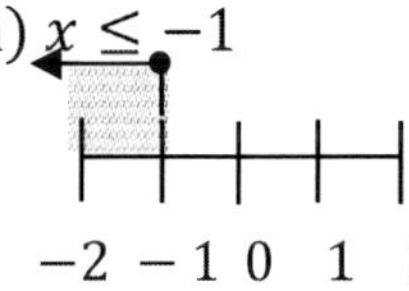

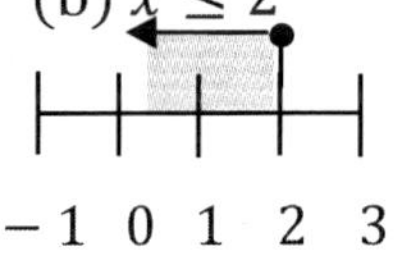

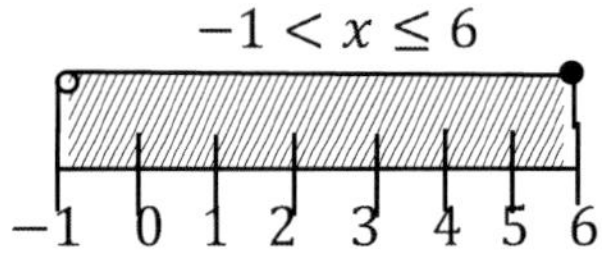

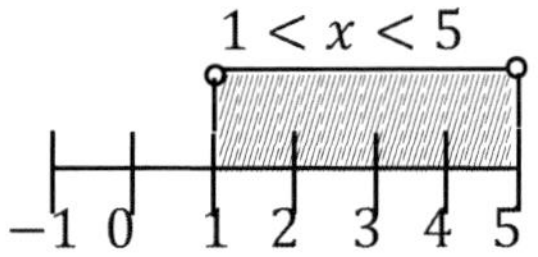

6. (a) $-1 \leq x \leq 3$ (b) $-2 \leq x \leq 1$ (c) $-1 < x < \frac{10}{3}$ (d) $-8 < x < -2$

7.

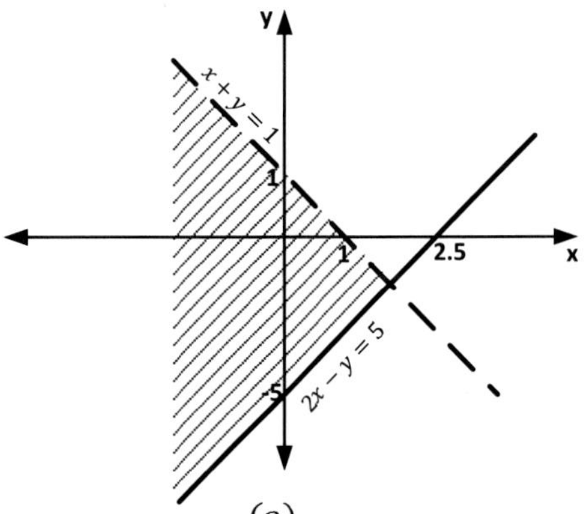

(a)

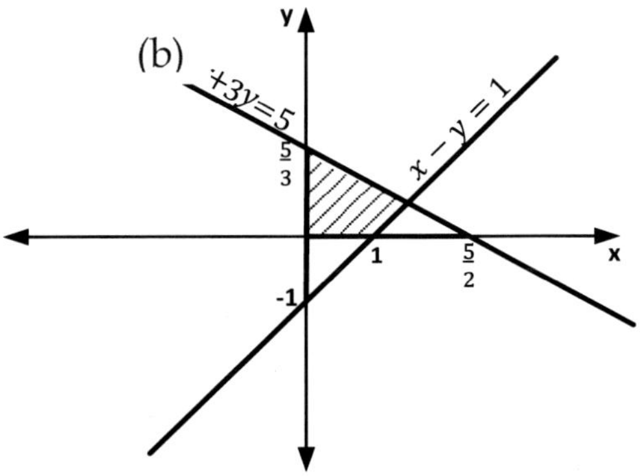

(b)

8. (a) $x > 3$ and $x < -4$ (b) $x > 2$ and $x < -8$ (c) $x > -2$ and $x < -1$
(d) $x < 2$ and $x > -\frac{1}{2}$ (e) $x < -\frac{5}{4}$ and $x > \frac{5}{6}$ (f) $1 < x < 7$ (g) $-2 < x < 1$

9. (a) $-3 < x < -4$ (b) $x < 2$ and $x < 2\frac{3}{4}$ (c) $-\frac{2}{5} \leq x \leq 11$ (d) $x \leq -10$ and $x \geq \frac{15}{34}$

10. (a) $\frac{1}{2} \leq x \leq 3$ or $-\frac{11}{2} \leq x \leq -5$ (b) $x < -\frac{15}{13}$ or $x > -\frac{5}{17}$
(c) $x \leq 3$ or $x \geq 3$ (d) the only real solution is $-1 < x < 0$

11.

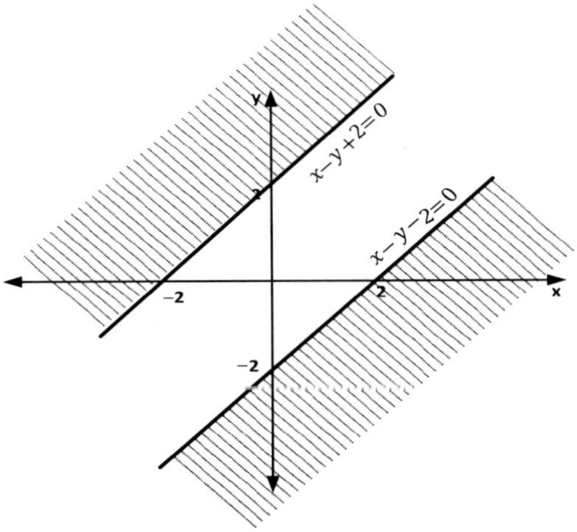

(a)

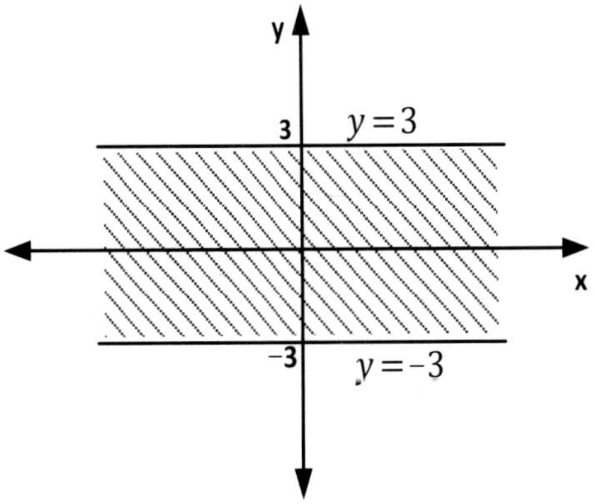

(b)

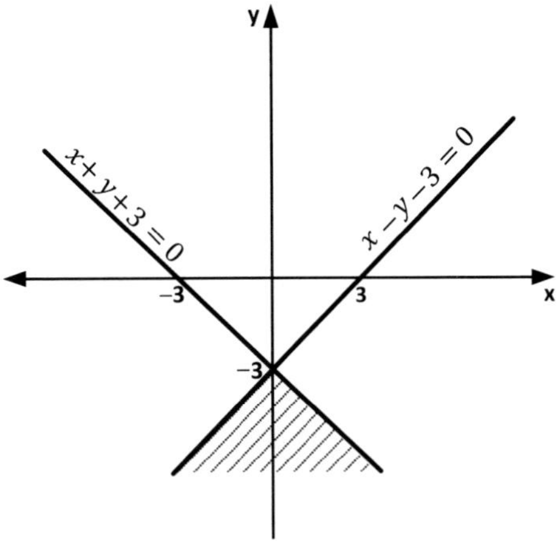

(c)

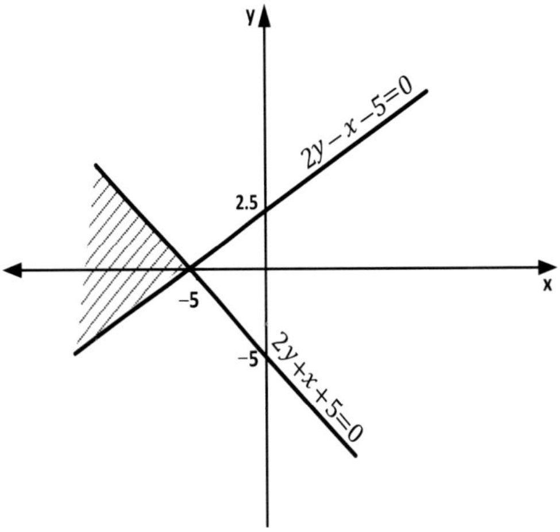

(d)

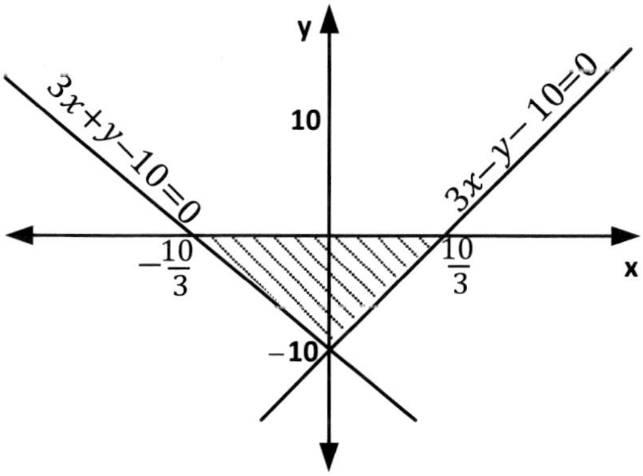
y
10
3x+y−10=0
3x−y−10=0
−10/3
10/3
x
−10
(e)

12.

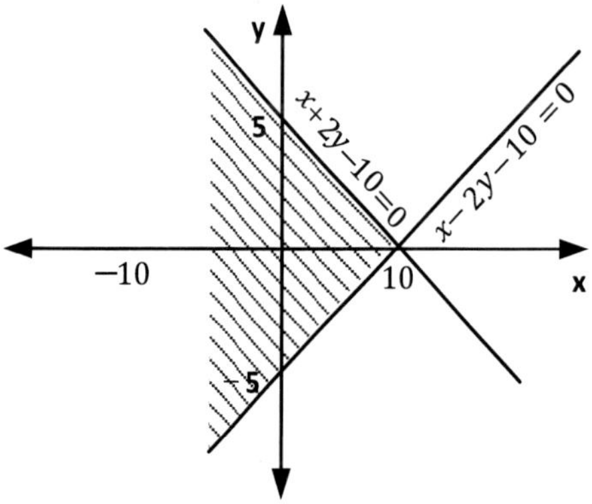
y
5
x+2y−10=0
x−2y−10=0
−10
10
x
−5
(a)

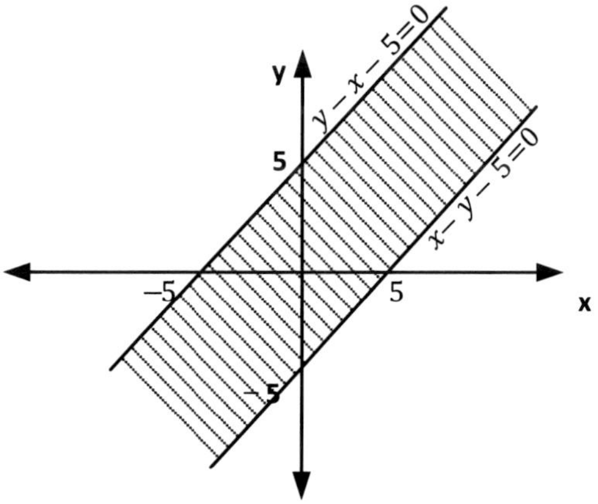
y
y−x−5=0
5
x−y−5=0
−5
5
x
−5
(b)

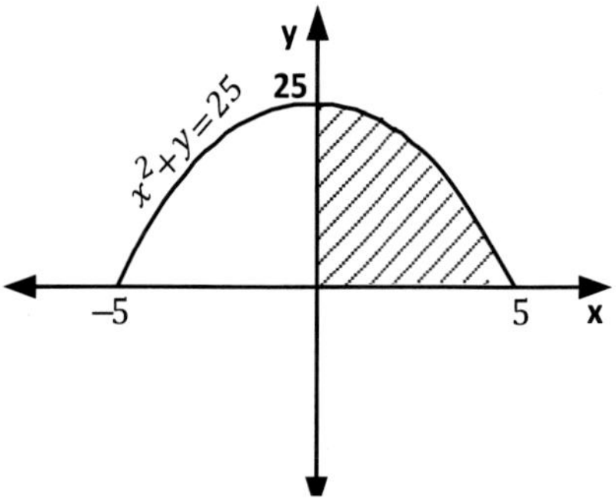
y
25
x²+y=25
−5
5
x
(c)

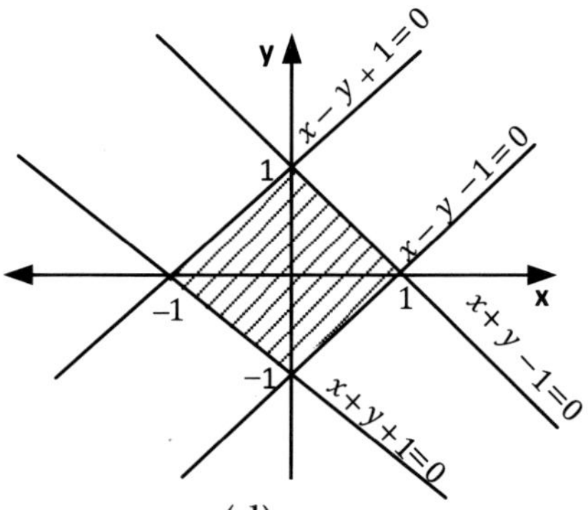
y
x−y+1=0
1
x−y−1=0
−1
1
x
x+y−1=0
−1
x+y+1=0
(d)

OBJECTIVE TEST 11

1C 2B 3D 4A 5D 6B 7A 8A 9A 10B

Exercise 11a

1. (a) 3 (b) 15 (c) $\frac{5}{8}$ (d) 1
2. (a) $6x^3 + 4x^2 + x + 7$ (b) $5x^3 + 2x^2 - x + 14$
   (c) $3x^3 - 8x + 6x + 5$ (d) $7x^3 - x^2 + 3x + 13$
3. (a) $4x^3 + 10x^2 - 6x + 6$ (b) $3x^3 - 9x^2 + 6x + 3$ (c) $16x^3 + 8x^2 - 12x + 16$
   (d) $7x^3 + x^2 + 9$ (e) $19x^3 - x^2 - 6x + 19$ (f) $-12x^3 + 2x^2 + 6x - 10$
4. $f(3) = 0$
5. $P = 3,\ Q = 2,\ R = 6$
6. Only one real zero exist at $x = 1.4$
7. $a = 6$
8. (a) $5x^6 + 23x^5 - 5x^4 - 6x^3 + 33x^2 - 20x + 6$
   (b) $2x^6 + 5x^5 - 15x^4 + 25x^3 - 4x^2 - 3x + 2$
   (c) $10x^3 - 9x^5 + 2x^4 + 37x^3 - 18x^2 + 14x + 12$
   (d) $12x^6 - 4x^5 - 13x^4 + 16x^3 - 22x^2 + 11x + 4$
   (e) $7x^6 + 28x^5 - 20x^4 + 19x^3 + 29x^2 + 23x + 8$
   (f) $-5x^6 + 32x^5 - 7x^4 - 43x^3 + 51x^2 - 34x - 6$
9. (a) 0 (b) 8 (c) 5 (d) -96
10. The remainder is 12
11. $k = 2\frac{1}{2}$ or $\frac{1}{2}$
12. $(x - 1)(x - 3)(x + 2)$
13. The zeroes of the polynomial are $-3, -2$, and 1
14. $f(-1) = 6x^3 - 13x^2 + 2x - 5 = 0$ (b) $(2x - 1), (x + 1)$, and $(3x + 5)$
15. $a = 14, b = -6$
16. $k = 2$
17. Remainder is -7

18. $(x+y)(x+3y)$

19. (a) $12x^3-4x^2+11x+18$ (b) $11x^3-4x^2+7x+10$ (c) $3x^3-4x^2+3x+6$
(d) $-3x^3+4x^2-3x-6$ (e) $16x^3-4x^2+13x+20$

20. (a) $x^4+7x^3+8x^2-3$ (b) $3x^4+7x^3+4x-11$
(c) $-3x^3+7x^3+8x^2-10x+21$ (d) $-3x^4+12x^3+20x^2-11x+20$

21. Quotient $= x+2$, Remainder $= 2$

22. (a) $f(0)=5$ (b) $f(2)=417$ (c) $f(-1)=-3$

Exercise 11b

1. (a) Non repeated linear factor; $\frac{1}{2(x-1)}-\frac{1}{2(x+1)}$
(b) Non repeated linear factor; $\frac{1}{4(x-3)}-\frac{1}{4(x+1)}$
(c) Non repeated with non-linear factor; $\frac{13}{6(x-1)}-\frac{10}{2x}+\frac{17x+4}{6(x^2+x+1)}$
(d) Non repeated with non-linear factor; $\frac{2}{3(x-1)}-\frac{2x+1}{3(x^2+x+1)}$
(e) Improper rational function; $1+x+\frac{7}{3(x-2)}-\frac{4}{3(x+1)}$

2. (a) $\frac{5}{x+2}-\frac{3}{x+1}$ (b) $\frac{1}{x+2}-\frac{4}{x-2}$ (c) $\frac{3}{x}-\frac{3}{x^2+1}$ (d) $x-\frac{x+1}{x^2+1}$

3. (a) $\frac{1}{5(x-2)}-\frac{1}{5(x+3)}$ (b) $\frac{17}{24(x+2)}-\frac{1}{30(x-1)}-\frac{17}{18(x+3)}+\frac{97}{360(x-6)}$
(c) $\frac{37}{7(x+5)}+\frac{12}{7(x-2)}$ (d) $\frac{221}{12(x-5)}+\frac{17}{24(x+1)}-\frac{73}{8(x-3)}$

4. (a) $\frac{1}{x}-\frac{1}{x^2+1}$ (b) $\frac{2x-5}{4(2x^2-5x-1)}-\frac{1}{4(x-1)}$ (c) $\frac{2x+1}{x^2-1}-\frac{1}{x-1}$ (d) $-\frac{2}{x}+\frac{2x^2+x+3}{x^3+1}$

5. (a) $\frac{3x^2-4x+5}{(x+1)(x-3)(2x-1)}$ (b) $\frac{3x^2+8x+3}{(x-1)(x^2+2x+5)}$ (c) $\frac{7x+2}{(2x-1)(x+1)^2}$

6. (a) $\frac{2}{x+1}+\frac{1}{(x+1)^2}-\frac{1}{x^2+1}$ (b) $\frac{12-16x}{25(x^2+1)}+\frac{16}{25(x+2)}+\frac{9}{5(x+2)^2}$

(c) $\frac{3}{x-3}-\frac{3x+1}{x^2+4}$ (d) $-\frac{2}{x+1}+\frac{4}{(x+2)^2}-\frac{1}{5(x+2)^2}$

7. (a) $\frac{1}{32(x+3}-\frac{1}{16(x-1)}+\frac{1}{32(x-5)}$ (b) $\frac{5}{x-1}-\frac{5}{x}-\frac{3}{x^2}$

(c) $2x+5+\frac{86}{9(x-2)}+\frac{7-68x}{9(x^2+3x+1)}$

OBJECTIVE TEST 12

1D 2C 3B 4C 5D 6A 7D 8C 9B 10C

Exercise 12

1. ...
2. ...
3. ...

4. (a) (i) 7 (ii) 1 (iii) 17 (b) 29

5. (i) 1 (ii) $\frac{11}{2}$ (iii) 1 (iv) $\frac{1}{2}$ (v) $\sqrt{3}$

6. (i) $\{-\frac{9}{2}, -2, 0, \frac{5}{2}\}$ (ii) $\{10, -5, -6, -5, 10\}$

(iii) $\{\frac{1}{4}, \frac{4}{9}, 0, 4, 16\}$ (iv) $\{3, 2\frac{17}{20}, \frac{9}{4}, 1, -3\}$

7.

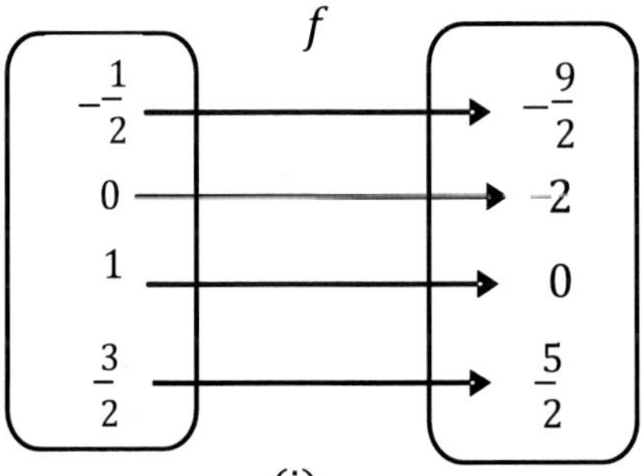

(i)

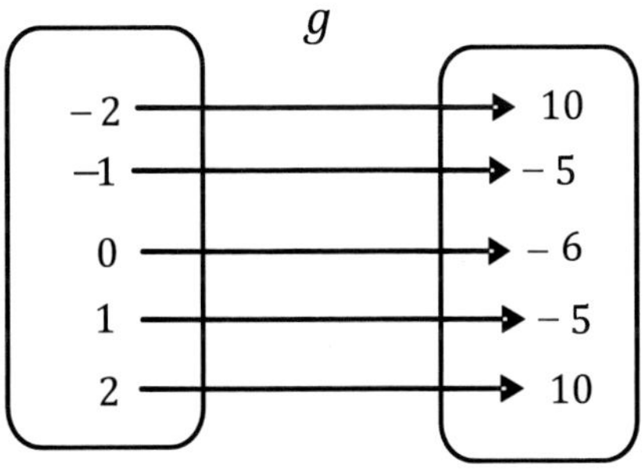

(ii)

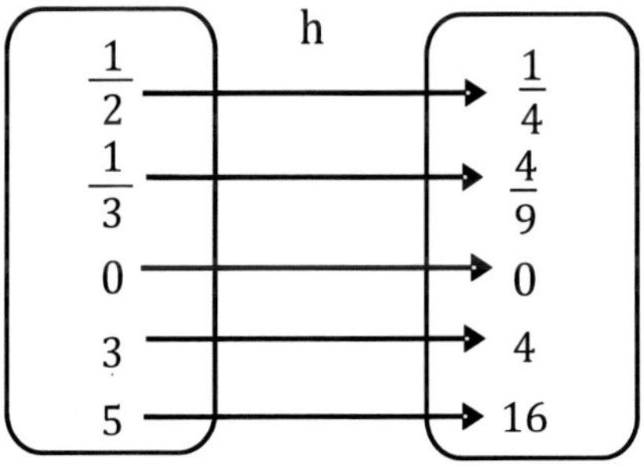

(iii)

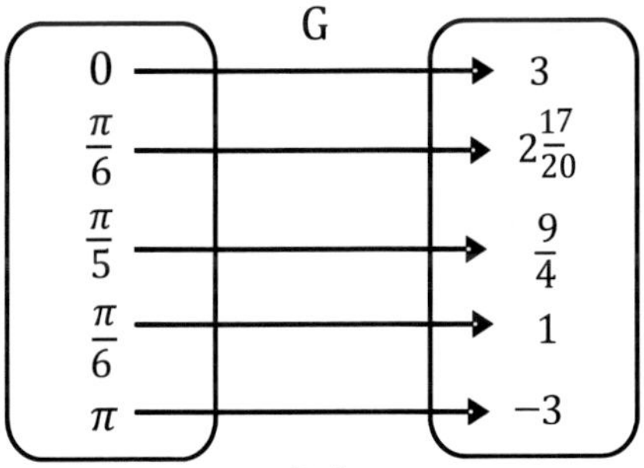

(iv)

8. $\{p, m\}$, $\{q, m\}$, $\{r, m\}$, $\{s, m\}$

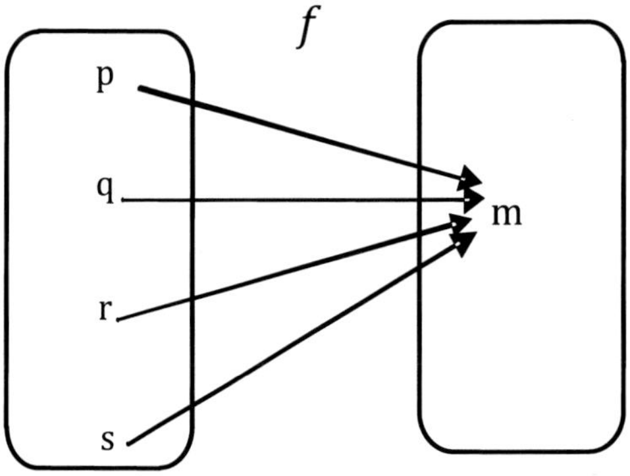

9. (i) $(x + 1)^2$ (ii) $x^2 + 1$

10. (i) x (ii) $Null$ (iii) (y, z)

11. f : is bijective mapping; g: is onto mapping; h: is constant mapping ;
i : is identity mapping; j: is bijective mapping; k: is not a mapping.

12. $x = -\frac{3}{10}$

13. (a) $gof = 9x^2 + 12x + 3$ and $fog = 3x^2 - 1$; $fog \neq gof$
(b) $gofoh = 9x^2 + 138x + 528$ and $fogoh = 9x^2 + 12x + 10$;
$fogoh \neq gofoh$

14. (a) $\frac{x^2-5x-16}{x+3}$ (b) $\frac{x^2-10x+24}{x-2}$

15. $(x + 125)^{\frac{1}{3}}$

16. (a) $16x - 3$ (b) $x^4 + 4x^3 + 4x^2 - 2$
(c) $5 - 8x - 4x^2$ (d) $16x^2 - 16x + 2$

17. (a) $\frac{1-3x}{2x-1}$ (b) $\frac{5-2x}{1-x}$ (c) $\frac{5x-14}{9-3x}$

OBJECTIVE TEST 13

1C 2C 3B 4C 5B 6D 7D 8B 9C 10B

Exercise 27

1. (a) 17 (b) 31 (c) −53 (d) −3
2. (a) 2 (b) 64 (c) $\frac{5}{6}$ (d) $\frac{3}{8}$
3. (i) 4 (ii) −1 (iii) 16 (iv) $\frac{35}{64}$
4. (a) (i) $2x + 5$ (ii) $-1 - 5x$ (iii) $x^4 - 2x^3 + 11x^2 - 10x + 24$ (iv) $2(3x + 8)$
(b) (i) $2(x + 1)$ (ii) $7 - 5x$ (iii) $x^4 - 3x^2 + 7$ (iv) $2(3x + 11)$

5.

(a)

+ (mod 3)	0	1
0	0	1
1	1	2

(b)

× (mod 3)	0	1	2
0	0	0	0
1	0	1	2
2	0	2	1

(c)

× (mod 3)	−1	0	1
−1	1	0	−1
0	0	0	0
1	−1	0	1

6. (a) Not closed because of 2

× (mod 6)	1	4
1	1	4
4	4	2

(b) Not closed because of 0

× (mod 6)	1	2	3
1	1	2	3
2	2	4	0
3	3	0	3

(c) Not closed because of 1

× (mod 6)	3	5	7
3	3	3	3
5	3	1	5
7	3	5	1

(d)Not closed because of 0

× (mod 6)	1	2	3	4
1	1	2	3	4
2	2	4	0	2
3	3	0	3	0
4	4	2	0	4

(e) It is closed

× (mod 6)	1	2	4
1	1	2	4
2	2	4	2
4	4	2	4

(f) It is closed

× (mod 6)	1	3	5
1	1	3	5
3	3	4	0
5	5	3	1

7. (i) It is closed (ii) It is closed (iii) It is closed (iv) It is closed

8. (i) It is not commutative (ii) It is commutative
(iii) It is commutative (iv) It is not commutative

9. (i) Not commutative (ii) Not commutative (iii) It is commutative
(iv) Not commutative (v)Not commutative (vi) It is commutative

10. (i) It is associative (ii) Not associative

11. (i) It has no identity element (ii) 0 (iii) 1 (iv) 1 (v) 1 (vi) ∅

1. (i) Identity element = 1 (ii) Identity element = 4
(iii) Identity element = 1

13. ...

14. ...

15.

+(mod 6)	0	1	2	3	4	5
0	0	1	2	3	4	5
1	1	2	3	4	5	0
2	2	3	4	5	0	1
3	3	4	5	0	1	2
4	4	5	0	1	2	3
5	5	0	1	2	3	4

(a) Identity element= 0 (b) (i) 2 (ii) 5 (iii) 1 (iv) 4 (v) 3

16. (a) Not distributive (b) Not distributive

17. (i) 26 (ii) −39 (iii) −4 (iv) −1

18. (i) 50 (ii) $x = 4$ or -1 (iii) $x = -18$ (iv) $x = 0$ or 23

OBJECTIVE TEST 14

1C 2B 3A 4D 5C 6D 7C 8D 9A 10C

Exercise 14

1. ...

2. ...

3. (i) 6 (ii) 10 (iii) 7 (iv) 4.69

4. (a) 3 (b) 3 (c) 3.1

5. (a)

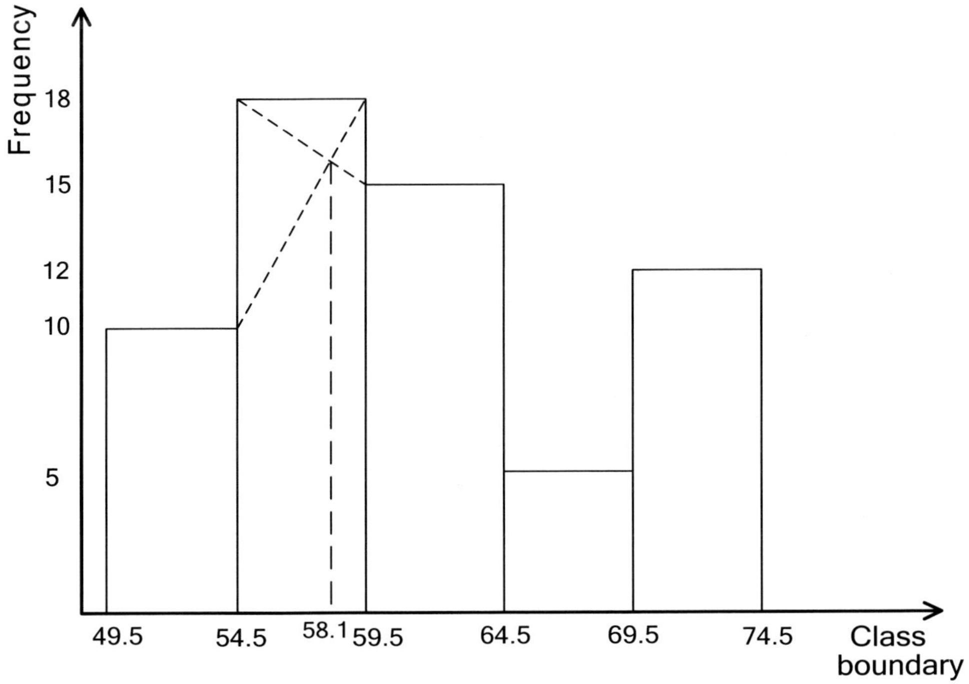

(b) (i) 61.25 (ii) 58.14 (iii) 60.17

6. (a)

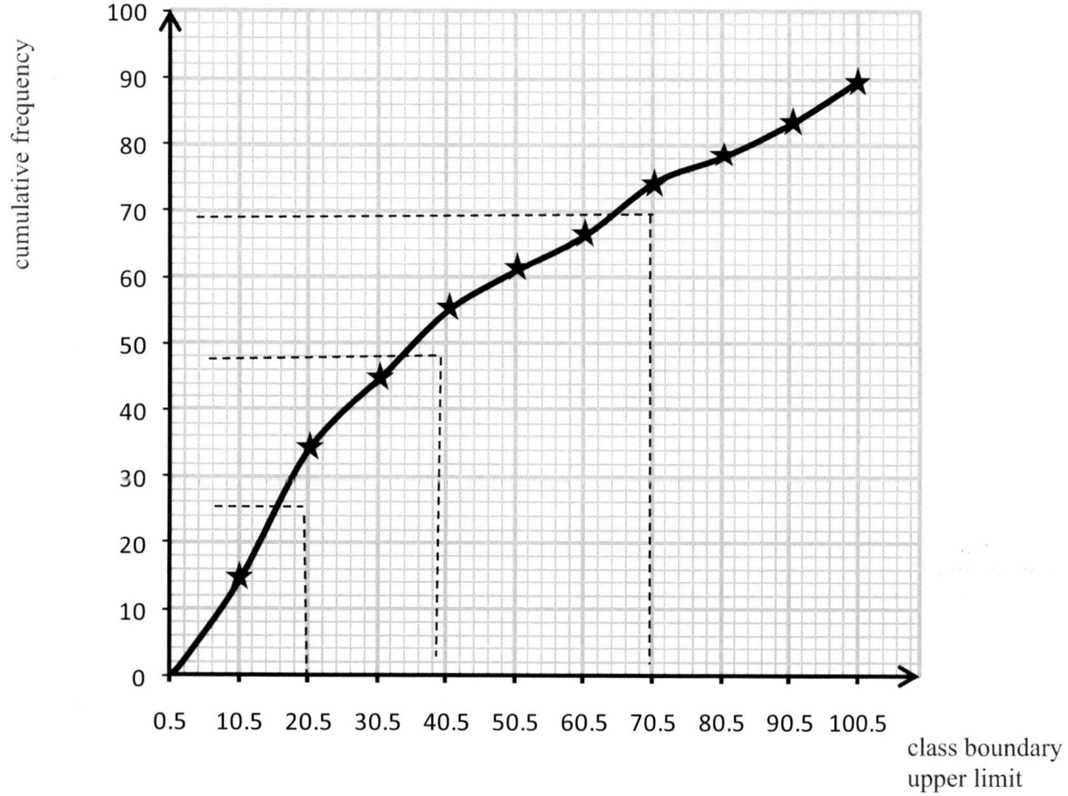

(c) Upper quartile $Q_3 = 63.5$; lower quartile $= Q_1 = 14.5$

7. (a)

Class marks	Frequency
21 – 30	2
31 – 40	7
41 – 50	9
51 – 60	10
61 – 70	10
71 – 80	3
81 – 90	2
91 – 100	2

(b) mean mark = 57.7 (c) (i) mode = 60.5 (ii) median = 55.5 (d) 37.8%

8. $x = 91\%$

9. $y = 47$

10.

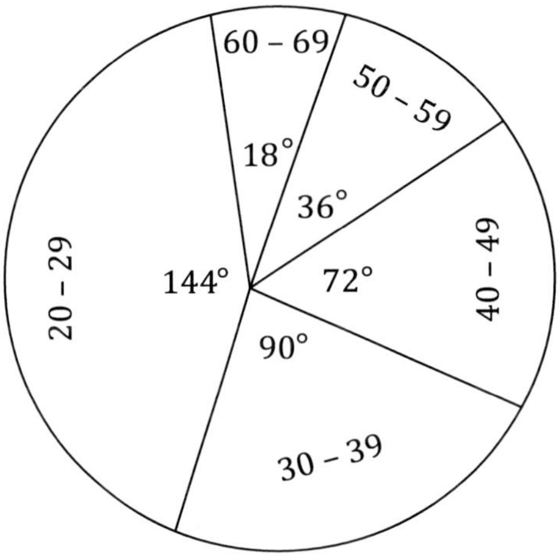

11. (a) modal class = 210 − 229; mode = 216.2
(b) mean height = 209.5

12. (a) mean = 5 (b) median = 5

13. $x = 42.92$kg

14. (i) mean = 4.49 (ii) standard deviation = 47

15. (i) mean = 48.03 (ii) 15.64 (iii) 399.3 (iv) 19.98

OBJECTIVE TEST 15

1D 2C 3C 4D 5A 6B 7C 8C 9C 10A

Exercise 15

1. ...
2. ...
3. (a) $S = \{HH, TT, HT, TH\}$ (b) (i) $\frac{1}{4}$ (ii) $\frac{1}{2}$ (iii) $\frac{1}{2}$ (iv) $\frac{1}{4}$
4. (a) $S = \{1, 2, 3, 4, 5, 6\}$ (b) (i) $\frac{1}{6}$ (ii) $\frac{1}{6}$ (iii) $\frac{1}{3}$ (iv) $\frac{1}{2}$

5. (a)

First throw / Second throw	1	2	3	4	5	6
1	1, 1	2,1	3, 1	4,1	5, 1	6, 1
2	1, 2	2, 2	3, 2	4,2	5, 2	6, 2
3	1, 3	2, 3	3, 3	4,3	5, 3	6, 3
4	1, 4	2, 4	3, 4	4,4	5, 4	6, 4
5	1, 5	2, 5	3, 5	4,5	5, 5	6, 5
6	1, 6	2, 6	3, 6	4,6	5, 6	6, 6

(b) (i) $\frac{1}{36}$ (ii) $\frac{5}{36}$ (iii) $\frac{1}{4}$ (iv) $\frac{1}{4}$

6. (a) $\frac{5}{8}$ (b) $\frac{3}{3}$

7. A and C, A and D, B and C, B and D

8. (a) (i) $\frac{1}{714}$ (ii) $\frac{1}{238}$ (iii) $\frac{3}{476}$ (iv) $\frac{11}{2380}$

(b) (i) $\frac{125}{46656}$ (ii) $\frac{5}{1296}$ (iii) $\frac{1}{144}$ (iv) $\frac{11}{1728}$

9. (i) $\frac{12}{25}$ (ii) $\frac{8}{25}$ (iii) $\frac{17}{25}$ (iv) $\frac{13}{25}$

10. (a) $\frac{1}{4}$ (b) $\frac{1}{4}$ (c) $\frac{1}{2}$ (d) $\frac{1}{2}$

11. (a) $S = \{0, 1, 2, 3, 4, 5, 6, 7, 8, 9, 10, \ldots 29\}$ (b) $\frac{1}{2}$ (c) $\frac{1}{2}$ (d) $\frac{3}{10}$

12. $x = 27$

13. Number of Apples introduced = 14

14. (i) $\frac{1}{3}$ (ii) $\frac{5}{12}$ (iii) $\frac{1}{4}$ (iv) $\frac{3}{4}$

15. (a) (i) $\frac{1}{36}$ (ii) $\frac{1}{24}$ (iii) $\frac{7}{24}$ (iv) $\frac{31}{72}$ (b) (i) $\frac{1}{66}$ (ii) $\frac{1}{22}$ (iii) $\frac{7}{22}$ (iv) $\frac{25}{66}$

16. (a) $\frac{2}{15}$ (b) $\frac{7}{15}$ (c) $\frac{2}{5}$ (d) $\frac{4}{15}$

17. (i) $\frac{1}{42}$ (ii) $\frac{5}{84}$ (iii) $\frac{1}{7}$ (iv) $\frac{15}{84}$ (v) $\frac{37}{84}$ (vi) $\frac{5}{14}$

18. (a) $\frac{1}{12}$ (b) $\frac{1}{4}$

OBJECTIVE TEST 16

1A 2D 3D 4C 5C 6A 7B 8B 9B 10C

Exercise 16

1. ...
2. ...
3. (a) combination (b) combination (c) permutation (d) combination (e) permutation (f) combination
4. (a) 5040 (b) 6720 (c) 120 (d) 498960 (e) 10080 (f) 90720 (g) 120 (h) 498960 (i) 360
5. 27720
6. (i) 1 (ii) 1 (iii) 120 (iv) 3628800 (v) 480 (vi) 6720 (vii) $\frac{1}{5}$ (viii) 241920 (ix) 1440 (x) $\frac{45}{2}$
7. 5040
8. 120

10. 220
11. (a) 56 (b) 1
12. 1024
13. 56
14. 5005
15. 66
16. (a) 24 (b) 6720 (c) 60 (d) $\frac{1}{6}$ (e) $\frac{5}{2}$
17. 4
18. 1140
19. 6840

OBJECTIVE TEST 17

1C 2 B 3 A 4 B 5D 6 C 7 A 8D 9B 10A

Exercise 17

1. (a) i (b) -1 (c) i (d) $-i$ (e) i
2. (a) $-3i$ (b) $5i$ (c) $0i$ (d) $0i$
3. (a) 8 (b) 4 (c) $1-i$ (d) $5+11i$
4. (a) 0 (b) 2
5. (a) $-14i-10$ (b) $12-8i$ (c) $-1-i$
6. (a) $1-i$ (b) $-33i$ (c) $-1+i$ (d) $-3i-5$ (e) $2-i$
7. (a) 5 (b) $\sqrt{5}$ (c) 5 (d) 1 (e) 1
8. (a) $a=-\frac{1}{2}, b=-\frac{3}{2}$
9. $\frac{-13}{5}$
10. (a) $-5i$ (b) $5i$ (c) $-28i-48$
11. $-9i+9$
12. $k=0$
13. (a) i (b) $\frac{18-i}{25}$
14. (a) $6+8i$ (b) 10

15. (a) $z=5(cos53.13^0+isin53.13^0)$ (b) $z=5(cos\,306.87^0+isin306.87^0)$
(c) $z=(cos\,270^0+isin270^0)$ same as $z=\left(cos\frac{3\pi}{2}+isin\frac{3\pi}{2}\right)$
16. (a) -1 (b) $\frac{5}{\sqrt{2}}+\frac{5}{\sqrt{2}}i$

OBJECTIVE TEST 18

1A 2B 3B 4A 5D 6A 7A 8D 9A 10C

Exercise 18

1. (i) $\begin{pmatrix}28 & 16\\21 & 15\end{pmatrix}$ (ii) $\begin{pmatrix}38 & 25\\27 & 38\end{pmatrix}$ (iii) $\begin{pmatrix}-9 & -21\\-22 & 14\end{pmatrix}$ (iv) $\begin{pmatrix}2 & 7\\7 & 4\end{pmatrix}$

2. (i) $\begin{pmatrix}-79 & -14\\16 & -49\end{pmatrix}$ (ii) $\begin{pmatrix}15 & 21\\37 & -9\end{pmatrix}$ (iii) $\begin{pmatrix}22 & 28\\20 & -16\end{pmatrix}$

   (iv) $\begin{pmatrix}60 & 27\\68 & 69\end{pmatrix}$ (v) $\begin{pmatrix}69 & 36\\51 & 60\end{pmatrix}$ (vi) $\begin{pmatrix}184 & 150\\112 & 12\end{pmatrix}$

3. (i) $|P| = 12$ (ii) $|Q| = -76$ (iii) $|R| = 192$ (iv) $|PQ| = -912$
   (v) $|QR| = -14592$ (vi) $|PR| = 2304$

4. (i) $P^T = \begin{pmatrix}3 & 1\\0 & 4\end{pmatrix}$ (ii) $Q^T = \begin{pmatrix}5 & 8\\7 & -4\end{pmatrix}$ (iii) $R^T = \begin{pmatrix}20 & 12\\9 & 15\end{pmatrix}$

   (iv) $P^TQ^T = \begin{pmatrix}22 & 4\\28 & -16\end{pmatrix}$ (v) $(PQ)^T = \begin{pmatrix}15 & 37\\21 & -9\end{pmatrix}$ (vi) $\begin{pmatrix}252 & 180\\64 & 24\end{pmatrix}$

5. (i) $P^{-1} = \frac{1}{2}\begin{pmatrix}252 & 180\\64 & 24\end{pmatrix}$ (ii) $Q^{-1} = \frac{1}{-76}\begin{pmatrix}-4 & -7\\-8 & 5\end{pmatrix}$

   (iii) $R^{-1} = \frac{1}{192}\begin{pmatrix}15 & -9\\-12 & 20\end{pmatrix}$ (iv) $(PQ)^{-1} = \frac{1}{912}\begin{pmatrix}9 & 21\\37 & -15\end{pmatrix}$

   (v) $(PR)^{-1} = \frac{1}{2304}\begin{pmatrix}69 & -27\\-68 & 60\end{pmatrix}$ (vi) $(QR)^{-1} = \frac{1}{14592}\begin{pmatrix}-12 & 150\\112 & -184\end{pmatrix}$

6. (i) 4 (ii) 9 (iii) −3 (iv) 2 (v) 15 (vi) −3 (vii) 2 (viii) 26

7. (i) $\begin{pmatrix}1 & 7\\-3 & 8\end{pmatrix}$ (ii) $\begin{pmatrix}3 & 7 & 4\\1 & 9 & 14\end{pmatrix}$ (iii) Not possible (iv) $\begin{pmatrix}1 & -1\\7 & 2\end{pmatrix}$

   (v) $\begin{pmatrix}-1 & 1\\-7 & -2\end{pmatrix}$ (vi) $\begin{pmatrix}7 & -1 & -2\\7 & 5 & 2\end{pmatrix}$

8. (i) $AB = \begin{pmatrix} -14 & 49 & 38 \\ -1 & 5 & 4 \\ -5 & 13 & 10 \end{pmatrix}$, $BA = \begin{pmatrix} -24 & -15 & -18 \\ 6 & 1 & 5 \\ 26 & -9 & 21 \end{pmatrix}$ hence $AB \neq BA$

9. (i) $\begin{pmatrix} -1 & 5 & 16 \\ 21 & 14 & 1 \\ 0 & 4 & 4 \end{pmatrix}$ (ii) $\begin{pmatrix} -3 & 39 & -24 \\ 156 & 87 & 96 \\ 26 & 10 & 57 \end{pmatrix}$ (iii) $\begin{pmatrix} 73 & 14 & 2 \\ 95 & -2 & -102 \\ 35 & 8 & 29 \end{pmatrix}$

(iv) $\begin{pmatrix} 3 & 7 & 2 \\ -3 & 64 & 63 \\ -8 & 35 & 42 \end{pmatrix}$ (v) $\begin{pmatrix} -79 & 10 & -24 \\ 5 & 48 & 33 \\ 2 & 3 & 7 \end{pmatrix}$ (vi) $\begin{pmatrix} 4 & 6 & 11 \\ -4 & 37 & 128 \\ -9 & 17 & 67 \end{pmatrix}$

(vii) $\begin{pmatrix} 46 & 40 & 41 \\ 587 & 319 & 101 \\ 354 & 167 & -2 \end{pmatrix}$

10. (i) $(15 \quad 31)$ (ii) (23) (iii) (247) (iv) impossible

11. (i) $\begin{pmatrix} 2 & 5 & 1 \\ -1 & 4 & 3 \\ 0 & 3 & 6 \end{pmatrix}$ (ii) $\begin{pmatrix} 4 & 43 & 25 \\ -8 & 29 & 35 \\ -3 & 36 & 58 \end{pmatrix}$

12. (i) $\begin{pmatrix} 1 & 2 \\ -2 & 5 \end{pmatrix}$ (ii) $\begin{pmatrix} 7 & 8 \\ 3 & 2 \end{pmatrix}$ (iii) $\begin{pmatrix} 13 & 12 \\ 1 & -6 \end{pmatrix}$ (iv) $\begin{pmatrix} -9 & 54 \\ -1 & 16 \end{pmatrix}$

(v) $\begin{pmatrix} 13 & 12 \\ 1 & -6 \end{pmatrix}$ (vi) $\begin{pmatrix} 38 & 25 \\ 27 & 38 \end{pmatrix}$

13. (i) $A(B+C) = AB + AC = \begin{pmatrix} -3 & 4 & 8 \\ -8 & -8 & -14 \\ -8 & 5 & -10 \end{pmatrix}$

(ii) $A(BC) = (AB)C = \begin{pmatrix} -30 & -11 & -46 \\ -8 & -12 & -30 \\ -54 & -22 & -164 \end{pmatrix}$

(iii) $AB = \begin{pmatrix} -7 & -7 & 5 \\ -2 & 0 & -2 \\ -8 & -14 & -2 \end{pmatrix}$ and $BA = \begin{pmatrix} 4 & -8 & 3 \\ 2 & -16 & 6 \\ 12 & -22 & 3 \end{pmatrix}$; hence $AB \neq BA$

14. (i) $x = 0, -1, 4$ (ii)) $x = 5, -9$

15. (i) -17 (ii) -33 (iii) -77 (iv) 48 (v) 0

(vi) 0 (vii) 0 (viii) $-7e^{2x}$ (ix) $5x^3$

16. (i) -15 (ii) -290 (iii) 0 (iv) 277 (v) 56 (vi) 640 (vii) 567

17. (a) (i) $\begin{pmatrix} 1 & 2 \\ -3 & 4 \end{pmatrix}$ (ii) $\begin{pmatrix} -2 & 2 \\ 12 & 5 \end{pmatrix}$ (iii) $\begin{pmatrix} 22 & 12 \\ 54 & 14 \end{pmatrix}$ (iv) $\begin{pmatrix} -8 & 4 \\ -3 & 44 \end{pmatrix}$

(b) (i) $\begin{pmatrix} 1 & -2 \\ 3 & 44 \end{pmatrix}$ (ii) $\begin{pmatrix} -2 & -2 \\ -12 & 5 \end{pmatrix}$ (iii) $\begin{pmatrix} 22 & -12 \\ -54 & 14 \end{pmatrix}$ (iv) $\begin{pmatrix} -8 & -4 \\ 3 & 44 \end{pmatrix}$

(c) (i) $\begin{pmatrix} 1 & 3 \\ -2 & 4 \end{pmatrix}$ (ii) $\begin{pmatrix} -2 & -12 \\ -2 & 5 \end{pmatrix}$ (iii) $\begin{pmatrix} 22 & -54 \\ -12 & 14 \end{pmatrix}$ (iv) $\begin{pmatrix} -8 & 3 \\ -4 & 44 \end{pmatrix}$

(d) (i) $\frac{1}{10}\begin{pmatrix} 1 & 3 \\ -2 & 4 \end{pmatrix}$ (ii) $\frac{1}{34}\begin{pmatrix} 2 & 12 \\ 2 & -5 \end{pmatrix}$ (iii) $\frac{1}{170}\begin{pmatrix} -11 & 27 \\ 6 & 7 \end{pmatrix}$ (iv) $\frac{1}{340}\begin{pmatrix} -44 & 3 \\ -4 & 8 \end{pmatrix}$

18. (i) $\begin{pmatrix} -70 & 26 & 144 \\ -104 & 0 & 130 \\ -10 & -13 & 19 \end{pmatrix}$ (ii) $\begin{pmatrix} -70 & -26 & 144 \\ 104 & 0 & -130 \\ -10 & 13 & 19 \end{pmatrix}$

19. (i) $\frac{1}{4}\begin{pmatrix} 4 & 3 & -6 \\ 0 & -1 & 2 \\ -4 & 1 & 2 \end{pmatrix}$ (ii) $\frac{1}{290}\begin{pmatrix} 58 & 0 & 0 \\ 0 & -10 & 60 \\ 0 & 25 & -5 \end{pmatrix}$ (iii) $\frac{1}{153}\begin{pmatrix} 15 & 15 & -1 \\ -108 & 198 & -3 \\ 264 & -501 & 13 \end{pmatrix}$

(iv) $\frac{1}{661}\begin{pmatrix} 26 & -24 & -17 \\ -28 & -25 & 120 \\ 45 & 111 & -4 \end{pmatrix}$ (v) $\frac{1}{53}\begin{pmatrix} -11 & 6 & 3 \\ 60 & 1 & -26 \\ 25 & -4 & -2 \end{pmatrix}$

(vi) $\frac{2}{17}\begin{pmatrix} -6 & \frac{5}{2} & -6 \\ 1 & 1 & 1 \\ 17 & -\frac{17}{4} & \frac{17}{2} \end{pmatrix}$ (vii) $\frac{20}{329}\begin{pmatrix} -\frac{57}{10} & 1 & \frac{13}{5} \\ -\frac{7}{2} & \frac{7}{2} & \frac{91}{10} \\ \frac{29}{4} & \frac{9}{2} & -\frac{19}{4} \end{pmatrix}$

(viii) It has no inverse (ix) It has no inverse

20. (i) $(x, y) = \left(\frac{7}{3}, -\frac{1}{3}\right)$ (ii) $(x, y) = (29, -48)$ (iii) $(x, y) = \left(-\frac{1}{8}, \frac{1}{11}\right)$

21. $(x, y, z) = \left(-\frac{11}{7}, \frac{6}{7}, 2\right)$

22. (i) $(x, y) = \left(\frac{2}{7}, \frac{3}{7}\right)$ (ii) $(x, y) = \left(2\frac{3}{11}, 3\frac{5}{11}\right)$

23. (i) $(x, y) = (4, -3)$ (ii) $(x, y) = \left(-\frac{56}{23}, \frac{-4}{23}\right)$

24. (i) $\begin{pmatrix} 1 & 2 & 2 \\ 5 & 3 & 1 \\ 1 & 1 & 4 \end{pmatrix}\begin{pmatrix} x \\ y \\ z \end{pmatrix} = \begin{pmatrix} -5 \\ 1 \\ 7 \end{pmatrix}$ (ii) $(x, y, z) = \left(\frac{69}{25}, -\frac{126}{25}, \frac{58}{25}\right)$

25. (i) $P = \begin{pmatrix} \frac{2}{11} & \frac{5}{22} \\ \frac{1}{11} & -\frac{3}{22} \end{pmatrix}$ (ii) $Q = \begin{pmatrix} 1 & \frac{3}{4} & -\frac{3}{2} \\ 0 & -\frac{1}{4} & \frac{1}{2} \\ -1 & \frac{1}{4} & \frac{1}{2} \end{pmatrix}$

26. (i) $A = \begin{pmatrix} 3 & 18 \\ 7 & -14 \end{pmatrix}$ (ii) $B = \begin{pmatrix} 40 & 30 \\ -5 & 17 \end{pmatrix}$ (iii) $\begin{pmatrix} 13 & 56 & 11 \\ 15 & 4 & 16 \\ -3 & -5 & 10 \end{pmatrix}$

(iv) $\begin{pmatrix} -85 & 43 & 132 \\ 105 & -38 & -160 \\ 122 & 90 & 58 \end{pmatrix}$ (v) $\begin{pmatrix} \frac{2}{3} & -\frac{4}{5} & 1 \\ -\frac{3}{7} & 0 & -1 \\ 1 & \frac{3}{8} & 0 \end{pmatrix}$

APPENDIX C

MATHEMATICAL SYMBOLS

The following symbols of mathematics as may be used in this book should be acquainted by the student(s):

SYMBOLS	MEANING
+	plus; positive; addition
-	minus; negative; subtraction.
×	product; multiply.
÷	division
±	plus or minus
=	equal to
≠	not equal to
≈	approximately
>	more than; greater than
<	less than
≥	greater than or equal to
≤	less than or equal to
∠ or ^	angle
△	triangle
∴	therefore
:	ratio
::	proportion
$\sqrt[n]{\square}$	n^{th} root
∝	varies as; proportional to
Σ	sigma; summation
'	compliment of
|AB|	the length of line AB
$\overline{AB}$	line segment A to B; length of line AB
₦	Naira
$	Dollar

$\in$	is an element of
$\notin$	is not an element of
n(A)	the number of element in set A
$\cup$	union of
$\cap$	intersection of
$\subset$	is a subset of
$\supset$	is a superset of
$\not\subset$	is not a subset of
$\not\supset$	is not a superset of
$\subseteq$	subset of or equal
$\supseteq$	superset of or equal
$a \wedge b$	a and b
$a \vee b$	a or b
$\rightarrow$	mapped on to
$\equiv$	is equivalent to
$\Leftrightarrow$	a statement of implication and its converse
$\Rightarrow$	implies that; if ... then
∞	infinity
{ }	brace or curly bracket
( )	open bracket
[ ]	square or closed bracket
n!	n factorial; factorial n $= 1 \times 2 \times 3 \times 4 \times \ldots \times n$
θ	theta
α	alpha
β	beta
π	pi
%	percentage
$^\circ$	degree
$\sqrt[3]{}$	cube root
$\sqrt[4]{}$	quad(fourth) root

INDEX

Made in United States
Cleveland, OH
06 November 2026

47787689R00247

Made in United States
Cleveland, OH
05 November 2026

47659897R00136